MULTIVARIATE STATISTICS

A complete list of the titles in this series appears at the end of this volume.

MULTIVARIATE STATISTICS

High-Dimensional and Large-Sample Approximations

YASUNORI FUJIKOSHI
Chuo University
Tokyo, Japan

VLADIMIR V. ULYANOV
Moscow State University
Moscow, Russia

RYOICHI SHIMIZU
Institute of Statistical Mathematics
Tokyo, Japan

WILEY

A JOHN WILEY & SONS, INC., PUBLICATION

Published by John Wiley & Sons, Inc., Hoboken, New Jersey.
Published simultaneously in Canada.

Library of Congress Cataloging-in-Publication Data:

Fujikoshi, Yasunori, 1942–
Multivariate statistics : high-dimensional and large-sample approximations / Yasunori Fujikoshi, Vladimir V. Ulyanov, Ryoichi Shimizu.
p. cm. — (Wiley series in probability and statistics)
Includes bibliographical references and index.
ISBN 978-0-470-41169-8 (cloth)
1. Multivariate analysis. 2. Approximation theory. I. Ulyanov, Vladimir V., 1953– II. Shimizu, Ryoichi, 1931– III. Title.
QA278.F84 2009
519.5'35—dc 22 2009017248

Contents

Preface

Many multivariate methods are based on large-sample approximations. These results can be found in the books in multivariate statistical analysis. See, for example, the books by Anderson (2003), Muirhead (1984) and Siotani et al. (1985). However, these approximations become increasingly inaccurate as dimension p of observations increases while sample size n remains fixed. On the other hand, in last years we encounter more and more problems in applications when p is comparable with n or even exceeds it. Some examples of high-dimensional data include curve data, spectra, images, and DNA micro-arrays.

Therefore, it becomes essential to revise the classical multivariate methods in order to make them useful in wide range of relations between p and n and to extend multivariate statistical theory in high-dimensional situations. One way of overcoming a weakness of classical large-sample approximations is to derive approximations under high-dimensional framework when $p/n \to c \in (0,1)$ or $(0,\infty)$.

Another problem related to multivariate approximations concerns their errors. Most results supply the so-called order estimates only. However, such estimates do not give information on actual errors for given values n, p and other parameters. Ideally, we wish to have computable error bounds, in addition to order estimates. It is made already for some multivariate statistics.

In multivariate methods it is important to reduce a set of original variables and canonical variables, so that we can make statistical inference more accurate and its interpretation more simple and effective. However, it is a difficult to choose an appropriate subset of variables, or an appropriate number of canonical variables. For such problem, a model selection approach is developed, in addition to traditional testing methods or sequential procedures.

Our book is focussing on high-dimensional and large-sample approximations. At the same time we describe many basic multivariate methods and derive the exact distributional results related to the methods too. For many approximations, its detailed derivation will take a lot of space. Therefore, we give mainly their outlines. In order to solve the above-mentioned problems, we consider in the book

(1) high-dimensional as well as large sample approximations for classical multivariate statistics,

(2) approximations for high-dimensional statistics,

(3) explicit error bounds for large-sample and high-dimensional approximations,

(4) selection of variables by model selection approach,

(5) basic multivariate methods and related exact distributions

This book is designed as a reference book for researchers interested in multivariate statistical analysis. However, we believe that it will be useful for graduate level courses as well, since it contains in first twelve chapters many basic facts and methods from multivariate analysis.

Broadly speaking, chapters 1 $\sim$ 12 deal with multivariate analysis focussing on (1), (2), (4) and (5). The last four chapters (Chapters 13 $\sim$ 16) concern with explicit error bounds for some large-sample and high-dimensional approximations. Chapter 1 gives basic properties of multivariate normal distributions and elliptical distributions. Sample covariance matrix and various sums of squares and products matrices have Wishart distributions, when their underlying distributions are normal. In Chapter 2 we describe properties of Wishart distributions. In Chapter 3 the Hoteling T^2 and the Lambda statistics are treated. We also study the likelihood ratio test for additional information when several mean vectors are compared. Definitions, inferences, and sampling distributions of several correlations (except canonical correlations) are discussed in Chapter 4. Covariance selection model which is related to partial correlations is discussed as well.

In Chapter 5 we summarize some methods and theories on asymptotic expansion of multivariate statistics. High-dimensional as well as large-sample approximations are discussed. In this chapter the reader will find the topics on Edgeworth, Cornish-Fisher and bootstrap approximations, and their validities. MANOVA problems are discussed in Chapter 6. The distributions of MANOVA tests and characteristic roots are treated. Multivariate regression and linear models are discussed in Chapter 7. We give C_p and AIC criteria for selection of the response variables as well as the explanatory variables. Classical and high-dimensional tests on covariance matrices are considered in Chapter 8.

In Chapter 9, discriminant analysis is studied. The concept of discriminant analysis is given, including a decision-theoretic approach and Fisher's method. Significance tests for discriminant functions and evaluation of probabilities of misclassifications are discussed. Further, the problems of selecting the canonical discriminant variables as well as the original variables are considered, based on model selection criteria. Principal component analysis and canonical correlation analysis are treated in Chapters 10 and 11, respectively. Some inferential problems on dimensionality are treated as the ones of selecting special types of covariance structures. Large-sample approximation is obtained for distributions of canonical correlations as special case of high-dimensional approximations. The growth curve model, which is a model for repeated measures data, is discussed in Chapter 12. Theoretically it can be considered as a multivariate linear model under a conditional set-up. Using this relation various inferential methods are derived, including the methods of defining a degree of polynomial in growth curve model.

Chapters 13 through 16 are concerned with explicit and computable error bounds for asymptotic approximations. We suggest a general approach to approximation of scale mixtures, including special cases of normal and chi-square mixtures and their multivariate extensions. In Chapter 14 we give the results on a location and scale mixture, the maximum of multivariate t- and F- variables, a mixture of F-distribution, and non-uniform error bounds. The applications of these basic results are discussed in Chapters 15 and 16. Error bounds are given in Chapter 15 for large-sample approximations of Hotelling's T_0^2 (Lawley and Hotelling criterion) and Lambda-statistics. In Chapter 16 we construct error bounds for large-sample and high-dimensional approximations of linear discriminant function. Furthermore, the estimators in profile analysis, growth curve analysis, and generalized linear model are treated.

We express our sincere thanks to Mr. Tetsuro Sakurai, Chuo University, Tokyo, for his kind help with numerical computations of some examples and the preparation of this book.

Our thanks also go to the authors, editors, and owners of copyrights for permission to reproduce the following materials: Table 7.3.1 (the chemical data introduced by Box and Youle, 1995 and examined by Rencher, 2002), Table 12.1.1 (the data on ramus height of 20 boys, taken from Elston and Grizzle, 1962), and Table 12.3.1 (the data of dental measurements on $n_1 = 11$ girls and $n_2 = 14$ boys, taken from Potthoff and Roy, 1964).

We believe and hope that the book will be useful in the future developments in multivariate analysis.

Yasunori Fujikoshi
Vladimir V. Ulyanov
Ryoichi Shimizu

Hiroshima, Moscow, Tokyo
November, 2009

Glossary of Notation and Abbreviations

R^p	p-dimensional Euclidean space $(\mathrm{R}^1 = \mathrm{R})$
$\boldsymbol{a}$, $\boldsymbol{a}'$	column and row vectors, respectively
$\boldsymbol{a} = (a_1, \ldots, a_p)'$	column vector with components $a_1, \ldots, a_p$
$\mathbf{1}_p = (1, \ldots, 1)'$	column vector consisting of p ones
$\mathrm{A}: p \times q$	matrix with p rows and q columns
(a_{ij})	matrix with elements a_{ij}'s
I_p	unit matrix of order p
A'	transposed matrix of A
$\mathrm{diag}\,(\theta_1, \ldots, \theta_p)$	diagonal matrix with diagonal elements $\theta_1, \ldots, \theta_p$
$\mathrm{diag}\,(\mathrm{A}_1, \ldots, \mathrm{A}_k)$	block-diagonal matrix with elements $\mathrm{A}_1, \ldots, \mathrm{A}_k$
$\lvert\mathrm{A}\rvert$	determinant of a matrix A
$\mathrm{tr}\ \mathrm{A}$	trace of a square matrix A
$\mathrm{A} \otimes \mathrm{B}$	direct product of matrices A and B
O	zero matrix consisting of 0s
$\mathrm{A} > \mathrm{B}$	the matrix $\mathrm{A} - \mathrm{B}$ is positive definite
$\mathrm{A} \geq \mathrm{B}$	the matrix $\mathrm{A} - \mathrm{B}$ is positive semi-definite
d.f.	degrees of freedom
i.i.d.	independent and identically distributed
$X, Y, \ldots$	random variable (r.v.) – italic
$\boldsymbol{X}, \boldsymbol{Y}, \ldots$	random vector (r.vec.) – slanted boldface
$\mathrm{A}, \mathrm{B}, \mathrm{C}, \ldots$	constant matrix — upright roman
$\mathbf{X}, \mathbf{Y}, \mathbf{S}, \ldots$	random matrix — wide boldface
$\mathrm{vec}\,(\mathbf{X})$	np-vector $(\boldsymbol{X}'_{(1)}, \ldots, \boldsymbol{X}'_{(p)})'$ obtained from the matrix $\mathbf{X} = (\boldsymbol{X}_{(1)}, \ldots, \boldsymbol{X}_{(p)})$
$\mathrm{E}(X), \mathrm{Var}(X)$	expectation and variance of random variable X
$\mathrm{E}(\boldsymbol{X}), \mathrm{Var}(\boldsymbol{X})$	expectation and covariance matrix of random vector $\boldsymbol{X}$
$\mathrm{E}(\mathbf{X}), \mathrm{Var}(\mathbf{X})$	expectation and covariance matrix of random matrices $\mathbf{X}$
$\mathrm{Cov}(\mathbf{X}, \mathbf{Y})$	covariance matrix of random metrices $\mathbf{X}$ and $\mathbf{Y}$
pdf	probability density function
cdf	cumulative distribution function
$\mathrm{N}(\mu, \sigma^2)$	the normal distribution with mean μ and variance σ^2
$\varphi(x)$	pdf of the standard normal distribution $\mathrm{N}(0, 1)$
$\Phi(x)$	cdf of the standard normal distribution $\mathrm{N}(0, 1)$

$G(\sigma, \lambda)$	the gamma distribution with scale and shape parameters σ and λ
$Be(p, q)$	the beta distribution
$t(n)$	the t-distribution with n d.f.
$\chi^2(p)$	the chi-square distribution with p d.f.
$\chi^2(p)$	the chi-square distribution with p d.f.
$\chi^2(p; \tau^2)$	the noncentral chi-square distribution with p d.f. and non-centrality parameter τ^2
$F(m, n)$	the F-distribution with (m, n) d.f.
$F(m, n; \tau^2)$	the noncentral F-distribution with (m, n) d.f. and noncetrality parameter τ^2
$\Lambda_p(m, n)$	the distribution of Wilks' lambda criterion of dimension p with (m, n) d.f.
$N_p(\boldsymbol{\mu}, \Sigma)$	p-variate normal distribution with mean vector $\boldsymbol{\mu}$ and covariance matrix Σ. $(N_1(\mu, \sigma^2) = N(\mu, \sigma^2))$
$W_p(\Sigma, n)$	Wishart distribution with n degrees of freedom and covariance matrix Σ
$ch_i(A)$	the ith largest characteristic root of A
ch.r.	characteristic root
LR	likelihood ratio, likelihood ratio test
ML, MLE	maximum likelihood, maximum likelihood estimator
PM	probability of misdiscrimination
MANOVA	multivariate analysis of variance
$\\|\boldsymbol{x}\\|$	the Eucledian norm of the vector $\boldsymbol{x} = (x_1, \ldots, x_p)'$: $\sqrt{\sum_{i=1}^p x_i^2}$
$\\|\mathbf{X}\\|$	maximum absolute value of latent roots of matrix $\mathbf{X}$
$\\|f(\boldsymbol{x})\\|_1$	the L_1-norm of real function f on R^p: $\int_{R^p} \|f(\boldsymbol{x})\| \, d\boldsymbol{x}$

Chapter 1

Multivariate Normal and Related Distributions

In multivariate analysis we treat ordered sets (random vectors) of several random variables as a whole, rather than individual varaibles separately. First we introduce some basic concepts that describe the behavior of random vectors together with their properties: moments, cumulants, disrtibution functions, density functions, and characteristic functions.

The uniqueness theorem of characteristic function and the continuity theorem are given without proofs, and then the multivariate normal, spherical, and elliptical distributions are defined. They are important in studying asymptotic expansions of various statistics in non-normal models.

1.1 Random Vectors

1.1.1 Mean Vector and Covariance Matrix

Multivariate analysis deals with issues related to the observation of correlated random variables on units of a selected random sample. We denote a set of p random variables $X_1, \ldots, X_p$ by a vector $\boldsymbol{X} = (X_1, \ldots, X_p)'$, which is called a *random vector*. The mean or expectation of $\boldsymbol{X}$ is defined to be the vector of expectations:

$$\mathrm{E}(\boldsymbol{X}) = \begin{pmatrix} \mathrm{E}(X_1) \\ \vdots \\ \mathrm{E}(X_p) \end{pmatrix}.$$

A typical set of multivariate random samples, $\{\boldsymbol{X}_1, \ldots, \boldsymbol{X}_n\}$, arises from taking measurements on a $p \times 1$ random vector $\boldsymbol{X}$ for each of n objects or people. It is convenient to express these observation vectors in matrix form,

$$\mathbf{X} = (\boldsymbol{X}_1, \ldots, \boldsymbol{X}_n)' = \begin{pmatrix} \boldsymbol{X}_1' \\ \boldsymbol{X}_2' \\ \vdots \\ \boldsymbol{X}_n' \end{pmatrix} = \begin{pmatrix} X_{11} & X_{12} & \cdots & X_{1p} \\ X_{21} & X_{22} & \cdots & X_{2p} \\ \vdots & \vdots & & \vdots \\ X_{n1} & X_{n2} & \cdots & X_{np} \end{pmatrix},$$

where the transpose of a vector $\boldsymbol{a}$ and a matrix A is denoted by $\boldsymbol{a}'$ and A$'$, respectively.

Let $\mathbf{X}$ be a matrix of random variables, which we call a *random matrix.* Here the rows of $\mathbf{X}$ may or may not be random observations of $\boldsymbol{X}$. More generally, the expectation of a random matrix $\mathbf{X} = (X_{ij})$ is defined by the matrix whose (i, j)th element is $\mathrm{E}(X_{ij})$: namely, $\mathrm{E}(\mathbf{X}) = (\mathrm{E}(X_{ij}))$.

Theorem 1.1.1 *Let* $\mathbf{X} = (X_{ij})$ *and* $\mathbf{Y} = (Y_{ij})$ *be* $n \times p$ *random matrices. If* A, B, *and* C *are* $m \times n, p \times q$, *and* $m \times q$ *matrices of constants, then:*

(1) $\mathrm{E}(\mathbf{X} + \mathbf{Y}) = \mathrm{E}(\mathbf{X}) + \mathrm{E}(\mathbf{Y})$.

(2) $\mathrm{E}(\mathrm{A}\mathbf{X}\mathrm{B} + \mathrm{C}) = \mathrm{A}\mathrm{E}(\mathbf{X})\mathrm{B} + \mathrm{C}$.

Proof. The (i, j)th element of the left-hand side of (1) can be expressed as

$$\mathrm{E}(X_{ij} + Y_{ij}) = \mathrm{E}(X_{ij}) + \mathrm{E}(Y_{ij}),$$

which is equal to the (i, j)th element of the right-hand side of (1). Similarly, the (i, j)th element of the left-hand side of (2) can be expressed as

$$\mathrm{E}\left(\sum_{k=1}^{n}\sum_{\ell=1}^{p} a_{ik}X_{k\ell}b_{\ell j} + c_{ij}\right) = \sum_{k=1}^{n}\sum_{\ell=1}^{p} a_{ik}\mathrm{E}(X_{k\ell})b_{\ell j} + c_{ij},$$

which is equal to the (i, j)th element of the right-hand side of (2). □

Definition 1.1.1 If a $p \times 1$ random vector $\boldsymbol{X} = (X_1, \ldots, X_p)'$ has mean $\boldsymbol{\mu} = (\mu_1, \ldots, \mu_p)'$, the covariance matrix of $\boldsymbol{X}$ is defined by

$$\Sigma \equiv \mathrm{Var}(\boldsymbol{X}) = \mathrm{E}[(\boldsymbol{X} - \boldsymbol{\mu})(\boldsymbol{X} - \boldsymbol{\mu})'].$$

Furthermore, if a $q \times 1$ random vector $\boldsymbol{Y} = (Y_1, \ldots, Y_q)'$ has mean $\boldsymbol{\eta} = (\eta_1, \ldots, \eta_p)'$, the covariance matrix of $\boldsymbol{X}$ and $\boldsymbol{Y}$ is defined by

$$\mathrm{Cov}(\boldsymbol{X}, \boldsymbol{Y}) = \mathrm{E}[(\boldsymbol{X} - \boldsymbol{\mu})(\boldsymbol{Y} - \boldsymbol{\eta})'].$$

In particular, $\mathrm{Cov}(\boldsymbol{X}, \boldsymbol{X}) = \mathrm{Var}(\boldsymbol{X})$.

The (i, j)th element of Σ is

$$\begin{aligned}\sigma_{ij} &= \mathrm{E}[(X_i - \mu_i)(X_j - \mu_j)] \\ &= \mathrm{Cov}(X_i, X_j),\end{aligned}$$

the covariance between X_i and X_j, and the (i, i)th element is

$$\begin{aligned}\sigma_{ii} &= \mathrm{E}[(X_i - \mu_i)^2] \\ &= \mathrm{Var}(X_i),\end{aligned}$$

the variance of X_i. The variance is denoted by $\sigma_i^2 = \sigma_{ii}$ with $\sigma_i \geq 0$. The σ_i is called the *standardized deviation.* The covariance matrix of $\boldsymbol{X}$ is symmetric (i.e., $\Sigma' = \Sigma$) and positive semidefinite, as in Theorem 1.1.2.

Theorem 1.1.2 *Let Σ be the covariance matrix of a $p \times 1$ random vector.*

(1) *Σ is positive semidefinite (nonnegative definite), that is, for any $p \times 1$ fixed vector $\boldsymbol{a} = (a_1, \ldots, a_p)'$,*

$$\boldsymbol{a}'\Sigma\boldsymbol{a} = \sum_{i=1}^{p}\sum_{j=1}^{p} \sigma_{ij} a_i a_j \geq 0.$$

(2) *Let B be a $q \times p$ constant matrix and $\boldsymbol{b}$ be a $q \times 1$ constant vector. Then the covariance matrix of $\boldsymbol{Y} = B\boldsymbol{X} + \boldsymbol{b}$ is*

$$\mathrm{Var}(\boldsymbol{Y}) = B\Sigma B'.$$

Proof. (1) For any constant vector $\boldsymbol{a}$, we have

$$\begin{aligned}\mathrm{Var}(\boldsymbol{a}'\boldsymbol{X}) &= \mathrm{E}[(\boldsymbol{a}'\boldsymbol{X} - \boldsymbol{a}'\boldsymbol{\mu})^2] \\ &= \mathrm{E}[\{\boldsymbol{a}'(\boldsymbol{X} - \boldsymbol{\mu})\}^2] \\ &= \mathrm{E}[\boldsymbol{a}'(\boldsymbol{X} - \boldsymbol{\mu}) \cdot (\boldsymbol{X} - \boldsymbol{\mu})'\boldsymbol{a}] = \boldsymbol{a}'\Sigma\boldsymbol{a}.\end{aligned}$$

The result required follows from $\mathrm{Var}(\boldsymbol{a}'\boldsymbol{X}) \geq 0$.

(2) Using $\boldsymbol{Y} - \mathrm{E}(\boldsymbol{Y}) = B(\boldsymbol{X} - \boldsymbol{\mu})$, we have

$$\mathrm{Var}(\boldsymbol{Y}) = \mathrm{E}[B(\boldsymbol{X} - \boldsymbol{\mu})\{B(\boldsymbol{X} - \boldsymbol{\mu})\}'] = B\Sigma B'. \qquad \square$$

The covariance matrix is said to be *positive definite* if

$$\boldsymbol{a}'\Sigma\boldsymbol{a} > 0 \quad \text{for any constant vector } \boldsymbol{a} \neq \boldsymbol{0}.$$

In general, the covariance matrix Σ is positive semidefinite. Therefore, it follows from [A.1.8] that the characteristic roots of Σ are nonnegative, and these are denoted by

$$\lambda_1 \geq \cdots \geq \lambda_p \geq 0.$$

For $i = 1, \ldots, p$, let γ_i be the characteristic vector corresponding to λ_i. We may assume without loss of generality that they are orthonormal: that is,

$$\gamma_i'\gamma_j = \begin{cases} 1 & \text{if} \quad i = j \\ 0 & \text{if} \quad i \neq j. \end{cases}$$

The characteristic roots and vectors satisfy

$$\Sigma\gamma_i = \lambda_i\gamma_i, \quad i = 1, \ldots, p,$$

which is expressed as

$$\Sigma\Gamma = \Gamma\Lambda, \tag{1.1.1}$$

where

$$\Lambda = \text{diag}\,(\lambda_1, \ldots, \lambda_p), \quad \Gamma = (\gamma_1, \ldots, \gamma_p).$$

Here the matrix Γ is normalized as $\Gamma'\Gamma = I_p$. Equation (1.1.1) implies that

$$\begin{aligned} \Sigma &= \Gamma\Lambda\Gamma' \\ &= \Gamma\Lambda^{1/2}\Lambda^{1/2}\Gamma' \\ &= \Gamma\Lambda^{1/2}\Gamma'\Gamma\Lambda^{1/2}\Gamma', \end{aligned}$$

where $\Lambda^{1/2} = \text{diag}\,(\sqrt{\lambda_1}, \ldots, \sqrt{\lambda_p})$. Defining $\Sigma^{1/2} = \Gamma\Lambda^{1/2}\Gamma'$ and putting $C = \Gamma\Lambda^{1/2}$, we can write

$$\Sigma = (\Sigma^{1/2})^2 = CC'.$$

Note that such expressions are used for any positive semidefinite matrix.

The covariance matrix Σ contains the variance of the p variables and the covariances between all pairs of variables. It is desirable to have a single measure for the overall multivariate scatter. One such is the generalized variance, defined as the *determinant* of the covariance matrix. It is expressed as

$$|\Sigma| = |\Gamma\Lambda\Gamma'| = |\Lambda| = \lambda_1 \cdots \lambda_p.$$

The other overall measure is

$$\begin{aligned} \text{tr}\ \Sigma &= \sigma_{11} + \cdots + \sigma_{pp} \\ &= \text{tr}\,\Gamma\Lambda\Gamma' = \text{tr}\,\Lambda\Gamma'\Gamma \\ &= \text{tr}\,\Lambda = \lambda_1 + \cdots + \lambda_p, \end{aligned}$$

which is called the *total variance.*

1.1.2 Characteristic Function and Distribution

Let $\boldsymbol{X}$ be a p-dimensional random vector. Suppose that the probability of the random point falling in any (measurable) set E in the p-dimensional Euclidean space R^p is expressed as

$$\Pr\{\boldsymbol{X} \in E\} = \int_E f(\boldsymbol{x})\,\mathrm{d}\boldsymbol{x},$$

where $\mathrm{d}\boldsymbol{x} = \mathrm{d}x_1 \cdots \mathrm{d}x_p$. Then the function $f(\boldsymbol{x})$ is called the *probability density function*, or simply, the *density* of $\boldsymbol{X}$. The characteristic function of $\boldsymbol{X}$ is defined by

$$C(\boldsymbol{t}) = \mathrm{E}[\mathrm{e}^{i\boldsymbol{t}'\boldsymbol{X}}],$$

where $i = \sqrt{-1}$, $\boldsymbol{t} = (t_1, \ldots, t_p)'$, and $-\infty < t_j < \infty, j = 1, \ldots, p$. It is well known that there exists, one-to-one correspondence between the distribution of $\boldsymbol{X}$ and its characteristic function. In particular, we have the following theorem.

Theorem 1.1.3 *If the $p \times 1$ random vector $\boldsymbol{X}$ has the density function $f(\boldsymbol{x})$ and the characteristic function $C(\boldsymbol{t})$, then*

$$f(\boldsymbol{x}) = \frac{1}{(2\pi)^p} \int_{-\infty}^{\infty} \cdots \int_{-\infty}^{\infty} \mathrm{e}^{-i\boldsymbol{t}'\boldsymbol{x}} C(\boldsymbol{t})\,\mathrm{d}t_1 \cdots \mathrm{d}t_p.$$

This shows that the characteristic function determines the density function uniquely. For proofs of this theorem and Theorems 1.1.4 and 1.1.5, see, for example, Lukacs (1970).

The characteristic function is also used to obtain various moments of $\boldsymbol{X}$. In fact, interchanging the integral formally with the differential, we have

$$\frac{\partial^m}{\partial t_1^{m_1} \cdots \partial t_p^{m_p}} C(\boldsymbol{t}) = \mathrm{E}\big[(iX_1)^{m_1} \cdots (iX_p)^{m_p} \mathrm{e}^{i\boldsymbol{t}'\boldsymbol{X}}\big],$$

where $m = m_1 + \cdots + m_p$, and putting $\boldsymbol{t} = \boldsymbol{0}$, we can expect to get the moment $\mathrm{E}[X_1^{m_1} \cdots X_p^{m_p}]$. The result is stated in the following theorem.

Theorem 1.1.4 *Suppose that $\mathrm{E}\big[|X_1|^{m_1} \cdots |X_p|^{m_p}\big]$ is finite; then the characteristic function is $(m_1, \ldots, m_p)$th times continuously differentiable, and*

$$\left.\frac{\partial^m}{\partial t_1^{m_1} \cdots \partial t_p^{m_p}} C(\boldsymbol{t})\right|_{\boldsymbol{t} = \boldsymbol{0}} = i^m \mathrm{E}\big[X_1^{m_1} \cdots X_p^{m_p}\big].$$

The characteristic function is useful for obtaining asymptotic distribution of a statistic when a parameter, typically the sample size, is large. The following result is known as the *continuity theorem*.

Theorem 1.1.5 *Let $\{F_n(\boldsymbol{x})\}$ be a sequence of distribution functions, and let $\{C_n(\boldsymbol{t})\}$ be the sequence of corresponding characteristic functions. A necessary and sufficient condition for $F_n(\boldsymbol{x})$ to converge to a distribution function $F(\boldsymbol{x})$ is that for every $\boldsymbol{t}$, $C_n(\boldsymbol{t})$ converges to a limit $C(\boldsymbol{t})$ that is continuous at $\boldsymbol{t} = \boldsymbol{0}$. When this condition is satisfied, the limit $C(\boldsymbol{t})$ is identical with the characteristic function of the limiting distribution function $F(\boldsymbol{x})$.*

In the distribution theory of multivariate statistics, the variable transformation method is often used, in addition to the characteristic function method. The method is based on the following well-known theorem.

Theorem 1.1.6 *Let $\boldsymbol{X}$ be a $p \times 1$ random vector having a density function $f(\boldsymbol{x})$ that is positive on a set $\mathcal{S} \subset R^p$. Suppose that the transformation $\boldsymbol{y} = \boldsymbol{y}(\boldsymbol{x}) = (y_1(\boldsymbol{x}), \ldots, y_p(\boldsymbol{x}))'$ is one-to-one of $\mathcal{S}$ onto $\mathcal{T}$, where $\mathcal{T}$ denotes the image of $\mathcal{S}$ under $\boldsymbol{y}$. Let $\boldsymbol{x} = \boldsymbol{x}(\boldsymbol{y})$ be the inverse transformation. Assume that partial derivatives $\partial x_i / \partial y_j$ $(i, j = 1, \ldots, p)$ exist and are continuous on $\mathcal{T}$. Then the density function of the random vector $\boldsymbol{Y} = \boldsymbol{y}(\boldsymbol{X})$ is given by*

$$g(\boldsymbol{y}) = f(\boldsymbol{x}(\boldsymbol{y}))|J(\boldsymbol{y} \to \boldsymbol{x})|,$$

where $J(\boldsymbol{y} \to \boldsymbol{x})$ is a determinant of the Jacobian matrix of the transformation from $\boldsymbol{y}$ to $\boldsymbol{x}$:

$$J(\boldsymbol{y} \to \boldsymbol{x}) = \begin{vmatrix} \dfrac{\partial x_1}{\partial y_1} & \cdots & \dfrac{\partial x_1}{\partial y_p} \\ \vdots & & \vdots \\ \dfrac{\partial x_p}{\partial y_1} & \cdots & \dfrac{\partial x_p}{\partial y_p} \end{vmatrix}.$$

1.2 Multivariate Normal Distribution

1.2.1 Bivariate Normal Distribution

Let X be distributed as a normal distribution with mean μ and variance σ^2 [i.e., $\mathrm{N}(\mu, \sigma^2)$]. Then, the probability of the event $a < X \le b$ is expressed as

$$\Pr\{a < X \le b\} = \int_a^b f(x)\,\mathrm{d}x,$$

where

$$f(x) = \frac{1}{\sqrt{2\pi}\sigma}\,\mathrm{e}^{-(x-\mu)^2/2\sigma^2}.$$

Here $-\infty < \mu < \infty$ and $\sigma > 0$. The function $f(x)$ is called the *probability density function (pdf)*, or simply the *density* of X, and is often denoted in full by $f(x\colon \mu, \sigma^2)$.

The distribution of two-dimensional variables (X_1, X_2) is defined by the density function $f(x_1, x_2)$, satisfying

$$\Pr\{a < X_1 \le b,\ c < X_2 \le d\} = \int_a^b \int_c^d f(x_1, x_2)\mathrm{d}x_1\,\mathrm{d}x_2$$

for any $a < b$ and $c < d$. For the bivariate normal distribution, its density function is defined by

$$f(x_1, x_2) = \frac{1}{2\pi\sigma_1\sigma_2\sqrt{1-\rho^2}} \exp\Big\{-\frac{1}{2(1-\rho^2)}Q\Big\}, \qquad (1.2.1)$$

where

$$Q(x_1, x_2) = \Big(\frac{x_1-\mu_1}{\sigma_1}\Big)^2 - 2\rho\frac{(x_1-\mu_1)(x_2-\mu_2)}{\sigma_1\sigma_2} + \Big(\frac{x_2-\mu_2}{\sigma_2}\Big)^2.$$

If (X_1, X_2) follows the bivariate normal distribution with density function (1.2.1), the mean and variance of X_j are μ_j and σ_j^2, respectively. The covariance and correlation between X_1 and X_2 are $\rho\sigma_1\sigma_2$ and ρ, respectively. The covariance matrix is

$$\Sigma = \begin{pmatrix} \sigma_1^2 & \rho\sigma_1\sigma_2 \\ \rho\sigma_1\sigma_2 & \sigma_2^2 \end{pmatrix}.$$

Furthermore, each of the distributions of X_1 and X_2 is normal. The conditional distribution of X_2 given $X_1 = x_1$ is also normal. The conditional means and conditional variance are

$$\mu_2 + \rho(\sigma_2/\sigma_1)\cdot(x_1-\mu_1) \quad \text{and} \quad \sigma_2^2(1-\rho^2),$$

respectively.

These properties are summarized as follows:

(1) $\mathrm{E}(X_1) = \mu_1,\ \mathrm{Var}(X_1) = \sigma_1^2,$
$\mathrm{E}(X_2) = \mu_1,\ \mathrm{Var}(X_2) = \sigma_2^2.$

(2) $X_1 \sim \mathrm{N}(\mu_1, \sigma_1^2), \quad X_2 \sim \mathrm{N}(\mu_2, \sigma_2^2).$

(3) $\mathrm{Cov}(X_1, X_2) = \sigma_1\sigma_2\rho,\ \mathrm{Corr}(X_1, X_2) = \rho.$

(4) $\mathrm{E}(X_2|X_1 = x_1) = \mu_2 + \rho(\sigma_2/\sigma_1)(x_1-\mu_1),$
$\mathrm{E}(X_1|X_2 = x_2) = \mu_1 + \rho(\sigma_1/\sigma_2)(x_2-\mu_2).$

(5) $\mathrm{Var}(X_2|X_1 = x_1) = \sigma_2^2(1-\rho^2), \quad \mathrm{Var}(X_1|X_2 = x_2) = \sigma_1^2(1-\rho^2).$

(6) $X_2|X_1 = x_1 \sim \mathrm{N}(\mu_2 + \rho(\sigma_2/\sigma_1)(x_1-\mu_1), \sigma_2^2(1-\rho^2)),$
$X_1|X_2 = x_2 \sim \mathrm{N}(\mu_1 + \rho(\sigma_1/\sigma_2)(x_2-\mu_2), \sigma_1^2(1-\rho^2)).$

For the derivation of these properties, it is essential that the exponent Q in the density function of the *bivariate normal distribution* can be put in the following quadratic form:

$$Q = (1-\rho^2)\Big(\frac{x_1-\mu_1}{\sigma_1}\Big)^2 + \Big\{\Big(\frac{x_2-\mu_2}{\sigma_2}\Big) - \rho\Big(\frac{x_1-\mu_1}{\sigma_1}\Big)\Big\}^2,$$

and hence the probability density function is expressed as

$$\begin{aligned} f(x_1,x_2) &= \frac{1}{\sqrt{2\pi}\,\sigma_1}\exp\Big\{-\frac{1}{2\sigma_1^2}\Big(\frac{x_1-\mu_1}{\sigma_1}\Big)^2\Big\} \\ &= \frac{1}{\sqrt{2\pi}\,\sigma_2\sqrt{1-\rho^2}}\exp\Big[\frac{-1}{2\sigma_2^2\sqrt{1-\rho^2}}\Big\{x_2-\mu_2-\rho\frac{\sigma_2}{\sigma_1}(x_1-\mu_1)\Big\}^2\Big]. \end{aligned}$$

Properties (1) through (6) can all be shown by using the decomposition above.

1.2.2 Definitions

The density function of a random variable Z with the standard normal distribution $\mathrm{N}(0,1)$ is

$$\varphi(z) \equiv \frac{1}{\sqrt{2\pi}}\mathrm{e}^{-z^2/2}, \quad -\infty < z < \infty. \tag{1.2.2}$$

A random variable X with a general normal distribution with mean μ and variance σ^2 is obtained by the linear transformation

$$X = \sigma Z + \mu.$$

In fact, starting from (1.2.2) we can show that the density function of X is given by

$$f(x;\mu,\sigma^2) = \frac{1}{\sqrt{2\pi}\,\sigma}\mathrm{e}^{-(x-\mu)^2/2\sigma^2}.$$

The approach is generalized as follows. The probability density function of $\boldsymbol{Z} = (Z_1,\ldots,Z_p)'$, where $Z_1,\ldots,Z_p$ are independent and identically distributed $\mathrm{N}(0,1)$, is given by

$$\prod_{i=1}^{p}\frac{1}{\sqrt{2\pi}}\mathrm{e}^{-z_i^2/2} = \Big(\frac{1}{\sqrt{2\pi}}\Big)^p\mathrm{e}^{-\boldsymbol{z}'\boldsymbol{z}/2}. \tag{1.2.3}$$

Consider the transformation

$$\boldsymbol{X} = \Sigma^{1/2}\boldsymbol{Z} + \boldsymbol{\mu}.$$

Then it is easy to see that

$$\mathrm{E}(\boldsymbol{X}) = \boldsymbol{\mu} \quad \text{and} \quad \mathrm{Cov}(\boldsymbol{X}) = \Sigma.$$

Furthermore, the Jacobian of transformation from $\boldsymbol{X}$ to $\boldsymbol{Z}$ is (see [A.3.1])

$$\frac{\partial \boldsymbol{z}}{\partial \boldsymbol{x}} = |\Sigma^{-1/2}| = |\Sigma|^{-1/2},$$

and the density function of $\boldsymbol{X}$ is given by

$$f(\boldsymbol{x}; \boldsymbol{\mu}, \Sigma) = (2\pi)^{-p/2}|\Sigma|^{-1/2} \exp\Big\{ -\frac{1}{2}(\boldsymbol{x} - \boldsymbol{\mu})'\Sigma^{-1}(\boldsymbol{x} - \boldsymbol{\mu})\Big\}, \qquad (1.2.4)$$

where $\boldsymbol{\mu} = (\mu_1, \ldots, \mu_p)'$, $-\infty < \mu_j < \infty$ and $\Sigma > 0$.

Definition 1.2.1 A p-dimensional random $\boldsymbol{X}$ is said to have a *p-variate normal distribution* with mean vector $\boldsymbol{\mu}$ and covariance matrix Σ if its density function is given by (1.2.4).

Theorem 1.2.1 *Let $\boldsymbol{\mu}$ be a p-dimensional fixed vector and Σ be a $p \times p$ positive definite matrix. Then the following two statements are equivalent.*

(1) $\boldsymbol{X} \sim \mathrm{N}_p(\boldsymbol{\mu}, \Sigma)$.

(2) $\boldsymbol{Z} \equiv \Sigma^{-1/2}(\boldsymbol{x} - \boldsymbol{\mu}) \sim \mathrm{N}_p(\boldsymbol{0}, \mathrm{I}_p)$.

Proof. If $f(\boldsymbol{x}; \boldsymbol{\mu}, \Sigma)$ is the probability density function of $\boldsymbol{X}$, the transformation $\boldsymbol{X} \to \boldsymbol{Z}$ gives

$$f(\boldsymbol{x}; \boldsymbol{\mu}, \Sigma) = |\Sigma|^{-1/2} f(\boldsymbol{z}; \boldsymbol{0}, \mathrm{I}_p).$$

with the Jacobian equal to $|\Sigma|^{-1/2}$, which means (1) $\Longrightarrow$ (2). A similar argument can apply to proving (2) $\Longrightarrow$ (1). □

Suppose that $\boldsymbol{X} \sim \mathrm{N}_p(\boldsymbol{\mu}, \Sigma)$. If Σ is singular, its density function is not defined. On the other hand, its characteristic function (see Theorem 1.2.2)

$$C_{\boldsymbol{X}}(\boldsymbol{t}) = i\boldsymbol{t}'\boldsymbol{\mu} - \frac{1}{2}\boldsymbol{t}'\Sigma\boldsymbol{t} \qquad (1.2.5)$$

is still valid. To have a natural definition of a singular normal distribution, consider the following decomposition of Σ with rank r by [A.1.11]:

$$\Sigma = \Gamma \begin{pmatrix} \mathrm{D}_\lambda & \mathrm{O} \\ \mathrm{O} & \mathrm{O} \end{pmatrix} \Gamma', \quad \Gamma = (\Gamma_1, \Gamma_2),$$

where Γ is a $p \times p$ orthogonal matrix, Γ_1 is an $a \times r$ matrix, and $\mathrm{D}_\lambda = \mathrm{diag}\,(\lambda_1, \ldots, \lambda_r)$, $\lambda_j > 0$, $j = 1, \ldots, r$. Putting $\mathrm{B} = \Gamma_1 \mathrm{D}_\lambda^{1/2}$, we have $\Sigma = \mathrm{BB}'$.

Note that B is not unique, since B can be replaced by BL, where L is any $r \times r$ orthogonal matrix. Using one such matrix, B, and $\boldsymbol{Z} \sim \mathrm{N}_r(\mathbf{0}, \mathrm{I}_r)$, consider

$$\tilde{\boldsymbol{X}} = \mathrm{B}\boldsymbol{Z} + \boldsymbol{\mu}. \tag{1.2.6}$$

Then the characteristic function of $\tilde{\boldsymbol{X}}$ is

$$\begin{aligned} C_{\tilde{\boldsymbol{X}}}(\boldsymbol{t}) &= \mathrm{E}[\exp\{i\boldsymbol{t}'(\mathrm{B}\boldsymbol{Z} + \boldsymbol{\mu})\}] \\ &= \exp\left\{i\boldsymbol{t}'\boldsymbol{\mu} - \frac{1}{2}\boldsymbol{t}'(\mathrm{BB}')\boldsymbol{t}\right\} = C_{\boldsymbol{X}}(\boldsymbol{t}). \end{aligned}$$

From this result a general p-variate normal distribution with mean $\boldsymbol{\mu}$ and covariance matrix Σ with rank $r \le p$ is defined as the distribution of $\mathrm{B}\boldsymbol{Z} + \boldsymbol{\mu}$, where $\Sigma = \mathrm{BB}'$ and $\boldsymbol{Z} \sim \mathrm{N}_r(\mathbf{0}, \mathrm{I}_r)$. In this book it is assumed that the covariance matrix of a multivariate normal variable is nonsingular. However, we note that most properties are also valid for *singular normal distributions*.

1.2.3 Some Properties

Theorem 1.2.2 *Let* $\boldsymbol{X}$ *be distributed as* $\mathrm{N}_p(\boldsymbol{\mu}, \Sigma)$. *Then:*

(1) $\mathrm{E}(\boldsymbol{X}) = \boldsymbol{\mu}, \mathrm{Var}(\boldsymbol{X}) = \Sigma$.

(2) *The characteristic function of* $\boldsymbol{X}$ *is*

$$C_{\boldsymbol{X}}(\boldsymbol{t}) = \exp\left(i\boldsymbol{\mu}'\boldsymbol{t} - \frac{1}{2}\boldsymbol{t}'\Sigma\boldsymbol{t}\right), \tag{1.2.7}$$

where $\boldsymbol{t} = (t_1, \ldots, t_p)'$ *and* t_j *is a real number.*

(3) *Let* B *be a* $q \times p$ *constant matrix with* $\mathrm{rank}(\mathrm{B}) = q$, *and* $\boldsymbol{b}$ *a* q *constant vector. Then* $\boldsymbol{Y} = \mathrm{B}\boldsymbol{X} + \boldsymbol{b} \sim \mathrm{N}_q(\mathrm{B}\boldsymbol{\mu} + \boldsymbol{b}, \mathrm{B}\Sigma\mathrm{B}')$.

Proof. From Theorem 1.2.1 we can write

$$\boldsymbol{X} = \Sigma^{1/2}\boldsymbol{Z} + \boldsymbol{\mu},$$

where $\boldsymbol{Z} = (Z_1, \ldots, Z_p)'$ and $Z_1, \ldots, Z_p \sim$ *i.i.d.* $\mathrm{N}(0, 1)$. Using this fact, it is easy to show (1). Writing $\boldsymbol{d} = (d_1, \ldots, d_p)' = \Sigma^{1/2}\boldsymbol{t}$, the characteristic function of $\boldsymbol{X}$ can be put in the form

$$\begin{aligned} C_{\boldsymbol{X}}(\boldsymbol{t}) &= \mathrm{E}[\mathrm{e}^{i\boldsymbol{t}'(\Sigma^{1/2}\boldsymbol{Z} + \boldsymbol{\mu})}] = \mathrm{e}^{i\boldsymbol{t}'\boldsymbol{\mu}}\mathrm{E}[\mathrm{e}^{i\boldsymbol{t}'\Sigma^{1/2}\boldsymbol{Z}}] = \mathrm{e}^{i\boldsymbol{t}'\boldsymbol{\mu}}\prod_{j=1}^{p}\mathrm{E}[\mathrm{e}^{id_jZ_j}] \\ &= \exp\left(i\boldsymbol{t}'\boldsymbol{\mu} - \frac{1}{2}\sum_{j=1}^{p}d_j^2\right), \end{aligned}$$

which coincides with (1.2.7). For a proof of (3), consider the characteristic function of $\boldsymbol{Y} = \mathrm{B}\boldsymbol{X} + \boldsymbol{b}$ given by

$$\begin{aligned}\mathrm{E}\big[\exp\{i\boldsymbol{t}'(\mathrm{B}\boldsymbol{X} + \boldsymbol{b})\}\big] &= \exp(i\boldsymbol{t}'\boldsymbol{b})\exp\{i(\mathrm{B}'\boldsymbol{t})'\boldsymbol{\mu} - \frac{1}{2}(\mathrm{B}'\boldsymbol{t})'\Sigma(\mathrm{B}'\boldsymbol{t})\} \\ &= \exp\{i\boldsymbol{t}'(\mathrm{B}\boldsymbol{\mu} + \boldsymbol{b}) - \frac{1}{2}\boldsymbol{t}'\mathrm{B}\Sigma\mathrm{B}'\boldsymbol{t}\}.\end{aligned}$$

Result (3) follows from (2) and the fact that the distribution is determined (see Theorem 1.1.3) by its characteristic function. □

Theorem 1.2.3 *For $i = 1, \ldots, n$, let $\boldsymbol{X}_i$ be a sequence of mutually independent random vectors following $\mathrm{N}_p(\boldsymbol{\mu}_i, \Sigma)$. Then, for any constant vector $\boldsymbol{a} = (a_1, \ldots, a_n)'$,*

$$\boldsymbol{Y} = a_1\boldsymbol{X}_1 + \cdots + a_n\boldsymbol{X}_n \sim \mathrm{N}_p(\boldsymbol{\eta}, \boldsymbol{a}'\boldsymbol{a}\Sigma),$$

where $\boldsymbol{\eta} = a_1\boldsymbol{\mu}_1 + \cdots + a_n\boldsymbol{\mu}_n$. In particular,

$$\bar{\boldsymbol{X}} = \frac{1}{n}\sum_{i=1}^{n}\boldsymbol{X}_i \sim \mathrm{N}_p(\boldsymbol{\mu}, \frac{1}{n}\Sigma).$$

Proof. Let $C_{\boldsymbol{X}_j}(\boldsymbol{t})$ be the characteristic function of $\boldsymbol{X}_j$. Then, using Theorem 1.2.2(2), the characteristic function of $\boldsymbol{Y}$ can be expressed as

$$\prod_{j=1}^{n} C_{\boldsymbol{X}_j}(a_j\boldsymbol{t}) = \exp\big(i\boldsymbol{\eta}'\boldsymbol{t} - \frac{1}{2}\boldsymbol{a}'\boldsymbol{a}\boldsymbol{t}'\Sigma\boldsymbol{t}\big),$$

which proves the result required. □

The central moments of $\boldsymbol{X}$ can be obtained by using Theorem 1.1.4 and the fact that the characteristic function of $\boldsymbol{X} - \boldsymbol{\mu}$ is $\exp(-\frac{1}{2}\boldsymbol{t}'\Sigma\boldsymbol{t})$. Some moments are given in the following theorem.

Theorem 1.2.4 *When $\boldsymbol{X} \sim \mathrm{N}_p(\boldsymbol{\mu}, \Sigma)$:*

(1) $\mathrm{E}[(X_i - \mu_i)(X_j - \mu_j)] = \sigma_{ij}$.

(2) $\mathrm{E}[(X_i - \mu_i)(X_j - \mu_j)(X_k - \mu_k)(X_\ell - \mu_\ell)] = \sigma_{ij}\sigma_{k\ell} + \sigma_{ik}\sigma_{j\ell} + \sigma_{i\ell}\sigma_{jk}$.

(3) $\mathrm{E}[(X_i - \mu_i)(X_j - \mu_j)(X_k - \mu_k)(X_\ell - \mu_\ell)(X_s - \mu_s)(X_t - \mu_t)]$

$$\begin{aligned}&= \sigma_{ij}\sigma_{k\ell}\sigma_{st} + \sigma_{ij}\sigma_{ks}\sigma_{\ell t} + \sigma_{ij}\sigma_{kt}\sigma_{\ell s} \\ &\quad + \sigma_{ik}\sigma_{j\ell}\sigma_{st} + \sigma_{ik}\sigma_{js}\sigma_{\ell t} + \sigma_{ik}\sigma_{jt}\sigma_{\ell s} \\ &\quad + \sigma_{i\ell}\sigma_{jk}\sigma_{st} + \sigma_{i\ell}\sigma_{js}\sigma_{kt} + \sigma_{i\ell}\sigma_{jt}\sigma_{kt} \\ &\quad + \sigma_{is}\sigma_{jk}\sigma_{\ell t} + \sigma_{is}\sigma_{j\ell}\sigma_{kt} + \sigma_{is}\sigma_{jt}\sigma_{k\ell} \\ &\quad + \sigma_{it}\sigma_{jk}\sigma_{\ell s} + \sigma_{it}\sigma_{j\ell}\sigma_{ks} + \sigma_{it}\sigma_{js}\sigma_{k\ell}.\end{aligned}$$

(4) *Every central moment of odd order is* 0.

Theorem 1.2.5 *Let* $\boldsymbol{X}$ *be* $\mathrm{N}_p(\boldsymbol{\mu}, \Sigma)$ *and decompose* $\boldsymbol{X}$, $\boldsymbol{\mu}$, *and* Σ *as*

$$\boldsymbol{X} = \begin{pmatrix} \boldsymbol{X}_1 \\ \boldsymbol{X}_2 \end{pmatrix}, \quad \boldsymbol{\mu} = \begin{pmatrix} \boldsymbol{\mu}_1 \\ \boldsymbol{\mu}_2 \end{pmatrix}, \quad \Sigma = \begin{pmatrix} \Sigma_{11} & \Sigma_{12} \\ \Sigma_{21} & \Sigma_{22} \end{pmatrix},$$

where $\boldsymbol{X}_i : p_i \times 1$, $\boldsymbol{\mu}_i : p_i \times 1$, *and* $\Sigma_{ij} : p_i \times p_j$. *Then the conditional distribution of* $\boldsymbol{X}_2$ *given* $\boldsymbol{X}_1$ *is normal with mean* $\mathrm{E}(\boldsymbol{X}_2|\boldsymbol{X}_1)$ *and covariance matrix* $\mathrm{Var}(\boldsymbol{X}_2|\boldsymbol{X}_1)$, *where*

$$\begin{aligned} \mathrm{E}(\boldsymbol{X}_2|\boldsymbol{X}_1) &= \boldsymbol{\mu}_2 + \Sigma_{21}\Sigma_{11}^{-1}(\boldsymbol{X}_1 - \boldsymbol{\mu}_1), \\ \mathrm{Var}(\boldsymbol{X}_2|\boldsymbol{X}_1) &= \Sigma_{22} - \Sigma_{21}\Sigma_{11}^{-1}\Sigma_{12} \equiv \Sigma_{22\cdot 1}. \end{aligned}$$

Proof. From Theorem 1.2.2 (3)

$$\boldsymbol{X}_1 \sim \mathrm{N}(\boldsymbol{\mu}_1, \Sigma_{11}).$$

Put

$$\mathrm{M} = \begin{pmatrix} \mathrm{I}_{p_1} & \mathrm{O} \\ -\Sigma_{21}\Sigma_{11}^{-1} & \mathrm{I}_{p_2} \end{pmatrix}.$$

Then

$$\mathrm{M}\Sigma\mathrm{M}' = \begin{pmatrix} \Sigma_{11} & \mathrm{O} \\ \mathrm{O} & \Sigma_{22\cdot 1}. \end{pmatrix}.$$

and hence $|\Sigma| = |\Sigma_{11}| \cdot |\Sigma_{22\cdot 1}|$. Furthermore, from [A.1.2],

$$\Sigma^{-1} = \begin{pmatrix} \Sigma_{11}^{-1} & \mathrm{O} \\ \mathrm{O} & \mathrm{O} \end{pmatrix} + \begin{pmatrix} -\Sigma_{11}^{-1}\Sigma_{12} \\ \mathrm{I}_{p_2} \end{pmatrix} \Sigma_{22\cdot 1}^{-1} \Big(-\Sigma_{21}\Sigma_{11}^{-1} \ \mathrm{I}_{p_2} \Big).$$

These imply that the density function of $\mathrm{N}_p(\boldsymbol{\mu}, \Sigma)$ can be decomposed as

$$\begin{aligned} f_p(\boldsymbol{x}; \boldsymbol{\mu}, \Sigma) &= f_{p_1}(\boldsymbol{x}_1; \boldsymbol{\mu}_1, \Sigma_{11}) \\ &= \times f_{p_2}(\boldsymbol{x}_2; \boldsymbol{\mu}_2 + \Sigma_{21}\Sigma_{11}^{-1}(\boldsymbol{x}_1 - \boldsymbol{\mu}_1), \Sigma_{22\cdot 1}). \end{aligned}$$

Using the decomposition, we can get the result required. □

Let $\boldsymbol{X}_1, \ldots, \boldsymbol{X}_n$ be independent p-dimensional normal vectors with means $\boldsymbol{\mu}_1, \ldots, \boldsymbol{\mu}_n$ and the same covariance matrix Σ. Put $\mathbf{X} = (\boldsymbol{X}_1, \ldots, \boldsymbol{X}_n)'$, and consider the transformation

$$\mathbf{Y} = (\boldsymbol{Y}_1, \ldots, \boldsymbol{Y}_n)' = \mathrm{H}\mathbf{X}, \tag{1.2.8}$$

where H is an $n \times n$ orthogonal matrix. Then $\mathbf{Y}$ has the same properties as $\mathbf{X}$ except that the mean of $\mathbf{Y}$ is changed to $\mathrm{E}(\mathbf{Y}) = \mathrm{HE}(\mathbf{X})$. More precisely, the result is stated in the following theorem.

Theorem 1.2.6 *Let* $\mathbf{X} = (\boldsymbol{X}_1, \ldots, \boldsymbol{X}_n)'$ *be an* $n \times p$ *random matrix whose rows are independent normal variates with mean* $\mathrm{E}(\mathbf{X}) = \mathrm{M}$ *and the same covariance matrix* Σ. *Let* $\mathbf{Y}$ *be the random matrix in* (1.2.8) *obtained by an* $n \times n$ *orthogonal transformation* H. *Then the rows of* $\mathbf{Y}$ *are independent normal variates with mean* $\mathrm{E}(\mathbf{Y}) = \mathrm{HM}$ *and the same covariance matrix* Σ.

Proof. The density function of $\mathbf{X}$ is expressed as

$$\prod_{i=1}^{n}(2\pi)^{-p/2}|\Sigma|^{-1/2}\,\mathrm{etr}\{-\frac{1}{2}\Sigma^{-1}(\boldsymbol{x}_i - \boldsymbol{\mu}_i)(\boldsymbol{x}_i - \boldsymbol{\mu}_i)'\}$$
$$= (2\pi)^{-pn/2}|\Sigma|^{-n/2}\,\mathrm{etr}\{-\frac{1}{2}\Sigma^{-1}(\mathrm{X}-\mathrm{M})'(\mathrm{X}-\mathrm{M})\},$$

where $\mathrm{etr}\,\bullet$ represents $\exp(\mathrm{tr}\,\bullet)$ and $\mathrm{X} = (\boldsymbol{x}_1, \ldots, \boldsymbol{x}_n)'$ is an $n \times p$ matrix. From the transformation (1.2.8) we have $\mathrm{X} = \mathrm{H}'\mathrm{Y}$, or equivalently

$$\begin{pmatrix} \boldsymbol{x}_{(1)} \\ \vdots \\ \boldsymbol{x}_{(p)} \end{pmatrix} = \mathrm{Diag}\,(\mathrm{H}', \ldots, \mathrm{H}') \begin{pmatrix} \boldsymbol{y}_{(1)} \\ \vdots \\ \boldsymbol{y}_{(p)} \end{pmatrix}, \tag{1.2.9}$$

where, for $j = 1, \ldots, p$, $\boldsymbol{x}_{(j)}$ and $\boldsymbol{y}_{(j)}$ denote the jth column vector of X and Y, respectively, that is, $\boldsymbol{x}_{(j)} = (x_{1j}, \ldots, x_{nj})'$ and $\boldsymbol{y}_{(j)} = (y_{1j}, \ldots, y_{nj})'$. Note, in particular, that the both sides of (1.2.9) are np-vectors. Then the Jacobian of the transformation is

$$\frac{\partial \mathrm{X}}{\partial \mathrm{Y}} = |\mathrm{H}'|^n = \pm 1.$$

Therefore the density function of $\mathbf{Y}$ is given by

$$(2\pi)^{-pn/2}|\Sigma|^{-n/2}\,\mathrm{etr}\{-\frac{1}{2}\Sigma^{-1}(\mathrm{H}'\mathrm{Y}-\mathrm{M})'(\mathrm{H}'\mathrm{Y}-\mathrm{M})\}$$
$$= (2\pi)^{-pn/2}|\Sigma|^{-n/2}\,\mathrm{etr}\{-\frac{1}{2}\Sigma^{-1}(\mathrm{Y}-\mathrm{HM})'(\mathrm{Y}-\mathrm{HM})\},$$

since

$$\begin{aligned}(\mathrm{H}'\mathrm{Y}-\mathrm{M})'(\mathrm{H}'\mathrm{Y}-\mathrm{M}) &= \{\mathrm{H}'(\mathrm{Y}-\mathrm{HM})\}'\{\mathrm{H}'(\mathrm{Y}-\mathrm{HM})\}' \\ &= (\mathrm{Y}-\mathrm{HM})'\mathrm{H}\mathrm{H}'(\mathrm{Y}-\mathrm{HM}) \\ &= (\mathrm{Y}-\mathrm{HM})'(\mathrm{Y}-\mathrm{HM}).\end{aligned}$$

This gives the result required. □

Definition 1.2.2 The distribution of an $n \times p$ random matrix $\mathbf{X} = (\boldsymbol{X}_{(1)}, \ldots, \boldsymbol{X}_{(p)})$ is said to be normal if the random vector defined by

$$\mathrm{vec}(\mathbf{X}) \equiv \begin{pmatrix} \boldsymbol{X}_{(1)} \\ \vdots \\ \boldsymbol{X}_{(p)} \end{pmatrix}$$

follows an np-variate normal distribution. If this is the case we simply write $\mathbf{X} \sim \mathrm{N}_{n\times p}(\Xi, \Psi)$ with $\Xi = \mathrm{E}(\mathbf{X})$ and $\Psi = \mathrm{Var}(\mathrm{vec}(\mathbf{X}))$.

By using the notation of *matrix normal distribution*, Theorem 1.2.6 can be stated as follows:

$$\mathbf{X} \sim \mathrm{N}_{n\times p}(\Xi, \Sigma \otimes \mathrm{I}_n) \Longrightarrow \mathbf{Y} = \mathrm{H}\mathbf{X} \sim \mathrm{N}_{n\times p}(\mathrm{H}\Xi, \Sigma \otimes \mathrm{I}_n),$$

where H is an orthogonal matrix. Here $\otimes$ denotes the Kronecker or direct product; that is, for matrices $\mathrm{A} = (a_{ij})$ and B, $\mathrm{A} \otimes \mathrm{B} = (a_{ij}\mathrm{B})$.

Theorem 1.2.7 *Let* $\mathbf{X} = (\boldsymbol{X}_1, \ldots, \boldsymbol{X}_n)'$ *be a* n*-vector consisting of mutually independent* $\boldsymbol{X}_i$*'s such that* $\boldsymbol{X}_i \sim \mathrm{N}_p(\boldsymbol{\xi}_i, \Sigma)$. *Let* B *be a* $q \times n$ *matrix with rank* q. *Then the distribution of* $\mathbf{Y} = \mathrm{B}\mathbf{X}$ *is normal with* $\mathrm{E}(\mathbf{Y}) = \mathrm{B}\Xi$ *and*

$$\mathrm{Cov}(\boldsymbol{Y}_{(i)}, \boldsymbol{Y}_{(j)}) = \sigma_{ij}\mathrm{BB}',$$

where $\Xi = (\boldsymbol{\xi}_1, \ldots, \boldsymbol{\xi}_n)'$, $\mathbf{Y} = (\boldsymbol{Y}_{(1)}, \ldots, \boldsymbol{Y}_{(p)})$, *and* $\mathrm{V} = \mathrm{BB}' = (v_{ij})$.

Proof. The density function of $\mathbf{X}$ is expressed as

$$f(\mathrm{X}) = 2^{pn/2}|\Sigma|^{-n/2}\,\mathrm{etr}\Big\{-\frac{1}{2}\Sigma^{-1}(\mathrm{X}-\Xi)'(\mathrm{X}-\Xi)\Big\}.$$

Let $\mathrm{T} = (t_{ij})$ be a $q \times p$ real matrix. Then the characteristic function of $\mathbf{Y}$ is

$$C_{\mathbf{Y}}(\mathrm{T}) = \int_{\mathrm{X}} \mathrm{etr}(i\mathrm{T}'\mathrm{BX}) f(\mathrm{X})\,\mathrm{dX}.$$

Note that

$$\begin{aligned}\mathrm{tr}\,\Sigma^{-1}(\mathrm{X} &- \Xi)'(\mathrm{X}-\Xi) - 2i\,\mathrm{tr}\,\mathrm{T}'\mathrm{BX} \\ &= \mathrm{tr}\,\Sigma^{-1}\{\mathrm{X} - (\Xi + i\mathrm{B}'\mathrm{T}\Sigma)\}'\{\mathrm{X} - (\Xi + i\mathrm{B}'\mathrm{T}\Sigma)\} \\ &\qquad -2i\ \mathrm{tr}\,\mathrm{T}'\mathrm{B}\Xi + \mathrm{tr}\,\Sigma\mathrm{T}'\mathrm{BB}'\mathrm{T}.\end{aligned}$$

Therefore, we have

$$C_{\mathbf{Y}}(\mathrm{T}) = \exp\Big\{i\ \mathrm{tr}\,\mathrm{T}'\mathrm{B}\Xi - \frac{1}{2}\mathrm{tr}\,\Sigma\mathrm{T}'\mathrm{BB}'\mathrm{T}\Big\}.$$

The required result is obtained by noting that

$$\mathrm{tr}\,\Sigma\mathrm{T}'\mathrm{BB}'\mathrm{T} = \mathrm{vec}(\mathrm{T})'[\Sigma \otimes \mathrm{BB}']\,\mathrm{vec}(\mathrm{T}). \qquad \square$$

1.3 Spherical and Elliptical Distributions

It is easily seen that the density function of a p-variate normal distribution $\mathrm{N}_p(\boldsymbol{\mu}, \Sigma)$ is constant on the ellipsoids

$$(\boldsymbol{x}-\boldsymbol{\mu})'\Sigma^{-1}(\boldsymbol{x}-\boldsymbol{\mu}) = k$$

for every positive value of k in a p-dimensional Euclidean space. A general class of distribution with this property is the class of elliptically countered distributions with density function of the form

$$\Lambda^{-1/2} g[(\boldsymbol{x}-\boldsymbol{\nu})'\Lambda^{-1}(\boldsymbol{x}-\boldsymbol{\nu})], \tag{1.3.1}$$

where Λ is a positive definite matrix, $g(\cdot) \geq 0$, and

$$\int_{-\infty}^{\infty} \cdots \int_{-\infty}^{\infty} g(\boldsymbol{z}'\boldsymbol{z}) \mathrm{d}z_1 \cdots \mathrm{d}z_p = 1. \tag{1.3.2}$$

Further, if $\boldsymbol{X}$ has the density function (1.3.1), then considering the transformed variate

$$\boldsymbol{Z} = \Lambda^{-1/2}(\boldsymbol{X} - \boldsymbol{\nu}), \tag{1.3.3}$$

we can see that the density function of $\boldsymbol{Z}$ is $g(\boldsymbol{z}'\boldsymbol{z})$, and the contours of constant density of $\boldsymbol{Z}$ are spheres centered at the origin. In other words, the distribution of $\boldsymbol{Z}$ belongs to the class of spherically contoured distributions. From (1.3.3) $\boldsymbol{X}$ is expressed in terms of $\boldsymbol{Z}$ as

$$\boldsymbol{X} = \boldsymbol{\nu} + \Lambda^{1/2}\boldsymbol{Z}. \tag{1.3.4}$$

Therefore, the class of elliptically contoured distributions is, in fact, derived from the smaller class of spherically contoured distributions through relation (1.3.4).

Let $\boldsymbol{Z} = (Z_1, \ldots, Z_p)'$ be distributed as $\mathrm{N}_p(\boldsymbol{0}, \mathrm{I}_p)$. Then $\boldsymbol{U} = \boldsymbol{Z}/\|\boldsymbol{Z}\|$ is a distribution on the sphere $\boldsymbol{u}'\boldsymbol{u} = 1$ satisfying
"The distributions of $\boldsymbol{U}$ and $\mathrm{H}\boldsymbol{U}$ are identical for all $\mathrm{H} \in \mathrm{O}(p)$",
where $\mathrm{O}(p)$ is the set of orthogonal matrices of order p. Such a distribution is called a *spherical distribution.* It is known [see Muirhead, 1982] that the spherical distribution on the unit sphere is determined uniquely.

Definition 1.3.1 A $p \times 1$ random vector $\boldsymbol{Z}$ is said to have a spherical distribution if $\boldsymbol{Z}$ and $\mathrm{H}\boldsymbol{Z}$ have the same distribution for all $p \times p$ orthogonal matrices H.

In the following we treat a spherical distribution with a density function.

Theorem 1.3.1 *The following statements are equivalent.*

(1) *$\boldsymbol{Z}$ has a spherical distribution.*

(2) *The density function of $\boldsymbol{Z}$ is expressed as $g(\boldsymbol{z}'\boldsymbol{z})$ for some scalar function $g(\cdot)$.*

(3) *The characteristic function of $\boldsymbol{Z}$ is of the form $\phi(\boldsymbol{t}'\boldsymbol{t})$ for some scalar function $\phi(\cdot)$.*

Proof. Let $f(\boldsymbol{z})$ be the density function of $\boldsymbol{Z}$. For any orthogonal matrix H, the density function of $\boldsymbol{X} = \mathrm{H}\boldsymbol{Z}$ is given by $f(\mathrm{H}'\boldsymbol{x})$. If (1) is true, $f(\boldsymbol{z}) = f(\mathrm{H}'\boldsymbol{z})$. Taking H whose first column is $(\boldsymbol{z}'\boldsymbol{z})^{-1/2}\boldsymbol{z}$, we can see that (2) is true, since $\mathrm{H}'\boldsymbol{z} = ((\boldsymbol{z}'\boldsymbol{z})^{1/2}, 0, \ldots, 0)'$. The converse is shown similarly. If (1) is true, the characteristic functions of $\boldsymbol{Z}$ and $\mathrm{H}\boldsymbol{Z}$ are the same, i.e. $C_{\boldsymbol{Z}}(\boldsymbol{t}) = C_{\mathrm{H}\boldsymbol{Z}}(\boldsymbol{t}) = C_{\boldsymbol{Z}}(\mathrm{H}'\boldsymbol{t})$. Therefore, we have (3). If (3) is true, it is easy to see that $C_{\boldsymbol{Z}}(\boldsymbol{t}) = C_{\mathrm{H}\boldsymbol{Z}}(\boldsymbol{t})$, and hence (1). □

Theorem 1.3.2 *Suppose that $\boldsymbol{X}$ has a p-variate spherical distribution with* $\Pr\{\boldsymbol{X} = \boldsymbol{0}\} = 0$. *Then*

$$R \equiv (\boldsymbol{X}'\boldsymbol{X})^{1/2} \quad \textit{and} \quad \boldsymbol{U}(\boldsymbol{X}) = R^{-1}\boldsymbol{X}$$

are mutually independent and the latter follows a spherical distribution on $\mathcal{S}_p = \{\boldsymbol{x} : \boldsymbol{x}'\boldsymbol{x} = 1\}$.

Proof. The second assertion follows from the identities

$$\boldsymbol{U}(\mathrm{H}\boldsymbol{X}) = (\boldsymbol{X}'\mathrm{H}'\mathrm{H}\boldsymbol{X})^{-1/2}\mathrm{H}\boldsymbol{X} = \mathrm{H}\boldsymbol{U}(\boldsymbol{X}),$$

which hold for any orthogonal matrix H. For the independence, see Muirhead (1982). □

Definition 1.3.2 Let $\boldsymbol{\mu}$ be a $p \times 1$ constant vector, and A be a $p \times p$ constant matrix. Random vector $\boldsymbol{X}$ is said to have an *elliptical distribution* with parameter $\boldsymbol{\mu}$ and $\Lambda = \mathrm{A}'\mathrm{A}$, if it can be put in the form $\boldsymbol{X} = \boldsymbol{\mu} + \mathrm{A}\boldsymbol{Z}$, where $\boldsymbol{Z}$ is a random vector following a spherical distribution. If this is the case, we write $\boldsymbol{X} \sim \mathrm{E}_p(\boldsymbol{\mu}, \Lambda)$.

Theorem 1.3.3 *Statements* (1) *through* (3) *are equivalent.*

(1) $\boldsymbol{X} \sim \mathrm{E}_p(\boldsymbol{\mu}, \Lambda)$.

(2) *The probability density function of $\boldsymbol{X}$ is of the form* $|\Lambda|^{-1/2}g((\boldsymbol{x} - \boldsymbol{\mu})'\Lambda^{-1}(\boldsymbol{x} - \boldsymbol{\mu}))$.

(3) *The characteristic function of $\boldsymbol{X}$ is of the form* $\exp(i\boldsymbol{t}'\boldsymbol{\mu})\phi(\boldsymbol{t}'\Lambda\boldsymbol{t})$.

Proof. By expression $\boldsymbol{X} = \boldsymbol{\mu} + R\mathrm{A}\boldsymbol{U}$, the (1), (2), and (3) can be stated in terms of $\boldsymbol{Z} = R\boldsymbol{U}$. The results follow from Theorem 1.3.1. □

The coordinate transformation in the case of p-variate is defined by

$$\begin{aligned} x_1 &= r\sin\theta_1 \sin\theta_2 \cdots \sin\theta_{p-2}\sin\theta_{p-1}, \\ x_2 &= r\sin\theta_1 \sin\theta_2 \cdots \sin\theta_{p-2}\cos\theta_{p-1}, \\ x_3 &= r\sin\theta_1 \sin\theta_2 \cdots \cos\theta_{p-2}, \\ &\vdots \\ x_{p-1} &= r\sin\theta_1 \cos\theta_2, \\ x_p &= r\cos\theta_1, \end{aligned} \tag{1.3.5}$$

where $r > 0$, $0 < \theta_i < \pi$ $(i = 1, \ldots, p-2)$, $0 < \theta_{p-1} < 2\pi$.

Theorem 1.3.4 *If $\boldsymbol{X}$ is $\mathrm{E}_p(\boldsymbol{0}, \mathrm{I}_p)$ with density function $c_p h(\boldsymbol{x}'\boldsymbol{x})$, the random variables R, $\Theta_1, \ldots, \Theta_{p-1}$ defined by (1.3.5) are independent. Further, the marginal density of Θ_k is*

$$\frac{\Gamma[\frac{1}{2}(p-k)]\cos^{p-k-1}\theta}{\{\Gamma(\frac{1}{2})\Gamma[\frac{1}{2}(p-k-1)]\}}, \quad k = 1, 2, \ldots, p-2,$$

and Θ_{p-1} is $1/(2\pi)$. The marginal density of R^2 is

$$f(y) = \frac{c_p \pi^{p/2}}{\Gamma(\frac{1}{2}p)} y^{p/2-1} h(y), \quad y > 0. \tag{1.3.6}$$

Proof. The Jacobian of the transformation from $x_1, \ldots, x_p$ to $r, \theta_1, \ldots, \theta_{p-1}$ given by (1.3.5) is $r^{p-1}\sin^{p-2}\theta_1 \sin^{p-3}\theta_2 \cdots \sin\theta_{p-2}$ (see Problem 1.7). It follows that the joint density function of R^2, $\Theta_1, \ldots, \Theta_{p-1}$ is

$$\frac{1}{2} c_p (r^2)^{p/2-1} \sin^{p-2}\theta_1 \sin^{p-3}\theta_2 \cdots \sin\theta_{p-2} h(r^2). \tag{1.3.7}$$

This shows that R^2, $\Theta_1, \ldots, \Theta_{p-1}$ are mutually independent and Θ_k has density function proportional to $\sin^{p-1-k}\theta_k$. Integration (1.3.7) with respect to $\theta_1, \ldots, \theta_{p-1}$ yields the factor $2\pi^{p/2}/\Gamma(\frac{1}{2}p)$, which is the surface area of a sphere of unit radius in R^p. Therefore, R^2 has the density function given by (1.3.6). The marginal density function of the Θ_i's is found by noting that

$$\int_{-\pi/2}^{\pi/2} \cos^{h-1}\theta \, \mathrm{d}\theta = \frac{\Gamma(\frac{1}{2}h)\Gamma(\frac{1}{2})}{\Gamma[\frac{1}{2}(h+1)]}.$$ □

When $\boldsymbol{X} \sim \mathrm{E}_p(\boldsymbol{\mu}, \Lambda)$, we can write

$$\boldsymbol{X} = \boldsymbol{\mu} + R\mathrm{A}\boldsymbol{U}, \tag{1.3.8}$$

where $R \geq 0$, $\boldsymbol{U}$ has a spherical distribution, and R and $\boldsymbol{U}$ are independent. Further from the uniqueness of a spherical distribution we may let

$$\boldsymbol{U} = \boldsymbol{Z}/\|\boldsymbol{Z}\|, \quad \boldsymbol{Z} \sim \mathrm{N}(\mathbf{0}, \mathrm{I}_p).$$

Using (1.3.8), we have the following theorem.

Theorem 1.3.5 *If* $\boldsymbol{X} \sim \mathrm{E}_p(\boldsymbol{\mu}, \Lambda)$, *then*

(1) $\mathrm{E}(\boldsymbol{X}) = \boldsymbol{\mu}$.

(2) $\mathrm{Var}(\boldsymbol{X}) = (1/p)\mathrm{E}(R^2)\Lambda = \Sigma$.

Further, using that the characteristic function in Theorem 1.3.3(3), we have

$$\mathrm{E}(R^2) = -2p\phi'(0).$$

Theorem 1.3.6 *Let* $\boldsymbol{X} \sim \mathrm{E}_p(\boldsymbol{\mu}, \Lambda)$, *and* B *be a* $q \times p$ *constant matrix with rank* q. *Then*

$$\boldsymbol{Y} = \mathrm{B}\boldsymbol{X} \sim \mathrm{E}_p(\mathrm{B}\boldsymbol{\mu}, \mathrm{B}\Lambda\mathrm{B}').$$

Proof. The characteristic function of $\boldsymbol{Y}$ is expressed as

$$C_{\boldsymbol{Y}}(\boldsymbol{t}) = \mathrm{E}[\mathrm{e}^{i\boldsymbol{t}'\mathrm{B}\boldsymbol{X}}] = \mathrm{e}^{i\boldsymbol{t}'\mathrm{B}\boldsymbol{\mu}}\phi(\boldsymbol{t}'\mathrm{B}\Lambda\mathrm{B}\boldsymbol{t}).$$

The desired result follows from the uniqueness of the characteristic function. □

Theorem 1.3.7 *Let* $\boldsymbol{X} \sim \mathrm{E}_p(\boldsymbol{\mu}, \Lambda)$ *and decompose* $\boldsymbol{X}$, $\boldsymbol{\mu}$, Σ *as*

$$\boldsymbol{X} = \begin{pmatrix} \boldsymbol{X}_1 \\ \boldsymbol{X}_2 \end{pmatrix}, \quad \boldsymbol{\mu} = \begin{pmatrix} \boldsymbol{\mu}_1 \\ \boldsymbol{\mu}_2 \end{pmatrix}, \quad \Sigma = \begin{pmatrix} \Sigma_{11} & \Sigma_{12} \\ \Sigma_{21} & \Sigma_{22} \end{pmatrix},$$

where $\boldsymbol{X}_i : p_i \times 1$, $\boldsymbol{\mu}_i : p_i \times 1$, $\Sigma_{ij} : p_i \times p_j$. *Then the conditional distribution of* $\boldsymbol{X}_1$ *given* $\boldsymbol{X}_2$ *is an elliptical distribution. The conditional mean and the conditional variance are given by*

$$\begin{aligned} \mathrm{E}(\boldsymbol{X}_1|\boldsymbol{X}_2) &= \boldsymbol{\mu}_1 + \Lambda_{12}\Lambda_{22}^{-1}(\boldsymbol{X}_2 - \boldsymbol{\mu}_2), \\ \mathrm{Var}(\boldsymbol{X}_1|\boldsymbol{X}_2) &= h(\boldsymbol{X}_2)\Lambda_{11\cdot 2}. \end{aligned}$$

where $\Lambda_{11\cdot 2} = \Lambda_{11} - \Lambda_{12}\Lambda_{22}^{-1}\Lambda_{21}$.

Proof. From Theorem 1.3.6 it follows that $\boldsymbol{X}_i \sim \mathrm{E}_{p_i}(\boldsymbol{\mu}_i, \Lambda_{ii})$, $i = 1, 2$. Note that

$$(\boldsymbol{x}-\boldsymbol{\mu})'\Lambda^{-1}(\boldsymbol{x}-\boldsymbol{\mu}) = a(\boldsymbol{x}_1|\boldsymbol{x}_2) + (\boldsymbol{x}_2-\boldsymbol{\mu}_2)'\Lambda_{22}^{-1}(\boldsymbol{x}_2-\boldsymbol{\mu}_2),$$

where

$$\begin{aligned} a(\boldsymbol{x}_1|\boldsymbol{x}_2) &= [\boldsymbol{x}_1-\boldsymbol{\mu}_1-\Lambda_{12}\Lambda_{22}^{-1}(\boldsymbol{x}_2-\boldsymbol{\mu}_2)]' \\ &\quad \times\Lambda_{22\cdot1}^{-1}[\boldsymbol{x}_1-\boldsymbol{\mu}_1-\Lambda_{12}\Lambda_{22}^{-1}(\boldsymbol{x}_2-\boldsymbol{\mu}_2)]. \end{aligned}$$

Therefore, the conditional density function of $\boldsymbol{X}_1$ given $\boldsymbol{X}_2$ is written as

$$\frac{g((\boldsymbol{x}-\boldsymbol{\mu})'\Lambda^{-1}(\boldsymbol{x}-\boldsymbol{\mu}))}{g((\boldsymbol{X}_2-\boldsymbol{\mu}_2)'\Lambda_{22}^{-1}(\boldsymbol{X}_2-\boldsymbol{\mu}_2))} = g_{1|2}(a(\boldsymbol{x}_1|\boldsymbol{X}_2)).$$

This proves the result required. □

Example 1.3.1 (Elliptical Distributions)

(1) Multivariate t-distribution. Let

$$\boldsymbol{X} = \boldsymbol{\mu} + \mathrm{A}\boldsymbol{Z}, \quad \boldsymbol{Z} = \frac{1}{S}\boldsymbol{Y},$$

where $mS^2 \sim \chi^2(m)$, $\boldsymbol{Y} \sim \mathrm{N}_p(\boldsymbol{0}, \mathrm{I}_p)$, and S and $\boldsymbol{Y}$ are independent. Then the pdf of $\boldsymbol{X}$ is given by

$$\frac{\Gamma(\frac{1}{2}(m+p))}{\Gamma(\frac{1}{2}m)m^{p/2}\pi^{p/2}} \mid \Lambda \mid^{-1/2} \left[1 + \frac{1}{m}(\boldsymbol{x}-\boldsymbol{\mu})'\Lambda^{-1}(\boldsymbol{x}-\boldsymbol{\mu})\right]^{-(m+p)/2}.$$

(2) The ϵ-contaminated p-variate normal. This is a mixture distribution of $\mathrm{N}_p(\boldsymbol{\mu}, \Lambda)$ and $\mathrm{N}_p(\boldsymbol{\mu}, c\Lambda)$ where the pdf is given by

$$\begin{aligned} &(1-\epsilon)(2\pi)^{-p/2} \mid \Lambda \mid^{-1/2} \exp\Big\{-\frac{1}{2}(\boldsymbol{x}-\boldsymbol{\mu})'\Lambda^{-1}(\boldsymbol{x}-\boldsymbol{\mu})\Big\} \\ &\quad +\epsilon(2\pi)^{-p/2} \mid c\Lambda \mid^{-1/2} \exp\Big\{-\frac{1}{2}(\boldsymbol{x}-\boldsymbol{\mu})'(c\Lambda)^{-1}(\boldsymbol{x}-\boldsymbol{\mu})\Big\}. \end{aligned}$$

1.4 Multivariate Cumulants

In general, the cumulants of a statistic play an important role in the derivation of asymptotic expansions for its distribution. In this section we consider expressing the cumulants of a random vector as a population distribution in terms of its moments.

First we consider the case of a univariate variable X. The characteristic function of X is defined as

$$C(t) = \mathrm{E}\big[\mathrm{e}^{itX}\big]$$

for all real numbers. In the following it is assumed for simplicity that all the moments exist. In general, if X has an rth moment, the results up to the r moments or cumulants are assured. Using

$$\mathrm{e}^{z} = 1 + z + \frac{1}{2!}(z)^2 + \frac{1}{3!}(z)^3 + \cdots,$$

the characteristic function can be expanded as

$$C(t) = \mathrm{E}\Big[\sum_{r=0}^{\infty} \frac{1}{r!}(itX)^r\Big] = \sum_{r=0}^{\infty} \mu_r \frac{(it)^r}{r!},$$

where μ_r is the rth moment of X, $\mu_r = \mathrm{E}[X^r]$. The *cumulants* of X are the coefficients κ_r in

$$\log C(t) = \sum_{r=1}^{\infty} \kappa_r \frac{(it)^r}{r!}.$$

The first four cumulants in terms of the moments are

$$\begin{aligned}
\kappa_1 &= \mu_1 = \mu, \\
\kappa_2 &= \mu_2 - \mu_1^2 = \sigma^2, \\
\kappa_3 &= \mu_3 - 3\mu_2\mu_1 + 2\mu_1^3, \\
\kappa_4 &= \mu_3 - 4\mu_3\mu_1 - 3\mu_2^2 + 12\mu_2\mu_1^2 - 6\mu_1^4.
\end{aligned}$$

The skewness $\kappa_{[3]}$ and kurtosis $\kappa_{[4]}$ of X are

$$\kappa_{[3]} = \kappa_3/(\kappa_2)^{3/2} \quad \text{and} \quad \kappa_{[4]} = \kappa_4/(\kappa_2)^2.$$

The r-th *moments* and *cumulants* of X_j in $\boldsymbol{X} = (X_1, \ldots, X_p)'$ are expressed as $\mu_r^j = \mathrm{E}[X_j^r]$ and κ_r^j. The *skewness* and *kurtosis* of X_j are $\kappa_{[3]}^j$ and $\kappa_{[4]}^j$, respectively.

Next we consider the cumulants of the joint distribution of $\boldsymbol{X} = (X_1, \ldots, X_p)'$. The characteristic function of $\boldsymbol{X}$ is defined as

$$C(\boldsymbol{t}) = \mathrm{E}\Big[\mathrm{e}^{i(t_1X_1+\cdots+t_pX_p)}\Big] = \mathrm{E}\Big[\mathrm{e}^{i\boldsymbol{t}'\boldsymbol{X}}\Big],$$

where t_j, $j = 1, \ldots, p$ are real numbers and $\boldsymbol{t} = (t_1, \ldots, t_p)'$. The characteristic function can be expanded as

$$\begin{aligned}
C(\boldsymbol{t}) &= \mathrm{E}\Big[\sum_{r=0}^{\infty} \frac{1}{r!}(i\boldsymbol{t}'\boldsymbol{X})^r\Big] \\
&= \sum_{r=0}^{\infty} \sum_{r_1+\cdots+r_p=r} \mu_{r_1\cdots r_p}^{1\cdots p} \times \frac{(it_1)^{r_1}\cdots(it_p)^{r_p}}{r_1!\cdots r_p!},
\end{aligned}$$

where $\mu^{1\cdots p}_{r_1\cdots r_p}$ is the moment of $\boldsymbol{X}$ defined by

$$\mu^{1\cdots p}_{r_1\cdots r_p} = \mathrm{E}\Big[X_1^{r_1}\cdots X_p^{r_p}\Big].$$

The *cumulants* of $\boldsymbol{X}$ are defined as the coefficients $\kappa^{1\cdots p}_{r_1\cdots r_p}$ in the expansion

$$\begin{aligned}\log C(\boldsymbol{t}) &= \log\Big\{1+\sum_{r=1}^{\infty}\sum_{r_1+\cdots+r_p=r}\mu^{1\cdots p}_{r_1\cdots r_p}\frac{(it_1)^{r_1}\cdots(it_p)^{r_p}}{r_1!\cdots r_p!}\Big\}\\ &= \sum_{r=1}^{\infty}\sum_{r_1+\cdots+r_p=r}\kappa^{1\cdots p}_{r_1\cdots r_p}\frac{(it_1)^{r_1}\cdots(it_p)^{r_p}}{r_1!\cdots r_p!}.\end{aligned}$$

In these notations, we write simply

$$\begin{aligned}&\mu^{12\cdots p}_{r_1 0\cdots 0} = \mu^{1}_{r_1}, \quad \mu^{12\cdots p}_{r_1 r_2\cdots 0} = \mu^{12}_{r_1 r_2},\\ &\kappa^{12\cdots p}_{r_1 0\cdots 0} = \kappa^{1}_{r_1}, \quad \kappa^{12\cdots p}_{r_1 r_2\cdots 0} = \kappa^{12}_{r_1 r_2}, \quad \text{etc.}\end{aligned}$$

If $\boldsymbol{X}$ is distributed as $\mathrm{E}_p(\boldsymbol{\mu},\Lambda)$, then from Theorem 1.3.3(3) we have

$$\log C(\boldsymbol{t}) = i\boldsymbol{t}'\boldsymbol{\mu} + \log\phi(\boldsymbol{t}'\Sigma\boldsymbol{t}).$$

Therefore,

$$\begin{aligned}&\kappa^{j}_{1} = \mu^{j}_{1},\\ &\kappa^{ij}_{11} = \sigma_{ij} = -2\phi'(0)\lambda_{ij}, \quad \Sigma = -2\phi'(0)\Lambda,\\ &\kappa^{r_1\cdots r_p}_{1\cdots p} = 0, \quad \text{if } r_1+\cdots+r_p (\geq 3) \text{ is odd},\\ &\kappa^{ijk\ell}_{1111} = \kappa(\sigma_{ij}\sigma_{k\ell}+\sigma_{ik}\sigma_{j\ell}+\sigma_{i\ell}\sigma_{jk}).\end{aligned}$$

The skewness and kurtosis of X_j are given by

$$\begin{aligned}\kappa^{j}_{[3]} &= \frac{\kappa^{j}_{3}}{(\kappa^{j}_{2})^{3/2}} = 0,\\ \kappa^{j}_{[4]} &= \frac{\kappa^{j}_{4}}{(\kappa^{j}_{2})^{2}} = 3\cdot\frac{[\phi''(0)-\phi'(0)^2]}{\phi'(0)^2}\\ &= 3\kappa.\end{aligned}$$

If, in particular, $\boldsymbol{X}$ follows $\mathrm{N}_p(\boldsymbol{\mu},\Sigma)$, the logarithm of its characteristc funcition takes most simple form:

$$\log C(\boldsymbol{t}) = i\boldsymbol{t}\boldsymbol{\mu} + \frac{1}{2}i^2\boldsymbol{t}'\Sigma\boldsymbol{t},$$

which immediately implies

$$\begin{aligned}&\kappa^{j}_{1} = \mu^{j}_{1},\\ &\kappa^{ij}_{11} = \sigma_{ij},\\ &\kappa^{1\cdots p}_{r_1\cdots r_p} = 0 \quad (r_1+\cdots+r_p \geq 3).\end{aligned}$$

The moments and cumulants are also denoted as follows:

$$\mu^{1\cdots p}_{r_1\cdots r_p} = \mu_{\underbrace{1\cdots 1}_{r_1}\cdots\underbrace{p\cdots p}_{r_p}} \quad \text{and} \quad \kappa^{1\cdots p}_{r_1\cdots r_p} = \kappa_{\underbrace{1\cdots 1}_{r_1}\cdots\underbrace{p\cdots p}_{r_p}}.$$

Then the characteristic function of $\boldsymbol{X}$ is expanded as

$$C(\boldsymbol{t}) = 1 + \sum_{r=1}^{\infty} \frac{i^r}{r!}\mu_{j_1\cdots j_r} t_{j_1}\cdots t_{j_r}$$

$$\log C(\boldsymbol{t}) = \sum_{r=1}^{\infty} \frac{i^r}{r!}\kappa_{j_1\cdots j_r} t_{j_1}\cdots t_{j_r}.$$

We have a formal relationship between moments and cumulants. As a rigorous statement, we have the following result.

Theorem 1.4.1 *If the sth moment of a random vector $\boldsymbol{X}$ exists, the characteristic function and the cumulant function of $\boldsymbol{X}$ can be expanded as*

$$C(\boldsymbol{t}) = 1 + \sum_{r=1}^{s} \frac{i^r}{r!}\mu_{j_1\cdots j_r} t_{j_1}\cdots t_{j_r} + \mathrm{o}(\|\boldsymbol{t}\|),$$

$$\log C(\boldsymbol{t}) = \sum_{r=1}^{s} \frac{i^r}{r!}\kappa_{j_1\cdots j_r} t_{j_1}\cdots t_{j_r} + \mathrm{o}(\|\boldsymbol{t}\|),$$

where $\|\boldsymbol{t}\| = \sqrt{t_1^2 + \cdots + t_p^2}$.

For *skewness* $\kappa_{[3]}$ and *kurtosis* $\kappa_{[4]}$ of a random vector $\boldsymbol{X}$, there are some definitions. Mardia (1970) defined

$$\kappa_3^{(1)} = \mathrm{E}\big[\{(\boldsymbol{X}-\boldsymbol{\mu})'\Sigma^{-1}(\boldsymbol{X}-\boldsymbol{\mu})\}^3\big],$$

$$\kappa_4^{(1)} = \mathrm{E}\big[\{(\boldsymbol{X}-\boldsymbol{\mu})'\Sigma^{-1}(\boldsymbol{X}-\boldsymbol{\mu})\}^2\big].$$

Let a random vector $\boldsymbol{X} = (X_1, \ldots, X_p)'$ have mean $\mathrm{E}(\boldsymbol{X}) = \boldsymbol{\mu} = (\mu_1, \ldots, \mu_p)'$ and covariance matrix $\mathrm{Cov}(\boldsymbol{X}) = \Sigma = (\sigma_{ij})$. Let the cumulants of $\boldsymbol{X} - \boldsymbol{\mu}$ be denoted by $\gamma_{j_1\cdots j_r}$ instead of $\kappa_{j_1\cdots j_r}$.

$$\begin{aligned}
\gamma_a &= 0,\\
\gamma_{ab} &= \sigma_{ab},\\
\gamma_{abc} &= E[(X_a-\mu_a)(X_b-\mu_b)(X_c-\mu_c)],\\
\gamma_{abcd} &= E[(X_a-\mu_a)(X_b-\mu_b)(X_c-\mu_c)(X_d-\mu_d)]\\
&\quad - (\sigma_{ab}\sigma_{cd} + \sigma_{ac}\sigma_{bd} + \sigma_{ad}\sigma_{bc}).
\end{aligned}$$

The quantities $\kappa_{[3]}^{(1)}$ and $\kappa_{[4]}^{(1)}$ can be expressed in terms of the cumulants of $\boldsymbol{X} - \boldsymbol{\mu}$ as

$$\kappa_{[3]}^{(1)} = \sum_{a,b,c,a',b',c'} \gamma_{abc}\gamma_{a'b'c'}\sigma^{aa'}\sigma^{bb'}\sigma^{cc'},$$
$$\kappa_{[4]}^{(1)} = \sum_{a,b,c,d} \gamma_{abcd}\sigma^{ab}\sigma^{cd},$$

where $\Sigma^{-1} = (\sigma^{ij})$. An alternative definition for skewness is

$$\kappa_{[3]}^{(2)} = \sum_{a,b,c,a',b',c'} \gamma_{abc}\gamma_{a'b'c'}\sigma^{ab}\sigma^{ca'}\sigma^{b'c'}.$$

For the standardized variate $\boldsymbol{Y} = (Y_1, \ldots, Y_p)' = \Sigma^{-1/2}(\boldsymbol{X} - \boldsymbol{\mu})$, we put

$$\mu_{i_1 \cdots i_k} = \mathrm{E}(Y_{i_1} \cdots Y_{i_k})$$

and denote the corresponding cumulants by $\kappa_{i_1 \cdots i_k}$. Then

$$\begin{aligned}
\mu_a &= \kappa_a \ (= 0), \\
\mu_{ab} &= \kappa_{ab} \ (= \delta_{ab}), \\
\mu_{abc} &= \kappa_{abc}, \\
\mu_{abcd} &= \kappa_{abcd} + \sum_{[3]} \delta_{ab}\delta_{cd}, \\
\mu_{abcde} &= \kappa_{abcde} + \sum_{[10]} \kappa_{abc}\delta_{de}, \\
\mu_{abcdef} &= \kappa_{abcdef} + \sum_{[15]} \kappa_{abcd}\delta_{ef} + \sum_{[10]} \kappa_{abc}\kappa_{def} + \sum_{[15]} \delta_{ab}\delta_{cd}\delta_{ef},
\end{aligned}$$

where $\sum_{[3]} \delta_{ab}\delta_{cd} = \delta_{ab}\delta_{cd} + \delta_{ac}\delta_{bd} + \delta_{ad}\delta_{bc}$. The multivariate skewness and kurtosis are expressed as

$$\kappa_{[3]}^{(1)} = \sum_{a,b,c} (\kappa_{abc})^2,$$
$$\kappa_{[3]}^{(2)} = \sum_{a} \Big(\sum_{b} \kappa_{abb} \Big)^2 = \sum_{a,b,c} \kappa_{aab}\kappa_{bcc},$$
$$\kappa_{[4]}^{(1)} = \sum_{a,b} \kappa_{aabb}.$$

Problems

1.1 Suppose that $\boldsymbol{X} = (X_1, X_2)'$ is distributed as a bivariate distribution with mean and variance given by

$$\boldsymbol{\mu} = \begin{pmatrix} \mu_1 \\ \mu_2 \end{pmatrix} \quad \text{and} \quad \Sigma = \begin{pmatrix} \sigma_1^2 & \rho\sigma_1\sigma_2 \\ \rho\sigma_1\sigma_2 & \sigma_2^2 \end{pmatrix},$$

respectively. The density function is a constant on ellipsoids

$$(\boldsymbol{x} - \boldsymbol{\mu})'\Sigma^{-1}(\boldsymbol{x} - \boldsymbol{\mu}) = c$$

for every positive value of c in two-dimensional Euclidian space. After transforming the coordinates by $(x_i - \mu_i)/\sigma_i = y_i$, $i = 1, 2$, show that the loci of constant density are ellipses defined by

$$\frac{1}{1-\rho^2}(y_1^2 - 2\rho y_1 y_2 + y_2^2) = k$$

for $k > 0$. Show that

(1) for $\rho > 0$, the principal axes are along the lines $y_1 = y_2, y_1 = -y_2$ with lengths $2\sqrt{k(1+\rho)}$ and $2\sqrt{k(1-\rho)}$, respectively, and that

(2) for $\rho < 0$, the principal axes are along the lines $y_1 = -y_2, y_1 = y_2 < 0$ with lengths $2\sqrt{k(1-\rho)}$ and $2\sqrt{k(1+\rho)}$, respectively.

Hint: Consider an orthogonal transformation

$$\begin{pmatrix} z_1 \\ z_2 \end{pmatrix} = \begin{pmatrix} \cos\theta & -\sin\theta \\ \sin\theta & \cos\theta \end{pmatrix} \begin{pmatrix} y_1 \\ y_2 \end{pmatrix}.$$

By choosing an appropriate θ, we can write the ellipse as

$$\frac{z_1^2}{\lambda_1} + \frac{z_2^2}{\lambda_2} = k',$$

where λ_1 and λ_2 are solutions of $(1-\lambda)^2 - \rho^2 = 0$: that is $\lambda_i = 1 \pm |\rho|$.

1.2 Let $\boldsymbol{Y}$ be a random vector defined by

$$\boldsymbol{Y} = \mathrm{X}\boldsymbol{\beta} + \mathrm{Z}\boldsymbol{B} + \boldsymbol{\varepsilon},$$

where X: $p \times q$ and Z: $p \times r$ are fixed matrices, $\boldsymbol{\beta}$: $q \times 1$ is a parameter vector, $\boldsymbol{B}$ and $\boldsymbol{\varepsilon}$ are random vectors having $\mathrm{N}_r(\mathbf{0}, \Delta)$ and $\mathrm{N}_p(\mathbf{0}, \sigma^2\mathrm{I}_p)$, respectively, and $\boldsymbol{B}$ and $\boldsymbol{\varepsilon}$ are independent. Then show the following.

(a) $\boldsymbol{y} \sim \mathrm{N}_p(\mathrm{X}\boldsymbol{\beta}, \mathrm{W})$, $\quad \mathrm{W} = \mathrm{Z}\Delta\mathrm{Z}' + \sigma^2\mathrm{I}_p$.

(b) $\begin{pmatrix} \boldsymbol{Y} \\ \boldsymbol{B} \end{pmatrix} \sim \mathrm{N}_{p+r}\left(\begin{pmatrix} \mathrm{X}\boldsymbol{\beta} \\ \mathbf{0} \end{pmatrix}, \begin{pmatrix} \mathrm{W} & \mathrm{Z}\Delta \\ \Delta\mathrm{Z}' & \Delta \end{pmatrix}\right)$.

(c) $\mathrm{E}(\boldsymbol{B}\,|\,\boldsymbol{Y}) = \Delta\mathrm{Z}\mathrm{W}^{-1}(\boldsymbol{Y} - \mathrm{X}\boldsymbol{\beta})$.

(d) $\boldsymbol{Y}|\boldsymbol{B} \sim \mathrm{N}_p(\mathrm{X}\boldsymbol{\beta} + \mathrm{Z}\boldsymbol{B}, \sigma^2\mathrm{I}_p)$.

1.3 Show that Theorem 1.2.2 holds for a singular multivariate normal variable.

Hint: Suppose that $\boldsymbol{X} \sim \mathrm{N}(\boldsymbol{\mu}, \Sigma)$ and Σ is singular with rank r. Then use the fact that $\boldsymbol{X}$ is expressed as $\boldsymbol{X} = \mathrm{B}\boldsymbol{Z}$, where $\mathrm{BB}' = \Sigma$ and $\boldsymbol{Z} \sim \mathrm{N}_r(\mathbf{0}, \mathrm{I}_r)$.

1.4 If $\boldsymbol{X} \sim \mathrm{N}_p(\boldsymbol{\mu}, \Sigma)$, the characteristic function of $\boldsymbol{X}'\mathrm{A}\boldsymbol{X}$ is given by

$$\begin{aligned}
\mathrm{E}\left[e^{it\boldsymbol{X}'\mathrm{A}\boldsymbol{X}}\right] &= |\mathrm{I}_p - 2it\mathrm{A}|^{-1/2}\exp\left(it\boldsymbol{\mu}'\mathrm{A}(\mathrm{I}_p - 2it\mathrm{A})\boldsymbol{\mu}\right) \\
&\qquad -\frac{1}{2}(\boldsymbol{x} - \boldsymbol{\mu})'\Sigma^{-1}(\boldsymbol{x} - \boldsymbol{\mu}) + it\boldsymbol{x}'\mathrm{A}\boldsymbol{x} \\
&= -\frac{1}{2}\left\{\boldsymbol{x} - (\Sigma^{-1} - 2it\mathrm{A})^{-1}\Sigma^{-1}\boldsymbol{\mu}\right\}(\Sigma^{-1} - 2it\mathrm{A}) \\
&\qquad \times\left\{\boldsymbol{x} - (\Sigma^{-1} - 2it\mathrm{A})^{-1}\Sigma^{-1}\boldsymbol{\mu}\right\} - \frac{1}{2}\boldsymbol{\mu}'\Sigma^{-1}\boldsymbol{\mu} \\
&\qquad + \frac{1}{2}\boldsymbol{\mu}'\Sigma^{-1}(\Sigma^{-1} - 2it\mathrm{A})^{-1}\Sigma^{-1}\boldsymbol{\mu}.
\end{aligned}$$

1.5 Let $\mathbf{X} = (\boldsymbol{X}_1, \ldots, \boldsymbol{X}_n)'$ be a random matrix such that $\boldsymbol{X}_i, i = 1, \ldots, n$ are independent and $\boldsymbol{X}_i \sim \mathrm{N}_p(\boldsymbol{\mu}_i, \Sigma)$. Show that the density function of $\mathbf{X}$ is written as

$$f(\mathrm{X}) = |(2\pi)\Sigma|^{-n/2}\,\mathrm{etr}\left\{-\frac{1}{2}\Sigma^{-1}(\mathrm{X} - \mathrm{M})'(\mathrm{X} - \mathrm{M})\right\},$$

where $\mathrm{X} = (\boldsymbol{x}_1, \ldots \boldsymbol{x}_n)'$ and $\mathrm{M} = (\boldsymbol{\mu}_1, \ldots, \boldsymbol{\mu}_n)'$ are $p \times n$ matrices, and the characteristic function of $\mathbf{X}$ is given by

$$C_{\mathbf{X}}(\mathrm{T}) = \mathrm{etr}\left\{i\mathrm{T}'\mathrm{M} - \frac{1}{2}\mathrm{T}'\Sigma\mathrm{T}\right\},$$

where $\mathrm{T} : p \times n$ is any real matrix.

Hint: The density function of $\mathbf{X}$ can be obtained by computing

$$f(\mathrm{X}) = \prod_{j=1}^{n} |(2\pi)\Sigma|^{-1/2}\,\mathrm{etr}\left\{-\frac{1}{2}\Sigma^{-1}(\boldsymbol{x}_j - \boldsymbol{\mu}_j)(\boldsymbol{x}_j - \boldsymbol{\mu}_j)'\right\}.$$

Note that for any $p \times n$ matrix $\mathrm{T} = (t_{ij})$, we have

$$\sum_{j=1}^{p}\sum_{k=1}^{n} t_{jk}x_{jk} = \mathrm{tr}\,\mathrm{TX}$$

and

$$
\begin{aligned}
-\frac{1}{2}\operatorname{tr}\Sigma^{-1}(\mathrm{X} - \mathrm{M})'(\mathrm{X}-\mathrm{M}) + i \operatorname{tr}\mathrm{TX} \\
= -\frac{1}{2}\operatorname{tr}\Sigma^{-1}\{\mathrm{X}-(\mathrm{M}+i\mathrm{T}\Sigma)\}'\{\mathrm{X}-(\mathrm{M}+i\mathrm{T}\Sigma)\} \\
+ i \operatorname{tr}\mathrm{T}'\mathrm{M} - \frac{1}{2}\operatorname{tr}\mathrm{T}'\Sigma\mathrm{T}.
\end{aligned}
$$

1.6 Suppose that $\boldsymbol{X}$ is distributed as $\mathrm{N}_p(\mathbf{0}, \mathrm{I}_p)$. Then, show that the mean of $||\boldsymbol{X}||^{-(p-1)}$ exists.

Hint: Use the coordinate transformation in Problem 1.9.

1.7 Let the density function of $\boldsymbol{X} = (X_1, \ldots, X_p)'$ be defined by

$$
f(\boldsymbol{x}) = \frac{\Gamma(\frac{1}{2}p+1)}{|\Sigma|^{1/2}\{(p+2)\pi\}^{p/2}}, \quad \text{if}(\boldsymbol{x}-\boldsymbol{\mu})'\Sigma^{-1}(\boldsymbol{x}-\boldsymbol{\mu}) \le p+2,
$$

and $f(\boldsymbol{x}) = 0$, otherwise. Then, show that $\mathrm{E}(\boldsymbol{X}) = \boldsymbol{\mu}$, and $\mathrm{Var}(\boldsymbol{X}) = \Sigma$.

1.8 (a) The volume of p-dimensional unit ball is given by

$$
\int_{\sum_{i=1}^p z_i^2 \le 1} \cdots \int \mathrm{d}z_1 \cdots \mathrm{d}z_p = \frac{\pi^{p/2}}{\Gamma((p+2)/2)}
$$

(b) the volume of the ellipsoid $(\boldsymbol{x}-\boldsymbol{\mu})'\Sigma^{-1}(\boldsymbol{x}-\boldsymbol{\mu}) = c^2$ $(c > 0)$ is

$$
\frac{1}{\Gamma((p+1)/2)}|\Sigma|^{1/2}c^p\pi^{p/2}.
$$

1.9 (a) Let (x_1, x_2, x_3) be written in polar coordinates as

$$
x_1 = r\sin\theta_1\sin\theta_2, \quad x_2 = r\sin\theta_1\cos\theta_2, \quad x_3 = r\cos\theta_1,
$$

where $r > 0, 0 < \theta_1 < \pi$, and $0 < \theta_2 < 2\pi$. Show that the Jacobian of the transformation is given by

$$
\frac{\partial(x_1, x_2, x_3)}{\partial(r, \theta_1, \theta_2)} = r^2\sin\theta_1.
$$

(b) The coordinate transformation in the case of p-variate is defined by

$$
\begin{aligned}
x_1 &= r\sin\theta_1\sin\theta_2\cdots\sin\theta_{p-2}\sin\theta_{p-1}, \\
x_2 &= r\sin\theta_1\sin\theta_2\cdots\sin\theta_{p-2}\cos\theta_{p-1}, \\
x_3 &= r\sin\theta_1\sin\theta_2\cdots\cos\theta_{p-2}, \\
&\ \vdots \\
x_{p-1} &= r\sin\theta_1\cos\theta_2, \\
x_p &= r\cos\theta_1,
\end{aligned}
$$

where $r > 0, 0 < \theta_i < \pi\ (i = 1, \ldots, p-2), 0 < \theta_{p-1} < 2\pi$. Show that the Jacobian of the transformation is given by

$$J = r^{p-1} \sin^{p-2}\theta_1 \sin^{p-3}\theta_2 \cdots \sin\theta_{p-2}.$$

1.10 Show that the volume of the ball B_a with radius a in R^p is

$$V = \frac{(a\sqrt{\pi})^p}{\Gamma((p+2)/2)} = \begin{cases} a^p \frac{\pi^{\frac{p}{2}} 2^{\frac{p}{2}}}{2\cdot 4\cdots p}, & (p : \text{even}) \\ a^p \frac{\pi^{\frac{p-1}{2}} 2^{\frac{p+1}{2}}}{1\cdot 3\cdots p}, & (p : \text{odd}) \end{cases}$$

Hint: The result follows from

$$\int_0^{\pi} \cdots \int_0^{\pi} \int_0^{2\pi} J \mathrm{d}\theta_1 \cdots \mathrm{d}\theta_{p-2} \mathrm{d}\theta_{p-1} = \frac{2\pi^{p/2}}{\Gamma(p/2)} \cdot \frac{a^p}{p}.$$

1.11 Suppose that a p-dimensional random vector $\boldsymbol{X} = (X_1, \ldots, X_p)'$ has the uniform distribution on the unit ball $S = \{\boldsymbol{x} = (x_1, \ldots, x_p)' : \|\boldsymbol{x}\| \le 1\}$ in R^p. Show that

$$\Pr\{\|\boldsymbol{X}\| \le a\} = \frac{\int_{B_a} \mathrm{d}\boldsymbol{x}}{\int_S \mathrm{d}\boldsymbol{x}} = a^p, \quad |a| < 1.$$

1.12 For the multivariate t-distribution (see Example 1.3.1) with the density, show that $\mathrm{E}(\boldsymbol{X}) = \boldsymbol{\mu}$ and $\mathrm{Var}(\boldsymbol{X}) = [m/(m-2)]\Lambda$.

1.13 Show that the first four cumulants are given, in terms of the moments, by

$$\begin{aligned} \kappa_1 &= \mu_1 = \mu, \\ \kappa_2 &= \mu_2 - \mu_1^2 = \sigma^2, \\ \kappa_3 &= \mu_3 - 3\mu_2\mu_1 + 2\mu_1^3, \\ \kappa_4 &= \mu_3 - 4\mu_3\mu_1 - 3\mu_2^2 + 12\mu_2\mu_1^2 - 6\mu_1^4. \end{aligned}$$

1.14 Let $\boldsymbol{X}$ be distributed as $\mathrm{N}_p(\boldsymbol{\mu}, \Sigma)$. Then, for a $p \times 1$ constant vector $\boldsymbol{a}$, show that

$$\mathrm{E}[\{\boldsymbol{a}'(\boldsymbol{X} - \boldsymbol{\mu})\}^j] = \begin{cases} \dfrac{(2m)!}{2^m m!}(\boldsymbol{a}'\Sigma\boldsymbol{a})^m & \text{if } j = 2m \text{ (even)}, \\ 0 & \text{if } j = 2m-1 \text{ (odd)}. \end{cases}$$

1.15 For $i = i = 1, \ldots, n$,

(1) let X_i be mutually independent random variables such that $X_i \sim \mathrm{N}(\mu_i, \sigma^2)$. What is the distribution of $\boldsymbol{X} = (X_1, \ldots, X_n)$? Find the mean vector and covariance matrix of $\boldsymbol{X}$. Similarly,

(2) let $\boldsymbol{X}_i$ be mutually independent random vectors such that $\boldsymbol{X}_i \sim \mathrm{N}_p(\boldsymbol{\mu}_i, \Sigma)$. What is the distribution of $\mathbf{X} = (\boldsymbol{X}_1, \ldots, \boldsymbol{X}_n)'$? Find the mean vector and covariance matrix of $\mathrm{vec}(\mathbf{X})$.

1.16 Suppose that $\boldsymbol{X}$ is distributed as a singular multivariate normal distribution $\mathrm{N}(\mathbf{0}, \Sigma)$ with $\mathrm{rank}(\Sigma) = k$.

(1) Using [A.1.12], show that we can express $\Sigma = \Gamma_1 \Lambda_1 \Gamma_1'$, where $\Lambda_1 = \mathrm{diag}\,(\lambda_1, \ldots, \lambda_k), \lambda_1 \geq \cdots \geq \lambda_k > 0$ and Γ_1 satisfies $\Gamma_1'\Gamma_1 = \mathrm{I}_k$.

(2) Let $\Sigma^+ = \Gamma_1 \Lambda_1^{-1} \Gamma_1$. Then, show that Σ^+ is the Moore–Penrose inverse matrix, that is, Σ^+ satisfies the following four conditions: (i) $\Sigma\Sigma^+\Sigma = \Sigma$, (ii) $\Sigma^+\Sigma\Sigma^+ = \Sigma^+$, (iii) $(\Sigma^+\Sigma)' = \Sigma^+\Sigma$, (iv) $(\Sigma\Sigma^+)' = \Sigma\Sigma^+$.

(3) Show that $\boldsymbol{X}'\boldsymbol{X} \sim \chi^2(k)$.

Chapter 2

Wishart Distribution

The Wishart distribution is a multivariate generalization of the chi-square distribution. The distributions of sample covariance matrix and various sums of squares and products matrices are of Wishart, provided the underlying distribution is normal.

In this book we use the generalized definition of Wishart distribution that includes both nonsingular and singular cases, with the probability density function for nonsingular case.

We provide some basic properties of the Wishart distribution; sampling distributions; Wishartness and independence of quadratic forms; asymptotic distributions.

2.1 Definition

Let $\boldsymbol{X}$ be a p-dimensional random vector that is distributed as $\mathrm{N}_p(\boldsymbol{\mu}, \Sigma)$ and $\boldsymbol{X}_1, \ldots, \boldsymbol{X}_n$ be a random sample of $\boldsymbol{X}$ with size n. Then the sample mean vector and covariance matrices are defined by

$$\bar{\boldsymbol{X}} = \frac{1}{n}\sum_{i=1}^{n} \boldsymbol{X}_i \quad \text{and} \quad \mathbf{S} = \frac{1}{n-1}\sum_{i=1}^{n}(\boldsymbol{X}_i - \bar{\boldsymbol{X}})(\boldsymbol{X}_i - \bar{\boldsymbol{X}})',$$

respectively. From Theorem 1.2.3 we know that $\bar{\boldsymbol{X}} \sim \mathrm{N}(\boldsymbol{\mu}, (1/n)\Sigma)$. Now we consider an orthogonal transformation

$$\begin{aligned}\mathbf{Z} = (\boldsymbol{Z}_1, \ldots, \boldsymbol{Z}_n)' &= \mathrm{H}'(\boldsymbol{X}_1, \ldots, \boldsymbol{X}_n)' \\ &= \mathrm{H}'\mathbf{X},\end{aligned}$$

where H is an $n \times n$ orthogonal matrix. Then from Theorem 1.2.6, $\boldsymbol{Z}_1, \ldots, \boldsymbol{Z}_n$ are mutually independent and each is distributed as $\mathrm{N}_p(\cdot, \Sigma)$. The mean of $\mathbf{Z}$

is $\mathrm{E}(\mathbf{Z}) = \mathrm{H}'E(\mathbf{X})$. If we choose the first column of H as $(1/\sqrt{n})\mathbf{1}_n$, then

$$\boldsymbol{Z}_1 \sim \mathrm{N}_p(\sqrt{n}\,\boldsymbol{\mu}, \Sigma), \quad \boldsymbol{Z}_2, \ldots, \boldsymbol{Z}_n \sim \text{i.i.d. } \mathrm{N}_p(\mathbf{0}, \Sigma).$$

Furthermore,

$$\bar{\boldsymbol{X}} = \frac{1}{\sqrt{n}}\boldsymbol{Z}_1, \quad \mathbf{W} = (n-1)\mathbf{S} = \sum_{i=2}^{n} \boldsymbol{Z}_i \boldsymbol{Z}_i'.$$

The expression of $\mathbf{W}$ shows that $\mathbf{W}$ is distributed as a Wishart distribution $\mathrm{W}_p(n-1, \Sigma)$. In general, a Wishart distribution is defined as follows.

Definition 2.1.1 If a $p \times p$ random matrix $\mathbf{W}$ is expressed as

$$\mathbf{W} = \sum_{j=1}^{n} \boldsymbol{Z}_j \boldsymbol{Z}_j',$$

where $\boldsymbol{Z}_j \sim \mathrm{N}_p(\boldsymbol{\mu}_j, \Sigma)$ and $\boldsymbol{Z}_1, \ldots, \boldsymbol{Z}_n$ are independent, $\mathbf{W}$ is said to have a *noncentral Wishart distribution* with n degrees of freedom, covariance matrix Σ, and noncentrality matrix $\Delta = \boldsymbol{\mu}_1\boldsymbol{\mu}_1' + \cdots + \boldsymbol{\mu}_n\boldsymbol{\mu}_n'$. We write that $\mathbf{W} \sim \mathrm{W}_p(n, \Sigma; \Delta)$. In the special case $\Delta = \mathrm{O}$, $\mathbf{W}$ is said to have a *Wishart distribution*, denoted by $\mathbf{W} \sim \mathrm{W}_p(n, \Sigma)$.

The noncentrality matrix is also defined as $\Delta = \Sigma^{-1}\mathrm{M}'\mathrm{M}$, not $\mathrm{M}'\mathrm{M}$, where $\mathrm{M} = (\boldsymbol{\mu}_1, \ldots, \boldsymbol{\mu}_n)'$. Using the $\boldsymbol{Z}_j$ in the definition of a Wishart distribution $\mathrm{W}_p(n, \Sigma)$, let $\mathbf{Z} = (\boldsymbol{Z}_1, \ldots, \boldsymbol{Z}_n)'$. Then $\mathbf{W} = \mathbf{Z}'\mathbf{Z}$. When $p = 1$, $W/\sigma^2 \sim \chi^2(n)$, where $\Sigma = \sigma^2$.

The Wishart distribution $\mathrm{W}_p(n, \Sigma)$ is defined as the distribution of $\mathbf{W} = \mathbf{Z}'\mathbf{Z}$, where the rows of $Z : n \times p$ are independent identically distributed as $\mathrm{N}_p(\mathbf{0}, \Sigma)$. When $n < p$, $\mathbf{W}$ is called *singular* and the $\mathrm{W}_p(n, \Sigma)$ distribution does not have a density function. When $n \geq p$, we derive the density function of $\mathbf{W}$ by using the following theorem due to Hsu (1940).

Theorem 2.1.1 *Let* $\mathbf{X}$ *be an* $n \times p$ *random matrix where* $n \geq p$. *If the density function of* $\mathbf{X}$ *has the form* $f(\mathrm{X}) = g(\mathrm{X}'\mathrm{X})$, *the density function of* $\mathbf{V} = \mathbf{X}'\mathbf{X}$ *is*

$$h(\mathrm{V}) = \frac{\pi^{pn/2}}{\Gamma_p(n/2)} |\mathrm{V}|^{(n-p-1)/2} g(\mathrm{V}),$$

where $\Gamma_p(a) = \pi^{p(p-1)/4} \prod_{i=1}^{p} \Gamma[a - (i-1)/2]$.

Proof. First we make the transformation $\mathbf{X} = \mathbf{ZT}$, where $\mathbf{T}$ is a $p \times p$ upper triangular matrix with positive diagonal elements $T_{ii} > 0$ and $\mathbf{Z}$ is an $n \times p$ column-orthogonal matrix (i.e., $\mathbf{Z}'\mathbf{Z} = \mathrm{I}_p$). Using [A.1.15], the transformation is one-to-one. Its Jacobian is $\mathrm{dX} = [2^p \pi^{pn/2}/\Gamma(n/2)] \prod_{i=1}^{p} t_{ii}^{n-i} \mathrm{dT} \cdot \mathrm{d(H)}$, where $\mathrm{dX} = \prod_{i=1}^{n}\prod_{j=1}^{p} \mathrm{d}x_{ij}$, $\mathrm{dT} = \prod_{i<j} \mathrm{d}t_{ij}$, and $\mathrm{d(H)}$ is the invariant measure defined over the totality of p orthogonal frames in n-space normalized so that $\int_{\mathrm{H'H=I}_p} \mathrm{d(H)} = 1$. This Jacobian was given by James (1954). Hence, the joint density function of $\mathbf{T}$ and $\mathbf{Z}$ is

$$\frac{2^p\pi^{pn/2}}{\Gamma_p(n/2)}\prod_{i=1}^{p} t_{ii}^{n-i} g(\mathrm{T'T}),$$

which is also the marginal density function of $\mathbf{T}$, since the joint density function does not depend on $\mathbf{Z}$. Now we make the transformation from $\mathbf{T}$ to $\mathbf{V} = \mathbf{T'T}$ whose Jacobian is, by [A.3.6], $J(\mathrm{T} \to \mathrm{V}) = 2^p \prod_{i=1}^{p} t_{ii}^{p-i+1}$ and hence $d\mathrm{T} = 2^{-p}\prod_{i=1}^{p} t_{ii}^{-p+i-1} d\mathrm{V}$. The final result is obtained by noting that $\prod_{i=1}^{p} t_{ii} = |\mathrm{V}|^{1/2}$. □

We consider the characteristic function of the noncentral Wishart distribution. Let $\mathbf{W} \sim \mathrm{W}_p(n, \Sigma; \Omega)$. Then without loss of generality we may write $\mathbf{W}$ as

$$\mathbf{W} = \sum_{i=1}^{n} \boldsymbol{X}_i\boldsymbol{X}_i',$$

where the random vectors $\boldsymbol{X}_i$ are mutually independent, each following the normal distribution $\mathrm{N}_p(\boldsymbol{\mu}_i, \Sigma)$ and $\Omega = \sum_{i=1}^{n} \boldsymbol{\mu}_i\boldsymbol{\mu}_i'$. The characteristic function of $\mathbf{W}$ is the expectation of

$$\begin{aligned}
&\exp\{i(t_{11}W_{11} + \cdots + t_{pp}W_{pp} + t_{12}W_{12} + \cdots + t_{p-1,p}W_{p-1,p})\}\\
&\quad = \exp(i \operatorname{tr} \mathrm{T}\mathbf{W})\\
&\quad = \exp(i\boldsymbol{X}_1'\mathrm{T}\boldsymbol{X}_1)\cdots\exp(i\boldsymbol{X}_n'\mathrm{T}\boldsymbol{X}_n),
\end{aligned}$$

where $\mathrm{T} = ((1+\delta_{ij})t_{ij}/2)$ is a symmetric matrix and δ_{ij} is the Kroneker delta [i.e., $\delta_{ij} = 1$, $\delta_{ij} = 0 (i \neq j)$]. Note that if $\boldsymbol{X} \sim \mathrm{N}_p(\boldsymbol{\mu}, \Sigma)$,

$$\mathrm{E}[\exp(i\boldsymbol{X}'\mathrm{T}\boldsymbol{X})] = |\mathrm{I} - 2i\Sigma\mathrm{T}|^{-1/2}\exp\{i\boldsymbol{\mu}'\mathrm{T}(\mathrm{I} - 2i\Sigma\mathrm{T})^{-1}\boldsymbol{\mu}\}.$$

The result is obtained by using

$$\begin{aligned}
&(\boldsymbol{X} - \boldsymbol{\mu})'\Sigma^{-1}(\boldsymbol{X} - \boldsymbol{\mu}) - 2i\boldsymbol{X}'\mathrm{T}\boldsymbol{X}\\
&\quad = (\boldsymbol{X} - \mathrm{A}\Sigma^{-1}\boldsymbol{\mu})'\mathrm{A}^{-1}(\boldsymbol{X} - \mathrm{A}\Sigma^{-1}\boldsymbol{\mu}) + \boldsymbol{\mu}'(\Sigma^{-1} - \Sigma^{-1}\mathrm{A}\Sigma^{-1})\boldsymbol{\mu},
\end{aligned}$$

where $\mathrm{A} = (\Sigma^{-1} - 2i\mathrm{T})^{-1}$. Therefore, the characteristic function of $\mathbf{W}$ is expressed as

$$C_{\mathbf{W}}(\mathrm{T}) = |\mathrm{I}_p - 2i\Sigma\mathrm{T}|^{-n/2}\exp\left\{i\sum_{j=1}^{n}\boldsymbol{\mu}_j'\mathrm{T}(\mathrm{I}_p - 2i\Sigma\mathrm{T})^{-1}\boldsymbol{\mu}_j\right\},$$

which gives the following result.

Theorem 2.1.2 *Let* $\mathbf{W}$ *be a noncentral Wishart matrix having* $\mathrm{W}_p(n, \Sigma; \Omega)$. *Then the characteristic function of* $\mathbf{W}$ *is given by*

$$C_{\mathbf{W}}(\mathrm{T}) = |\mathrm{I}_p - 2i\Sigma\mathrm{T}|^{-n/2}\,\mathrm{etr}\{i\Omega\mathrm{T}(\mathrm{I}_p - 2i\Sigma\mathrm{T})^{-1}\},$$

where $\mathrm{etr(A)} = \exp(\mathrm{tr\,A})$. *In particular, if* $\mathbf{W} \sim \mathrm{W}_p(n, \Sigma)$,

$$C_{\mathbf{W}}(\mathrm{T}) = |\mathrm{I}_p - 2i\mathrm{T}|^{-n/2}.$$

Theorem 2.1.3 *If* $\mathbf{W}$ *is* $\mathrm{W}_p(n, \Sigma)$ $(n \geq p)$, *the density function of* $\mathbf{W}$ *is*

$$\frac{1}{2^{pn/2}\Gamma_p(\frac{1}{2}n)|\Sigma|^{n/2}}\,\mathrm{etr}\Big(-\frac{1}{2}\Sigma^{-1}\mathrm{W}\Big)|\mathrm{W}|^{(n-p-1)/2}, \quad \mathrm{W} > \mathrm{O}, \tag{2.1.1}$$

where $\mathrm{etr(A)} = \exp(\mathrm{tr\,A})$.

Proof. Starting from the definition of $\mathbf{W}$, in Theorem 2.1.2 we have seen that the characteristic function of $\mathbf{W}$ is $|\mathrm{I}_p - 2i\mathrm{T}|^{-n/2}$. On the other hand, using the density function (2.1.1) we can obtain the characteristic function of $\mathbf{W}$ as $|\mathrm{I}_p - 2i\mathrm{T}|^{-n/2}$. The result follows from Theorem 1.1.3 on one-to-one correspondence between the characteristic function and its density function. □

2.2 Some Basic Properties

Theorem 2.2.1

(1) *If* $\mathbf{W}_1 \sim \mathrm{W}_p(n_1, \Sigma; \Delta_1)$, $\mathbf{W}_2 \sim \mathrm{W}_p(n_2, \Sigma; \Delta_2)$ *and they are independent, then* $\mathbf{W}_1 + \mathbf{W}_2 \sim \mathrm{W}_p(n_1 + n_2, \Sigma; \Delta_1 + \Delta_2)$.

(2) *If* $\mathbf{W} \sim \mathrm{W}_p(n, \Sigma; \Delta)$ *and* B *is a* $q \times p$ *matrix with rank* q, *then* $\mathrm{B}\mathbf{W}\mathrm{B}' \sim \mathrm{W}_q(n, \mathrm{B}\Sigma\mathrm{B}'; \mathrm{B}\Delta\mathrm{B}')$.

Proof. Result (1) is obtained easily from the definition of Wishart distribution. We can write $\mathbf{W}$ as

$$\mathbf{W} = \sum_{i=1}^{n} \boldsymbol{Z}_i\boldsymbol{Z}_i', \quad \boldsymbol{Z}_i \sim \mathrm{N}_p(\boldsymbol{\mu}_i, \Sigma)$$

and hence

$$\mathrm{B}\mathbf{W}\mathrm{B}' = \sum_{i=1}^{n}(\mathrm{B}\boldsymbol{Z}_i)(\mathrm{B}\boldsymbol{Z}_i)'.$$

Result (2) follows from the fact that $\mathrm{B}\boldsymbol{Z}_i \sim \mathrm{N}_q(\mathrm{B}\boldsymbol{\mu}_i, \mathrm{B}\Sigma\mathrm{B}')$.

Theorem 2.2.2 *Let* $\mathbf{W} \sim \mathrm{W}_p(n, \Sigma)$ *and* $\mathbf{W}$ *and* Σ *be partitioned as*

$$\mathbf{W} = \begin{pmatrix} \mathbf{W}_{11} & \mathbf{W}_{12} \\ \mathbf{W}_{21} & \mathbf{W}_{22} \end{pmatrix}, \quad \Sigma = \begin{pmatrix} \Sigma_{11} & \Sigma_{12} \\ \Sigma_{21} & \Sigma_{22} \end{pmatrix},$$

respectively, where $\mathbf{W}_{ij} : p_i \times p_j$ *and* $\Sigma_{ij} : p_i \times p_j$. *Then:*

(1) $\mathbf{W}_{ii} \sim \mathrm{W}_{p_i}(n, \Sigma_{ii})$, $i = 1, 2$.

(2) $\mathbf{W}_{11\cdot 2} = \mathbf{W}_{11} - \mathbf{W}_{12}\mathbf{W}_{22}^{-1}\mathbf{W}_{21} \sim \mathrm{W}_{p_1}(n - p_2, \Sigma_{11\cdot 2})$.

(3) $\mathbf{W}_{11\cdot 2}$ *and* $\{\mathbf{W}_{21}, \mathbf{W}_{22}\}$ *are independent.*

(4) *The conditional distribution of* $\mathbf{W}_{12} = (\boldsymbol{B}_1, \ldots. \boldsymbol{B}_{p_2})$ *is normal with mean* $\mathrm{E}(\mathbf{W}_{12}|\mathbf{W}_{22}) = \Sigma_{12}\Sigma_{22}^{-1}\mathbf{W}_{22}$ *and*

$$\mathrm{Cov}(\boldsymbol{B}_i, \boldsymbol{B}_j|\mathbf{W}_{22}) = (\mathbf{W}_{22})_{ij}\Sigma_{11\cdot 2},$$

where $(\mathrm{A})_{ij}$ *denotes the* (i, j)*th element of* A.

Proof. We may write $\mathbf{W}$ as

$$\mathbf{W} = \sum_{i=1}^{n} \boldsymbol{X}_i \boldsymbol{X}_i' = \sum_{i=1}^{n} \begin{pmatrix} \boldsymbol{X}_{i1} \\ \boldsymbol{X}_{i2} \end{pmatrix} \begin{pmatrix} \boldsymbol{X}_{i1} \\ \boldsymbol{X}_{i2} \end{pmatrix}',$$

where $\boldsymbol{X}_1, \ldots, \boldsymbol{X}_n \sim$ i.i.d. $\mathrm{N}_p(\mathbf{0}, \Sigma)$, and $\boldsymbol{X}_{i1} : p_i \times 1$. Then

$$\mathbf{W}_{11} = \sum_{i=1}^{n} \boldsymbol{X}_{i1}\boldsymbol{X}_{i1}', \quad \boldsymbol{X}_{i1}, \ldots, \boldsymbol{X}_{in} \sim \text{i.i.d. } \mathrm{N}_p(\mathbf{0}, \Sigma_{11}).$$

This shows (1). Let

$$\mathbf{X} = (\boldsymbol{X}_1, \ldots, \boldsymbol{X}_n)' = (\mathbf{X}_1 \ \mathbf{X}_2), \quad \mathbf{X}_1 : n \times p_1;$$

that is, $\mathbf{X}_1 = (\boldsymbol{X}_{11}, \ldots, \boldsymbol{X}_{1n})'$ and $\mathbf{X}_2 = (\boldsymbol{X}_{21}, \ldots, \boldsymbol{X}_{2n})'$. Consider the distribution of

$$\mathbf{W}_{11\cdot 2} = \mathbf{X}_1'\{\mathrm{I}_n - \mathbf{X}_2(\mathbf{X}_2'\mathbf{X}_2)^{-1}\mathbf{X}_2\}\mathbf{X}_1$$

given $\mathbf{X}_2$. Then $\boldsymbol{X}_{11}, \ldots, \boldsymbol{X}_{n1}$ are independent and

$$\boldsymbol{X}_{i1}|\boldsymbol{X}_{i2} \sim \mathrm{N}_{p_1}(\Sigma_{12}\Sigma_{22}^{-1}\boldsymbol{X}_{i2}, \Sigma_{11\cdot 2}).$$

Let $\mathbf{H} = (\mathbf{H}_1 \ \mathbf{H}_2)$ be the orthogonal matrix such that $\mathbf{H}_1 = \mathbf{X}_2(\mathbf{X}_2'\mathbf{X}_2)^{-1/2}$, and consider

$$\mathbf{Z} = (\boldsymbol{Z}_1, \ldots, \boldsymbol{Z}_n)' = \mathbf{H}'\mathbf{X}_1.$$

From Theorem 1.2.6 it is seen that $\boldsymbol{Z}_{p_2+1}, \ldots, \boldsymbol{Z}_n \sim$ i.i.d. $\mathrm{N}_{p_2}(\mathbf{0}, \Sigma_{11\cdot 2})$. Further,

$$\mathbf{W}_{11\cdot 2} = \mathbf{X}_1'\mathbf{X}_1 - \mathbf{X}_1\mathbf{H}_1\mathbf{H}_1'\mathbf{X}_1' = \sum_{i=p_2+1}^{n} \boldsymbol{Z}_i\boldsymbol{Z}_i'.$$

This implies that $\mathbf{W}_{11\cdot 2} \sim \mathrm{W}_{p_1}(n-p_2, \Sigma_{11\cdot 2})$, and that $\mathbf{W}_{11\cdot 2}$ does not depend on $\mathbf{X}_2$. The latter shows that $\mathbf{W}_{11\cdot 2}$ and $\mathbf{W}_{22}$ are independent. Further, we can express

$$\mathbf{W}_{11\cdot 2} = (\mathbf{X}_1'\mathbf{H}_2)\cdot(\mathbf{X}_1'\mathbf{H}_2)', \quad \mathbf{W}_{12} = \mathbf{X}_1\mathbf{H}_1 \cdot (\mathbf{X}_2'\mathbf{X}_2)^{1/2},$$

and hence the conditional distributions of $\mathbf{W}_{11\cdot 2}$ and $\mathbf{W}_{12}$ given $\mathbf{W}_{22}$ are independent. These imply (3). For a proof of (4), since $\mathbf{W}_{22} = \mathbf{X}_1'\mathbf{X}_2$, it is easily seen that the conditional distribution of $\mathbf{W}_{12}$ given $\mathbf{X}_2$ is normal with

$$\mathrm{E}(\mathbf{W}_{12}|\mathbf{X}_2) = \Sigma_{12}\Sigma_{22}^{-1}\mathbf{X}_2'\mathbf{X}_2 = \Sigma_{12}\Sigma_{22}^{-1}\mathbf{W}_{22}.$$

To compute its covariance matrix, put

$$\mathbf{X}_2 = (\boldsymbol{U}_1, \ldots, \boldsymbol{U}_{p_2}), \quad \boldsymbol{U}_i = (U_{1i}, \ldots, U_{ni})'.$$

Then using

$$\boldsymbol{B}_i = U_{1i}X_{11} + U_{2i}X_{21} + \cdots + U_{ni}X_{n1}, \quad i = 1, \ldots, p_2,$$

we obtain the desired result. □

From Theorem 2.2.2 we have:

(4)′ the conditional distribution of $\mathbf{W}_{12}\mathbf{W}_{22}^{-1/2} = (\tilde{\boldsymbol{B}}_1, \ldots, \tilde{\boldsymbol{B}}_{p_2})$ given $\mathbf{W}_{22}$ is normal with mean $\mathrm{E}(\mathbf{W}_{12}\mathbf{W}_{22}^{-1/2}|\mathbf{W}_{22}) = \Sigma_{12}\Sigma_{22}^{-1}\mathbf{W}_{22}^{1/2}$ and covariance matrix

$$\mathrm{Var}(\tilde{\boldsymbol{B}}_i|\mathbf{W}_{22}) = \Sigma_{11\cdot 2}, \quad \mathrm{Cov}(\tilde{\boldsymbol{B}}_i, \tilde{\boldsymbol{B}}_j|\mathbf{W}_{22}) = \mathrm{O}, \quad i \neq j.$$

Theorem 2.2.3 *If $\Sigma_{12} = \mathrm{O}$ in Theorem 2.2.2, then:*

(1) $\mathbf{W}_{11\cdot 2}$, $\mathbf{W}_{12}\mathbf{W}_{22}^{-1/2}$, *and* $\mathbf{W}_{22}$ *are mutually independent.*

(2) $\mathbf{W}_{11\cdot 2} \sim \mathrm{W}_{p_1}(n - p_2, \Sigma_{11})$, $\mathbf{W}_{22} \sim \mathrm{W}_{p_2}(n, \Sigma_{22})$, *and the columns of* $\mathbf{W}_{12}\mathbf{W}_{22}^{-1/2}$ *are independent and identically distributed as* $\mathrm{N}_{p_1}(\mathbf{0}, \Sigma_{11})$, *and hence* $\mathbf{W}_{12}\mathbf{W}_{22}^{-1}\mathbf{W}_{21} \sim \mathrm{W}_{p_1}(n, \Sigma_{11})$.

Proof. For a proof of (1), from Theorem 2.2.2(3) it is sufficient to show that $\mathbf{W}_{12}\mathbf{W}_{22}^{-1/2}$ is independent of $\mathbf{W}_{22}$. This follows from the fact that the conditional distribution of $\mathbf{W}_{12}\mathbf{W}_{22}^{-1/2}$ given $\mathbf{W}_{22}$ does not depend on $\mathbf{W}_{22}$. The second result follows the results of Theorem 2.2.2 and (4)′ in the case of $\Sigma_{12} = \mathrm{O}$. □

For an extension of Theorem 2.2.3 to the case when $\mathbf{W}$ has a noncentral Wishart distribution, see Kabe (1964) and Problem 2.8.

Theorem 2.2.4 *If* $\mathbf{W} \sim \mathrm{W}_p(n, \Sigma)$, *then*

$$\mathrm{E}[|\mathbf{W}|^h] = 2^{ph}|\Sigma|^h \prod_{i=1}^{p} \frac{\Gamma[\frac{1}{2}(n-i+1)+h]}{\Gamma[\frac{1}{2}(n-i+1)]}, \quad \Re(h) > -\frac{1}{2}n + \frac{1}{2}(p-1).$$

Proof. Let $C(p,n;\Sigma) = [2^{pn/2}\Gamma_p(\frac{1}{2}n)|\Sigma|^{n/2}]^{-1}$. Then, using Theorem 2.1.3

$$\begin{aligned}\mathrm{E}[|\mathbf{W}|^h] &= C(p,n;\Sigma)\int_{\mathrm{W}} |\mathrm{W}|^{(n+2h-p-1)/2} \times \operatorname{etr}\left(-\frac{1}{2}\Sigma^{-1}\mathrm{V}\right) d\mathrm{V} \\ &= \frac{C(p,n;\Sigma)}{C(p,n+2h;\Sigma)}\end{aligned}$$

which is equal to the required expression. Note that the above integral is extended to a complex number such that $\Re(n+2h) > p-1$. □

The first two *moments of Wishart* and *inverse Wishart matrices* are given in the following Theorems 2.2.5 through 2.2.8. For higher moments, see, for example Gupta and Nagar (2000) and Mathai et al. (1995).

Theorem 2.2.5 *Let* $\mathbf{W} \sim \mathrm{W}_p(n,\Sigma)$ *and let* A *be a* $p \times p$ *constant matrix. Then*

(1) $\mathrm{E}(\mathbf{W}) = n\Sigma$.

(2) $\mathrm{E}(\mathbf{WAW}) = n\Sigma\mathrm{A}'\Sigma + n\,\mathrm{tr}\,(\Sigma\mathrm{A})\Sigma + n^2\Sigma\mathrm{A}\Sigma$.

Theorem 2.2.6 *Let* $\mathbf{W} \sim \mathrm{W}_p(n,\Sigma)$ *and let* A *and* B *be* $p \times p$ *constant matrices. Then*

(1) $\mathrm{E}(\mathrm{tr}\,\mathrm{A}\mathbf{W}) = n\,\mathrm{tr}\,\mathrm{A}\Sigma$.

(2) $\mathrm{E}[\mathrm{tr}\,(\mathrm{A}\mathbf{W})\,\mathrm{tr}\,(\mathrm{B}\mathbf{W})] = n\,\mathrm{tr}\,(\mathrm{A}\Sigma\mathrm{B}\Sigma) + n\,\mathrm{tr}\,(\mathrm{A}'\Sigma\mathrm{B}\Sigma) + n^2\,\mathrm{tr}\,(\mathrm{A}\Sigma)\,\mathrm{tr}\,(\mathrm{B}\Sigma)$.

(3) $\mathrm{E}[\mathrm{tr}\,(\mathrm{A}\mathbf{W}\mathrm{B}\mathbf{W})] = n\,\mathrm{tr}\,(\mathrm{A}\Sigma\mathrm{B}'\Sigma) + n\,\mathrm{tr}\,(\mathrm{A}\Sigma)\,\mathrm{tr}\,(\mathrm{B}\Sigma) + n^2\,\mathrm{tr}\,(\mathrm{A}\Sigma\mathrm{B}\Sigma)$.

Theorem 2.2.7 *Let* $\mathbf{W} \sim \mathrm{W}_p(n,\Sigma)$, *and let* A *be a* $p \times p$ *constant matrix. Then*

(1) $\mathrm{E}(\mathbf{W}^{-1}) = \dfrac{1}{n-p-1}\Sigma, \quad n-p-1 > 0.$

(2) $\mathrm{E}(\mathbf{W}^{-1}\mathrm{A}\mathbf{W}^{-1}) = c_1\Sigma^{-1}\mathrm{A}\Sigma^{-1} + c_2[\Sigma^{-1}\mathrm{A}'\Sigma^{-1} + \mathrm{tr}\,(\mathrm{A}\Sigma^{-1})\Sigma^{-1}]$,

where $c_1 = (n-p-2)c_2$ *and* $c_2 = \{(n-p)(n-p-1)(n-p-3)\}^{-1}$.

Theorem 2.2.8 *Let* $\mathbf{W} \sim \mathrm{W}_p(n, \Sigma)$, *and let* A *and* B *be* $p \times p$ *constant matrices. Let* c_1 *and* c_2 *be the constants in Theorem 2.2.7. Then:*

$$\text{(1)}\quad \mathrm{E}[\,\mathrm{tr}\,(\mathrm{A}\mathbf{W}^{-1})] = \frac{1}{n-p-1}\,\mathrm{tr}\,(\mathrm{A}\Sigma), \quad n-p-1>0.$$

$$\text{(2)}\quad \mathrm{E}[(\,\mathrm{tr}\,\mathrm{A}\mathbf{W}^{-1})(\,\mathrm{tr}\,\mathrm{B}\mathbf{W}^{-1})] = c_1\,\mathrm{tr}\,(\mathrm{A}\Sigma^{-1})\,\mathrm{tr}\,(\mathrm{B}\Sigma^{-1}) + c_2[\,\mathrm{tr}\,(\mathrm{B}\Sigma^{-1}\mathrm{A}\Sigma^{-1} + \mathrm{tr}\,(\mathrm{B}\Sigma^{-1}\mathrm{A}'\Sigma^{-1})].$$

$$\text{(3)}\quad \mathrm{E}[\,\mathrm{tr}\,(\mathrm{A}\mathbf{W}^{-1}\mathrm{B}\mathbf{W}^{-1})] = c_1\,\mathrm{tr}\,(\mathrm{A}\Sigma^{-1}\mathrm{B}\Sigma^{-1}) + c_2[\,\mathrm{tr}\,(\mathrm{A}\Sigma^{-1}\mathrm{B}'\Sigma^{-1}) + \mathrm{tr}\,(\mathrm{A}\Sigma^{-1})\,\mathrm{tr}\,(\mathrm{B}\Sigma^{-1})].$$

Theorem 2.2.9 *Let* $\mathbf{W}$ *be a Wishart matrix having* $\mathrm{W}_p(n, \Sigma)$ *and* B *be a* $p \times q$ *constant matrix with rank* q. *Put*

$$\mathbf{Q} = (\mathrm{B}'\mathbf{W}^{-1}\mathrm{B})^{-1}\mathrm{B}'\mathbf{W}^{-1}\Sigma\mathbf{W}^{-1}\mathrm{B}(\mathrm{B}'\mathbf{W}^{-1}\mathrm{B})^{-1}.$$

Then

$$\mathrm{E}(\mathbf{Q}) = \frac{n-1}{n-(p-q)-1}(\mathrm{B}'\Sigma^{-1}\mathrm{B})^{-1}.$$

Proof. We may obtain the expectation of

$$\mathbf{Q} = (\mathrm{C}'\mathbf{W}^{-1}\mathrm{C})^{-1}\mathrm{C}'\mathbf{W}^{-2}\mathrm{C}(\mathrm{C}'\mathbf{W}^{-1}\mathrm{C})^{-1},$$

when $\mathbf{W} \sim \mathrm{W}_p(n, \mathrm{I}_p)$, where $\mathrm{C} = \Sigma^{-1/2}\mathrm{B}$. Let $\mathrm{H} = (\mathrm{H}_1\ \mathrm{H}_2)$ be an orthogonal matrix such that $\mathrm{H}_1 = \mathrm{C}(\mathrm{C}'\mathrm{C})^{-1/2}$. Then $\mathbf{V} = \mathrm{H}'\mathbf{W}\mathrm{H} \sim \mathrm{W}_p(m, \mathrm{I}_p)$. We decompose $\mathbf{V}$ as

$$\mathbf{V} = \begin{pmatrix} \mathbf{V}_{11} & \mathbf{V}_{12} \\ \mathbf{V}_{21} & \mathbf{V}_{22} \end{pmatrix}, \quad \mathbf{V}_{12} : q \times p,$$

and $\mathbf{Q}$ is expressed as

$$\mathbf{Q} = (\mathrm{B}'\Sigma^{-1}\mathrm{B})^{-1/2}\{\mathrm{I}_q + \mathbf{V}_{12}\mathbf{V}_{22}^{-2}\mathbf{V}_{21}\}(\mathrm{B}'\Sigma^{-1}\mathrm{B})^{-1/2}.$$

The required result is obtained by using Theorem 2.2.3. □

2.3 Functions of Wishart Matrices

Some properties of a Wishart matrix can be obtained by decomposing it to a triangular matrix. Let $\mathbf{W}$ be a $p \times p$ positive definite random matrix. Then, using [A.1.13] there exists a lower triangular matrix $\mathbf{T}$ such that

$$\mathbf{W} = \mathbf{T}\mathbf{T}', \quad \mathbf{T} = \begin{pmatrix} T_{11} & & & \\ T_{21} & T_{22} & & \mathrm{O} \\ \vdots & \vdots & \ddots & \\ T_{p1} & T_{p2} & \cdots & T_{pp} \end{pmatrix}.$$

Theorem 2.3.1 *Let* $\mathbf{W}$ *be* $\mathrm{W}_p(n, \Sigma)$, *where* $n \geq p$, *and put* $\mathbf{W} = \mathbf{TT}'$, *where* $\mathbf{T}$ *is a lower triangular matrix with positive diagonal elements. Then the pdf of* $\mathbf{T}$ *is given by*

$$f(\mathrm{T}) = \frac{2^p}{K} \prod_{i=1}^{p} t_{ii}^{n-i} \operatorname{etr}\Big(-\frac{1}{2}\Sigma^{-1}\mathrm{TT}' \Big),$$

where $K = 2^{np/2}|\Sigma|^{n/2}\Gamma_p(n/2)$.

Proof. The density of $\mathbf{W}$ is given (Theorem 2.1.3) by

$$\frac{1}{K} \operatorname{etr}\Big(-\frac{1}{2}\Sigma^{-1}\mathrm{W} \Big) |\mathrm{W}|^{(n-p-1)/2}.$$

Put $\mathrm{W} = \mathrm{TT}'$, where T is a lower triangular matrix.

$$\operatorname{tr} \Sigma^{-1}\mathrm{W} = \operatorname{tr} \Sigma^{-1}\mathrm{TT}',$$

$$|\mathrm{W}| = |\mathrm{TT}'| = |\mathrm{T}|^2 = \prod_{i=1}^{k} t_{ii}^2.$$

The required result is proved by seeing that the Jacobian is [see, e.g. Muirhead, 1982; Siotani et al., 1985]

$$\left| \frac{\partial w_{ij}}{\partial t_{k\ell}} \right| = 2^p \prod_{i=1}^{p} t_{ii}^{p-i+1}. \qquad \square$$

Corollary 2.3.1 *If* $\Sigma = \mathrm{I}_p$, *the elements* $T_{ij}(p \geq i \geq j \geq 1)$ *are all independent,* $T_{ii}^2 \sim \chi^2(n-i+1)(i = 1, \ldots, p)$, *and* $T_{ij} \sim N(0,1)(p \geq i > j \geq 1)$.

Proof. Since $\Sigma = \mathrm{I}_p$,

$$\operatorname{tr} \Sigma^{-1}\mathrm{TT}' = \operatorname{tr} \mathrm{TT}' = \sum_{i \geq j}^{p} t_{ij}^2.$$

Therefore, the joint density of $T_{ij}(i \geq j)$ can be written in the form

$$\prod_{i=1}^{p} \left\{ \frac{1}{2^{n-i+1}\Gamma[\frac{1}{2}(n-i+1)]} \exp\left(-\frac{1}{2}t_{ii}^2 \right) (t_{ij}^2)^{(n-i-1)/2} \right\}$$
$$\times \prod_{i>j}^{p} \left\{ \frac{1}{\sqrt{2\pi}} \exp\left(-\frac{1}{2}t_{ij}^2 \right) \right\},$$

which is the product of the marginal density functions for the elements of $\mathbf{T}$ stated in the corollary. $\square$

The result in Corollary 2.3.1, due to Bartlett (1933), is called *Bartlett's decomposition.*

Let $\mathbf{S}$ be the sample covariance matrix based on a sample of size n from $\mathrm{N}_p(\boldsymbol{\mu}, \Sigma)$. The sample generalized variance is defined by $|\mathbf{S}|$. Then since $(n-1)\mathbf{S} \sim \mathrm{W}_p(n-1, \Sigma)$, the distribution of $|\mathbf{S}|$ will be obtained based on the following theorem.

Theorem 2.3.2 *If $\mathbf{W} \sim \mathrm{W}(n, \Sigma)$, where $n \geq p$, then $|\mathbf{W}|/|\Sigma|$ has the same distribution as that of $\prod_{i=1}^p \chi_{n-i+1}^2$, where for $i = 1, \ldots, p$, χ_{n-i+1}^2 denote mutually independent random variables, each following the chi-square distribution with n-i+1 degrees of free.*

Proof. Note that

$$\frac{|\mathbf{W}|}{|\Sigma|} = |\Sigma^{-1/2}\mathbf{W}\Sigma^{-1/2}|$$

and $\Sigma^{-1/2}\mathbf{W}\Sigma^{-1/2} \sim \mathrm{W}_p(n, \mathrm{I}_p)$. Let T be a lower triangular matrix such that $\Sigma^{-1/2}\mathbf{W}\Sigma^{-1/2} = \mathbf{TT}'$. Then the result follows from Corollary 2.3.1 by taking $\mathbf{TT}' = \prod_{i=1}^p T_{ii}^2 = \prod_{i=1}^p \chi_{n-i+1}^2$. □

Theorem 2.3.3 *If $\mathbf{W}$ is distributed as $\mathrm{W}_p(n, \Sigma)$ and M is a $q \times p$ matrix of rank q, then $(\mathrm{M}\mathbf{W}^{-1}\mathrm{M}')^{-1}$ is distributed as $\mathrm{W}_q(n-p+q, (\mathrm{M}\Sigma^{-1}\mathrm{M}')^{-1})$.*

Proof. Put $\tilde{\mathbf{W}} = \Sigma^{-1/2}\mathbf{W}\Sigma^{-1/2}$, which is distributed [Theorem 2.2.1 (2)] as $\mathrm{W}_p(n, \mathrm{I}_p)$. Then

$$(\mathrm{M}\mathbf{W}^{-1}\mathrm{M}')^{-1} = (\mathrm{R}\tilde{\mathbf{W}}^{-1}\mathrm{R}')^{-1},$$

where $\mathrm{R} = \mathrm{M}\Sigma^{-1/2}$. Since $(\mathrm{RR}')^{-1} = (\mathrm{M}\Sigma^{-1}\mathrm{M}')^{-1}$, it is sufficient to show that $(\mathrm{R}\tilde{W}^{-1}\mathrm{R}')^{-1} \sim \mathrm{W}_q(n-p+q, (\mathrm{RR}')^{-1})$. Consider an orthogonal matrix $\mathrm{H} = (\mathrm{H}_1\ \mathrm{H}_2)$, whose first q columns H_1 are $\mathrm{H}_1 = \mathrm{R}'(\mathrm{RR}')^{-1/2}$. Then $\mathbf{V} = \mathrm{H}'\tilde{\mathbf{W}}\mathrm{H} \sim \mathrm{W}_p(n, \mathrm{I}_p)$ and let $\mathbf{V}$ partition as

$$\mathbf{V} = \begin{pmatrix} \mathbf{V}_{11} & \mathbf{V}_{12} \\ \mathbf{V}_{21} & \mathbf{V}_{22} \end{pmatrix}, \quad \mathbf{V}_{12} : q \times (p-q).$$

Noting that the first $q \times q$ submatrix of $\mathbf{V}^{-1}$ is expressed (see [A.1.2]) as $\mathbf{V}_{11\cdot 2} = \mathbf{V}_{11} - \mathbf{V}_{12}\mathbf{V}_{22}^{-1}\mathbf{V}_{21}$,

$$\begin{aligned} (\mathrm{R}\tilde{\mathbf{W}}^{-1}\mathrm{R}')^{-1} &= \{(\mathrm{RH})\mathbf{V}^{-1}(\mathrm{RH})'\}^{-1} \\ &= (\mathrm{RR}')^{-1/2}\{(\mathrm{I}\ \mathrm{O})\mathbf{V}^{-1}(\mathrm{I}\ \mathrm{O})'\}^{-1}(\mathrm{RR}')^{-1/2} \\ &= (\mathrm{RR}')^{-1/2}\mathbf{V}_{11\cdot 2}(\mathrm{RR}')^{-1/2}. \end{aligned}$$

Therefore, the required result is obtained by using Theorems 2.2.2(2) and 2.2.1(2). □

2.4 Cochran's Theorem

Cochran's theorem [Cochran, 1934] is useful in proving that certain vector quadratic forms $\boldsymbol{X}'\mathrm{A}_1\boldsymbol{X}, \ldots, \boldsymbol{X}'\mathrm{A}_k\boldsymbol{X}$ are independently distributed as central or noncentral chi-squares distributions, where $\boldsymbol{X} \sim \mathrm{N}_p(\boldsymbol{\mu}, \sigma^2\mathrm{I}_n)$ and A_i's are $n\times n$ symmetric matrices. There are various versions of *Cochran's theorem*, which include the following:

Theorem 2.4.1 *Let* $\mathbf{X} = (\boldsymbol{X}_1, \ldots, \boldsymbol{X}_n)'$ *be an n-vector such that* $\boldsymbol{X}_1, \ldots, \boldsymbol{X}_n$ *are independent and* $\boldsymbol{X}_i \sim \mathrm{N}_p(\boldsymbol{\mu}_i, \Sigma)$, $i = 1, \ldots, n$. *Suppose that the random variable* $\mathbf{Q} = \mathbf{X}'\mathrm{A}\mathbf{X}$ *is decomposed into a sum of k quadratic forms* $Q_i(\mathbf{X}) = \mathbf{X}'\mathrm{A}_i\mathbf{X}$, $i = 1, \ldots, k$:

$$Q(\mathbf{X}) = Q_1(\mathbf{X}) + \cdots + Q_k(\mathbf{X}),$$

where A *and* A_i $(i = 1, \ldots, k)$ *are* $n\times n$ *symmetric matrices of ranks* r *and* r_i. *If* A *is an idempotent matrix, then* $Q_i(\mathbf{X})$, $i = 1, \ldots, k$ *are independent and each has the noncentral Wishart* $\mathrm{W}_p(r_i, \Sigma; \Omega_i)$ *if and only if* $r = r_1 + \cdots + r_k$, *where* $\Omega_i = \mathrm{M}'\mathrm{A}_i\mathrm{M}$ *and* $\mathrm{M} = (\boldsymbol{\mu}_1, \ldots, \boldsymbol{\mu}_n)'$.

As is well known, Theorem 2.4.1 is a statistical statement for a matrix decomposition

$$\mathrm{A} = \mathrm{A}_1 + \cdots + \mathrm{A}_k.$$

For its details and a proof of Theorem 2.4.1, see, for example, Siotani et al. (1985). Here, we state a simple multivariate version of Cochran's theorem, which will be used often.

Theorem 2.4.2 *Let* $\mathbf{X} = (\boldsymbol{X}_1, \ldots, \boldsymbol{X}_n)'$, *where* $\boldsymbol{X}_i \sim \mathrm{N}_p(\boldsymbol{\mu}_i, \Sigma)$, $i = 1, \ldots, n$ *and* $\boldsymbol{X}_1, \ldots, \boldsymbol{X}_n$ *are independent. Let* A, A_1, *and* A_2 *be* $n \times n$ *symmetric matrices and* B *be a* $q \times n$ *matrix with rank* q. *Then:*

(1) $\mathbf{X}'\mathrm{A}\mathbf{X} \sim \mathrm{W}_p(k, \Sigma; \Delta) \Longleftrightarrow \mathrm{A}^2 = \mathrm{A},\ \mathrm{tr}\,\mathrm{A} = k, \Delta = \mathrm{E}(\mathbf{X})'\mathrm{A}\mathrm{E}(\mathbf{X})$.

(2) $\mathbf{X}'\mathrm{A}_1\mathbf{X}$ *and* $\mathbf{X}'\mathrm{A}_2\mathbf{X}$ *are independent* $\Longleftrightarrow \mathrm{A}_1\mathrm{A}_2 = \mathrm{O}$.

(3) $\mathbf{X}'\mathrm{A}\mathbf{X}$ *and* $\mathrm{B}\mathbf{X}$ *are independent* $\Longleftrightarrow \mathrm{BA} = \mathrm{O}$.

Proof. For a $p \times 1$ constant vector $\boldsymbol{a}$, $\boldsymbol{Y} = \mathbf{X}\boldsymbol{a} \sim N_n(\boldsymbol{\eta}, \sigma^2\mathrm{I}_p)$, where $\boldsymbol{\eta} = (\boldsymbol{\mu}_1, \ldots, \boldsymbol{\mu}_n)'\boldsymbol{a}$ and $\sigma^2 = \boldsymbol{a}'\Sigma\boldsymbol{a}$. The necessity ($\Longrightarrow$) of each of (1), (2), and (3) follows from the results in the univariate case. Now we prove the sufficiency of (1). For an $n\times n$ symmetric matrix A, there exist an orthogonal matrix H and a diagonal matrix $\mathrm{L} = \mathrm{diag}\,(\ell_1, \ldots, \ell_p)$, $\ell_1 \geq \cdots \geq \ell_n$ such that $\mathrm{A} = \mathrm{HLH}'$. Since $\mathrm{A}^2 = \mathrm{A}$ and $\mathrm{tr}\,A = k$, $\ell_1 = \cdots = \ell_k = 1$ and $\ell_{k+1} = \cdots = \ell_p = 0$. Put

$$\mathbf{Z} = (\boldsymbol{Z}_1, \ldots, \boldsymbol{Z}_n)' = \mathrm{H}'\mathbf{X}.$$

Then, $\boldsymbol{Z}_i$'s are independent and normal with the same covariance matrix Σ. Further

$$\mathbf{X}'\mathrm{A}\mathbf{X} = \mathbf{Z}'\mathrm{L}\mathbf{Z} = \sum_{i=1}^{k} \boldsymbol{Z}_i \boldsymbol{Z}_i' \sim \mathrm{W}_p(k, \Sigma; \Omega),$$

where $\Omega = \sum_{i=1}^k \mathrm{E}(\boldsymbol{Z}_i)\mathrm{L}\mathrm{E}(\boldsymbol{Z}_i)'$. The proof is completed by seeing that $\Omega = \mathrm{E}(\mathbf{Z})'\mathrm{L}\mathrm{E}(\mathbf{Z}) = \mathrm{E}(\mathbf{X})'\mathrm{A}\mathrm{E}(\mathbf{X})$. Next we prove the sufficiency ($\Leftarrow$) of (2). We can write $\mathrm{A}_i = \mathrm{H}_i\mathrm{L}_i\mathrm{H}_i$, $i = 1, 2$, where $\mathrm{H}_i'\mathrm{H}_i = \mathrm{I}_{n_i}$ and $\mathrm{L}_i = \mathrm{diag}\,(\ell_{i1}, \ldots, \ell_{im_i})$ with $\ell_{ij} \neq 0$. Since $\mathrm{A}_1'\mathrm{A}_2 = \mathrm{O}$, we can construct an orthogonal matrix $\mathrm{H} = (\mathrm{H}_1, \mathrm{H}_2, \mathrm{H}_3)$. Put $\mathrm{H}_i'\mathbf{X} = \mathbf{Z}_i$, $i = 1, 2, 3$. Then $\mathbf{Z}_1$, $\mathbf{Z}_2$, and $\mathbf{Z}_3$ are independent. The result is obtained by noting that $\mathbf{X}'\mathrm{A}_1\mathbf{X} = \mathbf{Z}_1'\mathrm{L}_1\mathbf{Z}_1$ and $\mathbf{X}'\mathrm{A}_2\mathbf{X} = \mathbf{Z}_2'\mathrm{L}_2\mathbf{Z}_2$. Similarly, the sufficiency ($\Leftarrow$) of (3) is proved. □

As an example of the use of Theorem 2.4.2, let $\bar{\boldsymbol{X}}$ and $\mathbf{S}$ be the sample mean vector and sample covariance matrix based on a random sample $\mathbf{X} = (\boldsymbol{X}_1, \ldots, \boldsymbol{X}_n)'$ of size n from $\mathrm{N}_p(\boldsymbol{\mu}, \Sigma)$. Then

$$\bar{\boldsymbol{X}} = \mathbf{X}' \cdot \frac{1}{n}\mathbf{1}_n, \quad (n-1)\mathbf{S} = \mathbf{X}'(\mathrm{I}_n - \mathrm{P}_0)\mathbf{X},$$

where $\mathbf{1}_n = (1, \ldots, 1)'$ and $\mathrm{P}_0 = (1/n)\mathbf{1}_n\mathbf{1}_n'$. It is easy to see that $\mathrm{I}_n - \mathrm{P}_0$ is an idempotent matrix and $(\mathrm{I}_n - \mathrm{P}_0)\mathbf{1}_n = \mathbf{0}$. Further, $\mathrm{tr}\,(\mathrm{I}_n - \mathrm{P}_0) = n - 1$ and $\Omega = \mathrm{M}'(\mathrm{I}_n - \mathrm{P}_0)\mathrm{M} = \mathrm{O}$ when $\mathrm{M} = (\boldsymbol{\mu}, \ldots, \boldsymbol{\mu})' = \mathbf{1}_n\boldsymbol{\mu}'$. Therefore, $\bar{\boldsymbol{X}}$ and $\mathbf{S}$ are independent and $(n-1)\mathbf{S} \sim \mathrm{W}_p(n-1, \Sigma)$.

2.5 Asymptotic Distributions

Let $n\mathbf{S} \sim \mathrm{W}_p(n, \Sigma)$ and consider *asymptotic distribution* of

$$\mathbf{V} = \sqrt{n}(\mathbf{S} - \Sigma). \tag{2.5.1}$$

The characteristic function of $\mathbf{V}$ is expressed as

$$\begin{aligned} C_{\mathbf{V}}(\mathrm{T}) &= \mathrm{E}[e^{i\,\mathrm{tr}\,\mathrm{T}\mathbf{V}}] \\ &= \exp(-\sqrt{n}i\,\mathrm{tr}\,\mathrm{T}\Sigma) \times \mathrm{E}[e^{i\,\mathrm{tr}\,(1/\sqrt{n})\mathrm{T}\cdot n\mathbf{S}}], \end{aligned}$$

where $\mathrm{T} = (1 + \delta_{ij})t_{ij}/2$. Using Theorem 2.1.2, we have

$$C_{\mathbf{V}}(\mathrm{T}) = \exp(-\sqrt{n}i\,\mathrm{tr}\,\Sigma\mathrm{T}) \times \left|\mathrm{I}_p - \frac{2i}{\sqrt{n}}\Sigma\mathrm{T}\right|^{-n/2}.$$

For any symmetric matrix A,

$$-\log|\mathrm{I}_p - \varepsilon\mathrm{A}| = \varepsilon\,\mathrm{tr}\,\mathrm{A} + \frac{1}{2}\varepsilon^2\,\mathrm{tr}\,\mathrm{A}^2 + \frac{1}{3}\varepsilon^3\,\mathrm{tr}\,\mathrm{A}^3 + \cdots. \tag{2.5.2}$$

Therefore

$$\begin{aligned}\log C_{\mathbf{V}}(\mathrm{T}) &= -\sqrt{n}\, i \operatorname{tr} \mathrm{T}\Sigma - \frac{n}{2}\log\left|\mathrm{I}_p - \frac{2i}{\sqrt{n}}\Sigma\mathrm{T}\right| \\ &= -\operatorname{tr}(\Sigma\mathrm{T})^2 + \mathrm{O}\left(\frac{1}{\sqrt{n}}\right),\end{aligned}$$

and hence

$$C_{\mathbf{V}}(\mathrm{T}) = \exp\{-\operatorname{tr}(\Sigma\mathrm{T})^2\} + \mathrm{O}\left(\frac{1}{\sqrt{n}}\right).$$

The asymptotic covariance between v_{ij} and $v_{k\ell}$ is:

$$\text{The coefficient of } t_{ij}t_{k\ell} \text{ in } 2\operatorname{tr}(\Sigma\mathrm{T})^2 = \sigma_{ik}\sigma_{j\ell} + \sigma_{i\ell}\sigma_{jk}.$$

Theorem 2.5.1 *Let* $n\mathbf{S} \sim \mathrm{W}_p(n, \Sigma)$ *and put* $\mathbf{S} = (S_{ij})$. *Then the limiting distribution of* $\mathbf{V} = \sqrt{n}(\mathbf{S} - \Sigma)$ *[i.e., the limiting joint distribution of* $\sqrt{n}(S_{ij} - \sigma_{ij})$*] is the* $\{p(p+1)/2\}$*-variate normal distribution with mean zero and covariance matrix with elements*

$$\operatorname{Cov}(\sqrt{n}(S_{ij} - \sigma_{ij}), \sqrt{n}(S_{k\ell} - \sigma_{k\ell})) = \sigma_{ik}\sigma_{j\ell} + \sigma_{i\ell}\sigma_{jk}. \qquad (2.5.3)$$

For asymptotic distribution of a function of a Wishart matrix, we can use the following theorem [see Serfling, 1980, Anderson, 2003, and Theorem 4.2.3].

Theorem 2.5.2 *Let* $\{\boldsymbol{U}(n)\}$ *be a sequence of* m*-component random vectors and* $\boldsymbol{b}$ *be a fixed vector such that* $\sqrt{n}(\boldsymbol{U}(n) - \boldsymbol{b})$ *has the limiting distribution* $\mathrm{N}_m(\mathbf{0}, \Psi)$ *as* $n \to \infty$. *Let* $\boldsymbol{f}(\boldsymbol{u}) = (f_1(\boldsymbol{u}), \ldots, f_k(\boldsymbol{u}))$, $k \le m$ *be a vector-valued function of* $\boldsymbol{u}$ *such that each component* $f_i(\boldsymbol{u})$ *is continuously differentiable in a neighborhood of* $\boldsymbol{u} = \boldsymbol{b}$. *Then the limiting distribution of* $\sqrt{n}\{\boldsymbol{f}(\boldsymbol{U}(n)) - \boldsymbol{f}(\boldsymbol{b})\}$ *is* $\mathrm{N}_k(\mathbf{0}, \mathrm{A}\Psi\mathrm{A}')$, *where* $\mathrm{A} = (a_{ij})$ *is a* $k \times m$ *matrix with* $a_{ij} = \partial f_i(\boldsymbol{u})/\partial u_j|_{\boldsymbol{u}=\boldsymbol{b}}$.

Proof. The proof is based on the result that for every $\varepsilon > 0$ there exists a neighborhood $N_\varepsilon(\boldsymbol{b})$ such that

$$\left| f_j(\boldsymbol{u}) - f_j(\boldsymbol{b}) - \sum_{i=1}^{m} \partial f_j(\boldsymbol{u})/\partial u_i(\boldsymbol{u}_j - \boldsymbol{b}_j) \right| \le \varepsilon \|\boldsymbol{u} - \boldsymbol{b}\|$$

for all $\boldsymbol{u} \in N_\varepsilon(\boldsymbol{b})$. □

Now we consider asymptotic variance (A-Var) of the random variable $f(\mathbf{S})$, where $n\mathbf{S} \sim \mathrm{W}_p(n, \Sigma)$. Suppose that $f(\mathrm{S})$ is continuously differentiable in a neighborhood of $\mathrm{S} = \Sigma$. Then the asymptotic distribution of $f(\mathbf{S})$ is normal with mean $f(\Sigma)$ and the following variance:

$$\text{A-Var}\{f(\mathbf{S})\} = \frac{2}{n}\operatorname{tr}(\Gamma\Sigma)^2, \qquad (2.5.4)$$

where $\Gamma \doteq (\gamma_{ij})$,

$$\gamma_{ij} = \partial_{ij} f(\Sigma), \quad \partial_{ij} \equiv \frac{1}{2}(1+\delta_{ij})\frac{\partial}{\partial \sigma_{ij}}.$$

The expression (2.5.4) is as follows. From Theorems 2.5.1 and 2.5.2,

$$\begin{aligned} &\text{A-Var}\{f(\mathbf{S})\} \\ &\quad = \sum_{i,j,k,\ell} \frac{1}{2}(1+\delta_{ij})\frac{\partial f(\Sigma)}{\partial \sigma_{ij}} \cdot \frac{1}{2}(1+\delta_{k\ell})\frac{\partial f(\Sigma)}{\partial \sigma_{k\ell}} \times \frac{1}{n}(\sigma_{ik}\sigma_{j\ell} + \sigma_{i\ell}\sigma_{jk}) \\ &\quad = \frac{1}{n} \sum_{i,j,k,\ell} \gamma_{ij}\gamma_{k\ell}(\sigma_{ik}\sigma_{j\ell} + \sigma_{i\ell}\sigma_{jk}) \\ &\quad = \frac{2}{n} \operatorname{tr}(\Gamma\Sigma)^2. \end{aligned}$$

Let R be a sample correlation coefficient based on a random sample of size $N = n+1$ from a normal population $\mathrm{N}_2(\boldsymbol{\mu}, \Sigma)$. Then we may write

$$R = S_{12}/\sqrt{S_{11}S_{22}} = V_{12}/\sqrt{V_{11}V_{22}},$$

where $\mathbf{S} = (S_{ij})$, $\mathbf{V} = (V_{ij})$, and $n\mathbf{S} = \mathbf{V} \sim \mathrm{W}_2(n, \Sigma)$. Then by Theorem 2.5.2, R is asymptotically normal with mean $\rho = \sigma_{12}/\sqrt{\sigma_{11}\sigma_{22}}$ and variance calculated by Theorem 2.5.2 on (2.5.4). In the following we obtain the asymptotic variance in two ways.

Since R is invariant under the transformation $\tilde{S}_{ij} = S_{ij}/\sqrt{\sigma_{ii}\sigma_{jj}}$, without loss of generality we may assume that

$$\Sigma = \begin{pmatrix} 1 & \rho \\ \rho & 1 \end{pmatrix} \equiv \Sigma_0.$$

For simplicity, let

$$\boldsymbol{U}(n) = \frac{1}{n}\begin{pmatrix} V_{11} \\ V_{22} \\ V_{12} \end{pmatrix} = \frac{1}{n}\begin{pmatrix} U_1 \\ U_2 \\ U_3 \end{pmatrix}, \quad \boldsymbol{b} = \begin{pmatrix} 1 \\ 1 \\ 1 \end{pmatrix},$$

and hence $R = U_3 U_1^{-1/2} U_2^{-1/2}$. Then, by Theorem 2.5.1 the random vector $\sqrt{n}\big(\boldsymbol{U}(n) - \boldsymbol{b}\big)$ has a limiting normal distribution with mean $\mathbf{0}$ and covariance matrix

$$\Phi = \begin{pmatrix} 2 & 2\rho^2 & 2\rho \\ 2\rho^2 & 2 & 2\rho \\ 2\rho & 2\rho & 1+\rho^2 \end{pmatrix}.$$

The function $r = u_3 u_1^{-1/2} u_2^{-1/2}$ satisfies the conditions of the theorem and

$$\begin{aligned} \boldsymbol{a}' &= \left(\frac{\partial r}{\partial u_1}, \frac{\partial r}{\partial u_2}, \frac{\partial r}{\partial u_3}, \right)\Bigg|_{\boldsymbol{u}=\boldsymbol{b}} \\ &= \left(-\frac{1}{2}\rho, -\frac{1}{2}\rho, 1 \right). \end{aligned}$$

Therefore, the variance of the limiting distribution of $\sqrt{n}(R-\rho)$ is

$$\boldsymbol{a}'\Phi\boldsymbol{a} = (1-\rho^2)^2,$$

and the asymptotic variance of R is $(1-\rho^2)^2/n$.

Next we consider the formula (2.5.4). The elements of Γ are

$$\begin{aligned}
\gamma_{11} &= \frac{\partial}{\partial\sigma_{11}}\sigma_{12}\sigma_{11}^{-1/2}\sigma_{22}^{-1/2}\bigg|_{\Sigma=\Sigma_0} = -\frac{1}{2}\rho, \\
\gamma_{22} &= \frac{\partial}{\partial\sigma_{22}}\sigma_{12}\sigma_{11}^{-1/2}\sigma_{22}^{-1/2}\bigg|_{\Sigma=\Sigma_0} = -\frac{1}{2}\rho, \\
\gamma_{12} = \gamma_{21} &= \frac{1}{2}\frac{\partial}{\partial\sigma_{12}}\sigma_{12}\sigma_{11}^{-1/2}\sigma_{22}^{-1/2}\bigg|_{\Sigma=\Sigma_0} = -\frac{1}{2}.
\end{aligned}$$

These yield the asymptotic variance

$$\frac{2}{n}\,\mathrm{tr}\,(\Gamma\Sigma_0)^2 = \frac{1}{n}(1-\rho^2)^2.$$

Problems

2.1 (Noncentral chi-square distribution) Assuming that $Z_k\ (k=1,\ldots,n)$ are independent and distributed as $\mathrm{N}(\mu_k,1)$, show that the density function of $X=\sum_{k=1}^n Z_k^2$ is given by

$$f(x) = \mathrm{e}^{-\delta^2/2}\sum_{k=0}^{\infty}\frac{1}{k!}\left(\frac{\delta^2}{2}\right)^k \frac{x^{(n+2k)/2-1}\mathrm{e}^{-x/2}}{\Gamma((n+2k)/2)2^{(n+2k)/2}}, \quad x>0.$$

This distribution, $\chi^2(n;\delta^2)$, is called the *noncentral chi-square distribution* with degrees of freedom n and *noncentrality parameter* δ^2.

Hint: If $Z\sim\mathrm{N}(\mu,1)$, the characteristic function is

$$\begin{aligned}
C_{Z^2}(t) &= \mathrm{E}[\exp(itZ^2)] \\
&= (1-2it)^{-1/2}\exp\Big\{-\frac{1}{2}\mu^2+\frac{1}{2}(1-2it)^{-1}\mu^2\Big\}.
\end{aligned}$$

Therefore, the characteristic function of $X=\sum_{k=1}^n Z_k^2$ can be put in the following form:

$$\begin{aligned} C_X(t) &= \prod_{k=1}^{n} \mathrm{E}[\exp(itZ_k^2)] \\ &= \prod_{k=1}^{n} (1-2it)^{-1/2} \exp\left\{ -\frac{1}{2}\mu_k^2 + \frac{1}{2}(1-2it)^{-1}\mu_k^2 \right\} \\ &= (1-2it)^{-1/2} \exp\left\{ -\frac{1}{2}\delta^2 + \frac{1}{2}(1-2it)^{-1}\delta^2 \right\} \\ &= \mathrm{e}^{-\delta^2/2} \sum_{k=0}^{\infty} \frac{1}{k!}\left(\frac{1}{2}\delta^2\right)^k (1-2it)^{-(n+2k)/2}. \end{aligned}$$

Inverting this characteristic function, we obtain the probability density function.

2.2 Let K be a random variable with Poisson distribution $\mathrm{Po}(\frac{1}{2}\delta^2)$. Then check that the density function and characteristic function of X with $\chi^2(n;\delta^2)$ are written as

$$\begin{aligned} f_X(x) &= \mathrm{E}\left[\frac{x^{(n+2K)/2-1}\mathrm{e}^{-x/2}}{\Gamma((n+2K)/2)2^{(n+2k)/2}}\right], \\ C_X(t) &= \mathrm{E}\left[(1-2it)^{-(n+2K)/2}\right], \end{aligned}$$

respectively. The density function is also written in terms of a hypergeometric function as

$$f_X(x) = \frac{x^{n/2-1}\mathrm{e}^{-x/2}}{\Gamma(n/2)2^{n/2}}\mathrm{e}^{-\delta/2}{}_0F_1\left(\frac{1}{2}n; \frac{1}{4}\delta x\right),$$

where

$${}_0F_1(b;z) = \sum_{k=0}^{\infty} \frac{1}{(b)_k}\frac{z^k}{k!}, \quad (b)_k = b(b+1)\cdots(b+k-1).$$

2.3 Let $\mathbf{W} \sim \mathrm{W}_p(n, \mathrm{I}_p)$. Show that $\mathrm{tr}\,\mathbf{W}$ has the chi-aquare distribution with np degrees of freedom.

2.4 Let $\mathbf{W} \sim \mathrm{W}_p(n, \Sigma)$. Show that $\boldsymbol{a}'\mathbf{W}\boldsymbol{a}$ and $\boldsymbol{b}'\mathbf{W}\boldsymbol{b}$ are independent if and only if $\boldsymbol{a}'\Sigma\boldsymbol{b} = 0$.

2.5 Let $\mathbf{W}$ be distributed as $\mathrm{W}_p(n, \Sigma)$. Then, show that the pdf of $\mathbf{U} = \mathbf{W}^{-1}$ is given by

$$\frac{1}{2^{pn/2}\Gamma_p(\frac{1}{2}n)|\Sigma|^{n/2}}|\mathrm{U}|^{-(n+p+1)/2}\,\mathrm{etr}\left(-\frac{1}{2}\Sigma^{-1}\mathrm{U}^{-1}\right).$$

Hint: Use [A.3.5].

2.6 Suppose that $\mathbf{W}$ is distribution as a Wishart distribution $\mathrm{W}_p(n, \Sigma)$, and let $\mathbf{W}$ and Σ partition as

$$\mathbf{W} = \begin{pmatrix} \mathbf{W}_{11} & \mathbf{W}_{12} \\ \mathbf{W}_{21} & \mathbf{W}_{22} \end{pmatrix}, \quad \Sigma = \begin{pmatrix} \Sigma_{11} & \Sigma_{12} \\ \Sigma_{21} & \Sigma_{22} \end{pmatrix}.$$

Put

$$\mathbf{U} = \begin{pmatrix} \mathbf{U}_{11} & \mathbf{U}_{12} \\ \mathbf{U}_{21} & \mathbf{U}_{22} \end{pmatrix},$$

where $\mathbf{U}_{11} = \mathbf{W}_{11\cdot 2} = \mathbf{W}_{11} - \mathbf{W}_{12}\mathbf{W}_{22}^{-1}\mathbf{W}_{21}$, $\mathbf{U}_{12} = \mathbf{U}_{21}' = \mathbf{W}_{11}\mathbf{W}_{22}^{-1/2}$ and $\mathbf{U}_{22} = \mathbf{W}_{22}$.

(1) Show that the transformation from $\mathbf{W}$ and $\mathbf{U}$ is one-to-one.

(2) When $\Sigma = \mathrm{I}_p$, obtain the joint pdf of $\mathbf{U}$.

Hint: Use Theorem 2.2.3, or obtain the Jacobian of the inverse transformation.

2.7 Let $\boldsymbol{X}_{i1}, \ldots, \boldsymbol{X}_{in_i} \sim$ i.i.d.$\mathrm{N}_p(\boldsymbol{\mu}_i, \Sigma)$, $i = 1, 2$, and let

$$\bar{\boldsymbol{X}}_i = \frac{1}{n_i}\sum_{j=1}^{n_i} \boldsymbol{X}_{ij}, \quad i = 1, 2, \qquad \bar{\boldsymbol{X}} = \frac{1}{n}\{n_1\bar{\boldsymbol{X}}_1 + n_2\bar{\boldsymbol{X}}_2\},$$

where $n = n_1 + n_2$. Examine that

$$\mathbf{B} = \sum_{i=1}^{2} n_i(\bar{\boldsymbol{X}}_i - \bar{\boldsymbol{X}})(\bar{\boldsymbol{X}}_i - \bar{\boldsymbol{X}})' \sim \mathrm{W}_p(1, \Sigma; \Omega),$$

by the following two methods, where $\Omega = (n_1n_2/n)(\boldsymbol{\mu}_1 - \boldsymbol{\mu}_2)(\boldsymbol{\mu}_1 - \boldsymbol{\mu}_2)'$.
Method 1: Use $\mathbf{B} = (n_1n_2/n)(\bar{\boldsymbol{X}}_1 - \bar{\boldsymbol{X}}_2)(\bar{\boldsymbol{X}}_1 - \bar{\boldsymbol{X}}_2)'$.
Method 2: Express $\mathbf{B} = \mathbf{X}'\mathrm{P}_b\mathbf{X}$, and use Cochran theorem, where $\mathbf{X} = (\boldsymbol{X}_{11}, \ldots, \boldsymbol{X}_{1n_1}, \boldsymbol{X}_{21}, \ldots, \boldsymbol{X}_{2n_2})$,

$$\mathrm{P}_b = \begin{pmatrix} \frac{1}{n_1}\mathrm{G}_{n_1} & \mathrm{O} \\ \mathrm{O} & \frac{1}{n_2}\mathrm{G}_{n_2} \end{pmatrix} - \frac{1}{n}\mathrm{G}_n,$$

and G_m is the $m \times m$ matrix whose elements are all one.

2.8 Let $\mathbf{W} \sim \mathrm{W}_p(n, \mathrm{I}_p; \Omega)$ and $\mathbf{W}$ and Ω be partitioned as

$$\mathbf{W} = \begin{pmatrix} \mathbf{W}_{11} & \mathbf{W}_{12} \\ \mathbf{W}_{21} & \mathbf{W}_{22} \end{pmatrix}, \quad \Omega = \begin{pmatrix} \Omega_{11} & \mathrm{O} \\ \mathrm{O} & \mathrm{O} \end{pmatrix},$$

respectively, where $\mathbf{W}_{ij} : p_i \times p_j$ and $\Omega_{11} : p_1 \times p_1$. Then show that:

(1) $\mathbf{W}_{22\cdot 1}$, $\mathbf{W}_{21}\mathbf{W}_{11}^{-1/2}$, and $\mathbf{W}_{11}$ are mutually independent.

(2) $\mathbf{W}_{22\cdot 1} \sim \mathrm{W}_{p_2}(n - p_1, \mathrm{I}_{p_2})$, $\mathbf{W}_{11} \sim \mathrm{W}_{p_1}(n, \Omega_{11})$, and the columns of $\mathbf{W}_{12}\mathbf{W}_{22}^{-1/2}$ are independent and identically distributed as $\mathrm{N}_{p_2}(\mathbf{0}, \mathrm{I}_{p_2})$, and hence $\mathbf{W}_{21}\mathbf{W}_{11}^{-1}\mathbf{W}_{12} \sim \mathrm{W}_{p_2}(n, \mathrm{I}_{p_1})$.

2.9 Suppose that $n\mathbf{S}$ is distributed as a Wishart distribution $\mathrm{W}_p(n, \Sigma)$. Obtain the asymptotic variances of $\sqrt{n}(\operatorname{tr}\mathbf{S} - \operatorname{tr}\Sigma)$ and $\sqrt{n}(\operatorname{tr}\mathbf{S}^2 - \operatorname{tr}\Sigma^2)$.

2.10 (Continuation) Show that the asymptotic distribution of $|\mathbf{S}|$ is normal $\mathrm{N}(|\Sigma|, (2p/n)|\Sigma|^2)$ or, what is the same thing, that the asymptotic distribution of $\sqrt{n}(|\mathbf{S}|/|\Sigma| - 1)$ is $\mathrm{N}(0, 2p)$.

2.11 Let $\mathbf{W}$ have a Wishart distribution $\mathrm{W}_p(n, \Sigma)$, and let $\mathbf{W}_i$ and Σ_i be the matrices consisting of the first i rows and columns of $\mathbf{W}$ and Σ, respectively with $\mathbf{W}_0 = 1$ and $\Sigma_0 = 1$. Show that

$$V_i = \frac{|\mathbf{W}_i|}{|\mathbf{W}_{i-1}|}\left(\frac{|\Sigma_i|}{|\Sigma_{i-1}|}\right)^{-1} \sim \chi^2(n - i + 1),$$

and that $V_1, \ldots, V_p$ are independent.

2.12 Suppose that $n\mathbf{S}$ is distributed as a Wishart distribution $\mathrm{W}_p(n, \Sigma)$. Then show that

$$\begin{pmatrix} \sqrt{n}\,(\operatorname{tr}\mathbf{S} - \operatorname{tr}\Sigma) \\ \sqrt{n}\,(\operatorname{tr}\mathbf{S}^2 - tr\Sigma^2) \end{pmatrix} \xrightarrow{d} \mathrm{N}\left(\begin{pmatrix} 0 \\ 0 \end{pmatrix}, \begin{pmatrix} 2\operatorname{tr}\Sigma^2 & 4\operatorname{tr}\Sigma^3 \\ 4\operatorname{tr}\Sigma^3 & 8\operatorname{tr}\Sigma^4 \end{pmatrix}\right).$$

Chapter 3

Hotelling's T^2 and Lambda Statistics

In this chapter we treat the *Hotelling* T^2 and Lambda statistics. They are statistics measuring the differences between two mean vectors and among several mean vectors, respectively. The exact distributions are given. Their decompositions are studied. We also study the likelihood ratio test for additional information when several mean vectors are compared.

3.1 T^2 Statistic

3.1.1 Distribution of the T^2 Statistic

Let $\bar{\boldsymbol{X}}$ and $\mathbf{S}$ be the sample mean vector and the sample covariance matrix based on a sample of size n from $\mathrm{N}_p(\boldsymbol{\mu}, \Sigma)$. Then it is important to know the distribution of

$$T_{(1)}^2 = n\bar{\boldsymbol{X}}'S^{-1}\bar{\boldsymbol{X}},$$

which is a test statistic for $\boldsymbol{\mu} = \mathbf{0}$. It is also important to know the distribution of the statistic obtained from $T_{(1)}^2$ by replacing $\bar{\boldsymbol{X}}$ by $\bar{\boldsymbol{X}} - \boldsymbol{\mu}$. Note that the latter distribution is obtained as the distribution of $T_{(1)}^2$ with $\boldsymbol{\mu} = \mathbf{0}$ (i.e., the *null distribution* of $T_{(1)}^2$).

For comparison of two mean vectors $\boldsymbol{\mu}_1$ and $\boldsymbol{\mu}_2$, let $\bar{\boldsymbol{X}}_i$ and $\mathbf{S}_i$ $(i = 1, 2)$ be the sample mean vectors and the sample covariance matrices based on samples of sizes n_i from $\mathrm{N}_p(\boldsymbol{\mu}_i, \Sigma)$, $i = 1, 2$. Let $\mathbf{S}$ be a pooled sample covariance matrix defined by

$$\mathbf{S} = \frac{1}{n-2}\{(n_1 - 1)\mathbf{S}_1 + (n_2 - 1)\mathbf{S}_2\},$$

where $n = n_1 + n_2$. Then it is also important to know the distribution of

$$T^2_{(2)} = \frac{n_1 n_2}{n}(\bar{\boldsymbol{X}}_1 - \bar{\boldsymbol{X}}_2)'\mathbf{S}^{-1}(\bar{\boldsymbol{X}}_1 - \bar{\boldsymbol{X}}_2).$$

The statistics $T^2_{(1)}$ and $T^2_{(2)}$ are called T^2 statistic, which was introduced by Hotelling (1936). 43 The distributions of these statistics are treated as

$$T^2 = \boldsymbol{Z}'\left(\frac{1}{m}\mathbf{W}\right)^{-1}\boldsymbol{Z}, \tag{3.1.1}$$

where $\boldsymbol{Z} \sim \mathrm{N}_p(\boldsymbol{\nu}, \Sigma)$, $\mathbf{W} \sim \mathrm{W}_p(m, \Sigma)$, and $\boldsymbol{Z}$ and $\mathbf{W}$ are independent. The distribution of T^2 is called the T^2 *distribution*. For $T^2_{(1)}$, $m = n - 1$, and for $T^2_{(2)}$, $m = n_1 + n_2 - 2$. We denote the distribution of T^2 by $T(p, m; \delta^2)$, where $\delta^2 = \boldsymbol{\nu}'\Sigma^{-1}\boldsymbol{\nu}$, and denote $T(p, m; 0)$ simply by $T(p, m)$.

Theorem 3.1.1 *Let T^2 be the random variable defined by* (3.1.1). *Then*

$$\frac{1}{m}T^2 \cdot \frac{m - p + 1}{p} \sim F(p, m - p + 1; \delta^2),$$

where $F(a, b; \delta^2)$ denotes a noncentral F-distribution with degrees of freedom a, b and noncentrality parameter δ^2.

Proof. When we consider the distribution of T^2, without loss of generality we may assume that $\Sigma = \mathrm{I}_p$ and $\boldsymbol{Z} \sim \mathrm{N}_p(\boldsymbol{\mu}, \mathrm{I}_p), \boldsymbol{\mu} = \Sigma^{-1}\boldsymbol{\nu}$. Consider the conditional distribution of T^2 given $\boldsymbol{Z}$. The conditional distribution of $\mathbf{W}$ given $\boldsymbol{Z}$ is $\mathrm{W}_p(m, \mathrm{I}_p)$ since $\mathbf{W}$ is independent of $\boldsymbol{Z}$. Let H be an orthogonal matrix whose first column is $\boldsymbol{Z}(\boldsymbol{Z}'\boldsymbol{Z})^{-1/2}$. Then the conditional distribution of $\mathbf{V} = \mathrm{H}'\mathbf{W}\mathrm{H}$ is $W_p(m, \mathrm{I}_p)$, which does not depend on $\boldsymbol{Z}$. Therefore, $\mathbf{V} \sim \mathrm{W}_p(m, \mathrm{I}_p)$ and is independent of $\boldsymbol{Z}$. Decompose $\mathbf{V}$ as

$$V = \begin{pmatrix} V_{11} & \boldsymbol{V}'_{21} \\ \boldsymbol{V}_{21} & \mathbf{V}_{22} \end{pmatrix}.$$

Note that the $(1, 1)$ element of $\mathbf{V}^{-1}$ is $V^{-1}_{11\cdot 2}$, where $V_{11\cdot 2} = V_{11} - \boldsymbol{V}'_{21}\mathbf{V}^{-1}_{22}\boldsymbol{V}_{21}$. Then

$$\begin{aligned} \frac{1}{m}\mathrm{T}^2 &= (\mathrm{H}\boldsymbol{Z})'(\mathrm{H}'\mathbf{W}\mathrm{H})^{-1}(\mathrm{H}'\boldsymbol{Z}) \\ &= ((\boldsymbol{Z}'\boldsymbol{Z})^{1/2}, 0, \ldots, 0)\mathbf{V}^{-1}((\boldsymbol{Z}'\boldsymbol{Z})^{1/2}, 0, \ldots, 0)' \\ &= \boldsymbol{Z}'\boldsymbol{Z} \cdot V^{-1}_{11\cdot 2}. \end{aligned}$$

The required result follows from the fact that $\boldsymbol{Z}'\boldsymbol{Z} \sim \chi^2(p; \delta^2)$ and $V_{11\cdot 2} \sim \chi^2(n - (p - 1))$ (see Theorem 2.2.2). □

The T^2 *statistic* $T^2_{(2)}$ in the *two-sample case* is closely related to the *Mahalanobis distance* between two population, defined by

$$D^2 = (\bar{\boldsymbol{X}}_1 - \bar{\boldsymbol{X}}_2)'\mathbf{S}^{-1}(\bar{\boldsymbol{X}}_1 - \bar{\boldsymbol{X}}_2). \tag{3.1.2}$$

In fact, $T^2_{(2)}$ is just a constant times D^2; that is,

$$T^2_{(2)} = cD^2, \quad c = n_1 n_2/(n_1 + n_2).$$

The population Mahalanobis distance is defined by

$$\Delta^2 = (\boldsymbol{\mu}_1 - \boldsymbol{\mu}_2)'\Sigma^{-1}(\boldsymbol{\mu}_1 - \boldsymbol{\mu}_2).$$

Theorem 3.1.2 *Suppose that D^2 is the sample Mahalanobis distance between two normal populations $\mathrm{N}_p(\boldsymbol{\mu}, \Sigma)$, $i = 1, 2$. Then:*

(1) $\mathrm{E}(D^2) = \dfrac{m}{m - p - 1}\left(\Delta^2 + \dfrac{p}{c}\right)$.

(2) $\mathrm{Var}(D^2) = \dfrac{2m^2}{(m-p-1)^2(m-p-3)}\left[\dfrac{p(m-1)}{c^2} + \dfrac{2(m-1)}{c}\Delta^2 + \Delta^4\right]$,

where $m = n_1 + n_2 - 2$.

Proof. The results are obtained by direct computation based on

$$\begin{aligned} D^2 = \frac{1}{c}\mathrm{T}^2 &= \frac{mp}{c(m-p-1)}F(p, m-p+1; c\Delta^2) \\ &= \frac{p}{c} \cdot \frac{\chi^2(p; c\Delta^2)}{\chi^2(m-p+1)}, \end{aligned}$$

which follows from Theorem 3.1.1. □

3.1.2 Decomposition of T^2 and D^2

Let $\boldsymbol{Z} \sim \mathrm{N}_p(\boldsymbol{\mu}, \Sigma)$, $m\mathbf{S} \sim \mathrm{W}_p(m, \Sigma)$, and $\boldsymbol{Z}$ and $\mathbf{S}$ be independent. Let $\boldsymbol{\mu}$, Σ, $\boldsymbol{Z}$, and $\mathbf{S}$ be

$$\boldsymbol{\mu} = \begin{pmatrix} \boldsymbol{\mu}_1 \\ \boldsymbol{\mu}_2 \end{pmatrix}, \quad \Sigma = \begin{pmatrix} \Sigma_{11} & \Sigma_{12} \\ \Sigma_{21} & \Sigma_{22} \end{pmatrix},$$
$$\boldsymbol{Z} = \begin{pmatrix} \boldsymbol{Z}_1 \\ \boldsymbol{Z}_2 \end{pmatrix}, \quad \mathbf{S} = \begin{pmatrix} \mathbf{S}_{11} & \mathbf{S}_{12} \\ \mathbf{S}_{21} & \mathbf{S}_{22} \end{pmatrix},$$

where $\boldsymbol{\mu}_i$: $p_i \times 1$, Σ_{ij}: $p_i \times p_j$, $\boldsymbol{Z}_i$: $p_i \times 1$, and $\mathbf{S}_{ij}$: $p_i \times p_j$. We also use the notations:

$$\begin{aligned} &\mathcal{B} = \Sigma_{21}\Sigma_{11}^{-1}, \quad \mathbf{B} = \mathbf{S}_{21}\mathbf{S}_{11}^{-1}, \\ &\Sigma_{22\cdot 1} = \Sigma_{22} - \Sigma_{21}\Sigma_{11}^{-1}\Sigma_{12} = \Sigma_{22} - \mathcal{B}\Sigma_{11}\mathcal{B}', \\ &\mathbf{S}_{22\cdot 1} = \mathbf{S}_{22} - \mathbf{S}_{21}\mathbf{S}_{11}^{-1}\mathbf{S}_{12} = \mathbf{S}_{22} - \mathbf{B}\mathbf{S}_{11}\mathbf{B}'. \end{aligned}$$

Now we consider the following decompositions of

$$\Delta^2 = \boldsymbol{\mu}'\Sigma^{-1}\boldsymbol{\mu} \quad \text{and} \quad \mathrm{T}^2 = \boldsymbol{Z}'\mathbf{S}^{-1}\boldsymbol{Z}$$

Theorem 3.1.3 *Let* $\Delta^2 = \boldsymbol{\mu}'\Sigma\boldsymbol{\mu}$, $\Delta_1 = \boldsymbol{\mu}_1'\Sigma_{11}^{-1}\boldsymbol{\mu}_1$, $T^2 = \boldsymbol{Z}'\mathbf{S}^{-1}\boldsymbol{Z}$ *and* $T_1^2 = \boldsymbol{Z}_1'\mathbf{S}_{11}^{-1}\boldsymbol{Z}_1$. *Then*

$$\Delta^2 = \Delta_1^2 + \Delta_{2\cdot1}^2, \quad T^2 = T_1^2 + T_{2\cdot1}^2,$$

where

$$\Delta_{2\cdot1}^2 = \boldsymbol{\mu}_{2\cdot1}'\Sigma_{22\cdot1}^{-1}\boldsymbol{\mu}_{2\cdot1}, \quad \boldsymbol{\mu}_{2\cdot1} = \boldsymbol{\mu}_2 - \mathcal{B}\boldsymbol{\mu}_1,$$
$$T_{2\cdot1}^2 = \boldsymbol{Z}_{2\cdot1}'\mathbf{S}_{22\cdot1}^{-1}\boldsymbol{Z}_{2\cdot1}, \quad \boldsymbol{Z}_{2\cdot1} = \boldsymbol{Z}_2 - \mathbf{B}\boldsymbol{Z}_1.$$

Proof. The results are obtained by applying [A.1.2] to Σ^{-1} and $\mathbf{S}^{-1}$. □

The quantity

$$U = \frac{T^2 - \mathrm{T}_1^2}{n + T_1^2} = \frac{T_{2\cdot1}^2}{n + T_1^2} \tag{3.1.3}$$

is called *Rao's U-statistic.* We now consider the sampling distribution of $U = T_{2\cdot1}^2/(n + T_1^2)$, where $T_{2\cdot1}^2$ and T_1^2 may be defined in terms of $\mathbf{V} = n\mathbf{S} \sim \mathrm{W}_p(n, \Sigma)$ as

$$T_{2\cdot1}^2 = n\boldsymbol{Z}_{2\cdot1}'\mathbf{V}_{22\cdot1}^{-1}\boldsymbol{Z}_{2\cdot1}, \quad T_1^2 = n\boldsymbol{Z}_1'\mathbf{V}_{11}\boldsymbol{Z}_1,$$

where $\boldsymbol{Z}_{2\cdot1} = \boldsymbol{Z}_2 - \mathbf{B}\boldsymbol{Z}_1$ and $\mathbf{V}$ is partitioned similarly to $\mathbf{S}$. First, we note that $\mathbf{V}_{22\cdot1} \sim \mathrm{W}_{p_2}(n - p_1, \Sigma_{22\cdot1})$; $\mathbf{V}_{22\cdot1}$ and $(\mathbf{B}, \mathbf{V}_{11})$ are independent. Let $\mathbf{V}$ denote, $\mathbf{V} = \mathbf{X}'\mathbf{X} = (\mathbf{X}_1\ \mathbf{X}_2)'(\mathbf{X}_1\ \mathbf{X}_2)$, where $\mathbf{X}_1 : n \times p_1$ and the rows of $\mathbf{X}$ are independently distributed as $\mathrm{N}_p(\mathbf{0}, \Sigma)$. The conditional distribution of $\mathbf{X}_2$ given $\mathbf{X}_1$ is normal such that the rows are independent with the same covariance matrix $\Sigma_{22\cdot1}$ and $\mathrm{E}(\mathbf{X}_2|\mathbf{X}_1) = \mathbf{X}_1\mathcal{B}$. Note that the conditional distribution of $\mathbf{B}(\mathbf{X}_1'\mathbf{X}_1)^{-1/2} = \mathbf{X}_2\mathbf{X}_1(\mathbf{X}_1'\mathbf{X}_1)^{-1/2}$ is normal such that its rows are independent with the same covariance matrix $\Sigma_{22\cdot1}$ and $\mathrm{E}(\mathbf{B}(\mathbf{X}_1'\mathbf{X}_1)^{-1/2}|\mathbf{X}_1) = \mathcal{B}(\mathbf{X}_1'\mathbf{X}_1)^{1/2}$. Therefore, we check that the conditional distribution $\boldsymbol{Z}_{2\cdot1} = \boldsymbol{Z}_2 - \mathbf{B}(\mathbf{X}_1'\mathbf{X}_1)^{1/2}(\mathbf{X}_1'\mathbf{X}_1)^{-1/2}\boldsymbol{Z}_1$ given that $(\boldsymbol{Z}_1, \mathbf{X}_1)$ is normal with mean

$$\mathrm{E}(\boldsymbol{Z}_{2\cdot1}|\boldsymbol{Z}_1, \mathbf{X}_1) = \boldsymbol{\mu}_2 + \mathcal{B}(\boldsymbol{Z}_1 - \boldsymbol{\mu}_1) - \mathcal{B}\boldsymbol{Z}_1 = \boldsymbol{\mu}_{2\cdot1}$$

and covariance matrix

$$\begin{aligned}\mathrm{Var}(\boldsymbol{Z}_{2\cdot1}|\boldsymbol{Z}_1, \mathbf{X}_1) &= \Sigma_{22\cdot1} + \boldsymbol{Z}_1'(\mathbf{X}_1'\mathbf{X}_1)^{-1}\boldsymbol{Z}_1 \cdot \Sigma_{22\cdot1} \\ &= \left(1 + \frac{1}{n}T_1^2\right)\Sigma_{22\cdot1}.\end{aligned}$$

These imply the following result.

Theorem 3.1.4 *Let* $\boldsymbol{Z} \sim \mathrm{N}_p(\boldsymbol{\mu}, \Sigma)$, $\mathbf{W} \sim \mathrm{W}_p(n, \Sigma)$, *and* $\boldsymbol{Z}$ *and* $\mathbf{W}$ *be independently distributed. Using the same notation as in Theorem 3.1.2, the conditional distribution of*

$$\frac{n-p+1}{p_2}U = \frac{n-p+1}{p_2} \cdot \frac{T_{2\cdot 1}^2}{n+T_1^2}$$

given that T_1^2 *is the noncentral F-distribution with* $(p_2, n-p+1)$ *degrees of freedom and noncentrality parameter*

$$\lambda^2 = \frac{n}{n+T_1^2}\Delta_{2\cdot 1}^2.$$

In particular, if $\Delta_{2\cdot 1}^2 = 0$ *(i.e.,* $\Delta^2 = \Delta_1^2$*), then* $\{(n-p+1)/p_2\}U$ *is distributed independent of* T_1^2 *as the central F-distribution with* $(p_2, n-p+1)$ *degrees of freedom.*

Theorem 3.1.4 can be applied to Hotelling's T^2-statistic $T_{(2)}^2$ in a two-sample case. Now we write $\boldsymbol{\mu}^{(i)}$ and $\bar{\boldsymbol{X}}^{(i)}$ instead of $\boldsymbol{\mu}_i$ and $\bar{\boldsymbol{X}}_i$. Then the Hotelling statistic is

$$T^2 = c(\bar{\boldsymbol{X}}^{(1)} - \bar{\boldsymbol{X}}^{(2)})'\mathbf{S}^{-1}(\bar{\boldsymbol{X}}^{(1)} - \bar{\boldsymbol{X}}^{(2)}) = cD^2.$$

Here $c = \sqrt{n_1 n_2 / n}$, $m\mathbf{S} \sim \mathrm{W}_p(m, \Sigma)$, and $m = n - 2$. In the test of $\boldsymbol{\mu}^{(1)} = \boldsymbol{\mu}^{(2)}$, we are interesting in deciding whether $\boldsymbol{X}_2$ has additional information in the presence of $\boldsymbol{X}_1$. This is can be examined by testing the hypothesis

$$\boldsymbol{\mu}_{2\cdot 1}^{(1)} = \boldsymbol{\mu}_{2\cdot 1}^{(2)},$$

where $\boldsymbol{\mu}_{2\cdot 1}^{(i)} = \boldsymbol{\mu}_2^{(i)} - \Sigma_{21}\Sigma_{11}^{-1}\boldsymbol{\mu}_1^{(i)}$, $i = 1, 2$. Then a reasonable test statistic is

$$U = c\frac{D^2 - D_1^2}{n + cD_1^2}, \tag{3.1.4}$$

which is shown to be a likelihood ratio test. The null distribution of $\{(m - p+1)/p_2\}U$ is an F-distribution with $(p_2, m-p+1)$ degrees of freedom.

Example 3.1.1 We consider the data used by Fisher (1936), which are cited in Mardia et al. (1979). Measurements were made on 50 specimens of flowers from each of two species of iris, *setosa* and *versicolors*, found growing together in the same colony. We denote these two populations by Π_1 and Π_2. Four flower measurements were taken: X_1, sepal length; X_2, sepal width; X_3, petal length; and X_4, petal width. Let the samples from Π_1 and Π_2 be denoted by $\boldsymbol{x}_j^{(1)}$, $j = 1, \ldots, n_1$ and $\boldsymbol{x}_j^{(2)}$, $j = 1, \ldots, n_2$, where $n_1 = n_2 = 50$. The sample mean vectors and covariance matrices are defined by

$$\bar{\boldsymbol{x}}^{(i)} = \frac{1}{n_i}\sum_{j=1}^{n_i} \boldsymbol{x}_j^{(i)}, \quad \mathrm{S}^{(i)} = \frac{1}{n_i - 1}\sum_{j=1}^{n_i}(\boldsymbol{x}_j^{(i)} - \bar{\boldsymbol{x}}^{(i)})(\boldsymbol{x}_j^{(i)} - \bar{\boldsymbol{x}}^{(i)})',$$

and the pooled sample covariance matrix is defined by

$$\mathrm{S} = \frac{1}{n-2}\{(n_1-1)\mathrm{S}^{(1)} + (n_2-1)\mathrm{S}^{(2)}\}, \qquad n = n_1 + n_2.$$

The sample mean vectors $\bar{\boldsymbol{x}}^{(i)}$ observed are

	X_1	X_2	X_3	X_4
Π_1	5.006	3.428	1.462	0.246
Π_2	5.936	2.770	4.260	1.326

The sample covariance matrices (upper triangular elements are correlation coefficients) are found as

Π_1	X_1	X_2	X_3	X_4
X_1	0.124	0.743	0.267	0.278
X_2	0.099	0.144	0.178	0.233
X_3	0.016	0.012	0.030	0.332
X_4	0.010	0.009	0.006	0.011

and

Π_2	X_1	X_2	X_3	X_4
X_1	0.266	0.526	0.754	0.546
X_2	0.085	0.098	0.561	0.664
X_3	0.183	0.083	0.221	0.787
X_4	0.056	0.041	0.073	0.039

The pooled sample covariance matrix is given by

Π	X_1	X_2	X_3	X_4
X_1	0.266	0.526	0.754	0.546
X_2	0.085	0.098	0.561	0.664
X_3	0.183	0.083	0.221	0.787
X_4	0.056	0.041	0.073	0.039

We assume that the populations Π_1 and Π_2 are $\mathrm{N}_p(\boldsymbol{\mu}_1, \Sigma)$ and $\mathrm{N}_p(\boldsymbol{\mu}_2, \Sigma)$, respectively. For testing $H : \boldsymbol{\mu}_1 = \boldsymbol{\mu}_2$, we have

$$\begin{aligned}\mathrm{T}^2 &= \frac{n_1 n_2}{n}(\bar{\boldsymbol{x}}^{(1)} - \bar{\boldsymbol{x}}^{(2)})'\mathrm{S}^{-1}(\bar{\boldsymbol{x}}^{(1)} - \bar{\boldsymbol{x}}^{(2)}) \\ &= \frac{n_1 n_2}{n} D^2 = 2580.84\end{aligned}$$

and

$$F = \frac{1}{m}T^2 \cdot \frac{m-p+1}{p}$$
$$= \frac{1}{98} \times 2580.84 \times \frac{98-4+1}{4} = 625.46.$$

The p-value is 0.000, and we reject the hypothesis. Next we consider whether X_1 has additional information in the presence of $\omega = \{X_2, X_3, X_4\}$. Let the Mahalanobis distances be based on $\Omega = \{X_1, \ldots, X_4\}$ and ω by D_Ω and D_ω, respectively. Then

$$D_\Omega = 10.16 \quad \text{and} \quad D_\omega = 10.12.$$

The test statistic is given by

$$T_{(\Omega-\omega)\cdot\omega} = \frac{m-p+1}{p-k_\omega} \frac{c(D_\Omega^2 - D_\omega^2)}{n-2+cD_\omega^2}$$

which is distributed as an F-distribution with $p - k_\omega$ and $m - p + 1$, where $c = \sqrt{n_1 n_2 / n}$ and k_ω is the number of variables in ω. The value of $T_{(\Omega-\omega)\cdot\omega}$ is 0.72, the p-value is 0.328, and hence X_1 has no additional information in the presence of $\{X_2, X_3, X_4\}$.

3.2 Lambda Statistic

3.2.1 Motivation of the Lambda Statistic

Let n_i samples

$$\boldsymbol{X}_{i1}, \ldots, \boldsymbol{X}_{in_i}, \quad i = 1, \ldots, k \tag{3.2.1}$$

be available from k populations with mean vector $\boldsymbol{\mu}_i$ and covariance matrix Σ, and denote the sample mean vectors and the sample covariance matrices by

$$\bar{\boldsymbol{X}}_i = \frac{1}{n_i}\sum_{j=1}^{n_i} \boldsymbol{X}_{ij} \quad \text{and} \quad \mathbf{S}_i = \frac{1}{n_i - 1}\sum_{j=1}^{n_i}(\boldsymbol{X}_{ij} - \bar{\boldsymbol{X}}_i)(\boldsymbol{X}_{ij} - \bar{\boldsymbol{X}}_i)'.$$

Further, let $\bar{\boldsymbol{X}}$ and $\mathbf{S}$ be the overall mean vector and the pooled sample covariance matrix, defined by

$$\bar{\boldsymbol{X}} = \frac{1}{n}\sum_{j=1}^{k} n_i \bar{\boldsymbol{X}}_i \quad (n = n_1 + \cdots + n_k) \quad \text{and} \quad \mathbf{S} = \frac{1}{n-k}\sum_{j=1}^{n_i}(n_i - 1)\mathbf{S}_i,$$

respectively. For testing

$$H_0 \colon \boldsymbol{\mu}_1 = \cdots = \boldsymbol{\mu}_k, \tag{3.2.2}$$

we have two basic matrices given by

$$\mathbf{S}_b = \sum_{i=1}^{k} n_i(\bar{\boldsymbol{X}}_i - \bar{\boldsymbol{X}})(\bar{\boldsymbol{X}}_i - \bar{\boldsymbol{X}})', \quad \mathbf{S}_w = \sum_{i=1}^{k}(n_i - 1)\mathbf{S}_i. \tag{3.2.3}$$

The matrices $\mathbf{S}_b$ and $\mathbf{S}_w$ are called *between-group and within-group matrices of sums of squares and products*, respectively. Note that

$$\begin{aligned}\mathbf{S}_t &= \sum_{i=1}^{k}\sum_{j=1}^{n_i}(\boldsymbol{X}_{ij} - \bar{\boldsymbol{X}})(\boldsymbol{X}_{ij} - \bar{\boldsymbol{X}})' \\ &= \mathbf{S}_b + \mathbf{S}_w.\end{aligned}$$

Then a typical test is based on a monotone function of $|\mathbf{S}_w|/|\mathbf{S}_w + \mathbf{S}_b|$ which is called the *lambda statistic*. If, in particular, the populations are all normal, $\mathrm{N}_p(\boldsymbol{\mu}_i, \Sigma)$, it reduces to the likelihood ratio test.

Theorem 3.2.1 *Let the samples* (3.2.1) *be from a p-variate normal population* $\mathrm{N}_p(\boldsymbol{\mu}_i, \Sigma), i = 1, \ldots, k$. *Then the likelihood ratio criterion* λ *for testing the hypothesis* (3.2.2) *can be expressed as*

$$-2\log\lambda = -n\log\frac{|\mathbf{S}_w|}{|\mathbf{S}_w + \mathbf{S}_b|}.$$

Proof. Let $g(\boldsymbol{\mu}_1, \ldots, \boldsymbol{\mu}_k, \Sigma)$ be -2 log likelihood for samples (3.2.1). Then

$$\begin{aligned}g(\boldsymbol{\mu}_1, \ldots, \boldsymbol{\mu}_k, \Sigma) &= n\log|\Sigma| + np\log(2\pi) \\ &\quad + \sum_{i=1}^{k}\sum_{j=1}^{n_i} \operatorname{tr}\Sigma^{-1}(\boldsymbol{x}_{ij} - \boldsymbol{\mu}_i)(\boldsymbol{x}_{ij} - \boldsymbol{\mu}_i)'.\end{aligned}$$

Note that

$$\begin{aligned}&\operatorname{tr}\Sigma^{-1}(\boldsymbol{x}_{ij} - \boldsymbol{\mu}_i)(\boldsymbol{x}_{ij} - \boldsymbol{\mu}_i)' \\ &\quad = \operatorname{tr}\Sigma^{-1}\mathrm{S}_w + \sum_{i=1}^{k} n_i \operatorname{tr}\Sigma^{-1}(\bar{\boldsymbol{x}}_i - \boldsymbol{\mu}_i)(\bar{\boldsymbol{x}}_i - \boldsymbol{\mu}_i)',\end{aligned}$$

which has the minimum at $\boldsymbol{\mu}_i = \hat{\boldsymbol{\mu}}_i = \bar{\boldsymbol{x}}_i, i = 1, \ldots, k$. Therefore,

$$\begin{aligned}\min_{\boldsymbol{\mu}_i, \Sigma} g(\boldsymbol{\mu}_1, \ldots, \boldsymbol{\mu}_k, \Sigma) &= \min_{\Sigma} g(\hat{\boldsymbol{\mu}}_1, \ldots, \hat{\boldsymbol{\mu}}_k, \Sigma) \\ &= \min_{\Sigma}\{n\log|\Sigma| + \operatorname{tr}\Sigma^{-1}\mathrm{S}_w\} + np\log(2\pi).\end{aligned}$$

The minimum for Σ of the last expression is attained at $\Sigma = \hat{\Sigma} = (1/n)\mathrm{S}_w$ by using [A.2.10], and hence

$$\min_{\boldsymbol{\mu}_i, \Sigma} g(\boldsymbol{\mu}_1, \ldots, \boldsymbol{\mu}_k, \Sigma) = n\log\left|\frac{1}{n}\mathrm{S}_w\right| + np(\log(2\pi) + 1).$$

The $-2\log$ likelihood under H_0 is expressed as $g_0(\boldsymbol{\mu}, \Sigma) = g(\boldsymbol{\mu}, \ldots, \boldsymbol{\mu}, \Sigma)$. Similarly, its minimum is attained at

$$\hat{\boldsymbol{\mu}} = \bar{\boldsymbol{y}}, \quad \Sigma = \hat{\Sigma} = \frac{1}{n} S_t = \frac{1}{n}(\mathrm{S}_w + \mathrm{S}_b)$$

and

$$\min_{\boldsymbol{\mu}_i, \Sigma} g_0(\boldsymbol{\mu}, \Sigma) = n \log \left| \frac{1}{n}(\mathrm{S}_w + \mathrm{S}_b) \right| + np(\log(2\pi) + 1).$$

Therefore,

$$\begin{aligned} -2\log\lambda &= \min g_0(\boldsymbol{\mu}, \Sigma) - \min g(\boldsymbol{\mu}_1, \ldots, \boldsymbol{\mu}_k, \Sigma) \\ &= -n \log \frac{|\mathrm{S}_w|}{|\mathrm{S}_w + \mathrm{S}_b|}. \end{aligned}$$ □

3.2.2 Distribution of the Lambda Statistic

Theorem 3.2.2 *Let* $\mathbf{S}_b$ *and* $\mathbf{S}_w$ *be the between group and within group matrices of sums of squares and products given by* (3.2.3), *based on* k p*-variate normal populations* $\mathrm{N}_p(\boldsymbol{\mu}_i, \Sigma)$. *Then* $\mathbf{S}_b$ *and* $\mathbf{S}_w$ *are independently distributed as* $\mathrm{W}_p(k-1, \Sigma; \Omega)$ *and* $\mathrm{W}_p(n-k, \Sigma)$, *respectively, where* $\Omega = \sum_{i=1}^k n_i(\boldsymbol{\mu}_i - \bar{\boldsymbol{\mu}})(\boldsymbol{\mu}_i - \bar{\boldsymbol{\mu}})'$ *and* $\bar{\boldsymbol{\mu}} = (1/n)\sum_{i=1}^k n_i \boldsymbol{\mu}_i$.

Proof. We have seen that $(n_i - 1)\mathbf{S}_i \sim \mathrm{W}_p(n_i - 1, \Sigma)$. Further, the $\mathbf{S}_i$ are independent, and hence $\mathbf{S}_w = \sum_{i=1}^k (n_i - 1)\mathbf{S}_i \sim \mathrm{W}_p(n-k, \Sigma)$. Put $\boldsymbol{X}_i = \sqrt{n_i}\,\bar{\boldsymbol{Y}}_i$, and then

$$\begin{aligned} \mathbf{S}_b &= \sum_{i=1}^k n_i \bar{\boldsymbol{Y}}_i \bar{\boldsymbol{Y}}_i' - n\bar{\boldsymbol{Y}}\bar{\boldsymbol{Y}}' \\ &= \sum_{i=1}^k \boldsymbol{X}_i \boldsymbol{X}_i' - \tilde{\boldsymbol{X}}\tilde{\boldsymbol{X}}', \end{aligned}$$

where $\tilde{\boldsymbol{X}} = \sum_{i=1}^k \sqrt{n_i/n}\,\boldsymbol{X}_i$. Let H be a $k \times k$ orthogonal matrix whose first column is $(\sqrt{n_1/n}, \ldots, \sqrt{n_k/n})'$, and consider the transformation

$$\mathbf{Z} = \begin{pmatrix} \boldsymbol{Z}_1' \\ \vdots \\ \boldsymbol{Z}_k' \end{pmatrix} = \mathrm{H}' \begin{pmatrix} \boldsymbol{X}_1' \\ \vdots \\ \boldsymbol{X}_k' \end{pmatrix} = \mathrm{H}'\mathbf{X}.$$

Then $\sum_{i=1}^k \boldsymbol{Z}_i \boldsymbol{Z}_i' = \sum_{i=1}^k \boldsymbol{X}_i \boldsymbol{X}_i'$ and $\boldsymbol{Z}_1 = \tilde{\boldsymbol{X}}$, and

$$\mathrm{S}_b = \sum_{i=1}^k \boldsymbol{Z}_i \boldsymbol{Z}_i' - \boldsymbol{Z}_1 \boldsymbol{Z}_1' = \sum_{i=2}^k \boldsymbol{Z}_i \boldsymbol{Z}_i'.$$

By Theorem 1.2.6, the $\boldsymbol{Z}_i$'s are independently and normally distributed with the same covariance matrix Σ. Therefore, $\mathbf{S}_b \sim \mathrm{W}_p(k-1, \Sigma; \Psi)$. The non-centrality matrix Ψ is

$$\begin{aligned}\Psi &= \sum_{i=2}^{k} \mathrm{E}(\boldsymbol{Z}_i)\mathrm{E}(\boldsymbol{Z}_i)' \\ &= \mathrm{E}(\mathbf{Z})'\mathrm{E}(\mathbf{Z}) - \mathrm{E}(\boldsymbol{Z}_1)\mathrm{E}(\boldsymbol{Z}_1)' \\ &= \mathrm{E}(\mathbf{X})'\mathrm{E}(\mathbf{X}) - n\bar{\boldsymbol{\mu}}\bar{\boldsymbol{\mu}}' \\ &= \sum_{i=1}^{k} n_i \boldsymbol{\mu}_i \boldsymbol{\mu}_i' - n\bar{\boldsymbol{\mu}}\bar{\boldsymbol{\mu}}' = \Omega. \qquad \square\end{aligned}$$

The null distribution of $|\mathbf{S}_w|/|\mathbf{S}_w + \mathbf{S}_b|$ is called lambda distribution, which is formally defined as follows:

Definition 3.2.1 Let $\mathbf{B}$ and $\mathbf{W}$ be independent random matrices following the Wishart distribution $\mathrm{W}_p(q, \Sigma)$ and $\mathrm{W}_p(n, \Sigma)$, respectively, with $n \geq p$. Then the distribution of

$$\Lambda = \frac{|\mathbf{W}|}{|\mathbf{B} + \mathbf{W}|}$$

is said to be the p-dimensional *Lambda distribution* with (q, n)-degrees of freedom, and is denoted by $\Lambda_p(q, n)$. In what follows, we write $\Lambda_{p;q,n}$ to denote a random variable with this distribution:

$$\Lambda_{p;q,n} \sim \Lambda_p(q, n)$$

.

When $p = 1$,

$$\Lambda_{1;q,n} = \frac{\chi_n^2}{\chi_n^2 + \chi_q^2}.$$

This implies that $\Lambda_1(q, n)$ is distributed as a beta distribution $\mathrm{Be}(n, q)$ and

$$\frac{1 - \Lambda_{1;q,n}}{\Lambda_{1;q,n}} \cdot \frac{n}{q} = \frac{\chi_q^2/q}{\chi_n^2/n} \sim F(q, n).$$

The distribution of $\Lambda_p(1, n)$ is also simplified. In this case we can write

$$\Lambda_{p;1,n} = \frac{|\mathbf{W}|}{|\mathbf{W} + \boldsymbol{X}\boldsymbol{X}'|},$$

where $\mathbf{W}$ and $\boldsymbol{X}$ are independent and distributed as $\mathrm{W}_p(n, \mathrm{I}_p)$ and $\mathrm{N}_p(\mathbf{0}, \mathrm{I}_p)$, respectively. Note

$$|\mathbf{W} + \boldsymbol{X}\boldsymbol{X}'| = |\mathbf{W}|(1 + \boldsymbol{X}'\mathbf{W}^{-1}\boldsymbol{X}),$$

which is obtained by computing

$$\begin{vmatrix} 1 & \boldsymbol{X}' \\ -\boldsymbol{X} & \mathbf{W} \end{vmatrix}$$

in two ways (see [A.1.1] (2)). Using the formula above,

$$\Lambda_{p;1,n} = (1 + \boldsymbol{X}'\mathbf{W}^{-1}\boldsymbol{X})^{-1}$$

and hence

$$\frac{1 - \Lambda_{p;1,n}}{\Lambda_{p:1,n}} = (1 + \boldsymbol{X}'\mathbf{W}^{-1}\boldsymbol{X})^{-1}.$$

The distribution of $\boldsymbol{X}'\mathbf{W}^{-1}\boldsymbol{X}$ is given in Theorem 3.1.1. Summarizing these results, we have the following theorem.

Theorem 3.2.3 *For the Lambda distribution, it holds that:*

(1) $\dfrac{1 - \Lambda_{1;q,n}}{\Lambda_{1;q,n}} \cdot \dfrac{n}{q} \sim F(q, n).$

(2) $\dfrac{1 - \Lambda_{p;1,n}}{\Lambda_{p;1,n}} \cdot \dfrac{n + 1 - p}{p} \sim F(p, n + 1 - p).$

The distribution of $\Lambda_p(q, n) = |\mathbf{W}|/|\mathbf{W}+\mathbf{B}|$ is defined by two independent $p \times p$ matrices $\mathbf{B}$ and $\mathbf{W}$, where $\mathbf{B} \sim \mathrm{W}_p(q, \mathrm{I}_p)$ and $\mathbf{W} \sim \mathrm{W}_p(n, \mathrm{I}_p)$. Next we shall see that it may be defined based on two $q \times q$ Wishart matrices. The result was given by Wakaki et al. (2003).

Theorem 3.2.4 *Let* $\mathbf{B} = \mathbf{X}'\mathbf{X}$ *and* $\mathbf{W}$ *be independently distributed as* $\mathrm{W}_p(q, \mathrm{I}_p; \mathrm{M}'\mathrm{M})$ *and* $\mathrm{W}_p(n, \mathrm{I}_p)$*, respectively. Here* X *is a* $q \times p$ *random matrix whose elements are independent normal variates with* $\mathrm{E}(\mathbf{X}) = \mathrm{M}$ *and the common variance* 1*. Put*

$$\mathbf{U} = \mathbf{X}\mathbf{X}' \quad \textit{and} \quad \mathbf{V} = \mathbf{U}^{1/2}(\mathbf{X}\mathbf{W}^{-1}\mathbf{X}')^{-1}\mathbf{U}^{1/2}.$$

Then:

(1) $\mathbf{U}$ *and* $\mathbf{V}$ *are independently distributed as* $\mathrm{W}_q(p, \mathrm{I}_q; \mathrm{M}'\mathrm{M})$ *and* $\mathrm{W}_q(n - p + q, \mathrm{I}_q)$*, respectively.*

(2) $\dfrac{|\mathbf{W}|}{|\mathbf{W} + \mathbf{B}|} = \dfrac{|\mathbf{V}|}{|\mathbf{V} + \mathbf{U}|}.$

Proof. By the definition of Wishart distribution, it follows that $\mathbf{U} \sim \mathrm{W}_q(p, \mathrm{I}_q; \mathrm{M}'\mathrm{M})$. On the other hand, Theorem 2.3.3 asserts that the conditional distribution of $(\mathbf{X}'\mathbf{W}^{-1}\mathbf{X})^{-1}$ given $\mathbf{X}$ is $\mathrm{W}_q(n - p + q, (\mathbf{X}'\mathbf{X})^{-1})$.

Therefore, $\mathbf{V} \sim \mathrm{W}_q(n-p+q, \mathrm{I}_q)$ and hence $\mathbf{V}$ is independent of $\mathbf{U}$. For the second result, note that

$$\frac{|\mathbf{W}|}{|\mathbf{W}+\mathbf{B}|} = \{(1+\ell_1)\cdots(1+\ell_p)\}^{-1},$$

where $\ell_1 \geq \cdots \geq \ell_1 \geq 0$. The nonzero characteristic roots of $\mathbf{UV}^{-1}$ are the same as those of $\mathbf{XW}^{-1}\mathbf{X}'$ and hence $\mathbf{BW}^{-1}$. This completes the proof. □

Theorem 3.2.5 *For the Lambda distribution* $\Lambda_p(q,n)$,

$$\Lambda_p(q,n) = \Lambda_q(p, n-p+q)$$

holds.

Proof. This is a direct consequence of Theorem 3.2.4. □

3.3 Test for Additional Information

In the test of $H_0\colon \boldsymbol{\mu}^{(1)} = \cdots = \boldsymbol{\mu}^{(k)}$ in (3.2.2), we are interested in whether a subset of variables $X_1, \ldots, X_p$ is sufficient, or is the set of remainder variables has no additional information or redundant. The distributions of tests based on $\mathbf{W}(\mathbf{W}+\mathbf{B})^{-1}$ depend on the parameters $\boldsymbol{\mu}^{(1)}, \ldots, \boldsymbol{\mu}^{(k)}$ and Σ through $\Sigma^{-1}\Omega$, where Ω is the noncentrality matrix of $\mathbf{B}$, defined by

$$\Omega = \sum_{j=1}^{k} n_j(\boldsymbol{\mu}^{(j)} - \bar{\boldsymbol{\mu}})(\boldsymbol{\mu}^{(j)} - \bar{\boldsymbol{\mu}})' = \Xi\Xi',$$

where $\bar{\boldsymbol{\mu}} = (1/n)\sum_{j=1}^{k} n_j \boldsymbol{\mu}^{(j)}$ and $\Xi = (n_1(\boldsymbol{\mu}^{(1)} - \bar{\boldsymbol{\mu}}), \ldots, n_k(\boldsymbol{\mu}^{(k)} - \bar{\boldsymbol{\mu}}))$. Note that Ω denotes departures from the null hypothesis H_0. A single measure for the overall departures is

$$\delta^2 = \mathrm{tr}\,\Sigma^{-1}\Omega, \tag{3.3.1}$$

which may be considered as an information quantity on departures from the null hypothesis H_0. Without loss of generality we consider the sufficiency of a subvector $\boldsymbol{X}_1 = (X_1, \ldots, X_q)'$ of $\boldsymbol{X}$, or redundancy of the remainder vector $\boldsymbol{X}_2 = (X_{q+1}, \ldots, X_p)'$. Let $\boldsymbol{\mu}^{(j)}$, $\bar{\boldsymbol{\mu}}$, and Ω partition in the same manner as the partition of $\boldsymbol{X}$:

$$\boldsymbol{X} = \begin{pmatrix} \boldsymbol{X}_1 \\ \boldsymbol{X}_2 \end{pmatrix}, \quad \boldsymbol{\mu}^{(j)} = \begin{pmatrix} \boldsymbol{\mu}_1^{(j)} \\ \boldsymbol{\mu}_2^{(j)} \end{pmatrix}, \quad \bar{\boldsymbol{\mu}} = \begin{pmatrix} \bar{\boldsymbol{\mu}}_1 \\ \bar{\boldsymbol{\mu}}_2 \end{pmatrix},$$

$$\Omega = \begin{pmatrix} \Omega_{11} & \Omega_{12} \\ \Omega_{21} & \Omega_{22} \end{pmatrix}.$$

Then the information quantity for $\boldsymbol{X}_1$ on departures from the null hypothesis is expressed as

$$\delta_1^2 = \operatorname{tr} \Sigma_{11}^{-1} \Omega_{11}.$$

Then we say that $\boldsymbol{X}_1$ is sufficient or $\boldsymbol{X}_2$ is redundant if $\delta^2 = \delta_1^2$; it is called a sufficiency of $\boldsymbol{X}_1$. Using [A.1.2], we can write

$$\delta^2 = \delta_1^2 + \operatorname{tr} \Sigma_{22\cdot 1}^{-1} \sum_{j=1}^{k} n_j \tilde{\boldsymbol{\mu}}_{2\cdot 1}^{(j)} \tilde{\boldsymbol{\mu}}_{2\cdot 1}^{(j)\prime},$$

where $\tilde{\boldsymbol{\mu}}_{2\cdot 1}^{(j)} = \boldsymbol{\mu}_2^{(j)} - \bar{\boldsymbol{\mu}}_2 - \Sigma_{21}\Sigma_{11}^{-1}(\boldsymbol{\mu}_1^{(j)} - \bar{\boldsymbol{\mu}}_1)$. Therefore, the sufficiency condition of $\boldsymbol{X}_1$ is expressed as $\tilde{\boldsymbol{\mu}}_{2\cdot 1}^{(j)} = \mathbf{0}$, $j = 1, \ldots, q$, which is equivalent to

$$H_{2\cdot 1}\colon \boldsymbol{\mu}_{2\cdot 1}^{(1)} = \cdots = \boldsymbol{\mu}_{2\cdot 1}^{(q)}, \tag{3.3.2}$$

where $\boldsymbol{\mu}_{2\cdot 1}^{(j)} = \boldsymbol{\mu}_2^{(j)} - \Sigma_{21}\Sigma_{11}^{-1}\boldsymbol{\mu}_1^{(j)}$, $j = 1, \ldots, q$. The hypothesis $H_{2\cdot 1}$ was introduced by Rao (1948, 1970). To obtain a likelihood ratio for $H_{2\cdot 1}$, we partition the observation matrix as

$$\begin{aligned} \mathbf{X} &= (\boldsymbol{X}_{11}, \ldots, \boldsymbol{X}_{1n_1}, \ldots, \boldsymbol{X}_{k1}, \ldots, \boldsymbol{X}_{kn_k})' \\ &= \begin{pmatrix} \mathbf{X}_1 & \mathbf{X}_2 \end{pmatrix}, \quad \mathbf{X}_1\colon n \times k. \end{aligned}$$

Then the conditional distribution of $\mathbf{X}_2$ given $\mathbf{X}_1$ is normal such that the rows of $\mathbf{X}_2$ are independently distributed with covariance matrix $\Sigma_{22\cdot 1} = \Sigma_{22} - \Sigma_{21}\Sigma_{11}^{-1}\Sigma_{12}$, and the conditional mean is given by

$$\mathrm{E}(\mathbf{X}_2 | \mathbf{X}_1) = \mathrm{A}\Theta + \mathbf{X}_1 \Sigma_{11}^{-1} \Sigma_{12}, \tag{3.3.3}$$

where

$$\mathrm{A} = \begin{pmatrix} \mathbf{1}_{n_1} & & \mathrm{O} \\ & \ddots & \\ \mathrm{O} & & \mathbf{1}_{n_k} \end{pmatrix}, \quad \Theta' = (\boldsymbol{\mu}_{2\cdot 1}^{(1)}, \ldots, \boldsymbol{\mu}_{2\cdot 1}^{(q)}).$$

We use the notation $\mathbf{W}$ and $\mathbf{T}$ for $\mathbf{S}_w$ and $\mathbf{S}_t$, respectively, and partition $\mathbf{W}$ and $\mathbf{T}$ as

$$\mathbf{W} = \begin{pmatrix} \mathbf{W}_{11} & \mathbf{W}_{12} \\ \mathbf{W}_{21} & \mathbf{W}_{22} \end{pmatrix}, \quad \mathbf{T} = \begin{pmatrix} \mathbf{T}_{11} & \mathbf{T}_{12} \\ \mathbf{T}_{21} & \mathbf{T}_{22} \end{pmatrix}, \tag{3.3.4}$$

where $\mathbf{W}_{12}\colon q \times (p-q)$ and $\mathbf{T}_{12}\colon q \times (p-q)$. Then, using [A.2.11], we have that the maximum likelihood estimations of $\Sigma_{22\cdot 1}$ and Σ_{11} under $H_{2\cdot 1}$ are

$$\begin{aligned} n\hat{\Sigma}_{22\cdot 1} &= \mathbf{X}_2' \left\{ \mathrm{I}_n - \mathrm{P}_0 - \mathrm{P}_{(\mathrm{I}_n - \mathrm{P}_0)\mathbf{X}_1} \right\} \mathbf{X}_2 \\ &= \mathbf{T}_{22} - \mathbf{T}_{21}\mathbf{T}_{11}^{-1}\mathbf{T}_{12} = \mathbf{T}_{22\cdot 1}, \\ n\hat{\Sigma}_{11} &= \mathbf{X}_1'(\mathrm{I}_n - \mathrm{P}_{\mathrm{A}})\mathbf{X}_1 \\ &= \mathbf{W}_{11}, \end{aligned}$$

where $\mathrm{P}_0 = (1/n)\mathbf{1}_n\mathbf{1}_n'$ and for an $n \times m$ matrix B with $\mathrm{rank}(B) = m$, $\mathrm{P}_\mathrm{B} = \mathrm{B}(\mathrm{B}'\mathrm{B})^{-1}\mathrm{B}'$. Therefore,

$$-2\log\max_{H_{2\cdot1}} f(\mathrm{X};\boldsymbol{\mu}^{(1)},\ldots,\boldsymbol{\mu}^{(k)},\Sigma) = n\log\left|\frac{1}{n}\mathrm{T}_{22\cdot1}\right| + n\log\left|\frac{1}{n}\mathrm{W}_{11}\right| + np\{\log 2\pi + 1\}.$$

On the other hand, under no restriction

$$-2\log\max f(\mathrm{X};\boldsymbol{\mu}^{(1)},\ldots,\boldsymbol{\mu}^{(k)},\Sigma) = n\log\left|\frac{1}{n}\mathrm{W}\right| + np\{\log 2\pi + 1\}.$$

These imply that the likelihood ratio test criterion for $H_{2\cdot1}$ is given by

$$\lambda = \left(\frac{|\mathbf{W}|}{|\mathbf{W}_{11}|\cdot|\mathbf{T}_{22\cdot1}|}\right)^{n/2} = \left(\frac{|\mathbf{W}_{22\cdot1}|}{|\mathbf{T}_{22\cdot1}|}\right)^{n/2}.$$

Theorem 3.3.1 *Suppose that* (3.2.1) *is a set of samples from* $\mathrm{N}_p(\boldsymbol{\mu}^{(j)},\Sigma)$, $j = 1,\ldots,k$. *Then the likelihood ratio criterion* λ *for the hypothesis* $H_{2\cdot1}$ *in* (3.3.2) *is given by*

$$\lambda = \left(\frac{|\mathbf{W}_{22\cdot1}|}{|\mathbf{T}_{22\cdot1}|}\right)^{n/2},$$

where $\mathbf{W} = \mathbf{S}_w$ *and* $\mathbf{T} = \mathbf{S}_t$ *are given by* (3.3.4). *Further, under* $H_{2\cdot1}$,

$$\frac{|\mathbf{W}_{22\cdot1}|}{|\mathbf{T}_{22\cdot1}|} \sim \Lambda_{p-q}(k-1, n-q-k).$$

Proof. We consider the conditional distribution of $\mathbf{W}_{22\cdot1}$ and $\mathbf{T}_{22\cdot1}$ given $\mathbf{X}_1$. The matrix $\mathbf{W}$ is expressed as $\mathbf{X}'(\mathrm{I}_n - \mathrm{P}_\mathrm{A})\mathbf{X}$. Note that, in general, for a matrix A, $\mathrm{P}_\mathrm{A} = \mathrm{A}(\mathrm{A}'\mathrm{A})^{-1}\mathrm{A}'$ denotes the projection matrix on the space spanned by the column vectors of A. Hence

$$\mathbf{W}_{22\cdot1} = \mathbf{X}_2'\mathbf{Q}_1\mathbf{X}_2, \quad \mathbf{Q}_1 = \mathrm{I}_n - \mathrm{P}_\mathrm{A} - \mathrm{P}_{(\mathrm{I}_n - \mathrm{P}_\mathrm{A})\mathbf{X}_1}.$$

It is easy to see that $\mathbf{Q}_1^2 = \mathbf{Q}_1$, $\mathrm{rank}(\mathbf{Q}_1) = \mathrm{tr}\,\mathbf{Q}_1 = n-q-k$, $\mathbf{Q}_1\mathrm{A} = \mathrm{O}$, $\mathbf{Q}_1\mathbf{X}_1 = \mathrm{O}$, and

$$\mathrm{E}(\mathbf{X}_2|\mathbf{X}_1)'\mathbf{Q}_1\mathrm{E}(\mathbf{X}_2|\mathbf{X}_1) = \mathrm{O}.$$

This implies that $\mathbf{W}_{22\cdot1}|\mathbf{X}_1 \sim \mathrm{W}_{p-k}(n-q-k, \Sigma_{22\cdot1})$ and hence $\mathbf{W}_{22\cdot1} \sim \mathrm{W}_{p-k}(n-q-k, \Sigma_{22\cdot1})$. For $\mathbf{T}_{22\cdot1}$, we have

$$\mathbf{T}_{22\cdot1} = \mathbf{X}_2'\mathbf{Q}_2\mathbf{X}_2, \quad \mathbf{Q}_2 = \mathrm{I}_n - \mathrm{P}_0 - \mathrm{P}_{(\mathrm{I}_n - \mathrm{P}_0)\mathbf{X}_1},$$

and hence

$$\mathbf{S}_h = \mathbf{T}_{22\cdot1} - \mathbf{W}_{22\cdot1} = \mathbf{X}_2'(\mathbf{Q}_2 - \mathbf{Q}_1)\mathbf{X}_2.$$

Similarly $\mathbf{Q}_2$ is idempotent. Using $\mathrm{P}_0\mathrm{P}_\mathrm{A} = \mathrm{P}_\mathrm{A}\mathrm{P}_0 = \mathrm{P}_0$, we have $\mathbf{Q}_1\mathbf{Q}_2 = \mathbf{Q}_2\mathbf{Q}_1 = \mathbf{Q}_1$, and hence

$$(\mathbf{Q}_2 - \mathbf{Q}_1)^2 = \mathbf{Q}_2 - \mathbf{Q}_1, \qquad \mathbf{Q}_1 \cdot (\mathbf{Q}_2 - \mathbf{Q}_1) = \mathrm{O}.$$

Further, under $H_{2\cdot 1}$,

$$\mathrm{E}(\mathbf{X}_2|\mathbf{X}_1)'(\mathbf{Q}_2 - \mathbf{Q}_1)\mathrm{E}(\mathbf{X}_2|\mathbf{X}_1) = \mathrm{O}.$$

The required result is obtained by using the Cochran Theorem in Theorem 2.2.3. □

3.3.1 Decomposition of the Lambda Statistic

Theorem 3.3.2 *Let* $\mathbf{B}$ *and* $\mathbf{W}$ *be independent,* $\mathbf{B} \sim \mathrm{W}_p(q, \Sigma)$*, and* $\mathbf{W} \sim \mathrm{W}_p(n, \Sigma)$ *with* $n \geq p$*. Put* $\mathbf{T} = \mathbf{W} + \mathbf{B}$*, and partition* $\mathbf{W}$ *and* $\mathbf{T}$ *as*

$$\mathbf{W} = \begin{pmatrix} \mathbf{W}_{11} & \mathbf{W}_{12} \\ \mathbf{W}_{21} & \mathbf{W}_{22} \end{pmatrix}, \quad \mathbf{T} = \begin{pmatrix} \mathbf{T}_{11} & \mathbf{T}_{12} \\ \mathbf{T}_{21} & \mathbf{T}_{22} \end{pmatrix},$$

where $\mathbf{W}_{ij}$: $p_i \times p_j$ *and* $\mathbf{T}_{ij}$: $p_i \times p_j$*. Then* $\Lambda = |\mathbf{W}|/|\mathbf{T}|$ *can be decomposed as*

$$\Lambda = \frac{|\mathbf{W}|}{|\mathbf{T}|} = \frac{|\mathbf{W}_{11}|}{|\mathbf{T}_{11}|} \cdot \frac{|\mathbf{W}_{22\cdot 1}|}{|\mathbf{T}_{22\cdot 1}|} = \Lambda^{(1)} \cdot \Lambda^{(2)}.$$

Further, it holds that:

(1) $\Lambda^{(1)} \sim \Lambda_{p_1}(q, n), \quad \Lambda^{(2)} \sim \Lambda_{p_2}(q, n - p_1)$.

(2) $\Lambda^{(1)}$ *and* $\Lambda^{(2)}$ *are independent.*

Proof. From Theorem 2.2.2 it follows that

$$\begin{aligned} &\mathbf{W}_{11} \sim \mathrm{W}_{p_1}(n, \Sigma_{11}), \quad \mathbf{W}_{22\cdot 1} \sim \mathrm{W}_{p_2}(n - p_1, \Sigma_{22\cdot 1}), \\ &\mathbf{T}_{11} \sim \mathrm{W}_{p_2}(n, \Sigma_{11}), \quad \mathbf{T}_{22\cdot 1} \sim \mathrm{W}_{p_2}(n + q - p_1, \Sigma_{22\cdot 1}). \end{aligned}$$

$\mathbf{W}_{22\cdot 1}$ is independent of $\mathbf{W}_{11}$, and $\mathbf{T}_{22\cdot 1}$ is independent of $\mathbf{T}_{11}$. Let $\mathbf{B}$ be partitioned in the same way as in $\mathbf{W}$ or $\mathbf{T}$. Then

$$\Lambda^{(1)} = \frac{|\mathbf{W}_{11}|}{|\mathbf{T}_{11}|} = \frac{|\mathbf{W}_{11}|}{|\mathbf{W}_{11} + \mathbf{B}_{11}|} \sim \Lambda_{p_1}(q, n),$$

and $\Lambda^{(2)}$ is independent of $\Lambda^{(1)}$. In the following we show that $\mathbf{S}_h = \mathbf{T}_{22\cdot 1} - \mathbf{W}_{22\cdot 1} \sim \mathrm{W}_{p_2}(q, \Sigma_{22\cdot 1})$ and $\mathbf{S}_h$ is independent of $\mathbf{W}_{22\cdot 1}$. We can write $\mathbf{B}$ and $\mathbf{W}$ as $\mathbf{B} = \mathbf{X}'\mathbf{X}$ and $\mathbf{W} = \mathbf{U}'\mathbf{U}$, respectively, where the rows of $\mathbf{X}$: $q \times p$ and

$\mathbf{U}$: $n \times p$ are independently distributed as $\mathrm{N}_p(\mathbf{0}, I_p)$. Let $\mathbf{X}$ and $\mathbf{U}$ partition as $\mathbf{X} = (\mathbf{X}_1\ \mathbf{X}_2)$, $\mathbf{X}_i$: $q \times p_i$ and $\mathbf{U} = (\mathbf{U}_1\ \mathbf{U}_2)$, $\mathbf{U}_i$: $n \times p_i$. Then

$$\mathbf{B} = \begin{pmatrix} \mathbf{X}_1'\mathbf{X}_1 & \mathbf{X}_1'\mathbf{X}_2 \\ \mathbf{X}_2'\mathbf{X}_1 & \mathbf{X}_2'\mathbf{X}_2 \end{pmatrix}, \quad \mathbf{W} = \begin{pmatrix} \mathbf{U}_1'\mathbf{U}_1 & \mathbf{U}_1'\mathbf{U}_2 \\ \mathbf{U}_2'\mathbf{U}_1 & \mathbf{U}_2'\mathbf{U}_2 \end{pmatrix},$$

and $\mathbf{W}_{22\cdot 1} = \mathbf{U}_2'(\mathrm{I}_n - \mathbf{U}_1(\mathbf{U}_1'\mathbf{U}_1)^{-1}\mathbf{U}_1')\mathbf{U}_2$. Noting that $(\mathrm{I}_q + \mathbf{X}_1\mathbf{W}_{11}^{-1}\mathbf{X}_1')^{-1} = \mathrm{I}_q - \mathbf{X}_1(\mathbf{X}_1'\mathbf{X}_1 + \mathbf{W}_{11})^{-1}\mathbf{X}_1'$ (use [A.1.4]),

$$\begin{aligned} \mathbf{S}_h = {} & (\mathbf{X}_2 - \mathbf{X}_1\mathbf{W}_{11}^{-1}\mathbf{W}_{12})' \\ & \times (\mathrm{I}_q + \mathbf{X}_1\mathbf{W}_{11}^{-1}\mathbf{X}_1')^{-1}(\mathbf{X}_2 - \mathbf{X}_1\mathbf{W}_{11}^{-1}\mathbf{W}_{12}). \end{aligned}$$

Let

$$\mathbf{Z} = \begin{pmatrix} \mathbf{X} \\ \mathbf{U} \end{pmatrix} = \begin{pmatrix} \mathbf{X}_1 & \mathbf{X}_2 \\ \mathbf{U}_1 & \mathbf{U}_2 \end{pmatrix} = (\mathbf{Z}_1\ \mathbf{Z}_2).$$

We can write $\mathbf{V}_{22\cdot 1} = \mathbf{Z}_2'\mathbf{F}\mathbf{Z}_2$, $\mathbf{S}_h = \mathbf{Z}_2'\mathbf{G}\mathbf{Z}_2$, where

$$\begin{aligned} \mathbf{F} &= \begin{pmatrix} \mathrm{O} & \mathrm{O} \\ \mathrm{O} & \mathrm{I}_n - \mathbf{U}_1(\mathbf{U}_1'\mathbf{U}_1)^{-1}\mathbf{U}_1' \end{pmatrix}, \\ \mathbf{G} &= (\mathrm{I}_q - \mathbf{X}_1\mathbf{W}_{11}^{-1}\mathbf{U}_1')'(\mathrm{I}_q + \mathbf{X}_1\mathbf{W}_{11}^{-1}\mathbf{X}_1')^{-1}(\mathrm{I}_q - \mathbf{X}_1\mathbf{W}_{11}^{-1}\mathbf{U}_1'). \end{aligned}$$

Given $\mathbf{Z}_1$, the rows of the $\mathbf{Z}_2$ are independently distributed as a p_2-variate normal distribution with the common covariance matrix $\Sigma_{22\cdot 1}$ and means $\mathrm{E}(\mathbf{Z}_2|\mathbf{Z}_1) = \mathbf{Z}_1\Sigma_{11}^{-1}\Sigma_{12}$. Note that $\mathbf{F}^2 = \mathbf{F}$, $\mathbf{G}^2 = \mathbf{G}$ and $\mathbf{F}\mathbf{G} = \mathrm{O}$. Therefore, from Theorem 2.2.3, $\mathbf{S}_h \sim \mathrm{W}_{p_2}(q, \Sigma_{11\cdot 2})$ and $\mathbf{S}_h$ is independent of $W_{22\cdot 1}$. This completes the proof. □

In Theorem 3.3.2, let $p_2 = p - 1$, and write

$$\begin{aligned} \mathbf{W}_{11} &\to \mathbf{W}_{(1\cdots p-1)(1\cdots p-1)}, & W_{22\cdot 1} &\to W_{pp\cdot 1\cdots p-1}, \\ \mathbf{T}_{11} &\to \mathbf{T}_{(1\cdots p-1)(1\cdots p-1)}, & T_{22\cdot 1} &\to T_{pp\cdot 1\cdots p-1}. \end{aligned}$$

Then

$$\begin{aligned} \Lambda &= \frac{|\mathbf{W}_{(1\cdots p-1)(1\cdots p-1)}|}{|\mathbf{T}_{(1\cdots p-1)(1\cdots p-1)}|} \cdot \frac{W_{pp\cdot 1\cdots p-1}}{T_{pp\cdot 1\cdots p-1}}, \\ \frac{W_{pp\cdot 1\cdots p-1}}{T_{pp\cdot 1\cdots p-1}} &\sim \frac{\chi^2(n-(p-1))}{\chi^2(n-(p-1)) + \chi^2(q)} = \mathrm{Be}\left(\frac{1}{2}\{n-(p-1)\}, \frac{1}{2}q\right). \end{aligned}$$

Applying a similar decomposition to $|\mathbf{W}_{(1\cdots p-1)(1\cdots p-1)}|/|\mathbf{T}_{(1\cdots p-1)(1\cdots p-1)}|$, we obtain the decomposition theorem of Λ:

Theorem 3.3.3 *Suppose that* $\Lambda = |\mathbf{W}|/|\mathbf{T}|$ *is distributed as a lambda distribution* $\Lambda_p(q, n)$, $\mathbf{W} \sim \mathrm{W}_p(n, \Sigma)$, $\mathbf{T} \sim \mathrm{W}_p(n+q, \Sigma)$, *and* $\mathbf{W}$ *and* $\mathbf{T} - \mathbf{W}$ *are independent. Then the distribution of* Λ *is the distribution of the product*

$\prod_{i=1}^{p} V_i$, *where* $V_1, \ldots, V_p$ *are independent and* $V_i \sim \mathrm{Be}(\frac{1}{2}(n-i+1), \frac{1}{2}q)$. *The variables* V_i *may be defined by*

$$V_1 = \frac{W_{11}}{T_{11}}, \quad V_i = \frac{W_{ii\cdot 1\cdots i-1}}{T_{ii\cdot 1\cdots i-1}}, \quad i = 2, \ldots, p,$$

where W_{11} *and* T_{11} *are the* $(1,1)$ *elements of* $\mathbf{W}$ *and* $\mathbf{T}$, *respectively. Let the first* $i \times i$ *matrices of* $\mathbf{W}$ *and* $\mathbf{T}$ *be denoted by*

$$\begin{pmatrix} \mathbf{W}_{11}^{(i)} & \boldsymbol{W}_{12}^{(i)} \\ \boldsymbol{W}_{12}^{(i)\prime} & W_{ii} \end{pmatrix} \quad \textit{and} \quad \begin{pmatrix} \mathbf{T}_{11}^{(i)} & \boldsymbol{T}_{12}^{(i)} \\ \boldsymbol{T}_{12}^{(i)\prime} & T_{ii} \end{pmatrix},$$

respectively. Then,

$$W_{ii\cdot 1\cdots i-1} = W_{ii} - \boldsymbol{W}_{12}^{(i)\prime}\{\mathbf{W}_{11}^{(i)}\}^{-1}\boldsymbol{W}_{12}^{(i)},$$
$$T_{ii\cdot 1\cdots i-1} = T_{ii} - \boldsymbol{T}_{12}^{(i)\prime}\{\mathbf{T}_{11}^{(i)}\}^{-1}\boldsymbol{T}_{12}^{(i)}.$$

Theorem 3.3.4 *When* $p = 2$:

(1) *The distribution of* $\Lambda_{2;q,n}$ *is the same as that of* Y^2, *where* $Y \sim \mathrm{Be}(n, q)$.

(2) $\dfrac{1-\sqrt{\Lambda_{2;q,n}}}{\sqrt{\Lambda_{2;q,n}}} \cdot \dfrac{2(n-1)}{2q} \sim F(2q, 2(n-1))$.

(3) $\dfrac{1-\sqrt{\Lambda_{2;q,n}}}{\sqrt{\Lambda_{2;q,n}}} \cdot \dfrac{2(n+1-p)}{2p} \sim F(2p, 2(n+1-p))$.

Proof. From Theorem 3.3.3 we may write

$$\Lambda_{2;q,n} = V_1 V_2, \quad V_1 \sim \mathrm{Be}\left(\frac{1}{2}n, \frac{q}{2}\right), \quad V_2 \sim \mathrm{Be}\left(\frac{1}{2}(n-1), \frac{q}{2}\right),$$

and V_1 and V_2 are independent. The hth moment of $\Lambda_{2;q,n}$ is

$$\mathrm{E}[\Lambda_{2;q,n}^h] = \frac{\Gamma[\frac{1}{2}n+h]\Gamma[\frac{1}{2}(n+q)]}{\Gamma[\frac{1}{2}n]\Gamma[\frac{1}{2}(n+q)+h]} \times \frac{\Gamma[\frac{1}{2}(n-1)+h]\Gamma[\frac{1}{2}(n+q-1)]}{\Gamma[\frac{1}{2}(n-1)]\Gamma[\frac{1}{2}(n+q-1)+h]}.$$

Using the duplication formula

$$\Gamma\left(\alpha+\frac{1}{2}\right)\Gamma(\alpha+1) = \frac{\sqrt{\pi}\,\Gamma(2\alpha+1)}{2^{2\alpha}},$$

$$\mathrm{E}[\Lambda_{2;q,n}^h] = \frac{\Gamma[q+n-1]\Gamma[n-1+h]}{\Gamma[q+n-3+2h]\Gamma[n+1]},$$

which is equal to $\mathrm{E}[Y^{2h}]$. This proves (1). From (1) we can write

$$\sqrt{\Lambda_{2;q,n}} = \frac{\chi_{2n}^2}{\chi_{2n}^2 + \chi_{2q}^2},$$

which gives (2). Result (3) follows from (2) and Theorem 3.2.5. □

Problems

3.1 (t-Distribution) Let X and Y be independent random variables following $\mathrm{N}(0,1)$ and $\chi^2(n)$, respectively. Then show that the density function is of $T = X/\sqrt{Y/n}$ is given by

$$f_T(t) = \frac{\Gamma((n+1)/2)}{\sqrt{n\pi}\,\Gamma(n/2)}\Big(1+\frac{1}{n}t^2\Big)^{-(n+1)/2}.$$

This distribution is called a *t-distribution* with n degrees of freedom.

Hint: The joint density function of (X, Y) is

$$f(x,y) = \frac{1}{\sqrt{n}}\exp\Big(-\frac{x^2}{2}\Big)\cdot\frac{1}{2^{n/2}}y^{n/2-1}\exp\Big(-\frac{y}{2}\Big).$$

Consider the transformation $(X,Y)\to(T,S); T = X/\sqrt{Y/n},\ S = Y$, and let

$$t = \frac{x}{\sqrt{y/n}}, \qquad s = y.$$

The Jacobian of the transformation is $\sqrt{s/n}$, and we have the joint density function $g(t,s)$ of (T,S). The result is obtained by integrating $g(t,s)$ with respect to s.

3.2 (Noncentral t-distribution) In problem 2.1, let the variable Z follow the normal distribution $\mathrm{N}(\lambda,1)$. Then, show that the density function $T = Z/\sqrt{Y/n}$ is given by

$$f(t) = \exp\Big(-\frac{1}{2}\lambda^2\Big)\frac{n^{n/2}}{\sqrt{\pi}\,\Gamma(n/2)}\sum_{k=0}^{\infty}\frac{\Gamma((n+k+1)/2)}{k!}\frac{(\sqrt{2}\,t\lambda)^k}{(n+t^2)^{(n+k+1)/2}}.$$

This distribution is called a *noncentral t-distribution* with n degrees of freedom and noncentrality parameter λ^2, which is denoted as $t(n;\lambda)$.

Hint: The conditional distribution of T given the condition $Y = y$ is the normal distribution $\mathrm{N}(\lambda(y/n)^{-1/2}, (y/n)^{-1})$. Therefore, the density function of T is expressed as

$$\begin{aligned}
f(t) &= \mathrm{E}_Y\left[\frac{\sqrt{Y/n}}{\sqrt{2\pi}}\exp\left\{-\frac{Y}{2n}\Big(t-\frac{\lambda}{\sqrt{Y/n}}\Big)^2\right\}\right] \\
&= \int_0^\infty \frac{\sqrt{y/n}}{\sqrt{2\pi}}\exp\left\{-\frac{y}{2n}\Big(t-\frac{\lambda}{\sqrt{y/n}}\Big)^2\right\}\cdot\frac{y^{n/2-1}\mathrm{e}^{-y/2}}{\Gamma(n/2)2^{n/2}}\,dy \\
&= \mathrm{e}^{-\lambda^2/2}\int_0^\infty \frac{\sqrt{y/n}}{\sqrt{2\pi}}\exp\left\{-\frac{1}{2}(1+t^2/n)y\right\}\exp\big(\sqrt{y/n}\,\lambda t\big) \\
&\qquad\times\frac{y^{n/2-1}\mathrm{e}^{-y/2}}{\Gamma(n/2)2^{n/2}}\,dy.
\end{aligned}$$

The result is obtained by expanding $\exp\left(\sqrt{y/n}\,\lambda t\right)$ in a power series and integrating term by term.

3.3 (F-distribution) Let X and Y be independent random variables following $\chi^2(m)$ and $\chi^2(n)$, respectively. Then show that the density function of $V = X/m/Y/n$ is

$$f_V(v) = \frac{\Gamma\big((m+n)/2\big)}{\Gamma\big(m/2\big)\Gamma\big(n/2\big)}\left(\frac{m}{n}\right)^{m/2} v^{m/2-1}\left(1+\frac{m}{n}v\right)^{-(m+n)/2}, \qquad v > 0.$$

The distribution is called an F-distribution with (m, n) degrees of freedom and is denoted by $\mathrm{F}(m, n)$.

Hint: The density function of (X, Y) is

$$f(x, y) = \frac{x^{m/2-1}\exp\big(-x/2\big)}{2^{m/2}\Gamma\big(m/2\big)}\cdot\frac{y^{n/2-1}\exp\big(-y/2\big)}{2^{n/2}\Gamma\big(n/2\big)}.$$

Consider the transformation from (X, Y) to (V, U) : $V = (X/m)/(Y/n)$, $U = Y$, and set

$$v = \frac{x/m}{y/n}, \qquad u = y.$$

The transformation is one-to-one and its Jacobian is $(m/n)u$. Therefore, the density function of (V, U) is

$$\begin{aligned} g(v, u) &= f\Big(\frac{m}{n}vu, u\Big)\Big|\frac{m}{n}\Big| \\ &= \frac{m^{m/2}n^{-n/2}v^{m/2-1}u^{(m+n)/2-1}\exp\big\{-\big((m/n)v+1\big)u/2\big\}}{2^{m/2}2^{n/2}\Gamma\big(m/2\big)\Gamma\big(n/2\big)}. \end{aligned}$$

The density function of V is obtained by integrating out $g(v, u)$ with respect to u: $g(v) = \int_0^\infty g(v, u)\,\mathrm{d}u$.

3.4 (Noncentral F-distribution) In Problem 2.5, let X be distributed as the noncental chi-squared distribution $\chi^2(m; \delta^2)$. Then show that the density function of V is given by

$$f(x) = \mathrm{e}^{-\delta^2/2}\sum_{k=0}^{\infty}\frac{(\delta^2/2)^k}{k!}\cdot\frac{(m/n)^{k+m/2}}{\mathrm{B}(k+m/2, n/2)}\cdot\frac{x^{m/2+k-1}}{\big(1+(m/n)x\big)^{k+(m+n)/2}},$$

where $\mathrm{B}(a, b) = \Gamma(a)\Gamma(b)/\Gamma(a+b)$. This distribution is called a *noncentral* F-*distribution* with (m, n) degrees of freedom and noncentrality parameter δ^2, which is denoted by $\mathrm{F}(m, n; \delta^2)$. Also, the density function is expressed as $f_{m,n}(x; \delta^2) = \mathrm{E}_K[f_{m+2K,n}(x)]$, where $K \sim \mathrm{Po}(\frac{1}{2}\delta^2)$.

Hint: The result is obtained by the same method as that used in deriving the F-distribution.

3.5 Consider the null distribution of $T^2_{2\cdot 1}$ appearing in Rao's U-statistic in (3.1.3):

$$U = \frac{T^2 - \mathrm{T}_1^2}{n + T_1^2} = \frac{T^2_{2\cdot 1}}{n + T_1^2},$$

which gives the expression

$$T^2_{2\cdot 1} = U \cdot (n + \mathrm{T}_1^2).$$

Show that the distribution of $T^2_{2\cdot 1}$ can be regarded as a scale mixture of the F-distribution by chcking that U and T_1^2 are mutually independent.

Note: Approximations to the scale mixture of the F-distribution will be discussed in Chapter 14].

3.6 Let $\bar{\boldsymbol{X}}$ and $\mathbf{S}$ be the sample mean vector and sample covariance matrix based on a random sample of size $N = n + 1$ from $\mathrm{N}_p(\boldsymbol{\mu}, \Sigma)$. Prove the following three propositions:

(1) $$\max_{\boldsymbol{a}} \frac{N\{\boldsymbol{a}'(\bar{\boldsymbol{X}} - \boldsymbol{\mu})\}}{\boldsymbol{a}'\mathbf{S}\boldsymbol{a}} = T^2 \equiv N(\bar{\boldsymbol{X}} - \boldsymbol{\mu})'\mathbf{S}^{-1}(\bar{\boldsymbol{X}} - \boldsymbol{\mu}).$$

(2) $$\frac{1}{n}T^2 \cdot \frac{n-p+1}{p} \sim \mathrm{F}(p, n-p+1).$$

(3) $$\mathrm{P}\{\boldsymbol{a}'\bar{\boldsymbol{X}} - c\sqrt{\boldsymbol{a}'\mathbf{S}\boldsymbol{a}} \leq \boldsymbol{a}'\boldsymbol{\mu} \leq \boldsymbol{a}'\bar{\boldsymbol{X}} + c\sqrt{\boldsymbol{a}'\mathbf{S}\boldsymbol{a}}, \ \text{for any} \ \boldsymbol{a}\} \geq 1 - \alpha,$$

where $$c^2 = \frac{np(n-p+1)}{n+1}\mathrm{F}(p, n-p+1; \alpha).$$

3.7 Let

$$T^2 = \boldsymbol{Z}'\left(\frac{1}{m}\mathbf{W}\right)^{-1}\boldsymbol{Z} \quad \text{and} \quad L = \frac{|\mathbf{W}|}{|\mathbf{W} + \boldsymbol{Z}\boldsymbol{Z}'|},$$

where $\boldsymbol{Z} \sim \mathrm{N}_p(\boldsymbol{\nu}, \Sigma)$, $m\mathbf{W} \sim \mathrm{W}_p(m, \Sigma)$, and $\boldsymbol{Z}$ and $\mathbf{W}$ are independent. Then, show that the distribution of $(1 - L)/L$ is the same as one of $(1/m)T^2$.

Hint: Use $|\mathbf{W} + \boldsymbol{Z}\boldsymbol{Z}'| = |\mathbf{W}| \cdot (1 + \boldsymbol{Z}'\mathbf{W}^{-1}\boldsymbol{Z})$.

3.8 Consider testing hypothesis $\mathrm{H}_0 : \mathrm{D}'\boldsymbol{\mu} = \mathbf{0}$ on the basis of a random sample of size $N = n+1$ from $\mathrm{N}_p(\boldsymbol{\mu}, \Sigma)$, where Σ is unknown, and D is a given matrix of size $p \times r$ with rank r. Let $\bar{\boldsymbol{X}}$ and $\mathbf{S}$ be the sample mean vector and sample covariance matrix, respectively.

(1) Show that the hypothesis H_0 is equivalent to '$\boldsymbol{\mu} = \mathrm{Q}\boldsymbol{\theta}$', where Q is a $p \times q\,(= p - r)$ matrix with rank q satisfying $\mathrm{D}'\mathrm{Q} = \mathrm{O}$, and $\boldsymbol{\theta}$ is an unknown vector. Further, show (see Sections 12.2.1 and 12.4.3) that the LR test is based on

$$T^2 = N(\mathrm{D}'\bar{\mathrm{X}})'(\mathrm{D}'\mathbf{S}\mathrm{D})^{-1}\mathrm{D}'\bar{\mathrm{X}}.$$

(2) Show that under the hypothesis, $(1/n)T^2 \cdot [(n-r+1)/r]$ is ditributed as $\mathrm{F}(r, n-r+1)$.

3.9 Let $\Lambda = |\mathbf{W}|/|\mathbf{T}|$ be distributed as a lambda distribution $\Lambda_p(q, n)$, and cosider the decomposition $\Lambda = V_1 V_2 \cdots V_p$ given in Theorem 3.3.3. Let $\Lambda(X_1, \ldots, X_i)$ be the Λ corresponding to the first i variables $X_1, \ldots, X_i$, and put

$$\Lambda(X_i | X_1, \ldots, X_{i-1}) = \Lambda(X_1, \ldots, X_i)/\Lambda(X_1, \ldots, X_{i-1}), \ i = 2, \ldots, p.$$

Then, show that $V_1 = \Lambda(X_1)$, $V_i = \Lambda(X_i | X_1, \ldots, X_{i-1}), i = 2, \ldots, p$.

Chapter 4

Correlation Coefficients

In this chapter we define several correlation coefficients between variables or random vectors in connection with linear regression models. Their sampling counterparts are obtained by the least squares and maximum likelihood methods. The sampling distributions and statistical inferences of these correlations are discussed. A covariance selection model that is related to partial correlations is discussed as well. The canonical correlation, which is the correlation between random vectors, will be discussed later in Chapter 11.

4.1 Ordinary Correlation Coefficients

4.1.1 Population Correlation

The correlation between two variables X_1 and X_2 is defined by

$$\rho = \rho(X_1, X_2) = \frac{\mathrm{Cov}(X_1, X_2)}{\sqrt{\mathrm{Var}(X_1)\mathrm{Var}(X_2)}}. \tag{4.1.1}$$

Let the mean vector, covariance matrix, and correlation matrix of $\boldsymbol{X} = (X_1, X_2)'$ be denoted by

$$\boldsymbol{\mu} = \begin{pmatrix} \mu_1 \\ \mu_2 \end{pmatrix}, \quad \Sigma = \begin{pmatrix} \sigma_{11} & \sigma_{12} \\ \sigma_{21} & \sigma_{22} \end{pmatrix}, \quad \mathrm{R} = \begin{pmatrix} 1 & \rho \\ \rho & 1 \end{pmatrix},$$

and let $\sigma_{11} = \sigma_1^2$ $(\sigma_1 \geq 0)$ and $\sigma_{22} = \sigma_2^2$ $(\sigma_2 \geq 0)$. Then the correlation between X_1 and X_2 is denoted by

$$\rho = \frac{\sigma_{12}}{\sigma_1 \sigma_2}.$$

The advantage of the correlation is that it is independent of the scale. More precisely, consider the transformation

$$Z_1 = aX_1 + b, \quad Z_2 = cX_2 + d,$$

where a, b, c, and d are constants and $a > 0$, $c > 0$. Then

$$\begin{aligned} &\mathrm{Var}(Z_1) = a^2\mathrm{Var}(X_1), \quad \mathrm{Var}(Z_2) = c^2\mathrm{Var}(X_2), \\ &\mathrm{Cov}(Z_1, Z_2) = ac\mathrm{Cov}(X_1, X_2). \end{aligned}$$

Therefore,

$$\rho(aX_1 + b, cX_2 + d) = \rho(X_1, X_2),$$

which shows that ρ is invariant under location and scale transformation.

It is well know that

$$|\rho| \le 1, \tag{4.1.2}$$

which is also shown from a consequence of Theorem 4.1.1.

Consider predicting X_1 by a linear function of X_2 (i.e., $\alpha + \beta X_2$). Then we have the following result.

Theorem 4.1.1

$$\min_{\alpha,\beta} \mathrm{E}[(X_1 - \alpha - \beta X_2)^2] = \sigma_1^2(1 - \rho^2).$$

The equality holds when $\beta = \rho\sigma_2/\sigma_1 \equiv b$ *and* $\alpha = \mu_1 - b\mu_2$.

Proof. Consider a reduction,

$$X_1 - \alpha - \beta X_2 = \{X_1 - \mu_1 - \beta(X_2 - \mu_2)\} + (\mu_1 - \alpha - \beta\mu_2).$$

Then

$$\begin{aligned} &\mathrm{E}[(X_1 - \alpha - \beta X_2)^2] \\ &\quad = \mathrm{E}[\{X_1 - \mu_1 - \beta(X_2 - \mu_2)\}^2] + (\mu_1 - \alpha - \beta\mu_2)^2 \\ &\quad = \sigma_{11} - 2\beta\sigma_{21} + \beta\sigma_{22} + (\mu_1 - \alpha - \beta\mu_2)^2 \\ &\quad = \sigma_{22}(\beta - \sigma_{21}/\sigma_{22})^2 + \sigma_{11} - \sigma_{21}^2/\sigma_{22} \\ &\qquad + (\mu_1 - \alpha - \beta\mu_2)^2 \\ &\quad \ge \sigma_{11}(1 - \rho^2). \end{aligned}$$

The equality condition is easily obtained. □

From Theorem 4.1.1 we have an optimum predictor

$$\hat{X}_1 = \mu_1 + b(X_2 - \mu_2), \quad b = \rho\sigma_2/\sigma_1, \tag{4.1.3}$$

which is called the *best linear predictor*. It satisfies

$$\mathrm{E}[(X_1 - \hat{X}_1)^2] = \sigma_1^2(1 - \rho^2). \tag{4.1.4}$$

This shows that $|\rho| \le 1$, since the left-hand side is nonnegative.

4.1.2 Sample Correlation

Let $\boldsymbol{X}_1 = (X_{11}, X_{12})', \ldots, \boldsymbol{X}_N = (X_{N1}, X_{N2})$ be a random sample drawn from a population with mean vector $\boldsymbol{\mu}$ and covariance matrix Σ. Let the sample mean vector and the sample covariance matrix be denoted by

$$\bar{\boldsymbol{X}} = \begin{pmatrix} \bar{X}_1 \\ \bar{X}_2 \end{pmatrix}, \quad \mathbf{S} = \begin{pmatrix} S_{11} & S_{12} \\ S_{21} & S_{22} \end{pmatrix}.$$

Then the sample correlation between X_1 and X_2 is defined by

$$R = R(X_1, X_2) = \frac{S_{12}}{\sqrt{S_{11}S_{22}}} = \frac{S_{12}}{S_1 S_2},$$

where $S_{ii} = S_i^2$, $S_i \geq 0$, $i = 1, 2$.

It is easy to see that:

(1) $|R| \leq 1$.

(2) $R(X_1, X_2) = R(aX_1 + b, cX_2 + d)$, where $a > 0$, $c > 0$.

The best linear predictor in the sample version is defined by

$$\hat{X}_1 = \bar{X}_1 + B(X_2 - \bar{X}_2), \quad B = RS_2/S_1.$$

The predictor can also be obtained by considering

$$\min_{\alpha,\beta} \sum_{i=1}^{N} (X_{i1} - \alpha - \beta X_{i2})^2.$$

Theorem 4.1.2 *Let R be the sample correlation coefficient of a sample of size $N = n + 1$ drawn from a bivariate normal distribution with correlation ρ. If $\rho = 0$, then*

$$\sqrt{n-1}\frac{R}{\sqrt{1-R^2}}$$

has the t-distribution with $n - 1$ degrees of freedom.

Proof. Since $n\mathbf{S} \sim \mathrm{W}_2(n, \Sigma)$, without loss of generality we may write

$$R = \frac{\sum_{i=1}^n X_i Y_i}{\sqrt{\sum_{i=1}^n X_i^2 \sum_{i=1}^n Y_i^2}} = \frac{\boldsymbol{X}'\boldsymbol{Y}}{\sqrt{\|\boldsymbol{X}\|^2 \cdot \|\boldsymbol{Y}\|^2}} = \frac{\boldsymbol{A}'\boldsymbol{Y}}{\sqrt{\|\boldsymbol{Y}\|^2}},$$

where $\boldsymbol{X} = (X_1, \ldots, X_n)'$, $\boldsymbol{Y} = (Y_1, \ldots, Y_n)'$, $\boldsymbol{A} = (1/\|\boldsymbol{X}\|)\boldsymbol{X}$, and

$$\begin{pmatrix} X_1 \\ Y_1 \end{pmatrix}, \ldots, \begin{pmatrix} X_n \\ Y_n \end{pmatrix} \sim \text{i.i.d. } \mathrm{N}_2\left(\begin{pmatrix} 0 \\ 0 \end{pmatrix}, \begin{pmatrix} 1 & \rho \\ \rho & 1 \end{pmatrix}\right).$$

Under the assumption $\rho = 0$, $\boldsymbol{Y}$ is independent of $\boldsymbol{X}$. The conditional distribution of $\boldsymbol{Y}$ given $\boldsymbol{X}$ is $\mathrm{N}_n(\mathbf{0}, \mathrm{I}_n)$. For any orthogonal random matrix $\mathbf{H}$ depending on $\boldsymbol{X}$ only, the distributions of $\boldsymbol{Y}$ and $\mathbf{H}'\boldsymbol{Y}$ are the same. We choose $\mathbf{H}$ as its first column $\boldsymbol{A}$. Then, by writing $\mathbf{H}'\boldsymbol{Y}$ as $\boldsymbol{Y}$, we have

$$\frac{R}{\sqrt{1-R^2}} = \frac{Y_1}{\sqrt{\sum_{i=1}^{2} Y_i^2}}$$

which gives our conclusion. □

It is known [see Kariya and Eaton, 1977] that the normality assumption is not important as long as one set of these samples has a spherical distribution.

From Theorem 4.1.2 we can write

$$\frac{\sqrt{n-1}\,R}{(1-R^2)^{1/2}} = \frac{Z_1}{\sqrt{\chi^2(n-1)/(n-1)}},$$

where $Z_1 \sim \mathrm{N}(0,1)$, and Z_1 and $\chi^2(n-1)$ are independent. This shows that $R^2 = \chi_1^2/(\chi_1^2 + \chi_{n-1}^2)$, $\chi_1^2 = Z^2$, and hence R^2 is distributed as a beta distribution with parameters $1/2$ and $(n-1)/2$. Therefore, the density function of R is given by

$$\frac{\Gamma(n/2)}{\pi^{1/2}\Gamma[(n-1)/2]}(1-r^2)^{(n-3)/2}, \quad -1 < r < 1.$$

The sample correlation is the MLE of ρ when the population is normal. This is shown by checking that:

(1) There exists the MLE of Σ.

(2) There exists a one-to-one correspondence between Σ and $(\sigma_{11}, \sigma_{22}, \rho)$.

The second condition can be relaxed as follows:

(2)′ $\rho = \rho(\Sigma)$ is a mapping from $\mathcal{S} = \{\Sigma\colon \Sigma > O\}$ to $(-1, 1)$.

Such a general result was given by Zehna (1966) as follows.

Lemma 4.1.1 *Let $L(\boldsymbol{\theta})$ be the likelihood function of a sample. Suppose that the MLE $\hat{\boldsymbol{\theta}}$ of $\boldsymbol{\theta}$ exists so that $L(\hat{\boldsymbol{\theta}}) \geq L(\boldsymbol{\theta})$ for all $\boldsymbol{\theta} \in \Omega$. Let $g(\boldsymbol{\theta})$ be an arbitrary mapping of Ω to Ω^*, where Ω^* is the range of $g(\boldsymbol{\theta})$. Then $g(\hat{\boldsymbol{\theta}})$ is a maximum (induced) likelihood estimator of $g(\boldsymbol{\theta})$.*

Proof. See Zehna (1966) or Siotani et al. (1985). □

To obtain the distribution of R, we start with $\mathbf{V} = n\mathbf{S} \sim \mathrm{W}_2(n, \Sigma)$. Let

$$\mathbf{V} = \begin{pmatrix} V_{11} & V_{12} \\ V_{21} & V_{22} \end{pmatrix}.$$

Then the density function of $\mathbf{V}$ (i.e., the joint density function of V_{11}, V_{12}, and V_{22}) is

$$\frac{\mathrm{etr}(-\frac{1}{2}\Sigma^{-1}\mathrm{V})|\mathrm{V}|^{(n-3)/2}}{2^n|\Sigma|^{n/2}\Gamma_2(\frac{1}{2}n)}.$$

Since V is invariant under the transformation

$$\mathrm{V} \to \begin{pmatrix} \sigma_1^{-1} & 0 \\ 0 & \sigma_2^{-1} \end{pmatrix} \mathrm{V} \begin{pmatrix} \sigma_1^{-1} & 0 \\ 0 & \sigma_2^{-1} \end{pmatrix},$$

we can assume without loss of generality that

$$\Sigma = \begin{pmatrix} 1 & \rho \\ \rho & 1 \end{pmatrix}.$$

Then

$$\Sigma^{-1} = \frac{1}{1-\rho^2}\begin{pmatrix} 1 & -\rho \\ -\rho & 1 \end{pmatrix},$$

so that

$$\delta = \frac{1}{2}\,\mathrm{tr}\,\Sigma^{-1}\mathrm{V} = \frac{1}{2(1-\rho^2)}(v_{11} + v_{22} - 2\rho v_{12}).$$

Therefore, the joint density function of V_{11}, V_{12}, and V_{22} is

$$\frac{\mathrm{e}^{-\delta}(v_{11}v_{22} - v_{12}^2)^{(n-3)/2}}{2^n(1-\rho^2)^{n/2}\Gamma(\frac{1}{2}n)}.$$

Consider the transformation $w_1 = \sqrt{v_{11}}$ and $w_2 = \sqrt{v_{22}}$. Noting that the Jacobian of its inverse transformation is $2^2 w_1 w_2$, the distribution of R is expressed as

$$\begin{aligned} f(r) &= 4 \cdot \frac{1}{2^n\Gamma(\frac{1}{2}n)}(1-\rho^2)^{-n/2}(1-r^2)^{(n-3)/2} \\ &\quad \times \int_0^\infty \int_0^\infty (w_1w_2)^{n-1} \exp\left\{-\frac{w_1^2 - 2\rho r w_1 w_2 + w_2^2}{2(1-\rho^2)}\right\} dw_1\, dw_2. \end{aligned}$$

By expanding $\exp\{\rho w_1 w_2/(1-\rho^2)\}$ in a power series and integrating term by term, we obtain

$$f(r) = \frac{1}{\Gamma(\frac{1}{2}n)}(1-\rho^2)^{n/2}(1-r^2)^{(n-3)/2}\sum_{j=0}^\infty \frac{(2\rho r)^j}{j!}\left\{\Gamma\left(\frac{n+j}{2}\right)\right\}^2.$$

Since, by the duplicate formula $\Gamma(a)\Gamma(a+1) = \sqrt{\pi}\,\Gamma(2a)/2^{2a-1}$,

$$\Gamma_2\Big(\frac{1}{2}n\Big) = \sqrt{\pi}\,\Gamma\Big(\frac{1}{2}n\Big)\Gamma\Big(\frac{1}{2}(n-1)\Big) = 2^{-n+2}\pi\Gamma(n-1),$$

we finally obtain the density function of R as

$$f(r) = \frac{2^{n-2}}{\pi\Gamma(n-1)}(1-\rho^2)^{n/2}(1-r^2)^{(n-3)/2}\sum_{j=0}^{\infty}\frac{(2\rho r)^j}{j!}\left\{\Gamma\Big(\frac{n+j}{2}\Big)\right\}^2.$$

Theorem 4.1.3 *Let R be a sample correlation coefficient based on a random sample of size $N = n+1$ drawn from a bivariate normal population with correlation ρ. Then R is distributed asymptotically as $\mathrm{N}(\rho, (1-\rho^2)^2/n)$, or the limiting distribution of $\sqrt{n}(R-\rho)$ is $\mathrm{N}(\rho, (1-\rho^2)^2)$.*

Proof. See Section 4.2.5. □

Based on Theorem 4.1.3, it holds asymptotically that with probability $1-\alpha$,

$$r - \frac{1}{\sqrt{n}}z_\alpha(1-\rho^2) \le \rho \le r + \frac{1}{\sqrt{n}}z_\alpha(1-\rho^2),$$

where z_α is the upper α-point of $\mathrm{N}(0,1)$. A conventional confidence interval is constructed by substituting r for the ρ in the endpoints. However, such a confidence interval will not be accurate. It is natural to look for a transformation of R such that the transformed variable converges in law to the normality faster than that of R itself and with a variance independent of ρ. Let $g(r)$ be an analytic function of r at $r = \rho$. Then by Theorem 2.5.2, $\sqrt{n}\{g(R) - g(\rho)\}$ is asymptotically normal with 0 and variance

$$\left\{\frac{dg(r)}{dr}\bigg|_{r=\rho}\right\}^2 (1-\rho^2)^2.$$

We wish to determine the function $g(\cdot)$ so that this variance is a constant, say, 1. Then we have a differential equation for $g(r)$ to satisfy

$$g'(\rho) = \frac{1}{1-\rho^2} = \frac{1}{2}\left(\frac{1}{1+\rho} + \frac{1}{1-\rho}\right).$$

Solving the equation, the transformation is given as

$$g(\rho) = \frac{1}{2}\log\left(\frac{1+\rho}{1-\rho}\right) = \tanh^{-1}\rho \equiv \xi. \tag{4.1.5}$$

Let

$$Z = \frac{1}{2}\log\left(\frac{1+R}{1-R}\right) = \tanh^{-1} R, \tag{4.1.6}$$

which was introduced by Fisher (1921) and is called *Fisher's z-transformation.*

Theorem 4.1.4 *Let Z and ξ be defined by (4.1.6) and (4.1.5), respectively. Then the limiting distribution of $\sqrt{n}(Z-\rho)$ is $\mathrm{N}(0,1)$.*

Konishi (1978) gave an interpretation of Fisher's z-transformation with an asymptotic expansion of the distribution function of R.

4.2 Multiple Correlation Coefficient

4.2.1 Population Multiple Correlation

In this section we consider a measure for the relationship between X_1 and $\boldsymbol{X}_2 = (X_2, \ldots, X_p)'$. Put $\boldsymbol{X} = (X_1, \boldsymbol{X}_2')'$, and let the mean vector and the covariance matrix of $\boldsymbol{X}$ decompose as

$$\boldsymbol{\mu} = \begin{pmatrix} \mu_1 \\ \boldsymbol{\mu}_2 \end{pmatrix}, \quad \Sigma = \begin{pmatrix} \sigma_{11} & \boldsymbol{\sigma}_{21}' \\ \boldsymbol{\sigma}_{21} & \Sigma_{22} \end{pmatrix}.$$

The multiple correlation coefficient can be characterized in various ways. Consider predicting X_1 by a linear predictor $\alpha + \boldsymbol{\beta}'\boldsymbol{X}_2$ of $\boldsymbol{X}_2$. Then the multiple correlation is the maximum correlation between X_1 and any linear function $\alpha + \boldsymbol{\beta}'\boldsymbol{X}_2$ of $\boldsymbol{X}_2$. The quantity is given as in the following.

Definition 4.2.1 The multiple correlation coefficient between X_1 and $\boldsymbol{X}_2 = (X_2, \ldots, X_p)'$, denoted by $\rho_{1(2\cdots p)}$, is defined by

$$\rho_{1(2\cdots p)} = \sqrt{\frac{\boldsymbol{\sigma}_{21}'\Sigma_{22}^{-1}\boldsymbol{\sigma}_{21}}{\sigma_{11}}}.$$

Theorem 4.2.1 *For linear predictor $\alpha + \boldsymbol{\beta}'\boldsymbol{X}_2$, it holds that:*

(1) $\min_{\alpha,\boldsymbol{\beta}} \mathrm{E}[(X_1 - \alpha - \boldsymbol{\beta}'\boldsymbol{X}_2)^2] = \sigma_1^2(1 - \rho_{1(2\cdots p)}^2)$.

(2) $\max_{\alpha,\boldsymbol{\beta}} \rho(X_1, \alpha + \boldsymbol{\beta}'\boldsymbol{X}_2) = \max_{\boldsymbol{\beta}} \rho(X_1, \boldsymbol{\beta}'\boldsymbol{X}_2) = \rho_{1(2\cdots p)}$.

The equality in (1) holds when

$$\boldsymbol{\beta} = \Sigma_{22}^{-1}\boldsymbol{\sigma}_{21} \equiv \boldsymbol{b}, \quad \alpha = \mu_1 - \boldsymbol{b}'\boldsymbol{\mu}_2 \equiv a.$$

The equality in (2) holds when $\boldsymbol{\beta} = c\boldsymbol{b}$, c is a proportional constant, and α is any constant.

Proof. The first result is proved by extending the proof of Theorem 4.1.1. In fact,

$$
\begin{aligned}
&\mathrm{E}[(X_1 - \alpha - \boldsymbol{\beta}' \boldsymbol{X}_2)^2] \\
&\quad = \mathrm{E}[\{X_1 - \mu_1 - \boldsymbol{\beta}'(\boldsymbol{X}_2 - \boldsymbol{\mu}_2)\}]^2 + (\mu_1 - \alpha - \boldsymbol{\beta}' \boldsymbol{\mu}_2)^2 \\
&\quad \geq \sigma_{11} - 2\boldsymbol{\sigma}_{21}' \boldsymbol{\beta} + \boldsymbol{\beta}' \Sigma_{22} \boldsymbol{\beta}.
\end{aligned}
$$

The equality in the last reduction folds when $\alpha = \mu_1 - \alpha - \boldsymbol{\beta}' \boldsymbol{\mu}_2$. Further, we have

$$
\begin{aligned}
&\sigma_{11} - 2\boldsymbol{\sigma}_{21}' \boldsymbol{\beta} + \boldsymbol{\beta}' \Sigma_{22} \boldsymbol{\beta} \\
&\quad = (\boldsymbol{\beta} - \Sigma_{22} \boldsymbol{\sigma}_{21})' \Sigma_{22} (\boldsymbol{\beta} - \Sigma_{22} \boldsymbol{\sigma}_{21}) + \sigma_{11} - \boldsymbol{\sigma}_{21}' \Sigma_{22} \boldsymbol{\sigma}_{21} \\
&\quad \geq \sigma_{11}(1 - \rho_{1(2\cdots p)}^2).
\end{aligned}
$$

The equality in the last reduction holds when $\boldsymbol{\beta} = \Sigma_{22}^{-1} \boldsymbol{\sigma}_{21}$. For the second result, it is easy to see that

$$
\begin{aligned}
\rho(X_1, \alpha + \boldsymbol{\beta}' \boldsymbol{X}_2) &= \rho(X_1, \boldsymbol{\beta}' \boldsymbol{X}_2) \\
&= \boldsymbol{\sigma}_{21}' \boldsymbol{\beta} / [\sigma_{11} (\boldsymbol{\beta}' \Sigma_{22} \boldsymbol{\beta})^{1/2}] = \phi(\boldsymbol{\beta}).
\end{aligned}
$$

By [A.2.2] we have

$$
\max_{\boldsymbol{\beta}} \phi^2(\boldsymbol{\beta}) = \frac{1}{\sigma_{11}} \boldsymbol{\sigma}_{21}' \Sigma^{-1} \boldsymbol{\sigma}_{21} = \rho_{1(2\cdots p)}^2,
$$

and the equality holds when $\boldsymbol{\beta} = c\Sigma^{-1} \boldsymbol{\sigma}_{21}$ (c is a proportional constant). This completes the proof. □

From Theorem 4.2.1 we have an optimum predictor

$$
\hat{X}_1 = \ell(\boldsymbol{X}_2) = \boldsymbol{\mu}_2 + \boldsymbol{b}'(\boldsymbol{X}_2 - \boldsymbol{\mu}_2), \quad \boldsymbol{b} = \Sigma_{22}^{-1} \boldsymbol{\sigma}_{21}, \tag{4.2.1}
$$

which is called the *best linear predictor*, and

$$
\begin{aligned}
\mathrm{E}[(X_1 - \hat{X}_1)^2] &= \sigma_{11}(1 - \rho_{1(2\cdots p)}^2), \\
\rho(X_1, \hat{X}_1) &= \rho_{1(2\cdots p)}.
\end{aligned} \tag{4.2.2}
$$

In the linear regression, $\ell(\boldsymbol{X}_2)$ given by (4.2.2) may be interpreted as the best predictor of X_1 in terms of $\boldsymbol{X}_2$ in the sense that it minimizes the mean square error $\mathrm{E}[(X_1 - \ell(\boldsymbol{X}_2))^2]$. The power of prediction may be measured by $\rho_{1(2\cdots p)}$.

4.2.2 Sample Multiple Correlation

Let

$$\boldsymbol{X}_1 = \begin{pmatrix} X_{11} \\ \boldsymbol{X}_{12} \end{pmatrix}, \ldots, \boldsymbol{X}_N = \begin{pmatrix} X_{N1} \\ \boldsymbol{X}_{N2} \end{pmatrix}$$

be a random sample drawn from a population with mean vector $\boldsymbol{\mu}$ and covariance matrix Σ. Let the sample mean vector, the sample covariance matrix, and the sample correlation matrix be denoted by

$$\bar{\boldsymbol{X}} = \begin{pmatrix} \bar{X}_1 \\ \bar{\boldsymbol{X}}_2 \end{pmatrix}, \quad \mathbf{S} = \begin{pmatrix} S_{11} & \boldsymbol{S}'_{21} \\ \boldsymbol{S}_{21} & \mathbf{S}_{22} \end{pmatrix}, \quad \mathbf{R} = \begin{pmatrix} 1 & \boldsymbol{R}'_{21} \\ \boldsymbol{R}_{21} & \mathbf{R}_{22} \end{pmatrix}.$$

The sample multiple correlation between X_1 and $\boldsymbol{X}_2$ is defined by

$$R_{1(2\cdots p)} = \sqrt{\frac{\boldsymbol{S}'_{21}\mathbf{S}_{22}^{-1}\boldsymbol{S}_{21}}{S_{11}}} = \sqrt{\boldsymbol{R}'_{21}\mathbf{R}_{22}^{-1}\boldsymbol{R}_{21}}, \tag{4.2.3}$$

which is the one obtained from Definition 4.2.1 by substituting $\mathbf{S}$ to Σ. The sample multiple correlation is the maximum sample correlation between X_1 and the linear function $\alpha + \boldsymbol{\beta}'\boldsymbol{X}_2$ of $\boldsymbol{X}_2$, that is,

$$\max_{\alpha,\boldsymbol{\beta}} r(X_1, \alpha + \boldsymbol{\beta}'\boldsymbol{X}_2) = \max_{\boldsymbol{\beta}} r(X_1, \boldsymbol{\beta}'\boldsymbol{X}_2) = R_{1(2\cdots p)}.$$

The maximum is attained at $\boldsymbol{\beta} = \hat{\boldsymbol{\beta}} = \mathbf{S}_{22}^{-1}\boldsymbol{S}_{21}$. Similarly, consider predicting X_1 by the linear function $\alpha + \boldsymbol{\beta}'\boldsymbol{X}_2$ of $\boldsymbol{X}_2$. Then we have

$$\min_{\alpha,\boldsymbol{\beta}} \frac{1}{n}\sum_{i=1}^{N}(X_{i1} - \alpha - \boldsymbol{\beta}'\boldsymbol{X}_{i2})^2 = S_{11}(1 - R_{1(2\cdots p)}^2).$$

The equality holds at $\boldsymbol{\beta} = \hat{\boldsymbol{\beta}}$, $\alpha = \hat{\alpha}$, where

$$\hat{\boldsymbol{\beta}} = \mathbf{S}_{22}^{-1}\boldsymbol{S}_{21} \equiv \boldsymbol{B}, \quad \hat{\alpha} = \bar{X}_1 - \boldsymbol{B}'\bar{\boldsymbol{X}}_2.$$

The best predictor is

$$\ell(\boldsymbol{X}_2) = \bar{X}_1 + \boldsymbol{S}'_{21}\mathbf{S}_{22}^{-1}(\boldsymbol{X}_2 - \bar{\boldsymbol{X}}_2).$$

When the population is a p-variate normal distribution $\mathrm{N}_p(\boldsymbol{\mu}, \Sigma)$, the sample multiple correlation is the MLE of $\rho_{1(2\cdots p)}$. This is obtained from [A.2.2].

For simplicity, put $\bar{\rho} \equiv \rho_{1(2\cdots p)}$ and $\bar{R} = R_{1(2\cdots p)}$. We first consider the distribution of $\bar{R}$ under $\bar{\rho} = 0$. Put $\mathbf{V} = n\mathbf{S}$; then $\mathbf{V} \sim \mathrm{W}_p(n, \Sigma)$. If we partition $\mathbf{V}$ similarly to Σ or $\mathbf{S}$ as

$$\mathbf{V} = \begin{pmatrix} V_{11} & \boldsymbol{V}'_{21} \\ \boldsymbol{V}_{21} & \mathbf{V}_{22} \end{pmatrix},$$

and the sample multiple correlation is given by

$$\bar{R}^2 = \frac{\boldsymbol{V}'_{21}\mathbf{V}_{22}^{-1}\boldsymbol{V}_{21}}{V_{11}} = \frac{\boldsymbol{V}'_{21}\mathbf{V}_{22}^{-1}\boldsymbol{V}_{21}}{V_{11\cdot 2} + \boldsymbol{V}'_{21}\mathbf{V}_{22}^{-1}\boldsymbol{V}_{21}},$$

where $V_{11\cdot 2} = V_{11} - \boldsymbol{V}'_{21}\mathbf{V}_{22}^{-1}\boldsymbol{V}_{21}$. Note that

$$\bar{\rho} = 0 \quad \Leftrightarrow \quad \boldsymbol{\sigma}_{21} = \mathbf{0}.$$

From Theorem 2.2.3 we know that

$$V_{11\cdot 2}/\sigma_{11} \sim \chi^2(n-(p-1)) \quad \text{and} \quad \boldsymbol{V}'_{21}\mathbf{V}_{22}^{-1}\boldsymbol{V}_{21}/\sigma_{11} \sim \chi^2(p-1),$$

and that they are independent, which implies that

$$\bar{R}^2 = \frac{\chi^2_{p-1}}{\chi^2_{p-1} + \chi^2_{n-(p-1)}},$$

which, in turn, gives the following theorem.

Theorem 4.2.2 *Let $\bar{R}$ be the sample multiple correlation between X_1 and $\boldsymbol{X}_2 = (X_2, \ldots, X_p)'$ based on a sample of size $N = n+1$ of $\boldsymbol{X} = (X_1, \boldsymbol{X}'_2)'$ whose distribution is $\mathrm{N}_p(\boldsymbol{\mu}, \Sigma)$. If the population multiple correlation $\bar{\rho}$ is zero, then*

$$\frac{n-(p-1)}{p-1} \cdot \frac{\bar{R}^2}{1-\bar{R}^2}$$

has an F-distribution with $(p-1, n-(p-1))$ degrees of freedom.

Next we consider the distribution of $\bar{R}^2$ when $\bar{\rho}$ is nonzero. We know that

$$\frac{\bar{R}^2}{1-\bar{R}^2} = \frac{\boldsymbol{V}'_{21}\mathbf{V}_{22}^{-1}\boldsymbol{V}_{21}}{V_{11\cdot 2}}.$$

From Theorem: 2.2.2 we have

(1) $V_{11\cdot 2}$ and $\boldsymbol{V}'_{21}\mathbf{V}_{22}^{-1}\boldsymbol{V}_{21}$ are independent,
(2) $V_{11\cdot 2}/\sigma_{11\cdot 2} \sim \chi^2(n-p+1)$, and
(3) the conditional distribution of $\boldsymbol{V}'_{21}\mathbf{V}_{22}^{-1}\boldsymbol{V}_{21}/\sigma_{11\cdot 2}$ given $\mathbf{V}_{22}$ is $\chi^2(p-1;\xi^2)$, where $\xi^2 = \boldsymbol{\sigma}'_{21}\Sigma_{22}^{-1}\mathbf{V}_{22}\Sigma_{22}^{-1}\boldsymbol{\sigma}_{21}/\sigma_{11\cdot 2}$.

Therefore,

$$Z \equiv \frac{n-(p-1)}{p-1} \cdot \frac{\bar{R}^2}{1-\bar{R}^2} = \frac{\chi^2_{p-1}(\xi^2)/(p-1)}{\chi^2_{n-p+1}/(n-p+1)},$$

which is the noncentral F-distribution with $(p-1, n-p+1)$ degrees of freedom and noncentrality parameter ξ^2. The conditional density of Z given ξ^2 is given (see Problem 4.) by

$$e^{-\xi^2/2}{}_1F_1\left(\frac{1}{2}n; \frac{1}{2}(p-1); \frac{(1/2)m\xi^2 z}{1+mz}\right) m^{(p-1)/2}$$
$$\times \frac{\Gamma(\frac{1}{2}n) z^{(p-3)/2}}{\Gamma[\frac{1}{2}(p-1)]\Gamma[\frac{1}{2}(n-p+1)](1+mz)^{n/2}},$$

where $m = (p-1)/(n-p+1)$ and ${}_1F_1$ denotes a hypergeometric function. The general hypergeometric function (or series) is

$$ {}_aF_b(a_1, \ldots, a_p; b_1, \ldots, b_q; z) = \sum_{k=0}^{\infty} \frac{(a_1)_k \cdots (a_p)_k}{(b_1)_k \cdots (b_q)_k} \cdot \frac{z^k}{k!}, \tag{4.2.4}$$

where $(a)_k = a(a+1)\cdots(a+k-1)$. By changing variables from Z to $\bar{R}^2$, the conditional distribution of $\bar{R}^2$ given $\mathbf{W}_{22}$ is

$$f(\bar{r}^2|\xi^2) = e^{-\xi^2/2}{}_1F_1\left(\frac{1}{2}n; \frac{1}{2}(p-1); \frac{1}{2}\xi^2\bar{r}^2\right) \cdot \frac{\Gamma(\frac{1}{2}n)}{\Gamma[\frac{1}{2}(p-1)]\Gamma[\frac{1}{2}(n-p+1)]}$$
$$\times (\bar{r}^2)^{(n-3)/2}(1-\bar{r}^2)^{(n-p-1)/2}, \qquad 0 < \bar{r}^2 < 1.$$

The (unconditional) density function is obtained by taking the expectation of $f(\bar{r}^2|\xi^2)$ with respect to ξ^2. Now $\boldsymbol{\sigma}_{21}'\Sigma_{22}^{-1}\mathbf{V}_{22}\Sigma_{22}^{-1}\boldsymbol{\sigma}_{21} \sim \mathrm{W}_1(n, \boldsymbol{\sigma}_{21}'\Sigma_{22}^{-1}\boldsymbol{\sigma}_{21})$, and hence

$$U = \frac{\boldsymbol{\sigma}_{21}'\Sigma_{22}^{-1}\mathbf{V}_{22}\Sigma_{22}^{-1}\boldsymbol{\sigma}_{21}}{\boldsymbol{\sigma}_{21}'\Sigma_{22}^{-1}\boldsymbol{\sigma}_{21}} \sim \chi^2(n).$$

We can write $\xi^2 = \delta^2 U$, where $\delta^2 = \boldsymbol{\sigma}_{21}'\Sigma_{22}^{-1}\boldsymbol{\sigma}_{21}/\sigma_{11\cdot 2}$. Replacing ξ^2 in $f(\bar{r}^2|\xi^2)$ by $\xi^2 = \delta^2 U$, and taking expectation with respect to U, we have the following theorem.

Theorem 4.2.3 *Under Theorem 4.2.2, if $\bar{\rho} \neq 0$, the density function of $\bar{R}^2$ is*

$$\frac{\Gamma(\frac{1}{2}n)}{\Gamma[\frac{1}{2}(p-1)]\Gamma[\frac{1}{2}(n-p+1)]}(\bar{r}^2)^{(n-3)/2}(1-\bar{r}^2)^{(n-p-1)/2}$$
$$\times (1-\bar{\rho}^2)^{n/2}{}_1F_1\left(\frac{1}{2}n; \frac{1}{2}n; \bar{\rho}^2\bar{r}^2\right), \qquad 0 < \bar{r}^2 < 1.$$

Example 4.2.1 Consider the marks from an examination on five subjects for 88 students given in Mardia et al. (1979). The five subjects are: X_1, mechanics; X_2, vector; X_3, algebra; X_4, analysis; and X_5, statistics. The correlation coefficients between these subjects are given in Table 4.2.1. Based on the

Table 4.2.1 Correlation matrix of five subjects

	X_1	X_2	X_3	R_4	R_5
X_1	1.000	–			
X_2	0.533	1.000	–		
X_3	0.547	0.610	1.000	–	
X_4	0.409	0.485	0.711	1.000	–
X_5	0.389	0.437	0.665	0.607	1.000

X_1	X_2	X_3	X_4	X_5
0.602	0.658	0.821	0.735	0.692

correlation matrix, we obtain multiple correlations between any subject and the other subjects as follows (see Table 4.2.1). The maximum value is 0.821, which is the multiple correlation between X_3 and the reminder variables.

4.3 Partial Correlation

4.3.1 Population Partial Correlation

Consider the correlation ρ_{12} between two random variables X_1 and X_2. If X_1 and X_2 are to be considered in conjunction with $p-2$ other variables $X_3, \ldots, X_p$, we may regard ρ_{12} as a mixture of a direct correlation between X_1 and X_2 and an indirect portion due to the presence of other variables correlating with X_1 and X_2. The partial correlation measuring the direct portion of the total correlation is defined as a correlation between X_1 and X_2 after the removal by linear regression of effects due to other variables. Let $\boldsymbol{X} = (X_1, X_2, \boldsymbol{X}_3')'$, and let the corresponding partitions of the mean vector $\boldsymbol{\mu}$ and the covariance matrix of $\boldsymbol{X}$ be

$$\boldsymbol{\mu} = \begin{pmatrix} \mu_1 \\ \mu_2 \\ \boldsymbol{\mu}_3 \end{pmatrix}, \qquad \Sigma = \begin{pmatrix} \sigma_{11} & \sigma_{12} & \boldsymbol{\sigma}_{31}' \\ \sigma_{21} & \sigma_{22} & \boldsymbol{\sigma}_{32}' \\ \boldsymbol{\sigma}_{31} & \boldsymbol{\sigma}_{32} & \Sigma_{33} \end{pmatrix}. \tag{4.3.1}$$

The best linear predictor of $X_i (i=1,2)$ by a linear function of $\boldsymbol{X}_3$ is

$$\ell_i(\boldsymbol{X}_3) = \mu_i + \boldsymbol{\sigma}_{i3}' \Sigma_{33}^{-1} (\boldsymbol{X}_3 - \boldsymbol{\mu}_3), \quad i = 1, 2.$$

The residuals

$$X_i^* = X_i - \ell_i(\boldsymbol{X}_3), \quad i = 1, 2, \tag{4.3.2}$$

are regarded as the portions of X_i remaining after removing the effects due to the presence of $\boldsymbol{X}_3$. Then

$$\boldsymbol{X}^* = \begin{pmatrix} X_1^* \\ X_2^* \end{pmatrix} = \begin{pmatrix} X_1 - \mu_1 \\ X_2 - \mu_2 \end{pmatrix} - \begin{pmatrix} \boldsymbol{\sigma}_{31}' \\ \boldsymbol{\sigma}_{32}' \end{pmatrix} \Sigma_{33}^{-1}(\boldsymbol{X}_3 - \boldsymbol{\mu}_3)$$

$$= \begin{pmatrix} 1 & 0 & -\boldsymbol{\sigma}_{31}'\Sigma_{33}^{-1} \\ 0 & 1 & -\boldsymbol{\sigma}_{32}'\Sigma_{33}^{-1} \end{pmatrix} \begin{pmatrix} X_1 - \mu_1 \\ X_2 - \mu_2 \\ \boldsymbol{X}_3 - \boldsymbol{\mu}_3 \end{pmatrix}$$

and hence

$$\mathrm{Var}(\boldsymbol{X}^*) = \begin{pmatrix} \sigma_{11} & \sigma_{12} \\ \sigma_{21} & \sigma_{22} \end{pmatrix} - \begin{pmatrix} \boldsymbol{\sigma}_{31}' \\ \boldsymbol{\sigma}_{32}' \end{pmatrix} \Sigma_{33}^{-1}(\boldsymbol{\sigma}_{31}\ \boldsymbol{\sigma}_{32})$$

$$\equiv \Sigma_{(12)(12)\cdot 3\cdots p}.$$

We denote

$$\Sigma_{(12)(12)\cdot 3\cdots p} = \begin{pmatrix} \sigma_{11\cdot 3\cdots p} & \sigma_{12\cdot 3\cdots p} \\ \sigma_{21\cdot 3\cdots p} & \sigma_{22\cdot 3\cdots p} \end{pmatrix}. \tag{4.3.3}$$

Definition 4.3.1 The partial correlation between X_1 and X_2 after the removal of effects due to $\boldsymbol{X}_3$ is

$$\rho_{12\cdot 3\cdots p} = \frac{\sigma_{12\cdot 3\cdots p}}{\sqrt{\sigma_{11\cdot 3\cdots p}\sigma_{22\cdot 3\cdots p}}},$$

which is also called the *partial correlation* (*coefficient*) between X_1 and X_2 given $\boldsymbol{X}_3$.

The partial correlation may be defined in terms of the correlation matrix

$$\mathrm{R} = (\rho_{ij}) = \begin{pmatrix} \rho_{11} & \rho_{12} & \boldsymbol{\rho}_{31}' \\ \rho_{21} & \rho_{22} & \boldsymbol{\rho}_{32}' \\ \boldsymbol{\rho}_{31} & \boldsymbol{\rho}_{32} & \mathrm{R}_{33} \end{pmatrix} \tag{4.3.4}$$

by substituting R for Σ. That is,

$$\rho_{12\cdot 3\cdots p} = \frac{\rho_{12}^*}{\sqrt{\rho_{11}^*\rho_{22}^*}},$$

where

$$\mathrm{R}_{(12)(12)\cdot 3\cdots p} = \begin{pmatrix} \rho_{11}^* & \rho_{12}^* \\ \rho_{21}^* & \rho_{22}^* \end{pmatrix} = \begin{pmatrix} \rho_{11} & \rho_{12} \\ \rho_{21} & \rho_{22} \end{pmatrix} - \begin{pmatrix} \boldsymbol{\rho}_{31}' \\ \boldsymbol{\rho}_{32}' \end{pmatrix} \mathrm{R}_{33}^{-1}(\boldsymbol{\rho}_{31}\ \boldsymbol{\rho}_{32}).$$

In particular, when $p = 3$,

$$\rho_{12\cdot 3} = \frac{\rho_{12} - \rho_{13}\rho_{23}}{\sqrt{(1 - \rho_{13}^2)(1 - \rho_{23}^2)}}.$$

4.3.2 Sample Partial Correlation

Let $\boldsymbol{X}$, $\mathbf{S}$, and $\mathbf{R}$ be versions of $\boldsymbol{\mu}$, Σ, and R based on a sample of size $N = n + 1$ from a p-variate population with mean $\boldsymbol{\mu}$ and covariance matrix Σ. We partition $\boldsymbol{X}$, $\mathbf{S}$, and $\mathbf{R}$ in the same manner as in the partition of $\boldsymbol{\mu}$, Σ, and R. Similarly, let

$$\begin{aligned}
\mathbf{S}_{(12)(12)\cdot 3\cdots p} &= \begin{pmatrix} S_{11} & S_{12} \\ S_{21} & S_{22} \end{pmatrix} - \begin{pmatrix} \boldsymbol{S}'_{31} \\ \boldsymbol{S}'_{32} \end{pmatrix} \mathbf{S}_{33}^{-1} (\boldsymbol{S}_{31}\ \boldsymbol{S}_{32}) \\
&= \begin{pmatrix} S^*_{11} & S^*_{12} \\ S^*_{21} & S^*_{22} \end{pmatrix}, \\
\mathbf{R}_{(12)(12)\cdot 3\cdots p} &= \begin{pmatrix} R_{11} & R_{12} \\ R_{21} & R_{22} \end{pmatrix} - \begin{pmatrix} \boldsymbol{R}'_{31} \\ \boldsymbol{R}'_{32} \end{pmatrix} \mathbf{R}_{33}^{-1} (\boldsymbol{R}_{31}\ \boldsymbol{R}_{32}) \\
&= \begin{pmatrix} R^*_{11} & R^*_{12} \\ R^*_{21} & R^*_{22} \end{pmatrix}.
\end{aligned}$$

Then the sample partial correlation $R_{12\cdot 3\cdots p}$ between X_1 and X_2 given $\boldsymbol{X}_3$ is defined from Definition 4.3.1 by substituting $\mathbf{S}$ or $\mathbf{R}$ for Σ. That is,

$$R_{12\cdot 3\cdots p} = \frac{S^*_{12}}{\sqrt{S^*_{11} S^*_{22}}} = \frac{R^*_{12}}{\sqrt{R^*_{11} R^*_{22}}}. \tag{4.3.5}$$

For the definition we may consider a regression approach by the best linear predictors of X_1 and X_2 based on $\boldsymbol{X}_3$, which are given by

$$\begin{aligned}
\hat{X}_1 &= \bar{X}_1 + \boldsymbol{S}'_{31}\mathbf{S}_{33}^{-1}(\boldsymbol{X}_3 - \bar{\boldsymbol{X}}_3), \\
\hat{X}_2 &= \bar{X}_2 + \boldsymbol{S}'_{32}\mathbf{S}_{33}^{-1}(\boldsymbol{X}_3 - \bar{\boldsymbol{X}}_3).
\end{aligned}$$

The residuals are

$$X_1^* = X_1 - \hat{X}_1, \quad X_2^* = X_1 - \hat{X}_2.$$

Then we can see that

$$R_{12\cdot 3\cdots p} = R(X_1^*, X_2^*).$$

We note that $R_{12\cdot 3\cdots p}$ is the MLE of $\rho_{12\cdot 3\cdots p}$ when the population is normal. For this, we have that there is a one-to-one correspondence between Σ and $(\Sigma_{(12)(12)\cdot 3\cdots p}, (\boldsymbol{\sigma}_{31}, \boldsymbol{\sigma}_{32}), \Sigma_{33})$, and the MLE of $\Sigma_{(12)(12)\cdot 3\cdots p}$ is $(n/N)\mathbf{S}_{(12)(12)\cdot 3\cdots p}$. Further, the partial correlation $\rho_{12\cdot 3\cdots p}$ is defined through $\Sigma_{(12)(12)\cdot 3\cdots p}$ [i.e., $\rho_{12\cdot 3\cdots p} = g(\Sigma_{(12)(12)\cdot 3\cdots p})$]. Therefore, from Lemma 4.1.1 the MLE of $\rho_{12\cdot 3\cdots p}$ is obtained by $g((n/N)\mathbf{S}_{(12)(12)\cdot 3\cdots p})$, which is

$$g(\mathbf{S}_{(12)(12)\cdot 3\cdots p}) = R_{12\cdot 3\cdots p}.$$

As seen in (4.3.5), the sample partial correlation $R_{12\cdot 3\cdots p}$ is regarded as a simple correlation defined from $\mathbf{S}_{(12)(12)\cdot 3\cdots p}$ instead of $\mathbf{S}_{(12)(12)}$. The distributions of $\mathbf{S}_{(12)(12)}$ and $\mathbf{S}_{(12)(12)\cdot 3\cdots p}$ are $\mathrm{W}_2(n, \Sigma_{(12)(12)})$ and $\mathrm{W}_2(n - (p-2), \Sigma_{(12)(12)\cdot 3\cdots p})$, respectively. Therefore, there is nothing new to be added to the results obtained for the simple correlation coefficient except for the replacement of n with $n-q$. In particular, the significance of $\rho_{12\cdot 3\cdots p}$ can be tested, based on

$$\sqrt{n-(p-2)-1}\frac{\bar{R}_{12\cdot 3\cdots p}}{\sqrt{1-\bar{R}^2_{12\cdot 3\cdots p}}} \sim \mathrm{t}(n-(p-2)-1)$$

under $\rho_{12\cdot 3\cdots p} = 0$.

4.3.3 Covariance Selection Model

For the relationship among p variables $X_1, \ldots, X_p$, it is interesting to examine whether two variables are independent, or conditionally independent, given the remainder variables. We denote the independence of X_i and X_j by $X_i \perp X_j$, and the conditional independence of X_i and X_j given $\boldsymbol{X}_r$ by $X_i \perp X_j | \boldsymbol{X}_r$, where $\boldsymbol{X}_r$ is a subvector of $\boldsymbol{X}$ obtained by deleting X_i and X_j.

In this section we assume that $\boldsymbol{X} = (X_1, \ldots, X_p)'$ follow $\mathrm{N}_p(\boldsymbol{\mu}, \Sigma)$. Then we know that:

(a) $X_1 \perp X_2 \Leftrightarrow \sigma_{12} = 0 \quad (\rho_{12} = 0)$.

(b) $X_1 \perp X_2 \,|\, \boldsymbol{X}_3 \Leftrightarrow \sigma_{12\cdot 3\cdots p} = 0 \quad (\rho_{12\cdot 3\cdots p} = 0)$.

More generally:

(1) $X_i \perp X_j \Leftrightarrow \sigma_{ij} = 0 \quad (\rho_{ij} = 0)$.

(2) $X_i \perp X_j \,|\, \boldsymbol{X}_r \Leftrightarrow \sigma_{ij\cdot \mathrm{rest}} = 0 \quad (\rho_{12\cdot \mathrm{rest}} = 0)$.

Here $\sigma_{ij\cdot \mathrm{rest}}(\rho_{12\cdot \mathrm{rest}})$ is the residual covariance (correlation) between X_i and X_j after removing the effects due to the remaining $(p-2)$ variables, or the conditional covariance (correlation) given the remaining p variables.

It is shown (see the following theorem) that statement (2) is equivalent to $\sigma^{ij} = 0$ $(\rho^{ij} = 0)$, where $\Sigma^{-1} = (\sigma^{ij})$, and $\mathrm{R}^{-1} = (\rho^{ij})$. A covariance matrix with $\sigma^{ij} = 0$ for a set of (i, j) called a *covariance selection model* was formulated by Dempster (1972).

Theorem 4.3.1 *Let* $\mathrm{R} = (\rho_{ij})$ *be the correlation matrix of* $\boldsymbol{X} = (X_1, X_2, \boldsymbol{X}_3')'$.

Then the partial correlation between X_1 *and* X_2 *given* $\boldsymbol{X}_3$ *is expressed as*

$$\rho_{12\cdot 3\cdots p} = -\frac{\rho^{12}}{\sqrt{\rho^{11}\rho^{22}}},$$

where $\mathrm{R}^{-1} = (\rho^{ij})$.

Proof. Let Q be the 2×2 matrix constructed by the first two rows and columns of R^{-1}. Then, from [A.1.2] we have

$$\begin{aligned} \mathrm{Q} &= \mathrm{R}^{-1}_{(12)(12)\cdot 3\cdots p} \\ &= \begin{pmatrix} \rho_{11\cdot 3\cdots p} & \rho_{12\cdot 3\cdots p} \\ \rho_{21\cdot 3\cdots p} & \rho_{22\cdot 3\cdots p} \end{pmatrix}^{-1}, \end{aligned}$$

and hence

$$\rho^{11} = \frac{\rho_{22\cdot 3\cdots p}}{|\mathrm{R}_{(12)(12)\cdot 3\cdots p}|}, \quad \rho^{22} = \frac{\rho_{11\cdot 3\cdots p}}{|\mathrm{R}_{(12)(12)\cdot 3\cdots p}|}, \quad \rho^{12} = \rho^{21} = \frac{-\rho_{12\cdot 3\cdots p}}{|\mathrm{R}_{(12)(12)\cdot 3\cdots p}|}.$$

These complete the proof. □

From Theorem 4.3.1 we have

$$\rho_{12\cdot 3\cdots p} = 0 \iff \rho^{12} = 0. \tag{4.3.6}$$

The result (4.3.6) is extended for the partial correlation between X_i and X_j given the remainder variables; that is,

$$\rho_{ij\cdot \mathrm{rest}} = 0 \iff \rho^{ij} = 0. \tag{4.3.7}$$

As a covariance selection model, we consider a model M such that all the partial correlations between every component of $\boldsymbol{X}_2$ and of $\boldsymbol{X}_3$ are zero, where $\boldsymbol{X} = (\boldsymbol{X}_1', \boldsymbol{X}_2', \boldsymbol{X}_3')'$ and $\boldsymbol{X}_i$: $p_i \times 1$. In accordance with the partition of $\boldsymbol{X}$, the covariance matrix Σ and its inverse $\Lambda = \Sigma^{-1}$ can be put in the form

$$\Sigma = \begin{pmatrix} \Sigma_{11} & \Sigma_{12} & \Sigma_{13} \\ \Sigma_{21} & \Sigma_{22} & \Sigma_{23} \\ \Sigma_{31} & \Sigma_{32} & \Sigma_{33} \end{pmatrix} \quad \text{and} \quad \Lambda = \begin{pmatrix} \Lambda_{11} & \Lambda_{12} & \Lambda_{13} \\ \Lambda_{21} & \Lambda_{22} & \Lambda_{23} \\ \Lambda_{31} & \Lambda_{32} & \Lambda_{33} \end{pmatrix}, \tag{4.3.8}$$

respectively, where Σ_{ij}: $p_i \times p_j$ and Λ_{ij}: $p_i \times p_j$. Our model is M: $\Lambda_{23} = \mathrm{O}$, which is equivalent to

$$M\colon \Sigma_{23\cdot 1} \equiv \Sigma_{23} - \Sigma_{21}\Sigma_{11}^{-1}\Sigma_{13} = \mathrm{O}. \tag{4.3.9}$$

This equivalence follows from

$$\Lambda_{(23)(23)} \equiv \begin{pmatrix} \Lambda_{22} & \Lambda_{23} \\ \Lambda_{32} & \Lambda_{33} \end{pmatrix} = \Sigma^{-1}_{(23)(23)\cdot 1} \equiv \begin{pmatrix} \Sigma_{22\cdot 1} & \Sigma_{23\cdot 1} \\ \Sigma_{32\cdot 1} & \Sigma_{33\cdot 1} \end{pmatrix}^{-1},$$

which is a consequence of [A.1.2] and Problem 4.7. Here $\Sigma_{22\cdot 1} = \Sigma_{22} - \Sigma_{21}\Sigma_{11}^{-1}\Sigma_{12}$, and so on. Similar notation will be used in the following whenever there is fear of confusion. The model is equivalent (see Chapter 11) to the redundancy condition of $\boldsymbol{X}_2$ in canonical correlation analysis between $(\boldsymbol{X}_1', \boldsymbol{X}_2')'$ and $\boldsymbol{X}_3$.

The likelihood ratio criterion is derived by using the fact that M is a model for the conditional independence between $\boldsymbol{X}_2$ and $\boldsymbol{X}_3$ given $\boldsymbol{X}_1$. It is a monotone function (see, Problem 4.8) of

$$L = \frac{|\mathbf{S}|}{|\mathbf{S}_{11}| \cdot |\mathbf{S}_{22\cdot 1}| \cdot |\mathbf{S}_{33\cdot 1}|} = \frac{|\mathbf{S}_{(23)(23)\cdot 1}|}{|\mathbf{S}_{22\cdot 1}| \cdot |\mathbf{S}_{33\cdot 1}|}. \tag{4.3.10}$$

The statistic can be written as

$$L = \frac{|\mathbf{W}_{(23)(23)}|}{|\mathbf{W}_{22}| \cdot |\mathbf{W}_{33}|} = \frac{|\mathbf{W}_{33\cdot 2}|}{|\mathbf{W}_{33}|}$$

in terms of

$$\mathbf{W} = \begin{pmatrix} \mathbf{W}_{22} & \mathbf{W}_{23} \\ \mathbf{W}_{32} & \mathbf{W}_{33} \end{pmatrix} = n \begin{pmatrix} \mathbf{S}_{22\cdot 1} & \mathbf{S}_{23\cdot 1} \\ \mathbf{S}_{32\cdot 1} & \mathbf{S}_{33\cdot 1} \end{pmatrix},$$

where $\mathbf{W}_{33\cdot 2} = \mathbf{W}_{33} - \mathbf{W}_{32}\mathbf{W}_{22}^{-1}\mathbf{W}_{23}$. Using Theorems 2.2.2 and 2.2.3, we know that $\mathbf{W} \sim \mathrm{W}_{p_2+p_3}(n - p_1, \Sigma_{(23)(23)\cdot 1})$. We also know that, under the model M, two variables $\mathbf{W}_{33\cdot 2}$ and $\mathbf{W}_{32}\mathbf{W}_{22}^{-1}\mathbf{W}_{23}$ are independently distributed as $\mathrm{W}_{p_3}(n - p_1 - p_3, \Sigma_{33\cdot 12})$ and $\mathrm{W}_{p_3}(p_2, \Sigma_{33\cdot 12})$, respectively. Here we use

$$\Sigma_{33\cdot 12} = \Sigma_{33\cdot 1} - \Sigma_{32\cdot 1}\Sigma_{22\cdot 1}^{-1}\Sigma_{23\cdot 1},$$

which is shown in Problem 4.1. Therefore, under M,

$$L \sim \Lambda_{p_3}(p_2, n - p_1 - p_2).$$

Let $\mathbf{R}$ be the sample correlation matrix, and partition it in the same way as those of $\mathbf{S}$. Then the statistic L is expressed as

$$L = \frac{|\mathbf{R}_{(23)(23)\cdot 1}|}{|\mathbf{R}_{22\cdot 1}| \cdot |\mathbf{R}_{33\cdot 1}|}$$

in terms of the correlation matrix $\mathbf{R}$, since L is invariant under the transformation from $\mathbf{S}$ to $\mathbf{R} = \mathbf{DSD}$, where $\mathbf{D} = \mathrm{diag}\,(1/\sqrt{S_{11}}, \ldots, 1/\sqrt{S_{pp}})$ and $\mathbf{S} = (S_{ij})$.

The above mentioned result is summarized as follows.

Theorem 4.3.2 *Suppose that the covariance matrix Σ of a random vector $\boldsymbol{X} = (\boldsymbol{X}_1', \boldsymbol{X}_2', \boldsymbol{X}_3')'$, $\boldsymbol{X}_i : p_i \times 1$ and its inverse Λ are partitioned as in* (4.3.8). *Let $\mathbf{S}$ be the sample covariance matrix based on a sample of size $N = n + 1$ from $\mathrm{N}_p(\boldsymbol{\mu}, \Sigma)$. Then, the likelihood statistic for $M : \Lambda_{23} = \mathrm{O}$ is given by L in* (4.3.10) *and its null distribution is $\Lambda_{p_3}(p_2, n - p_1 - p_2)$.*

Example 4.3.1 We examine the relationship between the five mathematics subjects considered in Example 4.2.1. The partial correlation coefficients between x_i and x_j given the remaining three variables are given in Table 4.3.1. From the table we may suggest a covariance selection model:

$$M: \rho^{14} = \rho^{24} = \rho^{15} = \rho^{25} = 0,$$

which is equivalent to

$$X_1 \perp X_4 | (X_2, X_3, X_5), \qquad X_2 \perp X_4 | (X_1, X_3, X_5),$$

$$X_1 \perp X_5 | (X_2, X_3, X_4), \qquad X_2 \perp X_5 | (X_1, X_3, X_4).$$

Table 4.3.1 Partial correlation matrix of five subjects

X_1	–				
X_2	0.33	–			
X_3	0.23	0.28	–		
X_4	0.00	0.08	0.43	–	
X_5	0.02	0.02	0.36	0.25	–

The covariance selection model means that all the partial correlations between each of $\{X_1, X_2\}$ and of $\{X_4, X_5\}$ given the reminder variables are zero. The likelihood ratio test is

$$L = \frac{|\mathbf{R}_{(1245)(1245)\cdot 3}|}{|\mathbf{R}_{(12)(12)\cdot 3}| \cdot |\mathbf{R}_{(45)(45)\cdot 3}|},$$

whose null distribution is $\Lambda_2(2, 87-1-2) = \Lambda_2(2, 84)$. Therefore, writing $L = \Lambda_{2;2,84}$ to make its distributional property clear, that is $\Lambda_{2;2,84} \sim \Lambda_2(2, 84)$, we obtain from Theorem 3.3.4,

$$F = \frac{1 - \sqrt{\Lambda_{2;2,84}}}{\sqrt{\Lambda_{2;2,84}}} \cdot \frac{2(87-1)}{2 \cdot 2} \sim F(4, 172).$$

The matrices in L are computed as

$$\mathrm{R}_{(1245)(1245)\cdot 3} = \begin{pmatrix} 1.000 & 0.533 & 0.409 & 0.389 \\ 0.533 & 1.000 & 0.485 & 0.437 \\ 0.409 & 0.485 & 1.000 & 0.607 \\ 0.389 & 0.437 & 0.607 & 1.000 \end{pmatrix}$$

$$- \begin{pmatrix} 0.547 \\ 0.610 \\ 0.711 \\ 0.665 \end{pmatrix} (0.547 \quad 0.610 \quad 0.711 \quad 0.665)$$

$$= \begin{pmatrix} 0.701 & 0.220 & 0.021 & 0.026 \\ 0.220 & 0.628 & 0.052 & 0.031 \\ 0.021 & 0.052 & 0.495 & 0.135 \\ 0.026 & 0.031 & 0.135 & 0.558 \end{pmatrix}.$$

Further, $|\mathrm{R}_{(1245)(1245)\cdot 3}| = 0.100$, and

$$|\mathrm{R}_{(12)(12)\cdot 3}| = \begin{vmatrix} 0.701 & 0.220 \\ 0.220 & 0.628 \end{vmatrix} = 0.392,$$

$$|\mathrm{R}_{(45)(45)\cdot 3}| = \begin{vmatrix} 0.495 & 0.135 \\ 0.135 & 0.558 \end{vmatrix} = 0.258.$$

Therefore. $L = 0.990, F = 0.219$, and its p-value is 0.927. This shows that the covariance selection model M is not rejected.

Problems

4.1 Let S be a $p \times p$ positive definite matrix, and partition it as

$$\mathrm{S} = \begin{pmatrix} \mathrm{S}_{11} & \mathrm{S}_{12} & \mathrm{S}_{13} \\ \mathrm{S}_{21} & \mathrm{S}_{22} & \mathrm{S}_{23} \\ \mathrm{S}_{31} & \mathrm{S}_{32} & \mathrm{S}_{33} \end{pmatrix}, \quad \mathrm{S}_{ij}\colon p_i \times p_j.$$

Prove the following:

$$\mathrm{S}_{3\cdot 12} \equiv \mathrm{S}_{33} - \mathrm{S}_{3(12)}\mathrm{S}_{(12)(12)}^{-1}\mathrm{S}_{(12)3} = \mathrm{S}_{33\cdot 1} - \mathrm{S}_{31\cdot 1}\mathrm{S}_{22\cdot 1}^{-1}\mathrm{S}_{23\cdot 1},$$

where $\mathrm{S}_{3(12)} = (\mathrm{S}_{31}\ \mathrm{S}_{32})$, $\mathrm{S}_{33\cdot 1} = \mathrm{S}_{33} - \mathrm{S}_{31}\mathrm{S}_{11}^{-1}\mathrm{S}_{13}$, and so on.

Hint: Let $\boldsymbol{X} = (\boldsymbol{X}_1', \boldsymbol{X}_2', \boldsymbol{X}_3')'$ be a random vector having a normal distribution $\mathrm{N}_p(\mathbf{0}, \Sigma)$. Then the result is shown from the fact that the conditional covariance matrix of $\boldsymbol{X}_3$ given $\boldsymbol{X}_1$ and $\boldsymbol{X}_2$ is the same as the conditional covariance matrix computed from the conditional distribution of $(\boldsymbol{X}_2', \boldsymbol{X}_3')'$ given $\boldsymbol{X}_1$. This fact is obtained from the fact that the conditional density function of $\boldsymbol{X}_3$ given $\boldsymbol{X}_1$ and $\boldsymbol{X}_2$ is expressed as

$$\begin{aligned} f(\boldsymbol{x}_3|\boldsymbol{x}_1, \boldsymbol{x}_2) &\equiv f(\boldsymbol{x}_1, \boldsymbol{x}_2, \boldsymbol{x}_3)/f(\boldsymbol{x}_1, \boldsymbol{x}_2) = \frac{f(\boldsymbol{x}_1, \boldsymbol{x}_2, \boldsymbol{x}_3)/f(\boldsymbol{x}_1)}{f(\boldsymbol{x}_1, \boldsymbol{x}_2)/f(\boldsymbol{x}_1)} \\ &= f(\boldsymbol{x}_2, \boldsymbol{x}_3|\boldsymbol{x}_1)/f(\boldsymbol{x}_2|\boldsymbol{x}_1). \end{aligned}$$

4.2 Let $h(\mathbf{S})$ be the sample multiple correlation between X_1 and $\boldsymbol{X}_2 = (X_2, \ldots, X_p)$, based on a sample of size $N = n + 1$, where $\mathbf{S}$ is the sample covariance matrix. Then show that $h(\mathbf{S})$ is invariant under the transformation from $\mathbf{S}$ to $\tilde{\mathbf{S}} = \mathrm{DSD}$, where $\mathrm{D} = \mathrm{diag}\,(d_1, \ldots, d_p)$ with $d_i > 0$.

4.3 In a bivariate normal sample, the regression coefficient defined by $B = V_{12}/V_{11} = R(V_{22}/V_{11})^{1/2}$ is the MLE of the population regression coefficient $\beta = \rho\sigma_2/\sigma_1$, where

$$\begin{pmatrix} X \\ Y \end{pmatrix} \sim \mathrm{N}_2(\boldsymbol{\mu}, \Sigma), \quad \boldsymbol{\mu} = \begin{pmatrix} \mu_1 \\ \mu_2 \end{pmatrix}, \quad \Sigma = \begin{pmatrix} \sigma_1 & \rho\sigma_1\sigma_2 \\ \rho\sigma_1\sigma_2 & \sigma_2 \end{pmatrix},$$

and the joint distribution of V_{11}, V_{12}, and V_{22} is $\mathrm{W}_2(n, \Sigma)$.

(a) Show that the pdf of B is

$$f(b) = \frac{\Gamma((n+1)/2)}{\sqrt{\pi}\,\Gamma(n/2)} |\Sigma|^{1/2n} [\sigma_1^{n-1}(\sigma_2^2 - 2\rho\sigma_1\sigma_2 b + \sigma_1^2 b^2)]^{-1/2(n+1)}.$$

(b) Prove that $\mathrm{E}(B) = \beta$ and $\mathrm{Var}(B) = (\sigma_2^2/\sigma_1^2)(1-\rho^2)/(n-2)$.

4.4 (Continuation) Show that the random variable

$$T_1 = \{\sqrt{n}\,\sigma_1/\sigma_2(1-\rho^2)^{1/2}\}(B - \beta)$$

is distributed as Student's t-distribution with n degrees of freedom (d.f.).

4.5 (Continuation) Prove that the random variable T defined by

$$T = \sqrt{n-1}\{V_{11}/V_{22}(1-R^2)\}^{1/2}(B-\beta)$$

has Student's t-distribution with $n-1$ d.f. [*Note*: This is due to Bartlett (1933), and we can obtain a test and confidence interval for β by using this distribution.]

4.6 Let

$$\boldsymbol{X}_1 = \begin{pmatrix} X_{11} \\ \boldsymbol{X}_{12} \end{pmatrix}, \ldots, \boldsymbol{X}_N = \begin{pmatrix} X_{N1} \\ \boldsymbol{X}_{N2} \end{pmatrix}$$

be a random sample of size $N = n + 1$ drawn from a population. Let the sample mean vector and the sample covariance matrix be denoted by

$$\bar{\boldsymbol{X}} = \begin{pmatrix} \bar{X}_1 \\ \bar{\boldsymbol{X}}_2 \end{pmatrix}, \quad \mathbf{S} = \begin{pmatrix} S_{11} & \boldsymbol{S}_{21}' \\ \boldsymbol{S}_{21} & \mathbf{S}_{22} \end{pmatrix}.$$

Then, observe that

(1) $$r^2(X_1, \alpha + \boldsymbol{\beta}'\boldsymbol{X}_2) = \frac{(\boldsymbol{\beta}'\boldsymbol{S}_{21})^2}{S_{11}\boldsymbol{\beta}'\mathbf{S}_{22}\boldsymbol{\beta}}.$$

(2) $$\frac{1}{n}\sum_{i=1}^{N}(X_{i1} - \alpha - \boldsymbol{\beta}'\boldsymbol{X}_{i2})^2 = (\bar{X}_1 - \alpha - \boldsymbol{\beta}'\bar{\boldsymbol{X}}_2)^2 + S_{11\cdot 2} + (\boldsymbol{\beta} - \mathbf{S}_{22}^{-1}\boldsymbol{S}_{21})'\mathbf{S}_{22}(\boldsymbol{\beta} - \mathbf{S}_{22}^{-1}\boldsymbol{S}_{21}).$$

Using these results, prove that

(3) $\max_{\alpha,\boldsymbol{\beta}} r(X_1, \alpha + \boldsymbol{\beta}'\boldsymbol{X}_2) = R_{1(2\cdots p)}$.

(4) $\min_{\alpha,\boldsymbol{\beta}} \frac{1}{n}\sum_{i=1}^{N}(X_{i1} - \alpha - \boldsymbol{\beta}'\boldsymbol{X}_{i2})^2 = S_{11}(1 - R^2_{1(2\cdots p)})$.

4.7 For a symmetric and nonsigular matrix A, let

$$\mathrm{A}^{-1} \equiv \begin{pmatrix} \mathrm{A}_{11} & \mathrm{A}_{12} \\ \mathrm{A}_{21} & \mathrm{A}_{22} \end{pmatrix}^{-1} = \begin{pmatrix} \mathrm{B}_{11} & \mathrm{B}_{12} \\ \mathrm{B}_{21} & \mathrm{B}_{22} \end{pmatrix} \equiv \mathrm{B}.$$

Then, show that $\mathrm{B}_{12} = \mathrm{O} \Leftrightarrow \mathrm{A}_{12} = \mathrm{O}$.

4.8 Show that the statistic L given by (4.3.10) is an LR test statistic for testing $M : \Lambda_{23} = \mathrm{O} \Leftrightarrow \Sigma_{23\cdot 1} = \mathrm{O}$, along the following lines. Let f_1 and $f_{2|1}$ be the likelihoods of $\boldsymbol{X}_1$ and $\boldsymbol{X}_2$ given $\boldsymbol{X}_1$, respectively. Then, observe that

$$\begin{aligned} -2\log f_1 &= N\log|\Sigma| + Np_1\log 2\pi \\ &\quad + \mathrm{tr}\,\Sigma^{-1}\{n\mathbf{S}_{11} + N(\bar{\boldsymbol{x}}_1 - \boldsymbol{\mu}_1)(\bar{\boldsymbol{x}}_1 - \boldsymbol{\mu}_1)'\} \\ &\equiv g_1(\boldsymbol{\mu}_1, \Sigma_{11}), \\ -2\log f_{2|1} &= N\log|\Sigma_{22\cdot 1}| + N\log 2\pi \\ &\quad + \mathrm{tr}\,\Sigma_{22\cdot 1}^{-1}\{n\mathbf{S}_{22\cdot 1} + (\mathcal{B} - \Sigma_{21}\Sigma_{11}^{-1})\mathbf{S}_{11}(\mathcal{B} - \Sigma_{21}\Sigma_{11}^{-1})' \\ &\quad + N(\bar{\boldsymbol{x}}_2 - \mathcal{B}\bar{\boldsymbol{X}}_1 - \boldsymbol{\mu}_{2\cdot 1})(\bar{\boldsymbol{x}}_2 - \mathcal{B}\bar{\boldsymbol{X}}_1 - \boldsymbol{\mu}_{2\cdot 1})'\} \\ &= g_{2|1}(\boldsymbol{\mu}_{2\cdot 1}, \mathcal{B}, \Sigma_{22\cdot 1}), \end{aligned}$$

where $\mathcal{B} = \Sigma_{21}\Sigma_{11}^{-1}$ and $\boldsymbol{\mu}_{2\cdot 1} = \boldsymbol{\mu}_2 - \mathcal{B}\boldsymbol{\mu}_1$. Using these results, show that

$$\begin{aligned} \max\{-2\log f_{(12)}\} &= \min g_1(\boldsymbol{\mu}_1, \Sigma_{11}) + \min g_{2|1}(\boldsymbol{\mu}_{2\cdot 1}, \mathcal{B}, \Sigma_{22\cdot 1}) \\ &= N\log|(n/N)\mathbf{S}_{11}| + +Np_1\log 2\pi \\ &\quad + N\log|(n/N)\mathbf{S}_{22\cdot 1}| + Np_2\log 2\pi. \end{aligned}$$

Let $f_{(123)}$ be the likelihood of $\boldsymbol{X} = (\boldsymbol{X}_1', \boldsymbol{X}_2', \boldsymbol{X}_3')'$. Then, noting that under M,

$$f_{(123)} = f_1 \cdot f_{(23)|1} = f_1 \cdot f_{2|1} \cdot f_{3|1},$$

obtain the conclusion.

4.9 Show that

$$1 - \rho^2_{1(2,\ldots,p)} = (1 - \rho^2_{12})(1 - \rho^2_{12\cdot 3})\cdots(1 - \rho^2_{12\cdot 34\cdots p}).$$

Chapter 5

Asymptotic Expansions for Multivariate Basic Statistics

In statistical inference it is of fundamental importance to obtain the sampling distribution of statistics. However, we often encounter situations where the exact distribution cannot be obtained in closed form, or even if it is obtained, it might be of little use because of its complexity. One practical way of getting around the problem is to provide reasonable approximations of the distribution function, along with extra information on their possible errors. In this chapter, we summarize some methods and theories on asymptotic expansions of multivariate statistics. Both high-dimensional and large-sample approximations are discussed. These include topics on the Edgeworth, Cornish–Fisher, and bootstrap approximations, and their validity. Perturbation methods on the characteristic roots and vectors of some random matrices are also discussed.

5.1 Edgeworth Expansion and its Validity

Let $\{F_n(x)\}$ be a set of distribution functions indexed by a parameter n: typically, sample size. Suppose, for example, that $F_n(x)$ is approximated by the first k-terms of asymptotic expansion:

$$G_{k,n}(x) = G(x) + \sum_{j=1}^{k-1} n^{-j/2} p_j(x) g(x), \tag{5.1.1}$$

where $G(x)$ is the limiting distribution function of $F_n(x)$, $g(x)$ is the density function of $G(x)$, and the $p_j(x)$ are suitable polynomials. For a wide class of statistics it is known that the error $R_{k,n}(x) = F_n(x) - G_{k,n}(x)$ satisfies the order condition [see Bhattacharya and Ghosh, 1978]

$$R_{k,n}(x) = \mathrm{O}(n^{-k/2}) \quad \text{uniformly in } x. \tag{5.1.2}$$

In this chapter we are concerned with either the methods of finding the first few terms of the asymptotic expansion or the validity of the expansion in the sense that the order condition (5.1.2) holds true. In this section we focus on a class of statistics having valid asymptotic expansions. The problem of obtaining its error bound is discussed in Chapters 13 through 16.

Let $\{\boldsymbol{Z}_j\}$, $j = 1, 2, \ldots$ be a sequence of *i.i.d.* random m-vectors from a distribution with mean $\boldsymbol{\xi}$ and covariance matrix $\Psi > \mathrm{O}$. Consider the distribution of

$$\boldsymbol{V}_n = \sqrt{n}(\bar{\boldsymbol{Z}}_n - \boldsymbol{\xi}), \tag{5.1.3}$$

where $\bar{\boldsymbol{Z}}_n = (1/n)\sum_{i=1}^n \boldsymbol{Z}_i$. Then it follows that the limiting distribution of $\boldsymbol{V}_n$ is $\mathrm{N}_p(\mathbf{0}, \Psi)$. The result, called the *multivariate central limit theorem*, is shown as follows. Let the characteristic function of $\boldsymbol{Z}_1$ be denoted by

$$C_{\boldsymbol{Z}_1}(\boldsymbol{t}) = \mathrm{E}[\mathrm{e}^{i\boldsymbol{t}'\boldsymbol{Z}_1}],$$

where $\boldsymbol{t} = (t_1, \ldots, t_m)'$ and the t_j are real numbers. Then the characteristic function of $\boldsymbol{V}_n$ is written as

$$\begin{aligned} C_{\boldsymbol{V}_n}(\boldsymbol{t}) &= \mathrm{E}[\mathrm{e}^{i\boldsymbol{t}'\boldsymbol{V}_n}] \\ &= \exp(-\sqrt{n}\,i\boldsymbol{t}'\boldsymbol{\xi})\mathrm{E}[\exp\{i\sum_{j=1}^{n} \boldsymbol{t}'\boldsymbol{Z}_j/\sqrt{n}\}] \\ &= \exp(-\sqrt{n}\,i\boldsymbol{t}'\boldsymbol{\xi})[C_{\boldsymbol{Z}_1}((1/\sqrt{n})\boldsymbol{t})]^n, \end{aligned}$$

and hence

$$\log C_{\boldsymbol{V}_n}(\boldsymbol{t}) = -\sqrt{n}\,i\boldsymbol{t}'\boldsymbol{\xi} + n\log C_{\boldsymbol{Z}_1}((1/\sqrt{n})\boldsymbol{t}). \tag{5.1.4}$$

Since $\boldsymbol{Z}_1$ has the second moments, we have (see Theorem 1.4.1)

$$\log C_{\boldsymbol{Z}_1}(\boldsymbol{t}) = i\boldsymbol{t}'\boldsymbol{\xi} - \frac{1}{2}\boldsymbol{t}'\Psi\boldsymbol{t} + \mathrm{o}(\|\boldsymbol{t}\|^2), \tag{5.1.5}$$

where $\|\boldsymbol{t}\| = \sqrt{t_1^2 + \cdots + t_m^2}$. Using (5.1.4) and (5.1.5), it is seen that

$$\lim_{n\to\infty} C_{\boldsymbol{V}_n}(\boldsymbol{t}) = \exp\left(-\frac{1}{2}\boldsymbol{t}'\Psi\boldsymbol{t}\right),$$

and hence from Theorem 1.1.5 we get the multivariate central limit theorem.

One important approach is to use an asymptotic expansion for the distribution of $\boldsymbol{V}_n$ in powers of $n^{-1/2}$. It is suggested that such an expansion will be obtained by expanding $C_{\boldsymbol{V}_n}(\boldsymbol{t})$ in powers of $n^{-1/2}$ and inverting. It is known that if $\boldsymbol{Z}_1$ has to sth moments with $s \geq 3$, then from Theorem 1.4.1 we can expand

$$\log C_{\boldsymbol{Z}_1}(\boldsymbol{t}) = i\boldsymbol{t}'\boldsymbol{\xi} + \frac{1}{2}i^2\boldsymbol{t}'\Psi\boldsymbol{t} + \sum_{r=3}^{s}\frac{i^r}{r!}\kappa_{j_1\cdots j_r}t_{j_1}\cdots t_{j_r} + \mathrm{o}(\|\boldsymbol{t}\|^s), \tag{5.1.6}$$

where $\kappa_{j_1\cdots j_r}$ is the cumulant of $\boldsymbol{Z}_1$. Substituting the expansion in (5.1.4), we obtain

$$\log C_{\boldsymbol{V}_n}(\boldsymbol{t}) = -\frac{1}{2}\boldsymbol{t}'\Psi\boldsymbol{t} + \sum_{r=3}^{s} n^{-(r-2)/2}\frac{i^r}{r!}\kappa_{j_1\cdots j_r}t_{j_1}\cdots t_{j_r} + \mathrm{o}(n^{(s-2)/2}) \tag{5.1.7}$$

and hence

$$C_{\boldsymbol{V}_n}(\boldsymbol{t}) = \exp\left(-\frac{1}{2}i^2\boldsymbol{t}'\Psi\boldsymbol{t}\right)\left[1 + \sum_{r=3}^{s} n^{-(r-2)/2}\frac{i^r}{r!}\tilde{\kappa}_{j_1\cdots j_r}t_{j_1}\cdots t_{j_r}\right] + \mathrm{o}(n^{-(s-2)/2}), \tag{5.1.8}$$

where $\tilde{\kappa}_{j_1j_2j_3} = \kappa_{j_1j_2j_3}$ and the remainder coefficients $\tilde{\kappa}_{j_1\cdots j_r}$ can be determined in terms of $\kappa_{j_1\cdots j_q}$ with $q \leq r$. Inverting (5.1.8), it is expected that there exists an asymptotic expansion of the density function of $\boldsymbol{V}_n$ given by

$$\phi_{s,n}(\boldsymbol{u}) = \phi(\boldsymbol{u})\left[1 + \sum_{j=1}^{s-2} a_j(\boldsymbol{u})n^{-j/2}\right], \tag{5.1.9}$$

where $\phi(\boldsymbol{u})$ is the density function of $\mathrm{N}_m(\boldsymbol{0}, \Psi)$.

The expansion $\phi_{s,n}(\boldsymbol{u})$ is called the Edgeworth expansion (E-expansion) of order $\mathrm{o}(n^{-(s-2)/2})$ for the distribution of $\boldsymbol{V}_n$. In general, the term *Edgeworth expansion* is used for a sequence $\{\boldsymbol{V}_n\}, n = 1, 2, \ldots$ of random m-vectors that possesses an expansion as in (5.1.9), with a general density function, not necessarily $\phi(\boldsymbol{u})$.

The E-expansion for the distribution of $\boldsymbol{V}_n$ defined by (5.1.3) has a long history. The first development was done by Edgeworth (1905), who gave a formal asymptotic expansion in the one-dimensional case. Cramér (1928) proved the validity of this formal expansion under certain regularity conditions. The following theorem on the existence of a valid asymptotic expansion for the distribution of $\boldsymbol{V}_n$ is due to Bhattacharya (1971).

Theorem 5.1.1 *Let $\{\boldsymbol{Z}_j\}$, $j = 1, 2, \ldots$ be a sequence of i.i.d. random p-vectors from a distribution with mean vector $\boldsymbol{\xi}$ and covariance matrix $\Psi >$ O. Suppose further that for some integer $s \geq 3$, $\boldsymbol{Z}_1$ satisfies the following assumptions:*

A1. $\mathrm{E}[\|\boldsymbol{Z}_1\|^s] < \infty$, *where* $\| * \|$ *is the Euclidean norm.*

A2. $\boldsymbol{Z}_1$ *(or its distribution) obeys Cramér's condition:*

$$\limsup_{\|\boldsymbol{t}\|\to\infty} \left| \mathrm{E}[\exp(i\boldsymbol{t}'\boldsymbol{Z}_1)] \right| < 1.$$

Then for the density function of $\boldsymbol{V}_n$ *defined by* (5.1.3),

$$\sup_{B\in\mathcal{B}} \left| \Pr\{\boldsymbol{V}_n \in B\} - \int_B \psi_{s,n}(\boldsymbol{u})\,\mathrm{d}\boldsymbol{u} \right| = \mathrm{o}(n^{-1/2(s-2)}),$$

where the supremum is over all convex Borel sets in R^m, $\phi_{s,n}$ *is defined by* (5.1.9), *and* $a_j(\boldsymbol{u})$ *is a polynomial in* $\boldsymbol{u}$ *whose coefficients are continuous functions of moments of* $\boldsymbol{Z}_1$ *up to the order* $j+2$.

It is more useful to consider a real-valued function $h(\bar{\boldsymbol{Z}}_n)$ of the sample mean $\bar{\boldsymbol{Z}}_n = (1/n)\sum_{j=1}^n \boldsymbol{Z}_j$, where $\boldsymbol{Z}_1, \boldsymbol{Z}_2, \ldots$ are i.i.d. random vectors with mean $\boldsymbol{\xi}$ and covariance matrix $\Psi > \mathrm{O}$. The random vectors may be defined as

$$\begin{aligned} \boldsymbol{Z}_j &= (Z_{j1}, \ldots, Z_{jm})' \\ &= (f_1(\boldsymbol{X}_j), \ldots, f_m(\boldsymbol{X}_j))', \; j = 1, 2, \ldots, \end{aligned} \tag{5.1.10}$$

where $f_1, \ldots, f_m$ are real-valued functions on R^p. In fact, many test statistics and estimators can be so expressed, since all functions of sample moments are of the form $h(\bar{\boldsymbol{Z}}_n)$. For example, we shall see that a statistic $T = g(\mathbf{S})$ is of the form $h(\bar{\boldsymbol{Z}}_n)$, where $\mathbf{S} = (S_{ij})$ is a $p \times p$ sample covariance matrix defined by

$$\mathbf{S} = \frac{1}{n}\sum_{j=1}^n (\boldsymbol{X}_j - \bar{\boldsymbol{X}})(\boldsymbol{X}_j - \bar{\boldsymbol{X}})' = \frac{1}{n}\sum_{j=1}^n \boldsymbol{X}_j\boldsymbol{X}_j' - \bar{\boldsymbol{X}}\bar{\boldsymbol{X}}'.$$

For simplicity, assume that $p = 2$, and let

$$\boldsymbol{X}_j = (X_{j1}, X_{j2})', \qquad j = 1, 2, \ldots.$$

We define $\boldsymbol{Z}_j = (Z_{j1}, \ldots, Z_{j5})'$, $j = 1, 2, \ldots$ as

$$Z_{j1} = X_{j1}, \quad Z_{j2} = X_{j2}, \quad Z_{j3} = X_{j1}^2, \quad Z_{j4} = X_{j2}^2, \quad Z_{j5} = X_{j1}X_{j2}.$$

Then, putting $\bar{Z}_{n\ell} = (1/n)\sum_{j=1}^n Z_{j\ell}$, we have

$$S_{11} = \bar{Z}_{n3} - \bar{Z}_{n1}^2, \quad S_{22} = \bar{Z}_{n4} - \bar{Z}_{n2}^2, \quad S_{12} = \bar{Z}_{n5} - \bar{Z}_{n1}\bar{Z}_{n2}.$$

These imply that T is a function of $\bar{\boldsymbol{Z}}_n = (\bar{Z}_{n1}, \ldots, \bar{Z}_{n5})'$.

Now we consider the distribution of

$$W_n = \sqrt{n}\{h(\bar{\boldsymbol{Z}}_n) - h(\boldsymbol{\xi})\}. \tag{5.1.11}$$

Assume that $h(\boldsymbol{z})$ is continuously differentiable in a neighborhood of $\boldsymbol{z} = \boldsymbol{\xi}$ and

$$\sigma^2 = \boldsymbol{b}'\Psi\boldsymbol{b} > 0,$$

where

$$\begin{aligned}\boldsymbol{b} &= (b_1, \ldots, b_p)' \\ &= (\partial h(\boldsymbol{z}_n)/\partial z_1, \ldots, \partial h(\boldsymbol{z}_n)/\partial z_p)'\Big|_{\boldsymbol{z}=\boldsymbol{\xi}} \equiv \partial h(\boldsymbol{z})/\partial \boldsymbol{z}\Big|_{\boldsymbol{z}=\boldsymbol{\xi}}.\end{aligned}$$

Then, according to Theorem 2.5.2, the limiting distribution of W_n is normal with mean zero and variance σ^2.

Precisely, the procedure is described as follows. Suppose that we have the characteristic function of W_n in the expanded form

$$C_n(t) = \mathrm{E}\big[\exp(i\,t\,W_n)\big] = C_{s,n}(t) + \mathrm{o}(n^{-(s-2)/2}),$$

where

$$C_{s,n}(t) = \exp\Big(-\frac{1}{2}\sigma^2 t^2\Big)\Big[1 + \sum_{j=1}^{s-2} p_j(it)n^{-j/2}\Big]; \tag{5.1.12}$$

$p_j(t)$ is a polynomial in t whose coefficients are independent of n. Then the E-expansion $\phi_{s,n}(u)$ is determined by inverting $C_{s,n}(it)$, that is,

$$\phi_{s,n}(u) = \frac{1}{2\pi}\int_{-\infty}^{\infty} \exp(-i\,t\,u)C_{s,n}(t)dt.$$

The terms in the inversion are evaluated by using the formulas

$$\begin{aligned}&\frac{1}{2\pi}\int_{-\infty}^{\infty} \exp(-i\,t\,u)\exp(-\sigma^2t^2/2)\,\mathrm{d}t = \phi_{\sigma^2}(u),\\ &\frac{1}{2\pi}\int_{-\infty}^{\infty} (-i\,t)^j \exp(-i\,t\,u)\exp(-\sigma^2t^2/2)\,\mathrm{d}t = \phi_{\sigma^2}^{(j)}(u),\end{aligned}$$

where $\phi_{\sigma^2}(u)$ is the density function of $\mathrm{N}(0, \sigma^2)$ and $\phi_{\sigma^2}^{(j)}(u)$ is the jth derivative of $\phi_{\sigma^2}(u)$. The inversion of (5.1.12) can be written as

$$\begin{aligned}\phi_{s,n}(u) &= \Big[1 + \sum_{j=1}^{s-2} n^{-j/2}p_j\Big(-\frac{d}{du}\Big)\Big]\phi_{\sigma^2}(u)\\ &= \phi_{\sigma^2}(u)\Big[1 + \sum_{j=1}^{s-2} n^{-j/2}q_j(u)\Big],\end{aligned} \tag{5.1.13}$$

where $\phi_{\sigma^2}(u)$ is the density function of $\mathrm{N}(0, \sigma^2)$ and $q_j(u)$ is a polynomial in u whose coefficients do not depend on n and is determined by the relation

$$\phi_{\sigma^2}(u)q_j(u) = p_j\Big(\frac{-\mathrm{d}}{\mathrm{d}u}\Big)\phi_{\sigma^2}(u).$$

Now we consider a general and powerful method for obtaining an asymptotic expansion of $C(t)$. First we use the approximation of W_n obtained by the Taylor expansion of $h(\bar{\boldsymbol{Z}}_n) = h(\boldsymbol{\xi} + \boldsymbol{V}_n/\sqrt{n})$; that is,

$$\hat{W}_{s,n}(\boldsymbol{V}_n) = \sum_j b_j V_{jn} + \frac{1}{2\sqrt{n}} \sum_{j_1,j_2} b_{j_1 j_2} V_{j_1 n} V_{j_2 n} + \cdots + \frac{1}{(s-1)!} n^{-(s-2)/2} \sum_{j_1,\ldots,j_{s-1}} b_{j_1 \cdots j_{s-1}} V_{j_1 n} \cdots V_{j_{s-1} n},$$

where $\boldsymbol{V}_n = (V_{n1}, \ldots, V_{np})'$, and

$$b_{j_1 \cdots j_\nu} = \partial^\nu h(\boldsymbol{z})/(\partial z_{j_1} \cdots \partial z_{j_\nu})|_{\boldsymbol{z}=\boldsymbol{\xi}}, \quad 1 \le j_1, \ldots, j_\nu \le p.$$

Next, using this approximation, we obtain the jth cumulant of $\tilde{W}_n$, which is given by

$$\kappa_{j,n} = \hat{\kappa}_{j,n} + \mathrm{o}(n^{-(s-2)/2}).$$

Then the expression

$$\exp\left\{ it\tilde{\kappa}_{1,n} + \frac{1}{2}(it)^2(\tilde{\kappa}_{2,n} - \sigma^2) + \sum_{j=3}^{s} \frac{1}{j!}(it)^j \hat{\kappa}_{j,n} \right\} \exp\left(-\frac{1}{2}\sigma^2 t^2 \right) \quad (5.1.14)$$

provides an approximation to the characteristic function of $\tilde{W}_n$ (or W_n). Expanding the first exponent, one may reduce (5.1.14) to (5.1.12).

The problem of the existence of a valid E-expansion of W_n was conjectured by Wallace (1958) and treated with limited success by several authors; see, for example, Chambers (1967), Sargan (1976), Phillips (1977, 1977a), Chibisov (1972), Bhattacharya (1985) and Bai and Rao (1991). The following theorem was developed by Bhattacharya and Ghosh (1978).

Theorem 5.1.2 *Suppose that for some integer* $s \ge 3$, $\boldsymbol{Z}_1$ *satisfies* A1 (Theorem 5.1.1), *while* $h(\boldsymbol{z})$ *satisfies* A3 *and* A4 *below.*

A3. *All the derivatives of* $h(\boldsymbol{z})$ *of order* s *and less are continuous in a neighborhood of* $\boldsymbol{z} = \boldsymbol{\xi}$.

A4. $\boldsymbol{a} = \left.\frac{\partial h(\boldsymbol{z})}{\partial \boldsymbol{z}}\right|_{\boldsymbol{z}=\boldsymbol{\xi}} \neq \boldsymbol{0}.$

If, in addition, (1) *the distribution of* $\boldsymbol{X}_1$ *in* (5.1.10) *has a nonzero absolutely continuous component (with respect to Lebesgue measure on* R^p*) and* (2) *the density of this component is strictly positive on some nonempty open set* U *on which* $f_1, \ldots, f_m$ *are continuously differentiable and* $1, f_1, \ldots, f_m$ *are* *linearly independent (as elements of the vector space of a continuous function on U) then*

$$\sup_B \left| P(W_n \in B) - \int_B \phi_{s,n}(u)\,\mathrm{d}u \right| = \mathrm{o}(n^{-(s-2)/2}), \tag{5.1.15}$$

where the supremum is over all measurable convex sets B in R^1 and $\phi_{s,n}(u)$ is given by (5.1.13).

(b) *If, instead of* (a), $\boldsymbol{Z}_1$ *satisfies* A2, *then the relation* (5.1.15) *holds uniformly over every class $\mathcal{B}$ of Borel sets satisfying*

$$\sup_{B\in\mathcal{B}} \int_{(\partial B)^\varepsilon} \phi_{\sigma^2}(v)\,\mathrm{d}v = \mathrm{O}(\varepsilon), \qquad \varepsilon \downarrow 0.$$

Here ∂B is the boundary of B and $(\partial B)^\varepsilon$ is the ε-neighborhood of B.

Note: Theorem 5.1.2 holds true for a vector-valued function $\boldsymbol{h}(\boldsymbol{z}) = (h_1(\boldsymbol{z}), \ldots, h_k(\boldsymbol{z}))'$ provided that $\mathrm{M} = \mathrm{G}'\Psi\mathrm{G}$ is nonsingular, where

$$\mathrm{G}: p \times k = \left(\left.\frac{\partial h_1(\boldsymbol{z})}{\partial \boldsymbol{z}}\right|_{\boldsymbol{z}=\boldsymbol{\xi}}, \ \ldots, \ \left.\frac{\partial h_k(\boldsymbol{z})}{\partial \boldsymbol{z}}\right|_{\boldsymbol{z}=\boldsymbol{\xi}} \right).$$

Next we consider the distribution of W_n when $\boldsymbol{a} = \partial h(\boldsymbol{z})/\partial \boldsymbol{z}|_{\boldsymbol{z}=\boldsymbol{\xi}} = 0$. This occurs when we are dealing with the null distributions of certain test statistics. For this case, let

$$Y_n = 2n\{h(\bar{\boldsymbol{Z}}_n) - h(\boldsymbol{\xi})\}.$$

Then again using the Taylor series of $h(\bar{\boldsymbol{z}}_n)$ around the point $\bar{\boldsymbol{z}}_n = \boldsymbol{\xi}$ up to $(s-1)$th order, we obtain an approximation to y_n as follows:

$$\begin{aligned}\hat{Y}_{s-1,n}(\boldsymbol{V}_n) &= \sum_{j_1,j_2} b_{j_1j_2} V_{j_1n} V_{j_2n} \\ &+ \frac{2}{3!} n^{-1/2} \sum_{j_1,j_2,j_3} b_{j_1j_2j_3} V_{j_1n} V_{j_2n} V_{j_3n} + \cdots \\ &+ \frac{2}{(s-1)!} n^{-1/2(s-3)} \sum_{j_1,\ldots,j_{s-1}} b_{j_1\cdots j_{s-1}} V_{j_1n} \cdots V_{j_{s-1}n}.\end{aligned}$$

We observe that since the leading term of Y_n is quadratic in V_{jn}, the limiting distribution of Y_n is a central shi-square distribution with q degrees of freedom $(\chi^2(q))$, where q is the rank of $\mathrm{B} = (b_{ij})$ under the following assumption:

A5. $\boldsymbol{b} = \boldsymbol{0}$ and $\mathrm{B}\Psi\mathrm{B} = \mathrm{B}$.

Chandra and Ghosh (1979) have established the existence of a valid asymptotic expansion for the distribution of Y_n under appropriate conditions. Thus in this case, the asymptotic expansion for the distribution of Y_n is obtained by formally inverting the exact characteristic function $\mathrm{E}[\exp(itY_n)]$ or the approximate characteristic function; that is,

$$\mathrm{E}[\exp\{it\hat{Y}_{s-1,n}(\boldsymbol{V}_n)\}], \quad \int \exp\{it\hat{y}_{s-1,n}(\boldsymbol{u})\phi_{s-1,n}(\boldsymbol{u})\}\,\mathrm{d}\boldsymbol{u},$$

and so on.

5.2 Sample Mean Vector and Covariance Matrix

Many multivariate tests are based on the sample mean vector $\bar{\boldsymbol{X}}$ and the sample covariance matrix $\mathbf{S}$, based on a sample of size n drawn from a population with mean $\boldsymbol{\mu}$ and covariance matrix Σ. Asymptotic results on the distribution of $\bar{\boldsymbol{X}}$ and $\mathbf{S}$ are basic for asymptotic expansions of various statistics. Under the assumption that the samples have been drawn for a normal population,

$$\bar{\boldsymbol{X}} \sim \mathrm{N}_p(\boldsymbol{\mu}, \Sigma), \; (n-1)\mathbf{S} \sim \mathrm{W}_p(n-1, \Sigma),$$

and they are independent. In this section we consider the case when the population is not necessarily normal. We consider asymptotic expansions of the distributions of the standardized statistics

$$\boldsymbol{Z} = \sqrt{n}\,\Sigma^{-1/2}(\bar{\boldsymbol{X}} - \boldsymbol{\mu}), \quad \mathbf{V} = \sqrt{n}\,\Sigma^{-1/2}(\mathbf{S} - \mathrm{I}_p)\Sigma^{-1/2}.$$

Let

$$\boldsymbol{Y} = \Sigma^{-1/2}(\boldsymbol{X} - \boldsymbol{\mu}), \; \boldsymbol{Y}_i = \Sigma^{-1/2}(\boldsymbol{X}_i - \boldsymbol{\mu}), \qquad i = 1, \ldots, n. \tag{5.2.1}$$

Then the standardized statistics are expressed as

$$\boldsymbol{Z} = \sqrt{n}\,\bar{\boldsymbol{Y}}, \quad \mathbf{V} = \sqrt{n}(\mathbf{S} - \mathrm{I}_p), \tag{5.2.2}$$

where $\mathbf{S} = (1/(n-1))\sum_{i=1}^{n}(\boldsymbol{Y}_i - \bar{\boldsymbol{Y}})(\boldsymbol{Y}_i - \bar{\boldsymbol{Y}})'$. In this section we denote the moments of $\boldsymbol{Y}$ by

$$\mathrm{E}\big[Y_{i_1} \cdots Y_{i_k}\big] = \mu_{i_1 \cdots i_k},$$

and the corresponding cumulants as $\kappa_{i_1 \cdots i_k}$. Then $\mathrm{E}(\boldsymbol{Y}) = \mathbf{0}, \mathrm{Var}(\boldsymbol{Y}) = \mathrm{I}_p$, and

$$\kappa_a = 0, \; \kappa_{ab} = \delta_{ab}, \; \kappa_{abc} = \mu_{abc},$$

$$\kappa_{abcd} = \mu_{abcd} - (\delta_{ab}\delta_{cd} + \delta_{ac}\delta_{bd} + \delta_{ad}\delta_{bc}).$$

We use the following moment notation:

$$\begin{aligned}
m_{j;k;\ell} &= m_{j;k\ell} = \kappa_{jk\ell}, \\
m_{ab;cd} &= \kappa_{abcd} + \sigma_{ac}\sigma_{bd} + \sigma_{ad}\sigma_{bc}, \\
m_{j;ab;cd} &= \kappa_{jabcd} + \sum_{[4]} \kappa_{jac}\sigma_{bd}, \\
m_{ab;cd;ef} &= \kappa_{abcdef} + \sum_{[12]} \kappa_{abce}\sigma_{bd} + \sum_{[4]} \kappa_{ace}\kappa_{bdf} + \sum_{[8]} \sigma_{ac}\sigma_{be}\sigma_{df}.
\end{aligned} \tag{5.2.3}$$

The multivariate skewness and kurtosis are expressed as

$$\begin{aligned}
\kappa_{[3]}^{(1)} &= \sum_{a,b,c} (\kappa_{abc})^2, \\
\kappa_{[3]}^{(2)} &= \sum_{a} (\sum_{b} \kappa_{abb})^2 = \sum_{a,b,c} \kappa_{aab}\kappa_{bcc}, \\
\kappa_{[4]}^{(1)} &= \sum_{a,b} \kappa_{aabb}.
\end{aligned}$$

Asymptotic expansion for the distribution of $\boldsymbol{Z}$ has been studied extensively. The formula is obtained from the following lemma.

Lemma 5.2.1 *Suppose that $\boldsymbol{Y}$ has a fourth moment. Then the characteristic function of $\boldsymbol{Z}$ can be expanded as*

$$\begin{aligned}
C_{\boldsymbol{Z}}(\boldsymbol{t}) = \exp\left(\frac{i^2}{2}\boldsymbol{t}'\boldsymbol{t}\right)\Bigg[1 &+ \frac{i^3}{6\sqrt{n}} \sum_{a,b,c} \kappa_{abc}t_a t_b t_c \\
&+ \frac{i^4}{24n} \sum_{a,b,c,d} \kappa_{abcd}t_a t_b t_c t_d \\
&+ \frac{i^6}{72n} \sum_{a,b,c,d,e,f} \kappa_{abc}\kappa_{def}t_a t_b t_c t_d t_e t_f \Bigg] + \mathrm{o}(n^{-1}),
\end{aligned} \tag{5.2.4}$$

where $\boldsymbol{t} = (t_1, \ldots, t_p)'$ and the t_j are real numbers.

Proof. The result is obtained from (5.1.6) and (5.1.7) with $\Psi = \mathrm{I}_p$. □

Inverting the characteristic function above and using Theorem 5.1.2 we have the following asymptotic expansion.

Theorem 5.2.1 *Suppose that the distribution of $\boldsymbol{Y}$ satisfies* Cramér*'s condition with a fourth order moment. Let $\boldsymbol{Z} = \sqrt{n}\,\bar{\boldsymbol{Y}}$ be defined as in* (5.2.2). *Then we have*

$$\sup_B \left| \Pr\{\boldsymbol{Z} \in B\} - \int_B \phi_{4,n}(\boldsymbol{x})\,\mathrm{d}\boldsymbol{x} \right| = \mathrm{o}(n^{-1}),$$

where the supremum is over all Borel sets B, and

$$\phi_{4,n}(\boldsymbol{z}) = \phi_p(\boldsymbol{z})\left[1 + \frac{1}{6\sqrt{n}}\sum_{a,b,c}\kappa_{abc}H_{abc}(\boldsymbol{z}) + \frac{1}{24n}\sum_{a,b,c,d}\kappa_{abcd}H_{abcd}(\boldsymbol{z}) + \frac{1}{72n}\sum_{a,b,c,d,e,f}\kappa_{abc}\kappa_{def}H_{abcdef}(\boldsymbol{z})\right]. \tag{5.2.5}$$

Here $\phi_p(\boldsymbol{z})$ *is the density function of* $\mathrm{N}_p(\mathbf{0}, \mathrm{I}_p)$*:*

$$\phi_p(\boldsymbol{z}) = \left(\frac{1}{\sqrt{2\pi}}\right)^p \exp\left(-\frac{1}{2}\boldsymbol{z}'\boldsymbol{z}\right),$$

and $H_a(\boldsymbol{z}), H_{ab}(\boldsymbol{z}), \ldots$ *are multivariate Hermite polynomials defined by*

$$H_{a_1\cdots a_j}(\boldsymbol{z})\phi_p(\boldsymbol{z}) = (-1)^{a_1+\cdots+a_j}\frac{\partial^j}{\partial z_{a_1}\cdots\partial z_{a_j}}\phi_p(\boldsymbol{z}). \tag{5.2.6}$$

Next we consider asymptotic expansions of the joint distribution as well as the joint characteristic function of $\boldsymbol{Z}$ and $\mathbf{V}$ up to the order $\mathrm{O}(n^{-1/2})$. Let

$$\tilde{\mathbf{S}} = n^{-1}\sum_{j=1}^{n}\boldsymbol{X}_j\boldsymbol{X}_j', \quad \tilde{\mathbf{V}} = \sqrt{n}(\tilde{\mathbf{S}} - \mathrm{I}_p). \tag{5.2.7}$$

Then it is easy to obtain an expansion formula of $C_{(\boldsymbol{Z},\tilde{\mathbf{V}})}(\boldsymbol{t}, \mathrm{T})$. On the other hand, note that

$$\mathbf{V} = \tilde{\mathbf{V}} + n^{-1/2}(\mathrm{I}_p - \boldsymbol{Z}\boldsymbol{Z}') + \mathrm{O}_p(n^{-1}).$$

This implies that

$$\begin{aligned} C_{(\boldsymbol{Z},\mathbf{V})}(\boldsymbol{t}, \mathrm{T}) &= \mathrm{E}\Bigg[\exp\Big(\sum_j it_jZ_j + \sum_{a\le b} it_{ab}\tilde{V}_{ab}\Big) \\ &\quad \times \Big\{1 + \frac{i}{\sqrt{n}}\Big(\sum_{a\le b} t_{ab}\delta_{ab} - \sum_{a\le b} t_{ab}Z_aZ_b\Big) + \mathrm{o}(n^{-1/2})\Big\}\Bigg] \\ &= \left[1 + \frac{i}{\sqrt{n}}\sum_{a\le b} t_{ab}\left(\delta_{ab} + \frac{\partial^2}{\partial t_a\partial t_b}\right) + \mathrm{o}(n^{-1/2})\right]C_{(\boldsymbol{Z},\tilde{\mathbf{V}})}(\boldsymbol{t}, \mathrm{T}). \end{aligned}$$

Using this relation we have the following expression for the joint characteristic function of $\boldsymbol{Z}$ and $\mathbf{V}$:

Lemma 5.2.2 *Suppose that $\boldsymbol{x}$ has a sixth moment. Then the characteristic function of $(\boldsymbol{Z}, \mathbf{V})$ can be expanded as*

$$C_{(\boldsymbol{Z},\mathbf{V})}(\boldsymbol{t}, \mathrm{T}) = \exp\left(\frac{i^2}{2}A_0\right)\left[1 + \frac{i^3}{6\sqrt{n}}A_1 + \mathrm{o}(n^{-1/2})\right], \tag{5.2.8}$$

where $\mathrm{T} = (\frac{1}{2}(1+\delta_{ab})t_{ab})$, $t_{ab} = t_{ba}$,

$$\begin{aligned} A_0 &= \sum_j t_j^2 + 2\sum_{j,a\le b} \kappa_{jab}t_j t_{ab} + \sum_{a\le b, c\le d} m_{ab;cd}t_{ab}t_{cd}, \\ A_1 &= \sum_{j,k,\ell} \kappa_{jk\ell}t_j t_k t_\ell + 3\sum_{j,k,a\le b} \kappa_{jkab}t_j t_k t_{ab} \\ &\quad + 3\sum_{j,a\le b,c\le d} m_{j;ab;cd}t_j t_{ab}t_{cd} + \sum_{a\le b,c\le d,e\le f} m_{ab;cd;ef}t_{ab}t_{cd}t_{ef}, \end{aligned}$$

and the coefficients $m_{j:k:\ell}, \ldots$ are as given in (5.2.3).

From Lemma 5.2.2 we have that $\boldsymbol{Z}$ and $\mathbf{V}$ are asymptotically independent if and only if $\kappa_{jab} = 0$ for all j and $a \le b$. The characteristic function of $\mathbf{V}$ is obtained from Lemma 5.2.2 with $\boldsymbol{t} = \mathbf{0}$ as

$$\begin{aligned} C_{(\boldsymbol{Z},\mathbf{V})}(\boldsymbol{t}, \mathrm{T}) &= \exp\left(\frac{i^2}{2}\sum_{a\le b,c\le d} m_{ab;cd}t_{ab}t_{cd}\right) \\ &\quad \times \left[1 + \frac{i^3}{6\sqrt{n}}\sum_{a\le b,c\le d,e\le f} m_{ab;cd;ef}t_{ab}t_{cd}t_{ef} + \mathrm{o}(n^{-1/2})\right]. \end{aligned} \tag{5.2.9}$$

The density function of the limiting distribution of $\mathbf{V}$ is given by

$$\begin{aligned} &\left(\frac{1}{2\pi}\right)^{p(p+1)/2}\int \exp\left(-i\sum_{a\le b} t_{ab}v_{ab} - \frac{1}{2}\sum_{a\le b,c\le d} m_{ab;cd}t_{ab}t_{cd}\right)\mathrm{dT} \\ &= \left(\frac{1}{2\pi}\right)^{p(p+1)/4}|\Lambda|^{1/2}\exp\left(-\frac{1}{2}\sum_{a\le b,c\le d}\lambda_{ab;cd}v_{ab}v_{cd}\right), \end{aligned} \tag{5.2.10}$$

where $\Lambda = (\lambda_{ab;cd})$ is the $\frac{1}{2}p(p+1) \times \frac{1}{2}p(p+1)$ matrix defined by

$$\Lambda^{-1} = (m_{ab;cd}), \qquad a \le b, c \le d. \tag{5.2.11}$$

We can obtain an asymptotic expansion of the density function of $\mathbf{V}$ by inverting the term of $\mathrm{O}(n^{-1/2})$. The inversion is computed by differentiating both sides of (5.2.10) with respect to v_{ab}.

Now, we consider the case when Σ is not I_p, assuming normality. Let $\mathbf{S}$ be the sample covariance matrix based on a sample of size $N = n + 1$ from $\mathrm{N}_p(\boldsymbol{\mu}, \Sigma)$. Then $n\mathbf{S} \sim \mathrm{W}_p(n, \Sigma)$, and let

$$\mathbf{U} = (U_{ij}) = \sqrt{n}(\mathbf{S} - \Sigma). \tag{5.2.12}$$

Theorem 5.2.2 *Let* $\mathbf{U}$ *be a random matrix defined by* (5.2.12). *Then the density function of* $\mathbf{U}$ *can be expanded as*

$$\begin{aligned} f(\mathrm{U}) &= [\pi^{p(p+1)/4} 2^{p(p+3)/4} |\Sigma|^{p(p+1)/2}]^{-1} \operatorname{etr}\left\{ -\frac{1}{4}(\Sigma^{-1}\mathrm{U})^2 \right\} \\ &\quad \times \left[1 + \frac{1}{\sqrt{n}} q_1(\mathrm{U}) + \frac{1}{n} q_2(\mathrm{U}) + \mathrm{o}(n^{-1}) \right], \end{aligned}$$

where

$$\begin{aligned} q_1(\mathrm{U}) &= -\frac{1}{2}(p+1)\operatorname{tr}(\Sigma^{-1}\mathrm{U}) + \frac{1}{6}\operatorname{tr}(\Sigma^{-1}\mathrm{U})^3, \\ q_2(\mathrm{U}) &= \frac{1}{2}\{q_1(\mathrm{U})\}^2 - \frac{1}{24}p(2p^2 + 3p - 1) \\ &\quad + \frac{1}{4}(p+1)\operatorname{tr}(\Sigma^{-1}\mathrm{U})^2 - \frac{1}{8}\operatorname{tr}(\Sigma^{-1}\mathrm{U})^4. \end{aligned}$$

Proof. Let $\mathbf{V} = (1/\sqrt{2})\Sigma^{-1/2}\mathbf{U}\Sigma^{-1/2}$, and $f(\mathrm{V})$ be the density function of $\mathbf{V}$. Since $J(\mathrm{V} \to \mathrm{U}) = 2^{-p(p+1)/2}|\Sigma|^{-(p+1)/2}$ (see [A.3.5]), the density function of $\mathbf{U}$ is

$$2^{-p(p+1)/2}|\Sigma|^{-(p+1)/2} g((1/\sqrt{2})\Sigma^{-1/2}\mathrm{U}\Sigma^{-1/2}),$$

so we obtain an asymptotic expansion of the density function of $\mathbf{V}$. Let $\mathrm{T} = ((1 + \delta_{ij})t_{ij}/2)$, where $t_{ij} = t_{ji}$ and δ_{ij} is the Kronecker delta. Then the characteristic function of $\mathbf{V}$ is

$$\begin{aligned} C_n(\mathrm{T}) &= \mathrm{E}[\operatorname{etr}(i\mathrm{T}\mathbf{V})] \\ &= \operatorname{etr}(-i\sqrt{n/2}\mathrm{T})\mathrm{E}[\operatorname{etr}(i\sqrt{n/2}\Sigma^{-1/2}\mathrm{T}\Sigma^{-1/2}\mathbf{S})] \\ &= \left|\mathrm{I}_p - \sqrt{n/2}\, i\mathrm{T}\right|^{-n/2} \operatorname{etr}(-i\sqrt{n/2}\mathrm{T}). \end{aligned}$$

Therefore, for any fixed T, we can expand in the form

$$C_n(\mathrm{T}) = \hat{C}_n(\mathrm{T}) + \mathrm{o}(n^{-1}),$$

where

$$\hat{C}_n(\mathrm{T}) = \operatorname{etr}\left(-\frac{1}{2}\mathrm{T}^2\right)\left[1 + \frac{1}{3}\sqrt{\frac{2}{n}}i^3 \operatorname{tr}\mathrm{T}^3 + \frac{1}{n}\left\{\frac{1}{2}i^4 \operatorname{tr}\mathrm{T}^4 + \frac{1}{9}i^6(\operatorname{tr}\mathrm{T}^3)^2\right\}\right].$$

Inverting $\hat{C}_n(\mathrm{T})$, we obtain an asymptotic expansion of the density function $g(\mathrm{V})$ of $\mathbf{V}$, which is given by

$$g(\mathrm{V}) = (2\pi)^{-p(p+1)/2} \int \mathrm{etr}(i\mathrm{TV})\hat{C}_n(\mathrm{T})\,d\mathrm{T}.$$

The inversion can be calculated by the formula in Problem 5.9.

Next, we consider the distribution of

$$W_n = \sqrt{n}\{h(\mathbf{S}) - h(\Sigma)\}, \tag{5.2.13}$$

which possesses a valid asymptotic expansion under assumptions A3 and A4. Following the general theory discussed in Section 5.2, consider a stochastic approximation to W_n up to the order n^{-1}, which is obtained by a Taylor series for $f(\mathbf{S}) = h(\Sigma + (1/\sqrt{n})\mathbf{U})$ about $\mathrm{S} = \Sigma$,

$$\hat{W}_n = \mathrm{tr}\,(\mathrm{A}\mathbf{U}) + n^{-1/2}q_1(\mathbf{U}) + n^{-1}q_2(\mathbf{U}),$$

where $\mathbf{U}$ is given by (5.2.12), and

$$\begin{aligned}
\mathrm{A}\colon p \times p &= \left(\frac{1}{2}(1+\delta_{ab})\partial h(\mathrm{S})/\partial s_{ab}\,\Big|_{\mathrm{S}=\Sigma}\right),\\
q_1(\mathrm{U}) &= \frac{1}{2}\sum_{a\geq b}\sum_{c\geq d} u_{ab}u_{cd}\frac{\partial^2 h(\mathrm{S})}{\partial s_{ab}\partial s_{cd}}\Big|_{\mathrm{S}=\Sigma},\\
q_2(\mathrm{U}) &= \frac{1}{6}\sum_{a\geq b}\sum_{c\geq d}\sum_{e\geq f} u_{ab}u_{cd}u_{ef}\frac{\partial^3 h(\mathrm{S})}{\partial s_{ab}\partial s_{cd}\partial s_{ef}}\Big|_{\mathrm{S}=\Sigma}.
\end{aligned}$$

To evaluate the characteristic function of W_n, we calculate the first three cumulants of $\hat{W}_n$ by using formulas on moments of $\mathbf{U}$ in Theorem 2.2.5 and 2.2.6; that is,

$$\begin{aligned}
\kappa_1(\hat{W}_n) &= \mathrm{E}[\hat{W}_n] = n^{-1/2}g_1 + \mathrm{O}(n^{-3/2}),\\
\kappa_2(\hat{W}_n) &= \mathrm{E}[\hat{W}_n^2] - \{\mathrm{E}[\hat{W}_n]\}^2 = \sigma^2 + \mathrm{O}(n^{-1}),\\
\kappa_3(\hat{W}_n) &= \mathrm{E}[\hat{W}_n^3] - 3\,\mathrm{E}[\hat{W}_n^2]\,\mathrm{E}[\hat{W}_n] + 2\{\mathrm{E}(\hat{W}_n)\}^3\\
&= 6n^{-1/2}g_3 + \mathrm{O}(n^{-3/2}),
\end{aligned}$$

where

$$\begin{aligned}
g_1 &= \frac{1}{2}\sum_{a\geq b}\sum_{c\geq d}(\sigma_{ac}\sigma_{bd} + \sigma_{ad}\sigma_{bc})\frac{\partial^2 h(\mathrm{S})}{\partial s_{ab}\partial s_{cd}}\Big|_{\mathrm{S}=\Sigma},\\
\sigma^2 &= 2\,\mathrm{tr}\,(\mathrm{A}\Sigma)^2,\\
g_3 &= \frac{4}{3}\mathrm{tr}\,(\mathrm{A}\Sigma)^3 + \sum_{a\geq b}\sum_{c\geq d}(\Sigma\mathrm{A}\Sigma)_{ab}(\Sigma\mathrm{A}\Sigma)_{cd}\frac{\partial^2 h(\mathrm{S})}{\partial s_{ab}\partial s_{cd}}\Big|_{\mathrm{S}=\Sigma},
\end{aligned}$$

and $(\mathrm{A})_{ab}$ denotes the (a, b) element of a matrix A. Hence the characteristic function of W_n is obtained in the form

$$\begin{aligned}
C_n(t) &= \mathrm{E}\big[\exp\,(itW_n)\big] \\
&= \exp\left\{it\,\kappa_1(W_n) + \frac{1}{2}(it)^2\kappa_2(W_n) + \frac{1}{6}(it)^3\kappa_3(W_n) + \cdots\right\} \\
&= \exp\left(-\frac{1}{2}\sigma^2 t^2\right)\left[1 + n^{-1/2}\{itg_1 + (it)^3 g_3\} + \mathrm{o}(n^{-1/2})\right],
\end{aligned}$$

and hence the characteristic function of W_n/σ is given by

$$\exp\left(-\frac{1}{2}t^2\right)\left[1 + n^{-1/2}\{it\,\sigma^{-1}g_1 + (it)^3\sigma^{-3}g_3\} + \mathrm{o}(n^{-1/2})\right]. \tag{5.2.14}$$

Inverting (5.2.14), the density function $f(x)$ and the distribution function $F(x)$ of W_n/σ are expanded as

$$f(x) = \phi(x)\left[1 + n^{-1/2}\{\sigma^{-1}g_1 H_1(x) + \sigma^{-3}g_3 H_3(x)\}\right] + \mathrm{o}(n^{-1/2}), \tag{5.2.15}$$

$$F(x) = \Phi(x) + n^{-1/2}\phi(x)\{\sigma^{-1}g_1 + \sigma^{-3}g_3 H_2(x)\} + \mathrm{o}(n^{-1/2}), \tag{5.2.16}$$

where H_j are the Hermite polynomials defined by

$$H_j(x)\phi(x) = (-1)^j (d/dx)^j \phi(x), \qquad j = 1, 2, \ldots. \tag{5.2.17}$$

5.3 T² Statistic

5.3.1 Outlines of Two Methods

Let $\boldsymbol{X}_1, \ldots, \boldsymbol{X}_n$ be a random sample from a p-dimensional population with mean vector $\boldsymbol{\mu}$ and covariance matrix Σ. Then we consider the distribution of T^2 statistic defined by

$$T^2 = n(\bar{\boldsymbol{X}} - \boldsymbol{\mu})'\mathrm{S}^{-1}(\bar{\boldsymbol{X}} - \boldsymbol{\mu}), \tag{5.3.1}$$

where

$$\bar{\boldsymbol{X}} = \frac{1}{n}\sum_{i=1}^{n}\boldsymbol{X}_j, \quad \mathrm{S} = \frac{1}{n-1}\sum_{j=1}^{n}(\boldsymbol{X}_j - \bar{\boldsymbol{X}})(\boldsymbol{X}_j - \bar{\boldsymbol{X}})'.$$

By considering the transformation $\boldsymbol{X}_j \to \Sigma^{-1/2}(\boldsymbol{X}_j - \boldsymbol{\mu})$, without loss of generality we may consider the distribution of

$$T^2 = n\bar{\boldsymbol{X}}'\mathrm{S}^{-1}\bar{\boldsymbol{X}}, \tag{5.3.2}$$

assuming that the moments and cumulants of $\boldsymbol{X}$ are those of $\Sigma^{-1/2}(\boldsymbol{X}-\boldsymbol{\mu})$. So $\mathrm{E}(\boldsymbol{X})=\mathbf{0}$, $\mathrm{Var}(\boldsymbol{X})=\mathrm{I}_p$, and the cumulants are as given in (5.2.1). Let

$$\boldsymbol{Z}=\sqrt{n}\,\bar{\boldsymbol{X}},\quad \mathbf{V}=\sqrt{n}(\mathbf{S}-\mathrm{I}_p).$$

In this section we give outlines of two methods for deriving an asymptotic expansion of T^2.

Method I. First we expand T^2 as

$$\begin{aligned}\mathrm{T}^2 &= \boldsymbol{Z}'(\mathrm{I}_p+n^{-1/2}\mathbf{V})^{-1}\boldsymbol{Z}\\ &= \boldsymbol{Z}\boldsymbol{Z}'-\frac{1}{\sqrt{n}}\boldsymbol{Z}'\mathbf{V}\boldsymbol{Z}+\frac{1}{n}\boldsymbol{Z}'\mathbf{V}^2\boldsymbol{Z}+\mathrm{o}(n^{-1}).\end{aligned}\tag{5.3.3}$$

Then the characteristic function of T^2 can be expanded as

$$C_{T^2}(t)=C_0(t)+\frac{1}{\sqrt{n}}C_1(t)+\frac{1}{n}C_2(t)+\mathrm{o}(n^{-1}),\tag{5.3.4}$$

where

$$\begin{aligned}C_0(t)&=\mathrm{E}\left[\exp(it\boldsymbol{Z}'\boldsymbol{Z})\right],\\ C_1(t)&=\mathrm{E}\left[-it\boldsymbol{Z}'\mathbf{V}\boldsymbol{Z}\exp(it\boldsymbol{Z}'\boldsymbol{Z})\right],\\ C_2(t)&=\mathrm{E}\left[\left\{it\boldsymbol{Z}'\mathbf{V}^2\boldsymbol{Z}+\frac{1}{2}(-it\boldsymbol{Z}'\mathbf{V}\boldsymbol{Z})^2\right\}\exp(it\boldsymbol{Z}'\boldsymbol{Z})\right].\end{aligned}\tag{5.3.5}$$

After much computation, the characteristic function of T^2 can be expressed as

$$C_{T^2}(t)=\varphi^{p/2}\left[1+\frac{1}{n}\sum_{j=0}^{3}b_j\varphi^j+\mathrm{o}(n^{-1})\right],\tag{5.3.6}$$

whose inversion gives a desired result, where $\varphi=(1-2it)^{-1}$. The computation of $C_0(t)$ is carried out by using an asymptotic expansion of the distribution of $\boldsymbol{Z}$ up to the order $\mathrm{O}(n^{-1})$. Similarly, $C_1(t)$ and $C_2(t)$ can be computed by using asymptotic expansions of the joint distribution of $\boldsymbol{Z}$ and $\mathbf{V}$ up to the orders $\mathrm{O}(n^{-1/2})$ and $\mathrm{O}(1)$, respectively.

The validity of the method can be checked from the results in Section 5.2. In fact, let

$$\begin{aligned}\boldsymbol{Y}&=(X_1,\cdots,X_p,X_1^2,X_1X_2,\cdots,X_p^2)'\\ &=(Y_1,\ldots,Y_m)',\quad m=\frac{1}{2}p(p+3),\end{aligned}$$

and $\boldsymbol{Y}_j, j=1,\ldots,n$, be the corresponding ones to $\boldsymbol{X}_j$. Then we can see that T^2 is of the form

$$T^2=n\{h(\bar{\boldsymbol{Y}})-h(\boldsymbol{\nu})\}(1-n^{-1})^{1/2},\tag{5.3.7}$$

where $\bar{\boldsymbol{Y}} = (1/n)\sum_{j=1}^{n} \boldsymbol{Y}_j, \boldsymbol{\nu} = \mathrm{E}(\boldsymbol{Y})$ and $h(\boldsymbol{y})$ is a smooth vector function in a neighborhood of $\boldsymbol{y} = \boldsymbol{\nu}$. In this case,

$$\mathrm{grad}(h(\boldsymbol{\nu})) \equiv ((\partial/\partial x_1)h(\boldsymbol{x})|_{\boldsymbol{x}=\boldsymbol{\nu}}, \ldots, (\partial/\partial x_m)h(\boldsymbol{x})|_{\boldsymbol{x}=\boldsymbol{\nu}})' = 0,$$

and we need to apply the validity condition as in Chandra and Ghosh (1979).

Method II. This method uses a decomposition of $T^2 = \boldsymbol{U}'\boldsymbol{U}$, where $\boldsymbol{U}$ is a multivariate t statistic defined by

$$\boldsymbol{U} = \sqrt{n}\,\mathbf{S}^{-1/2}\bar{\boldsymbol{X}}. \tag{5.3.8}$$

Similarly, it is seen that $\boldsymbol{U}$ is of the form

$$\boldsymbol{U} = \sqrt{n}\{\boldsymbol{h}(\bar{\boldsymbol{Y}}) - \boldsymbol{h}(\boldsymbol{\nu})\}(1 - n^{-1})^{1/2}. \tag{5.3.9}$$

In this case the gradient matrix $(\partial H_i/\partial y_j)$ at $\boldsymbol{y} = \boldsymbol{\nu}$ is nonsingular. Using a multivariate version of Theorem 5.1.2, we have a valid expansion for $\boldsymbol{U}$ under appropriate regularity conditions. To obtain an asymptotic expansion up to the order $\mathrm{O}(n^{-1})$, it is sufficient to satisfy:

C1. $\mathrm{E}(\| \boldsymbol{Y} \|^4) < \infty$, [i.e., $\mathrm{E}(\| \boldsymbol{X} \|^8) < \infty$].

C2-1. $\boldsymbol{X}$ has an absolutely continuous component.

The condition C2-1 may be changed to:

C2-2. $\boldsymbol{Y}$ satisfies Cramér's condition.

The distribution of $\boldsymbol{U}$ can be expanded as

$$\begin{aligned} g(\boldsymbol{u}) &= \phi_p(\boldsymbol{u})\left\{1 + \frac{1}{\sqrt{n}}p_1(\boldsymbol{u}) + \frac{1}{n}p_2(\boldsymbol{u})\right\} + \mathrm{o}(n^{-1}) \\ &= \phi_{4,n}(\boldsymbol{u}) + \mathrm{o}(n^{-1}), \end{aligned} \tag{5.3.10}$$

where $\phi_p(\boldsymbol{u})$ is the density of $\mathrm{N}_p(0, I_p)$. The derivation is based on a perturbation expansion of $\boldsymbol{U}$ given by

$$\begin{aligned} \boldsymbol{U} &= \left(\mathrm{I}_p + \frac{1}{\sqrt{n}}\mathbf{V}\right)^{-1/2}\boldsymbol{Z} \\ &= \left(\mathrm{I}_p - \frac{1}{2\sqrt{n}}\mathbf{V} + \frac{3}{8n}\mathbf{V}^2 + \mathrm{O}_p(n^{-3/2})\right)\boldsymbol{Z}. \end{aligned} \tag{5.3.11}$$

Next the characteristic function of T^2 can be computed as

$$\begin{aligned} C_{T^2}(t) &= \mathrm{E}\{\exp(itT^2)\} \\ &= \int \exp(it\boldsymbol{u}'\boldsymbol{u})\phi_{4,n}(\boldsymbol{u})d\boldsymbol{u} + \mathrm{o}(n^{-1}), \end{aligned} \tag{5.3.12}$$

which has an expression as in (5.3.4).

5.3.2 Multivariate t-Statistic

We consider an asymptotic expansion of the distribution function of a multivariate t-statistic $\boldsymbol{U}$ in (5.3.9) with a remainder term $\mathrm{o}(n^{-1})$. Let $\nu_{i_1\cdots i_k} = \mathrm{E}(U_{i_1}\cdots U_{i_k})$. Similarly, the corresponding cumulants of $\boldsymbol{U}$ are denoted by $\gamma_{i_1\cdots i_k}$. Then in Section 5.2 and Bhattacharya and Ghosh (1978) it is known that the cumulants of $\boldsymbol{U}$ take the following form:

$$\begin{aligned}
\gamma_i &= \frac{1}{\sqrt{n}}q_i + \mathrm{O}(n^{-3/2}),\\
\gamma_{ij} &= \delta_{ij} + \frac{2}{n}q_{ij} + \mathrm{O}(n^{-2}),\\
\gamma_{ijk} &= \frac{6}{\sqrt{n}}q_{ijk} + \mathrm{O}(n^{-3/2}),\\
\gamma_{ijk\ell} &= \frac{24}{n}q_{ijk\ell} + \mathrm{O}(n^{-2}),
\end{aligned}$$

and in general, for $j \geq 2$,

$$\gamma_{i_1\cdots i_j} = \mathrm{O}(n^{-(j-2)/2}).$$

The explicit expressions for q_i, q_{ij}, $\ldots$ can be computed from the definition by using the perturbation expansion of $\boldsymbol{U}$ in (5.3.11). They are expressed as

$$\begin{aligned}
q_i &= -\frac{1}{2}\sum_a \kappa_{iaa},\\
q_{ij} &= \frac{3}{8}\sum_{a,b}\kappa_{ija}\kappa_{abb} + \frac{1}{2}\sum_{a,b}\kappa_{iab}\kappa_{jab} + \frac{1}{2}(p+1)\delta_{ij},\\
q_{ijk} &= -\frac{1}{3}\kappa_{ijk},\\
q_{ijkl} &= -\frac{1}{12}\kappa_{ijkl} + \frac{1}{2}\sum_a \kappa_{aij}\kappa_{akl} + \frac{1}{4}\delta_{ij}\delta_{kj}.
\end{aligned} \tag{5.3.13}$$

Theorem 5.3.1 *Under assumptions C1 and C2-1 of Section 5.3.1 the distribution of a multivariate t-statistic* $\boldsymbol{U}$ *can be expanded as*

$$\sup_B \left| \Pr\{\boldsymbol{U} \in B\} - \int_B \phi_{4,n}(\boldsymbol{u})\,\mathrm{d}\boldsymbol{u} \right| = \mathrm{o}(n^{-1}), \tag{5.3.14}$$

the supremum is over all measurable convex sets B *in* R^m *for C2-1 and every class* $\mathcal{B}$ *of Borel sets satisfying a boundary condition (see Theorem 5.1.2).*

The density function $\phi_{4,n}(\boldsymbol{u})$ is given by

$$\begin{aligned}\phi_{2,n}(\boldsymbol{u}) = \phi_p(\boldsymbol{u}) \Bigg[1 &+ \frac{1}{\sqrt{n}} \left\{ \sum_i q_i H_i(\boldsymbol{x}) + \sum_{i,j,k} q_{ijk} H_{ijk}(\boldsymbol{x}) \right\} \\ &+ \frac{1}{n} \left\{ \sum_{i,j} (q_{ij} + \frac{1}{2} q_i q_j) H_{ij}(\boldsymbol{u}) + \sum_{i,j,k,\ell} (q_{ijk\ell} + q_i q_{jk\ell}) H_{ijk\ell}(\boldsymbol{u}) \right. \\ &+ \left. \frac{1}{2} \sum_{i,j,k,\ell,e,f} q_{ijk} q_{\ell e f} H_{ijk\ell e f}(\boldsymbol{u}) \right\} \Bigg]. \end{aligned} \tag{5.3.15}$$

Here the coefficients q_i, q_{ij}, q_{ijk}, and $q_{ijk\ell}$ are as given in (5.3.13).

In the special case $p = 1$, the density function (5.3.14) can be simplified as

$$\begin{aligned}\phi_{4,n}(u) = \phi_1(u) \Bigg[1 &- \frac{1}{6\sqrt{n}} \kappa_3 (2u^3 - 3u) \\ &+ \frac{1}{n} \left\{ \frac{1}{18} \kappa_3^2 (u^6 - 3u^4 - 9u^2 + 3) \right. \\ &- \left. \frac{1}{12} \kappa_4 (u^4 - 6u^2 + 3) + \frac{1}{4} (u^4 - 2u^2 - 1) \right\} \Bigg],\end{aligned}$$

where $\kappa_3 = \mathrm{E}(Y^3)$ and $\kappa_4 = \mathrm{E}(Y^4) - 3$. This result is essentially the same as a previous result [see, e.g., Hall, 1987] for the distribution function of U.

Note that the characteristic function of $\boldsymbol{U}$ can be expressed as

$$\begin{aligned}C_{\boldsymbol{U}}(\boldsymbol{t}) = \mathrm{E} \Bigg[\exp(i\boldsymbol{t}'\boldsymbol{Z}) \Bigg\{ 1 &+ \frac{-i}{2\sqrt{n}} \boldsymbol{t}'\mathbf{V}\boldsymbol{Z} \\ &+ \frac{3i}{8n} \boldsymbol{t}'\mathbf{V}^2\boldsymbol{Z} + \frac{1}{2} \left(\frac{-i}{2\sqrt{n}} \boldsymbol{t}'\mathbf{V}\boldsymbol{Z} \right)^2 \Bigg\} \Bigg] + \mathrm{o}(n^{-1}), \end{aligned} \tag{5.3.16}$$

which is evaluated based on an Edgeworth expansion of $(\boldsymbol{U}, V)$. By evaluating each term and inverting the resulting expansion, we can obtain Theorem 5.3.1.

It may be noted that the final results depend on fourth and less than the fourth moments of $\boldsymbol{X}$. So it is expected that condition C1 can be relaxed as

C1$'$. $\mathrm{E}(\|\boldsymbol{X}\|^2) < \infty$ [i.e., $\mathrm{E}(\|\boldsymbol{Y}\|^4) < \infty$].

For $p = 1$, the relaxation has been proved in Hall (1979). For related discussions, see Bhattacharya and Ghosh (1988) and Babu and Bai (1993).

5.3.3 Asymptotic Expansions

It is easily seen that T^2 has a valid expansion from Theorem 5.3.1 . In fact, taking $B = B_x = \{\boldsymbol{u}; \boldsymbol{u}'\boldsymbol{u} \le x\}$ in Theorem 5.3.1, the validity follows from

$$P(T^2 \le x) = \int_{B_x} \phi_{4,n}(\boldsymbol{u})\,\mathrm{d}\boldsymbol{u} + \mathrm{o}(n^{-1}), \tag{5.3.17}$$

uniformly for all positive real numbers x. A powerful method for getting a simplified asymptotic expansion is to expand the characteristic function in power series up to the order n^{-1} and then invert it term by term. Following method II, we can write

$$\begin{aligned} C_{T^2}(t) &= \mathrm{E}\{\exp(itT^2)\} \\ &= \int \exp(it\boldsymbol{u}'\boldsymbol{u})\phi_{4,n}(\boldsymbol{u})\,\mathrm{d}\boldsymbol{u} + \mathrm{o}(n^{-1}). \end{aligned}$$

Considering the transformation from $\boldsymbol{u}$ to $\boldsymbol{x} = \varphi^{-1}\boldsymbol{u}$, we have

$$\begin{aligned} C_{T^2}(t) = \varphi^p \mathrm{E}_{\boldsymbol{X}} &\left[1 + \frac{1}{\sqrt{n}} \left\{ \sum_i q_i H_i(\varphi \boldsymbol{X}) + \sum_{i,j,k} q_{ijk} H_{ijk}(\varphi \boldsymbol{X}) \right\} \right. \\ &+ \frac{1}{n} \left\{ \sum_{i,j} (q_{ij} + \frac{1}{2} q_i q_j H_{ij}(\varphi \boldsymbol{X}) \right. \\ &+ \sum_{i,j,k,\ell} (q_{ijk\ell} + q_i q_{jk\ell} H_{ijk\ell}(\varphi \boldsymbol{X}) \\ &+ \left. \left. \frac{1}{2} \sum_{i,j,k,l,e,f} q_{ijk} q_{\ell ef} H_{ijk\ell ef}(\varphi \boldsymbol{X}) \right\} \right] + \mathrm{o}(n^{-1}), \end{aligned} \tag{5.3.18}$$

where $\varphi = (1 - 2it)^{-1/2}$, and the expectation is taken under the normal random vector $\boldsymbol{X}$ whose elements are independently distributed as $\mathrm{N}(0, 1)$.

Lemma 5.3.1 *Suppose that* $\boldsymbol{X} = (X_1, \ldots, X_p)'$ *and the* X_i *are independently distributed as* $\mathrm{N}(0, 1)$. *Then*

$$\begin{aligned} \mathrm{E}[H_i(\varphi \boldsymbol{X})] &= 0, \\ \mathrm{E}[H_{ijk}(\varphi \boldsymbol{X})] &= 0, \\ \mathrm{E}[H_{ij}(\varphi \boldsymbol{X})] &= (\varphi^2 - 1)\delta_{ij}, \\ \mathrm{E}[H_{ijkl}(\varphi \boldsymbol{X})] &= (\varphi^2 - 1)^2 \sum_{[3]} \delta_{ij}\delta_{kl}, \\ \mathrm{E}[H_{ijklef}(\varphi \boldsymbol{X})] &= (\varphi^2 - 1)^3 \sum_{[15]} \delta_{ij}\delta_{kl}\delta_{ef}, \end{aligned}$$

where $\sum_{[m]}$ denotes a sum of m similar terms determined by suitable permutations of the indices.

Substituting the results in Lemma 5.3.1 in (5.3.18), we have

$$C_{T^2}(t) = \varphi^p \Big[1 + \frac{1}{n}\{a_1(\varphi^2 - 1) + a_2(\varphi^2 - 1)^2 + a_3(\varphi^2 - 1)^3\} \Big] + \mathrm{o}(n^{-1}), \tag{5.3.19}$$

where

$$\begin{aligned} a_1 &= \sum_i q_{ii} + \frac{1}{2}\sum_i q_i^2 \\ &= \frac{1}{2}p(p+1) + \frac{1}{2}(\kappa_3^{(1)})^2 + \frac{1}{2}(\kappa_3^{(2)})^2, \\ a_2 &= 3\sum_{i,j} q_{iijj} + 3\sum_{i,j} q_i q_{ijj} \\ &= \frac{1}{4}p(p+2) + (\kappa_3^{(1)})^2 + (\kappa_3^{(2)})^2 - \frac{1}{4}\kappa_4^{(1)}, \\ a_3 &= \frac{9}{2}\sum_{i,j,k} q_{iij}q_{jkk} + 3\sum_{i,j,k}(q_{ijk})^2 \\ &= \frac{1}{3}(\kappa_3^{(1)})^2 + \frac{1}{2}(\kappa_3^{(2)})^2. \end{aligned}$$

Inverting the characteristic function above formally, we obtain the following theorem.

Theorem 5.3.2 *Under the conditions C1 and C2-1 (or C2-2) the distribution function of T^2 can be expanded as*

$$\begin{aligned} \Pr\{T^2 \le x\} &= G_p(x) + \frac{1}{n}\sum_{j=0}^{3} b_j G_{p+2j}(x) + \mathrm{o}(n^{-1}) \\ &= G_p(x) - \frac{2x}{np}g_p(x)\Big\{ b_1 + b_2 + b_3 \\ &\quad + \frac{(b_2+b_3)x}{p+2} + \frac{b_3 x^2}{(p+2)(p+4)} \Big\} + \mathrm{o}(n^{-1}), \end{aligned} \tag{5.3.20}$$

uniformly for all positive real numbers x, where $G_p(x)$ and $g_p(x)$ are the distribution and density functions of a chi-square variate with p degrees of *freedom, respectively. The coefficients b_j are given by*

$$
\begin{aligned}
b_0 &= -\frac{1}{4}p^2 + \frac{1}{6}(\kappa_3^{(1)})^2 - \frac{1}{4}\kappa_4^{(1)}, \\
b_1 &= -\frac{1}{2}p - \frac{1}{2}(\kappa_3^{(1)})^2 + \frac{1}{2}\kappa_4^{(1)}, \\
b_2 &= \frac{1}{4}p(p+2) - \frac{1}{2}(\kappa_3^{(2)})^2 - \frac{1}{4}\kappa_4^{(1)}, \\
b_3 &= \frac{1}{3}(\kappa_3^{(1)})^2 + \frac{1}{2}(\kappa_3^{(2)})^2.
\end{aligned}
$$

The result was obtained independently by Kano (1995) and Fujikoshi (1997a) based on Methods I and II, respectively.

In the special case $p = 1$, the coefficients b_i are given by

$$
\begin{aligned}
b_0 &= \frac{1}{6}\kappa_3^2 - \frac{1}{4}\kappa_4 - \frac{1}{4}, \\
b_1 &= -\frac{1}{2}\kappa_3^2 + \frac{1}{2}\kappa_4, \\
b_2 &= -\frac{1}{2}\kappa_3^2 - \frac{1}{4}\kappa_4 + \frac{3}{4}, \\
b_3 &= \frac{5}{6}\kappa_3^2,
\end{aligned}
$$

where $\kappa_3 = \mathrm{E}(Y^3)$ and $\kappa_4 = \mathrm{E}(Y^4) - 3$.

If $\boldsymbol{Y}$ has an elliptical distribution $E_p(\mu, \Lambda)$ with the characteristic function $\exp(i\boldsymbol{t}'\boldsymbol{\mu})\psi(\boldsymbol{t}'\Lambda\boldsymbol{t})$, the coefficients are expressed as

$$
\begin{aligned}
b_0 &= -\frac{1}{4}p\{p + (p+2)\kappa\}, \\
b_1 &= -\frac{1}{2}p\{1 - (p+2)\kappa\}, \\
b_2 &= \frac{1}{4}p(p+2)(1-\kappa), \\
b_3 &= 0,
\end{aligned}
$$

where κ is the kurtosis parameter defined by $\kappa = \psi^{(2)}(0)/(\psi'(0))^2 - 1$. This result was obtained by Iwashita (1997).

5.4 Statistics with a Class of Moments

5.4.1 Large-Sample Expansions

Consider the Lambda distribution defined by

$$
\Lambda = \frac{|\mathbf{S}_e|}{|\mathbf{S}_e + \mathbf{S}_h|} \sim \Lambda_p(q, n), \tag{5.4.1}
$$

where $\mathbf{S}_h$ and $\mathbf{S}_e$ are independently distributed as $\mathrm{W}_p(q, \Sigma)$ and $\mathrm{W}_p(n, \Sigma)$, respectively. The Λ appears, for example, as a likelihood ratio test for testing the equality of mean vectors $\boldsymbol{\mu}_i$, $i = 1, \ldots, q+1$, based on an N_i sample from $\mathrm{N}_p(\boldsymbol{\mu}, \Sigma)$. In this case, $n = N - (q+1)$ and $N = N_1 + \cdots + N_{q+1}$. The likelihood ratio criterion is $\lambda = |\Lambda|^{N/2}$. From Theorem 3.3.3 the distribution of Λ is the same as

$$\Lambda = \beta_1 \cdots \beta_p, \tag{5.4.2}$$

where $\beta_1, \ldots, \beta_p$ are independent and $\beta_j \sim \mathrm{Be}(\frac{1}{2}(n-j+1), \frac{1}{2}q)$, and hence the hth moment of Λ is given by

$$\begin{aligned} \mathrm{E}[\Lambda^h] &= \prod_{j=1}^{p} \frac{\Gamma[\frac{1}{2}(n-j+1)+h]\Gamma[\frac{1}{2}(n+q-j+1)]}{\Gamma[\frac{1}{2}(n-j+1)]\Gamma[\frac{1}{2}(n+q-j+1)+h]} \\ &= K \prod_{j=1}^{p} \frac{\Gamma[\frac{1}{2}(n-j+1)+h]}{\Gamma[\frac{1}{2}(n+q-j+1)+h]}, \end{aligned} \tag{5.4.3}$$

where $K = \prod_{j=1}^{p} \Gamma[\frac{1}{2}(n+q-j+1)]/\Gamma[\frac{1}{2}(n-j+1)]$. Related to the distribution of $\lambda = \Lambda^{N/2}$, we consider a statistic W $(0 \le W \le 1)$ which has moments of the form

$$\mathrm{E}(W^h) = K \left(\frac{\prod_{k=1}^{b} y_k^{y_k}}{\prod_{j=1}^{a} x_j^{x_j}} \right)^h \frac{\prod_{j=1}^{a} \Gamma[x_j(1+h)+\xi_j]}{\prod_{k=1}^{b} \Gamma[y_k(1+h)+\eta_k]}, \quad h = 0, 1, \ldots, \tag{5.4.4}$$

where K is a constant such that $\mathrm{E}(W^0) = 1$ and

$$\sum_{j=1}^{a} x_j = \sum_{k=1}^{b} y_k. \tag{5.4.5}$$

The moments are expressed as

$$\mathrm{E}(W^h) = \left(\frac{\prod_{k=1}^{b} y_k^{y_k}}{\prod_{j=1}^{a} x_j^{x_j}} \right)^h \frac{\prod_{j=1}^{a} \Gamma[x_j(1+h)+\xi_j] \prod_{k=1}^{b} \Gamma[y_k+\eta_k]}{\prod_{j=1}^{a} \Gamma[x_j+\xi_j] \prod_{k=1}^{b} \Gamma[y_k(1+h)+\eta_k]}. \tag{5.4.6}$$

Let

$$V = -2\rho \log W, \tag{5.4.7}$$

where ρ is a constant such that $0 < \rho < 1$ and $\alpha_j = (1-\rho)x_j$, $\beta_k = (1-\rho)y_k$. Assume that:

C0: α_j and β_k have limits as x_j and y_k tend to infinity.

In most applications,

C1: $x_j = c_j N, \quad i = 1, \ldots, a; \quad y_k = d_k N, \quad k = 1, \ldots, b,$ (5.4.8) where c_j and d_k are nonzero constants such that $\sum_{j=1}^{a} c_j = \sum_{k=1}^{b} d_k$, and N is a sample size or a number growing with the sample size.

Let $C(t)$ be the characteristic function of V. Then the cumulant generating function $\log C(t)$ is

$$\log C(t) = g(t) - g(0), \tag{5.4.9}$$

where

$$\begin{aligned} g(t) = (2it) &\left(\sum_{j=1}^{a} x_j \log x_j - \sum_{k=1}^{b} y_k \log y_k \right) \\ &+ \sum_{j=1}^{a} \log \Gamma[\rho x_j(1-2it) + \alpha_j + \xi_j] \\ &- \sum_{k=1}^{b} \log \Gamma[\rho y_k(1-2it) + \beta_k + \eta_k]. \end{aligned}$$

To obtain an expansion for $\log C(t)$, we use the following generalized version of Stirling's formula for the gamma function, which is due to Barnes (1899):

$$\begin{aligned} \log \Gamma(z+h) = \log \sqrt{2\pi} + &\left(z + h - \frac{1}{2} \right) \log z - z \\ &- \sum_{r=1}^{m} (-1)^r \frac{B_{r+1}}{r(r+1)z^r} + R_{m+1}(z), \end{aligned} \tag{5.4.10}$$

where $R_{m+1}(z) = \mathrm{O}\left(z^{-(m+1)}\right)$ and $B_r(h)$ is the Bernoulli polynomial of degree r defined by

$$\frac{\tau e^{h\tau}}{e^{\tau}} = \sum_{r=0}^{\infty} \frac{\tau^r}{r!} B_r(h). \tag{5.4.11}$$

Some of them are

$$\begin{aligned} &B_0(h) = 1, \quad B_1(h) = h - \frac{1}{2}, \quad B_2(h) = h^2 - h + \frac{1}{6}, \\ &B_3(h) = h^3 - \frac{3}{2}h^2 + \frac{1}{2}h, \quad B_4(h) = h^4 - 2h^3 + h^2 - \frac{1}{30}, \\ &B_5(h) = h^5 - \frac{5}{2}h^4 + \frac{5}{3}h^3 - \frac{1}{6}h. \end{aligned} \tag{5.4.12}$$

Expanding each of the gamma functions in (5.4.10) by taking $z = \rho x_j(1 - 2it), \rho y_k(1 - 2it)$, and $h = \alpha_j + \xi_j, \beta_k + \eta_k$, we obtain

$$\log C(t) = -\frac{f}{2} \log(1-2it) + \sum_{\ell=1}^{r} \omega_r \left\{ (1-2it)^{-\ell} - 1 \right\} + R_{r+1}, \tag{5.4.13}$$

where

$$
\begin{aligned}
f &= -2\left\{\sum_{j=1}^{a}\xi_j - \sum_{k=1}^{b}\eta_k - \frac{1}{2}(a-b)\right\},\\
\omega_r &= \frac{(-1)^r}{r(r+1)}\left\{\sum_{j=1}^{a}\frac{B_{r+1}(\alpha_j+\xi_j)}{(\rho x_j)^r} - \sum_{k=1}^{b}\frac{B_{r+1}(\beta_k+\eta_k)}{(\rho y_k)^r}\right\},\\
R_{r+1} &= \sum_{j=1}^{a}\mathrm{O}\left(x_j^{-(r+1)}\right) + \sum_{k=1}^{b}\mathrm{O}\left(y_k^{-(r+1)}\right).
\end{aligned}
\tag{5.4.14}
$$

Using (5.4.13), we obtain the following result due to Box (1949).

Theorem 5.4.1 *Suppose that the moments of a random variable W $(0 < W \le 1)$ are given by (5.4.4) with $x_j = c_j N$ and $y_k = d_k N$ satisfying C1 in (5.4.8). Then the distribution function of the random variable $V = -2\rho\log W$ can be expanded as*

$$
\begin{aligned}
\Pr\{V \le x\} &= G_f(x) + \frac{\gamma_2}{m^2}\{G_{f+4}(x) - G_f(x)\}\\
&+ \frac{\gamma_3}{m^3}\{G_{f+6}(x) - G_f(x)\} + \frac{1}{m^4}\Big[\gamma_4\{G_{f+8}(x) - G_f(x)\}\\
&+ \frac{1}{2}\gamma_2^2\{G_{f+8}(x) - 2G_{f+4}(x) + G_f(x)\}\Big] + \mathrm{O}\left(m^{-5}\right),
\end{aligned}
\tag{5.4.15}
$$

where $G_f(x)$ is the distribution function of the central chi-square distribution with f degrees of freedom $(\chi^2(f))$, f is given by (5.4.12), and

$$
\rho = 1 - \gamma/N, \quad m = N\rho = N - \gamma. \tag{5.4.16}
$$

The coefficients γ and $\gamma_j (j = 2, 3, 4)$ are given by

$$
\begin{aligned}
\gamma &= \frac{1}{f}\left\{\sum_{j=1}^{a}c_j^{-1}B_2(\xi_j) - \sum_{k=1}^{b}d_k^{-1}B_2(\eta_k)\right\},\\
\gamma_2 &= -\frac{1}{6}\left\{\frac{3}{2}\gamma^2 f + \sum_{j=1}^{a}c_j^{-2}B_3(\xi_j) - \sum_{k=1}^{b}d_k^{-2}B_3(\eta_k)\right\},\\
\gamma_3 &= \frac{1}{12}\left\{-2\gamma^3 f - 24\gamma\gamma_2\sum_{j=1}^{a}c_j^{-3}B_4(\xi_j) - \sum_{k=1}^{b}d_k^{-3}B_4(\eta_k)\right\},\\
\gamma_4 &= -\frac{1}{20}\left\{\frac{5}{2}\gamma^4 f + 60\gamma^2\gamma_2 + 60\gamma\gamma_3\right.\\
&= -\frac{1}{20}\left\{+\sum_{j=1}^{a}c_j^{-4}B_5(\xi_j) - \sum_{k=1}^{b}d_k^{-4}B_5(\eta_k)\right\}.
\end{aligned}
\tag{5.4.17}
$$

Proof. First, we observe that the ρ given by (5.4.16) satisfies the condition C0 and that $\omega_j = \mathrm{O}(N^{-j})$. We also note that ρ is given so that $\omega_1 = 0$. This is from

$$\begin{aligned}
\omega_1 &= -\frac{1}{2}\left\{\sum_{j=1}^{a}(\rho x_j)^{-1}B_2(\alpha_j+\xi_j) - \sum_{k=1}^{b}(\rho y_k)^{-1}B_2(\beta_k+\eta_k)\right\} \\
&= -\frac{1}{2}\left[\sum_{j=1}^{a}(\rho x_j)^{-1}\{B_2(\xi_j)+\alpha_j(\alpha_j+2\xi_j-1)\}\right. \\
&\quad \left. - \sum_{k=1}^{b}(\rho y_k)^{-1}\{B_2(\eta_k)+\beta_k(\beta_k+2\eta_k-1)\}\right] \\
&= -\frac{1}{2\rho}\left[N^{-1}\left\{\sum_{j=1}^{a}c_j^{-1}B_2(\xi_j) - \sum_{k=1}^{b}d_k^{-1}B_2(\eta_k)\right\} - (1-\rho)f\right].
\end{aligned}$$

Then it follows from (5.4.11) that the characteristic function of $v = -2\rho\log w$ can be written as

$$\begin{aligned}
C(t) = (1-2it)^{-f/2}&\left[1+\sum_{r=2}^{4}\omega_r\{(1-2it)^{-j}-1\}\right. \\
&\left. +\frac{1}{2}\omega_2^2\{(1-2it)^{-j}-1\}^2\right] + \mathrm{O}\left(m^{-5}\right).
\end{aligned} \tag{5.4.19}$$

Hence the result (5.4.14) follows by inverting (5.4.19) and by noting that $\omega_j = m^{-j}\gamma_j$, $j = 2, 3, 4$.

Remark 5.4.1 Chandra and Ghosh (1979) proved the validity of the formal expansion (5.4.14) under suitable regularity conditions on the density function $f(\boldsymbol{x}:\boldsymbol{\theta})$ and the subset ω of the parameter space Ω.

Example 5.4.1 We consider the asymptotic approximations to Λ whose distribution occurs when n is sufficiently large. From (5.4.3), we observe that the hth moment of $W = \Lambda^{n/2}$ is

$$\mathrm{E}[W^h] = \mathrm{E}[\Lambda^{nh/2}] = K\frac{\prod_{j=1}^{p}\Gamma[\frac{1}{2}n(1+h)+\frac{1}{2}(1-j)]}{\prod_{j=1}^{p}\Gamma[\frac{1}{2}n(1+h)+\frac{1}{2}(q+1-j)]}, \tag{5.4.20}$$

which is of the form (5.4.3) in Theorem 5.4.1 with the correspondence

$$\begin{aligned}
a &= p, \quad x_q = \frac{1}{2}n, \quad \xi_q = \frac{1}{2}(1-q), \quad c_q = \frac{1}{2}, \\
b &= p, \quad y_k = \frac{1}{2}n, \quad \eta_k = \frac{1}{2}(q+1-k), \quad d_k = \frac{1}{2}.
\end{aligned}$$

The necessary parameters for the asymptotic expansion are calculated as follows:

$$
\begin{aligned}
f &= -2\left\{\sum_{j=1}^{p}(1-j)/2 - \sum_{k=1}^{p}(q+1-k)/2 - (p-p)/2\right\} = pq, \\
\gamma &= \frac{1}{pq}\left[\sum_{j=1}^{p} 2\{(1-j)^2/4 - (1-j)/2 + 1/6\}\right. \\
&\quad \left. - \sum_{k=1}^{p} 2\{(q+1-k)^2/4 - (q+1-k)/2 + 1/6\}\right] = (p+1-q)/2, \\
\rho &= 1 - \gamma/n, \quad m = n\rho = n - \gamma, \\
\gamma_2 &= pq(p^2+q^2-5)/48, \\
\gamma_3 &= pq\{3p^4 + 2q^4 + 10p^2q^2 - 50(p^2+q^2) + 159\}/1920,
\end{aligned}
$$

and $\gamma_3 = 0$. Thus we obtain, by (5.4.15), an asymptotic expansion for the distribution of

$$
V = -2\rho \log \Lambda^{n/2} = -\left\{n - \frac{1}{2}(p-q+1)\right\} \log \Lambda \tag{5.4.21}
$$

in the following form:

$$
\begin{aligned}
\Pr\{V \le x\} &= G_{pq}(x) + \frac{\gamma_2}{m^2}[G_{pq+4}(x) - G_{pq}(x)] \\
&\quad + \frac{1}{m^4}[\gamma_4\{G_{pq+8}(x) - G_{pq}(x)\} \\
&\qquad + \frac{1}{2}\gamma_2^2\{G_{pq+8}(x) - 2G_{pq+4}(x) + G_{pq}\}] + \mathrm{O}(m^{-5}).
\end{aligned}
$$

The Other Approximations. Rao (1951) proposed an F approximation of the distribution of another function of U, where U is distributed as $\Lambda_p(q,n)$. The approximation is to consider

$$
\frac{1 - U^{1/s}}{U^{1/s}} \cdot \frac{ms - r}{pq} \tag{5.4.22}
$$

as F approximation with pq and $ks - r$ degrees of freedom, where

$$
\begin{aligned}
m &= n - \frac{1}{2}(p-q-1), \quad r = \frac{1}{2}pq - 1, \\
s &= \sqrt{p^2q^2 - 4}/\sqrt{p^2+q^2-5}.
\end{aligned}
$$

For $p = 1$ or 2 (or $q = 1$ or 2) the F-distribution is exactly as given in Theorem 3.2.3 and 3.3.4. If $ms - r$ is not an integer, interpolation between two integer values can be used.

Much work has been done on approximation employing the saddle point method. Srivastava and Yau (1989) and Bulter and Huzurbazar (1992) proposed, among other things, saddle point approximations for the tail probability of $\Lambda_p(q,n)$.

5.4.2 High-Dimensional Expansions

In general, large-sample approximations as in Theorem 5.4.1 worsen when the dimensions are large. As an approach to covering the weak point, we consider high-dimensional approximations where the sample size and dimensions are both large.

Corresponding to a statistic $W^{2/N}$, we consider a statistic $\tilde{W}$ with hth moments

$$\mathrm{E}(\tilde{W}^h) = K\left(\frac{\prod_{k=1}^b y_k^{y_k}}{\prod_{j=1}^a x_j^{x_j}}\right)^h \frac{\prod_{j=1}^a \Gamma[x_j+\xi_j+h]}{\prod_{k=1}^b \Gamma[y_k+\eta_k+h]}, \quad h=0,1,\ldots, \tag{5.4.23}$$

where K is a constant such that $\mathrm{E}(\tilde{W}^0)=1$ and

$$\sum_{j=1}^a x_j = \sum_{k=1}^b y_k. \tag{5.4.24}$$

The moments are expressed as

$$\mathrm{E}(\tilde{W}^h) = \left(\frac{\prod_{k=1}^b y_k^{y_k}}{\prod_{j=1}^a x_j^{x_j}}\right)^h \frac{\prod_{j=1}^a \Gamma[x_j+\xi_j+h]\prod_{k=1}^b \Gamma[y_k+\eta_k]}{\prod_{j=1}^a \Gamma[x_j+\xi_j]\prod_{k=1}^b \Gamma[y_k+\eta_k+h]}. \tag{5.4.25}$$

Set

$$\tilde{V} = -\log \tilde{W} \tag{5.4.26}$$

and let $\tilde{C}(t)$ be the characteristic function of $\tilde{V}$. Then the cumulant generating function is

$$\begin{aligned}\log \tilde{C}(t) = {} & it\left(\sum_{j=1}^a x_j \log x_j - \sum_{k=1}^b y_k \log y_k\right) \\ & + \sum_{j=1}^a \{\log\Gamma[x_j+\xi_j] - \log\Gamma[x_j+\xi_j-it]\} \\ & - \sum_{k=1}^b \{\log\Gamma[y_k+\eta_k-it] - \log\Gamma[y_k+\eta_k]\}.\end{aligned}$$

Now we use the Taylor expansion formula:

$$\log\Gamma(a+b) = \log\Gamma(a) + \sum_{k=1}^\infty \frac{1}{k!}\psi^{(k-1)}(a)b^k,$$

where ψ is the digamma function defined by $\psi(z) = (d/dz)\log\Gamma(z)$. It is known that the polygamma function $\psi^{(s)}(a) = d^s\psi(z)/dz^s|_{z=a}$ can be expressed as

$$\psi^{(s)}(a) = \begin{cases} -C + \sum_{k=0}^{\infty}\left(\frac{1}{1+k} - \frac{1}{k+a}\right), & s = 0, \\ \sum_{k=0}^{\infty}\frac{(-1)^{s+1}s!}{(k+a)^{s+1}}, & s = 1, 2, \ldots, \end{cases}$$

where C is the Euler constant. Then we have

$$\log\tilde{C}(t) = \sum_{s=1}^{\infty}\frac{\kappa^{(s)}}{s!}(it)^s, \tag{5.4.27}$$

where

$$\kappa^{(1)} = \sum_{j=1}^{a}x_j\log x_j - \sum_{k=1}^{b}y_k\log y_k - \sum_{j=1}^{a}\psi^{(0)}(x_j+\xi_j) + \sum_{k=1}^{b}\psi^{(0)}(y_k+\eta_k), \tag{5.4.28}$$

$$\kappa^{(s)} = (-1)^s\left\{\sum_{j=1}^{a}\psi^{(s-1)}(x_j+\xi_j) - \sum_{k=1}^{b}\psi^{(s-1)}(y_k+\eta_k)\right\}, \quad s \geq 2. \tag{5.4.29}$$

Let Z be the standardized statistic defined by

$$Z = \frac{\tilde{V} - \kappa^{(1)}}{(\kappa^{(2)})^{1/2}}, \tag{5.4.30}$$

and let $\tilde{\kappa}^{(s)}$ be the standardized cumulants defined by $\tilde{\kappa}^{(s)} = \kappa^{(s)}/(\kappa^{(s)})^{s/2}$. The characteristic function of Z can be expanded as

$$\begin{aligned} C_Z(t) = \mathrm{E}[\exp(itZ)] &= \exp\left\{-\frac{t^2}{2} + \sum_{s=3}^{\infty}\frac{\tilde{\kappa}^{(s)}}{s!}(it)^s\right\} \\ &= \exp\left(-\frac{t^2}{2}\right)\left\{1 + \sum_{k=1}^{\infty}\frac{1}{k!}(it)^{3k}\left(\sum_{s=0}^{\infty}\frac{\tilde{\kappa}^{(s+3)}}{(s+3)!}(it)^s\right)^k\right\} \\ &= \exp\left(-\frac{t^2}{2}\right)\left\{1 + \sum_{k=1}^{\infty}\frac{1}{k!}(it)^{3k}\sum_{j=0}^{\infty}\gamma_{k,j}(it)^j\right\}, \end{aligned}$$

where

$$\gamma_{k,j} = \sum_{s_1+\cdots+s_k=j}\frac{\tilde{\kappa}^{(s_1+3)}\cdots\tilde{\kappa}^{(s_k+3)}}{(s_1+3)!\cdots(s_k+3)!}. \tag{5.4.31}$$

If $\gamma_{k,j} = \mathrm{O}(m^{-(j+k)})$ for some parameter, we can get an approximation for $C_Z(t)$. Therefore, let

$$C_{Z,s}(t) = \exp\Big(-\frac{t^2}{2}\Big)\Big\{1 + \sum_{k=1}^{s} \frac{1}{k!}(it)^{3k} \sum_{j=0}^{s-k} \gamma_{k,j}(it)^j\Big\}; \tag{5.4.32}$$

then

$$C_Z(t) = C_{Z,s}(t) + \mathrm{O}(m^{-(s+1)}), \qquad m \to \infty.$$

Inverting the characteristic function formally, we obtain the Edgeworth expansion of the distribution function of Z up to order $\mathrm{O}(m^{-s})$ as

$$Q_s(x) = \Phi(x) - \varphi(x)\Big\{\sum_{k=1}^{s} \frac{1}{k!} \sum_{j=0}^{s-k} \gamma_{k,j} H_{3k+j-1}(x)\Big\}, \tag{5.4.33}$$

where Φ and φ are the distribution and density functions of the standard normal distribution, respectively; $\gamma_{k,j}$ is given by (5.4.32); and $H_r(x)$ is the rth-order Hermite polynomial, defined by

$$\Big(\frac{d}{dx}\Big)^r \exp\Big(-\frac{x^2}{2}\Big) = (-1)^r H_r(x) \exp\Big(-\frac{x^2}{2}\Big).$$

The validity of such asymptotic methods has been shown for some likelihood ratio statistics by Wakaki (2009), Kato et al. (2009), and others. The validity was shown under the high-dimensional framework $p/n \to c \in (0, 1)$. It may be noted that the error estimates were done with computable error bounds.

As the special cases of (5.4.33), we give the first-, second and third-order asymptotic expansions:

$$\begin{aligned}
Q_0(x) &= \Phi(x), \\
Q_1(x) &= \Phi(x) - \frac{1}{6}\phi(x)\tilde{\kappa}^{(3)}(x^2 - 1), \qquad\qquad (5.4.34)\\
Q_2(x) &= \Phi(x) - \varphi(x)\Big\{\frac{1}{6}\tilde{\kappa}^{(3)}(x^2 - 1) + \frac{1}{24}\tilde{\kappa}^{(4)}(x^3 - 2x) \\
&\qquad + \frac{1}{36}(\tilde{\kappa}^{(3)})^2(x^5 - 10x^3 + 15x)\Big\}.
\end{aligned}$$

Note that the cumulants $\tilde{\kappa}^{(s)}$ involve the digamma and polygamma functions. However, these functions may be computed by for example, Mathematica Version 5.2.

Example 5.4.2 We consider the asymptotic approximation to Λ whose distribution is $\Lambda_p(q, n)$ under a high-dimensional framework: q is fixed and

$p/n \to c \in (0,1)$. Using the fact that $\Lambda \sim \Lambda_q(p, n-p+q)$ and that the hth moments of $\tilde{W} = \Lambda$ are expressed as

$$\mathrm{E}[\tilde{W}^h] = \mathrm{E}[\Lambda^h] = K \prod_{j=1}^{q} \frac{\Gamma[\frac{1}{2}(n-p+j)+h]}{\Gamma[\frac{1}{2}(n+j)+h]},$$

where $K = \prod_{j=1}^{q} \Gamma[\frac{1}{2}(n+j)]/\Gamma[\frac{1}{2}(n-p+j)]$. This expression is of the form (5.4.23) with correspondence

$$\begin{aligned} a &= q, \quad x_j = \frac{1}{2}n, \quad \xi_j = \frac{1}{2}(j-p), \\ b &= q, \quad y_k = \frac{1}{2}n, \quad \eta_j = \frac{1}{2}j. \end{aligned}$$

Therefore, from (5.4.28) and(5.4.29) the cumulants of $\tilde{V} = -\log \Lambda$ are given by

$$\kappa^{(s)} = (-1)^s \sum_{j=1}^{q} \left\{ \psi^{(s-1)}\left(\frac{n-p+j}{2}\right) - \psi^{(s-1)}\left(\frac{n+j}{2}\right) \right\}, \quad s = 1, 2, \ldots.$$

The results can also be obtained by expanding the cumulant generating function by directly:

$$\begin{aligned} \log C_{\tilde{V}}(t) &= \log \mathrm{E}[\exp(-it \log \Lambda)] \\ &= \log \mathrm{E}[\Lambda^{-it}] \\ &= \sum_{j=1}^{q} \left\{ \log \Gamma\left[\frac{1}{2}(n-p+j) - it\right] - \log \Gamma\left[\frac{1}{2}(n-p+j)\right] \right. \\ &\qquad \left. - \left(\log \Gamma\left[\frac{1}{2}(n+j) - it\right] - \log \Gamma\left[\frac{1}{2}(n+j)\right] \right) \right\}. \end{aligned}$$

We can use $Q_0, Q_1, Q_2, \ldots$ as the approximation for the distribution function of $Z = [-\log \Lambda - \kappa^{(1)}]/\kappa^{(2)}$. Wakaki (2009) has shown that under a high-dimensional framework,

$$\Pr\{Z \le x\} = Q_s(x) + \mathrm{O}(m^{-s-1}),$$

with a computable error bound, where $m = n - p$.

5.5 Perturbation Method

The perturbation method and Taylor expansions are often used to obtain asymptotic expansions of various statistics. Let $\mathbf{S}$ be the sample covariance matrix based on a sample of size n from a p-dimensional population, and let

$$\mathbf{S} = \Sigma + \frac{1}{\sqrt{n}}\mathbf{V} \quad \text{or} \quad \mathbf{V} = \sqrt{n}(\mathbf{S} - \Sigma).$$

Then we have obtained an asymptotic expansion of the distribution of $W = \sqrt{n}\{f(\mathbf{S}) - f(\Sigma)\}$. For a derivation it was essential to use an approximation based on perturbation method or Taylor expansions. In this section we consider a perturbation expansions of the characteristic roots or vectors of $\mathbf{S}$ and the other matrices. Note that the characteristic roots of $\mathbf{S}$ are invariant under the orthogonal transformation (i.e., $\mathbf{S} \to \mathbf{HSH}$, where $\mathbf{H}$ is an orthogonal matrix). So, when we consider the distribution of the characteristic roots of $\mathbf{S}$, we may assume that

$$\Sigma = \mathrm{diag}\,(\lambda_1, \ldots, \lambda_p),$$

where $\lambda_1 \geq \cdots \geq \lambda_p > 0$ are the characteristic roots of Σ.

Suppose that a $p \times p$ symmetric matrix Y has a perturbation expansion

$$\begin{aligned} \mathrm{Y} &= \mathrm{D}_\lambda + \varepsilon \mathrm{V}^{(1)} + \varepsilon^2 \mathrm{V}^{(2)} + \ldots \\ &= \mathrm{D}_\lambda + \varepsilon \mathrm{V}, \qquad \mathrm{V} = \mathrm{V}^{(1)} + \varepsilon \mathrm{V}^{(2)} + \cdots, \end{aligned} \tag{5.5.1}$$

where $\mathrm{V}^{(i)}$ are symmetric matrices, and

$$\mathrm{D}_\lambda = \mathrm{diag}\,(\lambda_1, \ldots, \lambda_p), \quad \lambda_1 \geq \cdots \geq \lambda_p.$$

Let ℓ_α be the αth characteristic roots of Y and the corresponding orthonormal characteristic vectors; that is,

$$\mathrm{Y}\boldsymbol{h}_\alpha = \ell_\alpha \boldsymbol{h}_\alpha, \quad \boldsymbol{h}_\alpha' \boldsymbol{h}_\beta = \delta_{\alpha\beta}. \tag{5.5.2}$$

Putting $\mathrm{L} = \mathrm{diag}\,(\ell_1, \ldots, \ell_p)$ and $\mathrm{H} = (\boldsymbol{h}_1, \ldots, \boldsymbol{h}_p)$, we have

$$\mathrm{YH} = \mathrm{HL}, \quad \mathrm{H}'\mathrm{H} = \mathrm{I}_p.$$

Case 1: Simple Root. Suppose that λ_α is simple; that is, λ_α is different from the other $p-1$ roots.

Theorem 5.5.1 *Suppose that a $p \times p$ symmetric matrix Y has a perturbation expansion around $\mathrm{Y} = \mathrm{D}$ as in (5.5.1). Let ℓ_α be the αth characteristic roots of Y and $\boldsymbol{h}_\alpha$ be the corresponding characteristic vector*

$$\boldsymbol{h}_\alpha = (h_{1\alpha}, \ldots, h_{p\alpha})', \quad \boldsymbol{h}_\alpha' \boldsymbol{h}_\alpha = 1, \quad h_{\alpha\alpha} > 0.$$

If λ_α is distinct from the other $p-1$ roots, then

$$\ell_\alpha = \lambda_\alpha + \varepsilon v_{\alpha\alpha}^{(1)} + \varepsilon^2 \left\{ v_{\alpha\alpha}^{(2)} + \sum_{\beta\neq\alpha}^{p} \lambda_{\alpha\beta}(v_{\alpha\beta}^{(1)})^2 \right\} + \mathrm{O}(\varepsilon^3),$$

$$h_{j\alpha} = -\lambda_{j\alpha}\left[\varepsilon v_{j\alpha}^{(1)} + \varepsilon^2 \left\{ \lambda_{j\alpha} v_{j\alpha}^{(1)} v_{\alpha\alpha}^{(1)} + \sum_{\beta\neq\alpha}^{p} \lambda_{\alpha\beta} v_{j\beta}^{(1)} v_{\beta\beta}^{(1)} + v_{j\alpha}^{(2)} \right\}\right]$$
$$+ \mathrm{O}(\varepsilon^3), \quad j \neq \alpha,$$

$$h_{\alpha\alpha} = 1 + \varepsilon^2 \left\{ -\frac{1}{2} \sum_{\beta\neq\alpha}^{p} \lambda_{\alpha\beta}^2 (v_{\alpha\beta}^{(1)})^2 \right\} + \mathrm{O}(\varepsilon^3),$$

where $\boldsymbol{e}_\alpha = (0,\ldots,0,1,0,\ldots,0)'$, $\mathrm{V}^{(j)} = (v_{\alpha\beta}^{(j)})$, *and* $\lambda_{\alpha\beta} = (\lambda_\alpha - \lambda_\beta)^{-1} (\alpha \neq \beta)$.

Proof. We look for solutions such that

$$\ell_\alpha = \lambda_\alpha + \epsilon\lambda_\alpha^{(1)} + \epsilon^2\lambda_\alpha^{(2)} + \cdots,$$
$$\boldsymbol{h}_\alpha = \boldsymbol{e}_\alpha + \epsilon\boldsymbol{e}_\alpha^{(1)} + \epsilon^2\boldsymbol{e}_\alpha^{(2)} + \cdots.$$

Substituting the expansions into

$$\mathrm{Y}\boldsymbol{h}_\alpha = \ell_\alpha \boldsymbol{h}_\alpha$$

and equating the coefficients of ϵ^i $(i = 0, 1, 2, 3)$ on both sides, we can obtain the results required. □

For details on the derivation of Theorem 5.5.1, see Bellman (1960). The first few derivatives of ℓ_α and $\boldsymbol{f}_\alpha$ were obtained by Sugiura (1973).

Case 2: When the Root λ_α Has Multiplicity. Suppose that

$$\lambda_1 = \cdots = \lambda_{q_1} = \theta_1,$$
$$\lambda_{q_1+1} = \cdots = \lambda_{q_1+q_2} = \theta_2, \qquad (5.5.3)$$
$$\vdots \qquad (5.5.4)$$
$$\lambda_{p-q_r+1} = \cdots = \lambda_p = \theta_r.$$

where $\theta_1 > \cdots > \theta_r$. In this case a perturbation expansion for each ℓ_α is not available, but it is possible to determine asymptotically a symmetric matrix L_i of order q_i such that the roots $\ell_{q_1+\cdots+q_{i-1}+1}, \ldots, \ell_{q_1+\cdots+q_i}$ are equal to those of L_i. Corresponding to the portions, decompose V and $\mathrm{V}^{(j)}$ as

$$\mathrm{V} = \begin{pmatrix} \mathrm{V}_{11} & \cdots & \mathrm{V}_{1r} \\ \vdots & \vdots & \vdots \\ \mathrm{V}_{r1} & \cdots & \mathrm{V}_{rr} \end{pmatrix}, \quad \mathrm{V}^{(j)} = \begin{pmatrix} \mathrm{V}_{11}^{(j)} & \cdots & \mathrm{V}_{1r}^{(j)} \\ \vdots & \vdots & \vdots \\ \mathrm{V}_{r1}^{(j)} & \cdots & \mathrm{V}_{rr}^{(j)} \end{pmatrix}.$$

Here $V_{\alpha\beta}$ and $V_{\alpha\beta}^{(j)}$ are $q_\alpha \times q_\beta$ submatrices of V and V^j, respectively, and

$$V_{\alpha\beta} = V_{\alpha\beta}^{(1)} + \varepsilon V_{\alpha i\beta}^{(2)} + \cdots .$$

Put $\theta_{\alpha\beta} = (\theta_\alpha - \theta_\beta)^{-1}$ for $\alpha \neq \beta$.

Theorem 5.5.2 *In the perturbation expansion* (5.5.1), *assume that the diagonal matrix* $D = \text{diag}(\lambda_1, \ldots, \lambda_p)$ *has a multiplicity as in* (5.5.3). *Then the* $(q_1 + \cdots + q_{i-1} + j)(1 \leq j \leq q_i)$th *root of* Y *is equal to the* jth *root of* W_i, *given by*

$$W_i = \theta_i I_{q_i} + \varepsilon V_{ii} + \varepsilon^2 \sum_{\beta\neq i} \theta_{i\beta} V_{i\beta} V_{\beta i} + \varepsilon^3 \left\{ \sum_{\beta\neq i} \sum_{\gamma\neq i} \theta_{i\beta}\theta_{i\gamma} V_{i\beta} V_{\beta\gamma} V_{\gamma i} \right.$$

$$\left. - \frac{1}{2} V_{ii} \sum_{\beta\neq i} \theta_{i\beta}^2 V_{i\beta} V_{\beta i} - \frac{1}{2} \sum_{\beta\neq i} \theta_{i\beta}^2 \theta_{i\beta}^2 V_{i\beta} V_{\beta i} V_{ii} \right\} + O(\varepsilon^4).$$

Proof. Consider $q_i \times q_i$ matrices L_i and an orthogonal matrix such that

$$YH = H \begin{pmatrix} L_1 & \cdots & O \\ \vdots & \ddots & \vdots \\ O & \cdots & L_r \end{pmatrix}.$$

Then L_i and H can be determined in the form

$$L_i = \theta_i I_{q_i} + \varepsilon L_i^{(1)} + \varepsilon^2 L_i^{(2)} + \cdots ,$$
$$H = \Gamma + \varepsilon\text{-}H_i^{(1)} + \varepsilon^2\text{-}H_i^{(2)} + \cdots ,$$

where

$$\Gamma = \begin{pmatrix} \Gamma_1 & \cdots & O \\ \vdots & \ddots & \vdots \\ O & \cdots & \Gamma_r \end{pmatrix},$$

and Γ_i are $q_i \times q_i$ orthogonal matrices. Finally, the W_i are obtained as $W_i = \Gamma_i L_i \Gamma_i'$. □

Lawley (1956) obtained Theorem 5.5.2 in the special case of $q_1 = \cdots = q_{r-1} = 1$. For details in the general case, see Fujikoshi (1977a, 1977b).

In discriminant analysis and canonical correlation analysis we use the characteristic roots and vectors in a general characteristic problem as follows. Let A and B be $p \times p$ positive definite matrices. Consider a general characteristic problem

$$A\boldsymbol{g}_\alpha = \ell_\alpha B\boldsymbol{g}_\alpha, \tag{5.5.5}$$

where $\ell_1 > \cdots > \ell_p > 0$ and the characteristic vectors are normalized as $\boldsymbol{g}'_\alpha \mathrm{B} \boldsymbol{g}_\beta = \delta_{\alpha\beta}$. Suppose that A and B are expanded as

$$\mathrm{A} = \mathrm{D}_\lambda + \varepsilon \mathrm{A}^{(1)} + \varepsilon^2 \mathrm{A}^{(2)} + \cdots,$$
$$\mathrm{B} = \mathrm{I}_p + \varepsilon \mathrm{B}^{(1)} + \varepsilon^2 \mathrm{B}^{(2)} + \cdots.$$

Then, putting

$$\mathrm{Y} = \mathrm{B}^{-1/2}\mathrm{A}\mathrm{B}^{-1/2} \quad \text{and} \quad \boldsymbol{h}_\alpha = \mathrm{B}^{1/2}\boldsymbol{g}_\alpha,$$

the characteristic roots ℓ_α and the transformed characteristic vectors $\boldsymbol{h}_\alpha$ are those of Y:

$$\mathrm{Y}\boldsymbol{h}_\alpha = \ell_\alpha \boldsymbol{h}_\alpha.$$

Therefore, we can obtain expansions of ℓ_α by using Theorems 5.5.1 and 5.5.2. More precisely, let

$$\mathrm{A} = \mathrm{D}_\lambda + \varepsilon \mathrm{C}, \quad \mathrm{B} = \mathrm{I}_p + \varepsilon \mathrm{U}.$$

Then

$$(\mathrm{I}_p + \varepsilon \mathrm{U})^{-1/2} = \mathrm{I}_p - \frac{1}{2}\varepsilon \mathrm{U} + \frac{3}{8}\varepsilon^2 \mathrm{U}^2 - \cdots$$

and

$$\begin{aligned}
\mathrm{Y} &= \mathrm{B}^{-1/2}\mathrm{A}\mathrm{B}^{-1/2} \\
&= (\mathrm{I}_p + \varepsilon \mathrm{U})^{-1/2}(\mathrm{D}_\lambda + \varepsilon \mathrm{C})(I_p + \varepsilon \mathrm{U})^{-1/2} \\
&= \mathrm{D}_\lambda + \varepsilon\left(\mathrm{C} - \frac{1}{2}\mathrm{D}_\lambda \mathrm{U} - \frac{1}{2}\mathrm{U}\mathrm{D}_\lambda\right) \\
&\quad + \varepsilon^2\left(\frac{1}{4}\mathrm{U}\mathrm{D}_\lambda\mathrm{U} - \frac{1}{2}\mathrm{U}\mathrm{C} - \frac{1}{2}\mathrm{C}\mathrm{U} + \frac{3}{8}\mathrm{U}^2\mathrm{D}_\lambda + \frac{3}{8}\mathrm{D}_\lambda\mathrm{U}^2\right) + \cdots.
\end{aligned}$$

Using Theorem 5.5.1 we obtain perturbation expansions for ℓ_α and $\boldsymbol{g}_\alpha$ when λ_α is simple,

$$\ell_\alpha = \lambda_\alpha + \varepsilon \ell_\alpha^{(1)} + \varepsilon^2 \ell_\alpha^{(2)} + \cdots,$$
$$\boldsymbol{g}_\alpha = \boldsymbol{e}_\alpha + \varepsilon \boldsymbol{g}_\alpha^{(1)} + \varepsilon^2 \boldsymbol{g}_\alpha^{(2)} + \cdots,$$

where

$$\mathrm{C}_\alpha^{(i)} = \mathrm{A}^{(i)} - \lambda_\alpha \mathrm{B}^{(i)} \quad \text{and} \quad \boldsymbol{g}_\alpha^{(i)} = (g_{\alpha 1}^{(i)}, \ldots, g_{\alpha p}^{(i)})',$$

and

$$\begin{aligned}
\ell_\alpha^{(1)} &= c_{\alpha,\alpha\alpha}^{(1)},\\
\ell_\alpha^{(2)} &= -c_{\alpha,\alpha\alpha}^{(1)} b_{\alpha\alpha}^{(1)} - \sum_{\beta\neq\alpha} \lambda_{\beta\alpha} \left\{ c_{\alpha,\alpha\beta}^{(1)} \right\}^2 + c_{\alpha,\alpha\alpha}^{(2)},\\
g_{\alpha\alpha}^{(1)} &= -\frac{1}{2} b_{\alpha\alpha}^{(1)}, \quad g_{\alpha\beta}^{(1)} = -\lambda_{\beta\alpha} c_{\alpha,\beta\alpha}^{(1)}, \ \beta \neq \alpha,\\
g_{\alpha\alpha}^{(2)} &= \frac{3}{8} \left\{ b_{\alpha\alpha}^{(1)} \right\}^2 + \sum_{\beta\neq\alpha} \lambda_{\beta\alpha} c_{\alpha,\beta\alpha}^{(1)} b_{\alpha\alpha}^{(1)} - \frac{1}{2} \sum_{\beta\neq\alpha} \lambda_{\beta\alpha}^2 c_{\alpha,\beta\alpha}^{(1)} - \frac{1}{2} b_{\alpha\alpha}^{(2)},\\
g_{\alpha\beta}^{(2)} &= \lambda_{\beta\alpha} \left\{ c_{\alpha,\alpha\alpha}^{(1)} b_{\beta\alpha}^{(1)} + \frac{1}{2} c_{\alpha,\beta\alpha}^{(1)} b_{\alpha\alpha}^{(1)} \right.\\
&\quad \left. - \lambda_{\beta\alpha} c_{\alpha,\alpha\alpha}^{(1)} c_{\alpha,\beta\alpha}^{(1)} + \sum_{\gamma\neq\alpha} \lambda_{\gamma\alpha} c_{\alpha,\beta\gamma}^{(1)} c_{\alpha,\gamma\alpha}^{(1)} - c_{\alpha,\beta\alpha}^{(2)} \right\}, \qquad \beta \neq \alpha.
\end{aligned}$$

Chambers (1967) discussed perturbation methods of producing multivariate approximations and justification of the approximations.

5.6 Cornish–Fisher Expansions

5.6.1 Expansion Formulas

Let X be a univariate random variable with a continuous distribution function. For any number $0 < \alpha < 1$, there exists a real number x such that $F(x) = \alpha$, which is called the (lower) $100\alpha\%$ *point* of F. If F is strictly increasing on its support, the inverse function $F^{-1}(\cdot)$ of F is well defined and the $100\alpha\%$ point is uniquely determined by $F^{-1}(\alpha)$. For an integer $q > 1$ and for $k = 1, \ldots, q-1$, $F^{-1}(k/q)$ is called the kth *q-quantile*. The 2-quantile has the special name of *median* and 100-quantiles are called *percentiles*, which is synonymous with $100\alpha\%$ *point* and α can be any positive number less than 1 (i.e. $100 \times \alpha$ need not be an integer). We also speak of "quantile" without reference to particular values of α. In other words, the quantiles of the distribution F are those values given by the inverse function $F^{-1}(\cdot)$.

Remark 5.6.1 Even in the general case, where $F(x)$ is not necessarily continuous nor is it strictly increasing, we can define its inverse function by, for example,

$$F^{-1}(u) = \inf\{x; F(x) > u\}.$$

This is a right-continuous nondecreasing function defined on the interval $(0, 1)$ and $F(x_0) \geq u_0$ if $x_0 = F^{-1}(u_0)$.

In this section we study asymptotic approximations of the quantiles of a given distribution function $F_n(x)$ with a parameter $n = 1, 2, \ldots$. We assume that each F_n admits the Edgeworth expansion in the powers of $\epsilon = n^{-1/2}$ or n^{-1}:

$$\begin{aligned} F_n(x) &= G_{k,n}(x) + \mathrm{O}(\epsilon^k) \quad \text{with} \\ G_{k,n}(x) &= G(x) + \{\epsilon a_1(x) + \cdots + \epsilon^{k-1} a_{k-1}(x)\} g(x), \end{aligned} \tag{5.6.1}$$

where $g(x)$ is the density function of the limiting distribution function $G(x)$. Then we can approximate quantiles of F_n by the corresponding quantiles of $G_{k,n}$. Let x_α and x_α^* be the upper $100\alpha\%$ points of F_n and $G_{k,n}$, respectively; that is, they satisfy

$$F_n(x_\alpha) = G_{k,n}(x_\alpha^*) = 1 - \alpha. \tag{5.6.2}$$

The approximate quantile x_α^* based on the Edgeworth expansion is available in numerical form but cannot be expressed in explicit form. An important approach to the problem of approximating the quantiles of F_n is to use their asymptotic relation to those of G's. Let x and u be the corresponding quantiles of F_n and G, respectively. Then we have

$$F_n(x) = G(u). \tag{5.6.3}$$

Write $x(u)$ and $u(x)$ to denote the solutions of (5.6.3) for x in terms of u and u in terms of x, respectively [i.e. $u(x) = G^{-1}(F_n(x))$ and $x(u) = F_n^{-1}(G(u))$]. Then we can take advantage of the Edgeworth expansion (5.6.1) for F_n to obtain formal solutions $x(u)$ and $u(x)$ in the form,

$$x(u) = u + \epsilon b_1(u) + \epsilon^2 b_2(u) + \cdots \tag{5.6.4}$$

and

$$u(x) = x + \epsilon c_1(x) + \epsilon^2 c_2(x) + \cdots . \tag{5.6.5}$$

Cornish and Fisher (1937) and Fisher and Cornish (1946) obtained the first few terms of these expansions when G is the standard normal distribution function (i.e., $G = \Phi$). We call both (5.6.4) and (5.6.5) the *Cornish–Fisher expansions*. Hill and Davis (1968) gave a general algorithm for obtaining each term of the expansions. In practical applications we use the Cornish–Fisher expansions in the following form with $k = 1, 2$, or 3:

$$x_k(u) = u + \sum_{j=1}^{k-1} \epsilon^j b_j(u), \tag{5.6.6}$$

which is a reasonable approximation of the quantity of F_n. On the other hand, we have

$$u_k(x) = x + \sum_{j=1}^{k-1} \epsilon^j c_j(x), \tag{5.6.7}$$

which may be regarded as a transformation of approximation F_n of G.

Now we look for explicit expressions for $b_1(u)$ and $b_2(u)$. Substituting (5.6.4) or (5.6.6) with $k = 3$ to (5.6.3) and using (5.6.1), we have

$$\begin{aligned} F_n(x) &= F_n(u + \epsilon b_1 + \epsilon^2 b_2 + \cdots) \\ &= G(u + \epsilon b_1 + \epsilon^2 b_2) + g(u + \epsilon b_1 + \epsilon^2 b_2) \\ &\quad \times \{\epsilon a_1(u + \epsilon b_1) + \epsilon^2 a_2(u)\} + \mathrm{O}(\epsilon^3). \end{aligned}$$

Considering Taylor's expansions for G, g, and a_1, we obtain

$$\begin{aligned} F_n(x) &= G(u) + \epsilon g(u)\{b_1 + a_1(u)\} \\ &\quad + \epsilon^2 \left[g(u)b_2 + \frac{1}{2}g'(u)b_1^2 + g(u)a_1'(u)b_1 \right. \\ &\quad \left. + g(u)a_2(u) + g'(u)b_1a_1(u) \right] + \mathrm{O}(\epsilon^3), \end{aligned}$$

which should be $G(u)$. Therefore,

$$\begin{aligned} b_1 &= -a_1(u), \\ b_2 &= \frac{1}{2}\{g'(u)/g(u)\}a_1^2(u) - a_2(u) + a_1'(u)a_1(u). \end{aligned} \tag{5.6.8}$$

Theorem 5.6.1 *Assume that the distribution function G is analytic and the density function $g(x)$ and G is arbitrarily differentiable. Then the following relations for the x and u satisfying $F_n(x) = G(u)$ hold:*

$$x = u - \sum_{r=1}^{\infty} \frac{1}{r!} \left\{ -[g(u)]^{-1} d_u \right\}^{r-1} \left[\{z_n(u)\}^r / g(u) \right], \tag{5.6.9}$$

where $d_u \equiv d/du$ and $z_n(u) = F_n(u) - G(u)$.

Proof. See Hill and Davis (1968). □

In many statistical applications, $F_n(x)$ is known to have an asymptotic expansion of the form

$$F_n(x) = G(x) + g(x)\left[n^{-a}p_1(x) + n^{-2a}p_2(x) + \cdots \right], \tag{5.6.10}$$

where $p_r(x)$ may be polynomials in x and $a = 1/2$ or 1. Then the formulas (5.6.9) can be written as

$$x = u - \sum_{r=1}^{\infty} \frac{1}{r!} d_{(r)} \{q_n(u)\}^r, \tag{5.6.11}$$

where $q_n(u) = n^{-a}p_1(u) + n^{-2a}p_2(u) + \cdots$, $m(x) = -g'(x)/g(x)$,

$$\begin{aligned} d_{(1)} &= \text{the identity operator}, \\ d_{(r)} &= \{m(u) - d_u\}\{2m(u) - d_u\} \cdots \{(r-1)m(u) - d_u\}, \quad r = 2, 3, \ldots. \end{aligned}$$

The rth term in each of (5.6.11) is $\mathrm{O}(n^{-ra})$.

Example 5.6.1 Suppose that $F_n(x)$ is expressed as

$$F_n(x) = \Phi(x) - \frac{1}{\sqrt{n}}\left\{a_1\Phi^{(1)}(x) + a_3\Phi^{(3)}(x)\right\} + \frac{1}{n}\left\{b_2\Phi^{(2)}(x) + b_4\Phi^{(4)}(x) + b_6\Phi^{(6)}(x)\right\} + \mathrm{o}(n^{-1}),$$

where $\Phi(x)$ is the distribution function of $\mathrm{N}(0,1)$ and $\Phi^{(r)}(x)$ is the rth derivative of $\Phi(x)$. Since

$$\Phi^{(r)}(x) = \phi^{(r-1)}(x) = (-1)^{r-1}h_{r-1}(x)\phi(x),$$

where $h_r(x)$ is the Hermite polynomial of degree r and

$$h_0(x) = 1, \quad h_1(x) = x, \quad h_2(x) = x^2 - 1,$$
$$h_3(x) = x^3 - 3x, \quad h_4(x) = x^4 - 6x^2 + 3, \quad h_5(x) = x^5 - 10x^3 + 15x,$$

$q_n(x)$ is written as

$$q_n(x) = -\frac{1}{\sqrt{n}}\{a_1h_0(x) + a_3h_2(x)\} + \frac{1}{n}\{b_2h_1(x) + b_4h_3(x) + b_6h_5(x)\} + \mathrm{o}(n^{-1}).$$

Hence it follows from (5.6.11) that

$$x = u - q_n(u) - \frac{u}{n}\Big[a_1 + a_3h_2(u)\Big] \times \left[\frac{1}{2}a_1 + a_3\left\{\frac{1}{2}h_2(u) - 2\right\}\right] + \mathrm{o}(n^{-1}).$$

Example 5.6.2 Suppose that

$$F_n(x) = G_f(x) + \frac{f\gamma}{4n}[G_f(x) - 2G_{f+2}(x) + G_{f+4}(x)] + \frac{f}{96n^2}\sum_{j=0}^{4}(-1)^jc_jG_{f+2j}(x) + \mathrm{o}(n^{-2}),$$

where $G_f(x) = \Pr\{\chi_f^2 \le x\}$; that is, the distribution function of the chi-square distribution with f degrees of freedom, γ is a constant, and c_j are constants such that $\sum_{j=0}^{4}(-1)^jc_j = 0$. Then $G(u) = G_f(u)$,

$$g(u) = g_f(u) = \Big[\Gamma(f/2)2^{f/2}\Big]^{-1} u^{f/2-1}\exp(-u/2),$$

$$m(u) = -g_f'(u)/g_f(u) = \frac{1}{2} - \frac{1}{u}\left(\frac{f}{2} - 1\right).$$

Noting that $G_{f+2}(u) = -2g_{f+2}(u) + G_f(u)$ and $g_{f+2}(u) = (u/f)g_f(u)$, we can write

$$\begin{aligned} q_n(u) = -\frac{u(u-f-2)}{2n(f+2)} - &\frac{u}{48n^2(f+2)(f+4)(f+6)}[c_4u^3 \\ &+ (c_4 - c_3)(f+6)u^2 + (c_1 - c_0)(f+4)(f+6)u \\ &- c_0(f+2)(f+4)(f+6)] + \mathrm{o}(n^{-2}). \end{aligned}$$

Therefore, we obtain

$$\begin{aligned} x = u - q_n(u) - &[\gamma^2 u(u-f-2)/\{16n^2(f+2)^2\}] \\ &\times \{u^2 - 2(f+4)u + (f+2)^2\} + \mathrm{o}(n^{-2}). \end{aligned}$$

In the study of approximations for the quantile of F_n it is important to obtain good upper and lower bounds for $x = x(u)$ and $u = u(x)$, where u and x satisfy the equation (5.6.3). More precisely we want to find upper and lower bounds such that

$$\begin{aligned} \underline{x}_n(u) &\geq x(u) \geq \bar{x}_n(u), \\ \underline{u}_n(x) &\geq u(x) \geq \bar{u}_n(x). \end{aligned}$$

Such bounds were obtained for some special distributions by Wallace (1959).

5.6.2 Validity of Cornish–Fisher Expansions

The approximate quantile x_α^* above based on the Edgeworth expansion is available in numerical form but cannot be expressed in explicit form. For the validity of the Cornish–Fisher expansions, there are some results. Suppose that the remainder term, $R_{k,n}(x) = F_n(x) - G_{k,n}(x)$, is such that

$$|R_{k,n}| \leq \epsilon^n C_k. \tag{5.6.12}$$

Then

$$\big|F_n(x_\alpha^*) - (1-\alpha)\big| = \big|F_n(x_\alpha^*) - G_{k,n}(x_\alpha^*)\big| \leq \epsilon^n C_k. \tag{5.6.13}$$

This gives an error bound for the absolute differences between the probabilities based on the true quantiles and their approximations. Following Wallace (1958), we can state the validity of (5.6.7) as follows. Consider the absolute differences between the probabilities based on the true quantile $u(x)$ and the approximate quantile $u_{k,n}(x)$. Suppose that G is k times continuously differentiable, and write $(\mathrm{d}^j/\mathrm{d}x^j)G(x) = G^{(j)}(x)$. Then, by Taylor's theorem, we have

$$\begin{aligned} G(u_{k,n}(x)) &= G(x) + \sum_{j=1}^{k-1} \frac{1}{j!} G^{(j)}(x) \left[\sum_{\ell=1}^{k-1} \epsilon^{\ell} c_{\ell}(x) \right]^j \\ &\quad + \frac{1}{k!} G^{(k)}(\tilde{x}) \left[\sum_{\ell=1}^{k-1} \epsilon^{\ell} c_{\ell}(x) \right]^k \\ &= G_{k,n}(x) + \big(\text{terms of } k\text{th or higher order in } \epsilon\big) \\ &= G_{k,n}(x) + \tilde{R}_{k,n}(x), \end{aligned}$$

where $x \leq \tilde{x} \leq u_{k,n}(x)$ or $u_{k,n}(x) \leq \tilde{x} \leq x$. It is easily seen that there exists a constant $\tilde{C}_k$ such that

$$|\tilde{R}_{k,n}(x)| \leq \epsilon^k \tilde{C}_k$$

uniformly in x on every compact interval. Therefore, if F_n has the valid Edgeworth expansion $G_{k,n}(x)$ of (5.6.1) satisfying (5.6.2), we have

$$\begin{aligned} |F_n(x) - G(u_{k,n}(x))| &\leq |F_n(x) - G_{k,n}(x)| + |G_{k,n}(x) - G(u_{k,n}(x))| \\ &\leq (C_k + \tilde{C}_k)\epsilon^k \end{aligned}$$

uniformly in x on every compact interval. For further details, the reader is referred to Hall (1983). The other validity of the Cornish–Fisher expansions was obtained by considering the distribution function $\tilde{F}_{k,n}$ of

$$\tilde{X} = U + \sum_{j=1}^{k-1} \epsilon^j b_j(U),$$

where U is the standard normal variable. Takeuchi and Takemura (1988) showed that if $|F_n(x) - G_{k,n}(x)| = \mathrm{o}(\epsilon^{k-1})$, then $|F_n(x) - \tilde{F}_{k,n}| = \mathrm{o}(\epsilon^{k-1})$.

Remark 5.6.2 If, in general, a random variable X follows a distribution with the continuous distribution function G, the random variable $G(X)$ follows the uniform distribution on the unit interval $(0, 1)$. If, further, $F(\cdot)$ is a strictly monotone distribution function, the variable $F^{-1}(G(X))$ follows the distribution F.

We are also interested in error bounds for the absolute differences between the true quantiles and the approximate quantiles. For simplicity, we give error bounds for the first-order Cornish–Fisher expansion due to Ulyanov et al. (2006). Suppose that the distribution of U can be approximated as

$$F(x) \equiv \Pr\{U \leq x\} = G(x) + R_1(x), \tag{5.6.14}$$

when a positive number ϵ tends to 0, where R_1 is a remainder term with $\mathrm{O}(\epsilon)$. Here we assume that density function g of the distribution G is continuous on its support $D = \{x \in \mathrm{R}\colon g(x) > 0\}$.

Theorem 5.6.2 *Suppose that the distribution function of U can be put in the form (5.6.14). For the error term $R_1(x)$, we assume that there exists a constant c_1 such that*

$$|R_1(x)| \le d_1 \equiv c_1\epsilon.$$

Let x_α and u_α be the upper $100\alpha\%$ points of F and G, respectively; that is,

$$\Pr\{U \le x_\alpha\} = G(u_\alpha) = 1 - \alpha. \tag{5.6.15}$$

Then, for any α such that $1 > \alpha > d_1$ and $1 > \alpha + d_1$:

(1) $u_{\alpha+d_1} \le x_\alpha \le u_{\alpha-d_1}$.

(2) $|x_\alpha - u_\alpha| \le d_1/g(u_{(1)})$, *where*

$$g(u_{(1)}) = \min_{u\in[u_{\alpha+d_1}, u_{\alpha-d_1}]} g(u).$$

Proof. By the mean value theorem,

$$|G(x_\alpha) - G(u_\alpha)| \ge |x_\alpha - u_\alpha| \min_{0<\theta<1} g(u_\alpha + \theta(x_\alpha - u_\alpha)).$$

From (5.6.14) and the definition of x_α and u_α, we have

$$\begin{aligned} |G(x_\alpha) - G(u_\alpha)| &= |\Pr\{S \le x_\alpha\} - R(x_\alpha) - G(x_\alpha)| \\ &= |R(x_\alpha)| \le d_1. \end{aligned}$$

Therefore,

$$|x_\alpha - u_\alpha| \le d_1 \cdot \frac{1}{\min_{0<\theta<1}} g(u_\alpha + \theta(x_\alpha - g(u_\alpha)). \tag{5.6.16}$$

On the other hand, from (5.6.14) we have

$$\begin{aligned} G(x_\alpha) &= 1 - \alpha - R_1(\alpha) \\ &\le 1 - (\alpha - d_1) = G(u_{\alpha-d_1}). \end{aligned}$$

This implies that $x_\alpha \le u_{\alpha-d_1}$. Similarly, we have $u_{\alpha+d_1} \le x_\alpha$. Therefore, we have (1). Using (1) we have

$$\min_{u\in[u_{\alpha+d}, u_{\alpha-d}]} g(u) \le \min_{0<\theta<1} g(u_\alpha + \theta(x_\alpha - u_\alpha)).$$

Result (2) is obtained by substituting the result above to (5.6.16). □

5.7 Transformations for Improved Approximations

Suppose that a statistic U has an asymptotic normal or chi-square approximation as some parameter; typically, the sample size n tends to infinity. In this case it is interesting to construct improved normal or chi-square approximations for U.

First, consider the case in which the limiting distribution of T as n tends to infinity is normal. Suppose that T has an asymptotic expansion

$$\Pr\{T \leq x\} = \Phi(x) + \frac{1}{\sqrt{n}}\phi(x)(ax^2 + b) + \mathrm{o}(n^{-1/2}), \tag{5.7.1}$$

which is a typical form. In general, we are interesting a transformation that leads to the disappearance of the terms of $\mathrm{O}(n^{-1/2})$ in the asymptotic expansion. It is seen (see Problem 5.) that such a transformation is given by $T^* = T + (aT^2 + b)/\sqrt{n}$. However, this transformation is not monotone. The monotonicity of a transformation is important in keeping optimality of a statistic U and computing the probability $\Pr\{T(U) \geq x\}$. As a method of making T^* monotone, we may consider

$$\tilde{T} = T^* + \frac{1}{3n}a^2T^3. \tag{5.7.2}$$

The transformation $\tilde{T}$ is monotone, since $\tilde{T}' = (1 + aT)^2 > 0$. Further, the added term is $\mathrm{O}(n^{-1})$. Therefore,

$$\Pr\{\tilde{T} \leq x\} = \Phi(x) + \mathrm{o}(n^{-1}).$$

This idea was used by Hall (1992a). This transformation removes its skewness.

Next we consider the case when U is asymptotically distributed to $\chi^2(q)$. A common approach is to consider a transformation $T = T(U)$ based on a Bartlett correction. More precisely, assume that

$$\mathrm{E}(U) = q\{1 + n^{-1}b_1 + \mathrm{O}(n^{-2})\}. \tag{5.7.3}$$

Then the corrected statistic is defined by

$$T = (1 - n^{-1}b_1)U, \quad \text{or} \quad T = (1 + n^{-1}b_1)^{-1}U.$$

Then

$$\mathrm{E}(T) = q + \mathrm{O}(n^{-2}), \tag{5.7.4}$$

so T has a better chi-square approximation in the sense that

$$\mathrm{E}(U - \chi_2^2) = \mathrm{O}(n^{-1}), \quad \text{but,} \quad \mathrm{E}(T - \chi_q^2) = \mathrm{O}(n^{-2}).$$

For $-2\log$ likelihood ratio criteria, it is known that under the appropriate regularity conditions,

$$\Pr\{T \le x\} = G_q(x) + \mathrm{O}(n^{-2}). \tag{5.7.5}$$

For example, consider the null distribution of a likelihood ratio statistic $T_{\mathrm{LR}} = -n\log(|\mathbf{S}_e|/|\mathbf{S}_e + \mathbf{S}_h|)$ for testing a linear hypothesis in a multivariate linear model, where $\mathbf{S}_h$ and $\mathbf{S}_e$ are independently distributed as Wishart distributions $\mathrm{W}_p(q, \Sigma)$ and $\mathrm{W}_p(n, \Sigma)$, respectively. Then the distribution of T_{LR} is expanded (see Section 4.1) as

$$\begin{aligned}\Pr\{T_{\mathrm{LR}} \le x\} &= G_r(x) \\ &\quad + \frac{r}{4n}(q-p-1)\{G_r(x) - G_{r+2}(x)\} + \mathrm{O}(n^{-2}),\end{aligned} \tag{5.7.6}$$

where $r = pq$. Its Bartlett correction is

$$\tilde{T}_{\mathrm{LR}} = -n\left\{1 + \frac{1}{2n}(q-p-1)\right\}\log(|\mathbf{S}_e|/|\mathbf{S}_e + \mathbf{S}_h|)$$

or

$$\tilde{T}_{\mathrm{LR}} = -n\left\{1 - \frac{1}{2n}(q-p-1)\right\}^{-1}\log(|\mathbf{S}_e|/|\mathbf{S}_e + \mathbf{S}_h|),$$

which satisfy (5.7.5). On the other hand, for the Lawley Hotelling trace criterion $T_{\mathrm{LH}} = n\ \mathrm{tr}\,\mathbf{S}_h\mathbf{S}_e^{-1}$ and Bartlett Nanda Pillai trace criterion $T_{\mathrm{BNP}} = n\ \mathrm{tr}\,\mathbf{S}_h(\mathbf{S}_e + \mathbf{S}_h)^{-1}$, their Bartlett corrections are

$$\tilde{T}_{\mathrm{LH}} = (n-p-1)\,\mathrm{tr}\,\mathbf{S}_h\mathbf{S}_e^{-1}, \text{ and } \tilde{T}_{LH} = (n+q)\,\mathrm{tr}\,\mathbf{S}_h(\mathbf{S}_e + \mathbf{S}_h)^{-1}.$$

However, the terms of $\mathrm{O}(n^{-1})$ in their asymptotic expansions do not disappear. In fact,

$$\begin{aligned}\Pr\{\tilde{T}_{\mathrm{LH}} \le x\} &= G_r(x) + \frac{r(p+q+1)}{4(n-p-1)} \\ &\quad \times \{G_r(x) - 2G_{r+2}(x) + G_{r+4}(x)\} + \mathrm{O}(n^{-2}).\end{aligned} \tag{5.7.7}$$

It is easy to find polynomial transformations that lead to the disappearance of the terms of $\mathrm{O}(n^{-1})$ in the asymptotic expansion. The transformation is given by

$$T_{\mathrm{LH}}^* = T_{\mathrm{LH}} + \frac{p+q+1}{2(n-p-1)}\left(T_{\mathrm{LH}} - \frac{1}{r+2}T_{\mathrm{LH}}^2\right).$$

The transformation is not monotone. There are several attempts to find monotone transformations that imply disappearance of the terms of $\mathrm{O}(n^{-1})$ in their asymptotic expansions. For T_{LH}, Fujikoshi (1997) proposed that

$$\tilde{T}_0^2 = \frac{r+2}{p+q+1}\left\{n + \frac{1}{2}(q-p-1)\right\}\log\left(1 + \frac{p+q+1}{r+2}\,\mathrm{tr}\,\mathbf{S}_h\mathbf{S}_e^{-1}\right)$$

as such a transformation. The problem could be considered in the more general framework

$$\Pr\{T \le x\} = G_f(x) + \frac{1}{n}\sum_{j=1}^{k} a_j G_{f+2j}(x) + \mathrm{O}(n^{-2}), \tag{5.7.8}$$

where k is a positive integer and the coefficients a_j are typically constants not depending on $n\,(>0)$. For $-2\log$ likelihood ratio criteria, we have $k = 1$. For most test statistics with a chi-square approximation, $k \le 3$. A polynomial transformation under which the terms of $\mathrm{O}(n^{-1})$ in its asymptotic expansion disappear is given [e.g., see Cordeiro and Ferrari, 1991] by

$$T^* = T\left(1 - \frac{1}{n}\sum_{j=1}^{k} c_j T^{j-1}\right),$$

where $c_j = 2\sum_{\ell=j}^{k} a_\ell / \prod_{\ell=j}^{j-1}(f + 2\ell)$. Fujikoshi (1997) gave a monotone transformation for $k = 2, 3$. For a general k, Kakizawa (1996) and Fujisawa (1997) proposed independently different transformations. For the case $k = 3$, Cordeiro and Ferrari (1998) proposed an alternative transformation. The transformation due to Kakizawa (1996), whose idea is based on Hall's (1992) Hall (1992) approach for the case of normal approximations, is given by

$$\tilde{T} = T^* + n^{-2}A(T),$$

where

$$\begin{aligned} A(x) &= \frac{1}{4}\int_0^x \left\{\frac{d}{dx}\left(\sum_1^k c_i x^i\right)\right\}^2 dx \\ &= \frac{1}{4}\int_0^x \left(\sum_1^k c_i i x^{i-1}\right)^2 dx \\ &= \frac{1}{4}\sum_{i,j=1}^{k} \frac{c_i c_j i j}{i+j-1} x^{i+j-1}. \end{aligned}$$

This transformation is applicable for a general k. However, it is a $(2k-1)$th-order polynomial, though T^* is a kth-order polynomial. Assume that U has asymptotically chi-square distribution $\chi^2(q)$. Then, if we can find a monotone transformation $T = T(U)$ such that

$$\Pr\{T \le x\} = G_q(x) + \mathrm{O}(n^{-2}),$$

we say that T has an improved chi-square approximation. It may be noted that as in the above-mentioned results, there are many statistics having an improved chi-square approximation. However, the number of papers on explicit error bounds is relatively small compared to those on improved chi-square approximations. For such error bounds, see, e.g., Ulyanov and Fujikoshi (2001).

5.8 Bootstrap Approximations

The bootstrap method was introduced by Efron (1979) as a method for estimating the sampling distribution of a statistical distribution. The method has been applied to various statistical procedures and their evaluations related to point estimators, confidence intervals, and significant tests [see, e.g., Hinkley, 1988; Diciccio and Romano, 1988; Hall. 1988; Babu, 1989; Konishi, 1991].

In the following we consider a univariate case, but it is straightforward to extend it to a multivariate case. Let $X_1, \ldots, X_n$ be a random sample from a population with the distribution function F, and set

$$\mathrm{E}(X_1) = \mu = \mu(F), \quad \mathrm{Var}(X_1) = \sigma^2,$$
$$\mathrm{E}[\{(X_1 - \mu)/\sigma\}^i] = \alpha_i, \quad i = 3, 4, \ldots.$$

Let $\boldsymbol{x} = (x_1, \ldots, x_n)'$ be the value observed for $\boldsymbol{X} = (X_1, \ldots, X_n)'$. Our interest is to approximate the distribution function of $T = T(\boldsymbol{X}, F)$,

$$G(x) = G_T(x, F) = \Pr\{T(\boldsymbol{X}, F) \leq x\}. \tag{5.8.1}$$

A typical statistic $T(\boldsymbol{X}, F)$ may be, for example,

$$T_1 = T_1(\boldsymbol{X}, F) = \sqrt{n}(\bar{X}_n - \mu)/\sigma,$$
$$T_2 = T_2(\boldsymbol{X}, F) = \sqrt{n}(\bar{X}_n - \mu)/S,$$
$$T_3 = T_3(\boldsymbol{X}, F) = (\hat{\theta} - \theta)/\hat{\tau},$$

where

$$\bar{X} = \frac{1}{n}\sum_{j=1}^{n} X_j, \quad S^2 = \frac{1}{n}\sum_{j=1}^{n}(X_j - \bar{X})^2$$

and $\hat{\theta}$ and $\hat{\tau}^2$ are estimators of unknown parameter θ and τ^2, respectively.

The bootstrap approximation is based on a bootstrap resample drawn by sampling with replacement from the original sample $\mathcal{X} = \{x_1, \ldots, x_n\}$. More theoretically, let F_n be the empirical distribution function; that is,

$$F_n(x) = \frac{1}{n}\sharp\{x_i : x_i \leq x\},$$

where $\sharp\{\cdot\}$ denotes the number of elements of the set $\{\cdot\}$, and let $X_1^*, \ldots, X_n^*$ be a random sample from a population with the distribution function F_n. Note that the mean and variance of X_1^*, which are denoted by E^* and Var^*, respectively, are

$$\mathrm{E}^*(X_1^*) = \mathrm{E}_{F_n}(X_1^*|\mathcal{X}) = \bar{x} = \frac{1}{n}\sum_{j=1}^n x_j,$$

$$\mathrm{Var}^*(X_1^*) = \mathrm{Var}_{F_n}(X_1^*|\mathcal{X}) = S_x^2 = \frac{1}{n}\sum_{j=1}^n (x_j - \bar{x})^2.$$

Put $\boldsymbol{X}^* = (X_1^*, \cdots, X_n^*)'$ and

$$T^* = T(\boldsymbol{X}^*, F_n),$$

which is a bootstrap version of T. For the statistics above,

$$\begin{aligned} T_1^* &= T_1(\boldsymbol{X}^*, F_n) = \sqrt{n}(\bar{X}_n^* - \bar{x})/S, \\ T_2^* &= T_2(\boldsymbol{X}^*, F_n) = \sqrt{n}(\bar{X}_n^* - \bar{x})/S^*, \\ T_3^* &= T_3(\boldsymbol{X}^*, F_n) = (\hat{\theta}^* - \hat{\theta})/\hat{\tau}^*, \end{aligned}$$

where

$$\bar{X}^* = \frac{1}{n}\sum_{j=1}^n X_j^*, \quad (S^*)^2 = \frac{1}{n}\sum_{j=1}^n (X_j^* - \bar{X}^*)^2,$$

and $\hat{\theta}^*$ and $\hat{\tau}^*$ are bootstrap versions of these estimators obtained from $\hat{\theta}$ and $\hat{\tau}$ by replacing $\boldsymbol{X}$ by $\boldsymbol{X}^*$. Now we can write the bootstrap approximation for $G(x)$ by

$$\hat{G}(x) = \Pr{}^*\{T^* \le x|\mathcal{X}\}. \tag{5.8.2}$$

Here $\Pr^*$ denotes the (conditional) probability distribution of $\boldsymbol{X}^*$ given $\boldsymbol{X} = \boldsymbol{x}$ or $\mathcal{X}$. Note that the right-hand side of (5.8.2) is computed numerically for a bootstrap resample. In fact, we draw a sample of size n with replacement from $(x_1, \ldots, x_n)$. We denote it by $x_{11}^*, \ldots, x_{1n}^*$. We obtain the value t_1^* of T^* based on these observations. We repeat this B times, giving $t_1^*, \ldots, x_B^*$. Then we can approximate $\hat{G}$ by

$$\hat{G}(x) = \frac{1}{B}\sharp\{t_i^* : t_i^* \le x\}.$$

Suppose that $T = T(\boldsymbol{X}, F)$ has an Edgeworth expansion such that

$$G(x) = \Phi_{k,n}(x) + \mathrm{O}(n^{-k/2}), \tag{5.8.3}$$

where

$$\Phi_{k,n}(x) = \Phi(x) + \varphi(x)\sum_{j=1}^{k-1} n^{-j/2}p_j(x).$$

Such expansions are obtained (see, e.g., Sections 5.1 through 5.3) for a class of statistics under appropriate regularity conditions. The polynomials $p_1(x)$ and $p_2(x)$ are given as follows:

The case T_1:

$$p_1(x) = -\frac{1}{6}\alpha_3(x^2 - 1),$$
$$p_2(x) = -\frac{1}{72}\alpha_3^2(x^5 - 10x^3 + 15x) - \frac{1}{24}(\alpha_4 - 3)(x^3 - x).$$

The case T_2:

$$p_1(x) = \frac{1}{6}\alpha_3(2x^2 + 1),$$
$$p_2(x) = -\frac{1}{18}\alpha_3^2(x^5 + 2x^3 - 3x) + \frac{1}{12}(\alpha_4 - 3)(x^3 - 3x) - \frac{1}{4}(x^3 + 3x).$$

The case T_3: p_1 has the form

$$p_1(x) = \beta + \gamma x^2.$$

Then it is known [Singh, 1981; Babu and Singh, 1984; Bhattacharya, 1987] that

$$\hat{G}(x) = \hat{\Phi}_{k,n}(x) + \mathrm{O}(n^{-k/2}), \tag{5.8.4}$$

where $\hat{\Phi}_{k,n}(x)$ is defined from $\Phi_{k,n}$ by replacing the population moments of X_1 by the corresponding sample moments:

$$\hat{\mu} = \bar{x},\ \hat{\sigma} = s,$$
$$\hat{\alpha}_i = \sum_{j=1}^{n}\left(\frac{x_j - \bar{x}}{s_x}\right)^i, \quad i = 3, 4, \ldots.$$

The approximation $\hat{\Phi}_{k,n}(x)$ is called an *empirical Edgeworth expansion.*

When we approximate G by its bootstrap approximation $\hat{G}$, we are interested in various types of errors:

$$\begin{aligned} R_T(x) &= R_T(x|\boldsymbol{x}) = G(x) - \hat{G}(x), \\ B_T(x) &= \mathrm{E}\{R_T(x|\boldsymbol{X})\}, \\ M_T(x) &= \mathrm{E}[\{R_T(x|\boldsymbol{X})\}^2]. \end{aligned} \tag{5.8.5}$$

These are important measures for the accuracy of bootstrap approximations. In general, it holds that

$$\begin{aligned} |R_T(x)| &= |G(x) - \Phi_{k,n}(x) + \hat{\Phi}_{k,n}(x) - \hat{\Phi}_{k,n}(x) + \hat{\Phi}_{k,n}(x) - \hat{G}(x)| \\ &\le |G(x) - \Phi_{k,n}(x)| + |\hat{\Phi}_{k,n}(x) - \hat{G}(x)| + |\hat{\Phi}_{k,n}(x) - \Phi_{k,n}(x)| \\ &= \mathrm{O}(n^{-k/2}) + |\Phi_{k,n}(x) - \hat{\Phi}_{k,n}(x)|. \end{aligned}$$

Suppose that an uniform error bound is available for the Edgeworth approximation $\Phi_{k,n}$ of $G(x)$; that is,

$$|G(x) - \Phi_{k,n}(x)| \le C_k/n^{k/2}, \tag{5.8.6}$$

where C_k is a known function of the population moments of X_1. Then we have

$$|R_T(x)| \le \frac{1}{n^{k/2}}\{\hat{C}_k + C_k\} + |\Phi_{k,n}(x) - \hat{\Phi}_{k,n}(x)|, \tag{5.8.7}$$

where $\hat{C}_k$ is obtained by replacing the population moments of X_1 by the corresponding sample moment. For the statistic T_1, we have a Berry Esseen bound. The second term on the right-hand of (5.8.9) with $k = 1$ disappears. Therefore, we have

$$\sup_x |R_{T_1}(x)| \le \frac{0.77}{\sqrt{n}}\{\alpha_3 + \hat{\alpha}_3\}. \tag{5.8.8}$$

Further, under appropriate regularity conditions we have an error bound for the first-order Edgeworth expansion of the distribution of T_1. Therefore, we can write (5.8.9) with $k = 2$ as

$$\begin{aligned}\sup_x |R_{T_1}(x)| &\le \frac{1}{n}\{C_2 + \hat{C}_2\} + \frac{1}{6\sqrt{n}} \sup |(x^2 - 1)\varphi(x)|\hat{\alpha}_3 - \alpha_3| \\ &= \frac{1}{n}\{C_2 + \hat{C}_2\} + \frac{1}{6\sqrt{n}} \cdot \frac{1}{\sqrt{2\pi}}|\hat{\alpha}_3 - \alpha_3|. \end{aligned} \tag{5.8.9}$$

5.9 High-Dimensional Approximations

5.9.1 Limiting Spectral Distribution

Let $\ell_1 \ge \cdots \ge \ell_p$ be the characteristic roots of a $p \times p$ symmetric matrix $\mathbf{S}$ depending on sample size or its modification n. Then we consider the empirical distribution function on the roots defined by

$$F_n(x) = \frac{1}{p}\sharp\{\ell_i : \ell_i \le x\}, \tag{5.9.1}$$

which is called the *spectral distribution* of the matrix $\mathbf{S}$, where $\sharp\{\cdot\}$ denotes the number of elements of the set $\{\cdot\}$. If the empirical distribution function $F_n(x)$ converges to F as $p \to \infty$, F is called the *limiting spectral distribution* (LSD) *function* of $\mathbf{S}$.

Now we define $\mathbf{S}$ as

$$\mathbf{S} = \frac{1}{n}\sum_{j=1}^{n} \boldsymbol{X}_j \boldsymbol{X}_j', \tag{5.9.2}$$

where $\boldsymbol{X}_j = (X_{1j}, \ldots, X_{pj})'$ and X_{ij} are i.i.d. random variables with mean 0 and variance σ^2. In spectral analysis of a high-dimensional random matrix, the sample covariance matrix is defined by (5.9.2), not by the usual sample covariance matrix $\mathbf{S} = (1/n)\sum_{j=1}^{n}(\boldsymbol{X}_j - \bar{\boldsymbol{X}})(\boldsymbol{X}_j - \bar{\boldsymbol{X}})'$. This is not essential under the assumption of normality, but is essential under the assumption of nonnormality.

Much of the mathematical groundwork has already been laid out by research into the spectral theory of large-dimensional random matrices. The LSD of $\mathbf{S}$ was first obtained by Marčenko and Pastur (1967). Subsequent work was done by Jonsson (1982), Yin (1986), Bai et al. (1987), and others. For a review of LSD, see Bai (1999).

Theorem 5.9.1 *Let F_n be the spectral distribution of the sample covariance matrix defined by (5.9.2). Suppose that $p/n \to c \in (0, \infty)$. Then*

$$F_n \xrightarrow{\text{a.s.}} F,$$

where $\xrightarrow{\text{a.s.}}$ denotes convergence almost surely. If $0 < c < 1$,

$$f(x) = \frac{dF(x)}{dx} = \begin{cases} \dfrac{1}{2\pi c x \sigma^2}\sqrt{(x-a)(b-x)}, & a < x < b, \\ 0, & \text{otherwise}, \end{cases}$$

and the constants a and b are given by $a = (1-\sqrt{c})^2\sigma^2$ and $b = (1+\sqrt{c})^2\sigma^2$, respectively. If $1 < c < \infty$,

$$f(x) = \frac{dF(x)}{dx} = \begin{cases} \dfrac{1}{2\pi c x \sigma^2}\sqrt{(x-a)(b-x)}, & a < x < b, \\ 1 - 1/c, & x = 0, \\ 0, & \text{otherwise}, \end{cases}$$

The LSD is useful in obtaining the asymptotic behavior of a symmetric function of the characteristic roots of $\mathbf{S}$. In fact, we have

$$\begin{aligned} T_n &\equiv \frac{1}{p}\{g(\ell_1) + \cdots + g(\ell_p)\} \\ &= \int_0^\infty g(x)\mathrm{d}F_n(x). \end{aligned} \tag{5.9.3}$$

Then, under appropriate regularity conditions we can see that T_n converges to

$$\int_0^\infty g(x)\mathrm{d}F(x) \tag{5.9.4}$$

under a high-dimensional framework.

Outline of Derivation. Let us denote the spectral distribution of the matrix $\mathbf{S}$ by $F_n^{(\mathrm{S})}$. Then the hth moment of $F_n^{(\mathrm{S})}$ given $\mathbf{S}$ can be written as

$$\begin{aligned}\beta_h(\mathbf{S}) &= \int x^h dF_n^{(\mathrm{S})}(x)\,\mathrm{d}x \\ &= \frac{1}{p}(\ell_1^h + \cdots + \ell_p^h) \\ &= \frac{1}{p}\operatorname{tr}\mathbf{S}^h.\end{aligned}$$

To show that $F_n^{(\mathrm{S})}$ tends to a limit, say F, the moment convergence theorem is employed. This is to show that

$$\beta_h(\mathbf{S}) \to \beta_h = \int x^h dF(x)$$

and Carleman's condition $\sum_{h=1}^{\infty}\beta_{2k}^{-1/(2h)} = \infty$. These proofs are carried out by estimating the mean, variance, or higher moments of $p^{-1}\operatorname{tr}\mathbf{S}^h$.

The LSD has been studied for a ratio matrix of two matrices. Let $\mathbf{S}_1$ and $\mathbf{S}_2$ be independently distributed as Wishart distributions $\mathrm{W}_p(m, \mathrm{I}_p)$ and $\mathrm{W}_p(n, \mathrm{I}_p)$, respectively. Assume that

$$p/m \to c_1 > 0 \quad \text{and} \quad p/n \to c_2 \in \left(0, \frac{1}{2}\right). \tag{5.9.5}$$

Then the LSD of $\mathbf{F} = \mathbf{S}_1\mathbf{S}_2^{-1}$ was also studied in the literature [see Bai, 1999].

The largest characteristic root of $\mathbf{S}$ has been also studied. Geman (1980) proved the almost sure ($a.s.$) convergence:

$$n^{-1}\ell_1 \overset{a.s.}{\to} (1+\sqrt{c})^2,$$

that is, $\ell_1 \sim (\sqrt{n}+\sqrt{p})^2$. When $\mathbf{S}$ is distributed as $W_p(n, \mathrm{I}_p)$, the limiting distribution of ℓ_1 was derived by Johnstone (2001). The center and scaling constants are defined by μ_{np} and σ_{np}, respectively, where

$$\begin{aligned}\mu_{np} &= (\sqrt{n-1}+\sqrt{p})^2, \\ \sigma_{np} &= (\sqrt{n-1}+\sqrt{p})\left(\frac{1}{\sqrt{n-1}}+\frac{1}{\sqrt{p}}\right)^{1/3}.\end{aligned}$$

5.9.2 Central Limit Theorem

Let $\mathbf{S}$ be the sample covariance matrix based on a sample of size $N = n+1$ from $\mathrm{N}_p(\boldsymbol{\mu}, \Sigma)$. Then it is basic to study asymptotic distribution of $(\operatorname{tr}\mathbf{S}, \operatorname{tr}\mathbf{S}^2)$ under a high-dimensional framework.

The exact sense in which sample size and dimensionality go to infinity together is defined [Ledoit and Wolf, 2002] by the following assumptions.

Assumption 1 (Asymptotics) Dimensionality and sample size are two increasing integer functions $p = p_k$ and $n = n_k$ of index $k = 1, 2, \ldots$ such that $\lim_{k\to\infty} p_k = \infty$, $\lim_{k\to\infty} n_k = \infty$ and there exists $c \in (0, \infty)$ such that $\lim_{k\to\infty} p_k/n_k = c$.

Assumption 2 (Data-generating process) For each positive integer k, $\mathbf{X}_k$ is an $(n_k+1)\times p_k$ matrix of n_k+1 i.i.d. observations on a system of p_k random variables that are jointly normally distributed with mean vector $\boldsymbol{\mu}_k$ and covariance matrix Σ_k. Let $\lambda_{1,k}, \ldots, \lambda_{p_k,k}$ be the characteristic roots of the covariance matrix Σ_k. Assume that the average $\lambda = \sum_{i=1}^{p_k} \lambda_{i,k}/p_k$ and dispersion $\delta^2 = \sum_{i=1}^{p_k}(\lambda_{i,k} - \lambda)^2/p_k$ are independent of the index k. Furthermore, we require that $\lambda > 0$.

Assumption 3 (Higher moments) The averages of the third and fourth moments of the characteristic roots of the population covariance matrix

$$\sum_{i=1}^{p_k} \frac{(\lambda_{i,k})^j}{p_k}, \qquad j = 3, 4$$

converge to finite limits, respectively.

In the following, we omit dependence on k. Then assumptions 1 through 3 are stated as:

$$\begin{aligned}
&\text{Assumption 1:} \quad p/n \;\to\; c \in (0, \infty),\\
&\text{Assumption 2:} \quad \frac{1}{p}\operatorname{tr}\Sigma^k = \mathrm{O}(1), \; k = 1, 2,\\
&\text{Assumption 3:} \quad \frac{1}{p}\operatorname{tr}\Sigma^k = \mathrm{O}(1), \; k = 3, 4.
\end{aligned}$$

Let $\lambda_1 \geq \cdots \geq \lambda_p$ be the characteristic roots of Σ and let

$$\frac{1}{p}\sum_{i=1}^{p} \lambda_i = \alpha, \qquad \frac{1}{p}\sum_{i=1}^{p}(\lambda_i - \alpha)^2 = \delta^2.$$

Let $\mathbf{S}$ be a sample covariance matrix based on a sample of size $N = n+1$ from $\mathrm{N}_p(\boldsymbol{\mu}, \Sigma)$. Two fundamental results on asymptotic behavior of $\operatorname{tr}\mathbf{S}$ and $\operatorname{tr}\mathbf{S}^2$ are given as follows:

Theorem 5.9.2 (Law of large numbers) *Under assumptions 1 through 3,*

$$\begin{aligned}
&\frac{1}{p}\operatorname{tr}\mathbf{S} \xrightarrow{p} \alpha.\\
&\frac{1}{p}\operatorname{tr}\mathbf{S}^2 \xrightarrow{p} (1+c)\alpha + \delta^2.
\end{aligned}$$

Theorem 5.9.3 (Central limit theorem) *Under assumptions 1 and 2, if* $\delta^2 = 0$, *then*

$$n \times \begin{pmatrix} \dfrac{1}{p}\operatorname{tr}\mathbf{S} - \alpha \\ \dfrac{1}{p}\operatorname{tr}\mathbf{S}^2 - \dfrac{n+p+1}{n}\alpha^2 \end{pmatrix} \xrightarrow{d} \mathrm{N}\left(\begin{pmatrix} 0 \\ 0 \end{pmatrix}, \begin{pmatrix} \gamma_{11} & \gamma_{12} \\ \gamma_{21} & \gamma_{22} \end{pmatrix}\right),$$

where $\xrightarrow{d}$ *denotes convergence in distribution, and*

$$\gamma_{11} = 2c^{-1}\alpha^2, \quad \gamma_{12} = \gamma_{21} = 4(1 + c^{-1})\alpha^3,$$
$$\gamma_{22} = 4(2c^{-1} + 5 + 2c)\alpha^4.$$

For a proof of Theorem 5.9.3, see Yin and Krishnaiah (1983) or Ledoit and Wolf (2002). The asymptotic normality was shown by Jonsson (1982), who treated the joint distribution of $\operatorname{tr}\mathbf{S}^i, i = 1, \ldots, k$. The asymptotic covariance matrix with $k = 2$ was given by Ledoit and Wolf (2002). In the following we give an outline of the asymptotic normality along the lines of Jonsson (1982).

Outline of Asymptotic Normality. Without loss of generality we assume that $\lambda = 1$. Then let

$$\mathbf{W} = n\mathbf{S} = (W_{ij}) \quad \text{and} \quad W_{ij} = \sum_{k=1}^{n} X_{ki}X_{kj},$$

where X_{ij}, $i = 1, \ldots, n$, $j = 1, \ldots, p$ are independently and identically distributed according to the standard normal distribution. Put

$$Y_i = \frac{n^i}{(n+p)^i}\left[\operatorname{tr}(\mathbf{W}^i) - \mathrm{E}\left\{\operatorname{tr}(\mathbf{W}^i)\right\}\right], \quad i = 1, \ldots, k.$$

We consider asymptotic distribution of the k-dimensional random vector $\boldsymbol{Y} = (Y_1, \ldots, Y_k)'$ under A0, based on the moment method. Consider the mixed moments

$$\mathrm{E}(Y_1^{s_1} \cdots Y_k^{s_k}),$$

where $s_1, \ldots, s_k$ are nonnegative integers. Put

$$s = s_1 + \cdots + s_k.$$

Under the high-dimensional framework A0,

$$\mathrm{E}(Y_1^{s_1} \cdots Y_k^{s_k}) \to \begin{cases} 0, & s \text{ is odd}, \\ \sum \sigma_{m_1m_2} \cdots \sigma_{m_{s-1}m_s}, & s \text{ is even}, \end{cases}$$

where the notation Σ stands for the summation over all partitions of s_1 1's, ..., s_k k's into $(m_1, m_2), \ldots, (m_{s-1}, m_s)$ [see, e.g., Jonsson, 1982]. Here

σ_{ij} denotes the covariance of the ith element Z_i and the jth Z_j of $\boldsymbol{Z} = (Z_1, \ldots, Z_k)'$, which is distributed according to the k-variate normal distribution. This leads to

$$\begin{aligned}\mathrm{Var}(Y_1^{s_1} \cdots Y_k^{s_k}) &= \mathrm{E}(Y_1^{2s_1} \cdots Y_k^{2s_k}) - \{\mathrm{E}(Y_1^{s_1} \cdots Y_k^{s_k})\}^2 \\ &\to \sum \sigma_{m_1 m_2} \cdots \sigma_{m_{s-1} m_s} \sigma_{m_{s+1} m_{s+2}} \cdots \sigma_{m_{2s-1} m_{2s}} \\ &\quad - \left\{\sum \sigma_{m_1 m_2} \cdots \sigma_{m_{s-1} m_s}\right\}^2 .\end{aligned}$$

As a result,

$$Y_1^{s_1} \cdots Y_k^{s_k} \xrightarrow{p} \begin{cases} 0, & s \text{ is odd}, \\ \sum \sigma_{m_1 m_2} \cdots \sigma_{m_{s-1} m_s}, & s \text{ is even}. \end{cases} \tag{5.9.6}$$

We need to show that the moment sequence $\{\sum \sigma_{m_1 m_2} \cdots \sigma_{m_{s-1} m_s}\}$ determines the k-variate normal distribution uniquely. But this is obvious from the characteristic function. High-dimensional approximation of $\mathrm{E}[\mathrm{tr}(\mathbf{W}^i)/p]$ is given as

$$\mathrm{E}\left[\frac{1}{p}\mathrm{tr}(\mathbf{W}^i)\right] \to \sum_{j=1}^{i} c^{i-j} \sum_{\substack{n_1 + \cdots + n_j = i-j+1 \\ n_1 + \cdots + j n_j = i}} \frac{i!}{n_1! \cdots n_j! j!}.$$

5.9.3 Martingale Limit Theorem

Recently, high-dimensional asymptotic distributions of some test statistics have been obtained by expressing them as sums of martingale differences and using the martingale limit theorem as in Schott (2005), etc.

Definition 5.9.1 Let $\{X_n\}$ be a sequence of random variables and $\mathcal{F}_n = \{X_1, \ldots, X_n\}$. Then, the sequence $\{X_n\}$ is called a *martingale* if

(1) $\mathrm{E}|X_n| < \infty$.

(2) $\mathrm{E}(X_{n+1}|\mathcal{F}_n) = X_n$.

Especially, it is called a *martingale difference* if

(1) $\mathrm{E}|X_n| < \infty$.

(2) $\mathrm{E}(X_{n+1}|\mathcal{F}_n) = 0$.

Each X_n is also called a martingale difference.

As an example expressed as a sum of martingale differences, consider a test statistic for complete independence based on a sample of size $N = n + 1$ from $\mathrm{N}_p(\boldsymbol{\mu}, \Sigma)$. Let $\mathbf{S} = (S_{ij})$ be the sample covariance matrix and $\mathbf{R} = (R_{ij})$

and $\mathrm{R} = (\rho_{ij})$ be the sample and population correlation matrices. Consider the null distribution of a test statistic

$$T_{np} = \sum_{i=2}^{p} \sum_{j=1}^{i-1} R_{ij}^2 - \frac{p(p-1)}{2n} \tag{5.9.7}$$

for $H_0: \rho_{ij} = 0 (i > j)$. This statistic can be expressed as a sum of martingale differences under assumption 1. For details, see Chapter 8 or Schott (2005).

The central limit theorem for a sum of martingale differences is given in the following theorem. For a proof, see Hall and Heyde (1980).

Theorem 5.9.4 *Let $\{X_{n_k}\}$ be the martingale difference $\mathcal{F}_{n_k} = \{X_1, \ldots, X_{n_k}\}$ and*

$$\gamma^2 = \lim_{k \to \infty} \sum_{i=1}^{n_k} \mathrm{Var}(S_i).$$

Suppose that:

(1) *For all $\varepsilon > 0$.*

$$\sum_{i=1}^{n_k} \mathrm{E}\{X_i^2 I(|X_i| > \varepsilon) | \mathcal{F}_{i-1}\} \xrightarrow{p} 0,$$

(2) $\sum_{i=1}^{n_k} \mathrm{E}\{X_i^2 | \mathcal{F}_{i-1}\} \xrightarrow{p} \gamma^2$, *where* $\xrightarrow{p}$ *denotes convergence in probability.*

Then

$$\sum_{i=1}^{n_k} X_i \xrightarrow{d} \mathrm{N}(0, \gamma^2),$$

where $\xrightarrow{d}$ *denotes convergence in the distribution.*

5.9.4 Geometric Representation

Let $\{\boldsymbol{X}_1, \ldots, \boldsymbol{X}_n\}$ be a random sample of size n from a p-dimensional population, which can be regarded as a set of n-vectors or points in the p-dimensional space R^p. There has been attracted a considerable interest for data sets of high-dimension, i.e., the cases when p is large. In high-dimensional asymptotic theory it is assumed that either

(1) both p and n tend to infinity, or

(2) p tends to infinity while n remains fixed (high-dimensional low sample size– HDLSS).

In this geometric interpretation, it is basic to study asymptotic behavior of the following three statistics:

$\|\boldsymbol{X}_i\|$: the length of a p-dimensional vector,

$\|\boldsymbol{X}_i - \boldsymbol{X}_j\|$: the distance of two independent vectors, and
$\mathrm{ang}(\boldsymbol{X}_i, \boldsymbol{X}_j)$: the angle between two vectors at the mean vector.

where $\|\cdot\|$ and O_p are the Euclidean distance and the stochastic order in probability, respectively.

Hall et al. (2005) obtained asymptotic values of the three statistics for case (2): assuming that the underlying population is $\mathrm{N}_p(0, \mathrm{I}_p)$, these statistics satisfy the following:

$$\begin{aligned}
\|\boldsymbol{X}_i\| &= \sqrt{p} + \mathrm{O}_p(1), \quad i = 1, \ldots, n, \\
\|\boldsymbol{X}_i - \boldsymbol{X}_j\| &= \sqrt{2p} + \mathrm{O}_p(1), \quad i, j = 1, \ldots n, \ i \neq j, \\
\mathrm{ang}(\boldsymbol{X}_i, \boldsymbol{X}_j) &= \frac{1}{2}\pi + \mathrm{O}_p(p^{-1/2}), \quad i, j = 1, \ldots n, \ i \neq j,
\end{aligned}$$

This implies, in particular, that the observed data approach in probability to the vertices of a deterministic regular simplex. This property was extended to non-normal cases with some additional assumptions. The result was also extended to two sample problem and the performance of some discrimination rules was studied.

Problems

5.1 Suppose that the distribution of X is absolutely continuous; that is, the distribution function $F(x)$ of X can be expressed as

$$F(x) \equiv \Pr\{X \leq x\} = \int_{-\infty}^{x} f(u)\,\mathrm{d}u.$$

In general, it is known that for any $\varepsilon > 0$ there exists a step function $f_\varepsilon(x) = \sum_{j=1}^{p} b_j I_{(a_j, a_{j+1})}$ such that

$$\int_{-\infty}^{\infty} |f(x) - f_\varepsilon(x)|\,\mathrm{d}x < \varepsilon.$$

Using the result above, show that the characteristic function $C(t)$ of X satisfies $\lim_{|t|\to\infty} C(t) = 0$.

Hint:

$$\begin{aligned}
|C(t)| &= \left| \int_{-\infty}^{\infty} \mathrm{e}^{itx} f(x)\,\mathrm{d}x \right| \\
&\leq \left| \int_{-\infty}^{\infty} |f(x) - f_\varepsilon(x)|\,\mathrm{d}x \right| + \left| \int_{-\infty}^{\infty} \mathrm{e}^{itx} f_\varepsilon(x)\,\mathrm{d}x \right| \\
&\leq \varepsilon + \sum_{j=1}^{p} \left| \int_{a_{j+1}}^{a_j} \mathrm{e}^{itx} b_j\,\mathrm{d}x \right| \leq \varepsilon + 2\sum_{j=1}^{p} \left| \frac{b_j}{t} \right|.
\end{aligned}$$

5.2 If the distribution of X has an absolutely continuous component, X satisfies Cramér's condition.

Hint: From the decomposition theorem we can write $F(x) = p_1 F_a(x) + p_2 F_d(x) + p_3 F_s(x)$, where $p_1 > 0$ and $p_1 + p_2 + p_3 = 1$, F_a is absolutely continuous, F_d is a step function, and F_s is a continuous function.

5.3 Show that for any polynomial $p(x)$,

$$\frac{1}{2\pi}\int_{-\infty}^{\infty} \exp(-itu)p(it)\exp\left(-\frac{1}{2}t^2\right) dt = p(-d/dx)\phi(x).$$

Hint:

$$\begin{aligned} p(it) &= a_0 + a_1 \times (it) + \cdots + a_p \times (it)^p, \\ I(p(it)) &= a_0 + a_1 I(it) + \cdots + a_p I((it)^p). \end{aligned}$$

5.4 Derive the following expansions:

$$\begin{aligned} (\mathrm{I}_p + \varepsilon \mathrm{V})^{-1} &= \mathrm{I}_p - \varepsilon \mathrm{V} + \varepsilon^2 \mathrm{V}^2 - \varepsilon^3 \mathrm{V}^3 + \varepsilon^4 \mathrm{V}^4 - \varepsilon^5 \mathrm{V}^5 (\mathrm{I} + \varepsilon \mathrm{V})^{-1}, \\ (\mathrm{I}_p + \varepsilon \mathrm{V})^{1/2} &= \mathrm{I}_p + \frac{1}{2}\varepsilon \mathrm{V} - \frac{1}{8}\varepsilon^2 \mathrm{V}^2 + \frac{1}{16}\varepsilon^3 \mathrm{V}^3 + \cdots, \\ (\mathrm{I}_p + \varepsilon \mathrm{V})^{-1/2} &= \mathrm{I}_p - \frac{1}{2}\varepsilon \mathrm{V} + \frac{3}{8}\varepsilon^2 \mathrm{V}^2 - \frac{5}{16}\varepsilon^3 \mathrm{V}^3 + \cdots. \end{aligned}$$

5.5 Let

$$\mathrm{V} = \mathrm{V}^{(1)} + \varepsilon \mathrm{V}^{(2)} + \cdots = \begin{pmatrix} \mathrm{V}_{[11]} & \mathrm{V}_{[12]} \\ \mathrm{V}_{[21]} & \mathrm{V}_{[22]} \end{pmatrix},$$

where $\mathrm{V}_{[22]}$ is a $q_r \times q_r$ matrix and

$$\Delta = \begin{pmatrix} \theta_{1r}\mathrm{I}_{q_1} & \cdots & \mathrm{O} \\ \vdots & \ddots & \vdots \\ \mathrm{O} & \cdots & \theta_{r-1,r}\mathrm{I}_{q_r} \end{pmatrix}.$$

Then show that W_r in Theorem 5.5.2 can be written as

$$\begin{aligned} \mathrm{W}_r &= \theta_r \mathrm{I}_r + \varepsilon \mathrm{V}_{[22]} - \varepsilon^2 \mathrm{V}_{[21]}\Delta \mathrm{V}_{[12]} \\ &\quad + \varepsilon^3 \left\{ \mathrm{V}_{[21]}\Delta \mathrm{V}_{[11]}\Delta \mathrm{V}_{[12]} - \frac{1}{2}\mathrm{V}_{[21]}\Delta \mathrm{V}_{[12]}\mathrm{V}_{[22]} - \frac{1}{2}\mathrm{V}_{[21]}\Delta \mathrm{V}_{[12]} \right\} \\ &\quad + \mathrm{O}(\varepsilon^4). \end{aligned}$$

5.6 Let $n\mathbf{S} \sim \mathrm{W}_p(n, \Sigma)$ and let $\mathbf{U} = \sqrt{n}(\mathbf{S} - \Sigma)$. Show that for any $p \times p$

symmetric matrices A, B, and C such that they are mutually commutable,

$$\begin{aligned}
\mathrm{E}[\operatorname{etr}(it\mathrm{A}\mathbf{U})] &= \exp\{-t^2 \operatorname{tr}(\mathrm{A}\Sigma)^2\}\Big[1 + \frac{4}{3\sqrt{n}}(it)^3 \operatorname{tr}(\mathrm{A}\Sigma)^3 \\
&+ \frac{(it)^4}{n}\Big\{2\operatorname{tr}(\mathrm{A}\Sigma)^4 + \frac{8}{9}(it)^2(\operatorname{tr}(\mathrm{A}\Sigma)^3)^2\Big\} + \mathrm{O}(n^{-3/2})\Big], \\
\mathrm{E}[\{\operatorname{etr}(it\mathrm{A}\mathbf{U})\}\operatorname{tr}(\mathbf{BUCU})] &= \exp\{-t^2 \operatorname{tr}(\mathrm{A}\Sigma)^2\}[\operatorname{tr}(\mathrm{BC}\Sigma^2) \\
&+ \operatorname{tr}(\mathrm{B}\Sigma)\operatorname{tr}(\mathrm{C}\Sigma) + 4(it)^2 \operatorname{tr}(\mathrm{BC}A^2\Sigma^4) + \mathrm{O}(n^{-1/2})], \\
\mathrm{E}[\{\operatorname{etr}(it\mathrm{A}\mathbf{U})\}\operatorname{tr}(\mathbf{BU})\operatorname{tr}(\mathbf{CU})] &= \exp\{-t^2 \operatorname{tr}(\mathrm{A}\Sigma)^2\}[\operatorname{tr}(\mathrm{BC}\Sigma^2) \\
&+ 4(it)^2 \operatorname{tr}(\mathrm{BA}\Sigma^2)\operatorname{tr}(\mathrm{CA}\Sigma^2) + \mathrm{O}(n^{-1/2})].
\end{aligned}$$

5.7 Let $n\mathbf{S} \sim \mathrm{W}_p(n, \Sigma)$ and $\mathbf{U} = \sqrt{n}(\mathbf{S} - \Sigma)$. Let

$$\mathcal{U}_n = \{U : |\text{each element of } \Sigma^{-1/2}U\Sigma^{-1/2}| < k \cdot \log n\}.$$

Then show that $P(\mathcal{U}_n) = 1 + \mathrm{o}(n^{-k})$. [Anderson, 1973].

5.8 Assuming that $n\mathbf{S} \sim \mathrm{W}_p(n, \Sigma)$ and $\mathbf{U} = \sqrt{n}(\mathbf{S} - \Sigma)$, prove the following expressions:

$$\begin{aligned}
&\mathrm{E}(u_{ab}) = 0, \\
&\mathrm{E}(u_{ab}u_{cd}) = \sigma_{ac}\sigma_{bd} + \sigma_{ad}\sigma_{bc} \equiv \alpha(ab, cd), \\
&\mathrm{E}(u_{ab}u_{cd}u_{ef}) = n^{-1/2}\{\sigma_{ac}\sigma_{be}\sigma_{df} + \sigma_{ac}\sigma_{bf}\sigma_{de} + \sigma_{ad}\sigma_{be}\sigma_{cf} + \sigma_{ad}\sigma_{bf}\sigma_{ce} \\
&\qquad + \sigma_{ae}\sigma_{bc}\sigma_{df} + \sigma_{ae}\sigma_{bd}\sigma_{cf} + \sigma_{af}\sigma_{bc}\sigma_{de} + \sigma_{af}\sigma_{bd}\sigma_{ce}\} + \mathrm{O}(n^{-3/2}), \\
&\mathrm{E}(u_{ab}u_{cd}u_{ef}u_{gh}) = \alpha(ab, cd)\alpha(ef, gh) + \alpha(ab, ef)\alpha(cd, gh) \\
&\qquad + \alpha(ab, gh)\alpha(cd, ef) + \mathrm{O}(n^{-1}).
\end{aligned}$$

5.9 Put

$$I[g(\mathrm{T})] = (2\pi)^{-1/2p(p+1)} \int_{-\infty}^{\infty} \cdots \int_{-\infty}^{\infty} \operatorname{etr}\left(-i\mathrm{TU} - \frac{1}{2}\mathrm{T}^2\right) g(\mathrm{T}) \prod_{j \geq k} \mathrm{d}_{jk},$$

where $\mathrm{T} = ((1 + \delta_{jk})/2t_{jk})$ and U is a $p \times p$ symmetric matrix. Prove the following:

$$
\begin{aligned}
I[1] &= \left[\pi^{1/4p(p+1)}2^{1/2p}\right]^{-1} \operatorname{etr}\left(-\frac{1}{2}\mathrm{U}^2\right) \equiv \Psi(\mathrm{U}), \\
I[i^3 \operatorname{tr}(\mathrm{T}^3)] &= \left\{\operatorname{tr}(\mathrm{U}^3) - \frac{3}{2}(p+1)\operatorname{tr}(\mathrm{U})\right\}\Psi(\mathrm{U}), \\
I[i^4 \operatorname{tr}(\mathrm{T}^4)] &= \left\{\operatorname{tr}(\mathrm{U}^4) - [\operatorname{tr}(\mathrm{U})]^2 - (2p+3)\operatorname{tr}(\mathrm{U}^2)\right. \\
&\quad \left. + \frac{1}{4}p(2p^2+5p+5)\right\}\Psi(\mathrm{U}), \\
I[i^6\{\operatorname{tr}(\mathrm{T}^3)\}^2] &= \left\{[\operatorname{tr}(\mathrm{U}^3)]^2 - 3(p+1)\operatorname{tr}(\mathrm{U})\operatorname{tr}(\mathrm{U}^3) - 9\operatorname{tr}(\mathrm{U}^4)\right. \\
&\quad + \frac{9}{4}(p^2+2p+3)[\operatorname{tr}(\mathrm{U})]^2 + \frac{9}{2}(3p+4)\operatorname{tr}(\mathrm{U}^2) \\
&\quad \left. - \frac{3}{4}p(4p^2+9p+7)\right\}\Psi(\mathrm{U}).
\end{aligned}
$$

5.10 Let $\mathbf{Y}$ be a $p \times p$ symmetric random matrix with the pdf

$$
f(\mathrm{Y}) = \left[2^{1/2p}\pi^{1/4p(p+1)}\right]^{-1} \operatorname{etr}\left(-\frac{1}{2}\mathrm{Y}^2\right).
$$

Then show that for any symmetric matrices A and B,

$$
\begin{aligned}
\mathrm{E}[\{\operatorname{tr}(\mathrm{A}\mathbf{Y})\}\{\operatorname{tr}(\mathrm{B}\mathbf{Y})\}] &= \operatorname{tr}(\mathrm{AB}), \\
\mathrm{E}[\operatorname{tr}(\mathbf{AYBY})] &= \frac{1}{2}\operatorname{tr}(\mathrm{AB}) + \frac{1}{2}\{\operatorname{tr}(\mathrm{A})\}\{\operatorname{tr}(\mathrm{B})\}.
\end{aligned}
$$

Use these formulas to derive asymptotic expansion (5.2.14) of W_n/σ (for the definition of W_n, see (5.2.13).

5.11 Suppose that the statistics $\Lambda_{(i)}$, $i = 1, \ldots, d$ are independently distributed as $\Lambda_{a_i}(b_i, m_i)$, respectively. Let $m_i = c_i n = \mathrm{O}(n)$, $i = 1, \ldots, d$. Find np or ρ such that

$$
\Pr\left\{-n\rho \log \prod_{i=1}^{d} \Lambda_{(i)} \le x\right\} = \Pr\{\chi^2(f) \le x\} + \mathrm{O}(n^{-2}),
$$

where $f = \sum_{i=1}^{d} a_i b_i$.

Hint: A solution of ρ is given by

$$
\rho n = f\{\sum_{i=1}^{d} \frac{a_i b_i}{m_i + (1/2)(b_i - a_i - 1)}\}^{-1}.
$$

Chapter 6

MANOVA Models

In this chapter we first describe the one-way multivariate analysis of variance (MANOVA) model and then, a two-way MANOVA model with unbalanced data. Both of them are regarded as special cases of the multivariate linear model.

Three test criteria, including the LR test, are examined in details for linear hypotheses and for dimensionality. A theoretical comparison is made between powers of these tests. Further, their null distributions under nonnormality are studied employing both the asymptotic expansion and bootstrap methods. The distributions of characteristic roots related to the tests are discussed.

Finally, we take up testing problem in high-dimensional situations.

6.1 Multivariate One-Way Analysis of Variance

In this section we extend univariate one-way analysis of variance to multivariate analysis of variance (MANOVA), in which we measure p dependent variables on each experimental unit instead of just one variable. For multivariate one-way analysis of variance, we consider q treatments and assign n_i subjects to the ith treatment. It is assumed that all the n $(= n_1 + \cdots + n_q)$ observations are normally distributed with the common covariance matrix Σ.

Let $\boldsymbol{Y}_{i1}, \ldots, \boldsymbol{Y}_{in_i}$ be samples from the ith treatment group. The model for each observation vector is

$$\begin{aligned} \boldsymbol{Y}_{ij} &= \boldsymbol{\mu} + \boldsymbol{\alpha}_i + \boldsymbol{\varepsilon}_{ij} \\ &= \boldsymbol{\mu}_i + \boldsymbol{\varepsilon}_{ij}, \quad i = 1, \ldots, q, \quad j = 1, \ldots, n_i, \end{aligned} \tag{6.1.1}$$

where $\boldsymbol{\mu}$, $\boldsymbol{\alpha}_i$, and $\boldsymbol{\mu}_i$ are unknown parameter vectors, and the errors ε_{ij} are i.i.d. $\mathrm{N}_p(\mathbf{0}, \Sigma)$. To define components $\boldsymbol{\mu}$ and $\boldsymbol{\alpha}_i$ in the decomposition,

$$\boldsymbol{\mu}_i = \boldsymbol{\mu} + \boldsymbol{\alpha}_i, \quad i = 1, \ldots, q, \tag{6.1.2}$$

uniquely, we impose the restriction

$$\sum_{i=1}^{q} \boldsymbol{\alpha}_i = \mathbf{0}. \tag{6.1.3}$$

This is equivalent to defining $\boldsymbol{\mu} = (1/n)\sum_{i=1}^{q} \boldsymbol{\mu}_i$ and $\boldsymbol{\alpha}_i = \boldsymbol{\mu}_i - \boldsymbol{\mu}$. $\boldsymbol{\alpha}_i$ is called an *effect of the ith treatment.* We may use the restriction with weights,

$$\sum_{i=1}^{q} n_i \boldsymbol{\alpha}_i = \mathbf{0}. \tag{6.1.4}$$

In this case $\boldsymbol{\mu}$ should be defined by $\boldsymbol{\mu} = (1/n)\sum_{i=1}^{q} n_i \boldsymbol{\mu}_i$. Now we wish to test

$$H\colon \boldsymbol{\alpha}_1 = \cdots = \boldsymbol{\alpha}_q = \mathbf{0} \Leftrightarrow \boldsymbol{\mu}_1 = \cdots = \boldsymbol{\mu}_q. \tag{6.1.5}$$

We define $\bar{\boldsymbol{Y}}_{i\cdot}$ to be the sample mean vector of the ith treatment group and $\bar{\boldsymbol{Y}}_{\cdot\cdot}$ to be the overall sample mean vector; that is,

$$\bar{\boldsymbol{Y}}_{i\cdot} = \frac{1}{n_i}\sum_{j=1}^{n_i} \boldsymbol{Y}_{ij}, \quad \bar{\boldsymbol{Y}}_{\cdot\cdot} = \frac{1}{n}\sum_{i=1}^{q}\sum_{j=1}^{n_i} \boldsymbol{Y}_{ij}.$$

The sums of squares in the univariate case are extended as

$$\begin{aligned}
\mathbf{S}_\alpha &= \sum_{i=1}^{q} n_i(\bar{\boldsymbol{Y}}_{i\cdot} - \bar{\boldsymbol{Y}}_{\cdot\cdot})(\bar{\boldsymbol{Y}}_{i\cdot} - \bar{\boldsymbol{Y}}_{\cdot\cdot})' = \sum_{i=1}^{q} n_i \bar{\boldsymbol{Y}}_{i\cdot}\bar{\boldsymbol{Y}}_{i\cdot}' - n\bar{\boldsymbol{Y}}_{\cdot\cdot}\bar{\boldsymbol{Y}}_{\cdot\cdot}', \\
\mathbf{S}_t &= \sum_{i=1}^{q}\sum_{i=1}^{n_i}(\boldsymbol{Y}_{ij} - \bar{\boldsymbol{Y}}_{\cdot\cdot})(\boldsymbol{Y}_{ij} - \bar{\boldsymbol{Y}}_{\cdot\cdot})' = \sum_{i=1}^{q}\sum_{i=1}^{n_i} \boldsymbol{Y}_{ij}\boldsymbol{Y}_{ij}' - n\bar{\boldsymbol{Y}}_{\cdot\cdot}\bar{\boldsymbol{Y}}_{\cdot\cdot}', \\
\mathbf{S}_e &= \mathbf{S}_t - \mathbf{S}_\alpha.
\end{aligned} \tag{6.1.6}$$

Then

$$\begin{aligned}
\mathbf{S}_e &= \sum_{i=1}^{q}\sum_{i=1}^{n_i}(\boldsymbol{Y}_{ij} - \bar{\boldsymbol{Y}}_{i\cdot})(\boldsymbol{Y}_{ij} - \bar{\boldsymbol{Y}}_{i\cdot})' \\
&= \sum_{i=1}^{q}\left(\sum_{i=1}^{n_i} \boldsymbol{Y}_{ij}\boldsymbol{Y}_{ij}' - n_i\bar{\boldsymbol{Y}}_{i\cdot}\bar{\boldsymbol{Y}}_{i\cdot}'\right).
\end{aligned}$$

The $\mathbf{S}_\alpha$ and $\mathbf{S}_e$ are called matrices of sums of squares and products (SSP Matrices) due to treatments (between groups) and errors (within groups). In Chapter 3 we have seen that an LR test for H is

$$\Lambda_\alpha = \frac{|\mathbf{S}_e|}{|\mathbf{S}_e + \mathbf{S}_\alpha|},$$

and under H, Λ_α is distributed as $\Lambda_p(q-1, n-q)$. For the distribution of $\Lambda_p(k, n)$, using the following rule is suggested:

1. For the case p or $k \le 2$: Use the exact distribution in Theorems 3.2.3 and 3.3.4.

2. The other case:

  2.1. For $n \gg p$: Use large-sample asymptotic expansion in Example 5.4.1.

  2.2. For the other case: Use high-dimensional asymptotic expansion in Example 5.4.2.

From Theorem 3.2.3 we know that $\mathbf{S}_\alpha$ and $\mathbf{S}_e$ are independently distributed as $\mathrm{W}_p(q-1, \Sigma; \Omega)$ and $\mathrm{W}_p(n-q, \Sigma)$, respectively. Further, we have the decomposition

$$\mathbf{S}_t = \mathbf{S}_\alpha + \mathbf{S}_e.$$

These are summarized as a one-way MANOVA table as in the univariate case (Table 6.1.1). We note that the LR test does not depend on the restrictions

Table 6.1.1 One-way MANOVA table

	SSP matrix	d.f.
Treatments	$\mathbf{S}_\alpha$	$q-1$
Errors	$\mathbf{S}_e$	$n-q$
Total	$\mathbf{S}_t$	$n-1$

(6.1.3) or (6.1.4). However, the MLEs of the $\boldsymbol{\alpha}_i$ depend on the restrictions. A merit in the use of (6.1.4) is that the MLE of $\boldsymbol{\alpha}_i$ is a natural quantity $\hat{\boldsymbol{\alpha}}_i = \bar{\boldsymbol{Y}}_{i\cdot} - \bar{\boldsymbol{Y}}_{\cdot\cdot}$.

We shall see that the test statistic Λ_α can be obtained based on a general theory of the multivariate linear model (see Chapter 7). Let $\mathbf{Y}$ be the observation matrix defined by

$$\mathbf{Y} = (\boldsymbol{Y}_{11}, \ldots, \boldsymbol{Y}_{1n_1}, \ldots, \boldsymbol{Y}_{q1}, \ldots, \boldsymbol{Y}_{qn_q})'.$$

Then

$$\mathrm{E}(\mathbf{Y}) = (\mathbf{1}_n, \mathrm{G}) \begin{pmatrix} \boldsymbol{\mu}' \\ \mathcal{A} \end{pmatrix} = \mathrm{A}\Theta,$$

where $\mathcal{A} = (\boldsymbol{\alpha}_1, \ldots, \boldsymbol{\alpha}_q)'$ and

$$\mathrm{G} = \mathrm{Diag}\,(\mathbf{1}_{n_1}, \mathbf{1}_{n_2}, \ldots, \mathbf{1}_{n_g}) = \begin{pmatrix} \mathbf{1}_{n_1} & \mathbf{0} & \cdots & \mathbf{0} \\ \mathbf{0} & \mathbf{1}_{n_2} & \cdots & \mathbf{0} \\ \vdots & \vdots & \ddots & \vdots \\ \mathbf{0} & \mathbf{0} & \cdots & \mathbf{1}_{n_q} \end{pmatrix}.$$

Let ω and Ω be the range spaces generated by the column vectors of $\mathrm{E}[\mathbf{Y}]$ under the hypothesis H and model, respectively. Since each of the p columns of $\mathrm{E}[\mathbf{Y}]$ spans the same space, we can write

$$\begin{aligned} \Omega &= \{\boldsymbol{\eta} : \boldsymbol{\eta} = \mathbf{1}_n \mu + \mathrm{G}\boldsymbol{\alpha}, \quad \boldsymbol{\alpha} = (\alpha_1, \ldots, \alpha_q)', \\ &\qquad n_1\alpha_1 + \cdots + n_k\alpha_k = 0\} \\ &= \mathcal{R}[\mathrm{G}], \\ \omega &= \{\boldsymbol{\eta} : \boldsymbol{\eta} = \mathbf{1}_n \mu\} = \mathcal{R}[\mathbf{1}_n], \end{aligned}$$

where $\mathcal{R}[\mathrm{A}]$ represents range space generated by the columns of A. Further, let P_ω and P_Ω be the projection matrices (operators) to the spaces ω and Ω. They are given by

$$P_\omega = \frac{1}{n}\mathbf{1}_n\mathbf{1}_n', \quad P_\Omega = \mathrm{G}(\mathrm{G}'\mathrm{G})^{-1}\mathrm{G}'.$$

The SSP matrices may be defined by

$$\begin{aligned} \mathbf{S}_e &= \mathbf{Y}'(\mathrm{I}_n - \mathrm{P}_\Omega)\mathbf{Y}, \\ \mathbf{S}_\alpha &= \mathbf{Y}'(\mathrm{I}_n - \mathrm{P}_\omega)\mathbf{Y} - \mathbf{Y}'(\mathrm{I}_n - \mathrm{P}_\Omega)\mathbf{Y} \\ &= \mathbf{Y}'(\mathrm{P}_\Omega - \mathrm{P}_\omega)\mathbf{Y}. \end{aligned}$$

These expressions are useful in deriving the distributions of $\mathbf{S}_e$ and $\mathbf{S}_\alpha$. In fact, we can apply the multivariate Cochran theorem.

6.2 Multivariate Two-Way Analysis of Variance

In two-way analysis of variance we consider treatments for combinations of levels of two factors, A and B. Let the numbers of levels of factors A and B be a and b, respectively. Suppose that we assign n_{ij} subjects to treatment with the ith level of A and the jth level of B, and measured p variables on each subject. Let the observations in the (i, j)th treatment group or (i, j)th cell be denoted by

$$\boldsymbol{Y}_{ij1}, \boldsymbol{Y}_{ij2}, \ldots, \boldsymbol{Y}_{ijn_{ij}}.$$

The model for each observation is

$$\boldsymbol{Y}_{ijk} = \boldsymbol{\eta}_{ij} + \boldsymbol{\varepsilon}_{ijk}, \quad i = 1, \ldots, a, \quad j = 1, \ldots, b, \quad k = 1, \ldots, n_{ij}, \tag{6.2.1}$$

where $\boldsymbol{\eta}_{ij}$ is the mean vector of the (i, j)th cell, and the errors $\boldsymbol{\varepsilon}_{ijk}$ are i.i.d. $\mathrm{N}_p(\mathbf{0}, \Sigma)$. We decompose the cell mean vectors as

$$\boldsymbol{\eta}_{ij} = \boldsymbol{\mu} + \boldsymbol{\alpha}_i + \boldsymbol{\beta}_j + \boldsymbol{\gamma}_{ij}, \tag{6.2.2}$$

where $\boldsymbol{\mu}$ is a general mean vector, $\boldsymbol{\alpha}_i$ is an effect due to the ith level of the factor A, $\boldsymbol{\beta}_j$ is an effect due to the jth level of the factor B, and $\boldsymbol{\gamma}_{ij}$ represents an effect for the interaction of the ith level of A and the jth level of B. If nothing more is stated about the decomposition (6.2.2), the components of the decomposition are not defined uniquely. Usually, the following restrictions are imposed:

$$\sum_{i=1}^{a} \boldsymbol{\alpha}_i = \mathbf{0}, \quad \sum_{j=1}^{b} \boldsymbol{\beta}_j = \mathbf{0}, \quad \sum_{i=1}^{a} \boldsymbol{\gamma}_{ij} = \mathbf{0}, \quad \sum_{j=1}^{b} \boldsymbol{\gamma}_{ij} = \mathbf{0}. \tag{6.2.3}$$

This is equivalent to defining the effects as follows:

$$\begin{aligned} &\boldsymbol{\mu} = \bar{\boldsymbol{\eta}}_{\cdot\cdot}, \quad \boldsymbol{\alpha}_i = \bar{\boldsymbol{\eta}}_{i\cdot} - \bar{\boldsymbol{\eta}}_{\cdot\cdot}, \quad \boldsymbol{\beta}_j = \bar{\boldsymbol{\eta}}_{\cdot j} - \bar{\boldsymbol{\eta}}_{\cdot\cdot}, \\ &\boldsymbol{\gamma}_{ij} = \boldsymbol{\eta}_{ij} - \bar{\boldsymbol{\eta}}_{i\cdot} - \bar{\boldsymbol{\eta}}_{\cdot j} + \bar{\boldsymbol{\eta}}_{\cdot\cdot}, \end{aligned} \tag{6.2.4}$$

where $\bar{\boldsymbol{\eta}}_{\cdot\cdot} = (ab)^{-1} \sum_{i=1}^{a} \sum_{j=1}^{b} \boldsymbol{\eta}_{ij}$, $\bar{\boldsymbol{\eta}}_{i\cdot} = b^{-1} \sum_{j=1}^{b} \boldsymbol{\eta}_{ij}$, $\bar{\boldsymbol{\eta}}_{\cdot j} = a^{-1} \sum_{i=1}^{a} \boldsymbol{\eta}_{ij}$. The overbar and dot notation is also used in other places.

As alternative restrictions, we may impose

$$\sum_{i=1}^{a} n_{i\cdot} \boldsymbol{\alpha}_i = \mathbf{0}, \quad \sum_{j=1}^{b} n_{\cdot j} \boldsymbol{\beta}_j = \mathbf{0}, \quad \sum_{i=1}^{a} n_{ij} \boldsymbol{\gamma}_{ij} = \mathbf{0}, \quad \sum_{j=1}^{b} n_{ij} \boldsymbol{\gamma}_{ij} = \mathbf{0}, \tag{6.2.5}$$

where $n_{i\cdot} = \sum_{j=1}^{b} n_{ij}$ and $n_{\cdot j} = \sum_{i=1}^{a} n_{ij}$. For a discussion of various weight restrictions, see, for example, Fujikoshi (1993).

We now consider the following testing problems:

$$\begin{aligned} &H_{\gamma}\colon \boldsymbol{\gamma}_{ij} = \mathbf{0}, \quad i = 1, \ldots, a, \quad j = 1, \ldots, b, \\ &H_{\alpha}\colon \boldsymbol{\alpha}_i = \mathbf{0}, \quad i = 1, \ldots, a, \\ &H_{\beta}\colon \boldsymbol{\beta}_j = \mathbf{0}, \quad j = 1, \ldots, b. \end{aligned} \tag{6.2.6}$$

Related to these testing problems, we define the SSP matrices due to errors, interactions, factor A, and factor B as follows:

$$\begin{aligned}
\mathbf{S}_e &= \sum_{i=1}^{a}\sum_{j=1}^{b}\sum_{k=1}^{n_{ij}}(\boldsymbol{Y}_{ijk} - \bar{\boldsymbol{Y}}_{ij\cdot})(\boldsymbol{Y}_{ijk} - \bar{\boldsymbol{Y}}_{ij\cdot})', \\
\mathbf{S}_\gamma &= \sum_{i=1}^{a}\sum_{j=1}^{b} n_{ij}(\bar{\boldsymbol{Y}}_{ij\cdot} - \bar{\boldsymbol{Y}}_{i\cdot\cdot} - \bar{\boldsymbol{Y}}_{\cdot j\cdot} + \bar{\boldsymbol{Y}}_{\cdots})(\bar{\boldsymbol{Y}}_{ij\cdot} - \bar{\boldsymbol{Y}}_{i\cdot\cdot} - \bar{\boldsymbol{Y}}_{\cdot j\cdot} + \bar{\boldsymbol{Y}}_{\cdots})', \\
\mathbf{S}_\alpha &= \sum_{i=1}^{a} n_{i\cdot}(\bar{\boldsymbol{Y}}_{i\cdot\cdot} - \bar{\boldsymbol{Y}}_{\cdots})(\bar{\boldsymbol{Y}}_{i\cdot\cdot} - \bar{\boldsymbol{Y}}_{\cdots})', \qquad (6.2.7) \\
\mathbf{S}_\beta &= \sum_{j=1}^{b} n_{\cdot j}(\bar{\boldsymbol{Y}}_{\cdot j\cdot} - \bar{\boldsymbol{Y}}_{\cdots})(\bar{\boldsymbol{Y}}_{\cdot j\cdot} - \bar{\boldsymbol{Y}}_{\cdots})'.
\end{aligned}$$

Then if all the frequencies n_{ij} are equal, we can test H_γ, H_α, and H_β based on

$$\begin{aligned}
\Lambda_\gamma &= |\mathbf{S}_e|/|\mathbf{S}_e + \mathbf{S}_\gamma| \sim \Lambda_p((a-1)(b-1), m), \\
\Lambda_\alpha &= |\mathbf{S}_e|/|\mathbf{S}_e + \mathbf{S}_\alpha| \sim \Lambda_p(a-1, m), \qquad (6.2.8) \\
\Lambda_\beta &= |\mathbf{S}_e|/|\mathbf{S}_e + \mathbf{S}_\beta| \sim \Lambda_p(b-1, m),
\end{aligned}$$

where $m = n - ab$ and $n = \sum_{i=1}^{a}\sum_{j=1}^{b} n_{ij}$. These tests are LR tests. Further, let $\mathbf{W}_\alpha$, $\mathbf{W}_\beta$, $\mathbf{W}_\gamma$, and $\mathbf{W}_e$ be the SSP matrices due to H_α, H_β, H_γ, and the errors, respectively. Then, if all the frequencies n_{ij} are equal,

$$\mathbf{W}_\alpha = \mathbf{S}_\alpha, \quad \mathbf{W}_\beta = \mathbf{S}_\beta, \quad \mathbf{W}_\gamma = \mathbf{S}_\gamma, \quad \mathbf{W}_e = \mathbf{S}_e.$$

However, if all the frequencies n_{ij} are not equal, some modifications are necessary. To see these SSP matrices, let ω_α, ω_β, ω_γ, and Ω be the range spaces of $\mathrm{E}(\mathbf{Y})$ under H_α, H_β, H_γ, and the model (6.2.2), respectively. Similarly, let P_α, P_β, P_γ, and P_Ω be the projection operators on ω_α, ω_β, ω_γ, and Ω, respectively. In general, the SSP matrix due to the errors is

$$\mathbf{W}_e = \mathbf{Y}'(\mathrm{I}_n - \mathrm{P}_\Omega)\mathbf{Y}.$$

The SSP matrix due to a hypothesis H is

$$\begin{aligned}
\mathbf{W}_h &= \mathbf{Y}'(\mathrm{I}_n - \mathrm{P}_\omega)\mathbf{Y} - \mathbf{Y}'(\mathrm{I}_n - \mathrm{P}_\Omega)\mathbf{Y} \\
&= \mathbf{Y}'(\mathrm{P}_\Omega - \mathrm{P}_\omega)\mathbf{Y},
\end{aligned}$$

when ω is the range space of $\mathrm{E}(\mathbf{Y})$ under H.

The SSP matrices in an unbalanced case are obtained by extending the results [see, e.g., Arnold, 1981 or Fujikoshi, 1993] on the sums of squares in the univariate case as follows:

(1) $\mathbf{W}_e = \mathbf{S}_e$.

(2) $\mathbf{W}_\gamma$ does not depend on the restrictions (6.2.3) and (6.2.5).

(3) The SSP matrix $\mathbf{W}_\gamma$ is given by

$$\mathbf{W}_\gamma = \sum_{i=1}^{a}\sum_{j=1}^{b} n_{ij}(\bar{\boldsymbol{Y}}_{ij\cdot} - \hat{\boldsymbol{\mu}} - \hat{\boldsymbol{\alpha}}_i - \hat{\boldsymbol{\beta}}_j)(\bar{\boldsymbol{Y}}_{ij\cdot} - \hat{\boldsymbol{\mu}} - \hat{\boldsymbol{\alpha}}_i - \hat{\boldsymbol{\beta}}_j)',$$

where $\hat{\boldsymbol{\mu}} = \bar{\boldsymbol{Y}}_{\cdots}$ and $\hat{\boldsymbol{\alpha}}_i$ and $\hat{\boldsymbol{\beta}}_j$ are the solutions of

$$\begin{pmatrix} \mathrm{D}_a & \mathrm{N} \\ \mathrm{N}' & \mathrm{D}_b \end{pmatrix}\begin{pmatrix} \mathcal{A}' \\ \mathcal{B}' \end{pmatrix} = \begin{pmatrix} \mathrm{D}_a\bar{\mathbf{Y}}_a' \\ \mathrm{D}_b\bar{\mathbf{Y}}_b' \end{pmatrix}$$

$$\sum_{i=1}^{a} n_{i\cdot}\boldsymbol{\alpha}_i = \mathbf{0} \quad \text{and} \quad \sum_{j=1}^{b} n_{\cdot j}\boldsymbol{\beta}_j = \mathbf{0}.$$

Here $\mathrm{N} = (n_{ij}); a \times b$ matrix,

$$\mathrm{D}_a = \mathrm{diag}\,(n_{1\cdot}, \ldots, n_{a\cdot}), \quad \bar{\mathbf{Y}}_a = (\bar{\boldsymbol{Y}}_{1\cdot\cdot} - \bar{\boldsymbol{Y}}_{\cdots}, \ldots, \bar{\boldsymbol{Y}}_{a\cdot\cdot} - \bar{\boldsymbol{Y}}_{\cdots}),$$
$$\mathrm{D}_b = \mathrm{diag}\,(n_{\cdot 1}, \ldots, n_{\cdot b}), \quad \bar{\mathbf{Y}}_b = (\bar{\boldsymbol{Y}}_{\cdot 1\cdot} - \bar{\boldsymbol{Y}}_{\cdots}, \ldots, \bar{\boldsymbol{Y}}_{\cdot b\cdot} - \bar{\boldsymbol{Y}}_{\cdots}).$$

(4) The SSP matrices $\mathbf{W}_\alpha$ and $\mathbf{W}_\beta$ under the model (6.2.2) with the restriction (6.2.5) are given by

$$\mathbf{W}_\alpha = \sum_{i=1}^{a}\sum_{j=1}^{b} n_{ij}(\bar{\boldsymbol{Y}}_{ij\cdot} - \bar{\boldsymbol{Y}}_{\cdots})(\bar{\boldsymbol{Y}}_{ij\cdot} - \bar{\boldsymbol{Y}}_{\cdots}) - \mathbf{W}_\gamma - \mathbf{S}_\beta,$$
$$\mathbf{W}_\beta = \sum_{i=1}^{a}\sum_{j=1}^{b} n_{ij}(\bar{\boldsymbol{Y}}_{ij\cdot} - \bar{\boldsymbol{Y}}_{\cdots})(\bar{\boldsymbol{Y}}_{ij\cdot} - \bar{\boldsymbol{Y}}_{\cdots}) - \mathbf{W}_\gamma - \mathbf{S}_\alpha.$$

(5) The SSP matrices $\mathbf{W}_\alpha$ and $\mathbf{W}_\beta$ under the model (6.2.2) with the restriction (6.2.3) are obtained by the multivariate least squares method. For the details, see the justification below.

(6) For the case of proportional sampling (i.e., $n_{ij} = n_{i\cdot}n_{\cdot j}/n$) in the model (6.2.2) with the restriction (6.2.5), $\mathbf{W}_\gamma = \mathbf{S}_\gamma$, $\mathbf{W}_\alpha = \mathbf{S}_\alpha$, and $\mathbf{W}_\beta = \mathbf{S}_\beta$.

For a justification, we can expand a general multivariate linear theory as in the multivariate one-way analysis of variance model. In this case, let

$$\mathbf{Y} = (\boldsymbol{Y}_{111}, \ldots, \boldsymbol{Y}_{11n_{11}}, \boldsymbol{Y}_{121}, \ldots, \boldsymbol{Y}_{12n_{12}}, \ldots, \boldsymbol{Y}_{ab1}, \ldots, \boldsymbol{Y}_{abn_{ab}})'.$$

Then

$$\mathrm{E}(\mathbf{Y}) = \mathrm{G}\begin{pmatrix} \boldsymbol{\eta}_{11}' \\ \vdots \\ \boldsymbol{\eta}_{ab}' \end{pmatrix} = \mathrm{A}\Theta,$$

where

$$\begin{aligned} &\mathrm{G} = \mathrm{Diag}\,(\mathbf{1}_{n_{11}}, \mathbf{1}_{n_{12}}, \ldots, \mathbf{1}_{n_{ab}}), \\ &\Theta = (\boldsymbol{\mu}, \mathcal{A}', \mathcal{B}', \Gamma')', \quad \mathcal{A} = (\boldsymbol{\alpha}_1, \ldots, \boldsymbol{\alpha}_a)', \quad \mathcal{B} = (\boldsymbol{\beta}_1, \ldots, \boldsymbol{\beta}_b)', \\ &\Gamma = (\gamma_{11}, \ldots, \gamma_{1b}, \ldots, \gamma_{a1}, \ldots, \gamma_{ab})'. \end{aligned}$$

Further, the design matrix A is defined by

$$\begin{aligned} &\mathrm{A} = (\mathbf{1}_n,\ \mathrm{A}_1,\ \mathrm{A}_2,\ \mathrm{A}_3), \quad \mathrm{A}_1 = \mathrm{G}\cdot \mathrm{I}_a \otimes \mathbf{1}_b, \\ &\mathrm{A}_2 = \mathrm{G}\cdot \mathrm{I}_a \otimes \mathrm{I}_b, \quad \mathrm{A}_3 = \mathrm{G}. \end{aligned}$$

Note that

$$\begin{aligned} &\sum_{i=1}^{a}\sum_{j=1}^{b}\sum_{k=1}^{n_{ij}}(\boldsymbol{Y}_{ijk} - \boldsymbol{\eta}_{ij})(\boldsymbol{Y}_{ijk} - \boldsymbol{\eta}_{ij}) \\ &\qquad = \mathbf{S}_e + \sum_{i=1}^{a}\sum_{j=1}^{b} n_{ij}(\boldsymbol{Y}_{ij} - \bar{\boldsymbol{Y}}_{ij\cdot})(\boldsymbol{Y}_{ij} - \bar{\boldsymbol{Y}}_{ij\cdot})', \end{aligned}$$

which leads to $\mathbf{W}_e = \mathbf{S}_e$. Further, for $h = \alpha, \beta$, and γ,

$$\mathbf{W}_h = \min_{\omega_h}\sum_{i=1}^{a}\sum_{j=1}^{b}\sum_{k=1}^{n_{ij}}(\boldsymbol{Y}_{ijk} - \boldsymbol{\eta}_{ij})(\boldsymbol{Y}_{ijk} - \boldsymbol{\eta}_{ij}).$$

Therefore, we can see that $\mathbf{W}_\gamma$ is obtained by (3). For a justification of (4) and (5), note that the restriction (6.2.5) is expressed as

$$\mathbf{1}_n' A_1 \mathcal{A} = \mathrm{O},\ \mathbf{1}_n' A_2 \mathcal{B} = \mathrm{O},\ (A_1\ A_2)'\Gamma = \mathrm{O},$$

which implies

$$\omega_\alpha = \mathcal{R}[\mathbf{1}_n] + (\mathcal{R}[A_2] \cap \mathcal{R}[\mathbf{1}_n]^\perp) + (\mathcal{R}[G] \cap \mathcal{R}[A_1\ A_2]^\perp).$$

Therefore we have

$$\mathbf{W}_\alpha = \mathbf{Y}'\left\{(\mathrm{P}_G - \mathrm{P}_{\mathbf{1}_n}) - (\mathrm{P}_G - \mathrm{P}_{(A_1\ A_2)}) - (\mathrm{P}_{\mathrm{A}_2} - \mathrm{P}_{\mathbf{1}_n})\right\}\mathbf{Y},$$

which is equal to the right-hand side of $\mathbf{W}_\alpha$ in (4). Similarly the SSP matrix $\mathbf{W}_\beta$ is justified.

6.3 MANOVA Tests

6.3.1 Test Criteria

In a multivariate q-sample problem or one-way MANOVA model, let $\boldsymbol{Y}_{ij}$ be the jth sample observation ($j = 1, \ldots, n_i$) from the ith p- dimensional population Π_i ($i = 1, \ldots, q$) with mean vector $\boldsymbol{\mu}_i$ and common covariance matrix Σ, where the $\boldsymbol{\mu}_i$ and Σ are unknown. Let $\mathbf{S}_b$ and $\mathbf{S}_w$ be the between-group and the within-group SSP; that is,

$$\mathbf{S}_b = \sum_{i=1}^{q} n_i(\bar{\boldsymbol{Y}}_i - \bar{\boldsymbol{Y}})(\bar{\boldsymbol{Y}}_i - \bar{\boldsymbol{Y}})', \quad \mathbf{S}_w = \sum_{i=1}^{q}(n_i - 1)\mathbf{S}_i, \tag{6.3.1}$$

where $n = n_1 + \cdots + n_q$, $\bar{\boldsymbol{Y}}_i = n_i^{-1}\sum_{j=1}^{n_j} \boldsymbol{Y}_{ij}$, $\bar{\boldsymbol{Y}} = n^{-1}\sum_{i=1}^{q}\sum_{j=1}^{n_i} \boldsymbol{Y}_{ij}$, and $\mathbf{S}_i = (n_i - 1)^{-1}\sum_{j=1}^{n_i}(\boldsymbol{Y}_{ij} - \bar{\boldsymbol{Y}}_i)(\boldsymbol{Y}_{ij} - \bar{\boldsymbol{Y}}_i)'$. For testing,

$$H_0 : \boldsymbol{\mu}_1 = \cdots = \boldsymbol{\mu}_q \text{ vs. } H_1 : \boldsymbol{\mu}_i \neq \boldsymbol{\mu}_j \quad \text{for some } i \neq j, \tag{6.3.2}$$

we have seen that under normality, the likelihood ratio criterion is

$$\lambda = \{|\mathbf{S}_w|/|\mathbf{S}_w + \mathbf{S}_b|\}^{n/2}. \tag{6.3.3}$$

Then $\mathbf{S}_b$ and $\mathbf{S}_w$ are independently distributed as a noncentral Wishart distribution $\mathrm{W}_p(q-1, \Sigma; \Gamma)$ and a central Wishart distribution $\mathrm{W}_p(n-q, \Sigma)$, respectively, where $\Gamma = \sum_{i=1}^{q} n_i(\boldsymbol{\mu}_i - \bar{\boldsymbol{\mu}})(\boldsymbol{\mu}_i - \bar{\boldsymbol{\mu}})'$ with $\bar{\boldsymbol{\mu}} = (1/n)\sum_{i=1}^{q} n_i\boldsymbol{\mu}_i$.

In general, suppose that we have two SSP matrices, $\mathbf{S}_h$ and $\mathbf{S}_e$ due to the hypothesis and the errors, where $\mathbf{S}_h$ and $\mathbf{S}_e$ are distributed independently as a noncentral Wishart distribution $\mathrm{W}_p(q, \Sigma; \mathrm{MM}')$ and a central Wishart distribution $\mathrm{W}_p(n, \Sigma)$, respectively. For example, the matrix M in the q-sample problem is

$$\mathrm{M} = (\sqrt{n_1}(\boldsymbol{\mu}_1 - \bar{\boldsymbol{\mu}}), \ldots, \sqrt{n_1}(\boldsymbol{\mu}_q - \bar{\boldsymbol{\mu}})).$$

For the testing problem

$$H_0 : \mathrm{M} = \mathrm{O} \text{ vs. } H_1 : \mathrm{M} \neq \mathrm{O},$$

there are three major test statistics:

$$\begin{aligned}
&\text{Likelihood ratio statistic: } T_{\mathrm{LR}} = -n \ \log\{|\mathbf{S}_e|/|\mathbf{S}_e + \mathbf{S}_h|\}, \\
&\text{Lawley–Hotelling statistic: } T_{\mathrm{LH}} = n \ \mathrm{tr}\,\mathbf{S}_h\mathbf{S}_e^{-1}, \\
&\text{Bartlett–Nanda–Pillai statistic: } T_{\mathrm{BNP}} = n \ \mathrm{tr}\,\mathbf{S}_h(\mathbf{S}_e + \mathbf{S}_h)^{-1}.
\end{aligned} \tag{6.3.4}$$

Great interest has been expressed in asymptotic expansions of the null and nonnull distributions of these test statistics and power comparison of these tests. When we consider the distributions of the test statistics above, we may assume that:

(1) $\mathbf{S}_h \sim \mathrm{W}_p(q, \mathrm{I}_p; \Omega)$, $\mathbf{S}_e \sim \mathrm{W}_p(q, \mathrm{I}_p)$, where $\Omega = \mathrm{diag}(\omega_1, \ldots, \omega_p)$ and ω_i's are the characteristic roots of $\Sigma^{-1/2}\mathrm{MM}'\Sigma^{-1/2}$ or $\mathrm{M}'\Sigma^{-1}\mathrm{M}$.

(2) $\mathbf{S}_h$ and $\mathbf{S}_e$ are independent.

6.3.2 Large-Sample Approximations

In this section we consider asymptotic expansions of the test statistics in (6.3.4) under a large-sample framework:

$$\text{A1: } p \text{ and } q \text{ are fixed, } n \to \infty.$$

In general, though, the noncentrality matrix Ω depends on p, q, and n, usually $\Omega = \mathrm{O}(n)$, We assume that $\Omega = \mathrm{O}(1)$, and hence the powers of the tests correspond to those local alternatives. Their asymptotic expansions have been investigated extensively. Box (1949) derived an asymptotic expansion of the null distribution of T_{LR}. Asymptotic expansions of the null and nonnull distributions of T_{LR} were given by Siotani (1956, 1957) and Ito (1956, 1960). The moment of T_{LR} was obtained by Constantine (1963) in terms of the hypergeometric function ${}_1F_1$ of the matrix argument, which is a power series of zonal polynomials reported by James (1960, 1964). Sugiura and Fujikoshi (1969) derived asymptotic expansions of the nonnull distribution of T_{LR} by expanding the characteristic function, which is expressed in terms of the hypergeometric function ${}_1F_1$ of the matrix argument. Moreover, they obtained formulas of weighted sums of zonal polynomials. Fujikoshi (1970) derived asymptotic expansions of the null and nonnull distributions of T_{BNP} by proving formulas of new types of weighted sums of zonal polynomials. On the other hand, Muirhead (1970) obtained asymptotic expansions of T_{BNP} by using a partial differential equation for ${}_1F_1$.

Under condition A1, the null distributions of the test statistics tend to the chi-square distribution with $f = pq$ degrees of freedom ($\chi^2(pq)$). The test statistics with Bartlett corrections are given by

$$\begin{aligned} T_1 &= -n_1 \log\{|\mathbf{S}_e|/|\mathbf{S}_e + \mathbf{S}_h|\}, \quad n_1 = n + \frac{1}{2}(q - p - 1), \\ T_2 &= n_2 \,\mathrm{tr}\, \mathbf{S}_h \mathbf{S}_e^{-1}, \quad n_2 = n - p - 1, \\ T_3 &= n_3 \,\mathrm{tr}\, \mathbf{S}_h(\mathbf{S}_e + \mathbf{S}_h)^{-1}, \quad n_3 = n + q, \end{aligned} \tag{6.3.5}$$

so that under the hypothesis these test statistics satisfy

$$\mathrm{E}(T_i) = f + \mathrm{O}(n^{-2}), \quad i = 1, 2, 3.$$

The results on their asymptotic null and nonnull distributions are summarized as follows:

Theorem 6.3.1 *Under condition A1 and $\Omega = O(1)$, we have the following asymptotic expansions of the nonnull distributions of T_1, T_2, and T_3:*

$$\Pr\{T \le x\} = G_f(x;\delta^2) + \frac{1}{4n}\sum_{j=0}^{4} a_j G_{f+2j}(x;\delta^2) + O(n^{-2}), \tag{6.3.6}$$

where $G_f(x;\delta^2)$ is the distribution function of a noncentral chi-square distribution with $f = pq$ degrees of freedom and noncentrality parameter $\delta^2 = \operatorname{tr}\Omega$. The coefficients a_j of T_1, T_2, and T_3 are given from the a_j below by putting $\xi = -1$, 0, and 2, respectively.

$$\begin{aligned}
a_0 &= (\xi+1)f\gamma, \quad a_1 = -2(\xi+1)f\gamma, \\
a_2 &= (\xi+1)f\gamma - (3\xi+4)\operatorname{tr}\Omega + \operatorname{tr}\Omega^2, \\
a_3 &= 2(\xi+1)\operatorname{tr}\Omega - (\xi+2)\operatorname{tr}\Omega^2, \quad a_4 = (\xi+1)\operatorname{tr}\Omega^2,
\end{aligned}$$

where $\gamma = p + q + 1$ and $\omega_i = \operatorname{tr}\Omega^i$.

Corollary 6.3.1 *The null distributions of T_i, $i = 1, 2, 3$ are expanded as*

$$\begin{aligned}
\Pr\{T \le x\} &= G_f(x) + \frac{1}{4n} f\gamma(\xi+1) \\
&\quad \times \{G_f(x) - 2G_{f+2}(x) + G_{f+4}(x)\} + O(n^{-2}) \\
&= G_f(x) - \frac{1}{2n} g_f(x)\gamma(\xi+1)\cdot\frac{u(u-f-2)}{f+2} + O(n^{-2}),
\end{aligned}$$

with $\xi = -1$, 0, and -2 for T_1, T_2, and T_3, respectively.

Using a Cornish–Fisher formula, the upper α-point T is expanded in terms of the upper α-point u of $\chi^2(f)$ as

$$t = u + \frac{1}{2n}\gamma(\xi+1)\cdot\frac{u(u-f-2)}{f+2} + O(n^{-2}). \tag{6.3.7}$$

6.3.3 Comparison of Powers

The asymptotic expansions of the null and nonnull distributions of the three statistics allow us to compare the powers of the tests. The following result was obtained by Rothenberg (1977).

Theorem 6.3.2 *Under the local alternatives $\Omega = O(1)$, the powers of T_1, T_2, and T_3 with significance level α can be expanded as*

$$\beta_T = \beta_0 + \frac{1}{2n}(\xi+1)\eta(\Omega)g_{f+8}(u;\delta^2) + O(n^{-2}), \tag{6.3.8}$$

with $\xi = -1$, 0, *and* -2 *for* T_1, T_2, *and* T_3,*respectively, where* u *is the upper* α*-point of* $\chi^2(f)$,

$$\beta_0 = 1 - G_f(u;\delta^2) + \frac{1}{2n}\left\{\gamma\omega_1 g_{f+4}(u;\delta^2) - \omega_2 g_{f+6}(u;\delta^2)\right\},$$

$$\eta(\Omega) = \frac{\sigma_\omega^2}{\bar{\omega}^2} - \frac{(p-1)(p+2)}{pq+2},$$

$$\bar{\omega} = \frac{1}{p}\sum_{i=1}^{p}\omega_i, \quad \sigma_\omega^2 = \frac{1}{p}\sum_{i=1}^{p}(\omega_i - \bar{\omega})^2.$$

Further, if the value of η *is positive, the power of* T_{LH} *is greater than that of* T_{LR}*, which in turn is greater than that of* T_{BNP}*, neglecting terms of order* n^{-2}. *If the value of* γ *is negative, the order of the powers is reversed.*

Proof. Let t and u be the upper α points of T and $\chi^2(f)$, respectively. Then $t = u + c/n + \mathrm{O}(n^{-2})$, where $c = (1/2)\gamma(\xi+1)u(u-f-2)/(f+2)$. Using

$$G_{f+2}(x;\delta^2) = -2g_{f+2}(x;\delta^2) + G_f(x;\delta^2)$$

and

$$G_f\left(x + \frac{c}{n};\delta^2\right) = G_f(x;\delta^2) + \frac{c}{n}g_f(x;\delta^2) + \mathrm{O}(n^{-2}),$$

we can express the power function as

$$\begin{aligned}\beta_T &= \Pr\{T \geq t\}\\ &= 1 - G_f(u;\delta^2) + \frac{1}{2n}\sum_{j=1}^{4} b_j g_{f+2j}(u;\delta^2)\\ &\quad - \frac{c}{n}g_f(u;\delta^2) + \mathrm{O}(n^{-2}),\end{aligned}$$

where $b_j = \sum_{i=j}^{4} a_i$. This expression can be reduced to (6.3.8) by using

$$ug_f(u;\delta^2) = fg_{f+2}(u;\delta^2) + 2g_{f+4}(u;\delta^2). \qquad \square$$

For given p and q, there are two extremes for the value of $\eta(\Omega)$:

(1) If $\Omega = \omega \mathrm{I}_p$, $\eta(\Omega) = -(p-1)(p+2)/(pq+2) \leq 0$.

(2) If $\Omega = \mathrm{diag}\,(\omega_1, 0, \ldots, 0)$, $\eta(\Omega) = (p-1)\{1 - (p+2)/(pq+2)\} \geq 0$.

Theorem 6.3.2 was extended [see, Fujikoshi, 1988] to a class of test statistics $\sum_{j=1}^{p} h(\ell_j)$, where $\ell_1 > \cdots > \ell_p > 0$ are characteristic roots of $\mathbf{S}_h\mathbf{S}_e^{-1}$, and h satisfies: (1) $h(\ell)$ is a monotone increasing function for $\ell \geq 0$ with $h(0) = 0$, (2) $h'(0) = 1$, and (3) $h(\ell)$ has a continuous third-order derivative in a neighborhood of $\ell = 0$. The test statistics T_{LR}, T_{LH}, and T_{BNP} are expressed by putting $h(\ell) = \log(1+\ell)$, ℓ, and $\ell/(1+\ell)$, respectively. Similar results are obtained for some tests on dimensionality under MANOVA and canonical correlation models [see Fujikoshi, 1988a]. Berndt and Savin (1977) have noted that

$$\mathrm{tr}\,\mathbf{S}_h(\mathbf{S}_h+\mathbf{S}_e)^{-1} \le -\log|\mathbf{S}_e|/|\mathbf{S}_h+\mathbf{S}_e| \le \mathrm{tr}\,\mathbf{S}_h\mathbf{S}_e^{-1}$$

(see Problem 6.5).

6.3.4 High-Dimensional Approximations

Our interest is to consider asymptotic expansions of the null and nonnull distributions of the three test statistics under

$$\text{A2:}\quad q\text{: fix },\ p/n \to c \in (0,1). \tag{6.3.9}$$

This assumption implies that $m = n - p + q$ also tends to infinity. For the LR statistic, there is some work that has been done. It is known [see, e.g., Anderson, 2003] that $|\mathbf{S}_e|/|\mathbf{S}_e+\mathbf{S}_h|$ is distributed as a lambda distribution $\Lambda(p,q,n)$. Mudholkar and Trivedi (1980) have proposed a normal approximation. Their idea is to improve an approximation by a power transformation such that a leading term in an expansion for skewness of transformed statistic vanishes. Saddlepoint approximations have been proposed by Srivastava and Yau (1989). Tonda and Fujikoshi (2004) obtained an asymptotic expansion of the null distribution of the LR statistic by using the property that $\Lambda_p(q,n) = \Lambda_q(p, n-p+q)$.

A key idea is to express the test statistics in terms of two $q \times q$ matrices $\mathbf{B}$ and $\mathbf{W}$ instead of two $p \times p$ matrices $\mathbf{S}_h$ and $\mathbf{S}_e$. Such expressions have been discussed in Section 3.5. In fact, the three statistics can be represented as

(1) $$\frac{|\mathbf{S}_e|}{|\mathbf{S}_e+\mathbf{S}_h|} = \frac{|\mathbf{W}|}{|\mathbf{W}+\mathbf{B}|},$$

(2) $$\mathrm{tr}\ \mathbf{S}_h\mathbf{S}_e^{-1} = \mathrm{tr}\ \mathbf{B}\mathbf{W}^{-1},$$

(3) $$\mathrm{tr}\ \mathbf{S}_h(\mathbf{S}_e+\mathbf{S}_h)^{-1} = \mathrm{tr}\,\mathbf{B}(\mathbf{W}+\mathbf{B})^{-1}.$$

Here $\mathbf{B}$ and $\mathbf{W}$ are independent and follow a noncentral Wishart distribution $\mathrm{W}_q(p, \mathrm{I}_q; \Omega)$ and a central Wishart distribution $\mathrm{W}_q(m, \mathrm{I}_q)$, respectively, where $m = n - p + q$ and the noncentrality matrix Ω is given by

$$\Omega = \mathrm{M}'\Sigma^{-1}\mathrm{M}.$$

In the following we take up the null distributions only, but the nonnull distributions are obtained similarly. To make the statements sweeping, we write

$$
\begin{aligned}
T_{\mathrm{LR}} &= -\sqrt{p}\left(1+\frac{m}{p}\right)\left\{\log\frac{|\mathbf{S}_e|}{|\mathbf{S}_e+\mathbf{S}_h|}+q\log\left(1+\frac{p}{m}\right)\right\},\\
T_{\mathrm{LH}} &= \sqrt{p}\left(\frac{m}{p}\operatorname{tr}\mathbf{S}_h\mathbf{S}_e^{-1}-q\right),\\
T_{\mathrm{BNP}} &= \sqrt{p}\left(1+\frac{p}{m}\right)\left\{\left(1+\frac{m}{p}\right)\operatorname{tr}\mathbf{S}_h(\mathbf{S}_e+\mathbf{S}_h)^{-1}-q\right\}.
\end{aligned}
\tag{6.3.10}
$$

We write the three statistics above as T_G ($G =$ LR, LH, and BNP). The derivation of asymptotic expansions is based on the use of an expansion of T_G in terms of

$$
\mathbf{U} = \sqrt{p}\left(\frac{1}{p}\mathbf{B}-\mathrm{I}_q\right) \quad \text{and} \quad \mathbf{V} = \sqrt{m}\left(\frac{1}{m}\mathbf{W}-\mathrm{I}_q\right).
$$

The characteristic function of T_G can be evaluated as

$$
C_{T_G}(t) = \exp\left\{q(1+r)(it)^2\right\}\left[1+\frac{1}{\sqrt{p}}\left\{itb_1+(it)^3b_3\right\}\right]+\mathrm{O}(p^{-1}), \tag{6.3.11}
$$

where $r = p/m$, and

$$
\begin{aligned}
b_1 &= q(q+1)\left\{c(1+r)+r\right\},\\
b_3 &= 4q(1+r)\left\{c(1+r)+r\right\}+\frac{4}{3}q(1-r^2),
\end{aligned}
$$

with $c = -\frac{1}{2}p(m+p)^{-1}$, 0, and $-p(m+p)^{-1}$ for $G =$ LR, LH, and BNP, respectively. Inversion of (6.3.11) gives the following result [Wakaki et al., 2003]:

Theorem 6.3.3 *Let T_G be the transformed test statistics for $G =$ LR, LH, and BNP defined by* (6.3.10). *Then*

$$
\Pr\left\{\frac{1}{\sigma}T_G \le z\right\} = \Phi(z)-\phi(z)\frac{1}{\sqrt{p}}\left\{\frac{1}{\sigma}b_1+\frac{1}{\sigma^3}b_3(z^2-1)\right\} + \mathrm{O}(p^{-1}), \tag{6.3.12}
$$

where Φ and ϕ are the distribution and density functions of the standard normal distribution, $\sigma = \sqrt{2q(1+r)}$, and the b_j are as given in (6.3.11).

Applying the Cornish–Fisher formula to (6.3.12), we can obtain an asymptotic expansion for the upper percent point of the distribution of T_G, as in the following corollary.

Corollary 6.3.2 *Let z_α be the upper $100(1-\alpha)\%$ point of the standard normal distribution, and let*

$$t_G(\alpha) = z_\alpha + \frac{1}{\sqrt{p}}\left\{\frac{b_1}{\sigma} + (z_\alpha^2 - 1)\frac{b_3}{\sigma^3}\right\}. \qquad (6.3.13)$$

Then

$$\Pr\left\{\frac{1}{\sigma}T_G \leq t_G(\alpha)\right\} = 1 - \alpha + \mathrm{O}(p^{-1}).$$

The terms of $\mathrm{O}(p^{-1})$ in Theorem 6.3.3 and Corollary 6.3.2 have also been obtained. The numerical accuracy of the large-sample approximation (6.3.7) and the high-dimensional approximation (6.3.13) has been compared by Wakaki et al. (2003) and Tonda and Fujikoshi (2004). It may be noted that (6.3.13) is more accurate than (6.3.7) in wide range of (p, q, n).

Now we examine the relationship between large-sample ((n)-) asymptotics and high-dimensional ((n,p)-) asymptotics. For example, for the Lawley Hotelling trace criterion $\mathrm{tr}\,\mathbf{S}_h\mathbf{S}_e^{-1}$,

$$\frac{1}{\sqrt{p}}\{m\ \mathrm{tr}\,\mathbf{S}_h\mathbf{S}_e^{-1} - pq\} \xrightarrow{d} \mathrm{N}(1, 2q(1+r))\colon (n,p)\text{-asymptotic},$$

where $\xrightarrow{d}$ denotes convergence in distribution. On the other hand, first consider asymptotic distribution under the large-sample framework such that n tends to infinity and p and q are fixed, and then consider asymptotic distribution of the result when p tends to infinity. Then we have

$$\frac{1}{\sqrt{p}}\{m\ \mathrm{tr}\,\mathbf{S}_h\mathbf{S}_e^{-1} - pq\} \xrightarrow{d} \frac{1}{\sqrt{p}}\{\chi^2(pq) - pq\}\colon (n)\text{-asymptotic},$$
$$\frac{1}{\sqrt{p}}\{\chi^2(pq) - pq\} \xrightarrow{d} \mathrm{N}(1, 2q)\colon (p)\text{-asymptotic}.$$

The results of the two approaches differ in the variances $2q(1+r)$ and $2q$ of the asymptotic normality. So if $r = p/(n-p+q)$ is large, the difference is large. In other words, the existing (n)-asymptotics break down when p goes to infinity with n.

6.4 Approximations Under Nonnormality

6.4.1 Asymptotic Expansions

In previous sections we treated the distributions of MANOVA tests under the assumption that the distributions of observation vectors are normal. In this section we consider asymptotic expansions of the null distributions of the three test statistics $T_{\mathrm{LR}}, T_{\mathrm{HL}}$ and T_{BNP} in a one-way MANOVA model under nonnormality. Let $\mathbf{S}_b$ and $\mathbf{S}_w$ be the between-group SSP and the within-group SSP, respectively, in (6.3.1). The test statistics with Bartlett corrections are given by

$$\begin{aligned} T_1 &= -(n+d_1)\log\{|\mathbf{S}_w|/|\mathbf{S}_w+\mathbf{S}_b|\}, \quad d_1 = -\frac{1}{2}(p+q+2), \\ T_2 &= (n+d_2)\operatorname{tr}\mathbf{S}_b\mathbf{S}_w^{-1}, \quad d_2 = -(p+q+1), \\ T_3 &= (n+d_3)\operatorname{tr}\mathbf{S}_b(\mathbf{S}_w+\mathbf{S}_b)^{-1}, \quad d_3 = -1, \end{aligned} \tag{6.4.1}$$

where $n = n_1 + \cdots + n_q$.

Since the null distributions of the three test statistics are invariant under the transformation from $\boldsymbol{Y}_{ij}$ to $\Sigma^{-1/2}(\boldsymbol{Y}_{ij} - \mu_i)$, we may start by assuming that $\boldsymbol{Y}_{ij}(j = 1,\ldots,n_i; i = 1,\ldots,q)$ are i.i.d. samples of $\boldsymbol{Y}$ with $\mathrm{E}(\boldsymbol{Y}) = 0$ and $\mathrm{Cov}(\boldsymbol{Y}) = \mathrm{I}_p$. For $i = 1,\ldots,q$, let

$$\boldsymbol{Z}_i = \sqrt{n_i}\,\bar{\boldsymbol{Y}}_i, \quad \mathbf{V}_i = \sqrt{n_i}(\mathbf{S}_i - \mathrm{I}_p) \tag{6.4.2}$$

and

$$\mathbf{Z} = (\boldsymbol{Z}_1,\ldots,\boldsymbol{Z}_q)', \quad \mathbf{V} = \rho_1\mathbf{V}_1 + \cdots + \rho_q\mathbf{V}_q, \tag{6.4.3}$$

where $\rho_i = \sqrt{n_i/n}$ and $\boldsymbol{\rho} = (\rho_1,\cdots,\rho_q)'$. In Section 5.3 we treated the null distribution of the T^2 statistic in a one-sample problem under nonnormality. For its large-sample asymptotic expansion, conditions C1 and C2-1 (or C2-2) were assumed. In addition to the assumptions, we assume C3 on the sample sizes:

C1. $\mathrm{E}(\|\boldsymbol{Y}\|^8) < \infty$.

C2-1. $\boldsymbol{Y}$ has an absolutely continuous component.

C3. $\rho_i \to c_i \in (0,1)$ as $n \to \infty$, $i = 1,\ldots,q$.

We can express $\mathbf{S}_h$ and $\mathbf{S}_e$ as

$$\mathbf{S}_h = \sum_{i=1}^{q} n_i\bar{\boldsymbol{Y}}_i\bar{\boldsymbol{Y}}_i' - n\bar{\boldsymbol{Y}}\bar{\boldsymbol{Y}}' = \mathbf{Z}\Omega\mathbf{Z}' \tag{6.4.4}$$

and

$$\frac{1}{n}\mathbf{S}_e = \mathrm{I}_p + \frac{1}{\sqrt{n}}\mathbf{V} - \frac{q}{n}\mathrm{I}_p + \mathrm{O}_p(n^{-3/2}), \tag{6.4.5}$$

where $\Omega = \mathrm{I}_q - \boldsymbol{\rho}\boldsymbol{\rho}'$. Note that each of the three statistics is a smooth function of $\mathbf{Z} = (\boldsymbol{Z}_1,\ldots,\boldsymbol{Z}_q)'$ and $\mathbf{V}_1,\ldots,\mathbf{V}_q$. We have seen that each of the distributions of $(\boldsymbol{Z}_i, \mathbf{V}_i)$ has a valid expansion under C1 and C2-1. By an argument similar to the one in Hotelling T^2 statistic we can see that each distribution of the three statistics has a valid expansion up to the order n^{-1} under assumptions C1, C2-1, and C3.

Using (6.4.4) and (6.4.5), we can expand our test statistics T_i $(i = 1,2,3)$ as

$$T = T^{(0)} + \frac{1}{\sqrt{n}}T^{(1)} + \frac{1}{n}T^{(2)} + \mathrm{O}_p(n^{-3/2}), \tag{6.4.6}$$

where

$$\begin{aligned} T^{(0)} &= \operatorname{tr} \mathbf{Z}\Omega\mathbf{Z}', \quad T^{(1)} = -\operatorname{tr} \mathbf{V}\mathbf{Z}\Omega\mathbf{Z}', \\ T^{(2)} &= \operatorname{tr} \mathbf{Z}\Omega\mathbf{Z}'(\mathbf{V}^2 + r_1 \mathbf{I}_p) + r_2 \operatorname{tr} \{\mathbf{Z}\Omega\mathbf{Z}'\}^2, \end{aligned}$$

with the constants

$$(r_1, r_2) = (d_1, -1/2), \ (d_2, 0), \ (d_3, -1) \tag{6.4.7}$$

for T_1, T_2, and T_3, respectively. It may be noted that the leading term of $T^{(i)}$ is a homogeneous polynomial of degree $i + 2$ in the elements of $\mathbf{Z}$ and $\mathbf{V}$, but it is a polynomial of degree less than i in the elements of $\mathbf{V}$. This property should be related to the problem of relaxing the moment condition C1 to C1-1: $\mathrm{E}(\| \boldsymbol{Y} \|^4) < \infty$.

Using (6.4.6), the characteristic function of T can be expanded as

$$C_T(t) = (1 - 2it)^{-p(q-1)/2} \left[1 + \frac{1}{n} \sum_{j=0}^{3} b_j (1 - 2it)^{-j} + \mathrm{o}(n^{-1}) \right], \tag{6.4.8}$$

where

$$\begin{aligned} b_0 &= \frac{1}{4} p(q-1)\{q - p - 2 - 2r_1\} \\ &\quad - (a_2 \kappa_{[3]}^{(1)} + a_3 \kappa_{[3]}^{(2)}) + a_1 \kappa_{[4]}^{(1)}, \\ b_1 &= \frac{1}{2} p(q-1)\{-q + 1 + r_1 - r_2(p+q)\} \\ &\quad + 3(a_2 \kappa_{[3]}^{(1)} + a_3 \kappa_{[3]}^{(2)}) - 2a_1 \kappa_{[4]}^{(1)}, \\ b_2 &= \frac{1}{4} p(q-1)(p+q)(1 + 2r_2) \\ &\quad - 3(a_2 \kappa_{[3]}^{(1)} + a_3 \kappa_{[3]}^{(2)}) + a_1 \kappa_{[4]}^{(1)}, \\ b_3 &= a_2 \kappa_{[3]}^{(1)} + a_3 \kappa_{[3]}^{(2)}, \end{aligned} \tag{6.4.9}$$

with $a_1 = \frac{1}{8}(n_* - q^2 - 2q + 2)$, $a_2 = \frac{1}{12}(n_* - 3q + 2)$, $a_3 = \frac{1}{8}(n_* - q^2)$, and $n_* = \sum_{i=1}^{q} n/n_i$. The quantities $\kappa_{[3]}^{(1)}$, $\kappa_{[3]}^{(2)}$, and $\kappa_{[4]}^{(1)}$ are the multivariate skewness and kurtosis defined in Section 5.2. Inverting (6.4.8), we obtain the following result, which was derived by Fujikoshi (2002, 2002c). For an extension to multivariate linear hypothesis, see Wakaki et al. (2002).

Theorem 6.4.1 *Under assumptions C1, C2-1, and C3 the null distribution* *of* T $(= T_1, T_2, T_3)$ *can be expanded as*

$$\begin{aligned}\Pr\{T \le x\} &= G_f(x) + \frac{1}{n}\sum_{j=0}^{3} b_j G_{f+2j}(x) + \mathrm{o}(n^{-1}) \\ &= G_f(x) - \frac{2x}{nf} g_f(x)\Bigg\{b_1 + b_2 + b_3 \\ &\quad + \frac{(b_2+b_3)x}{f+2} + \frac{b_3 x^2}{(f+2)(f+4)}\Bigg\} + \mathrm{o}(n^{-1}), \end{aligned} \tag{6.4.10}$$

where $f = p(q-1)$, *and* G_f *is the distribution function of the central chi-square distribution with* f *degrees of freedom, and the coefficients* b_j *are given by* (6.4.9).

It may be noted that the final result depends on the cumulants up to fourth order. On the other hand, applying the Cornish–Fisher formula to (6.4.10), we can obtain an asymptotic expansion for the upper percentage point of the distribution of T, as in the following corollary.

Corollary 6.4.1 *Let* u *be the upper* $100(1-\alpha)\%$ *point of a* $\chi^2(f)$, *and let*

$$t = u + \frac{2u}{nf}\left\{b_1 + b_2 + b_3 + \frac{(b_2+b_3)u}{f+2} + \frac{b_3 u^2}{(f+2)(f+4)}\right\} \tag{6.4.11}$$

Then

$$\Pr\{T \le t\} = 1 - \alpha + \mathrm{O}(n^{-1}).$$

In the special case $p = 1$, we obtain an asymptotic expansion (6.4.10) with $f = q - 1$ and the following coefficients:

$$\begin{aligned} b_0 &= \frac{1}{4}(q-1)\{q - 3 - 2r_1 - 2r_2(q+1)\} - \tilde{a}_2\kappa_3^2 + a_1\kappa_4, \\ b_1 &= \frac{1}{2}(q-1)\{1 - q + r_1 + r_2(q+1)\} + 3\tilde{a}_2\kappa_3^2 - 2a_1\kappa_4, \\ b_2 &= \frac{1}{4}(q-1)(q+1) - 3\tilde{a}_2\kappa_3^2 + a_1\kappa_4, \\ b_3 &= \tilde{a}_2\kappa_3^2, \end{aligned} \tag{6.4.12}$$

where $\tilde{a}_2 = \frac{1}{24}(5\sum_{\alpha=1}^{q}\rho_\alpha^{-2} - 3q^2 - 6q + 4)$, and κ_3 and κ_4 are the standardized third- and fourth-order cumulants.

A Bartlett correction for T under nonnormality is defined by

$$\tilde{T} = \left(1 - \frac{c}{n}\right)T, \tag{6.4.13}$$

where the constant c is determined by

$$\mathrm{E}(T) = f\left\{1 + \frac{c}{n} + \mathrm{o}(n^{-1})\right\}.$$

From (6.4.10) it is seen that

$$\begin{aligned} c &= \frac{2}{f}\sum_{j=1}^{3} jb_j \\ &= p + 1 + r_1 + r_2(p+q) \\ &= 0 \quad \text{for the three cases (6.4.7)} . \end{aligned} \tag{6.4.14}$$

This means that the Bartlett corrections under nonnormality do not depend on the third- and fourth-order cumulants, and furthermore, they are the same as the Bartlett corrections under normality; that is,

$$\mathrm{E}(T_i) = f + \mathrm{o}(n^{-1}), \qquad i = 1, 2, 3. \tag{6.4.15}$$

For example, the Lawley Hotteling statistic T_2 with Bartlett correction in two-sample case is

$$T_2 = (n - p - 3)\frac{n_1 n_2}{n(n-2)}(\bar{\boldsymbol{Y}}^{(1)} - \bar{\boldsymbol{Y}}^{(2)})'\mathbf{S}^{-1}(\bar{\boldsymbol{Y}}^{(1)} - \bar{\boldsymbol{Y}}^{(2)}),$$

and under the null hypothesis, $\mathrm{E}(T_2) = p + \mathrm{o}(n^{-1})$.

On the other hand, consider Hotelling's T^2 statistic for the one-sample case given by

$$T^2 = n(\bar{\boldsymbol{X}} - \boldsymbol{\mu})'\mathbf{S}^{-1}(\bar{\boldsymbol{X}} - \boldsymbol{\mu}).$$

From Theorem 5.3.2 we have

$$\mathrm{E}(T^2) = p + \frac{1}{n}\{\kappa_3^{(1)} + \kappa_3^{(2)} + p(p+2)\} + \mathrm{o}(n^{-1}).$$

This shows that the Bartlett correction for T^2 depends on the third-order cumulants $\kappa_3^{(1)}$ and $\kappa_3^{(2)}$, but the correction for T_2 does not depend on them. In this sense we can say that the two-sample test is more robust than the one-sample test. This confirms results due to Chase and Bulgren (1971) and Everitt (1979) based on Monte Carlo investigation.

6.4.2 Bootstrap Tests

Fisher and Hall (1990) proposed the method of constructing bootstrap hypothesis tests using the example of one- and two-way ANOVA models. The aim in this section is to extend their idea to the construction of bootstrap tests in MANOVA models.

For simplicity we consider a one-way MANOVA model or multivariate q-sample problems. Assume that a random sample $\{\boldsymbol{Y}_{ij},\ 1 \le j \le n_i\}$ is gathered from a p-dimensional population Π_i for $i = 1, \dots, q$. The ith population has mean vector $\boldsymbol{\mu}_i$ and covariance matrix Σ_i, where the $\boldsymbol{\mu}_i$ and Σ_i are unknown.

Suppose that the Π_i have a common covariance matrix (i.e., $\Sigma_1 = \cdots = \Sigma_q$). However, without making the distributional assumptions about the Π's, we wish to test the null hypothesis

$$H_0 \colon \boldsymbol{\mu}_1 = \cdots = \boldsymbol{\mu}_q.$$

Put

$$n = \sum_{i=1}^{q} n_i, \quad \bar{\boldsymbol{Y}}_i = n_i^{-1} \sum_{j=1}^{n_j} \boldsymbol{Y}_{ij}, \quad \bar{\boldsymbol{Y}} = n^{-1} \sum_{i=1}^{q} \sum_{j=1}^{n_i} \boldsymbol{Y}_{ij}.$$

We define the sample covariance matrix based on a random sample $\{\boldsymbol{Y}_{ij},\ 1 \le j \le n_i\}$ from Π_i by

$$\mathbf{S}_i = \frac{1}{n_i} \sum_{j=1}^{n_i} (\boldsymbol{Y}_{ij} - \bar{\boldsymbol{Y}}_i)(\boldsymbol{Y}_{ij} - \bar{\boldsymbol{Y}}_i)',$$

not the usual unbiased estimator. Since the population covariance matrices are common, it is natural to use a statistic based on the between-group SSP $\mathbf{S}_b$ and the within-group SSP $\mathbf{S}_w$, which are defined in (6.3.1); that is,

$$\mathbf{S}_b = \sum_{i=1}^{q} n_i (\bar{\boldsymbol{Y}}_i - \bar{\boldsymbol{Y}})(\bar{\boldsymbol{Y}}_i - \bar{\boldsymbol{Y}})', \quad \mathbf{S}_w = \sum_{i=1}^{q} n_i \mathbf{S}_i.$$

As one such multivariate test, we take up the Lawley–Hotelling statistic given by

$$T = n \ \mathrm{tr}\, \mathbf{S}_b \mathbf{S}_w^{-1},$$

since our bootstrap tests are constructed similarly for the other test statistics.

Our concern is how to estimate the percentage point of T based on the bootstrap method and to examine its theoretical accuracy.

First we note that T is invariant under the transformation

$$\boldsymbol{Y}_{ij} \ \rightarrow \ \boldsymbol{Z}_{ij} = \Sigma_i^{-1/2} (\boldsymbol{Y}_{ij} - \boldsymbol{\mu}_i). \tag{6.4.16}$$

Put $\mathcal{Z} = \{\boldsymbol{Z}_{ij};\ 1 \le j \le n_i, 1 \le i \le q\}$. Let $\mathbf{S}_b(\mathcal{Z})$ and $\mathbf{S}_w(\mathcal{Z})$ be the random matrices obtained from $\mathbf{S}_b$ and $\mathbf{S}_w$, respectively, by substituting $\boldsymbol{Z}_{ij}$ for $\boldsymbol{Y}_{ij}$. Put

$$T_1 = n \ \mathrm{tr}\, \mathbf{S}_b(\mathcal{Z}) \mathbf{S}_w(\mathcal{Z})^{-1}. \tag{6.4.17}$$

Then the distribution of T_1 is the same as that of T, since $T_1 = T$. Using this property, two resampling schemes are proposed. In the following we describe these methods along the lines of Fisher and Hall (1990).

Resampling Scheme 1. Let $\mathcal{Y}^* = \{\boldsymbol{Y}_{i1}^*, \ldots, \boldsymbol{Y}_{in_i}^*\}$ be a resample, with replacement, from the sample $\mathcal{Y} = \{\boldsymbol{Y}_{i1}, \ldots, \boldsymbol{Y}_{in_i}\}$. Put

$$\boldsymbol{Z}_{ij}^* = \Sigma_i^{-1/2} (\boldsymbol{Y}_{ij}^* - \bar{\boldsymbol{Y}}_i), \quad \mathcal{Z}^* = \{\boldsymbol{Z}_{ij}^*;\ 1 \le j \le n_i, 1 \le i \le q\}, \tag{6.4.18}$$

and let

$$T_1^* = n \ \operatorname{tr} \mathbf{S}_b(\mathcal{Z}^*)\mathbf{S}_w(\mathcal{Z}^*)^{-1}. \tag{6.4.19}$$

An approximate test of H_0 with significance level α is to reject H_0 if $T > \hat{y}_1$, where $\hat{y}_1$ is defined by

$$\Pr\{T_1^* \le \hat{y}_1|\mathcal{Y}\} = 1 - \alpha.$$

Note that $\hat{y}_1$ is numerically computed (see Section 5.8) by resampling the value of T_1^*. Now we examine the error in the bootstrap approximation to the null distribution of T_1 $(= T)$. By modifying the derivation of Theorem 6.4.1, we know under the appropriate regularity conditions that the null distribution of T_1 can be expanded as

$$\Pr\{T_1 \le y\} = G_f(y) + n^{-1}p_1(y) + \mathrm{O}(n^{-2}), \tag{6.4.20}$$

where p_1 is a function depending on Σ and the third- and the fourth-order cumulants of each population, but not on n. Let $\hat{p}_1$ denote the version of p_1 when all population cumulants are replaced by their sample cumulants. Then

$$\begin{aligned}\Pr\{T_1^* \le y|\mathcal{Y}\} &= G_f(y) + n^{-1}\hat{p}_1(y) + \mathrm{O}_p(n^{-2}) \\ &= G_f(y) + n^{-1}p_1(y) + \mathrm{O}_p(n^{-3/2}).\end{aligned} \tag{6.4.21}$$

The last expression is obtained since $\hat{p}_1 = p_1 = \mathrm{O}_p(n^{-1/2})$. Therefore,

$$\Pr\{T_1^* \le y|\mathcal{Y}\} = \Pr(T_1 \le y) + \mathrm{O}_p(n^{-3/2}).$$

Resampling Scheme 2. This sampling scheme is the case where populations are assumed distributed identically around means. Let $\mathcal{W}$ denote the set of all n values of $\boldsymbol{W}_{ij} = \mathbf{S}_i^{-1/2}(\boldsymbol{X}_{ij}^* - \bar{\boldsymbol{X}}_i)$ (i.e., $\mathcal{W} = \{\boldsymbol{W}_{ij};\ j \le n_i, i \le q\}$) and let

$$\mathcal{W}^* = \{\boldsymbol{W}_{ij}^*;\ j \le n_i, i \le q\}$$

be an n-sample drawn with replacement from $\mathcal{W}$. Further, Put

$$T_2^* = n \ \operatorname{tr} \mathbf{S}_b(\mathcal{W}^*)\mathbf{S}_w(\mathcal{W}^*)^{-1}. \tag{6.4.22}$$

An approximate test of the hypothesis H_0 with significance level α is to reject H_0 if $T > \hat{y}_2$, where $\hat{y}_2$ is defined by

$$\Pr\{T_2^* \le \hat{y}_2|\mathcal{Y}\} = 1 - \alpha.$$

If populations are identically distributed around their means. Then, repeating the argument above for T_1^* for T_2^*, it is shown that

$$\Pr\{T_2^* \le y|\mathcal{Y}\} = \Pr\{T_1 \le y\} + \mathrm{O}_p(n^{-3/2}).$$

6.5 Distributions of Characteristic Roots

Let $\mathbf{S}_h$ and $\mathbf{S}_e$ be two $p \times p$ random matrices that appear in MANOVA problems. Suppose that $\mathbf{S}_h \sim \mathrm{W}_p(q, \Sigma; \mathrm{MM}')$, $\mathbf{S}_e \sim \mathrm{W}_p(n, \Sigma)$, and they are independent. We are interested in the distributions of the characteristic roots of $\mathbf{S}_h\mathbf{S}_e^{-1}$ or $\mathbf{S}_h(\mathbf{S}_h + \mathbf{S}_e)^{-1}$. In canonical discriminant analysis (see Section 9.5) the characteristic roots denote discrimination powers of the canonical discriminant functions. The smallest roots are used for tests of the dimensionality of the noncentrality matrix or canonical discriminant analysis. The largest roots are used for the construction of simultaneous confidence intervals (see Section 7.6).

When we consider the distribution of these characteristic roots, we may assume that $\Sigma = \mathrm{I}_p$ and the noncentrality matrix is $\Omega = \mathrm{diag}\,(\omega_1, \cdots, \omega_p)$, where $\omega_1 \geq \cdots \geq \omega_p \geq 0$ are the characteristic roots of $\Sigma^{-1/2}\mathrm{MM}'\Sigma^{-1/2}$. Let $1 > r_1 > \cdots > r_p > 0$ be the characteristic roots of $\mathbf{S}_h(\mathbf{S}_h + \mathbf{S}_e)^{-1}$, which are solutions of

$$|\mathbf{S}_h - r(\mathbf{S}_h + \mathbf{S}_e)| = 0.$$

Similarly, $\ell_1 > \cdots > \ell_p > 0$ are the characteristic roots of $\mathbf{S}_h\mathbf{S}_e^{-1}$, which are the solutions of

$$|\mathbf{S}_h - \ell\mathbf{S}_e| = 0.$$

In this section it is assumed without loss of generality that

$$\mathbf{S}_h \sim \mathrm{W}(q, \mathrm{I}_q; \Omega), \quad \mathbf{S}_e \sim \mathrm{W}_p(n, \mathrm{I}_p) \tag{6.5.1}$$

and $\mathbf{S}_h$ and $\mathbf{S}_e$ are independent, where $\Omega = \mathrm{diag}\,(\omega_1, \ldots, \omega_p)$.

6.5.1 Exact Distributions

The exact density function of the characteristic roots in MANOVA problems is very complicated, although it is expressed in terms of hypergeometric functions of matrix arguments. However, the density function under $\Omega = \mathrm{O}$ is expressed in a simple form.

Theorem 6.5.1 *Suppose that $\Omega = \mathrm{O}$. Then if $q \geq p$, the density functions of $1 > r_1 > \cdots > r_p > 0$ and $\ell_1 > \cdots > \ell_p > 0$ are*

$$\frac{\pi^{p^2/2}\Gamma_p((n+q)/2)}{\Gamma_p(n/2)\Gamma_p(q/2)\Gamma_p(p/2)} \prod_{i=1}^{p} r_i^{(q-p-1)/2} \prod_{i=1}^{p}(1-r_i)^{(n-p-1)/2} \prod_{i<j}(r_i - r_j),$$

$$\frac{\pi^{p^2/2}\Gamma_p((n+q)/2)}{\Gamma_p(n/2)\Gamma_p(q/2)\Gamma_p(p/2)} \prod_{i=1}^{p} \ell_i^{(q-p-1)/2} \prod_{i=1}^{p}(\ell_i+1)^{(n-p-1)/2} \prod_{i<j}(\ell_i - \ell_j),$$

respectively. The density functions for $p > q$ are obtained from the those for $p \leq q$ by changing

$$p \to q, \;\; q \to p, \;\; n \to n - p + q. \tag{6.5.2}$$

Proof. For notational simplicity, we write $\mathbf{S}_h = \mathbf{B}$ and $\mathbf{S}_e = \mathbf{W}$. Then the joint density function of $\mathbf{B}$ and $\mathbf{W}$ is

$$\frac{1}{2^{p(n+q)/2}\Gamma_p\left(\frac{1}{2}q\right)\Gamma_p\left(\frac{1}{2}n\right)}|\mathrm{B}|^{(q-p-1)/2}|\mathrm{W}|^{(n-p-1)/2}\mathrm{e}^{-(1/2)\,\mathrm{tr}\,(\mathrm{B}+\mathrm{W})}.$$

Consider the following transformation from (B, W) to (H, R):

$$\mathrm{BH} = (\mathrm{B}+\mathrm{W})\mathrm{HR},$$

where $\mathrm{R} = \mathrm{diag}\,(r_1, \ldots, r_p)$ and H is a nonsingular matrix. In this transformation, the column vectors of H are the characteristic vectors of $\mathrm{B}(\mathrm{B}+\mathrm{W})^{-1}$ or BW^{-1} with lengths $\mathrm{H}'(\mathrm{B}+\mathrm{W})\mathrm{H} = \mathrm{I}_p$. The diagonal elements of R are the characteristic roots of $\mathrm{B}(\mathrm{B}+\mathrm{W})^{-1}$. Then, from [A.3.7] and [A.3.5],

$$J(\mathrm{R}, \mathrm{H} \to \mathrm{B}, \mathrm{W}) = 2^p|\mathrm{H}|^{p+2}\prod_{i<j}(r_i - r_j),$$

$$J(\mathrm{H}^{-1} \to \mathrm{H}) = |\mathrm{H}|^{p+1}.$$

Therefore the joint density function of $\mathbf{R}$ and $\mathbf{E} = \mathbf{H}^{-1}$ is

$$\frac{2^{p(2-n-q)/2}}{\Gamma_p\left(\frac{1}{2}q\right)\Gamma_p\left(\frac{1}{2}n\right)}|\mathrm{E}'\mathrm{E}|^{(n+q-p)/2}\mathrm{e}^{-\frac{1}{2}\,\mathrm{tr}\,\mathrm{E}'\mathrm{E}}$$
$$\times \prod_{i=1}^{p} r_i^{(q-p-1)/2}\prod_{i=1}^{p}(1-r_i)^{(n-p-1)/2}\prod_{i<j}(r_i - r_j).$$

Integrating the joint density function with respect to E, the joint density function of $1 > r_1 > \cdots > r_p > 0$ is obtained. The joint density function of $\mathrm{L} = \mathrm{diag}\,(\ell_1, \ldots, \ell_p)$ is obtained from the joint density function of R by considering the transformation

$$\ell_i = \frac{r_i}{1-r_i}, \quad i = 1, \ldots, p.$$

The results for $q < p$ are obtained from (6.3.4). □

Using Theorem 6.5.1, the distribution function of the maximum root r_1 in the special case $p = 2$ is expressed as

$$\Pr\{r_1 \le x\} = I_x(q-1, n-1) - \frac{\sqrt{\pi}\,\Gamma(\frac{1}{2}(n+q-1))}{\Gamma(\frac{1}{2}q)\Gamma(\frac{1}{2}n)}$$
$$\times\, x^{(q-1)/2}(1-x)^{(n-1)/2} I_x\left(\frac{1}{2}(q-1), \frac{1}{2}(n-1)\right),$$

where $I_x(a, b)$ is the incomplete beta function defined by

$$I_x(a, b) = \int_0^x u^{a-1}(1-u)^{b-1}du.$$

6.5.2 Large-Sample Case

In this section we consider asymptotic distributions of the characteristic roots $\ell_1 > \cdots > \ell_p > 0$ of $\mathbf{S}_h\mathbf{S}_e^{-1}$ under a large-sample framework:

A1: p and q are fixed, $n \to \infty$.

The results are obtained by using a perturbation method. For the method it is basic to express S_h and S_e as

$$\begin{aligned}\frac{1}{n}\mathbf{S}_h &= \frac{1}{n}(\mathbf{X}_1 + \Omega^{1/2},\ \mathbf{X}_2)(\mathbf{X}_1 + \Omega^{1/2},\ \mathbf{X}_2)' \\ &= \Delta + \frac{1}{\sqrt{n}}\mathbf{X}^{(1)} + \frac{1}{n}\mathbf{X}^{(2)}, \qquad (6.5.3)\\ \frac{1}{n}\mathbf{S}_e &= I_p + \frac{1}{\sqrt{n}}\mathbf{V}, \qquad (6.5.4)\end{aligned}$$

where $\Delta = \mathrm{diag}\,(\delta_1, \ldots, \delta_p) = (1/n)\Omega$, $\mathbf{X} = (\mathbf{X}_1,\ \mathbf{X}_2)$ is a $p \times q$ random matrix whose elements are independently distributed as $\mathrm{N}(0, 1)$, $\mathbf{X}_1 : p \times p$, and

$$\mathbf{X}^{(1)} = \mathbf{X}_1\Delta^{1/2} + \Delta^{1/2}\mathbf{X}_1, \quad \mathbf{X}^{(2)} = \mathbf{X}'\mathbf{X}.$$

For the order of Ω, it is assumed that

B1: $\Omega = \mathrm{O}(n)$,

and hence $\Delta = \mathrm{O}(1)$. The characteristic roots $\ell_1 > \cdots > \ell_p > 0$ are the characteristic roots of $\mathbf{Y} = \mathbf{S}_e^{-1/2}\mathbf{S}_h\mathbf{S}_e^{-1/2}$, and $\mathbf{Y}$ is expanded as

$$\begin{aligned}\mathbf{Y} &= \left(I_p + \frac{1}{\sqrt{n}}\mathbf{V}\right)^{-1/2}\left(\Delta + \frac{1}{\sqrt{n}}\mathbf{X}^{(1)} + \frac{1}{n}\mathbf{X}^{(2)}\right) \\ &\quad \times \left(I_p + \frac{1}{\sqrt{n}}\mathbf{V}\right)^{-1/2} \\ &= \Delta + \frac{1}{\sqrt{n}}\mathbf{Y}^{(1)} + \frac{1}{n}\mathbf{Y}^{(2)} + \cdots, \qquad (6.5.5)\end{aligned}$$

where

$$\begin{aligned}\mathbf{Y}^{(1)} &= \mathbf{X}^{(1)} - \frac{1}{2}(\Delta\mathbf{V} + \mathbf{V}\Delta), \\ \mathbf{Y}^{(2)} &= \mathbf{X}^{(2)} - \frac{1}{2}\left(\mathbf{V}\mathbf{X}^{(1)} + \mathbf{X}^{(1)}\mathbf{V}\right) \\ &\quad + \frac{1}{4}\mathbf{V}\Delta\mathbf{V} + \frac{3}{8}(\Delta\mathbf{V}^2 + \mathbf{V}^2\Delta).\end{aligned}$$

Now it is possible to give perturbation expansions for the characteristic roots $\ell_1 > \cdots > \ell_p > 0$ based on the results in Section 5.5. For example, if ω_α is simple, the limiting distribution of $\sqrt{n}(\ell_\alpha - \delta_\alpha)$ is the same as that of

$$Y_{\alpha\alpha}^{(1)} = 2\sqrt{\delta_\alpha} X_{\alpha\alpha} - \delta_\alpha V_{\alpha\alpha},$$

where $\mathbf{V} = (V_{\alpha\beta})$, $\mathbf{X} = (X_{\alpha\beta})$, and $\mathbf{Y}^{(i)} = (Y_{\alpha\beta}^{(i)}), i = 1, 2$. Then it is easy to find the limiting distribution of $Y_{\alpha\alpha}^{(1)}$, so that the limiting distribution of $\sqrt{n}(\ell_\alpha - \delta_\alpha)$ is $\mathrm{N}(0, 4\delta_\alpha + 2\delta_\alpha^2)$. More generally, suppose that

B1: $\quad \mathrm{rank}(\Omega) = k$, i.e., $\omega_k > \omega_{k+1} = \cdots = \omega_p = 0$,
Ω has r different roots with multiplicities $q_1, \ldots, q_r$, i.e.,

$$\begin{aligned} \omega_1 = \cdots = \omega_{q_1} &= n\lambda_1, \\ \omega_{q_1+1} = \cdots = \omega_{q_1+q_2} &= n\lambda_2, \\ &\vdots \\ \omega_{p-q_r+1} = \cdots = \omega_p &= n\lambda_r = 0, \end{aligned} \tag{6.5.6}$$

where $\lambda_1 > \lambda_2 > \cdots > \lambda_r = 0$ and $\sum_{\alpha=1}^{r} q_\alpha = p$. Put

$$\begin{aligned} Z_i &= \left(\frac{n}{4\lambda_\alpha + 2\lambda_\alpha^2}\right)^{1/2} (\ell_i - \lambda_\alpha), \quad i \in J_\alpha, \quad \alpha = 1, \ldots, r-1, \\ Z_i &= n\ell_i, \quad i \in J_r, \end{aligned} \tag{6.5.7}$$

where J_α denotes the set of integers

$$\sum_{i=1}^{\alpha-1} q_i + 1, \ldots, \sum_{i=1}^{\alpha} q_i$$

with $q_0 = 0$.

Theorem 6.5.2 *Let $\mathbf{S}_h$ and $\mathbf{S}_e$ be independent Wishart matrices such that $\mathbf{S}_h \sim \mathrm{W}_p(q, \Sigma; MM')$ and $\mathbf{S}_e \sim \mathrm{W}_p(n, \Sigma)$. Under $q \ge p$, let $\ell_1 > \cdots > \ell_p$ be the characteristic roots of $\mathbf{S}_h\mathbf{S}_e^{-1}$ with $n \ge \max\{p, q\}$. Let $\omega_1 \ge \cdots \ge \omega_p$ be the characteristic roots of $MM'\Sigma^{-1}$, and put $\Omega = \mathrm{diag}(\omega_1, \ldots, \omega_p)$. Under assumptions A1 and B1, the normalized roots $(Z_1, \ldots, Z_p)$ in (6.5.7) are asymptotically distributed as*

$$\left\{\prod_{\alpha=1}^{r-1} f_1(\{z_i,\ i \in J_\alpha\}; q_\alpha)\right\} f_0(L_r; m, s), \tag{6.5.8}$$

where

$$f_1(\{z_i;\ i \in J_\alpha\}; q_\alpha) = \frac{\pi^{q_\alpha(q_\alpha-1)/4}}{2^{q_\alpha/2}\Gamma_{q_\alpha}\left(\frac{1}{2}q_\alpha\right)} \exp\left(-\frac{1}{2}\sum_{i\in J_\alpha} z_i^2\right) \times \prod_{i<j;i,j\in J_\alpha} (z_i - z_j), \tag{6.5.9}$$

$$f_0(\{z_i;\ i \in J_r\}; s, t) = \frac{\pi^{t^2/2}}{2^{st/2}\Gamma_t\left(\frac{1}{2}s\right)\Gamma_t\left(\frac{1}{2}t\right)} \exp\left(-\frac{1}{2}\sum_{i=1}^{t} z_{k+i}\right) \times \prod_{i=1}^{t} z_{k+i}^{(s-t-1)/2} \prod_{i<j}^{t} (z_{k+i} - z_{k+j}), \tag{6.5.10}$$

$s = q - k$, $t = p - k$, *and* $\sigma_\alpha^2 = 4\lambda_\alpha + 2\lambda_\alpha^2$. *For the results for* $p \geq q$, *only* p *and* q *has to change.*

Proof. Let $\mathbf{Y}^{(i)}$, $i = 1, 2$ be partitioned in $q_1, \ldots q_r$ columns and rows and denote their (j,k) block submatrices by $\mathbf{Y}_{jk}^{(i)}$, $j, k = 1, \ldots, r$. By using perturbation expansion (Theorem 5.5.2) of the characteristic roots with multiplicities we can see that the limiting distribution of $\sqrt{n}(\ell_i - \lambda_\alpha), i \in J_\alpha$ $(\alpha = 1, \ldots, r-1)$ is the same as that of the characteristic roots of $\mathbf{X}_{\alpha\alpha}^{(1)}$. Similarly, the limiting distribution of $Z_i = n\ell_i$, $i = k+1, \ldots, p$ is the same as that of the characteristic roots of $\mathbf{Y}_{rr}^{(2)}$. These imply the results required for $q \geq p$. The results for $p \geq q$ are obtained from those for $q \geq p$. The distribution of the characteristic roots $\ell_1 > \cdots > \ell_q > 0$ for $p \geq q$ becomes that for $q \geq p$ by changing (6.5.1) (see also Theorem 3.2.4). □

6.5.3 High-Dimensional Case

In this section we assume that $p > q$ and consider asymptotic distribution of the possibly nonzero roots $\ell_1 > \cdots > \ell_q > 0$ of $\mathbf{S}_h\mathbf{S}_e^{-1}$ under a high-dimensional framework:

$$\text{A2:}\quad q\text{: fix },\ c = p/n \to c_0 \in (0, 1).$$

A key idea is to use the fact that the possibly nonzero roots may be defined by two $q \times q$ random matrices, not two $p \times p$ matrices. In fact, the possibly nonzero roots are those of $\mathbf{BW}^{-1}$. Here $\mathbf{B}$ and $\mathbf{W}$ are distributed independently as a noncentral Wishart distribution $W_q(p, \mathrm{I}_q; \Omega)$ and a central Wishart distribution $W_q(m, \mathrm{I}_q)$, respectively, where $m = n - p + q$. The matrix Ω is a $q \times q$ diagonal matrix with the same nonzero diagonal elements $\omega_1 \geq \cdots \geq \omega_q \geq 0$ as those for $q \geq p$.

Now we give an outline for deriving high-dimensional asymptotic distributions. Let $\mathbf{U}$ and $\mathbf{V}$ be matrices defined by

$$\mathbf{U} = \frac{1}{\sqrt{p}}(\mathbf{B} - p\mathbf{I}_q - \Omega) \quad \text{and} \quad \mathbf{V} = \frac{1}{\sqrt{m}}(\mathbf{W} - m\mathbf{I}_q),$$

respectively. Note that we use the same notations in the large-sample case for Ω, $\mathbf{V}$, and so on, but they are slightly different. Then it is checked that $\mathbf{U}$ is asymptotically normal as $p \to \infty$, and $\mathbf{V}$ is asymptotically normal as $m \to \infty$. The random matrices $\mathbf{B}$ and $\mathbf{W}$ are expressed in terms of $\mathbf{U}$ and $\mathbf{V}$ as

$$\mathbf{B} = p\left(I_q + \frac{1}{p}\Omega + \sqrt{p}\,\mathbf{U}\right) = m\left(\Gamma + \frac{1}{\sqrt{m}}\sqrt{r}\,\mathbf{U}\right),$$
$$\mathbf{W} = m\left(I_q + \frac{1}{\sqrt{m}}\mathbf{V}\right),$$

where

$$\Gamma = \operatorname{diag}(\gamma_1, \ldots, \gamma_q), \quad \gamma_i = c + \frac{1}{m}\omega_i, \quad i = 1, \ldots, q. \tag{6.5.11}$$

Using the expressions above, we can express

$$\begin{aligned}\mathbf{Y} &= \mathbf{W}^{-1/2}\mathbf{B}\mathbf{W}^{-1/2} \\ &= \Gamma + \frac{1}{\sqrt{m}}\mathbf{Y}^{(1)} + \cdots,\end{aligned}$$

where

$$\mathbf{Y}^{(1)} = \sqrt{c}\,\mathbf{U} - \frac{1}{2}\mathbf{V}\Gamma - \frac{1}{2}\Gamma\mathbf{V}.$$

For simplicity, suppose that ω_α is simple. Then the limiting distribution of $\sqrt{m}(\ell_\alpha - \gamma_\alpha)$ is the same as that of

$$Y_{\alpha\alpha}^{(1)} = \sqrt{c}\,U_{\alpha\alpha} - \gamma_\alpha V_{\alpha\alpha},$$

where $\mathbf{U} = (U_{\alpha\beta})$ and $\mathbf{V} = (V_{\alpha\beta})$. Then $U_{\alpha\alpha}$ is asymptotically distributed as $\mathrm{N}(0, 2 + 4\omega_\alpha/p)$, $V_{\alpha\alpha}$ is asymptotically distributed as $\mathrm{N}(0, 2)$, and they are asymptotically independent. Therefore, $\sqrt{m}(\ell_\alpha - \gamma_\alpha)$ is asymptotically normal with mean 0 and variance

$$2c(1+c) + 4(1+c)\frac{\omega_\alpha}{m} + 2\left(\frac{\omega_\alpha}{m}\right)^2.$$

More generally, suppose that the roots $\omega_1 \geq \cdots \geq \omega_q \geq 0$ have the multiplicities (6.5.6) with the replacement of p by q. Put

$$Z_i = \frac{\sqrt{m}}{\sigma_\alpha}(\ell_i - \mu_\alpha), \quad i \in J_\alpha, \quad \alpha = 1, \ldots, r, \tag{6.5.12}$$

where for $\alpha = 1, \ldots, r$,

$$\begin{aligned} \mu_\alpha &= c + \frac{n}{m}\lambda_\alpha, \\ \sigma_\alpha^2 &= 2c(1+c) + 4(1+c)\frac{n}{m}\lambda_\alpha + 2\left(\frac{n}{m}\lambda_\alpha\right)^2. \end{aligned} \tag{6.5.13}$$

Theorem 6.5.3 *Let* $\mathbf{S}_h$ *and* $\mathbf{S}_e$ *be independent Wishart matrices such that* $\mathbf{S}_h \sim \mathrm{W}_p(q, \Sigma; MM')$ *and* $\mathbf{S}_e \sim \mathrm{W}_p(n, \Sigma)$. *Under* $p \geq q$, *let* $\ell_1 > \cdots > \ell_q$ *be the characteristic roots of* $\mathbf{S}_h\mathbf{S}_e^{-1}$ *with* $n \geq \max\{p, q\}$. *Let* $\omega_1 \geq \cdots \geq \omega_q$ *be the characteristic roots of* $MM'\Sigma^{-1}$, *and put* $\Omega = \mathrm{diag}\,(\omega_1, \ldots, \omega_q)$. *Under assumptions A2 and B1 (with p replaced by q), the normalized roots* $(Z_1, \ldots, Z_p)$ *in* (6.5.7) *are asymptotically distributed as*

$$\left\{\prod_{\alpha=1}^{r} f_1(\{z_i,\ i \in J_\alpha\}; q_\alpha)\right\}, \tag{6.5.14}$$

where $f_1(\{z_i;\ i \in J_\alpha\}; q_\alpha)$ *is given by* (6.5.9).

Theorem 6.5.3 was stated by Fujikoshi et al. (2008) when $\omega_1 > \cdots > \omega_{k+1} = \cdots = \omega_q = 0$. Now we compare (6.5.8) and (6.5.13). The asymptotic distribution of $Z_i = \sqrt{m}(\ell_i - \mu_\alpha)/\sigma_\alpha, i \in J_\alpha$ in (6.5.12) is the same as that of $Z_i = \sqrt{m}(\ell_i - \lambda_\alpha)/\sqrt{4\lambda_\alpha + 2\lambda_\alpha^2}, i \in J_\alpha$ for $\alpha = 1, \ldots, r-1$, but they are different for $\alpha = r$. It is interesting that normalization in the high-dimensional case under the large-sample case tends to that in the large-sample case for $\alpha = 1, \ldots, r-1$, since $\sqrt{m} \to \sqrt{n}$, $\mu_\alpha \to \lambda_\alpha$, and $\sigma_\alpha \to \sqrt{4\lambda_\alpha + 2\lambda_\alpha^2}$. This means that the high-dimensional approximation includes more information than the large-sample approximation.

6.6 Tests for Dimensionality

6.6.1 Three Test Criteria

In a multivariate q-sample problem, one of the aims is to test the hypothesis

$$H_0\colon \boldsymbol{\mu}_1 = \cdots = \boldsymbol{\mu}_q,$$

based on random samples $\boldsymbol{Y}_{ij}, j = 1, \ldots, n_i$ from p-dimensional normal population $\Pi_i; \mathrm{N}_p(\boldsymbol{\mu}, \Sigma), i = 1, \ldots, q$. The hypothesis is equivalent to

$$\begin{aligned} \Xi &\equiv (\boldsymbol{\mu}_1 - \boldsymbol{\mu}_q, \ldots, \boldsymbol{\mu}_{q-1} - \boldsymbol{\mu}_q) = \mathrm{O}, \\ \Leftrightarrow \tilde{\mathrm{M}} &\equiv (\boldsymbol{\mu}_1 - \bar{\boldsymbol{\mu}}, \ldots, \boldsymbol{\mu}_q - \bar{\boldsymbol{\mu}}) = \mathrm{O}, \\ \Leftrightarrow \Gamma &\equiv \sum_{i=1}^{q} n_i(\boldsymbol{\mu}_i - \bar{\boldsymbol{\mu}})(\boldsymbol{\mu}_i - \bar{\boldsymbol{\mu}})' = \mathrm{O}, \\ \Leftrightarrow \Omega &\equiv \Sigma^{-1/2}\Gamma\Sigma^{-1/2} = \mathrm{O}, \end{aligned}$$

where $\bar{\boldsymbol{\mu}} = (1/n)\sum_{i=1}^{q} n_i \boldsymbol{\mu}_i$ and $n = n_1 + \cdots + n_q$. If H_0 is rejected, one might be interested in testing

$$\begin{aligned} H_{0k}: \operatorname{rank}(\Xi) &= k, \\ \Leftrightarrow \operatorname{rank}(\Omega) &= k, \\ \Leftrightarrow \omega_k &> \omega_{k+1} = \cdots = \omega_p = 0, \end{aligned} \tag{6.6.1}$$

where $\omega_1 \geq \cdots \geq \omega_p \geq 0$ are the characteristic roots of Ω, and $0 \leq k < t = \min\{p, q-1\}$ since $\operatorname{rank}(\Omega) \leq t$. In canonical discriminant analysis (see Section 9.4) it is interesting to determine the number of discriminant functions necessary to describe group differences. The number is defined as $\operatorname{rank}(\Omega)$, which is called the *dimension* or dimensionality in canonical discriminant analysis.

Theorem 6.6.1 *Let λ be the likelihood ratio criterion for testing*

$$H_{0k}: \ \operatorname{rank}(\Omega) = k \text{ vs. } H_{1k}: \ \operatorname{rank}(\Omega) \geq k+1. \tag{6.6.2}$$

Then

$$-2\log\lambda = n \log \prod_{j=k+1}^{t} (1 + \ell_j), \tag{6.6.3}$$

where $\ell_1 > \cdots > \ell_t \geq 0$ are the possibly nonzero characteristic roots of $\mathbf{S}_b\mathbf{S}_w^{-1}$. Here $\mathbf{S}_b$ and $\mathbf{S}_w$ are the between-group and within-group SSPs defined in (6.6.1).

Proof. Let L be the likelihood of all the $\boldsymbol{Y}_{ij}$ observed. Then

$$\begin{aligned} -2\log L &= n\log|\Sigma| + np\log 2\pi \\ &\quad + \operatorname{tr}\Sigma^{-1}\left\{\sum_{i=1}^{k} n_i(\bar{\boldsymbol{y}}_i - \boldsymbol{\mu}_i)(\bar{\boldsymbol{y}}_i - \boldsymbol{\mu}_i)' + \mathrm{S}_e\right\}. \end{aligned}$$

We can see that

$$\begin{aligned} &\sum_{i=1}^{q} n_i(\bar{\boldsymbol{y}}_i - \boldsymbol{\mu}_i)(\bar{\boldsymbol{y}}_i - \boldsymbol{\mu}_i)' \\ &\quad = (\mathrm{Z} - \Xi)\mathrm{Q}^{-1}(\mathrm{Z} - \Xi)' + n(\bar{\boldsymbol{y}} - \bar{\boldsymbol{\theta}})(\bar{\boldsymbol{y}} - \bar{\boldsymbol{\mu}})', \end{aligned}$$

where $\mathrm{Z} = (\bar{\boldsymbol{y}}_1 - \bar{\boldsymbol{y}}_q, \ldots, \bar{\boldsymbol{y}}_{q-1} - \bar{\boldsymbol{y}}_q)$ and

$$\mathrm{Q}^{-1} = \operatorname{diag}(n_1, \ldots, n_{q-1}) - \frac{1}{n}\boldsymbol{n}_{q-1}\boldsymbol{n}_{q-1}'.$$

Then the likelihood of L_1 after maximizing L with respect to $\bar{\boldsymbol{\mu}}$ is

$$\begin{aligned} -2\log L_1 &= n\log|\Sigma| + np\log 2\pi \\ &\quad + \operatorname{tr}\Sigma^{-1}\left\{(\mathrm{Z} - \Xi)\mathrm{Q}^{-1}(\mathrm{Z} - \Xi)' + \mathrm{S}_e\right\}. \end{aligned}$$

The likelihood of L_2 after maximizing L_1 with respect to Σ is

$$\begin{aligned}-2\log L_2 &= n\log|(\mathrm{Z}-\Xi)\mathrm{Q}^{-1}(\mathrm{Z}-\Xi)'+\mathrm{S}_e|+np(\log 2\pi+1)\\ &= n\log|\mathrm{I}_p+\mathrm{S}_e^{-1/2}(\mathrm{Z}-\Xi)\mathrm{Q}^{-1}(\mathrm{Z}-\Xi)'\mathrm{S}_e^{-1/2}|\\ &\quad+n\log|(1/n)\mathrm{S}_e|+np(\log 2\pi+1).\end{aligned}$$

The final result is obtained by noting that

$$\mathrm{S}_b = \mathrm{Z}\mathrm{Q}^{-1}\mathrm{Z}'$$

and using [A.2.8] and [A.2.11]. □

Anderson (1951) derived the likelihood ratio test by Lagrange's method of undetermined multipliers. Fujikoshi (1974) gave another algebraic proof as in the approach above and derived it as well as the likelihood ratio criteria for testing the dimensions in canonical correlation analysis and the growth curve model.

For testing the hypothesis $H_{0k}: \omega_k > \omega_{k+1} = \cdots = \omega_p = 0$ vs. $H_{1k}: \omega_{k+1} > \omega_{k+1} = \cdots = \omega_p = 0$, the following test statistics are proposed as analogies for testing the overall hypothesis H_0:

$$\text{Likelihood ratio test: } T_{\mathrm{LR};k} = n\log\prod_{j=k+1}^{t}(1+\ell_j),$$

$$\text{Lawley–Hotelling statistic: } T_{\mathrm{LH};k} = n\sum_{j=k+1}^{t}\ell_j,$$

$$\text{Bartlett–Nanda–Pillai statistic: } T_{\mathrm{BNP};k} = n\sum_{j=1}^{t}\ell_j/(1+\ell_j),$$

where $\ell_1 > \cdots > \ell_t > 0$ are the possible nonzero roots of $\mathbf{S}_h\mathbf{S}_e^{-1}$. When we consider asymptotic distributions of the statistics above, we may assume that $\mathbf{S}_h$ and $\mathbf{S}_e$ are two independent random matrices having a noncentral Wishart distribution $\mathrm{W}_p(q, \mathrm{I}_p; \Omega)$ and a central Wishart distribution $\mathrm{W}_p(n, \mathrm{I}_p)$, respectively, where $\Omega = \mathrm{diag}(\omega_1, \ldots, \omega_p)$ and $\omega_1 \geq \cdots \geq \omega_p$.

6.6.2 Large-Sample and High-Dimensional Asymptotics

In this section we consider asymptotic null distributions of the three test statistics for dimensionality, assuming B1 and B2; that is,

$$\omega_i = \mathrm{O}(n),\ i = 1, \ldots, k \quad \text{and} \quad \mathrm{rank}(\Omega) = k. \tag{6.6.4}$$

Under the large-sample framework it follows from Theorem 6.5.1 that the limiting density function of $Z_{k+1}, \ldots, Z_t$, where $Z_i = n\ell_i, i = k+1, \ldots, t$, is

$$f_0(z_{k+1}, \ldots, z_t; s, t),$$

where $f_0(z_{k+1}, \ldots, z_t; s, t)$ is defined by (6.5.10) with $s = \max\{p, q\}$ and $t = \min\{p, q\}$. This implies that

$$\sum_{i=k+1}^{t} Z_i = n \sum_{i=k+1}^{t} \ell_i$$

is asymptotically distributed as $\chi^2(st)$. This result can be obtained from the following lemma.

Lemma 6.6.1 *Let* $\mathbf{W}$ *be distributed as a Wishart distribution* $\mathrm{W}_p(q, I_p; \mathrm{I}_p)$ *with* $q \geq p$, *and let* $d_1 > \cdots > d_p > 0$ *be the characteristic roots of* $\mathbf{W}$. *Then:*

(1) *The pdf of* $d_1 > \cdots > d_p > 0$ *is given by* $f_0(d_1, \ldots, t; q, p)$.

(2) *The distribution of* $\sum_{i=1}^{p} d_i$ *is* $\chi^2(pq)$.

Proof. The density function of $\mathbf{W}$ is written as

$$\frac{1}{2^{1/2pq}\Gamma_p\left(\frac{1}{2}q\right)} e^{-(1/2)\operatorname{tr}\mathrm{W}} |\mathrm{W}|^{(q-p-1)/2}$$
$$= \frac{1}{2^{1/2pq}\Gamma_p\left(\frac{1}{2}q\right)} e^{-(1/2)\sum_{i=1}^{p} d_i} \prod_{i=1}^{p} d_i^{(q-p-1)/2}.$$

The first result is obtained by considering the transformation from W to D, H) such that $\mathrm{D} = \operatorname{diag}(d_1, \ldots, d_p)$ and H is an orthogonal matrix, and using the Jacobian given in [A.3.7]. The second result is obtained from the fact that $\sum_{i=1}^{p} d_i = \operatorname{tr}\mathbf{W}$ and $\operatorname{tr}\mathbf{W} \sim \chi^2(pq)$. □

Noting that $n\ell_i = \mathrm{O}_p(1), i = k+1, \ldots, t$, it is seen that all three statistics $T_{\mathrm{LR};k}$, $T_{\mathrm{LH};k}$, and $T_{\mathrm{BNP};k}$ are asymptotically equivalent to $n\sum_{i=1k+1}^{t} n\ell_i$. These imply the following theorem.

Theorem 6.6.2 *Under the large-sample framework A1 and assumptions B1 and B2, the null distributions of* $T_{\mathrm{LR};k}$, $T_{\mathrm{LH};k}$, *and* $T_{\mathrm{BNP};k}$ *are asymptotically distributed as* $\chi^2(f_k)$, *where* $f_k = (p-k)(q-k)$.

The Bartlett corrections of these test statistics were studied by Lawley (1959), Fujikoshi (1977), and others. The statistics with corrections are given by

$$T_{1;k} = -n_1 \log\{|\mathbf{S}_e|/|\mathbf{S}_e + \mathbf{S}_h|\}, \quad n_1 = n + \frac{1}{2}(q-p-1) + \sum_{i=1}^{k}(\omega_i/n)^{-1},$$

$$T_{2;k} = n_2 \operatorname{tr} \mathbf{S}_h \mathbf{S}_e^{-1}, \quad n_2 = n - p - 1 + k + \sum_{i=1}^{k}(\omega_i/n), \tag{6.6.5}$$

$$T_{3;k} = n_3 \operatorname{tr} \mathbf{S}_h(\mathbf{S}_e + \mathbf{S}_h)^{-1}, \quad n_3 = n + q - k + \sum_{i=1}^{k}(\omega_i/n)^{-1},$$

so that under the hypothesis these test statistics satisfy

$$\mathrm{E}(T_{i;k}) = f_k + \mathrm{O}(n^{-2}), \quad i = 1, 2, 3.$$

For a practical use, we need to replace $\sum_{i=1}^{k}(\omega_i/n)^{-1}$ by $\sum_{i=1}^{k}(\ell_i/n)^{-1}$ or may neglect the term if it is small in comparison with n.

Next, we consider asymptotic null distributions of the three statistics under the high-dimensional framework. Then note that $t = q$, since $p > q$. Consider the transformed versions of LR, LH, and BNP statistics defined by

$$\tilde{T}_{\mathrm{LR}} = \sqrt{p}\left(1 + \frac{m}{p}\right)\left\{\log \prod_{j=k+1}^{q}(1+\ell_j) - (q-k)\log\left(1+\frac{p}{m}\right)\right\},$$

$$\tilde{T}_{\mathrm{LH}} = \sqrt{p}\left\{\frac{m}{p}\sum_{j=k+1}^{q}\ell_j - (q-k)\right\}, \tag{6.6.6}$$

$$\tilde{T}_{\mathrm{BNP}} = \sqrt{p}\left(1+\frac{p}{m}\right)\left\{\left(1+\frac{p}{m}\right)\sum_{j=k+1}^{q}\frac{\ell_j}{1+\ell_j} - (q-k)\right\}.$$

It has essentially been shown in Section 6.5.3 that the last $q-k$ characteristic roots $\ell_{k+1}, \ldots, \ell_q$ are the characteristic roots of

$$\mathbf{Q} = c\mathrm{I}_{q-k} + \frac{1}{\sqrt{m}}(\sqrt{c}\,\mathbf{U}_{22} - c\mathbf{V}_{22}) + \mathrm{O}_p(m^{-1}),$$

where

$$\mathbf{U} = \begin{pmatrix} \mathbf{U}_{11} & \mathbf{U}_{12} \\ \mathbf{U}_{21} & \mathbf{U}_{22} \end{pmatrix}, \quad \mathbf{V} = \begin{pmatrix} \mathbf{V}_{11} & \mathbf{V}_{12} \\ \mathbf{V}_{21} & \mathbf{V}_{22} \end{pmatrix},$$

and $\mathbf{U}_{22}$ and $\mathbf{V}_{22}$ are $(q-k)\times(q-k)$ matrices. This implies that

$$\sum_{i=k+1}^{q}\ell_i = c(q-k) + \frac{1}{\sqrt{m}}\operatorname{tr}(\sqrt{c}\,\mathbf{U}_{22} - c\mathbf{V}_{22}) + \mathrm{O}_p(m^{-1}). \tag{6.6.7}$$

Using the perturbation expansion above it is shown that the asymptotic distributions of all three statistics are the same as those of $\mathrm{tr}\,(U_{22} - \sqrt{r}\,V_{22})$. Furthermore, each of the diagonal elements of $\mathbf{U}_{22}$ and $\mathbf{V}_{22}$ is asymptotically distributed as $\mathrm{N}(0, 2)$, and they are asymptotically independent. Therefore, we obtain the following theorem.

Theorem 6.6.3 *Under the high-dimensional framework A2 and assumptions B1 and B2, the null distributions of* $\tilde{T}_{\mathrm{LR};k}$, $\tilde{T}_{\mathrm{LH};k}$, *and* $\tilde{T}_{\mathrm{BNP};k}$ *are asymptotically the same as* $\mathrm{N}(0, \sigma^2)$ *with the variance* $\sigma^2 = 2(q - k)(1 + c)$.

6.7 High-Dimensional Tests

Suppose that we have two SSP matrices $\mathbf{S}_h$ and $\mathbf{S}_e$ due to the hypothesis and the error, where $\mathbf{S}_h$ and $\mathbf{S}_e$ are independently distributed as a noncentral Wishart distribution $\mathrm{W}_p(q, \Sigma; \mathrm{MM}')$ and a central Wishart distribution $\mathrm{W}_p(n, \Sigma)$, respectively. These are related to the testing problem

$$H_0 \colon \mathrm{M} = \mathrm{O} \text{ vs. } H_1 : \mathrm{M} \neq \mathrm{O}.$$

When $n < p$, $\mathbf{S}_e$ becomes singular, and it will be impossible to use the classical statistics. For such cases, a nonexact test was first proposed by Dempster (1958, 1960)Dempster (1958)Dempster (1960) for one- and two-sample cases. The statistic is based on

$$T_D = (\,\mathrm{tr}\,\mathbf{S}_h)(\,\mathrm{tr}\,\mathbf{S}_e)^{-1}. \tag{6.7.1}$$

As an alternative statistic, it is proposed to use

$$T_B = (1/q)\,\mathrm{tr}\,\mathbf{S}_h - (1/n)\,\mathrm{tr}\,\mathbf{S}_e. \tag{6.7.2}$$

This statistic was considered by Bai and Saranadasa (1996) for the equality of two mean vectors and by Schott (2006) for the equality of several mean vectors. A motivation of this statistic is that T_B has mean 0 if and only if the null hypothesis holds, since $\mathrm{E}(\mathbf{S}_h) = q\Sigma + \mathrm{MM}'$ and $\mathrm{E}(\mathbf{S}_e) = n\Sigma$.

For high-dimensional asymptotics of T_D and T_B, suppose that

$$\begin{aligned}
&\text{A3:} \quad \frac{p}{n} \to c_0 \in (0, \infty), \\
&\text{C1:} \quad \frac{1}{p}\,\mathrm{tr}\,\Sigma^k = \mathrm{O}(1), \quad k = 1, 2, \\
&\text{C2:} \quad \frac{1}{p}\,\mathrm{tr}\,\Sigma^k \Omega = \mathrm{O}(1), \quad k = 1, 2,
\end{aligned} \tag{6.7.3}$$

where $\Omega = \Sigma^{-1/2}\mathrm{MM}'\Sigma^{-1/2}$.

Lemma 6.7.1 *Let*

$$U = \frac{1}{\sqrt{p}}\{\operatorname{tr}\mathbf{S}_h - q\operatorname{tr}\Sigma - \operatorname{tr}\Omega\Sigma\} \text{ and } V = \frac{1}{\sqrt{np}}\{\operatorname{tr}\mathbf{S}_e - n\operatorname{tr}\Sigma\}.$$

Assume that the conditions A3, C1, *and* C3 *are satisfied. Then* U *and* V *are asymptotically and independently distributed as* $\mathrm{N}(0, \sigma_U^2)$ *and* $\mathrm{N}(0, \sigma_V^2)$, *respectively, where*

$$\sigma_U^2 = 2qp^{-1}\operatorname{tr}\Sigma^2 + 4p^{-1}\operatorname{tr}\Sigma^2\tilde{\Omega} \quad \text{and} \quad \sigma_V^2 = 2p^{-1}\operatorname{tr}\Sigma^2.$$

Proof. Note that the characteristic functions $C_U(t)$ and $C_V(t)$ of U and V are expressed explicitly by using Theorem 2.1.3. The results are obtained by checking that $C_U(t)$ and $C_V(t)$ converge to $\exp(-\sigma_U^2/2)$ and $\exp(-\sigma_V^2/2)$, respectively. □

Using Lemma 6.7.1 we obtain asymptotic distributions of T_D and T_B, which are given in the following theorems.

Theorem 6.7.1 *Under assumptions A3, C1, and C2,*

$$\frac{\sqrt{p}}{\sigma_D}\left(n\frac{\operatorname{tr}\mathbf{S}_h}{\operatorname{tr}\mathbf{S}_e} - q - \frac{\operatorname{tr}\Sigma\tilde{\Omega}}{\operatorname{tr}\Sigma}\right) \xrightarrow{d} \mathrm{N}(0,1),$$

where

$$\sigma_D^2 = \left(\frac{1}{p}\operatorname{tr}\Sigma\right)^{-2}\left\{2q\left(\frac{1}{p}\operatorname{tr}\Sigma^2\right) + 4\left(\frac{1}{p}\operatorname{tr}\Sigma^2\Omega\right)\right\};$$

in particular, under the hypothesis

$$\tilde{T}_D \equiv \frac{\sqrt{p}}{\tilde{\sigma}_D}\left(n\frac{\operatorname{tr}\mathbf{S}_h}{\operatorname{tr}\mathbf{S}_e} - q\right) \xrightarrow{d} \mathrm{N}(0,1),$$

where

$$\tilde{\sigma}_D^2 = 2q\left(\frac{1}{p}\operatorname{tr}\Sigma\right)^{-2}\left(\frac{1}{p}\operatorname{tr}\Sigma^2\right).$$

Theorem 6.7.2 *Under assumptions A3, C1, and C2,*

$$\frac{q}{\sqrt{p}\,\sigma_U}\left(\frac{1}{q}\operatorname{tr}\mathbf{S}_h - \frac{1}{n}\operatorname{tr}\mathbf{S}_e - \operatorname{tr}\Sigma\Omega\right) \xrightarrow{d} \mathrm{N}(0,1),$$

where σ_U *is as given in Lemma* 6.7.1; *in particular, under the hypothesis*

$$\frac{q}{\sqrt{p}\,\tilde{\sigma}_U}\left(\frac{1}{q}\operatorname{tr}\mathbf{S}_h - \frac{1}{n}\operatorname{tr}\mathbf{S}_e\right) \xrightarrow{d} \mathrm{N}(0,1),$$

where

$$\tilde{\sigma}_U^2 = 2q\cdot\frac{1}{p}\operatorname{tr}\Sigma^2.$$

Theorem 6.7.1 was proved by Fujikoshi et al. (2004). Theorem 6.7.2 was proved by Bai and Saranadasa (1996) for the equality of two mean vectors and by Schott (2006) for the equality of several mean vectors. For a practical use, we need to estimate $\tilde{\sigma}_D^2$ and $\tilde{\sigma}_B^2$. Their (n, p)-consistency estimators are given by

$$\hat{\sigma}_D^2 = 2pq\frac{\operatorname{tr}\mathbf{S}_e^2 - n^{-1}(\operatorname{tr}\mathbf{S}_e)^2}{(\operatorname{tr}\mathbf{S}_e)^2},$$

$$\hat{\sigma}_U^2 = 2q \cdot \frac{1}{p}\left\{\operatorname{tr}\mathbf{S}_e^2 - n^{-1}(\operatorname{tr}\mathbf{S}_e)^2\right\},$$

respectively. For a review of this subsection, see Fujikoshi (2004).

Problems

6.1 For $i = 1, \ldots, q$, let $y_1^{(i)}, \ldots, y_{n_i}^{(i)}$ be a sample from the population Π_i, and let $\bar{y}^{(i)} = (1/n_i)\sum_{j=1}^{n_i} y_j^{(i)}$ be the sample mean. Let s_b^2 and s_w^2 be the between-group and within-group sums of squares, respectively:

$$s_b^2 = \sum_{i=1}^{q} n_i(\bar{y}^{(i)} - \bar{y})^2, \quad s_w^2 = \sum_{i=1}^{q}\sum_{j=1}^{n_i}(y_j^{(i)} - \bar{y}^{(i)})^2,$$

where $\bar{y} = (1/n)\sum_{i=1}^{q}\sum_{j=1}^{n_i} y_j^{(i)}$ and $n = \sum_{i=1}^{q} n_i$.

(a) Show that s_b^2 and s_w^2 can be expressed as

$$s_b^2 = \boldsymbol{y}'\mathrm{P}_b\boldsymbol{y}, \quad s_w^2 = \boldsymbol{y}'\mathrm{P}_w\boldsymbol{y},$$

where $\boldsymbol{y} = (y_1^{(1)}, \ldots, y_{n_1}^{(1)}, \ldots, y_1^{(q)}, \ldots, y_{n_q}^{(q)})'$,

$$\mathrm{P}_b = \operatorname{Diag}\left(\frac{1}{n_1}\mathrm{G}_{n_1}, \ldots, \frac{1}{n_q}\mathrm{G}_{n_q}\right) - \frac{1}{n}\mathrm{G}_n,$$

$$\mathrm{P}_w = \operatorname{Diag}\left(\mathrm{I}_{n_1} - \frac{1}{n_1}\mathrm{G}_{n_1}, \ldots, \mathrm{I}_{n_q} - \frac{1}{n_q}\mathrm{G}_{n_q}\right),$$

and G_m is the $m \times m$ matrix whose elements are all 1.

(b) Show that $\mathrm{P}_b^2 = \mathrm{P}_b$, $\mathrm{P}_w^2 = \mathrm{P}_w$, and $\mathrm{P}_b\mathrm{P}_w = \mathrm{O}$.

(c) Show that $\operatorname{rank}(\mathrm{P}_b) = \operatorname{tr}\mathrm{P}_b = q - 1$, and $\operatorname{rank}(\mathrm{P}_w) = \operatorname{tr}\mathrm{P}_w = n - q$.

6.2 (Continuation) Consider the case when the $vy_j^{(i)}$ are $p \times 1$ vectors (i.e., $\boldsymbol{y}_j^{(i)}$). Let

$$\mathrm{S}_b = \sum_{i=1}^{q} n_i(\bar{\boldsymbol{y}}^{(i)} - \bar{\boldsymbol{y}})(\bar{\boldsymbol{y}}^{(i)} - \bar{\boldsymbol{y}})',$$

$$\mathrm{S}_w = \sum_{i=1}^{q} \sum_{j=1}^{n_i} (\boldsymbol{y}_j^{(i)} - \bar{\boldsymbol{y}}^{(i)})(\boldsymbol{y}_j^{(i)} - \bar{\boldsymbol{y}}^{(i)})',$$

where $\bar{\boldsymbol{y}}^{(i)} = (1/n_i) \sum_{j=1}^{n_i} \boldsymbol{y}_j^{(i)}$ and $\bar{\boldsymbol{y}} = (1/n) \sum_{i=1}^{q} \sum_{j=1}^{n_i} \boldsymbol{y}_j^{(i)}$. Then show that

$$\mathrm{S}_b = \mathrm{Y}'\mathrm{P}_b\mathrm{Y}, \quad \mathrm{S}_w = \mathrm{Y}'\mathrm{P}_w\mathrm{Y},$$

where $\mathrm{Y} = (\boldsymbol{y}_1^{(1)}, \ldots, \boldsymbol{y}_{n_1}^{(1)}, \ldots, \boldsymbol{y}_1^{(q)}, \ldots, \boldsymbol{y}_{n_q}^{(q)})'$.

6.3 (Continuation) Suppose that $\boldsymbol{Y}_{i1}, \ldots, \boldsymbol{Y}_{in_i} \sim$ i.i.d. $\mathrm{N}_p(\boldsymbol{\mu}_i, \Sigma)$, that is, $\Pi_i : \mathrm{N}_p(\boldsymbol{\mu}_i, \Sigma)$. Using the multivariate Cochran theorem, show that $\mathbf{S}_b \sim \mathrm{W}_p(q-1, \Sigma; \Gamma)$, $\mathbf{S}_w \sim \mathrm{W}_p(n-q, \Sigma)$, and that they are independent, where $\Gamma = \sum_{i=1}^{q} n_i(\boldsymbol{\mu}_i - \bar{\boldsymbol{\mu}})(\boldsymbol{\mu}_i - \bar{\boldsymbol{\mu}})'$ with $\bar{\boldsymbol{\mu}} = (1/n)\sum_{i=1}^{q} n_i \boldsymbol{\mu}_i$.

6.4 (Continuation) Suppose that a p-variate $\boldsymbol{y}$ is decomposed as $\boldsymbol{y} = (\boldsymbol{y}_1', \boldsymbol{y}_2')'$, $\boldsymbol{y}_i : p_i \times 1$, where $\boldsymbol{y}_1$ and $\boldsymbol{y}_2$ are the main variate and covariate, respectively. Along the same lines, we decompose $\boldsymbol{\mu}_i$, $\mathbf{W} = \mathbf{S}_e$ and $\mathbf{T} = \mathbf{S}_e + \mathbf{S}_w$ as

$$\boldsymbol{\mu}_i = \begin{pmatrix} \boldsymbol{\mu}_{i1} \\ \boldsymbol{\mu}_{i2} \end{pmatrix}, \quad \mathbf{W} = \begin{pmatrix} \mathbf{W}_{11} & \mathbf{W}_{12} \\ \mathbf{W}_{21} & \mathbf{W}_{22} \end{pmatrix}, \quad \mathbf{T} = \begin{pmatrix} \mathbf{T}_{11} & \mathbf{T}_{12} \\ \mathbf{T}_{21} & \mathbf{T}_{22} \end{pmatrix}.$$

Assuming that $\boldsymbol{\mu}_{12} = \cdots = \boldsymbol{\mu}_{q2}$, consider a test of hypothesis $H_0 : \boldsymbol{\mu}_{11} = \cdots = \boldsymbol{\mu}_{q1}$, which is called the multivariate analysis of covariance test. Show that

(1) The likelihood ratio criterion λ is given by

$$\lambda^{2/n} = |\mathbf{W}_{11\cdot 2}|/|\mathbf{T}_{11\cdot 2}| = \frac{|\mathbf{W}|}{|\mathbf{T}|} \cdot \left(\frac{|\mathbf{W}_{11}|}{|\mathbf{T}_{11}|}\right)^{-1},$$

where $\mathbf{W}_{11\cdot 2} = \mathbf{W}_{11} - \mathbf{W}_{12}\mathbf{W}_{22}^{-1}\mathbf{W}_{21}$ and $\mathbf{T}_{11\cdot 2} = \mathbf{T}_{11} - \mathbf{T}_{12}\mathbf{T}_{22}^{-1}\mathbf{T}_{21}$.

(2) Under H_0, $\lambda^{2/n} \sim \Lambda_{p_1}(q-1, n-q-p_2)$.

6.5 Let B be a $p \times p$ positive semidefinite matrix and W a $p \times p$ positive definite matrix. Let $\ell_1 \geq \cdots \geq \ell_p \geq 0$ be the characteristic roots of BW^{-1}. Then prove that:

$$\text{(a)} \quad \operatorname{tr} \mathrm{B}(\mathrm{W}+\mathrm{B})^{-1} \leq -\log \frac{|\mathrm{W}|}{|\mathrm{W}+\mathrm{B}|} \leq \operatorname{tr} \mathrm{BW}^{-1}.$$

$$\text{(b)} \quad \sum_{i=1}^{p} \frac{\ell_i}{1+\ell_i} \leq \log \prod_{i=1}^{p} (1+\ell_i) \leq \sum_{i=1}^{p} \ell_i.$$

Hint: For $x \geq 0$, let $h_1(x) = x/(1+x)$, $h_2(x) = \log(1+x)$, $h_3(x) = x$. Show that $h_1(x) \leq h_2(x) \leq h_3(x)$.

6.6 Let $G_f(x;\delta^2)$ and $g_f(x;\delta^2)$ be the distribution and the density functions of $\chi^2(f;\delta^2)$. Show that

$$\begin{aligned} &(1) \quad G_{f+2}(x;\delta^2) = -2g_{f+2}(x;\delta^2) + G_f(x;\delta^2). \\ &(2) \quad xg_f(x;\delta^2) = fg_{f+2}(x;\delta^2) + 2g_{f+4}(x;\delta^2). \end{aligned}$$

6.7 Let $\mathbf{V} = (V_{ij})$ be a symmetric random matrix such that $V_{ii} \sim \mathrm{N}(0,1)$, $V_{ij} \sim \mathrm{N}(0, \frac{1}{2})$ $(i \neq j)$, and the V_{ij} $(i \leq j)$ are independent.

(a) Show that the joint density function of $\mathbf{V}$ with $\ell_1 > \cdots > \ell_p$ is

$$\frac{1}{2^{p/2}\pi^{p(p+1)/2}} \mathrm{e}^{-(1/2)\operatorname{tr} \mathrm{V}^2}.$$

(b) The density function of the characteristic roots shows that the joint density function of $\mathbf{V}$ is

$$\frac{\pi^{p^2/2}}{2^{p/2}\pi^{p(p+1)/2}\Gamma_p(\frac{1}{2}p)} \mathrm{e}^{-(\ell_1^2+\cdots+\ell_p^2)/2} \prod_{i<j} (\ell_i - \ell_j).$$

Hint: For a proof of (b), use the fact that $g(\mathrm{V})$ is a function of $\ell_1, \ldots, \ell_p$, and consider the orthogonal transformation $\mathrm{V} = \mathrm{HLH}'$, $\mathrm{L} = \operatorname{diag}(\ell_1, \ldots, \ell_p)$ whose Jacobian is given by [A.3.8].

6.8 (Continuation) Let $Z_i = \sqrt{n}/(4\lambda_\alpha + 2\lambda_\alpha^2)^{1/2}(\ell_i - \lambda_\alpha)$, $i \in J_\alpha$, $\alpha = 1, \ldots, r-1$ be the normalized random variables defined by (6.5.7). Obtain the limiting distribution of $\sum_{i \in J_\alpha} Z_i$.

6.9 The LR test for testing the dimensionality hypothesis (6.6.1) is based on $\prod_{j=k+1}^{t}(1+\ell_j)$ (see Theorem 6.6.1), where $\ell_1 > \cdots > \ell_t > 0$ are the nozero characteristic roots of $\mathbf{S}_b\mathbf{S}_w^{-1}$. This results in the case when Σ is unknown. Show that if Σ is known, the LR test is based on $\tilde{\ell}_{k+1} + \tilde{\ell}_t$, where $\tilde{\ell}_1 > \cdots > \tilde{\ell}_t > 0$ are the nonzero characteristic roots of $\mathbf{S}_b\Sigma^{-1}$.

6.10 Consider constructing a simultaneous confidence interval for $\sum_{i=1}^{q} \boldsymbol{a}'\boldsymbol{\mu}_i \cdot b_i$ in multivariate one-way model (6.1.1), where $\boldsymbol{a} = (a_1, \ldots, a_p)$, and $\boldsymbol{b} = (b_1, \ldots, b_q)'$, based on the following line:

(1) Let

$$T(\boldsymbol{b}) = \sum_{i=1}^{q} \boldsymbol{a}'(\bar{\mathbf{Y}}_i - \boldsymbol{\mu}_i)b_i \Big/ \sqrt{\sum_{i=1}^{q} (b_i^2/n_i)\boldsymbol{a}'\mathbf{S}_e\boldsymbol{a}/m},$$

where $m = n - q$. Then, show that for any $\boldsymbol{a}$, $\mathrm{T}(\boldsymbol{b})$ is ditributed as a t-distribution $\mathrm{t}(m)$

(2) Let $t_m(\alpha)$ be the α point of $\mathrm{t}(m)$, and consider k given values of $\boldsymbol{b}$, $\boldsymbol{b}_1, \ldots, \boldsymbol{b}_k$. Then, show that

$$\Pr\{|T(\boldsymbol{b}_1)| \le t_m(\alpha/2k), \ldots, |T(\boldsymbol{b}_1)| \le t_m(\alpha/2k)\} \ge 1 - \alpha.$$

(3) Show that a simultaneous confidence interval for $\sum_{i=1}^{q} \boldsymbol{a}'\boldsymbol{\mu}_i \cdot b_i$ for all $\boldsymbol{a}$ and $\boldsymbol{b} = \boldsymbol{b}_1, \ldots, \boldsymbol{b}_k$ is given by

$$\sum_{i=1}^{q} \boldsymbol{a}'\bar{\boldsymbol{Y}}_i b_i \pm t_m(\alpha/2k) \left[\left(\sum_{i=1}^{q} \frac{b_i^2}{n_i} \right) \frac{(\boldsymbol{a}'\mathbf{S}_e\boldsymbol{a})}{m} \right]^{1/2}.$$

6.11 (Continuation) Consider considering a simultaneous confidence interval for $\sum_{i=1}^{q} \boldsymbol{a}'\boldsymbol{\mu}_i \cdot b_i$ for all $\boldsymbol{a}$ and $\boldsymbol{b}$ such that $\sum_{i=1}^{q} b_i = 0$, based on the following line.

(1) Using

$$\sum_{i=1}^{q} b_i \boldsymbol{a}'(\bar{\boldsymbol{Y}}_i - \boldsymbol{\mu}_i) = \sum_{i=1}^{q} (b_i/\sqrt{n_i})\boldsymbol{a}'\sqrt{n_i}\left\{\bar{\boldsymbol{Y}}_i - \boldsymbol{\mu}_i - (\bar{\boldsymbol{Y}} - \bar{\boldsymbol{\mu}})\right\},$$

show that

$$\begin{aligned} \sup_{\boldsymbol{a},\boldsymbol{b}} \left\{ \frac{\sum_{i=1}^{q} b_i \boldsymbol{a}'(\bar{\boldsymbol{Y}}_i - \boldsymbol{\mu}_i)}{\sqrt{\left(\sum_{i=1}^{q} \frac{b_i^2}{n_i}\right) \boldsymbol{a}'\mathbf{S}_e\boldsymbol{a}}} \right\}^2 &\le \sup_{\boldsymbol{a}} \frac{\boldsymbol{a}'\tilde{\mathbf{S}}_h\boldsymbol{a}}{\boldsymbol{a}'\mathbf{S}_e\boldsymbol{a}} \\ &= ch_i(\tilde{\mathbf{S}}_h\mathbf{S}_e^{-1}), \end{aligned}$$

where $\tilde{\mathbf{S}}_h = \sum_{i=1}^{q} n_i\{\bar{\boldsymbol{Y}}_i - \bar{\boldsymbol{Y}} - (\boldsymbol{\mu}_i - \bar{\boldsymbol{\mu}})\}\{\bar{\boldsymbol{Y}}_i - \bar{\boldsymbol{Y}} - (\boldsymbol{\mu}_i - \bar{\boldsymbol{\mu}})\}'$.

(2) Let $\mathbf{B}$ and $\mathbf{W}$ be independently distributed as Whishart distributions $\mathrm{W}_p(q-1, \Sigma)$ and $\mathrm{W}_p(m, \Sigma)$. Let $\ell(\alpha)$ and $d(\alpha)$ be the upper α points of the largest characteristic roots of $\mathbf{B}\mathbf{W}^{-1}$ and $\mathbf{B}(\mathbf{B}+\mathbf{W})^{-1}$, respectively. Then, show that $\ell(\alpha) = d(\alpha)/(1 - d(\alpha))$.

(3) Show that a simultaneous confidence interval for $\sum_{i=1}^{q} \boldsymbol{a}'\boldsymbol{\mu}_i \cdot b_i$ for all $\boldsymbol{a}$ and $\boldsymbol{b}$ such that $\sum_{i=1}^{q} b_i = 0$ is given by

$$\sum_{i=1}^{q} \boldsymbol{a}'\bar{\boldsymbol{Y}}_i b_i \pm \left[\left(\sum_{i=1}^{q} \frac{b_i^2}{n_i} \right) (\boldsymbol{a}'\mathbf{S}_e\boldsymbol{a}) \frac{d(\alpha)}{1 - d(\alpha)} \right]^{1/2}.$$

Chapter 7

Multivariate Regression

In this chapter we first give some basic results on multivariate regression. Tests are discussed for additional information. Then C_p and AIC criteria are explained for the selection of response and explanatory variables, and their corrections are discussed. We also explain the method for selecting variables based on stepwise procedure. Next we treat principal component regression. We also treat the case when the number of explanatory variables is larger than that of sample size. Finally a theory is presented for testing a general linear hypothesis and constructing confidence intervals.

7.1 Multivariate Linear Regression Model

We consider the linear relationship between p response (dependent) variables and k explanatory (independent, predictor) variables. Let the dependent variables and explanatory variables be denoted by $\boldsymbol{Y} = (Y_1, \ldots, Y_p)'$ and $\boldsymbol{x} = (x_1, \ldots, x_k)'$, respectively. We use a linear model to relate $\boldsymbol{Y}$ to $\boldsymbol{x}$ and are concerned with the statistical inference of the parameters in the model and with the selection of variables. When $p = 1$, the model is called a *multiple regression model.* We treat the case when $\boldsymbol{x}$ is fixed, as is usually assumed in most regression models.

Suppose that $\boldsymbol{y}$ and $\boldsymbol{x}$ are measured for n subjects, and let the observations of the ith subject be denoted by

$$\boldsymbol{Y}_i = (Y_{i1}, \ldots, Y_{ip})', \ \ \boldsymbol{x}_i = (x_{i1}, \ldots, x_{ik})', \quad i = 1, \ldots, n.$$

The relationship between $\boldsymbol{Y}$ and $\boldsymbol{x}$ is expressed as

$$\boldsymbol{Y}_i = \boldsymbol{\beta}_0 + \boldsymbol{\beta}_1 x_{i1} + \cdots + \boldsymbol{\beta}_k x_{ik} + \boldsymbol{\varepsilon}_i, \quad i = 1, \ldots, n, \tag{7.1.1}$$

where the coefficients $\boldsymbol{\beta}_0, \ldots, \boldsymbol{\beta}_k$ are unknown p-dimensional parameter vectors. The error terms $\boldsymbol{\varepsilon}_1, \ldots, \boldsymbol{\varepsilon}_n$ are p-dimensional random vectors. It is assumed that they are mutually uncorrelated, and that

$$\mathrm{E}(\boldsymbol{\varepsilon}_i) = \mathbf{0}, \ \mathrm{Var}(\boldsymbol{\varepsilon}_i) = \Sigma, \quad i = 1, \ldots, n.$$

Further, the normality of $\boldsymbol{\varepsilon}_i$ is assumed in some statistical inference.

To express the multivariate linear regression model (7.1.1) in matrix notation, write

$$\mathbf{Y} = \begin{pmatrix} \boldsymbol{Y}_1' \\ \vdots \\ \boldsymbol{Y}_n' \end{pmatrix}, \quad \mathrm{X} = \begin{pmatrix} 1 & \boldsymbol{x}_1' \\ \vdots & \vdots \\ 1 & \boldsymbol{x}_n' \end{pmatrix} \quad \text{and} \quad \mathcal{E} = \begin{pmatrix} \boldsymbol{\varepsilon}_1' \\ \vdots \\ \boldsymbol{\varepsilon}_n' \end{pmatrix}.$$

Then the multivariate linear regression model is expressed as follows:

$$\mathbf{Y} = \mathrm{X}\mathcal{B} + \mathcal{E}, \tag{7.1.2}$$

where $\mathcal{B} = (\boldsymbol{\beta}_0, \boldsymbol{\beta}_1, \ldots, \boldsymbol{\beta}_k)'$ and our assumptions stated above are equivalent to the following:

1. $\mathrm{E}(\mathbf{Y}) = \mathrm{X}\mathcal{B}$.
2. $\mathrm{Var}(\boldsymbol{Y}_i) = \Sigma$ for all $i = 1, \ldots, n$.
3. $\mathrm{Cov}(\boldsymbol{Y}_i, \boldsymbol{Y}_j) = \mathrm{O}$ for all $i \neq j$.

We have yet another expression of the model, which is related to a univariate regression model. Let

$$\mathbf{Y} = (\boldsymbol{Y}_{(1)}, \ldots, \boldsymbol{Y}_{(p)}), \quad \mathcal{B} = (\boldsymbol{\beta}_{(1)}, \ldots, \boldsymbol{\beta}_{(p)}), \quad \mathcal{E} = (\boldsymbol{\varepsilon}_{(1)}, \ldots, \boldsymbol{\varepsilon}_{(p)}).$$

Then the expression (7.1.1) and (7.1.2) are equivalent to the following p univariate multiple regression models:

$$\boldsymbol{Y}_{(i)} = \mathrm{X}\boldsymbol{\beta}_{(i)} + \boldsymbol{\varepsilon}_{(i)}, \quad i = 1, \ldots, p. \tag{7.1.3}$$

Here we note that three p univariate multiple regression models are not independent. A stronger relationship is obtained by considering

$$\boldsymbol{Y} = (\boldsymbol{Y}_{(1)}', \ldots, \boldsymbol{Y}_{(p)}')'.$$

Then we have

$$\boldsymbol{Y} = \tilde{\mathrm{X}}\boldsymbol{\beta} + \boldsymbol{\varepsilon}, \tag{7.1.4}$$

where $\boldsymbol{\beta} = (\boldsymbol{\beta}_{(1)}', \ldots, \boldsymbol{\beta}_{(p)}')'$ȦC$\boldsymbol{\varepsilon} = (\boldsymbol{\varepsilon}_{(1)}', \ldots, \boldsymbol{\varepsilon}_{(p)}')'$, and

$$\tilde{\mathrm{X}} = \begin{pmatrix} \mathrm{X} & & \mathrm{O} \\ & \ddots & \\ \mathrm{O} & & \mathrm{X} \end{pmatrix} = \mathrm{I}_n \otimes \mathrm{X},$$

which differs from that in the usual univariate regression model. Here the notation $\otimes$ represents the Kroneker product for $\mathrm{A} = (a_{ij})$ and B, $\mathrm{A} \otimes \mathrm{B} = (a_{ij}\mathrm{B})$. The covariance matrix of ε is

$$\mathrm{Var}(\varepsilon) = \begin{pmatrix} \sigma_{11}\mathrm{I}_n & \cdots & \sigma_{1p}\mathrm{I}_n \\ \vdots & \ddots & \vdots \\ \sigma_{p1}\mathrm{I}_n & \cdots & \sigma_{pp}\mathrm{I}_n \end{pmatrix} = \Sigma \otimes \mathrm{I}_n.$$

7.2 Statistical Inference

The *sum of squares error* is expressed as

$$\begin{aligned} \sum_{i=1}^{n}\sum_{j=1}^{p} \varepsilon_{ij}^2 &= \mathrm{tr}\,(\mathrm{Y} - \mathrm{X}\mathcal{B})'(\mathrm{Y} - \mathrm{X}\mathcal{B}) \\ &= \sum_{i=1}^{p} (\boldsymbol{y}_{(i)} - \mathrm{X}\boldsymbol{\beta}_{(i)})'(\boldsymbol{y}_{(i)} - \mathrm{X}\boldsymbol{\beta}_{(i)}). \end{aligned}$$

Therefore, from the theory of the least squares in univariate regression, we have the least squares estimator of $\mathcal{B}$, given by

$$\hat{\boldsymbol{\beta}}_{(i)} = (\mathrm{X}'\mathrm{X})^{-1}\mathrm{X}'\boldsymbol{y}_{(i)},\ i = 1, \ldots, p \qquad [\text{i.e., } \hat{\mathcal{B}} = (\mathrm{X}'\mathrm{X})^{-1}\mathrm{X}'\mathrm{Y}].$$

The estimator satisfies the fact that for any $\Sigma > 0$,

$$\begin{aligned} \mathrm{tr}\,\Sigma^{-1}(\mathrm{Y} - \mathrm{X}\mathcal{B})'(\mathrm{Y} - \mathrm{X}\mathcal{B}) &\geq \mathrm{tr}\,\Sigma^{-1}(\mathrm{Y} - \mathrm{X}\hat{\mathcal{B}})'(\mathrm{Y} - \mathrm{X}\hat{\mathcal{B}}) \\ &= \mathrm{tr}\,\Sigma^{-1}\mathrm{Y}'(\mathrm{I}_n - \mathrm{P}_\mathrm{X})\mathrm{Y}, \end{aligned} \tag{7.2.1}$$

where $\mathrm{P}_\mathrm{X} = \mathrm{X}(\mathrm{X}'\mathrm{X})^{-1}\mathrm{X}'$. For a proof of the first inequality, let

$$\mathrm{Q} = (\mathrm{Y} - \mathrm{X}\mathcal{B})'(\mathrm{Y} - \mathrm{X}\mathcal{B}) - (\mathrm{Y} - \mathrm{X}\hat{\mathcal{B}})'(\mathrm{Y} - \mathrm{X}\hat{\mathcal{B}})$$

and it is enough to show that $\mathrm{tr}\,\Sigma^{-1}\mathrm{Q} \geq 0$, or Q is positive semidefinite since $\mathrm{tr}\,\Sigma^{-1}\mathrm{Q} = \mathrm{tr}\,\Sigma^{-1/2}\mathrm{Q}\Sigma^{-1/2}$. Consider the decomposition

$$\mathrm{Y} - \mathrm{X}\mathcal{B} = (\mathrm{Y} - \mathrm{X}\hat{\mathcal{B}}) + \mathrm{X}(\hat{\mathcal{B}} - \mathcal{B}).$$

Then $(\mathrm{Y} - \mathrm{X}\hat{\mathcal{B}})'\mathrm{X}(\hat{\mathcal{B}} - \mathcal{B}) = \{(\mathrm{I}_n - \mathrm{P}_\mathrm{X})\mathrm{Y}\}'\mathrm{X}(\hat{\mathcal{B}} - \mathcal{B}) = \mathrm{O}$. Therefore, we have

$$\mathrm{Q} = \{\mathrm{X}(\hat{\mathcal{B}} - \mathcal{B})\}'\{\mathrm{X}(\hat{\mathcal{B}} - \mathcal{B})\} \geq \mathrm{O}.$$

On the other hand, the likelihood function $\mathrm{L}(\Theta, \Sigma)$ of Y under normality satisfies

$$\begin{aligned} -2\log \mathrm{L}(\mathcal{B}, \Sigma) &= n \log|\Sigma| + \mathrm{tr}\,\Sigma^{-1}(\mathrm{Y} - \mathrm{X}\mathcal{B})'(\mathrm{Y} - \mathrm{X}\mathcal{B}) + np\log 2\pi \\ &\geq n\{\log|\Sigma| + \mathrm{tr}\,\Sigma^{-1}\mathrm{S}\} + np\log 2\pi, \end{aligned} \tag{7.2.2}$$

where $\mathrm{S} = n^{-1}\mathrm{Y}'(\mathrm{I}_n - \mathrm{P_A})\mathrm{Y}$. The equality holds when $\mathcal{B} = \hat{\mathcal{B}}$. Further, considering the minimization with respect to Σ, we obtain the MLEs given by

$$\hat{\mathcal{B}} = (\mathrm{X}'\mathrm{X})^{-1}\mathrm{X}'\mathbf{Y}, \quad \hat{\Sigma} = \mathbf{S} = \frac{1}{n}(\mathbf{Y} - \mathrm{X}\hat{\mathcal{B}})'(\mathbf{Y} - \mathrm{X}\hat{\mathcal{B}}). \tag{7.2.3}$$

Theorem 7.2.1 *For the estimators $\hat{\mathcal{B}}$ and $\hat{\Sigma}$ in (7.2.3), it holds that:*

(1) $\mathrm{E}(\hat{\mathcal{B}}) = \mathcal{B}$, $\mathrm{E}(\hat{\Sigma}) = \dfrac{1}{n(n-k+1)}\Sigma$, $\mathrm{Cov}(\hat{\boldsymbol{\beta}}_i, \hat{\boldsymbol{\beta}}_j) = v_{ij}\Sigma$, *where* $\mathrm{V} = (\mathrm{X}'\mathrm{X})^{-1} = (v_{ij})$.

(2) *If* $\mathbf{Y}$ *is normally distributed, then* $\hat{\mathcal{B}} \sim \mathrm{N}_{(k+1)\times p}(\mathcal{B}, \Sigma \otimes \mathrm{V})$, $n\hat{\Sigma} \sim \mathrm{W}_p(n-k+1, \Sigma)$, *and* $\hat{\mathcal{B}}$ *and* $n\hat{\Sigma}$ *are independent.*

Proof. It is easy to see that

$$\begin{aligned}\mathrm{E}(\hat{\mathcal{B}}) &= \mathrm{E}[(\mathrm{X}'\mathrm{X})^{-1}\mathrm{X}'\mathbf{Y}] \\ &= (\mathrm{X}'\mathrm{X})^{-1}\mathrm{X}'\mathrm{E}(\mathbf{Y}) = (\mathrm{X}'\mathrm{X})^{-1}\mathrm{X}'\mathrm{X}\mathcal{B} = \mathcal{B}.\end{aligned}$$

Using $n\hat{\Sigma} = \mathcal{E}'(\mathrm{I}_n - \mathrm{P_X})\mathcal{E}$, we have

$$\begin{aligned}\mathrm{E}(n\hat{\sigma}_{ij}) &= \mathrm{E}[\boldsymbol{\varepsilon}_{(i)}'(\mathrm{I}_n - \mathrm{P_X})\boldsymbol{\varepsilon}_{(j)}] = \mathrm{E}[\,\mathrm{tr}\,\boldsymbol{\varepsilon}_{(i)}'(\mathrm{I}_n - \mathrm{P_X})\boldsymbol{\varepsilon}_{(j)}] \\ &= \mathrm{E}[\,\mathrm{tr}\,(\mathrm{I}_n - \mathrm{P_X})\boldsymbol{\varepsilon}_{(j)}\boldsymbol{\varepsilon}_{(i)}'] = \sigma_{ij}\,\mathrm{tr}\,(\mathrm{I}_n - \mathrm{P_X}) \\ &= (n-k+1)\sigma_{ij}.\end{aligned}$$

Let $\boldsymbol{c}_i = (c_{i1}, \ldots, c_{in})'$ be the ith column vector of $\mathrm{X}(\mathrm{X}'\mathrm{X})^{-1}$. Then $\hat{\boldsymbol{\beta}}_i = \sum_{\ell=1}^n c_{i\ell}\boldsymbol{Y}_\ell$ and

$$\begin{aligned}\mathrm{Cov}(\hat{\boldsymbol{\beta}}_i, \hat{\boldsymbol{\beta}}_j) &= \sum_{\ell=1}^n\sum_{m=1}^n c_{i\ell}c_{jm}\mathrm{Cov}(\boldsymbol{Y}_i, \boldsymbol{Y}_j) = \sum_{\ell=1}^n c_{i\ell}c_{jm}\Sigma \\ &= \boldsymbol{c}_i'\boldsymbol{c}_j\Sigma = v_{ij}\Sigma.\end{aligned}$$

Using Theorem 1.2.7 we can see that the distribution is normal. Therefore, the first result of (2) follows from (1). The distribution of $n\hat{\Sigma}$ is obtained using the multivariate Cochran theorem (Theorem 2.4.2). □

From Theorem 7.2.1(1), $\hat{\Sigma}$ is not unbiased. A commonly used unbiased estimator is

$$\tilde{\Sigma} = \frac{1}{n-k}(\mathbf{Y} - \mathrm{X}\hat{\mathcal{B}})'(\mathbf{Y} - \mathrm{X}\hat{\mathcal{B}}).$$

The covariance matrices of $\hat{\boldsymbol{\beta}}_i$ and between $\hat{\boldsymbol{\beta}}_i$ and $\hat{\boldsymbol{\beta}}_j$ are estimated by

$$\hat{\mathrm{Var}}(\hat{\boldsymbol{\beta}}_i) = v_{ii}\tilde{\Sigma}, \quad \hat{\mathrm{Cov}}(\hat{\boldsymbol{\beta}}_i, \hat{\boldsymbol{\beta}}_j) = v_{ij}\tilde{\Sigma}, \quad i \neq j.$$

Now we consider testing the hypothesis that a subset of the explanatory variables x_j has no additional information in predicting $\boldsymbol{y}$ in the presence of the remaining explanatory variables. Without loss of generality we assume that $\{x_{j+1}, \ldots, x_k\}$ has no additional information in the presence of $\{x_1, \ldots, x_j\}$. In other words, that $\{x_1, \ldots, x_j\}$ is sufficient. The hypothesis is defined by

$$H_J\colon\ \boldsymbol{\beta}_{j+1} = \cdots = \boldsymbol{\beta}_k = \mathbf{0}.$$

Here J represents the set of indices of sufficient variables $\{x_1, \ldots, x_j\}$. The multivariate regression model under H_J is expressed as

$$M_J\colon\ \boldsymbol{Y}_i = \boldsymbol{\beta}_0 + \boldsymbol{\beta}_1 x_{i1} + \cdots + \boldsymbol{\beta}_j x_{ij} + \boldsymbol{\varepsilon}_i, \quad i = 1, \ldots, n, \tag{7.2.4}$$

or in matrix notation,

$$M_J\colon\ \mathbf{Y} = \mathrm{X}_J \mathcal{B}_J + \mathcal{E}, \tag{7.2.5}$$

where X_J and $\mathcal{B}'_J$ are the matrices comprising by the first $j+1$ columns of X and $\mathcal{B}'$, respectively.

The residual sum of squares and products matrix under model (7.1.1) or (7.1.2) is defined by

$$\begin{aligned}\mathbf{S}_\Omega &= (\mathbf{Y} - \mathrm{X}\hat{\mathcal{B}})'(\mathbf{Y} - \mathrm{X}\hat{\mathcal{B}})\\ &= \{(\mathrm{I}_n - \mathrm{P}_\mathrm{X})\mathbf{Y}\}'\{(\mathrm{I}_n - \mathrm{P}_\mathrm{X})\mathbf{Y}\}\\ &= \mathbf{Y}'(\mathrm{I}_n - \mathrm{P}_\mathrm{X})\mathbf{Y},\end{aligned}$$

where $\mathrm{P}_\mathrm{X} = \mathrm{X}(\mathrm{X}'\mathrm{X})^{-1}\mathrm{X}'$. The matrix $\mathbf{S}_\Omega$ is called the SSP (sum of squares and products) *matrix due to errors*, which is also denoted $\mathbf{S}_e$. Similarly, the residual SSP matrix under the hypothesis H or model (7.2.5) is defined by

$$\begin{aligned}\mathbf{S}_\omega &= (\mathbf{Y} - \mathrm{X}_J\hat{\mathcal{B}}_J)'(\mathbf{Y} - \mathrm{X}_J\hat{\mathcal{B}}_J)\\ &= \mathbf{Y}'(\mathrm{I}_n - \mathrm{P}_J)\mathbf{Y},\end{aligned}$$

where $\mathrm{P}_J = \mathrm{X}_J(\mathrm{X}'_J\mathrm{X}_J)^{-1}\mathrm{X}'_J$. Then the SSP matrix due to the hypothesis is defined by

$$\mathbf{S}_h = \mathbf{S}_\omega - \mathbf{S}_\Omega = \mathbf{Y}'(\mathrm{P}_\mathrm{X} - \mathrm{P}_J)\mathbf{Y}.$$

Using Cochran's multivariate theorem (Theorem 2.4.2), we have the following theorem.

Theorem 7.2.2 *Suppose that the error terms $\boldsymbol{\varepsilon}_1, \ldots, \boldsymbol{\varepsilon}_n$ in* (7.1.2) *are normal. Then:*

(1) $\mathbf{S}_e \sim \mathrm{W}_p(n - k - 1, \Sigma)$.

(2) *Under the hypothesis,* $\mathbf{S}_h \sim \mathrm{W}_p(k - j, \Sigma)$.

(3) $\mathbf{S}_e$ *and* $\mathbf{S}_h$ *are mutually independent.*

A test procedure is based on whether or not $\mathbf{S}_h$ is large compared with $\mathbf{S}_e$. The LR test is based on

$$|\mathbf{S}_e|/|\mathbf{S}_e + \mathbf{S}_h|. \tag{7.2.6}$$

In fact, let $L(\mathcal{B}, \Sigma)$ be the likelihood function of $\mathbf{Y}$. Then, using (7.2.2), we have

$$-2\log\max \mathrm{L}(\mathcal{B}, \Sigma) = n\log\left|\frac{1}{n}\mathbf{S}_e\right| + np(\log 2\pi + 1).$$

Similarly,

$$-2\log\max_{H_J} \mathrm{L}(\mathcal{B}, \Sigma) = n\log\left|\frac{1}{n}(\mathbf{S}_e + \mathbf{S}_h)\right| + np(\log 2\pi + 1).$$

Therefore, the LR criterion is given by

$$\lambda = (|\mathbf{S}_e|/|\mathbf{S}_e + \mathbf{S}_h|)^{n/2}.$$

From Theorem 7.2.2 under H_J,

$$\lambda^{2/n} = \frac{|\mathbf{S}_e|}{|\mathbf{S}_e + \mathbf{S}_h|} \sim \Lambda_p(k - j, n - k - 1).$$

For numerical computations of S_h and S_e, we first compute the sample covariance matrix or the SSP matrix, given by

$$\begin{aligned}
\mathrm{W} &= \sum_{i=1}^{n} \begin{pmatrix} \boldsymbol{y}_i - \bar{\boldsymbol{y}} \\ \boldsymbol{x}_i - \bar{\boldsymbol{x}} \end{pmatrix} \begin{pmatrix} \boldsymbol{y}_i - \bar{\boldsymbol{y}} \\ \boldsymbol{x}_i - \bar{\boldsymbol{x}} \end{pmatrix}' \\
&= \begin{pmatrix} \mathrm{W}_{yy} & \mathrm{W}_{yx} \\ \mathrm{W}_{xy} & \mathrm{W}_{xx} \end{pmatrix}.
\end{aligned}$$

Along the partition of $\boldsymbol{x} = (\boldsymbol{x}_1', \boldsymbol{x}_2')'$, $\boldsymbol{x}_1 = (x_1, \ldots, x_j)'$ and $\boldsymbol{x}_2 = (x_{j+1}, \ldots, x_k)$, we partition W as

$$\mathrm{W} = \begin{pmatrix} \mathrm{W}_{yy} & \mathrm{W}_{y1} & \mathrm{W}_{y2} \\ \mathrm{W}_{1y} & \mathrm{W}_{11} & \mathrm{W}_{12} \\ \mathrm{W}_{2y} & \mathrm{W}_{21} & \mathrm{W}_{22} \end{pmatrix}.$$

Then it holds that

$$\mathrm{S}_\Omega = \mathrm{W}_{yy} - \mathrm{W}_{yx}\mathrm{W}_{xx}^{-1}\mathrm{W}_{xy}, \quad \mathrm{S}_\omega = \mathrm{W}_{yy} - \mathrm{W}_{y1}\mathrm{W}_{11}^{-1}\mathrm{W}_{1y}. \tag{7.2.7}$$

There results can be shown by using the fact that the projection matrix $\mathrm{P_X}$ and P_J are expressed as $\mathrm{P_X} = \mathrm{P_0} + \mathrm{P}_{(\mathrm{I_n} - \mathrm{P_0})\mathrm{X}}$, $\mathrm{P}_J = \mathrm{P_0} + \mathrm{P}_{(\mathrm{I_n} - \mathrm{P_0})\mathrm{X}_J}$ where $\mathrm{P_0} = (1/n)\mathbf{1}_n\mathbf{1}_n'$.

When J is empty, the hypothesis becomes

$$H_0 \colon \boldsymbol{\beta}_1 = \cdots = \boldsymbol{\beta}_k = \mathbf{0}.$$

The hypothesis H_0 means that all the explanatory variables are not useful in predicting $\boldsymbol{y}$. The LR criterion for H_0 is

$$(|\mathbf{S}_e|/|\mathbf{S}_0|)^{n/2},$$

where

$$\mathbf{S}_0 = \mathbf{Y}'(\mathrm{I}_n - \mathrm{P}_0)\mathbf{Y} = \sum_{j=1}^{n}(\boldsymbol{Y}_j - \bar{\boldsymbol{Y}})(\boldsymbol{Y}_j - \bar{\boldsymbol{Y}})'.$$

We denote

$$\Lambda(x_1, \ldots, x_k) = |\mathbf{S}_e|/|\mathbf{S}_0|.$$

Similarly, we use $\Lambda(x_1)$, $\Lambda(x_1, x_2)$, and so on. For example, $\Lambda(x_1)$ is the LR test for $\boldsymbol{\beta}_1 = \mathbf{0}$ under the model when only x_1 is used. The statistic $\Lambda(x_1, x_2)$ is the LR test for $\boldsymbol{\beta}_1 = \boldsymbol{\beta}_2 = \mathbf{0}$ under the model when only x_1 and x_2 are used. Then consider the LR test for H_J under the model when all the variables $x_1, \ldots, x_k$ are used.

The LR test is given by (7.2.6), for which the LR test is expressed as

$$\begin{aligned} \frac{|\mathbf{S}_e|}{|\mathbf{S}_e + \mathbf{S}_h|} &= \frac{|\mathbf{S}_e|/|\mathbf{S}_0|}{|\mathbf{S}_e + \mathbf{S}_h|/|\mathbf{S}_0|} \\ &= \frac{\Lambda(x_1, \ldots, x_k)}{\Lambda(x_1, \ldots, x_j)} \\ &\equiv \Lambda(x_{j+1}, \ldots, x_k | x_1, \ldots, x_j). \end{aligned} \tag{7.2.8}$$

Now we consider testing a linear hypothesis

$$H: \ \mathrm{C}\mathcal{B} = \mathrm{O} \tag{7.2.9}$$

under the model (7.1.2), where $\mathrm{C} = (c_{ij})$ is a $c \times (k+1)$ matrix of rank c. The hypothesis H_J is a special case of H, since it is obtained from H by putting $\mathrm{C} = (\mathrm{O}, \mathrm{I}_{k-j})$. The hypothesis is expressed as

$$c_{i0}\boldsymbol{\beta}_0 + c_{i1}\boldsymbol{\beta}_1 + \cdots + c_{ik}\boldsymbol{\beta}_k = \mathbf{0}, \ i = 1, \ldots, c.$$

The testing problem has the following features:

(1) The rows of $\mathbf{Y}$ are independently and normally distributed with the same unknown covariance matrix Σ.

(2) The spaces spanned by each of the columns of $\mathrm{E}(\mathbf{X})$ under the model are the same, which is denoted by Ω.

(3) The spaces spanned by each of the columns of $\mathrm{E}(\mathbf{X})$ under the hypothesis are the same, which is denoted by ω.

The spaces Ω and ω are given by $\mathcal{R}[\mathrm{X}]$ and $\mathrm{N}[\mathrm{B}]\cap\Omega$ (see the proof of [A.1.25]), respectively, where $\mathrm{B} = \mathrm{C}(\mathrm{X}'\mathrm{X})^{-1}\mathrm{X}'$. Then, from [A.2.11] the MLS's of Σ under the model and the hypothesis are given by

$$\hat{\Sigma}_\Omega = (1/n)\mathbf{Y}'(\mathrm{I}_n - \mathrm{P}_\Omega)\mathbf{Y} = (1/n)\mathbf{S}_\Omega, \quad \text{and}$$
$$\hat{\Sigma}_\omega = (1/n)\mathbf{Y}'(\mathrm{I}_n - \mathrm{P}_\omega)\mathbf{Y} = (1/n)\mathbf{S}_\omega,$$

respectively. Further, the likelihood criterion λ is expressed as

$$\lambda^{2/n} = \frac{|\mathbf{S}_\Omega|}{|\mathbf{S}_\omega|} = \frac{|\mathbf{S}_\Omega|}{|\mathbf{S}_\Omega + \mathbf{S}_h|}, \tag{7.2.10}$$

where $\mathbf{S}_h = \mathbf{S}_\omega - \mathbf{S}_\Omega$. Using [A.1.25], we can express

$$\mathbf{S}_h = \mathrm{X}(\mathrm{X}'\mathrm{X})^{-1}\mathrm{C}'\{\mathrm{C}(\mathrm{X}'\mathrm{X})^{-1}\mathrm{C}'\}^{-1}\mathrm{C}(\mathrm{X}'\mathrm{X})^{-1}\mathrm{X}. \tag{7.2.11}$$

By the argument similar to the proof of Theorem 7.2.2, it holds that

(1) $\mathbf{S}_\Omega \sim \mathrm{W}_p(n-k-1, \Sigma)$.

(2) Under the hypothesis, $\mathbf{S}_h \sim \mathrm{W}_p(c, \Sigma)$.

(3) $\mathbf{S}_\Omega$ and $\mathbf{S}_h$ are mutually independent.

This implies that under H, $\lambda^{2/n} = |\mathbf{S}_\Omega|/|\mathbf{S}_\Omega + \mathbf{S}_h| \sim \Lambda_p(c, n-k-1)$.

7.3 Selection of Variables

In this chapter we consider the problem of selecting a model from a collection of candidate models specified by linear regression of $\boldsymbol{Y}$ on subvectors of $\boldsymbol{x}$. Here we are concerned with selection methods based on stepwise procedures and the construction of approximately unbiased estimators of a risk function or an underlying criterion function, in particular the C_p criterion [Mallows, 1973] and the AIC [Akaike, 1973].

Related to a selection of models, it may be noted that there are two ideas about the best model or a target model Geisser and Eddy (1979): (1) which of the models $M_1, \ldots, M_m$ best explain a given set of data? (2) which models $M_1, \ldots, M_m$ yield the best predictions for future observations from the same process which generated the given set of data? The C_p criterion and the AIC can be regarded as a realization of (2).

7.3.1 Stepwise Procedure

Forward Selection Procedure. At the first step, we consider test statistics $\Lambda(x_j)$ for each j. We choose the variable with minimum $\Lambda(x_j)$. This is searching for the variable x_j which best predicts the p-variate $\boldsymbol{Y}$.

Suppose that x_1 is the variable entered at the first step. Then at the second step, we seek the variable yielding the smallest partial Λ given by

$$\Lambda(x_j|x_1) = \frac{\Lambda(x_1, x_j)}{\Lambda(x_1)}$$

for $x_j \neq x_1$. In this step we are searching for the variable x_j that predicts best in the presence of x_1.

If we denote the second variable to ender by x_2, then at the third step we seek the x_j that minimizes

$$\Lambda(x_j|x_1, x_2) = \frac{\Lambda(x_1, x_2, x_j)}{\Lambda(x_1, x_2)},$$

where x_j is a candidate variable from among the $p - 2$ remaining variables $x_3, \ldots, x_p$.

The procedure continues until a step is reached at which the p-value of the minimum partial Λ exceeds a predetermined value (e.g., 0.05).

Backward Elimination Procedure. The backward elimination procedure begins with all x's included in the model and deletes one at a time using a partial Λ. At the first step, the partial Λ for each x_j is

$$\Lambda(x_j|x_1, \ldots, x_{j-1}, x_{j+1}, \ldots, x_k) = \frac{\Lambda(x_1, \ldots, x_k)}{\Lambda(x_1, \ldots, x_{j-1}, x_{j+1}, \ldots, x_k)}.$$

The variable with the largest Λ, which is the least important variable in the presence of the other variables, is deleted. At the second step, a partial Λ is calculated for each of the $p - 1$ remaining variables, and again the least important variable in the presence of the others is eliminated. More precisely, suppose that x_k is the variable deleted at the first step. Then, at the second step, we seek the variable yielding the largest partial Λ given by

$$\Lambda(x_j|x_1, \ldots, x_{j-1}, x_{j+1}, \ldots, x_{k-1}) = \frac{\Lambda(x_1, \ldots, x_{k-1})}{\Lambda(x_1, \ldots, x_{j-1}, x_{j+1}, \ldots, x_{k-1})}$$

for $x_j \neq x_k$. This procedure continues until a step is reached where the largest Λ is significant.

Stepwise Method. Forward selection may be extended as follows. Each time a variable enters, all the variables that have entered previously are checked by a partial Λ to see if the least significant is now redundant and can be deleted. For example, suppose that $x_1, x_2, \ldots, x_q$ have entered previously and x_{q+1} enter. Then we seek the variable x_j that maximizes

$$\Lambda(x_j|x_1, \ldots, x_{j-1}, x_{j+1}, \ldots, x_q) = \frac{\Lambda(x_1, \ldots, x_{q+1})}{\Lambda(x_1, \ldots, x_{j-1}, x_{j+1}, \ldots, x_{q+1})}$$

for $x_j \neq x_1$. If the largest p-value is larger than the predetermined p-value, the variable is deleted. Such a stepwise procedure is called.

Similarly backward elimination is extended, by checking all the variables that, have entered previously in the same way as for forward selection.

In each of the steps in a stepwise method we need to know p-values of the tests for significance of an entered variable or a deleted variable. These p-values are obtained by using the following results:

(1) Suppose that a set $\{x_1, \ldots, x_{j-1}\}$ has been entered, and we wish to test the significance of $x_a \notin \{x_1, \ldots, x_{j-1}\}$. An LR test statistic is given by

$$L_e = \Lambda(x_a | x_1, \ldots, x_{j-1})$$

whose null distribution is $\Lambda_p(1, n-j-1)$. Therefore, the p-value is obtained by using

$$\frac{1 - L_e}{L_e} \cdot \frac{n-j-1}{p} \sim \mathrm{F}(p, n-j-1).$$

(2) Suppose that a set $\{x_1, \ldots, x_j\}$ has been entered, and we wish to test a significance of $x_a \in \{x_1, \ldots, x_{j-1}\}$. An LR test statistic is given by

$$L_d = \Lambda(x_a | x_1, \ldots, x_{a-1}, x_{a+1}, \ldots, x_j)$$

whose null distribution is $\Lambda_p(1, n-j-1)$. Therefore, the p-value is obtained by using

$$\frac{1 - L_d}{L_d} \cdot \frac{n-j-1}{p} \sim \mathrm{F}(p, n-j-1).$$

7.3.2 C_p Criterion

We assume that the model (7.1.1) is a multivariate regression model when we use all available explanatory variables. Then it is called the *full* or *initial multivariate linear model* and is denoted by M_F:

$$M_F: \ \boldsymbol{Y}_i = \boldsymbol{\beta}_0 + \boldsymbol{\beta}_1 x_{i1} + \cdots + \boldsymbol{\beta}_k x_{ik} + \boldsymbol{\varepsilon}_i, \quad i = 1, \ldots, n, \tag{7.3.1}$$

where the coefficient vectors $\boldsymbol{\beta}_0, \ldots, \boldsymbol{\beta}_k$ are unknown and the error terms $\boldsymbol{\varepsilon}_1, \ldots, \boldsymbol{\varepsilon}_n$ are distributed independently as a p-variate normal distribution with mean $\mathbf{0}$ and covariance matrix Σ. Using matrix notation in Section 7.1, the full model is denoted as

$$M_F: \mathbf{Y} = \mathrm{X}\mathcal{B} + \mathcal{E}.$$

Suppose that the true model for $\boldsymbol{Y}_1, \ldots, \boldsymbol{Y}_n$ is

$$M_*: \ \boldsymbol{Y}_i = \boldsymbol{\eta}_i + \boldsymbol{\varepsilon}_i, \quad i = 1, \ldots, n, \tag{7.3.2}$$

where the error terms $\varepsilon_1, \ldots, \varepsilon_n$ are i.i.d. with mean $\mathbf{0}$ and covariance matrix Σ_0. The true model is expressed in matrix notation as

$$M_*: \mathbf{Y} = \boldsymbol{\eta}_0 + \mathcal{E}.$$

where $\boldsymbol{\eta}_0 = (\boldsymbol{\eta}_{10}, \ldots, \boldsymbol{\eta}_{n0})'$.

Now we consider a linear model using a subset of the k explanatory variables. A genetic candidate model can be expressed in terms of a subset $J \subset \{x_1, \ldots, x_k\}$. Without loss of generality, we consider the model based on the first j explanatory variables $x_1, \ldots, x_j$, which is denoted by

$$M_J\colon\ \boldsymbol{Y}_i = \boldsymbol{\beta}_0 + \boldsymbol{\beta}_1 x_{i1} + \cdots + \boldsymbol{\beta}_j x_{ij} + \boldsymbol{\varepsilon}_i, \quad i = 1, \ldots, n, \tag{7.3.3}$$

where the coefficient vectors $\boldsymbol{\beta}_0, \ldots, \boldsymbol{\beta}_j$ are unknown and the error terms $\varepsilon_1, \ldots, \varepsilon_n$ are independently distributed as a p-variate normal distribution with mean $\mathbf{0}$ and covariance matrix Σ. In matrix notation,

$$M_J\colon \mathbf{Y} = \mathrm{X}_J \mathcal{B}_J + \mathcal{E},$$

where X_J is the matrix consisting of the first $j+1$ columns of X, and $\mathcal{B}_J$ is the matrix consisting the first $j+1$ rows of $\mathcal{B}$. Let $\boldsymbol{x}_i$ partition as

$$\begin{aligned} \boldsymbol{x}_i' &= (x_{i1}, \ldots, x_{ij}, \ldots, x_{ik}) \\ &= (\boldsymbol{x}_{i1}', \boldsymbol{x}_{i2}'), \end{aligned}$$

then

$$\mathrm{X}_J = \begin{pmatrix} 1 & \boldsymbol{x}_{i1}' \\ \vdots & \vdots \\ 1 & \boldsymbol{x}_{in}' \end{pmatrix}, \quad \mathcal{B}_J = (\boldsymbol{\beta}_0, \boldsymbol{\beta}_1, \ldots, \boldsymbol{\beta}_j)'.$$

For selection of the best model from a collection of candidate models specified by the linear regression of $\boldsymbol{Y}$ on subvectors of $\boldsymbol{x}$, it is important to define a measure of goodness of a candidate model (i.e., to define what the target model is). To consider a measure of goodness, introduce new n observations $\boldsymbol{Z}_i$ of $\boldsymbol{Y}$ at $\boldsymbol{x}_i$, and let

$$\mathbf{Z} = \begin{pmatrix} \boldsymbol{Z}_1' \\ \vdots \\ \boldsymbol{Z}_n' \end{pmatrix}.$$

Here it is assumed that $\mathbf{Z}$ has the same distribution as $\mathbf{Y}$ and is independent of $\mathbf{Y}$. Now we consider a measure of goodness for a general candidate model M_J. Under M_J, $\mathbf{Y}$ is commonly predicted by

$$\hat{\mathbf{Y}}_J = \mathrm{X}_J(\mathrm{X}_J'\mathrm{X}_J)^{-1}\mathrm{X}_J'\mathbf{Y} = \mathrm{P}_{\mathrm{X}_J}\mathbf{Y},$$

where $\mathrm{P}_{\mathrm{X}_J} = \mathrm{X}_J(\mathrm{X}_J'\mathrm{X}_J)^{-1}\mathrm{X}_J'$. The residual SSP matrix is

$$\mathbf{S}_J = (\mathbf{Y} - \hat{\mathbf{Y}}_J)'(\mathbf{Y} - \hat{\mathbf{Y}}_J)' = \mathbf{Y}'(\mathrm{I}_n - \mathrm{P}_{\mathrm{X}_J})\mathbf{Y}.$$

The quantities in the full model corresponding to these are denoted by $\hat{\mathbf{Y}}_F$ and S_F. A multivariate version of the C_p criterion [Mallows, 1973, 1995] is defined by

$$\begin{aligned} C_p &= \operatorname{tr} \mathbf{S}_J \tilde{\Sigma}_F^{-1} + 2p(j+1) \\ &= (n-k-1) \operatorname{tr} \mathbf{S}_J \mathbf{S}_F^{-1} + 2p(j+1), \end{aligned} \tag{7.3.4}$$

where $\tilde{\Sigma}_F$ is the usual unbiased estimator of Σ under the full model M_F, and is given by $\tilde{\Sigma}_F = (n-k-1)^{-1}\mathbf{S}_F$. This criterion is considered to be an estimator of the risk function given by

$$\begin{aligned} R_p &= \mathrm{E}_{\mathbf{Y}}^*\mathrm{E}_{\mathbf{Z}}^*[\operatorname{tr} \Sigma_0^{-1}(\mathbf{Z} - \hat{\mathbf{Y}}_J)'(\mathbf{Z} - \hat{\mathbf{Y}}_J)] \\ &= \mathrm{E}_{\mathbf{Y}}^*\mathrm{E}_{\mathbf{Z}}^*[\operatorname{tr} \Sigma_0^{-1}(\boldsymbol{\eta} - \hat{\mathbf{Y}}_J)'(\boldsymbol{\eta} - \hat{\mathbf{Y}}_J)] + np \end{aligned} \tag{7.3.5}$$

where $\mathrm{E}_{\mathbf{Y}}^*$ and $\mathrm{E}_{\mathbf{Z}}^*$ denote the expectations with respect to the true model of $\mathbf{Y}$ and $\mathbf{Z}$. In fact, Mallows (1973) considered the first term of the second expression of (7.3.5) as a risk function. Therefore, his C_p criterion in the multivariate case corresponds to $C_p - np$.

Now we examine properties of C_p as an estimator of R_C and consider a corrected version that reflects its risk more faithfully. A key approach is to rewrite R_C as

$$R_C = \mathrm{E}_{\mathbf{Y}}^*[\operatorname{tr} \mathbf{S}_J \tilde{\Sigma}_F^{-1}] + b_C, \tag{7.3.6}$$

where

$$b_C = \mathrm{E}_{\mathbf{Y}}^*[\operatorname{tr} \Sigma_0^{-1}(\boldsymbol{\eta} - \hat{\mathbf{Y}}_J)'(\boldsymbol{\eta} - \hat{\mathbf{Y}}_J) - \operatorname{tr} \mathbf{S}_J \tilde{\Sigma}_F^{-1}] + np.$$

Then we can regard b_C as the bias term when we estimate R_C by $\operatorname{tr} \mathbf{S}_J \tilde{\Sigma}_F^{-1}$. Note that this naive estimator is obtained from

$$\operatorname{tr} \Sigma_0^{-1}(\mathbf{Z} - \hat{\mathbf{Y}}_J)'(\mathbf{Z} - \hat{\mathbf{Y}}_J)$$

by substituting $\tilde{\Sigma}_F$ and $\mathbf{Y}$ to Σ_0 and $\mathbf{Z}$, respectively. The C_p in (7.3.4) arises from estimating b_C by $2p(j+1)$.

Theorem 7.3.1 *Let b_C be the bias term in* (7.3.6) *when one estimates R_C by $\operatorname{tr} \mathbf{S}_J \tilde{\Sigma}_F^{-1}$. Suppose that the full model includes the true model M_*. Then*

$$b_C = 2p(j+1) + \frac{p+1}{n-k-p-2}\left[p(k-j) + \operatorname{tr}\Omega\right], \tag{7.3.7}$$

where $\Omega = \Sigma_0^{-1/2}\boldsymbol{\eta}_0'(\mathrm{P}_{\mathrm{X}} - \mathrm{P}_{\mathrm{X}_J})\boldsymbol{\eta}_0\Sigma_0^{-1/2}$.

Proof. Since M_F includes M_*, we may write

$$\mathrm{E}^*(\mathbf{Y}) = \boldsymbol{\eta}_0 = \mathrm{X}\mathcal{B}_0.$$

In the following we simply write E, suppressing the asterisk. Noting that

$$\begin{aligned}\boldsymbol{\eta}_0 - \hat{\mathbf{Y}}_J &= \mathrm{X}\mathcal{B}_0 - \mathrm{P}_{\mathrm{X}_J}(\mathbf{Y} - \mathrm{X}\mathcal{B}_0 + \mathrm{X}\mathcal{B}_0) \\ &= (\mathrm{I}_n - \mathrm{P}_{\mathrm{X}_J})\mathrm{X}\mathcal{B}_0 - \mathrm{P}_{\mathrm{X}_J}(\mathbf{Y} - \mathrm{X}\mathcal{B}_0),\end{aligned}$$

we have

$$\begin{aligned}&(\boldsymbol{\eta}_0 - \hat{\mathbf{Y}}_J)'(\boldsymbol{\eta}_0 - \hat{\mathbf{Y}}_J) \\ &= \Sigma_0^{-1/2}\boldsymbol{\eta}_0'(\mathrm{P}_{\mathrm{X}} - \mathrm{P}_{\mathrm{X}_J})\boldsymbol{\eta}_0\Sigma_0^{-1/2} + (\mathbf{Y} - \mathrm{X}\mathcal{B}_0)'\mathrm{P}_{\mathrm{X}_J}(\mathbf{Y} - \mathrm{X}\mathcal{B}_0).\end{aligned}$$

Also, we have from Theorem 2.4.2

$$(\mathbf{Y} - \mathrm{X}\mathcal{B}_0)'\mathrm{P}_{\mathrm{X}_J}(\mathbf{Y} - \mathrm{X}\mathcal{B}_0) \sim \mathrm{W}_p(j+1, \Sigma_0),$$

which, in turn, implies that

$$\mathrm{E}[\,\mathrm{tr}\,\Sigma_0^{-1}(\boldsymbol{\eta}_0 - \hat{\mathbf{Y}}_J)'(\boldsymbol{\eta}_0 - \hat{\mathbf{Y}}_J)] = p(j+1) + \mathrm{tr}\,\Omega. \tag{7.3.8}$$

On the other hand, $\mathbf{S}_F = \mathbf{Y}'(\mathrm{I}_n - \mathrm{P}_{\mathrm{X}})\mathbf{Y}$, and $\mathbf{S}_J - \mathbf{S}_F = \mathbf{Y}'(\mathrm{P}_{\mathrm{X}} - \mathrm{P}_{\mathrm{X}_J})\mathbf{Y}$. Using Theorem 2.4.2 gives

$$\mathbf{S}_F \sim \mathrm{W}_p(n-(k+1), \Sigma_0), \quad \mathbf{S}_J - \mathbf{S}_F \sim \mathrm{W}_p(k-j, \Sigma_0; \Sigma_0^{-1/2}\Omega\Sigma_0^{-1/2}),$$

and $\mathbf{S}_F$ and $(\mathbf{S}_J - \mathbf{S}_F)$ are independent. These imply that

$$\begin{aligned}\mathrm{E}(\,\mathrm{tr}\,\mathbf{S}_F^{-1}\cdot\mathbf{S}_J) &= \mathrm{E}[\,\mathrm{tr}\,\mathbf{S}_F^{-1}\{\mathbf{S}_F + (\mathbf{S}_J - \mathbf{S}_F)\}] \\ &= p + \frac{1}{n-k-p-2}\{\,\mathrm{tr}\,\Omega + (k-j)p\}.\end{aligned} \tag{7.3.9}$$

The results (7.3.8) and (7.3.9) imply (7.3.7). □

From (7.3.9) we have an unbiased estimator for $\mathrm{tr}\,\Omega$ given by

$$\mathrm{tr}\,\hat{\Omega} = (n-k-p-2)\{\,\mathrm{tr}\,\mathbf{S}_F^{-1}\mathbf{S}_J - p\} - (k-j)p. \tag{7.3.10}$$

A corrected estimator of b_C is obtained by substituting the unbiased estimator for $\mathrm{tr}\,\Omega$ in (7.3.7). This gives a modified C_p criterion [Fujikoshi and Satoh, 1997] given by

$$MC_p = (n-k-p-2)\,\mathrm{tr}\,\mathbf{S}_J\mathbf{S}_F^{-1} + 2\left\{p(j+1) + \frac{1}{2}p(p+1)\right\}. \tag{7.3.11}$$

Theorem 7.3.2 *Suppose that the full model M_F includes the true model M_*. Then:*

(1) *If $\Omega = \mathrm{O}(1)$ for large n, C_p is an asymptotically unbiased estimator of R_C.*

(2) *MC_p is an unbiased estimator of R_C.*

Proof. If $\Omega = \mathrm{O}(1)$, from (7.3.7) we have

$$b_C = 2p(j+1) + \mathrm{O}(n^{-1}),$$

and hence, from (7.3.6),

$$R_C = \mathrm{E}^*_{\mathbf{Y}}[(n-k-1)\,\mathrm{tr}\,\mathbf{S}_J\mathbf{S}_F^{-1}] + 2p(j+1) + \mathrm{O}(n^{-1}).$$

This shows the first result. Let $\hat{b}_C$ be the b_C in (7.3.7) with $\mathrm{tr}\,\hat{\Omega}$ in (7.3.10) instead of $\mathrm{tr}\,\Omega$. Then we have an unbiased estimator of R_C,

$$(n-k-1)\,\mathrm{tr}\,\mathbf{S}_J\mathbf{S}_F^{-1} + \hat{b}_C,$$

which is equal to MC_p. □

The relationship between the two criteria C_p and MC_p is close in the following sense:

$$MC_p = C_p - (p+1)(\,\mathrm{tr}\,\mathbf{S}_F^{-1}\mathbf{S}_J - p) \le C_p,$$

which is a consequence of the fact that $\mathbf{S}_J - \mathbf{S}_F$ is positive semidefinite. This shows that C_p has a tendency to overestimate the risk function R_C.

7.3.3 AIC Criterion

Next we consider the AIC criterion [Akaike, 1973] which is applicable for selections of various models as well as multivariate regression models. Let $\mathbf{Y}$ be a random matrix. Suppose that the true density function of $\mathbf{Y}$ is $g(\mathrm{Y})$. Consider a candidate model M such that

M: the density function of $\mathbf{Y}$ belongs to a class of density functions $\{f(\mathrm{Y};\Theta);\Theta\in\Omega\}$,

where Ω is the parameter space for unknown parameter matrix Θ. We consider the risk function [Akaike, 1973] for M defined by

$$R_A = \mathrm{E}^*_{\mathbf{Y}}\mathrm{E}^*_{\mathbf{Z}}[-2\log f(\mathbf{Z};\hat{\Theta})], \tag{7.3.12}$$

where $\hat{\Theta}$ is the MLE of Θ under M and $\mathbf{Z}$ is a future random matrix used in the definition of the risk R_C in (7.3.5). Akaike (1973) proposed

$$AIC = -2\log f(\mathbf{Y};\hat{\Theta}) + 2\times\dim(\Theta), \tag{7.3.13}$$

as an approximately unbiased estimator of R_A.

We note that the risk function R_A is regarded as the expected log-predictive function for a model M. Further, it may be noted that the problem of seeking a model with minimum R_A is equivalent to that of seeking a model with minimum Kullback–Leibler information. In fact, consider the Kullback–Leibler information on the distance from the density function $g(\mathrm{Z})$ to $\hat{f} = f(\mathrm{Z};\hat{\Theta})$, which is given by

$$\begin{aligned} D_{KL}(g, \hat{f}) &= \int g(\mathrm{Z}) \log \left\{ \frac{g(\mathrm{Z})}{f(\mathrm{Z};\hat{\Theta})} \right\} d\mathrm{Z} = \mathrm{E}^*_{\mathbf{Z}} \left[\log \left\{ \frac{g(\mathbf{Z})}{f(\mathbf{Z};\hat{\Theta})} \right\} \right] \\ &= -\mathrm{E}^*_{\mathbf{Z}}[\log f(\mathbf{Z};\hat{\Theta})] + \mathrm{E}^*_{\mathbf{Z}}[\log g(\mathbf{Z})]. \end{aligned}$$

Note that the second term of the last expression does not depend on model M, say c_0. Then we have the relationship

$$R_A = \mathrm{E}^*_{\mathbf{Y}}[D_{KL}(g, \hat{f})] + c_0.$$

The *AIC* in (7.3.13) is proposed by writing R_A as

$$R_A = \mathrm{E}^*_{\mathbf{Y}}[-2\log f(\mathbf{Y};\hat{\Theta})] + b_A,$$

where

$$b_A = \mathrm{E}^*_{\mathbf{Y}}\mathrm{E}^*_{\mathbf{Z}}\left[-2\log\left\{ \frac{f(\mathbf{Z};\hat{\Theta})}{f(\mathbf{Y};\hat{\Theta})} \right\} \right], \tag{7.3.14}$$

and finding an approximation for b_A.

Now we consider AIC and its correction, CAIC, for the multivariate regression model M_J in (7.3.3). The likelihood function of $\mathbf{Y}$ under M_J is

$$-2\log f(\mathrm{Y}, \Theta) = n \log|\Sigma| + \operatorname{tr} \Sigma^{-1}(\mathrm{Y} - \mathrm{X}_J\mathcal{B}_J)'(\mathrm{Y} - \mathrm{X}_J\mathcal{B}_J) + np\log(2\pi).$$

The MLEs of $\mathcal{B}_J$ and Σ are given (see Section 7.2) by

$$\hat{\mathcal{B}}_J = (\mathrm{X}_J'\mathrm{X}_J)^{-1}\mathrm{X}_J'\mathbf{Y}, \quad \hat{\Sigma}_J = \frac{1}{n}\mathbf{Y}'(\mathrm{I}_n - \mathrm{P}_{\mathrm{X}_J})\mathbf{Y}.$$

The number of independent parameters is $p(j+1) + \frac{1}{2}p(p+1)$. Therefore, using (7.3.13) we have

$$AIC = n\log|\hat{\Sigma}_J| + np(\log 2\pi + 1) + 2\left\{ p(j+1) + \frac{1}{2}p(p+1) \right\}. \tag{7.3.15}$$

A corrected AIC was proposed [Sugiura, 1978; Bedrick and Tsai, 1994] as

$$\begin{aligned} CAIC &= n\log|\hat{\Sigma}_J| + np(\log 2\pi + 1) \\ &\quad + \frac{2n}{n-j-p-2}\left\{ p(j+1) + \frac{1}{2}p(p+1) \right\}. \end{aligned} \tag{7.3.16}$$

Theorem 7.3.3 *Suppose that the model M_J includes the true model. Then for the AIC and CAIC in (7.3.15) and (7.3.16), it holds that:*

(1) *AIC is asymptotically unbiased: that is,*

$$\lim_{n\to\infty} \mathrm{E}(AIC) = R_A.$$

(2) *CAIC is an unbiased estimator of R_A if $n-j-p-1>0$.*

Proof. In general, we can write the bias term b_A as

$$\begin{aligned} b_A &= \mathrm{E}^*_{\mathbf{Y}}\mathrm{E}^*_{\mathbf{Z}}[\operatorname{tr}\hat{\Sigma}_J^{-1}(\mathbf{Z}-\mathrm{X}_J\hat{\mathcal{B}}_J)'(\mathbf{Z}-\mathrm{X}_J\hat{\mathcal{B}}_J)] - np \\ &= \mathrm{E}^*_{\mathbf{Y}}[n\operatorname{tr}\hat{\Sigma}_J^{-1}\{\Sigma_0 + (\boldsymbol{\eta}_0 - \mathrm{P}_{\mathrm{X}_J}\mathbf{Y})'(\boldsymbol{\eta}_0 - \mathrm{P}_{\mathrm{X}_J}\mathbf{Y})\}] - np \\ &= n\operatorname{tr}\mathrm{E}^*_{\mathbf{Y}}(\hat{\Sigma}_J^{-1})\mathrm{E}^*_{\mathbf{Y}}[\Sigma_0 + (\boldsymbol{\eta}_0 - \mathrm{P}_{\mathrm{X}_J}\mathbf{Y})'(\boldsymbol{\eta}_0 - \mathrm{P}_{\mathrm{X}_J}\mathbf{Y})\}] - np, \end{aligned}$$

since $\mathbf{S}_J$ and $\mathrm{P}_{\mathrm{X}_J}\mathbf{Y}$ are independent. From our assumption, $M_* \in M_J$, so $\mathrm{P}_{\mathrm{X}_J}\boldsymbol{\eta}_0 = \boldsymbol{\eta}_0$. We have $\mathbf{S}_J \sim \mathrm{W}_p(n-j-1, \Sigma_0)$ and

$$\begin{aligned} (\boldsymbol{\eta}_0 - \mathrm{P}_{\mathrm{X}_J}\mathbf{Y})'(\boldsymbol{\eta}_0 - \mathrm{P}_{\mathrm{X}_J}\mathbf{Y}) &= (\mathbf{Y}-\boldsymbol{\eta}_0)'\mathrm{P}_{\mathrm{X}_J}(\mathbf{Y}-\boldsymbol{\eta}_0) \\ &\sim \mathrm{W}_p(j+1, \Sigma_0). \end{aligned}$$

Therefore, using Theorems 2.2.6 and 2.2.7 we get

$$\begin{aligned} b_A &= \frac{n}{n-j-p-2}\{np + p(j+1)\} - np \\ &= \frac{2n}{n-j-p-2}\left\{p(j+1) + \frac{1}{2}p(p+1)\right\}, \end{aligned}$$

which implies the first and second results. □

Since $n/(n-j-p-2) > 1$,

$$AIC < CAIC.$$

This means that AIC is underestimating the risk. For corrections of AIC in some other models, see, e.g., Sugiura (1978), Hurvich and Tsai (1898)

7.3.4 Numerical Example

We consider the chemical reaction data in Table 7.3.1 introduced by Box and Youle (1995) and examined by Rencher (2002) as an example of a multivariate regression model. The explanatory (independent) variables are

$$x_1 = \text{temperature}, \quad x_2 = \text{concentration}, \quad x_3 = \text{time}.$$

as an approximately unbiased estimator of R_A.

We note that the risk function R_A is regarded as the expected log-predictive function for a model M. Further, it may be noted that the problem of seeking a model with minimum R_A is equivalent to that of seeking a model with minimum Kullback–Leibler information. In fact, consider the Kullback–Leibler information on the distance from the density function $g(\mathrm{Z})$ to $\hat{f} = f(\mathrm{Z}; \hat{\Theta})$, which is given by

$$\begin{aligned} D_{KL}(g, \hat{f}) &= \int g(\mathrm{Z}) \log \left\{ \frac{g(\mathrm{Z})}{f(\mathrm{Z}; \hat{\Theta})} \right\} d\mathrm{Z} = \mathrm{E}^*_{\mathbf{Z}} \left[\log \left\{ \frac{g(\mathbf{Z})}{f(\mathbf{Z}; \hat{\Theta})} \right\} \right] \\ &= -\mathrm{E}^*_{\mathbf{Z}}[\log f(\mathbf{Z}; \hat{\Theta})] + \mathrm{E}^*_{\mathbf{Z}}[\log g(\mathbf{Z})]. \end{aligned}$$

Note that the second term of the last expression does not depend on model M, say c_0. Then we have the relationship

$$R_A = \mathrm{E}^*_{\mathbf{Y}}[D_{KL}(g, \hat{f})] + c_0.$$

The *AIC* in (7.3.13) is proposed by writing R_A as

$$R_A = \mathrm{E}^*_{\mathbf{Y}}[-2 \log f(\mathbf{Y}; \hat{\Theta})] + b_A,$$

where

$$b_A = \mathrm{E}^*_{\mathbf{Y}} \mathrm{E}^*_{\mathbf{Z}} \left[-2 \log \left\{ \frac{f(\mathbf{Z}; \hat{\Theta})}{f(\mathbf{Y}; \hat{\Theta})} \right\} \right], \tag{7.3.14}$$

and finding an approximation for b_A.

Now we consider AIC and its correction, CAIC, for the multivariate regression model M_J in (7.3.3). The likelihood function of $\mathbf{Y}$ under M_J is

$$-2 \log f(\mathrm{Y}, \Theta) = n \log |\Sigma| + \mathrm{tr}\, \Sigma^{-1} (\mathrm{Y} - \mathrm{X}_J \mathcal{B}_J)'(\mathrm{Y} - \mathrm{X}_J \mathcal{B}_J) + np \log(2\pi).$$

The MLEs of $\mathcal{B}_J$ and Σ are given (see Section 7.2) by

$$\hat{\mathcal{B}}_J = (\mathrm{X}_J' \mathrm{X}_J)^{-1} \mathrm{X}_J' \mathbf{Y}, \quad \hat{\Sigma}_J = \frac{1}{n} \mathbf{Y}'(\mathrm{I}_n - \mathrm{P}_{\mathrm{X}_J}) \mathbf{Y}.$$

The number of independent parameters is $p(j+1) + \frac{1}{2}p(p+1)$. Therefore, using (7.3.13) we have

$$AIC = n \log |\hat{\Sigma}_J| + np(\log 2\pi + 1) + 2 \left\{ p(j+1) + \frac{1}{2} p(p+1) \right\}. \tag{7.3.15}$$

A corrected AIC was proposed [Sugiura, 1978; Bedrick and Tsai, 1994] as

$$\begin{aligned} CAIC &= n \log |\hat{\Sigma}_J| + np(\log 2\pi + 1) \\ &\quad + \frac{2n}{n - j - p - 2} \left\{ p(j+1) + \frac{1}{2} p(p+1) \right\}. \end{aligned} \tag{7.3.16}$$

Theorem 7.3.3 *Suppose that the model M_J includes the true model. Then for the AIC and CAIC in* (7.3.15) *and* (7.3.16), *it holds that:*

(1) *AIC is asymptotically unbiased: that is,*

$$\lim_{n\to\infty} \mathrm{E}(AIC) = R_A.$$

(2) *CAIC is an unbiased estimator of R_A if $n - j - p - 1 > 0$.*

Proof. In general, we can write the bias term b_A as

$$\begin{aligned} b_A &= \mathrm{E}_{\mathbf{Y}}^* \mathrm{E}_{\mathbf{Z}}^* [\mathrm{tr}\, \hat{\Sigma}_J^{-1} (\mathbf{Z} - \mathrm{X}_J \hat{\mathcal{B}}_J)'(\mathbf{Z} - \mathrm{X}_J \hat{\mathcal{B}}_J)] - np \\ &= \mathrm{E}_{\mathbf{Y}}^* [n\, \mathrm{tr}\, \hat{\Sigma}_J^{-1} \{\Sigma_0 + (\boldsymbol{\eta}_0 - \mathrm{P}_{\mathrm{X}_J} \mathbf{Y})'(\boldsymbol{\eta}_0 - \mathrm{P}_{\mathrm{X}_J} \mathbf{Y})\}] - np \\ &= n\, \mathrm{tr}\, \mathrm{E}_{\mathbf{Y}}^* (\hat{\Sigma}_J^{-1}) \mathrm{E}_{Y}^* [\Sigma_0 + (\boldsymbol{\eta}_0 - \mathrm{P}_{\mathrm{X}_J} \mathbf{Y})'(\boldsymbol{\eta}_0 - \mathrm{P}_{\mathrm{X}_J} \mathbf{Y})\}] - np, \end{aligned}$$

since $\mathbf{S}_J$ and $\mathrm{P}_{\mathrm{X}_J}\mathbf{Y}$ are independent. From our assumption, $M_* \in M_J$, so $\mathrm{P}_{\mathrm{X}_J}\boldsymbol{\eta}_0 = \boldsymbol{\eta}_0$. We have $\mathbf{S}_J \sim \mathrm{W}_p(n - j - 1, \Sigma_0)$ and

$$\begin{aligned} (\boldsymbol{\eta}_0 - \mathrm{P}_{\mathrm{X}_J} \mathbf{Y})'(\boldsymbol{\eta}_0 - \mathrm{P}_{\mathrm{X}_J} \mathbf{Y}) &= (\mathbf{Y} - \boldsymbol{\eta}_0)' \mathrm{P}_{\mathrm{X}_J} (\mathbf{Y} - \boldsymbol{\eta}_0) \\ &\sim \mathrm{W}_p(j + 1, \Sigma_0). \end{aligned}$$

Therefore, using Theorems 2.2.6 and 2.2.7 we get

$$\begin{aligned} b_A &= \frac{n}{n - j - p - 2} \{np + p(j + 1)\} - np \\ &= \frac{2n}{n - j - p - 2} \left\{ p(j + 1) + \frac{1}{2} p(p + 1) \right\}, \end{aligned}$$

which implies the first and second results. □

Since $n/(n - j - p - 2) > 1$,

$$AIC < CAIC.$$

This means that AIC is underestimating the risk. For corrections of AIC in some other models, see, e.g., Sugiura (1978), Hurvich and Tsai (1898)

7.3.4 Numerical Example

We consider the chemical reaction data in Table 7.3.1 introduced by Box and Youle (1995) and examined by Rencher (2002) as an example of a multivariate regression model. The explanatory (independent) variables are

$$x_1 = \text{temperature}, \quad x_2 = \text{concentration}, \quad x_3 = \text{time}.$$

The response (dependent) variables are

$$Y_1 = \text{percentage of unchanged starting material},$$
$$Y_2 = \text{percentage converted to the desired product},$$
$$Y_3 = \text{percentage of unwanted by-product}.$$

Here we examine the selection of the explanatory variables x_1, x_2, and x_3 by C_p MC_p, AIC, and $CAIC$. The S_F and S_J may be computed numerically by using (7.2.7). Table 7.3.2 shows that all three variables should be considered in multivariate regression analysis of (Y_1, Y_2, Y_3) to (x_1, x_2, x_3). On the other hand, we tried to select the explanatory variables in multiple regression analysis of Y_i to (x_1, x_2, x_3). Then, in multiple regression analysis of Y_2 to (x_1, x_2, x_3), it is suggested from Table 7.3.3 that x_2 may be deleted.

Table 7.3.1 Chemical Reaction Data

Experiment Number	Response Variables			Explanatory Variables		
	Y_1	Y_2	Y_3	x_1	x_2	x_3
1	41.5	45.9	11.2	162.0	23.0	3.0
2	33.8	53.3	11.2	162.0	23.0	8.0
3	27.7	57.5	12.7	162.0	30.0	5.0
4	21.7	58.8	16.0	162.0	30.0	8.0
5	19.9	60.6	16.2	172.0	25.0	5.0
6	15.0	58.0	22.6	172.0	25.0	8.0
7	12.2	58.6	24.5	172.0	30.0	5.0
8	4.3	52.4	38.0	172.0	30.0	8.0
9	19.3	56.9	21.3	167.0	27.5	6.5
10	6.4	55.4	30.8	177.0	27.5	6.5
11	37.6	46.9	14.7	157.0	27.5	6.5
12	18.0	57.3	22.2	167.0	32.5	6.5
13	26.3	55.0	18.3	167.0	22.5	6.5
14	9.9	58.9	28.0	167.0	27.5	9.5
15	25.0	50.3	22.1	167.0	27.5	3.5
16	14.1	61.1	23.0	177.0	20.0	6.5
17	15.2	62.9	20.7	177.0	20.0	6.5
18	15.9	60.0	22.1	160.0	34.0	7.5
19	19.6	60.6	19.3	160.0	34.0	7.5

7.4 Principal Component Regression

In this section we consider a multivariate regression model (7.1.1) in a situation where there are considerable relationships among explanatory variables.

Table 7.3.2 Values of C_p, MC_p, AIC, and $CAIC$

Model	C_p	Model	MC_p	Model	AIC	Model	$CAIC$
M_{123}	69.0	M_{123}	69.0	M_{123}	271.6	M_{123}	297.8
M_{12}	111.5	M_{12}	98.6	M_{12}	293.0	M_{12}	310.5
M_{13}	182.1	M_{13}	150.3	M_{13}	307.2	M_1	323.0
M_1	259.7	M_1	205.6	M_1	312.0	M_{13}	324.7
M_{23}	371.6	M_{23}	289.3	M_0	318.3	M_0	324.8
M_3	375.2	M_3	290.4	M_3	318.7	M_3	329.7
M_2	432.3	M_2	332.2	M_2	323.2	M_2	334.3
M_0	447.3	M_0	341.6	M_{23}	324.0	M_{23}	341.5

Table 7.3.3: Values of C_p, MC_p, AIC, and $CAIC$ when Y_2 is the only response variable

Model	C_p	Model	MC_p	Model	AIC	Model	$CAIC$
M_{13}	22.3	M_{13}	22.1	M_{13}	112.4	M_{13}	115.2
M_{123}	23.0	M_{123}	23.0	M_{123}	112.8	M_1	116.1
M_{12}	24.0	M_1	23.5	M_{12}	114.3	M_3	116.3
M_1	24.2	M_{12}	23.6	M_1	114.5	M_0	116.6
M_3	24.4	M_3	23.7	M_3	114.7	M_{12}	117.1
M_0	26.1	M_0	24.9	M_0	115.8	M_{123}	117.5
M_{23}	26.4	M_{23}	25.6	M_{23}	116.6	M_2	119.3
M_2	27.9	M_2	26.7	M_2	117.7	M_{23}	119.5

In such situations it might happen that the variances of some estimated regression coefficients are large, and the prediction will become unstable. As one of the methods of overcoming such faults, we may consider the selection of explanatory variables. As a second method we may consider using the principal components of the explanatory variables instead of the original explanatory variables.

Consider a multivariate regression model (7.1.1). Here we denote the observation matrix of the explanatory variables $\boldsymbol{x} = (x_1, \ldots, x_n)'$ by

$$\mathrm{X} = \begin{pmatrix} x_{11} & \cdots & x_{1k} \\ \vdots & \vdots & \vdots \\ x_{n1} & \cdots & x_{nk} \end{pmatrix} = \begin{pmatrix} \boldsymbol{x}_1' \\ \vdots \\ \boldsymbol{x}_n' \end{pmatrix}.$$

Let the sample mean vector be denoted by

$$\bar{\boldsymbol{x}} = \frac{1}{n}(\boldsymbol{x}_1 + \cdots + \boldsymbol{x}_n) = (\bar{x}_1, \ldots, \bar{x}_p)'$$

and the centralized observation matrix be denoted by

$$\tilde{\mathrm{X}} = \begin{pmatrix} x_{11} - \bar{x}_1 & \cdots & x_{1k} - \bar{x}_k \\ \vdots & \vdots & \vdots \\ x_{n1} - \bar{x}_1 & \cdots & x_{nk} - \bar{x}_k \end{pmatrix}. \tag{7.4.1}$$

Then the sample covariance matrix is expressed as

$$\mathrm{S} = \frac{1}{n-1}\sum_{j=1}^{n}(\boldsymbol{x}_j - \bar{\boldsymbol{x}})(\boldsymbol{x}_j - \bar{\boldsymbol{x}})' = \frac{1}{n-1}\tilde{\mathrm{X}}'\tilde{\mathrm{X}}.$$

The principal components of $\boldsymbol{x}$ are defined (see Chapter 11) in the following. Let $\ell_1 > \cdots > \ell_p > 0$ be the characteristic roots of S and $\boldsymbol{h}_1, \ldots, \boldsymbol{h}_p$ be the corresponding orthonormal characteristic vectors. These are the solutions of

$$\mathrm{S}\boldsymbol{h}_i = \ell_i \boldsymbol{h}_i, \quad i = 1, \ldots, k,$$

where

$$\ell_1 > \cdots > \ell_k > 0, \quad \boldsymbol{h}_i'\boldsymbol{h}_i = 1, \quad \boldsymbol{h}_i'\boldsymbol{h}_j = 0, \quad i \neq j.$$

Then the jth principal component is defined by

$$u_j = \boldsymbol{h}_j'(\boldsymbol{x} - \bar{\boldsymbol{x}}), \quad j = 1, \ldots, k.$$

The values of the jth principal component for the ith subject are

$$u_{ij} = \boldsymbol{h}_j'(\boldsymbol{x}_i - \bar{\boldsymbol{x}}) = h_{1j}(x_{i1} - \bar{x}_1) + \cdots + h_{kj}(x_{ik} - \bar{x}_k),$$

which are called the *principal component scores*. They are expressed as

$$\mathrm{U} = \begin{pmatrix} u_{11} & \cdots & u_{1k} \\ \vdots & \vdots & \vdots \\ u_{n1} & \vdots & u_{nk} \end{pmatrix} = \tilde{\mathrm{X}}(\boldsymbol{h}_1, \ldots, \boldsymbol{h}_k). \tag{7.4.2}$$

Now we have a multivariate principal components regression model

$$\tilde{M}_F: \ \boldsymbol{Y}_i = \boldsymbol{\alpha}_0 + \boldsymbol{\alpha}_1 u_{i1} + \cdots + \boldsymbol{\alpha}_k u_{ik} + \boldsymbol{\varepsilon}_i, \quad i = 1, \ldots, n, \tag{7.4.3}$$

where $\boldsymbol{\alpha}_0, \ldots, \boldsymbol{\alpha}_k$ are unknown parameter vectors and the error terms $\boldsymbol{\varepsilon}_1, \ldots, \boldsymbol{\varepsilon}_n$ are distributed independently with mean $\mathbf{0}$ and covariance matrix Σ. We are also interested in a multivariate principal components model based on the first j principal components: that is,

$$\tilde{M}_j: \ \boldsymbol{y}_i = \boldsymbol{\alpha}_0 + \boldsymbol{\alpha}_1 u_{i1} + \cdots + \boldsymbol{\alpha}_j u_{ij} + \boldsymbol{\varepsilon}_i, \quad i = 1, \ldots, n. \tag{7.4.4}$$

More generally, it is suggested that a multivariate principal components regression based on a subset of all the principal components be considered. For choosing a candidate multivariate principal components model, we can use C_p, MC_p, AIC, $CAIC$, stepwise procedure, and so on.

7.5 Selection of Response Variables

In a multivariate regression model (7.1.1) or (7.1.2), we consider testing the hypothesis $H_{2\cdot 1}$: $\boldsymbol{Y}_2 = (Y_{q+1}, \ldots, Y_p)'$ has no additional information in the presence of $\boldsymbol{Y}_1 = (Y_1, \ldots, Y_q)'$. It is assumed that $\boldsymbol{\varepsilon}_1, \ldots, \boldsymbol{\varepsilon}_n \sim i.i.d.\ \mathrm{N}_p(\mathbf{0}, \Sigma)$. We rewrite (7.1.2) as

$$\mathbf{Y} = \mathbf{1}_n \boldsymbol{\theta}_0' + \tilde{\mathrm{X}}\Theta + \mathcal{E}, \tag{7.5.1}$$

where $\tilde{\mathrm{X}}$ is the $n \times k$ matrix given by (7.4.1):

$$\boldsymbol{\theta}_0 = \boldsymbol{\beta}_0 + \bar{x}_1 \boldsymbol{\beta}_1 + \cdots + \bar{x}_k \boldsymbol{\beta}_k,$$

$$\Theta = \begin{pmatrix} \boldsymbol{\beta}_1' \\ \vdots \\ \boldsymbol{\beta}_k' \end{pmatrix} = \begin{pmatrix} \boldsymbol{\theta}_1' \\ \vdots \\ \boldsymbol{\theta}_k' \end{pmatrix}.$$

Now we start from a multivariate regression model (7.5.1) and consider formulating $H_{2\cdot 1}$. Along the partition of $\boldsymbol{Y}$ into $(\boldsymbol{Y}_1', \boldsymbol{Y}_2')$, let $\mathbf{Y}$, $\boldsymbol{\theta}_0$, Θ, and Σ partition as

$$\mathbf{Y} = (\mathbf{Y}_1 \ \mathbf{Y}_2), \quad \Theta = (\Theta_1 \ \Theta_2),$$

$$\boldsymbol{\theta}_0 = \begin{pmatrix} \boldsymbol{\theta}_{01} \\ \boldsymbol{\theta}_{02} \end{pmatrix}, \quad \Sigma = \begin{pmatrix} \Sigma_{11} & \Sigma_{12} \\ \Sigma_{21} & \Sigma_{22} \end{pmatrix}.$$

The conditional distribution of $\mathbf{Y}_2$ given $\mathbf{Y}_1$ is normal with mean

$$\begin{aligned}\mathrm{E}(\mathbf{Y}_2|\mathbf{Y}_1) &= \mathbf{1}_n\boldsymbol{\theta}_{02}' + \mathrm{X}\Theta_2 + (\mathbf{Y}_1 - \mathbf{1}_n\boldsymbol{\theta}_{01}' - \mathrm{X}\Theta_1)\Sigma_{11}^{-1}\Sigma_{12}\\ &= \mathbf{1}_n\tilde{\boldsymbol{\theta}}_{02}' + \mathrm{X}\tilde{\Theta}_2 + \mathbf{Y}_1\Sigma_{11}^{-1}\Sigma_{12} \end{aligned} \tag{7.5.2}$$

and the covariance matrix

$$\mathrm{Var}[\mathrm{vec}\,(\mathbf{Y}_2|\mathbf{Y}_1)] = \Sigma_{22\cdot 1} \otimes \mathrm{I}_n, \tag{7.5.3}$$

where $\Sigma_{22\cdot 1} = \Sigma_{22} - \Sigma_{21}\Sigma_{11}^{-1}\Sigma_{12}$, Now we define the hypothesis $H_{2\cdot 1}$ as

$$H_{2\cdot 1}\colon \Theta_2 = \Theta_1\Sigma_{11}^{-1}\Sigma_{12}. \tag{7.5.4}$$

The hypothesis $H_{2\cdot 1}$ means that the $\boldsymbol{Y}_2$ after removing the effects of $\boldsymbol{Y}_1$ does not depend on $\boldsymbol{x}$. In other words, the relationship between $\boldsymbol{Y}_2$ and $\boldsymbol{x}$ can be described by the relationship between $\boldsymbol{Y}_1$ and $\boldsymbol{x}$. In this sense $\boldsymbol{Y}_2$ is redundant in the relationship between $\boldsymbol{Y}$ and $\boldsymbol{x}$.

The hypothesis $H_{2\cdot 1}$ can be stated in terms of

$$\delta^2 = \mathrm{tr}\,\Theta\Sigma^{-1}\Theta' = \sum_{i=1}^{k} \boldsymbol{\theta}_i'\Sigma^{-1}\boldsymbol{\theta}_i$$

and

$$\delta_1^2 = \mathrm{tr}\,\Theta_1\Sigma^{-1}\Theta_1' = \sum_{i=1}^{k} \boldsymbol{\theta}_{i1}'\Sigma^{-1}\boldsymbol{\theta}_{i1},$$

where $\boldsymbol{\theta}_i = (\boldsymbol{\theta}_{i1}'\ \boldsymbol{\theta}_{i2}')'$, $i = 1, \ldots, k$, as

$$H_{2\cdot 1}\colon \delta^2 = \delta_1^2. \tag{7.5.5}$$

The equivalence is proved from the fact that

$$\delta^2 = \delta_1^2 + \mathrm{tr}\,\Sigma_{22\cdot 1}^{-1}(\Theta_2 - \Theta_1\Sigma_{11}^{-1}\Sigma_{12})(\Theta_2 - \Theta_1\Sigma_{11}^{-1}\Sigma_{12})',$$

which follows by using the inverse formula for Σ (see [A.1.2]).

Now we derive the LR criterion for testing the hypothesis $H_{2\cdot 1}$. Let $\mathbf{W}$ be the SSP matrix of $(\boldsymbol{Y}', \boldsymbol{x}')'$:

$$\begin{aligned}\mathbf{W} &= \sum_{i=1}^{n} \begin{pmatrix} \boldsymbol{Y}_i - \bar{\boldsymbol{Y}} \\ \boldsymbol{x}_i - \bar{\boldsymbol{x}} \end{pmatrix} \begin{pmatrix} \boldsymbol{Y}_i - \bar{\boldsymbol{Y}} \\ \boldsymbol{x}_i - \bar{\boldsymbol{x}} \end{pmatrix}'\\ &= \begin{pmatrix} \mathbf{W}_{yy} & \mathbf{W}_{yx} \\ \mathbf{W}_{xy} & \mathrm{W}_{xx} \end{pmatrix}. \end{aligned}$$

Along the partition of $\boldsymbol{Y} = (\boldsymbol{Y}_1', \boldsymbol{Y}_2')'$, we partition $\mathbf{W}$ as

$$\mathbf{W} = \begin{pmatrix} \mathbf{W}_{11} & \mathbf{W}_{12} & \mathbf{W}_{1x} \\ \mathbf{W}_{21} & \mathbf{W}_{22} & \mathbf{W}_{2x} \\ \mathbf{W}_{x1} & \mathbf{W}_{x2} & \mathrm{W}_{xx} \end{pmatrix}.$$

Theorem 7.5.1 *In the multivariate regression model* (7.5.1), *consider the hypothesis* $H_{2\cdot 1}$ *in* (7.5.4) (*i.e., no additional information on* $\boldsymbol{Y}_2$ *in the presence of* $\boldsymbol{Y}_1$). *Then the LR criterion* λ *is given by*

$$\lambda^{2/N} = \frac{|\mathbf{W}_{22\cdot 1x}|}{|\mathbf{W}_{22\cdot 1}|},$$

where $\mathbf{W}_{22\cdot 1x} = \mathbf{W}_{22\cdot x} - \mathbf{W}_{21\cdot x}\mathbf{W}_{11\cdot x}^{-1}W_{12\cdot x}$.

Proof. The density function of $\mathbf{Y} = (\mathbf{Y}_1\ \mathbf{Y}_2)$ can be expressed as the product of the marginal density function of $\mathbf{Y}_1$ and the conditional density function of $\mathbf{Y}_2$ given $\mathbf{Y}_1$. Since the marginal density functions of $\mathbf{Y}_1$ under $H_{2\cdot 1}$ and the alternatives are the same, we may obtain the LR criterion based on the conditional density function of $\mathbf{Y}_2$ given $\mathbf{Y}_1$. The rows of $\mathbf{Y}_2$ given $\mathbf{Y}_1$ are independent and normally distributed with the same covariance matrix $\Sigma_{22\cdot 1}$. The conditional mean of $\mathbf{Y}_2$ is given by (7.5.2), and under $H_{2\cdot 1}$,

$$\mathrm{E}(\mathbf{Y}_2|\mathbf{Y}_1) = \mathbf{1}_n\tilde{\boldsymbol{\theta}}_{20}' + \mathbf{Y}_1\Sigma_{11}^{-1}\Sigma_{12}.$$

Let $\Omega = \mathcal{R}[\mathbf{1}_n, \tilde{\mathrm{X}}, \mathbf{Y}_1]$ and $\omega = \mathcal{R}[\mathbf{1}_n, \mathbf{Y}_1]$, and let P_Ω and P_ω be the projection matrices to the spaces Ω and ω, respectively. Note that $\dim(\Omega) = 1 + k + q$ and $\dim(\omega) = 1 + q$. Then we know from (7.2.10) that

$$\lambda^{1/n} = \Lambda = |\mathbf{S}_\Omega|/|\mathbf{S}_\omega|,$$

where $\mathbf{S}_\Omega = \mathbf{Y}_1'(\mathrm{I}_n - \mathrm{P}_\Omega)\mathbf{Y}_1$ and $\mathbf{S}_\omega = \mathbf{Y}_1'(\mathrm{I}_n - \mathrm{P}_\omega)\mathbf{Y}_1$. Further, under $H_{2\cdot 1}$, $\Lambda \sim \Lambda_{p-q}(k, n-q-k-1)$. Our proof is completed when we show that $\mathbf{S}_\omega = \mathbf{W}_{22\cdot 1}$ and $\mathrm{S}_\Omega = \mathrm{W}_{22\cdot 1x}$. The first result is obtained by using the fact that

$$\omega = \mathcal{R}[\mathbf{1}_n] + \mathcal{R}[(\mathrm{I}_n - \mathrm{P}_0)\mathrm{Y}_1]$$

and $\mathrm{P}_\mathrm{A} = \mathrm{A}(\mathrm{A}'\mathrm{A})^{-1}\mathrm{A}'$ for an $n \times m$ matrix with $\mathrm{rank}(\mathrm{A}) = m$, where $\mathrm{P}_0 = (1/n)\mathbf{1}_n\mathbf{1}_n'$. Let $\tilde{\mathbf{Y}}_1 = (\mathrm{I}_n - \mathrm{P}_0)\mathbf{Y}_1$. Then we have

$$\begin{aligned}\Omega &= \mathcal{R}[\mathbf{1}_n] + \mathcal{R}[\tilde{\mathbf{Y}}_1, \tilde{\mathrm{X}}] \\ &= \mathcal{R}[\mathbf{1}_n] + \mathcal{R}[\tilde{\mathrm{X}}] + \mathcal{R}[(\mathrm{I}_n - \mathrm{P}_{\tilde{\mathrm{X}}})\tilde{\mathbf{Y}}_1],\end{aligned}$$

and hence

$$\mathbf{S}_\Omega = \mathbf{W}_{22\cdot 1x} = \mathbf{W}_{22\cdot x} - \mathbf{W}_{21\cdot x}\mathbf{W}_{11\cdot x}^{-1}\mathbf{W}_{12\cdot x}. \qquad \square$$

For the distribution of $\Lambda_{p-q}(k, n-q-k-1)$, see Sections 3.3 and 6.1. If one is interested in finding a set of response variables that has no additional information in the presence of other response variables, it is suggested that AIC be used for each $H_{2\cdot 1}$. Let $H_{2\cdot 1}$ be denoted by $M_{q+1,\cdots p(1\cdots q)}$. The MLEs of Σ_{11} and $\Sigma_{22\cdot 1}$ under $M_{q+1,\cdots p(1\cdots q)}$ are

$$\begin{aligned}n\hat{\Sigma}_{11} &= \mathbf{W}_{11} - \mathbf{W}_{1x}\mathbf{W}_{xx}^{-1}\mathbf{W}_{x1} = \mathbf{W}_{11\cdot x}, \\ n\hat{\Sigma}_{22\cdot 1} &= \mathbf{W}_{22} - \mathbf{W}_{21}\mathbf{W}_{11}^{-1}\mathbf{W}_{12} = \mathbf{W}_{22\cdot 1}.\end{aligned}$$

and the independent parameter under $M_{q+1,\cdots p(1\cdots q)}$ is $kq + p + \frac{1}{2}p(p+1)$. Therefore,

$$AIC = n\log|\mathbf{W}_{22\cdot 1}| + n\log|\mathbf{W}_{11\cdot x}| + np(\log 2\pi + 1) + 2\left\{kq + p + \frac{1}{2}p(p+1)\right\}. \tag{7.5.6}$$

In particular, let $M_{1\cdots p}$ be a model such that there are no redundant response variables (i.e., all the response variables are useful). The AIC is

$$AIC_F = n\log|\mathbf{W}_{yy\cdot x}| + np(\log 2\pi + 1) + 2\left\{(k+1)p + \frac{1}{2}p(p+1)\right\}$$

and

$$AIC - AIC_F = -n\log\frac{|\mathbf{W}_{22\cdot 1x}|}{|\mathbf{W}_{22\cdot 1}|} - 2k(p-q).$$

We consider selecting the response variables Y_1, Y_2, and Y_3 for the chemical reaction data in Table 7.3.1 by AIC. The values are given in Table 7.5. The result shows that Y_1 is redundant in the presence of Y_2 and Y_3.

Table 7.5.1 Selection of Y_1, Y_2, and Y_3 for the chemical reaction data

Model	AIC
M_1	436.3
M_{123}	440.5
M_{12}	441.6
M_{13}	441.6
M_{23}	453.4
M_3	471.3
M_2	486.2

7.6 General Linear Hypotheses and Confidence Intervals

In this section we consider general linear hypotheses and confidence intervals in a multivariate linear model which includes MANOVA models in Chapter 6 and multivariate regression models in this chapter. Let us consider a linear relationship between the p response variables $Y_1, \ldots, Y_p$ and the q explanatory or dummy variables $a_1, \ldots, a_q$. Suppose that there are observations of $\boldsymbol{Y} = (Y_1, \ldots, Y_p)'$ and $\boldsymbol{a} = (a_1, \ldots, a_q)'$ for n subjects, which are denoted by

$$\mathbf{Y} = \begin{pmatrix} \boldsymbol{Y}_1' \\ \vdots \\ \boldsymbol{Y}_n' \end{pmatrix}, \quad \mathrm{A} = \begin{pmatrix} \boldsymbol{a}_1' \\ \vdots \\ \boldsymbol{a}_n' \end{pmatrix}.$$

Then it is assumed that

$$\mathbf{Y} = \mathrm{A}\Theta + \mathcal{E}, \tag{7.6.1}$$

where A is an $n \times q$ known matrix with rank q and Θ is a $q \times p$ unknown parameter matrix. For the error matrix $\mathcal{E} = (\varepsilon_1, \ldots, \varepsilon_n)'$ we assume that:

(1) $\mathrm{E}(\varepsilon_j) = \mathbf{0}, \quad j = 1, \ldots, n.$

(2) $\mathrm{Var}(\varepsilon_j) = \Sigma, \quad j = 1, \ldots, n.$

(3) $\varepsilon_1, \ldots, \varepsilon_n$ are independent and normal.

The assumption is equivalent to

$$\varepsilon_1, \ldots, \varepsilon_n \sim \text{i.i.d. } \mathrm{N}_p(\mathbf{0}, \Sigma). \tag{7.6.2}$$

Here Σ is unknown.

Consider a testing problem for

$$H: \mathrm{C}\Theta\mathrm{D} = \mathrm{O} \quad \text{vs.} \quad K: \mathrm{C}\Theta\mathrm{D} \neq \mathrm{O}, \tag{7.6.3}$$

where C is a $c \times q$ known matrix with rank c and D is a $p \times d$ known matrix with rank q. For a motivation of C and D, let

$$\begin{aligned} &\boldsymbol{Y}_1, \ldots, \boldsymbol{Y}_{n_1} \sim \text{i.i.d. } \mathrm{N}_p(\boldsymbol{\theta}_1, \Sigma), \\ &\boldsymbol{Y}_{n_1+1}, \ldots, \boldsymbol{Y}_{n_1+n_2} \sim \text{i.i.d. } \mathrm{N}_p(\boldsymbol{\theta}_2, \Sigma), \\ &\boldsymbol{Y}_{n_1+n_2+1}, \ldots, \boldsymbol{Y}_{n_1+n_2+n_3} \sim \text{i.i.d. } \mathrm{N}_p(\boldsymbol{\theta}_3, \Sigma), \end{aligned}$$

where $n = n_1 + n_2 + n_3$. Then

$$\mathrm{A} = \begin{pmatrix} \mathbf{1}_{n_1} & \mathbf{0} & \mathbf{0} \\ \mathbf{0} & \mathbf{1}_{n_2} & \mathbf{0} \\ \mathbf{0} & \mathbf{0} & \mathbf{1}_{n_3} \end{pmatrix}, \quad \Theta = \begin{pmatrix} \boldsymbol{\theta}_1' \\ \boldsymbol{\theta}_2' \\ \boldsymbol{\theta}_3' \end{pmatrix}.$$

The hypothesis $\boldsymbol{\theta}_1 = \boldsymbol{\theta}_2 = \boldsymbol{\theta}_3$ is expressed as $\mathrm{C}\Theta = \mathrm{O}$, where

$$\mathrm{C} = \begin{pmatrix} 1 & 0 & -1 \\ 0 & 1 & -1 \end{pmatrix}.$$

Further, if one is interested in the significance of the first two components of $\boldsymbol{\theta}_i$, $i = 1, 2, 3$, we can express it as

$$\begin{pmatrix} 1 & 0 & -1 \\ 0 & 1 & -1 \end{pmatrix} \begin{pmatrix} \boldsymbol{\theta}_1' \\ \boldsymbol{\theta}_2' \\ \boldsymbol{\theta}_3' \end{pmatrix} \begin{pmatrix} 1 & 0 \\ 0 & 1 \\ 0 & 0 \end{pmatrix} = \mathrm{O}.$$

For the derivation of LR test of (7.6.3), we can use the following conventional approach: If $\mathbf{U} = \mathbf{Y}\mathrm{D}$, then the rows of $\mathbf{U}$ are independent and normally distributed with the identical covariance matrix $\mathrm{D}'\Sigma\mathrm{D}$, and

$$\mathrm{E}(\mathbf{U}) = \mathrm{A}\Xi, \tag{7.6.4}$$

where $\Xi = \Theta\mathrm{D}$. The testing problem (7.6.3) is expressed as

$$H : \mathrm{C}\Xi = \mathrm{O} \quad \text{vs.} \quad K : \mathrm{C}\Xi \neq \mathrm{O}. \tag{7.6.5}$$

Applying a general theory (see Section 7.2) for a testing problem (7.6.5) under (7.6.4), we have an LR statistic:

$$T = \frac{|\mathbf{S}_e|}{|\mathbf{S}_e + \mathbf{S}_h|}, \tag{7.6.6}$$

where

$$\begin{aligned} \mathbf{S}_e &= \mathbf{U}'(\mathrm{I}_n - \mathrm{P_A})\mathbf{U} \\ &= \mathrm{D}'\mathbf{Y}'(\mathrm{I}_n - \mathrm{P_A})\mathbf{Y}\mathrm{D} \end{aligned}$$

and

$$\begin{aligned} \mathbf{S}_h &= (\mathrm{C}\hat{\Xi})'\{\mathrm{C}(\mathrm{A}'\mathrm{A})^{-1}\mathrm{C}'\}^{-1}\mathrm{C}\hat{\Xi} \\ &= (\mathrm{C}\hat{\Theta}\mathrm{D})'\{\mathrm{C}(\mathrm{A}'\mathrm{A})^{-1}\mathrm{C}'\}^{-1}\mathrm{C}\hat{\Theta}\mathrm{D}, \end{aligned}$$

where $\hat{\Theta} = (\mathrm{A}'\mathrm{A})^{-1}\mathrm{A}'\mathbf{U} = (\mathrm{A}'\mathrm{A})^{-1}\mathrm{A}'\mathbf{Y}\mathrm{D}$ and $\hat{\Theta} = (\mathrm{A}'\mathrm{A})^{-1}\mathrm{A}'\mathbf{Y}$.

Theorem 7.6.1 *The statistic T in* (7.6.6) *is an LR statistic for testing problem* (7.6.3) *under* (7.6.1). *Further, under H_0, $T \sim \Lambda_d(c, n-q)$.*

Proof. Let $\mathrm{G} = (\mathrm{G}_1\ \mathrm{G}_2)$ be a $p \times p$ matrix such that $\mathrm{G}_1 = \mathrm{D}$, $\mathrm{G}_1'\mathrm{G}_2 = \mathrm{O}$, and $|\mathrm{G}| \neq 0$. Consider a transformation from $\mathbf{Y}$ to

$$(\mathbf{U}\ \mathbf{V}) = \mathbf{Y}(\mathrm{G}_1\ \mathrm{G}_2).$$

Then the rows of $(\mathbf{U}\ \mathbf{V})$ are independently normal with the same covariance matrix

$$\Psi = \mathrm{G}'\Sigma\mathrm{G} = \begin{pmatrix} \Psi_{11} & \Psi_{12} \\ \Psi_{21} & \Psi_{22} \end{pmatrix}, \quad \Psi_{12} : d \times (p-d),$$

and

$$\begin{aligned}\mathrm{E}[(\mathbf{U}\ \mathbf{V})] &= \mathrm{A}\Theta(\mathrm{G}_1\ \mathrm{G}_2)\\ &= \mathrm{A}(\Xi\ \Delta), \quad \Xi = \Theta\mathrm{G}_1,\ \Delta = \Theta\mathrm{G}_2.\end{aligned}$$

The conditional of $\mathbf{V}$ given $\mathbf{U}$ is normal. The rows of $\mathbf{V}$ given $\mathbf{U}$ are independently normal with the same covariance matrix $\Psi_{11\cdot 2}$, and

$$\begin{aligned}\mathrm{E}(\mathbf{V}|\mathbf{U}) &= \mathrm{A}\Delta + (\mathbf{U} - \mathrm{A}\Xi)\Gamma\\ &= \mathrm{A}\Delta^* + \mathbf{U}\Gamma,\end{aligned}$$

where $\Delta^* = \Delta - \Xi\Gamma$ and $\Gamma = \Psi_{11}^{-1}\Psi_{12}$. We see that the maximum likelihood of $\mathbf{V}$ given $\mathbf{U}$ does not depend on the hypothesis. Therefore, an LR statistic is obtained from the marginal distribution of $\mathbf{U}$, which implies the results required. □

Next we consider confidence intervals for Θ. First we consider a confidence region for

$$c_1\boldsymbol{\theta}_1 + \cdots + c_q\boldsymbol{\theta}_q = \Theta'\boldsymbol{c},$$

where $\boldsymbol{c} = (c_1, \ldots, c_q)'$ is a given vector and $\Theta = (\boldsymbol{\theta}_1, \ldots, \boldsymbol{\theta}_q)'$. From Theorem 7.2.1 it follows that

$$\begin{aligned}(\hat{\Theta} - \Theta)'\boldsymbol{c} &= \{(\mathrm{A}'\mathrm{A})^{-1}\mathrm{A}'(\mathbf{Y} - \mathrm{A}\Theta)\}'\boldsymbol{c}\\ &\sim \mathrm{N}_p(\mathbf{0}, \boldsymbol{c}'(\mathrm{A}'\mathrm{A})^{-1}\boldsymbol{c}\cdot\Sigma)\end{aligned}$$

since $\mathbf{S}_e \sim \mathrm{W}_p(n-q, \Sigma)$ (see Theorem 7.2.1) from Theorem 3.1.1. Therefore, a confidence region for $\Theta\boldsymbol{c}$ with confidence coefficient $1-\alpha$ is given by

$$\frac{(\hat{\Theta}'\boldsymbol{c} - \Theta'\boldsymbol{c})'\mathbf{S}_e^{-1}(\hat{\Theta}'\boldsymbol{c} - \Theta'\boldsymbol{c})}{\boldsymbol{c}'(\mathrm{X}'\mathrm{X})^{-1}\boldsymbol{c}} \cdot \frac{n-k-p+1}{p} \le \mathrm{F}(p, n-k-p+1; \alpha).$$

Next we give a simultaneous confidence interval of $\boldsymbol{a}'\mathrm{C}\Theta\mathrm{D}\boldsymbol{b}$ for any given $\boldsymbol{a} = (a_1, \ldots, a_c)'$ and $\boldsymbol{b} = (b_1, \ldots, b_d)$. A method is based on

$$\boldsymbol{a}'\mathrm{C}(\hat{\Theta} - \Theta)\mathrm{D}\boldsymbol{b} \sim \mathrm{N}(0, \boldsymbol{a}'\mathrm{C}(\mathrm{A}'\mathrm{A})^{-1}\mathrm{C}'\boldsymbol{a}\cdot\boldsymbol{b}'\mathrm{D}'\Sigma\mathrm{D}\boldsymbol{b})$$

or

$$\frac{\boldsymbol{a}'\mathrm{C}(\hat{\Theta} - \Theta)\mathrm{D}\boldsymbol{b}}{\sqrt{\boldsymbol{a}'\mathrm{C}(\mathrm{A}'\mathrm{A})^{-1}\mathrm{C}'\boldsymbol{a}}\sqrt{\boldsymbol{b}'\mathrm{D}\Sigma\mathrm{D}'\boldsymbol{b}}} \sim \mathrm{N}(0,1).$$

More precisely, we use its Studentized statistic, whose square is evaluated as

$$\begin{aligned}&\frac{\{\boldsymbol{a}'\mathrm{C}(\hat{\Theta} - \Theta)\mathrm{D}\boldsymbol{b}\}^2}{\boldsymbol{a}'\mathrm{C}(\mathrm{A}'\mathrm{A})^{-1}\mathrm{C}'\boldsymbol{a}\cdot\boldsymbol{b}'\mathrm{D}'\hat{\Sigma}\mathrm{D}\boldsymbol{b}}\\ &\qquad\le \frac{\boldsymbol{b}'\{\mathrm{C}(\hat{\Theta} - \Theta)\mathrm{D}\}'\{\mathrm{C}(\mathrm{A}'\mathrm{A})^{-1}\mathrm{C}'\}^{-1}\{\mathrm{C}(\hat{\Theta} - \Theta)\mathrm{D}\}\boldsymbol{b}}{\boldsymbol{b}'\mathrm{D}\hat{\Sigma}\mathrm{D}'\boldsymbol{b}}\\ &\qquad\le n\ell_1,\end{aligned}$$

where ℓ_1 is the maximum characteristic root of $\tilde{\mathbf{S}}_h \mathbf{S}_e^{-1}$ and

$$\begin{aligned}\tilde{\mathbf{S}}_h &= \{\mathrm{C}(\hat{\Theta} - \Theta)\mathrm{D}\}'\{\mathrm{C}(\mathrm{A}'\mathrm{A})^{-1}\mathrm{C}'\}^{-1}\{\mathrm{C}(\hat{\Theta} - \Theta)\mathrm{D}\},\\ \mathbf{S}_e &= \mathrm{D}'\mathbf{Y}'(\mathrm{I}_n - \mathrm{P}_{\mathrm{A}})\mathbf{Y}\mathrm{D}.\end{aligned}$$

Note that $\mathbf{S}_e$ and $\tilde{\mathbf{S}}_h$ are independent and

$$\tilde{\mathbf{S}}_h \sim \mathrm{W}_d(c, \mathrm{D}\Sigma\mathrm{D}), \quad \mathbf{S}_e \sim \mathrm{W}_d(n-k, \mathrm{D}\Sigma\mathrm{D}').$$

Let $\ell_1(\alpha)$ be the upper α point of ℓ_1. Then a simultaneous confidence interval is given by

$$\boldsymbol{a}'\mathrm{C}\hat{\Theta}\mathrm{D}\boldsymbol{b} \pm \{\ell_1(\alpha/2) \cdot \boldsymbol{a}'\mathrm{C}(\mathrm{A}'\mathrm{A})^{-1}\mathrm{C}'\boldsymbol{a} \cdot \boldsymbol{b}'\mathrm{D}'\mathbf{S}_e\mathrm{D}\boldsymbol{b}\}^{1/2}.$$

7.7 Penalized Regression Models

In this section we treat two methods applicable even when the number of explanatory variables is much larger than the number of independent samples. Consider the multivariate regression model (7.1.1) or (7.1.2). For notational simplicity, without loss of generality we may suppose that the observation matrix $\mathbf{Y}$ has been centralized or standardized as $\mathbf{Y} - \mathbf{1}_n \bar{\boldsymbol{Y}}'$, and observation matrix X has been centralized as

$$\begin{pmatrix} (x_{11} - \bar{x}_1)/\sqrt{s_{11}} & \cdots & (x_{1k} - \bar{x}_k)/\sqrt{s_{kk}} \\ \vdots & \vdots & \vdots \\ (x_{n1} - \bar{x}_1)/\sqrt{s_{11}} & \cdots & (x_{nk} - \bar{x}_k)/\sqrt{s_{kk}} \end{pmatrix}, \tag{7.7.1}$$

where $s_{ii} = (n-1)^{-1}\sum_{j=1}^n (x_{ij} - \bar{x}_j)^2$, $i = 1, \ldots, k$. Then we may start from the regression model

$$\mathbf{Y} = \mathrm{X}\mathcal{B} + \mathcal{E}, \tag{7.7.2}$$

where $\mathbf{Y} = (\boldsymbol{Y}_1, \ldots, \boldsymbol{Y}_n)$, $\mathrm{X} = (\boldsymbol{x}_1, \ldots, \boldsymbol{x}_n)'$, and $\mathcal{B} = (\boldsymbol{\beta}_1, \ldots, \boldsymbol{\beta}_k)'$. In general, if there is a linear relationship between the column vectors of the design matrix X, it is said that multicollinearity exists. We often have multicollinearity or near multicollinearity as the number of explanatory variables increases. For such cases, the $k \times k$ matrix $\mathrm{X}'\mathrm{X}$ becomes singular or nearly singular. Further, if the sample size n is less that k, $\mathrm{X}'\mathrm{X}$ becomes singular.

When near multicollinearity or near singularity exists, the usual least squares estimator $\hat{\mathcal{B}} = (\mathrm{X}'\mathrm{X})^{-1}\mathrm{X}'\mathbf{Y}$ will not be appropriate. In fact, consider the ordinary mean-squared error defined by

$$\begin{aligned}\mathrm{MSE}_{(1)}(\hat{\mathcal{B}}) &= \mathrm{E}[\sum_{j=1}^{k}(\hat{\boldsymbol{\beta}}_j - \boldsymbol{\beta}_j)'(\hat{\boldsymbol{\beta}}_j - \boldsymbol{\beta}_j)]\\ &= \mathrm{E}[\,\mathrm{tr}\,(\hat{\mathcal{B}} - \mathcal{B})'(\hat{\mathcal{B}} - \mathcal{B})],\end{aligned}$$

and its standardized version defined by

$$\begin{aligned}\mathrm{MSE}_{(2)}(\hat{\mathcal{B}}) &= \mathrm{E}\Big[\sum_{j=1}^{k}(\hat{\boldsymbol{\beta}}_j - \boldsymbol{\beta}_j)'\Sigma^{-1}(\hat{\boldsymbol{\beta}}_j - \boldsymbol{\beta}_j)\Big] \\ &= \mathrm{E}[\,\mathrm{tr}\,(\hat{\mathcal{B}} - \mathcal{B})'\Sigma^{-1}(\hat{\mathcal{B}} - \mathcal{B})].\end{aligned}$$

It is seen (see Problem 7.7) that

$$\mathrm{MSE}_{(1)}(\hat{\mathcal{B}}) = (\,\mathrm{tr}\,\Sigma)\,\mathrm{tr}\,(\mathrm{X}'\mathrm{X})^{-1}, \quad \mathrm{MSE}_{(2)}(\hat{\mathcal{B}}) = p\,\mathrm{tr}\,(\mathrm{X}'\mathrm{X})^{-1}.$$

Let $r_1 \geq \cdots \geq r_k$ be the characteristic roots of $\mathrm{X}'\mathrm{X}$. Then

$$\mathrm{tr}\,(\mathrm{X}'\mathrm{X})^{-1} = \sum_{j=1}^{k} r_j^{-1} \geq r_k^{-1}.$$

Therefore, we can see that as the design matrix X approaches collinearity, $r_k \to 0$, and $\mathrm{MSE}_{(i)}(\hat{\mathcal{B}}) \to \infty$.

As a method for coping effectively with the collinearlity problem, Hoerl and Kennard (1970) proposed using ridge regression estimators in the univariate case $p = 1$. Assuming that the design matrix X has been standardized as in (7.7.1), one of its multivariate versions is defined as

$$\hat{\mathcal{B}}_\lambda = (\mathrm{X}'\mathrm{X} + \lambda \mathrm{I}_k)^{-1}\mathrm{X}'\mathbf{Y}, \tag{7.7.3}$$

where λ is a constant chosen to be greater than zero. The multivariate ridge regression estimator may be defined as the one that minimizes

$$\mathrm{tr}\,(\mathbf{Y} - \mathrm{X}\mathcal{B})'(\mathbf{Y} - \mathrm{X}\mathcal{B}) + \lambda\,\mathrm{tr}\,\mathcal{B}'\mathcal{B}. \tag{7.7.4}$$

Using

$$(\mathrm{X}'\mathrm{X} + \lambda \mathrm{I}_k)^{-1} = (\mathbf{X}'\mathbf{X})^{-1} - (\mathbf{X}'\mathbf{X})^{-1}\{(\mathbf{X}'\mathbf{X})^{-1} + \lambda^{-1}\mathrm{I}_k\}^{-1}(\mathbf{X}'\mathbf{X})^{-1}$$

(see [A.1.3]), we can write

$$\hat{\mathcal{B}}_\lambda = [\mathrm{I}_k - \{\mathrm{I}_k + \lambda^{-1}\mathrm{X}'\mathrm{X}\}^{-1}]\hat{\mathcal{B}}.$$

Estimators of this kind are called shrinkage estimators.

The multivariate ridge estimator is not unbiased, and its bias is expressed as

$$\mathrm{E}(\hat{\mathcal{B}}_\lambda) - \mathcal{B}_\lambda = -\lambda(\mathrm{X}'\mathrm{X} + \lambda \mathrm{I}_k)^{-1}\mathcal{B}.$$

However, following the result in the case $p = 1$ due to Hoerl and Kennard (1970), there exists a value of λ such that $\mathrm{MSE}_{(i)}(\hat{\mathcal{B}}_\lambda)$ is smaller than $\mathrm{MSE}_{(i)}(\hat{\mathcal{B}})$. But the value depends on the unknown parameters λ and Σ.

Some methods of choosing or estimating λ have been proposed for the case $p = 1$ [for their details, see e.g., Srivastava, 2002]. A very common approach to the choice of λ is to use cross-validation, whose multivariate version is stated as follows: Let $\hat{\boldsymbol{Y}}_{(j)}$ be the predictive value of $\boldsymbol{Y}_j$ obtained without $\boldsymbol{Y}_j$ in the model using the multivariate ridge estimator, and consider the sum of squares of prediction errors defined by

$$R(\lambda) = \sum_{j=1}^{n} (\mathbf{Y}_j - \hat{\mathbf{Y}}_{(j)})'(\mathbf{Y}_j - \hat{\mathbf{Y}}_{(j)}).$$

We choose the value of λ minimizing $R(\lambda)$ by calculating $R(\lambda)$ for various values of λ.

Another method applicable even when $k \geq n$ is to use principal component regression in Section 7.7. We will see that the method is possible. In this case the design matrix $\boldsymbol{X}$ may be centralized or standardized. Depending on the normalizations, we consider the characteristic roots $\ell_1 \geq \cdots \geq \ell_k$ of the sample covariance matrix or the sample correlation matrix, and the corresponding orthonormal characteristic vectors $\boldsymbol{h}_1, \ldots, \boldsymbol{h}_k$. In this process, note that $\ell_{m+1} = \cdots = \ell_k = 0$, where $m = \min\{n-1, k\}$. So we may consider a design matrix based on m principal components:

$$\mathrm{U} = \mathrm{X}(\boldsymbol{h}_1, \ldots, \boldsymbol{h}_m).$$

A major issue in principal components regression is determining the number of components (i.e., a set of columns of U to use). We may use the methods of selection of variables in the multivariate regression model. We may also use the cross-validation method, which is computer intensive.

Problems

7.1 Consider a general linear hypothesis $\mathrm{C\Theta D} = \mathrm{O}$ on normal populations $\mathrm{N}_p(\boldsymbol{\theta}_i, \Sigma)$, where $\Theta = (\boldsymbol{\theta}_1, \ldots, \boldsymbol{\theta}_q)'$, C is a $c \times q$ known matrix with rank c, and D is a $p \times d$ known matrix with rank d. Choosing C and D appropriately, show that the general linear hypothesis includes the following hypotheses as special cases:

$$\begin{aligned}
&(1) \quad \boldsymbol{\theta}_1 = \cdots = \boldsymbol{\theta}_q. \\
&(2) \quad \boldsymbol{\theta}_i = \boldsymbol{\theta}_j, \quad i \neq j. \\
&(3) \quad \theta_{1\ell} = \cdots = \theta_{q\ell}. \\
&(4) \quad \theta_{i\ell} = \theta_{j\ell}, \quad i \neq j.
\end{aligned}$$

Here $\boldsymbol{\theta}_i = (\theta_{i1}, \ldots, \theta_{ip})'$, $i = 1, \ldots, q$.

7.2 Consider testing a linear hypothesis $H_*: \mathrm{C}\mathcal{B} = \mathrm{G}_0$ instead of $\mathrm{C}\mathcal{B} = \mathrm{O}$ in a multivariate linear model (7.1.2). Here G_0 is a $c \times p$ given matrix such that there exists a matrix $\mathcal{B}_0$ satisfying $\mathrm{C}\mathcal{B} = \mathrm{G}_0$. Let $\mathrm{Y}_0 = \mathrm{X}\mathcal{B}_0$ and $\mathbf{Y}_* = \mathbf{Y} - \mathrm{Y}_0$.

(1) Show that

$$\begin{aligned}
&\mathbf{Y}_*'(\mathrm{I}_n - \mathrm{P_X})\mathbf{Y}_* - \mathbf{Y}'(\mathrm{I}_n - \mathrm{P_X})\mathbf{Y} \equiv \mathbf{S}_e, \\
&\mathbf{Y}_*'\mathrm{X}(\mathrm{X}'\mathrm{X})^{-1}\mathrm{C}'\{\mathrm{C}(\mathrm{X}'\mathrm{X})^{-1}\mathrm{C}'\}^{-1}\mathrm{C}(\mathrm{X}'\mathrm{X})^{-1}\mathrm{X}'\mathbf{Y}_* \\
&= (\mathrm{C}\hat{\mathcal{B}} - \mathrm{G}_0)'\{\mathrm{C}(\mathrm{X}'\mathrm{X})^{-1}\mathrm{C}'\}^{-1}\mathrm{C}(\mathrm{X}'\mathrm{X})^{-1}\mathrm{X}'\mathbf{Y} \\
&\equiv \tilde{\mathbf{S}}_h,
\end{aligned}$$

where $\hat{\mathcal{B}} = (\mathrm{X}'\mathrm{X})^{-1}\mathrm{X}'\mathrm{Y}$.

(2) Show that under the hypothesis H_*,the statistics $\tilde{\mathbf{S}}_h$ and $\mathbf{S}_e$ are independent and distributed as $\mathrm{W}_p(c, \Sigma)$ and $\mathrm{W}_p(n-k-1, \Sigma)$, respectively.

(3) Show that an LR test statistic is given by $|\mathbf{S}_e|/|\mathbf{S}_e + \tilde{\mathbf{S}}_h|$.

7.3 Consider the regression model (7.3.3) for $p = 1$. Let AIC for $J = \{x_1, \ldots, x_{j-1}\}$ and $J = \{x_1, \ldots, x_j\}$ be denoted by $AIC(j-1)$ and $AIC(j)$, respectively. Then

$$AIC(j) = n \log s_j^2 + n(\log 2\pi + 1) + 2(j+2)$$

and

$$AIC(j-1) - AIC(j) = n \log \frac{s_{j-1}^2}{s_j^2} - 2,$$

where s_j^2 is the residual sums of squares. Then show that $AIC(j-1) - AIC(j)$ $\Leftrightarrow$

$$F = \frac{s_{j-1}^2 - s_j^2}{s_j^2/(n-j-1)} > (n-j-1)(\mathrm{e}^{2/n} - 1) = F_0.$$

Further,

$$\begin{aligned}
F_0 &= (n-j-1)\left[1 + \frac{2}{n} + \frac{1}{2}\left(\frac{2}{n}\right)^2 + \mathrm{O}(n^{-3})\right] \\
&= 2\left(1 - \frac{j}{n}\right) + \mathrm{O}(n^{-2}).
\end{aligned}$$

This result provides a relation between methods based on an F-test and AIC in comparison with the models $M_{\{x_1,\ldots,x_{j-1}\}}$ and $M_{\{x_1,\ldots,x_j\}}$.

7.4 (Continuation) Reexamine Problem 7.3.

Hint:

$$
\begin{aligned}
C_p(j-1) - C_p(j) &= \frac{s_{j-1}^2 - s_j^2}{s_F^2/(n-k+1)} - 2 \\
&= \frac{s_{j-1}^2 - s_j^2}{s_j^2/(n-j-1)} \cdot \frac{s_j^2/(n-j-1)}{s_F^2/(n-k+1)} - 2.
\end{aligned}
$$

Under $M_{\{x_1,\dots,x_j\}}$,

$$
\lim_{n\to\infty} \frac{s_j^2/(n-j-1)}{s_F^2/(n-k+1)} = \frac{\sigma^2}{\sigma^2} = 1.
$$

Therefore, "$C_p(j-1) - C_p(j) > 0$" is asymptotically equivalent to "$F_0 > 2$".

7.5 Let $AIC(j-1)$ and $AIC(j)$ be the AIC criteria for the multivariate regression model M_{j-1} and M_j based on the use of explanatory variables -$x_1, \dots, x_{j-1}$ and $x_1, \dots, x_j$, respectively.

(1) Show that

$$
AIC(j-1) - AIC(j) = n \log |\hat{\Sigma}_j| / |\hat{\Sigma}_{j-1}| - 2p,
$$

where $\hat{\Sigma}_{j-1}$ and $\hat{\Sigma}_j$ are the MLE's of Σ under M_{j-1} and M_j, respectively.

(2) Show that an LR test statistic for testing $\boldsymbol{\beta}_j = \mathbf{0}$ under M_j is given by $L = |\hat{\Sigma}_j|/|\hat{\Sigma}_{j-1}|$. Further, show that the null distribution of L is $\Lambda_p(1, n-j-1)$, and hence

$$
\frac{1-L}{L} \cdot \frac{n-j-1}{p} \sim \mathrm{F}(p, n-j-1).
$$

(3) Show that $AIC(j-1) - AIC(j) > 0 \Leftrightarrow$

$$
\frac{1-L}{L} \cdot \frac{n-j-1}{p} > F_0 = \frac{n-j-1}{p}(e^{2p/n} - 1),
$$

and $F_0 = 2\{1 + (1/n)(p-j-1) + O(n^{-2})\}$. This result provides a relation between methods based on an LR test and AIC in comparison with the model M_{j-1} and M_j.

7.6 Consider testing a general linear hypothesis $H_* : \mathrm{C}\mathcal{B}\mathrm{D} = \mathrm{G}_0$ instead of $\mathrm{C}\mathcal{B}\mathrm{D} = \mathrm{O}$ (see (7.6.3)) in multivariate linear model (7.1.2). Here G_0 is a $c \times d$ given matrix such that there exists a matrix $\mathcal{B}$ satisfying $\mathrm{C}\mathcal{B}\mathrm{D} = \mathrm{G}$. Expand the argument similar to the one as in Problem 7.2.

7.7 Let $\hat{\mathcal{B}} = (X'X)^{-1}X'\mathbf{Y}$ be the least squares estimator. Then, without the assumption of normality show that

(1) $\mathrm{E}[\operatorname{tr}(\hat{\mathcal{B}} - \mathcal{B})'\Sigma^{-1}(\hat{\mathcal{B}} - \mathcal{B})] = \operatorname{tr}\Sigma \operatorname{tr}(X'X)^{-1}$.

(2) $\mathrm{E}[\operatorname{tr}(\hat{\mathcal{B}} - \mathcal{B})'\Sigma^{-1}(\hat{\mathcal{B}} - \mathcal{B})] = p \operatorname{tr}(X'X)^{-1}$.

Chapter 8

Classical and High-Dimensional Tests for Covariance Matrices

In this chapter we are concerned primarily with tests of covariance matrices of multivariate normal populations. The LR tests are derived. The distributions of test statistics are handled asymptotically. Asymptotic expansions of the null distributions of test statistics are obtained under a high-dimensional framework as well as a large-sample framework. The testing problems are also discussed for situations where the dimension is larger than the sample size. High-dimensional test statistics and their asymptotic distributions are given.

8.1 Specified Covariance Matrix

8.1.1 Likelihood Ratio Test and Moments

Let $\mathbf{X} = (\boldsymbol{X}_1, \ldots, \boldsymbol{X}_N)$ be a sample of size $N = n + 1$ drawn from $\mathrm{N}_p(\boldsymbol{\mu}, \Sigma)$, and let

$$\bar{\boldsymbol{X}} = \frac{1}{N}\sum_{j=1}^{N} \boldsymbol{X}_j, \quad \mathbf{S} = \frac{1}{n}\mathbf{V}, \quad \mathbf{V} = \sum_{j=1}^{N}(\boldsymbol{X}_j - \bar{\boldsymbol{X}})(\boldsymbol{X}_j - \bar{\boldsymbol{X}})'. \tag{8.1.1}$$

We wish to test

$$H\colon \Sigma = \Sigma_0 \text{ vs. } K\colon \Sigma \neq \Sigma_0,$$

where Σ_0 is a specified positive definite matrix.

Theorem 8.1.1 *The likelihood ratio criterion for testing H: $\Sigma = \Sigma_0$ vs. K: $\Sigma \neq \Sigma_0$ is*

$$\lambda = \left(\frac{e}{N}\right)^{pN/2} \operatorname{etr}\left(-\frac{1}{2}\Sigma_0^{-1}\mathbf{V}\right) |\Sigma_0^{-1}\mathbf{V}|^{N/2}. \tag{8.1.2}$$

Proof. Let $L(\boldsymbol{\mu}, \Sigma)$ be the likelihood function of $\mathbf{X}$. The likelihood function after maximizing with respect to $\boldsymbol{\mu}$ is

$$L(\Sigma) = L(\bar{\boldsymbol{x}}, \Sigma) = (2\pi)^{-pN/2}|\Sigma|^{-N/2} \operatorname{etr}\left(-\frac{1}{2}\Sigma^{-1}\mathrm{V}\right).$$

From [A.2.10] we have

$$\max_{\Sigma} L(\Sigma) = L(N^{-1}\mathrm{V}) = (2\pi)^{-pN/2}|N^{-1}\mathrm{V}|^{-N/2} \operatorname{etr}\left(-\frac{1}{2}pN\right).$$

When H is true,

$$\max_{\Sigma=\Sigma_0} L(\Sigma_0) = (2\pi)^{-pN/2}|\Sigma_0|^{-N/2} \operatorname{etr}\left(-\frac{1}{2}\Sigma_0^{-1}\mathrm{V}\right).$$

These give the likelihood ratio criterion $\lambda = \max_{\Sigma=\mathrm{I}_p} L(\Sigma)/\max_{\Sigma} L(\Sigma)$ in (8.1.2). □

It is known [Sugiura and Nagao, 1968; Das Gupta, 1969] that the test based on the λ is biased, whereas if we replace N with n, the test becomes unbiased. Thus we use the test based on

$$\lambda^* = \left(\frac{e}{n}\right)^{pn/2} \operatorname{etr}\left(-\frac{1}{2}\Sigma_0^{-1}\mathbf{V}\right) |\Sigma_0^{-1}\mathbf{V}|^{n/2}. \tag{8.1.3}$$

Then the test rejects H if $\lambda^* \leq \lambda^*(\alpha)$, where $\lambda^*(\alpha)$ is the lower α-point of λ^*.

When we consider the distribution of λ^*, we may treat

$$\lambda^* = \left(\frac{e}{n}\right)^{pn/2} \operatorname{etr}\left(-\frac{1}{2}\mathbf{V}\right) |\mathbf{V}|^{n/2}, \tag{8.1.4}$$

where $\mathbf{V}$ is distributed as $\mathrm{W}_p(n, \Sigma_0^{-1/2}\Sigma\Sigma_0^{-1/2})$. In the following we denote $\Sigma_0^{-1/2}\Sigma\Sigma_0^{-1/2}$ simply by Σ. Therefore, the hypothesis H corresponds to $\Sigma = \mathrm{I}_p$.

Theorem 8.1.2 *If $\lambda^* = (e/n)^{pn/2} \operatorname{etr}\left(-\frac{1}{2}\mathbf{V}\right) |\mathbf{V}|^{n/2}$, where $\mathbf{V}$ is distributed as $\mathrm{W}_p(n, \Sigma)$, then*

$$\mathrm{E}[(\lambda^*)^h] = \left(\frac{2e}{n}\right)^{pnh/2} \frac{|\Sigma|^{nh/2}}{|\mathrm{I}_p + h\Sigma|^{(1+h)n/2}} \prod_{j=1}^{p} \frac{\Gamma[\frac{1}{2}(n-p+j+nh)]}{\Gamma[\frac{1}{2}(n-p+j)]}. \tag{8.1.5}$$

In particular, when H is true,

$$\mathrm{E}[(\lambda^*)^h] = \left(\frac{2e}{n}\right)^{pnh/2} (1+h)^{-pn(h+1)/2} \prod_{j=1}^{p} \frac{\Gamma[\frac{1}{2}(n-p+j+nh)]}{\Gamma[\frac{1}{2}(n-p+j)]}. \quad (8.1.6)$$

Proof. Since $\mathbf{V} \sim \mathrm{W}_p(n, \Sigma)$, using Theorem 2.1.3 we have

$$\begin{aligned}
\mathrm{E}[(\lambda^*)^h] &= \left(\frac{e}{n}\right)^{pnh/2} |\Sigma|^{-n/2} K(p,n) \\
&\quad \times \int_{\mathrm{V}>O} |\mathrm{V}|^{(nh+n-p-1)/2} \mathrm{etr}\left\{-\frac{1}{2}(h\mathrm{I}_p + \Sigma^{-1})\mathrm{V}\right\} d\mathrm{V} \\
&= \left(\frac{e}{n}\right)^{pnh/2} |\Sigma|^{-n/2} |h\mathrm{I}_p + \Sigma^{-1}|^{-n(h+1)/2} \frac{K(p,n)}{K(p, nh+n)},
\end{aligned}$$

where $K(p,n) = [2^{pn/2}\Gamma_p(n/2)]^{-1}$. This gives the result required. □

8.1.2 Asymptotic Expansions

Large-Sample Case. First we consider asymptotic approximations of the null distribution of λ^* in (8.1.4) under a large-sample framework:

$$\text{A1: } p \text{ is fixed}, \quad n \to \infty. \quad (8.1.7)$$

The form of moments given in Theorem 8.1.2 is not that of (5.1.2), so we cannot apply Theorem 5.1.1 to derive an asymptotic expansion of the null distribution of λ^*. However, it is possible to obtain it by a similar method. In fact, using Theorem 8.1.2, the characteristic function of $-2\log\lambda^*$ is written as

$$\begin{aligned}
C(t) &= \mathrm{E}[\exp\{-2it\log\lambda^*\}] = \mathrm{E}[(\lambda^*)^{-2it}] \\
&= \left(\frac{2e}{n}\right)^{-itpn} (1-2it)^{-pn(1-2it)/2} \prod_{j=1}^{p} \frac{\Gamma[\frac{1}{2}\{n(1-2it)-j+1\}]}{\Gamma[\frac{1}{2}(n-j+1)]}.
\end{aligned}$$

Then we use the expansion formula (5.4.10) for the gamma function to obtain

$$\begin{aligned}
\log C(t) &= -\frac{1}{4}p(p+1)\log(1-2it) \\
&\quad - \sum_{r=1}^{k} \frac{(-2)^r}{r(r+1)n^r} \left\{\sum_{s=1}^{p} B_{r+1}\left(\frac{1-s}{2}\right)\right\}((1-2it)^{-r}-1) + \mathrm{O}(n^{-k-1}),
\end{aligned}$$

where the $B_r(c)$ are Bernoulli polynomials. After some computation we obtain

$$C(t) = (1-2it)^{p(p+1)/4}\left[1 + \frac{1}{n}\xi_2\{(1-2it)^{-1} - 1\}\right.$$
$$\left. + \frac{1}{6n^2}\{(3\xi_2^2 - 4\xi_3)(1-2it)^{-2} - 6\xi_2^2(1-2it)^{-1} + (3\xi_2^2 + 4\xi_3)\}\right]$$
$$+ \mathrm{O}(n^{-3}), \tag{8.1.8}$$

where

$$\xi_2 = \frac{1}{24}p(2p^2+3p-1), \quad \xi_3 = -\frac{1}{32}p(p-1)(p+1)(p+2).$$

Let $\rho = 1 - c/n$. Then

$$\begin{aligned} 1 - 2it\rho &= (1-2it) + 2itc/n \\ &= (1-2it)\left[1 + \frac{c}{n}\{(1-2it)^{-1} - 1\}\right]. \end{aligned}$$

Therefore, the characteristic function of $-2\rho\log\lambda^*$ can be expanded to

$$C(\rho t) = (1-2it)^{p(p+1)/4}\left[1 - \frac{cp(p+1)}{4n}\{(1-2it)^{-1} - 1\}\right]$$
$$\times \left[1 + \frac{1}{n}\xi_2\{(1-2it)^{-1} - 1\}\right] + \{\text{terms of } \mathrm{O}(n^{-2})\} + \mathrm{O}(n^{-3}).$$

To delete the terms of $\mathrm{O}(n^{-1})$, we define ρ (or c) as

$$m = \rho n = n - \frac{2p^2+3p-1}{6(p+1)}. \tag{8.1.9}$$

Then, after some computation, we obtain

$$C(\rho t) = (1-2it)^{-f/2}\left[1 + \frac{\gamma}{m^2}\{(1-2it)^{-2} - 1\} + \mathrm{O}(m^{-3})\right],$$

where

$$f = \frac{1}{2}p(p+1), \quad \gamma = \frac{p(2p^4+6p^3+p^2-12p-13)}{288(p+1)}. \tag{8.1.10}$$

Inverting this expanded characteristic function, we have the following theorem.

Theorem 8.1.3 *Let λ^* be the modified LR criterion given by (8.1.3) for testing $H\colon \Sigma = \Sigma_0$ vs. $K\colon \Sigma \neq \Sigma_0$. Then an asymptotic expansion of the null* *distribution for large n is*

$$\Pr\{-2\rho\log\lambda^* \le x\} = G_f(x) + \frac{\gamma}{m^2}\{G_{f+4}(x) - G_f(x)\} + \mathrm{O}(m^{-3})$$
$$= G_f(x) - \frac{2\gamma}{m^2}g_f(x)\frac{x}{f}\left(1 + \frac{x}{f+2}\right) + \mathrm{O}(m^{-3}), \tag{8.1.11}$$

where G_f and g_f are the distribution and density functions of $\chi^2(f)$, and m, ρ, f, and γ are as defined in (8.1.9) *and* (8.1.10).

Using the Cornish–Fisher formula, the upper α-point t of $-2\rho\log\lambda$ can be expanded in terms of the upper α-point u of $\chi^2(f)$ as

$$t = u + \frac{2\gamma}{m^2}\frac{x}{f}\left(1 + \frac{x}{f+2}\right) + \mathrm{O}(m^{-3}). \tag{8.1.12}$$

We have seen that, for LR criteria for tests of covariance matrices, their asymptotic null distributions are of chi-squares, under the normality assumption. However, this property is known to be very sensitive to a departure from normality. If, instead, we assume a common elliptical distribution with kurtosis parameter κ, asymptotically valid tests can be constructed by properly estimating κ [see, e.g. Muirhead, 1982]. For bootstrap procedures for estimating the critical values, see Zhang and Boos (1992).

High-Dimensional Case. If p is large, the approximation formulas given by (8.1.11) and (8.1.12) will become worse. To overcome this weakness, we consider high dimensional approximation as in Section 5.4.2 under

$$\text{A2: } p \to \infty, \quad n \to \infty, \quad p/n \to c \in (0,1). \tag{8.1.13}$$

The characteristic function of $T = -(2/n)\log\lambda^*$ under H is given by

$$C_T(t) = \left(\frac{2e}{n}\right)^{-pit}\left(1 - \frac{2it}{n}\right)^{-p(n/2-it)}\prod_{j=1}^{p}\frac{\Gamma[\frac{1}{2}(n-p+j) - it]}{\Gamma[\frac{1}{2}(n-p+j)]},$$

where z^w for complex number z indicates the principal branch, that is, $z^w = \exp(w\log z) = \exp\{w(\log|z| + i\arg z)\}$. The sth cumulant of T is expressed as

$$\kappa_T^{(s)} = (-i)^{-s}\frac{\partial^s}{\partial t^s}\log C_T(t)\bigg|_{t=0}. \tag{8.1.14}$$

To obtain $\kappa_T^{(s)}$ we consider expanding $\log C_T(t)$ given by

$$\log C_T(t) = -pit\log\left(\frac{2e}{n}\right) - p(n/2 - it)\log\left(1 - \frac{2it}{n}\right)$$
$$+ \sum_{j=1}^{p}\left\{\log\Gamma\Big(\frac{1}{2}(n-p+j) - it\Big) - \log\Gamma\Big(\frac{1}{2}(n-p+j)\Big)\right\}.$$

Note that

$$-\log\left(1-\frac{2it}{n}\right)=\left(\frac{2it}{n}\right)+\frac{1}{2}\left(\frac{2it}{n}\right)^2+\frac{1}{3}\left(\frac{2it}{n}\right)^3+\cdots, \quad \text{and}$$

$$\log\Gamma\Big(\frac{1}{2}(n-p+j)-it\Big)-\log\Gamma\Big(\frac{1}{2}(n-p+j)\Big)$$
$$=\sum_{s=1}^{\infty}\psi^{(s-1)}\Big(\frac{1}{2}(n-p+j)\Big)(-it)^s,$$

where ψ is the digamma function defined by $\psi(z)=(d/dz)\log\Gamma(z)$. These give the following expressions for the cumulants:

$$\begin{aligned}\kappa_T^{(1)}&=p-p\log\left(\frac{2e}{n}\right)-\sum_{j=1}^{p}\psi\left(\frac{1}{2}(n-p+j)\right),\\ \kappa_T^{(s)}&=-p(s-2)!\Big(\frac{2}{n}\Big)^{s-1}\\ &\quad+(-1)^s\sum_{j=1}^{p}\psi^{(s-1)}\psi\left(\frac{1}{2}(n-p+j)\right),\ s\geq 2.\end{aligned}\tag{8.1.15}$$

Let $\tilde{T}$ and $\tilde{\kappa}_T^s$ be the standardized statistic and cumulants defined by

$$\tilde{T}=\frac{T-\kappa_T^{(1)}}{(\kappa_T^{(2)})^{1/2}}\tag{8.1.16}$$

and

$$\tilde{\kappa}_T^{(s)}=\kappa_T^{(s)}/(\kappa_T^{(2)})^{s/2},\tag{8.1.17}$$

respectively. Then, following a general method in Section 5.4.2, we can expand the characteristic function of $\tilde{T}$ as

$$\begin{aligned}C_{\tilde{T}}(t)&=\exp\Big\{-\frac{t^2}{2}+\sum_{s=3}^{\infty}\frac{\tilde{\kappa}^{(s)}}{s!}(it)^s\Big\}\\ &=\exp\left(-\frac{t^2}{2}\right)\Big\{1+\sum_{k=1}^{\infty}\frac{(it)^{3k}}{k!}\sum_{j=0}^{\infty}\gamma_{k,j}(it)^j\Big\},\end{aligned}\tag{8.1.18}$$

where

$$\gamma_{k,j}=\sum_{s_1+\cdots+s_k=j}\frac{\tilde{\kappa}_T^{(s_1+3)}\cdots\tilde{\kappa}_T^{(s_k+3)}}{(s_1+3)!\cdots(s_k+3)!}.\tag{8.1.19}$$

Under assumption (8.1.13) it is shown that $\gamma_{k,j}=\mathrm{O}_{j+k}^*$, where the O_j^* denote a term of jth order with respect to (p^{-1},n^{-1}). Therefore, we have an approximation of $C_{\tilde{T}}(t)$ up to order O_s^* given by

$$C_{\tilde{T},s}(t)=\exp\left(-\frac{t^2}{2}\right)\Big\{1+\sum_{k=1}^{s}\frac{(it)^{3k}}{k!}\sum_{j=0}^{s-k}\gamma_{k,j}(it)^j\Big\}.\tag{8.1.20}$$

Inversion of (8.1.20) gives an asymptotic approximation for the distribution function of $\tilde{T}$ up to order O_s^* as

$$Q_s(x) = \Phi(x) - \varphi(x)\Big\{\sum_{k=1}^{s}\frac{1}{k!}\sum_{j=0}^{s-k}\gamma_{k,j}H_{3k+j-1}(x)\Big\}. \qquad (8.1.21)$$

Theorem 8.1.4 *Let λ^* be the modified LR statistic given by* (8.1.3), *and let $\tilde{T}$ be the standardized statistic of $T = -(2/n)\log\lambda^*$ given by* (8.1.17). *Then under* (8.1.13) *the distribution function of $\tilde{T}$ is expanded as*

$$\Pr\{\tilde{T} \le x\} = Q_s(x) + \mathrm{O}_{s+1}^*, \qquad (8.1.22)$$

where Q_s is given by (8.1.19).

Wakaki (2009) gave a rigorous proof of the error term in (8.1.22) by deriving an error bound. For practical use, it is suggested to use $Q_0 = \Phi$, Q_1, or Q_2, which are written as

$$Q_1(x) = \Phi(x) - \frac{1}{6}\varphi(x)\tilde{\kappa}^{(3)}(x^2-1), \qquad (8.1.23)$$

$$\begin{aligned} Q_2(x) = \Phi(x) - \varphi(x)\Big\{&\frac{1}{6}\tilde{\kappa}^{(3)}(x^2-1) + \frac{1}{24}\tilde{\kappa}^{(4)}(x^3-2x) \\ &+ \frac{1}{36}(\tilde{\kappa}^{(3)})^2(x^5-10x^3+15x)\Big\}. \end{aligned} \qquad (8.1.24)$$

Note that the standardized cummulants $\tilde{\kappa}^{(s)}$ involve the polygamma functions, but they are computed numerically by, for example, Mathematica Version 5.2.

8.1.3 High-Dimensional Tests

For the case $p > n = N - 1$, we consider testing H: $\Sigma = \mathrm{I}_p$ vs. K: $\Sigma \neq \mathrm{I}_p$, since the identity I_p in the hypothesis can be replaced by any other positive definite matrix Σ_0 by multiplying the observations by $\Sigma_0^{-1/2}$. In this case the LR test cannot be used, since $|\mathbf{S}|$ becomes singular. A test statistic will be

$$T_1 = \frac{1}{p}\,\mathrm{tr}\,(\mathbf{S} - \mathrm{I}_p)^2. \qquad (8.1.25)$$

Under a large-sample framework A1 in (8.1.7), $\mathbf{S} \xrightarrow{p} \Sigma$; therefore, $T_1 \xrightarrow{p} (1/p)\,\mathrm{tr}\,(\Sigma - \mathrm{I}_p)^2$. This shows that T_1 is consistent. We call the consistency *n-consistency*, while the consistency under a high-dimensional framework:

$$\text{A2: } p \to \infty, \quad n \to \infty, \quad p/n \to c \in (0, \infty), \qquad (8.1.26)$$

is called (n,p)-*consistency*. We will see that the n-consistency of the test based on T_1 does not extend to (n,p)-asymptotics. To overcome its fault, Ledoit and Wolf (2002) proposed

$$T_2 = \frac{1}{p}\operatorname{tr}(\mathbf{S}-\mathrm{I}_p)^2 - \frac{p}{n}\left(\frac{1}{p}\operatorname{tr}\mathbf{S}\right)^2 + \frac{p}{n}. \tag{8.1.27}$$

As for (n,p)-consistency, assume A2 and

$$\begin{aligned} &\text{C1: } (1/p)\operatorname{tr}\Sigma^k = \mathrm{O}(1), \quad k=1,2,\\ &\text{C2: } (1/p)\operatorname{tr}\Sigma^k = \mathrm{O}(1), \quad k=3,4. \end{aligned} \tag{8.1.28}$$

Then, from Theorem 5.9.2,

$$\begin{aligned} T_1 &= \frac{1}{p}\operatorname{tr}\mathbf{S}^2 - \frac{2}{p}\operatorname{tr}\mathbf{S} + 1 \xrightarrow{p} (1+c)\alpha^2 + \delta^2 - 2\alpha + 1\\ &= c\alpha^2 + (\alpha-1)^2 + \delta^2, \end{aligned}$$

where

$$\alpha = \frac{1}{p}\operatorname{tr}\Sigma = \frac{1}{p}\sum_{i=1}^{p}\lambda_i,\ \delta^2 = \frac{1}{p}\sum_{i=1}^{p}(\lambda_i-\alpha)^2$$

and $\lambda_1 \geq \cdots \geq \lambda_p \geq 0$ are the characteristic roots of Σ. Note that the hypothesis is equivalent to $(\alpha-1)^2+\delta^2=0$. Based on the result above, it is shown that T_1 is not (n,p)-consistent. For a counter example, see Ledoit and Wolf (2002).

Under A1, $T_2 \xrightarrow{p} 1/p\operatorname{tr}(\Sigma-\mathrm{I}_p)^2$; therefore, the test of $\Sigma = \mathrm{I}_p$ based on T_2 is also n-consistent. For (n,p)-consistency, similarly, we have

$$T_2 = c\alpha^2 + (\alpha-1)^2 + \delta^2 - c\alpha^2 + c = c + (\alpha-1)^2 + \delta^2.$$

This shows that the test based on T_2 is (n,p)-consistent. Next we examine asymptotic null distributions of T_1 and T_2. Nagao (1973) shows that under A1, the limiting null distribution of T_1 is given by

$$\frac{np}{2}T_1 \xrightarrow{d} \chi^2(p(p+1)/2)$$

or, equivalently,

$$nT_1 - p \xrightarrow{d} \frac{2}{p}\chi^2_{p(p+1)/2} - p,$$

where Y_d denotes a random variable distributed as chi-square distribution with d degrees of freedom $(\chi^2(d))$. In general, it is easily seen that

$$Y_{p(p+1)/2+a} - p \xrightarrow{d} \mathrm{N}(1,4) \tag{8.1.29}$$

when p tends to infinity, where a is a constant not depending on p. Therefore,

$$\lim_{p\to\infty}\lim_{n\to\infty} nT_1 - p \xrightarrow{d} \mathrm{N}(1,4).$$

The (n,p)-limiting null distribution of T_1 is derived by using Theorem 5.9.3:

$$nT_2 - p \xrightarrow{d} \mathrm{N}(1, 4+8c). \tag{8.1.30}$$

This means that the existing n-asymptotic theory (where p is fixed) breaks down when p goes to infinity with n, including the case $p > n$.

On the other hand, it is shown that under a large-sample framework, the limiting null distribution of T_2 is the same as that of T_1, and

$$\lim_{p\to\infty}\lim_{n\to\infty} nT_2 - p \xrightarrow{d} \mathrm{N}(1,4).$$

The (n,p)-limiting null distribution of T_2 is derived using Theorem 5.9.3:

$$nT_2 - p \xrightarrow{d} \mathrm{N}(1,4) \tag{8.1.31}$$

Therefore, the n-asymptotic theory developed for T_1 is directly applicable to T_2, and it remains valid (for T_2 but not T_1) if p goes to infinity with n, even when $p > n$.

8.2 Sphericity

8.2.1 Likelihood Ratio Tests and Moments

In this section we consider the problem of testing the *sphericity hypothesis*:

$$H\colon \Sigma = \sigma^2 \mathrm{I}_p \text{ vs. } K\colon \Sigma \neq \sigma^2 \mathrm{I}_p,$$

based on a random sample of $\mathbf{X} = (\boldsymbol{X}_1, \ldots, \boldsymbol{X}_N)$ drawn from $\mathrm{N}_p(\boldsymbol{\mu}, \Sigma)$. Let $\bar{\boldsymbol{X}}$, $\mathbf{S}$, and $\mathbf{V}$ be as defined in (8.1.1). The testing problem is a combination of the following two testing problems:

$$\text{(1) } H\colon \Sigma = \sigma^2 \mathrm{I}_p \text{ vs. } H_1\colon \Sigma = \mathrm{diag}\,(\sigma_1^2, \ldots, \sigma_p^2) \equiv \mathrm{D}_{\sigma^2}.$$
$$\text{(2) } H_1\colon \Sigma = \mathrm{D}_{\sigma^2} \text{ vs. } K_1\colon \Sigma \neq \mathrm{D}_{\sigma^2}.$$

Testing problems (1) and (2) are special cases of problems for equality of covariance matrices (see Section 8.5) and for independence (see Section 8.4), respectively. We denote the likelihood ratio criteria for (1) and (2) by λ_1 and λ_2, respectively.

Theorem 8.2.1 *The likelihood ratio criterion for testing* H: $\sigma^2 \mathrm{I}_p$ vs. K: $\Sigma \neq \sigma^2 \mathrm{I}_p$, *is*

$$\lambda = \lambda_1 \lambda_2 = \left[|\mathrm{V}| \Big/ \left\{ \frac{1}{p} \operatorname{tr}(\mathrm{V}) \right\}^p \right]^{N/2}. \tag{8.2.1}$$

Proof. The criterion λ can be obtained from the relation $\lambda = \lambda_1 \lambda_2$. It is easy to see that

$$\lambda_1 = \Lambda_1^{N/2}, \quad \Lambda_1 = \frac{|\mathrm{V}|}{v_{11} v_{22} \cdots v_{pp}},$$
$$\lambda_2 = \Lambda_2^{N/2}, \quad \Lambda_2 = \frac{v_{11} v_{22} \cdots v_{pp}}{\{\frac{1}{p} \operatorname{tr}(\mathrm{V})\}^p}.$$

□

Put

$$\Lambda = \lambda^{2/N} = |\mathbf{V}| \Big/ \left\{ \frac{1}{p} \operatorname{tr}(\mathbf{V}) \right\}^p = \Lambda_1 \Lambda_2, \tag{8.2.2}$$

where $\Lambda_i = \lambda_i^{2/N}$, $i = 1, 2$. The unbiasedness of the modified LR criterion $\lambda^* = \Lambda^{n/2}$ was proved by Gleser (1966) and Sugiura and Nagao (1968).

Theorem 8.2.2 *The moments of* $\Lambda = \Lambda_1 \Lambda_2$ *under* H *are*

$$\begin{aligned} \mathrm{E}[\Lambda^h] &= \mathrm{E}[\Lambda_1^h] \mathrm{E}[\Lambda_2^h] \\ &= p^{ph} \frac{\Gamma[\frac{1}{2}pn]}{\Gamma[\frac{1}{2}p(n+2h)]} \prod_{i=1}^{p} \frac{\Gamma[\frac{1}{2}(n+2h+1-i)]}{\Gamma[\frac{1}{2}p(n+1-i)]}. \end{aligned} \tag{8.2.3}$$

Proof. The moments of Λ_1 are obtained as a special case of the LR criterion for equality of covariance matrices (see Section 8.5). The moments of Λ_2 are obtained as a special case of the LR criterion for independence (see Section 8.4). □

8.2.2 Asymptotic Expansions

Large-Sample Case. First we shall see that an asymptotic expansion of the null distribution of $\lambda^* = \Lambda^{n/2}$ (modified LR criterion) under a large-sample framework (8.1.7) is obtained as a special case of Box's theorem (Theorem 5.4.1). From Theorem 8.2.2 the moments of λ^* can be expressed as

$$\begin{aligned} \mathrm{E}[(\lambda^*)^h] &= \mathrm{E}[\Lambda^{nh/2} \\ &= K p^{ph/2} \frac{\prod_{i=1}^p \Gamma[\frac{1}{2}n(1+h) - \frac{1}{2}(i-1))]}{\Gamma[\frac{1}{2}pn(1+h)]}, \end{aligned} \tag{8.2.4}$$

where K is the constant such that $\mathrm{E}[(\lambda^*)^0] = 1$. Note that (8.2.4) is of the form (5.4.4). Therefore, applying Box's theorem we have the following theorem.

Theorem 8.2.3 *Let $\lambda^* = \Lambda^{n/2}$ be the modified LR criterion with Λ in (8.2.2) for testing H: $\Sigma = \sigma^2 \mathrm{I}_p$ vs. K: $\Sigma \neq \sigma^2 \mathrm{I}_p$. Then an asymptotic expansion of the null distribution of $-2\rho \log \lambda^*$ for large n is*

$$\begin{aligned}\Pr\{-2\rho \log \lambda^* \leq x\} & G_f(x) + \frac{\gamma}{m^2}\{G_{f+4}(x) - G_f(x)\} + \mathrm{O}(m^{-3}), \\ &= G_f(x) - \frac{2\gamma}{m^2} g_f(x) \frac{x}{f}\left(1 + \frac{x}{f+2}\right) + \mathrm{O}(m^{-3}), \end{aligned} \tag{8.2.5}$$

where G_f and g_f are the distribution and density functions of $\chi^2(f)$. The constants m, ρ, f, and γ are defined as follows:

$$\begin{aligned} m &= \rho n = n - \frac{1}{6p}(2p^2 + p + 2), \quad f = \frac{1}{2}(p+2)(p-1), \\ \gamma &= \frac{1}{288p^2}(p-1)(p-2)(p+2)(2p^3 + 6p^2 + 3p + 2). \end{aligned}$$

Using the Cornish Fisher formula, the upper α-point t of $-2\rho \log \lambda$ can be expanded in terms of the the upper α-point u of $\chi^2(f)$ as

$$t = u + \frac{2\gamma}{m^2} \frac{x}{f}\left(1 + \frac{x}{f+2}\right) + \mathrm{O}(m^{-3}). \tag{8.2.6}$$

High-Dimensional Case. An asymptotic expansion under a high-dimensional case (8.1.13) is also obtained. Its formal derivation is similar to that of T in the preceding section, although some modifications are required for a proof of its validity. Let us consider

$$T = -(2/n) \log \lambda^* = -\log \Lambda \tag{8.2.7}$$

instead of $-2 \log \lambda^*$. A crucial point in a formal derivation is to give the cumulants of T. Using Theorem 8.2.2, the characteristic function of T is expressed as

$$C_T(t) = p^{-itp} \frac{\Gamma[\frac{1}{2}pn]}{\Gamma[\frac{1}{2}pn - pit]} \prod_{j=1}^{p} \frac{\Gamma[\frac{1}{2}(n-p+j) - it]}{\Gamma[\frac{1}{2}(n-p+j)]}.$$

Expanding $\log C_T(t)$ with the use of (5.4.26), we can find the sth cumulants of T as

$$\begin{aligned} \kappa_T^{(1)} &= -p \log p + p\psi\left(\frac{1}{2}np\right) - \sum_{j=1}^{p} \psi\left(\frac{1}{2}(n-p+j)\right), \\ \kappa_T^{(s)} &= (-1)^{s-1}\left\{p^s \psi^{(s-1)}\left(\frac{1}{2}np\right)\right. \\ &\quad \left. - \sum_{j=1}^{p} \psi^{(s-1)}\left(\frac{1}{2}(n-p+j)\right)\right\}, \quad s = 2, 3, \ldots. \end{aligned} \tag{8.2.8}$$

Let T and $\kappa_T^{(s)}$ be normalized as

$$\tilde{T} = \frac{T - \kappa_T^{(1)}}{(\kappa_T^{(2)})^{1/2}}, \quad \tilde{\kappa}^{(s)} = \frac{\kappa_T^{(s)}}{(\kappa_T^{(2)})^{s/2}}, \quad s = 3, 4, \ldots. \tag{8.2.9}$$

Inversion of (8.1.20) with above $\kappa_T^{(s)}$ gives an asymptotic approximation for the distribution function of $\tilde{T}$ up to order O_s^* as

$$Q_s(x) = \Phi(x) - \varphi(x)\left\{\sum_{k=1}^{s} \frac{1}{k!} \sum_{j=0}^{s-k} \gamma_{k,j} H_{3k+j-1}(x)\right\}. \tag{8.2.10}$$

Theorem 8.2.4 *Let $\lambda^* = \Lambda^{n/2}$ be the modified LR statistic with Λ in* (8.2.2)*, and let $\tilde{T}$ be the standardized statistic of $T = -(2/n)\log\lambda^*$ given by* (8.2.9)*. Then under* (8.1.13)*, the distribution function of $\tilde{T}$ is expanded as*

$$\Pr\{\tilde{T} \le x\} = Q_s(x) + \mathrm{O}_{s+1}^*, \tag{8.2.11}$$

where Q_s is given by (8.1.19)*:*

$$Q_s(x) = \Phi(x) - \varphi(x)\left\{\sum_{k=1}^{s} \frac{1}{k!} \sum_{j=0}^{s-k} \gamma_{k,j} H_{3k+j-1}(x)\right\},$$

where

$$\gamma_{k,j} = \sum_{s_1+\cdots+s_k=j} \frac{\tilde{\kappa}_T^{(s_1+3)} \cdots \tilde{\kappa}_T^{(s_k+3)}}{(s_1+3)! \cdots (s_k+3)!}.$$

Wakaki (2009) gave a rigorous proof of the error term in (8.2.11) by deriving an error bound.

8.2.3 High-Dimensional Tests

Ledoit and Wolf (2002) examined a high-dimensional test for sphericity based on

$$\begin{aligned} U &= \frac{1}{p}\,\mathrm{tr}\left(\frac{\mathbf{S}}{(1/p)\,\mathrm{tr}\,\mathbf{S}} - \mathrm{I}_p\right)^2 \\ &= \frac{(1/p)\,\mathrm{tr}\,\mathbf{S}^2}{[(1/p)\,\mathrm{tr}\,\mathbf{S}]^2} - 1. \end{aligned} \tag{8.2.12}$$

It is easy to see that the sphericity test based on U is n-consistent. As for (n,p)-consistency, Theorem 5.9.2 implies that under assumptions 1 through 3 in Section 5.9.2, or (8.1.26) and (8.1.27),

$$U \xrightarrow{p} \frac{(1+c)\alpha^2 + \delta^2}{\alpha^2} - 1 = c + \frac{\delta^2}{\alpha^2}.$$

Since c can be approximated by the known quantity p/n, the power of this test to separate the null hypothesis of sphericity $\delta^2/\alpha^2 = 0$ from the alternative $\delta^2/\alpha^2 > 0$ converges to 1 as n and p go to infinity together; this constitutes an (n,p)-consistent test.

The limiting distribution of U when $p/n \to c \in (0,\infty)$ is called the (n,p)-*limiting distribution* of U. The (n,p)-distribution is obtained by using Theorem 5.9.3, and the result is given as follows.

Theorem 8.2.5 *Under the assumptions of Theorem 5.9.3, the hypothesis*

$$nU - p \xrightarrow{d} \mathrm{N}(1,4). \tag{8.2.13}$$

That is, the (n,p)-limiting distribution of $nU - p$ is $\mathrm{N}(1,4)$.

Further, as n goes to infinity while p remains fixed, the limiting distribution of U under the null is given by

$$\frac{np}{2}U \xrightarrow{d} Y_{p(p+1)/2-1} \tag{8.2.14}$$

or, equivalently,

$$nU - p \xrightarrow{d} \frac{2}{p}Y_{p(p+1)/2-1} - p, \tag{8.2.15}$$

where Y_d denotes a random variable distributed as chi-square distribution with d degrees of freedom. Similarly,

$$\frac{2}{p}Y_{p(p+1)/2-1} - p \xrightarrow{d} \mathrm{N}(1,4),$$

as p is large.

8.3 Intraclass Covariance Structure

8.3.1 Likelihood Ratio Tests and Moments

The interclass covariance structure of $\boldsymbol{X} = (X_1, \dots, X_p)'$ is defined by

$$\begin{aligned} \mathrm{Var}(X_i) &= \sigma^2, \quad i = 1, \dots, \\ \mathrm{Cov}(X_i, X_j) &= \sigma^2\rho, \quad i \neq j = 1, \dots, p, \end{aligned}$$

which is equivalent to

$$\mathrm{Var}(\boldsymbol{X}) = \sigma^2\{(1-\rho)\mathrm{I}_p + \rho\mathbf{1}_p\mathbf{1}_p'\} \equiv \Sigma_I.$$

Here $\sigma^2 >$ and $1 > \rho > -1/(p-1)$, since Σ_I is positive definite. We wish to test

$$H\colon \Sigma = \Sigma_I \text{ vs. } K\colon \Sigma \neq \Sigma_I,$$

based on a random sample of $\mathbf{X} = (\boldsymbol{X}_1, \dots, \boldsymbol{X}_N)$ drawn from $\mathrm{N}_p(\boldsymbol{\mu}, \Sigma)$. Let $\bar{\boldsymbol{X}}$, $\mathbf{S}$, and $\mathbf{V}$ be those defined in (8.1.1).

Theorem 8.3.1 *The likelihood ratio criterion for testing* $H: \Sigma = \Sigma_I = \sigma^2\{(1-\rho)\mathrm{I}_p + \rho \mathbf{1}_p \mathbf{1}_p'\}$ *vs.* $K: \Sigma \neq \Sigma_I$ *as*

$$\lambda = [(p-1)^{p-1}|\mathrm{V}|/uv^{p-1}]^{N/2}, \tag{8.3.1}$$

where u *and* v *are defined by*

$$v = \operatorname{tr} \mathrm{V} - u. \tag{8.3.2}$$

Proof. Let Ω be the parameter space of $(\boldsymbol{\mu}, \Sigma)$ and ω be the subset of Ω restricted by H. Let $L(\Omega)$ and $L(\omega)$ be the likelihoods of $\mathbf{X}$ in Ω and ω, respectively. As before, we have

$$\max_{\Omega} L(\Omega) = |(2\pi)\mathrm{V}/N|^{-N/2} \exp(-pN/2).$$

To compute $\max_\omega L(\omega)$, let

$$\tau_1 = \sigma^2\{1 + (p-1)\rho\}, \quad \tau_2 = \sigma^2(1-\rho) \tag{8.3.3}$$

and

$$u = \frac{1}{p}\mathbf{1}_p'\mathrm{V}\mathbf{1}_p, \quad v = \operatorname{tr}\mathrm{V} - u.$$

Then the $L(\omega)$ can be written as

$$L(\omega) = (2\pi)^{-pN/2}(\tau_1\tau_2^{p-1})^{-N/2} \exp\left\{-\frac{1}{2}\frac{1}{\tau_2}\left[(u+v) - \frac{\tau_1 - \tau_2}{\tau_1}u\right]\right\}$$
$$\times \operatorname{etr}\left\{-\frac{N}{2}(\bar{\boldsymbol{x}} - \boldsymbol{\mu})'\Sigma_I^{-1}(\bar{\boldsymbol{x}} - \boldsymbol{\mu})\right\},$$

where $\Sigma_I = \sigma^2\{(1-\rho)\mathrm{I}_p + \rho\mathbf{1}_p\mathbf{1}_p'\}$. The MLEs of $\boldsymbol{\mu}$, τ_1, and τ_2 are, respectively, given by

$$\hat{\boldsymbol{\mu}} = \bar{\boldsymbol{x}}, \quad \hat{\tau}_1 = \frac{1}{N}u, \quad \hat{\tau}_2 = \frac{v}{N(p-1)}.$$

Hence

$$\max_{\omega} = (2\pi)^{pN/2}N^{pN/2}\left\{u\left(\frac{v}{p-1}\right)^{p-1}\right\}^{-N/2} \exp(-pN/2). \qquad \square$$

Theorem 8.3.2 *The* h*th moment of* $\Lambda = (p-1)^{p-1}|\mathbf{V}|/\{UV^{p-1}\}$ *under* H *is obtained as*

$$\mathrm{E}[\Lambda^h] = (p-1)^{(p-1)h}\frac{\Gamma[\frac{1}{2}(p-1)n]}{\Gamma[\frac{1}{2}(p-1)(n+2h)]}\prod_{i=1}^{p}\frac{\Gamma[\frac{1}{2}(n+2h-i)]}{\Gamma[\frac{1}{2}(n-i)]}.$$

Proof. To compute the moments of Λ, note that if $\mathbf{V} \sim \mathrm{W}_p(n, \Sigma_I)$, then U and V are distributed independently and $U/\tau_1 \sim \chi^2(n)$ and $V/\tau_2 \sim \chi^2((p-1)n)$; see Problem 2.9.3. Then under H: $\Sigma = \Sigma_I$, we obtain

$$\begin{aligned}
\mathrm{E}[\Lambda^h] &= (p-1)^{(p-1)h}[K(p,n)/K(p,n+2h)]|\Sigma_I|^h \\
&\quad \times \int_{\mathrm{V}>\mathrm{O}} K(p,n+2h)|\Sigma_I|^{-(n+2h)/2}[u^{-h}v^{-(p-1)h}] \\
&\qquad \times |\mathrm{V}|^{(n+2h-p-1)/2} \operatorname{etr}\left\{-\frac{1}{2}\Sigma_I^{-1}\mathrm{V}\right\} d\mathrm{V} \\
&= (p-1)^{(p-1)h}[K(p,n)/K(p,n+2h)]\mathrm{E}[(U/\tau_1)^{-h}(V/\tau_2)^{-(p-1)h}],
\end{aligned}$$

where $K(p,m) = [2^{pm/2}\Gamma_p(m/2)]^{-1}$ and the expectation is taken with respect to $\mathrm{W}_p(n+2h, \Sigma_I)$. But this expectation can be evaluated based on a chi-square distribution. Since U and V are independently distributed,

$$\begin{aligned}
\mathrm{E}[(U/\tau_1)^{-h}] &= 2^{-h}\frac{\Gamma[\frac{1}{2}n]}{\Gamma[\frac{1}{2}(n+2h)]}, \\
\mathrm{E}[(V/\tau_2)^{-(p-1)h}] &= 2^{-(p-1)h}\frac{\Gamma[\frac{1}{2}(p-1)n]}{\Gamma[\frac{1}{2}(p-1)(n+2h)]}. \qquad \square
\end{aligned}$$

8.3.2 Asymptotic Expansions

Large-Sample Case. We consider the distribution of the modified LR criterion

$$\lambda^* = [(p-1)^{p-1}|\mathbf{V}|/\{UV^{p-1}\}]^{n/2} = \Lambda^{n/2}. \tag{8.3.4}$$

The hth moment of λ^* under H is given by

$$\mathrm{E}[(\lambda^*)^h] = K \cdot (p-1)^{(p-1)nh/2} \prod_{i=1}^{p} \frac{\Gamma[\frac{1}{2}(n+2nh-i)]}{\Gamma[\frac{1}{2}n(p-1)(1+h)]}.$$

Theorem 8.3.3 *Let $\lambda^* = \Lambda^{n/2}$ be the modified LR criterion with Λ in (8.3.4) for testing H: $\Sigma = \Sigma_I$ vs. K: $\Sigma \neq \Sigma_I$. Then an asymptotic expansion of the null distribution of $-2\rho\log\lambda^*$ for large n is*

$$\begin{aligned}
\Pr\{-2\rho\log\lambda^* \le x\}G_f(x) &+ \frac{\gamma}{m^2}\{G_{f+4}(x) - G_f(x)\} + \mathrm{O}(m^{-3}) \\
&= G_f(x) - \frac{2\gamma}{m^2}g_f(x)\frac{x}{f}\left(1 + \frac{x}{f+2}\right) + \mathrm{O}(m^{-3}),
\end{aligned} \tag{8.3.5}$$

where G_f and g_f are the distribution and density functions of $\chi^2(f)$. The *constants* m, ρ, f, *and* γ *are defined as follows:*

$$m = \rho n = n - \frac{p(2p^3 + p^2 - 4p - 3}{6(p-1)(p^2+p-4)}, \quad f = \frac{1}{2}(p+2)(p-1),$$
$$\gamma = \frac{1}{288}\frac{p}{(p-1)^2(p^2+p-4)}$$
$$\times (2p^7 + 2p^6 - 33p^5 + 8p^4 + 116p^3 - 66p^2 - 93p + 48).$$

High-Dimensional Case. The characteristic function of $T = -\log \Lambda$ is given by

$$\begin{aligned} C_T(t) &= \mathrm{E}[\exp(itT)] = \mathrm{E}[\Lambda^{-it}] \\ &= K(p-1)^{(p-1)it} \frac{\prod_{j=1}^{p-1} \Gamma[it + (n-j)/2]}{\Gamma[(p-1)(it + n/2)]}. \end{aligned}$$

Accordingly, the cumulant generating function of T can be put in the form

$$\begin{aligned} \log C_T(t) = &\log K + (p-1)it \log(p-1) \\ &+ \sum_{j=1}^{p-1} \log \Gamma\Big[it + \frac{1}{2}(n-j)\Big] - \log \Gamma\Big[(p-1)(\frac{1}{2}n + it)\Big]. \end{aligned}$$

Using Taylor's expansion formula:

$$\log \Gamma(a+b) = \log \Gamma(a) + \sum_{k=1}^{\infty} \frac{1}{k!} \psi^{(k-1)}(a) b^k,$$

we obtain

$$\begin{aligned} \log C_T(t) &= \log K + (p-1)it \log(p-1) \\ &\quad + \sum_{j=1}^{p-1} \Big\{ \log \Gamma\Big[\frac{1}{2}(n-j)\Big] + \sum_{k=1}^{\infty} \frac{1}{k!} \psi^{(k-1)}\Big(\frac{1}{2}(n-j)\Big)(it)^k \Big\} \\ &\quad - \log \Gamma\Big(\frac{1}{2}n(p-1)\Big) - \sum_{k=1}^{\infty} \frac{1}{k!} \psi^{(k-1)}\Big(\frac{1}{2}n(p-1)\Big)((p-1)it)^k \\ &= \sum_{s=1}^{\infty} \frac{1}{s!}(it)^s \kappa_T^{(s)}, \end{aligned} \tag{8.3.6}$$

where

$$\kappa_T^{(1)} = (p-1)\log(p-1) + \psi_{(p-1)}\Big(\frac{1}{2}(n-1)\Big) - \psi\Big(\frac{1}{2}n(p-1)\Big)(p-1),$$
$$\kappa_T^{(s)} = \psi_{(p-1)}^{(s-1)}\Big(\frac{1}{2}(n-1)\Big) - \psi^{(s-1)}\Big(\frac{1}{2}n(p-1)\Big)(p-1)^r, \ r \geq 3.$$

Here $\psi_{(p-1)}(a) = \sum_{j=1}^{p-1} \psi(a - \frac{1}{2}(j-1))$. Under a high-dimensional framework, it is shown [see, Kato et al., 2009] that $\kappa_T^{(2)} = \mathrm{O}_0^*$ and $\kappa_T^{(s)} = \mathrm{O}_{s-2}^*$, where O_j^* denotes a term of the jth order with respect to $(1/n, 1/p)$. Let T and $\kappa_T^{(s)}$ be normalized as

$$\tilde{T} = \frac{T - \kappa_T^{(1)}}{(\kappa_T^{(2)})^{1/2}}, \quad \tilde{\kappa}^{(s)} = \frac{\kappa_T^{(s)}}{(\kappa_T^{(2)})^{s/2}}, \quad s = 3, 4, \ldots. \tag{8.3.7}$$

Theorem 8.3.4 *Let* $\lambda^* = \Lambda^{n/2}$ *be the modified LR statistic with* Λ *in* (8.3.4), *and let* $\tilde{T}$ *be the standardized statistic of* $T = -(2/n)\log\lambda^*$ *given by* (8.1.30). *Then under* (8.1.13) *the distribution function of* $\tilde{T}$ *is expanded as*

$$\Pr\{\tilde{T} \le x\} = Q_s(x) + \mathrm{O}_{s+1}^*, \tag{8.3.8}$$

where Q_s *is given by* (8.1.19)*:*

$$Q_s(x) = \Phi(x) - \varphi(x)\left\{\sum_{k=1}^{s} \frac{1}{k!} \sum_{j=0}^{s-k} \gamma_{k,j} H_{3k+j-1}(x)\right\},$$

with

$$\gamma_{k,j} = \sum_{s_1+\cdots+s_k=j} \frac{\tilde{\kappa}_T^{(s_1+3)} \cdots \tilde{\kappa}_T^{(s_k+3)}}{(s_1+3)! \cdots (s_k+3)!}.$$

A rigorous proof of the error term in (8.3.8) was given by Kato et al. (2009). They also obtained a computable error bound.

We note that the testing problem for an intraclass covariance structure is closely related to that for sphericity. Let H be a $p \times (p-1)$ matrix satisfying $\mathrm{H}'\mathbf{1}_p = \mathbf{0}$. Then, if the covariance matrix of $\boldsymbol{X}$ has an intraclass covariance structure Σ_I, the covariance of $\mathrm{H}'\boldsymbol{X}$ is $\sigma^2(1-\rho)\mathrm{I}_p$. Using this property, it is possible to check the intraclass covariance structure when $p > n$, since a sphericity test is available for the case $p > n$.

8.3.3 Numerical Accuracy

In this section we examine the numerical accuracy of two types of approximations. One is an asymptotic expansion in terms of chi-square distributions under a large-sample framework (see Theorem 8.3.3). The other is an asymptotic expansion in terms of normal distributions under a high-dimensional framework (see Theorem 8.3.4).

A large-sample approximate test rejects the hypothesis H with significance level α if $-2\rho\log(\Lambda)^{n/2}$ is larger than the upper α-point of the chi-square distribution $\chi^2(f)$ with $f(= p(p+1)/2 - 2)$ degrees of freedom, and ρ

is the Bartlett correction factor defined by

$$\rho = 1 - \frac{p(2p^3 + p^2 - 4p - 3)}{6n(p-1)(p^2 + p - 4)}.$$

A high-dimensional approximate test rejects the hypothesis H with significance level α if $Z = [T - \kappa_T^{(1)}]/(\kappa_T^{(2)})^{1/2}$ is larger than the upper α-point of $\mathrm{N}(0,1)$, where $T = -\log \Lambda$. The normal approximation can be modified by Q_1in Theorem 8.3.4.

We list the estimated significance levels of these approximations by simulations. In the simulations, the parameters in the interclass correlation matrix are $\sigma^2 = 1, \rho = 1/2$, and $N = n + 1 = 20$. We list the estimated significance levels for nominal significance levels of 0.05, 0.50, 0.95 and 0.99 with 1,000,000 repetitions. Table 8.3.1 give high-dimensional approximations. The values in parentheses are the approximations by Q_1. Table 8.3.2 gives large-sample approximations. The accuracy of the high-dimensional approximation become better than that of the large-sample approximation as p increases in value. Further, the high-dimensional approximation is improved considerably when using the approximation Q_1.

Table 8.3.1 Actual probabilities of high-dimensional approx. for $N = 20$

p	0.05	0.50	0.95	0.99
5	0.0820 (0.0634)	0.4480 (0.4991)	0.9666 (0.9435)	0.9980 (0.9918)
11	0.0628 (0.0524)	0.4755 (0.5002)	0.9591 (0.9480)	0.9953 (0.9904)
17	0.0617 (0.0529)	0.4764 (0.4974)	0.9584 (0.9490)	0.9952 (0.9910)

Table 8.3.2 Actual probabilities of large-sample approx. for $N = 20$

p	0.05	0.50	0.95	0.99
5	0.0502	0.5036	0.9519	0.9905
11	0.0679	0.5785	0.9716	0.9956
17	0.2928	0.9130	0.9996	1.0000

8.4 Test for Independence

8.4.1 Likelihood Ratio Tests and Moments

Let $\boldsymbol{X} \sim \mathrm{N}_p(\boldsymbol{\mu}, \Sigma)$, where $\boldsymbol{\mu}$ and Σ are unknown. Suppose that $\boldsymbol{X}$ is partitioned into k vectors $\boldsymbol{X}_1 : p_1 \times 1, \ldots, \boldsymbol{X}_k : p_k \times 1$, where $p_1 + \cdots + p_k = p$.

Let the corresponding partitions of $\boldsymbol{\mu}$ and Σ be

$$\begin{aligned}\boldsymbol{\mu} &= (\boldsymbol{\mu}_1', \ldots, \boldsymbol{\mu}_k')', \quad \boldsymbol{\mu}_i : p_i \times 1, \\ \Sigma &= (\Sigma_{ij}), \quad \Sigma_{ij} : p_i \times p_j, \quad i, j = 1, \ldots, k.\end{aligned} \tag{8.4.1}$$

We wish to test the hypothesis that all the subvectors $\boldsymbol{X}_1, \ldots, \boldsymbol{X}_k$ are independently distributed that is, to test the hypothesis

$$H\colon \Sigma = D_\Sigma \equiv \text{Diag}\,(\Sigma_{11}, \ldots, \Sigma_{kk}) \text{ vs. } K\text{: not } H_0. \tag{8.4.2}$$

based on a random sample of size $N = n + 1$ drawn from $\text{N}_p(\boldsymbol{\mu}, \Sigma)$. Let $\bar{\boldsymbol{X}}$ and $\mathbf{S} = (1/n)\mathbf{V}$ be the sample mean vector and the sample covariance matrix, respectively. Further, let $\bar{\boldsymbol{X}}_i : p_i \times 1, i = 1, \ldots, k$, and $\mathbf{V}_{ij} : p_i \times p_j, i, j = 1, \ldots, k$ be the subvectors and the submatrices in the partitions of $\boldsymbol{X}$ and $\mathbf{V}$ corresponding to those of $\boldsymbol{\mu}$ and Σ, respectively, and let $D_{\mathbf{V}} = \text{Diag}\,(\mathbf{V}_{11}, \ldots, \mathbf{V}_{kk})$.

Theorem 8.4.1 *The likelihood ratio criterion is given by*

$$\lambda = \left[|\mathrm{V}| \Big/ \prod_{i=1}^{k} |\mathrm{V}_{ii}| \right]^{N/2} = |\mathrm{V}\mathrm{D}_{\mathrm{V}}^{-1}|^{N/2}, \tag{8.4.3}$$

where $\mathrm{D}_{\mathrm{V}} = \text{Diag}\,(\mathrm{V}_{11}, \ldots, \mathrm{V}_{kk})$ *and the test rejects* $H\colon \Sigma = \text{Diag}\,(\Sigma_{11}, \ldots, \Sigma_{kk})$ *for a smaller value of* λ. *This is obviously equivalent to the based on*

$$\Lambda = |\mathrm{V}\mathrm{D}_{\mathrm{V}}^{-1}|. \tag{8.4.4}$$

Proof. Let Ω be the parameter space of $(\boldsymbol{\mu}, \Sigma)$ and ω be the subspace of Ω restricted by H [ω be the space of $(\boldsymbol{\mu}, D_\Sigma)$]. Then we have, as before,

$$\max_{\Omega}\{L(\Omega)\} = |(2\pi)\mathrm{V}/N|^{-N/2} \exp(-pN/2)$$

and

$$\max_{\omega}\{L(\omega)\} = (2\pi)^{-N/2} \exp(-pN/2) \prod_{i=1}^{k} |\mathrm{V}_{ii}/N|. \qquad \square$$

Let $\mathbf{R} = \mathbf{D}_{\mathbf{V}}^{-1/2}\mathbf{V}\mathbf{D}_{\mathbf{V}}^{-1/2}$, where $\mathbf{D}_{\mathbf{V}} = \text{diag}\,(V_{11}, \ldots, V_{pp})$, and the V_{ii} be diagonal elements of $\mathbf{V}$; that is, $\mathbf{R}$ is the sample correlation matrix. Then the criterion λ^* can be expressed as

$$\lambda^* = |\mathbf{R}\mathbf{D}_{\mathbf{R}}^{-1}| = |\mathbf{R}| \Big/ \prod_{i=1}^{k} |\mathbf{R}_{ii}|, \tag{8.4.5}$$

where $\mathbf{R}_{ii} (i = 1, \ldots, k)$ is the submatrix of $\mathbf{R}$ corresponding to $\mathbf{V}_{ii}$.

Theorem 8.4.2 *Under the null hypothesis $H: \Sigma = \mathrm{D}_\Sigma$, the moments of $\Lambda = |\mathbf{V}\mathbf{D}_\mathbf{V}|^{-1}$ are given by*

$$\begin{aligned}\mathrm{E}[\Lambda^h] &= \prod_{j=p_1+1}^{p} \frac{\Gamma[\frac{1}{2}(n+1+2h-j)]}{\Gamma[\frac{1}{2}(n+1-j)]} \prod_{i=2}^{k}\prod_{r=1}^{p_i} \frac{\Gamma[\frac{1}{2}(n+1-r)]}{\Gamma[\frac{1}{2}(n+1+2h-r)]} \\ &= \prod_{i=2}^{k}\prod_{r=1}^{p_i} \frac{\Gamma[\frac{1}{2}(n+1+2h-\bar{p}_i-r)]\Gamma[\frac{1}{2}(n+1-r)]}{\Gamma[\frac{1}{2}(n+1-\bar{p}_i-r)]\Gamma[\frac{1}{2}(n+1+2h-r)]}, \end{aligned} \tag{8.4.6}$$

where $\bar{p}_i = p_1 + \cdots + p_{i-1}$.

Proof. First, we observe that λ^* is invariant under the group of transformations $\mathrm{V} \to \mathrm{D_A V D_A'}$, where $\mathrm{D_A} = \mathrm{Diag}(\mathrm{A}_1, \ldots, \mathrm{A}_k)$, $\mathrm{A}_i : p_i \times p_i$ and nonsingular. Hence we may assume without any loss of generality that $\mathrm{D}_\Sigma = \mathrm{I}_p$, so that $\mathbf{V} \sim \mathrm{W}_p(n, \mathrm{I}_p)$. Now

$$\begin{aligned}\mathrm{E}[\Lambda] &= \mathrm{E}[|\mathbf{V}|^h |\mathbf{D}_\mathbf{V}|^{-h}] \\ &= K(p,n) \int_{\mathrm{V}>\mathrm{O}} |\mathrm{V}|^{(n+2h-p-1)} \left(\prod_{i=1}^{k} |\mathrm{V}_{ii}|^{-h} \right) \mathrm{etr}\left(-\frac{1}{2}\mathrm{V}\right) d\mathrm{V} \\ &= \frac{K(p,n)}{K(p,n+2h)} \mathrm{E}\left[\prod_{i=1}^{k} |\mathbf{V}_{ii}|^{-h} \right],\end{aligned}$$

where the expectation is now taken with respect to $\mathbf{V} \sim \mathrm{W}_p(n+2h, \mathrm{I}_p)$. Then it follows from Theorem 2.2.1 that $\mathbf{V}_{ii}$ are distributed independently as $\mathrm{W}_{p_i}(n+2h, \mathrm{I}_{p_i})$. Hence

$$\mathrm{E}[\Lambda] = \frac{K(p,n)}{K(p,n+2h)} \prod_{i=1}^{k} \mathrm{E}[|\mathbf{V}_{ii}|^{-h}] = \frac{K(p,n)}{K(p,n+2h)} \prod_{i=1}^{k} \frac{K(p_i, n+2h)}{K(p_i, n)}.$$

□

8.4.2 Asymptotic Expansions

We can see that the moment of the LR criterion λ is of the form (5.4.4). Therefore, applying Box's theorem (Theorem 5.4.1) we have the following theorem.

Theorem 8.4.3 *When H is true, an asymptotic expansion of the distribution of $-2\rho\log\lambda$ for large n is*

$$\begin{aligned}\Pr\{-2\rho\log\lambda \le x\} G_f(x) &+ \frac{\gamma}{m^2}\{G_{f+4}(x) - G_f(x)\} + \mathrm{O}(m^{-3}), \\ &= G_f(x) - \frac{2\gamma}{m^2} g_f(x) \frac{x}{f}\left(1 + \frac{x}{f+2}\right) + \mathrm{O}(m^{-3}),\end{aligned} \tag{8.4.7}$$

where G_f and g_f are the distribution and density functions of $\chi^2(f)$. The constants m, ρ, f, and γ are defined as follows:

$$m = \rho N = N - \frac{1}{2}\left(\frac{3}{2} + \frac{1}{3}\frac{p^3 - \sum_{i=1}^k p_i^3}{p^2 - \sum_{i=1}^k p_i^2}\right), \quad f = \frac{1}{2}\left(p^2 - \sum_{i=1}^k p_i^2\right),$$

$$\gamma = \frac{1}{48}\left[\left(p^4 - \sum_{i=1}^k p_i^4\right) 6 \left(p^3 - \sum_{i=1}^k p_i^3\right) + 11 \left(p^2 - \sum_{i=1}^k p_i^2\right)\right] - \frac{1}{4}N^2(1-\rho)^2.$$

Similarly, it is possible to give a high-dimensional approximation. However, its validity has not been demonstrated by an independence test of a general partition of $\boldsymbol{X}$.

8.4.3 High-Dimensional Tests

If ρ_{ij} is the (i,j)th element of the population correlation matrix R, the hypothesis of complete independence can be written as H: $\rho_{ij} = 0 (i > j)$. A simple and intuitively plausible statistic for testing this hypothesis is the sum of squared R_{ij} for $i > j$. Since $\mathrm{E}(R_{ij}^2) = n^{-1}$ under H_0 when $i \neq j$,

$$\mathrm{E}\left(\sum_{i=2}^{p}\sum_{j=1}^{i-1} R_{ij}^2\right) = \frac{p(p-1)}{2n},$$

so a statistic with mean 0 under the null hypothesis is given by

$$T_{np} = \sum_{i=2}^{p}\sum_{j=1}^{i-1} R_{ij}^2 - \frac{p(p-1)}{2n}. \tag{8.4.8}$$

Schott (2005) studied the high-dimensional approximation of T_{np} under a high-dimensional framework; that is, p and n approach infinity in such a way that

$$\lim(p/n) = c \in (0, \infty). \tag{8.4.9}$$

If h, i, j, and k are distinct integers, then, under H,

$$\mathrm{E}(R_{ij}^4) = 3/\{n(n+2)\}, \quad \mathrm{E}(R_{ij}^2 R_{ik}^2) = \mathrm{E}(R_{hi}^2 R_{jk}^2) = n^{-2}, \tag{8.4.10}$$

and this leads to

$$\sigma_{np}^2 = \mathrm{Var}(T_{np}) = \frac{p(p-1)(n-1)}{n^2(n+2)},$$

so that T_{np}/σ_{np} will have mean 0 and variance 1 if the complete-independence hypothesis holds.

Note that, under condition (8.4.9),

$$\lim \sigma_{np}^2 = \lim \frac{p(p-1)(n-1)}{n^2(n+2)} = c^2.$$

Theorem 8.4.4 *Suppose that the sample correlation matrix* $\mathbf{R} = (R_{ij})$ *has been computed from a random sample from a multivariate normal distribution with the population correlation matrix* $\mathrm{R} = (\rho_{ij})$. *If* $\mathrm{R} = \mathrm{I}_p$ *and condition* (8.4.9) *holds,* T_{np} *converges in distribution to a normal random variable with mean* 0 *and variance* c^2.

The theorem was proved (Schoot, 2005) by the martingale limit theorem (see Section 5.9). In the following we give an outline of the derivation along the lines of Schott (2005). Note that the (i, j)th element of $\mathbf{S}$ can be put in the form $S_{ij} = n^{-1}\sigma_{ii}^{1/2}\sigma_{jj}^{1/2}\boldsymbol{Z}_i'\boldsymbol{Z}_j$, where $\boldsymbol{Z}_1, \ldots, \boldsymbol{Z}_p$ are independent and identically distributed as $\mathrm{N}_n(\mathbf{0}, \mathrm{I}_n)$. As a result, R_{ij} can be expressed as $R_{ij} = \boldsymbol{U}_i'\boldsymbol{U}_j$, where $\boldsymbol{U}_i = (\boldsymbol{Z}_i'\boldsymbol{Z}_i)^{-1/2}\boldsymbol{Z}_i$ and $\boldsymbol{U}_1, \ldots, \boldsymbol{U}_p$ are independently distributed, each having a uniform distribution on the surface of the n-sphere. For $\ell = 2, \ldots, p$, let

$$X_{n\ell} = T_{n\ell} - T_{n,\ell-1} = \sum_{i=1}^{\ell-1} R_{\ell i}^2 - \frac{\ell - 1}{n},$$

where $T_{n1} = 0$, so that $T_{np} = \sum_{\ell=2}^{p} X_{n\ell}$. If we defined the set $\mathcal{F}_{n,\ell-1} = \{\boldsymbol{U}_1, \ldots, \boldsymbol{U}_{\ell-1}\}$, then

$$\mathrm{E}(R_{\ell i}^2|\mathcal{F}_{n,\ell-1}) = \boldsymbol{U}_i'\mathrm{E}(\boldsymbol{U}_\ell\boldsymbol{U}_\ell')\boldsymbol{U}_i = \boldsymbol{U}_i'\mathrm{E}(n^{-1}\mathrm{I}_n)\boldsymbol{U}_i = \frac{1}{n},$$

so that $\mathrm{E}(X_{n\ell}|\mathcal{F}_{n,\ell-1}) = 0$. Consequently, for each n, $\{T_{n\ell}, \ell = 2, \ldots, p\}$ is a martingale and $X_{n2}, \ldots, X_{np}$ are martingale differences. By the martingale limit theorem (see Theorem 5.9.4) the theorem will follow if we can show that

$$\sum_{\ell} \mathrm{E}\{X_{n\ell}^2 I(|X_{n\ell}| > \varepsilon)|\mathcal{F}_{n,\ell-1}\} \to 0 \tag{8.4.11}$$

in probability for all $\varepsilon > 0$, and

$$\sum_{\ell} \mathrm{E}\{X_{n\ell}^2|\mathcal{F}_{n,\ell-1}\} \to \gamma^2 \tag{8.4.12}$$

in probability. Here $I(\cdot)$ denotes the indicator function. To check these conditions, see Schott (2005).

8.5 Tests for Equality of Covariance Matrices

8.5.1 Likelihood Ratio Test and Moments

Let $\bar{\boldsymbol{X}}_i$ and $\mathbf{S}_i = n_i\mathbf{V}_i$ be the sample mean vector and the sample covariance matrix based on a random sample of size $N_i = n_i + 1$ drawn from $\mathrm{N}(\boldsymbol{\mu}_i, \Sigma_i)$ for $i = 1, \ldots, k$. We wish to test

$$H\colon \Sigma_1 = \cdots = \Sigma_k \text{ vs. } K\colon \text{not } H.$$

Let $\mathbf{V} = \mathbf{V}_1 + \cdots + \mathbf{V}_k$. Then the LR criterion is given by the following theorem.

Theorem 8.5.1 *The likelihood ratio criterion of $H\colon \Sigma_1 = \cdots = \Sigma_k$ is*

$$\lambda = \frac{N^{pN/2}}{\prod_{i=1}^k N_i^{pN_i/2}} \frac{\prod_{i=1}^k |\mathrm{V}_i|^{N_i/2}}{|\mathrm{V}|^{N/2}}. \tag{8.5.1}$$

Proof. The LR of $\{\boldsymbol{\mu}_i, \Sigma_i, i = 1, \ldots, k\}$ is given by

$$\begin{aligned}
L(\Omega) &= \prod_{i=1}^k L_i(\Omega_i) \\
&= (2\pi)^{pN/2} \prod_{i=1}^k |\Sigma_i|^{-N_i/2} \operatorname{etr}\left\{-\frac{1}{2}\Sigma_i^{-1}[V_i + N_i(\bar{\boldsymbol{x}}_i - \boldsymbol{\mu}_i)(\bar{\boldsymbol{x}}_i - \boldsymbol{\mu}_i)']\right\},
\end{aligned}$$

where Ω is the parameter space $\{(\boldsymbol{\mu}_i, \Sigma_i), i = 1, \ldots, k\}$ and $\Omega_i = \{\boldsymbol{\mu}_i, \Sigma_i\}$. Hence we obtain

$$\max_{\Omega}\{L(\Omega)\} = (2\pi)^{pN/2} \prod_{i=1}^k N_i^{pN_i/2} |\mathrm{V}_i|^{-N_i/2} \exp(-pN_i/2).$$

When H is true, the LF reduces to

$$L(\omega) = |(2\pi)\Sigma|^{-N/2} \operatorname{etr}\left\{-\frac{1}{2}\Sigma^{-1}\left[\mathrm{V} + \sum_{i=1}^k N_i(\bar{\boldsymbol{x}}_i - \boldsymbol{\mu}_i)(\bar{\boldsymbol{x}}_i - \boldsymbol{\mu}_i)'\right]\right\},$$

where $\omega = \{\boldsymbol{\mu}_1, \ldots, \boldsymbol{\mu}_k, \Sigma\}$, the parameter space under H, from which we obtain

$$\max\{L(\omega)\} = |(2\pi)\mathrm{V}|^{-N/2} N^{pN/2} \exp(-pN/2). \qquad \square$$

The test with rejection region $[\lambda < c]$ is, however, not unbiased. But as in the test of $\Sigma = \Sigma_0$, this becomes unbiased if we replace the N_i and N in

λ_1 with $n_i = N_i - 1$ and $n = N - k$, respectively [see Perlman, 1980]. Then we use the following modified LR criterion:

$$\lambda^* = \prod_{i=1}^{k}(n/n_i)^{pn_i/2} \cdot W, \quad W = \prod_{i=1}^{k}|\mathbf{V}_i|^{n_i/2}/|\mathbf{V}|^{n/2}, \tag{8.5.2}$$

and we reject H whenever $\lambda^* < c^*$, where c^* is chosen so that the test has a preassigned significance level.

Theorem 8.5.2 *The* hth *moment of* λ^* *under* H *is obtained as*

$$\mathrm{E}[(\lambda^*)^h] = K\left(\frac{\prod_{j=1}^{p}(n/2)^{n/2}}{\prod_{i=1}^{k}\prod_{\ell=1}^{p}(n_i/2)^{n_i/2}}\right)\frac{\prod_{i=1}^{k}\prod_{\ell=1}^{p}\Gamma[\frac{1}{2}(n_i+n_ih-\ell+1)]}{\prod_{j=1}^{p}\Gamma[\frac{1}{2}(n+nh-j+1)]}. \tag{8.5.3}$$

Proof. We observe that under H, $\mathbf{V}_i$ are independently distributed as $\mathrm{W}_p(n_i, \Sigma)$, $i = 1, \ldots, k$ and $\mathbf{V} \sim \mathrm{W}_p(n, \Sigma)$. Since w_1 is invariant under any nonsingular transformation $\mathrm{V}_i \to A\mathrm{V}_iA'$ and $\mathrm{V} \to A\mathrm{V}A'$, we may assume without any loss of generality that $\Sigma = \mathrm{I}_p$. Hence the moments of W can be computed as follows:

$$\begin{aligned}
\mathrm{E}[W^h] &= \mathrm{E}\left[\prod_{i=1}^{k}|\mathbf{V}_i|^{n_ih/2}|\mathbf{V}|^{-nh/2}\right] \\
&= \left(\frac{K(p,n_i)}{K(p,n_ih+n_i)}\right)\int_{\mathrm{V}_i>\mathrm{O}}\!\!\cdots\int |\mathrm{V}|^{-nh/2} \\
&\quad\times\left[\prod_{i=1}^{k}K(p,n_ih+n_i)|\mathrm{V}_i|^{(n_ih+n_i-p-1)/2}\,\mathrm{etr}\left(-\frac{1}{2}\mathrm{V}_i\right)d\mathrm{V}_i\right],
\end{aligned}$$

where $K(p,m) = [2^{mp/2}\Gamma_p(m/2)]^{-1}$, as before. Since the integral in the expression above is the expectation of $|\mathbf{V}|^{nh/2}$, it can, alternatively, be computed based on the distribution of $\mathbf{V}$, which is $\mathrm{W}_p(nh+n, \mathrm{I}_p)$. It follows now that

$$\mathrm{E}[W^h] = \prod_{i=1}^{k}\frac{K(p,n_i)}{K(p,n_ih+n_i)}\,\frac{K(p,nh+n)}{K(p,n)}.$$

Thus we obtain

$$\mathrm{E}[W^h] = \prod_{j=1}^{p}\frac{\Gamma[\frac{1}{2}(n-j+1)]}{\Gamma[\frac{1}{2}(n+nh-j+1)]}\left(\prod_{i=1}^{k}\frac{\Gamma[\frac{1}{2}(n_i+n_ih-j+1)]}{\Gamma[\frac{1}{2}(n_i-j+1)]}\right). \qquad \square$$

8.5.2 Asymptotic Expansions

Large-Sample Case. We can see that the moment of the modified LR criterion λ^* is of the form (5.4.4) under the condition

$$n_i/n \to r_i \in (0,1), \quad i = 1, \ldots, k. \tag{8.5.4}$$

Therefore, applying Box's theorem, we have the following theorem.

Theorem 8.5.3 *When H is true, an asymptotic expansion of the distribution of $-2\rho \log \lambda^*$ under (8.5.4) is*

$$\begin{aligned} \Pr\{-2\rho \log \lambda \le x\} G_f(x) &+ \frac{\gamma}{m^2}\{G_{f+4}(x) - G_f(x)\} + \mathrm{O}(m^{-3}), \\ = G_f(x) &- \frac{2\gamma}{m^2} g_f(x) \frac{x}{f}\left(1 + \frac{x}{f+2}\right) + \mathrm{O}(m^{-3}), \end{aligned} \tag{8.5.5}$$

where G_f and g_f are the distribution and density functions of $\chi^2(f)$. The constants m, ρ, f, and γ are defined as follows:

$$\begin{aligned} m &= \rho n = n - \frac{2p^2 + 3p - 1}{6(p+1)(k-1)}\left(\sum_{i=1}^{k} \frac{n}{n_i} - 1\right), \\ f &= \frac{1}{2} p(p+1)(k-1), \\ \gamma &= \frac{p(p+1)}{48}\left\{(p-1)(p+2)\left(\sum_{i=1}^{k} \frac{n^2}{n_i^2} - 1\right) - 6(k-1)n^2(1-\rho)^2\right\}. \end{aligned}$$

High-Dimensional Case. We consider the distribution of

$$W = (\det \mathbf{V})^{-n/2} \prod_{i=1}^{r} (\det \mathbf{V}_i)^{n_i/2}, \tag{8.5.6}$$

whose hth moment is given (see Theorem 8.5.2) by

$$\mathrm{E}[W^h] = \prod_{j=1}^{p} \left\{ \frac{\Gamma[\frac{1}{2}(n-p+j)]}{\Gamma[\frac{1}{2}(n-p+j)+h]} \prod_{i=1}^{k} \frac{\Gamma[\frac{1}{2}(n_i-p+j)+h]}{\Gamma[\frac{1}{2}(n_i-p+j)]} \right\}.$$

Hence the characteristic function and the sth cumulant of $T = -\log W$ are given by

$$C_T(t) = \mathrm{E}[W^{-it}] = \prod_{j=1}^{p} \left\{ \frac{\Gamma[\frac{1}{2}(n-p+j)]}{\Gamma[\frac{1}{2}(n-p+j)-it]} \prod_{i=1}^{k} \frac{\Gamma[\frac{1}{2}(n_i-p+j)-it]}{\Gamma[\frac{1}{2}(n_i-p+j)]} \right\}$$

and

$$\begin{aligned}\kappa_T^{(s)} &= (-i)^{-s} \left.\frac{\partial^s}{\partial t^s} \log C_T(t)\right|_{t=0} \\ &= (-1)^s \sum_{j=1}^{p} \left\{ -\psi^{(s-1)}\left(\frac{1}{2}(n-p+j)\right) + \sum_{i=1}^{r} \psi^{(s-1)}\left(\frac{1}{2}(n_i-p+j)\right) \right\},\end{aligned} \tag{8.5.7}$$

respectively, where $\psi^{(s)}$ is the polygamma function.

Suppose that

$$p/n_i \to c_i \in (0,1), \quad i = 1, \ldots, k. \tag{8.5.8}$$

Let T and $\kappa_T^{(s)}$ be normalized as

$$\tilde{T} = \frac{T - \kappa_T^{(1)}}{(\kappa_T^{(2)})^{1/2}}, \quad \tilde{\kappa}^{(s)} = \frac{\kappa_T^{(s)}}{(\kappa_T^{(2)})^{s/2}}, \quad s = 3, 4, \ldots. \tag{8.5.9}$$

Theorem 8.5.4 *Let W be the modified LR statistic in (8.5.6), and let $\tilde{T}$ be the standardized statistic of $T = -\log W$ given by (8.5.9). Then under (8.5.8) the distribution function of $\tilde{T}$ is expanded as*

$$\Pr\{\tilde{T} \le x\} = Q_s(x) + \mathrm{O}_{s+1}^*, \tag{8.5.10}$$

where Q_s is given by (8.1.19), that is,

$$Q_s(x) = \Phi(x) - \varphi(x)\left\{ \sum_{k=1}^{s} \frac{1}{k!} \sum_{j=0}^{s-k} \gamma_{k,j} H_{3k+j-1}(x) \right\},$$

with

$$\gamma_{k,j} = \sum_{s_1+\cdots+s_k=j} \frac{\tilde{\kappa}_T^{(s_1+3)} \cdots \tilde{\kappa}_T^{(s_k+3)}}{(s_1+3)! \cdots (s_k+3)!}.$$

A rigorous proof of the error term in (8.5.10) was given by Wakaki (2009). He also obtained a computable error bound.

8.5.3 High-Dimensional Tests

Since the sample covariance matrix $\mathbf{S}_i$ is singular if $n_i < p$, the LR test is valid only if $n_i \ge p$ for $i = 1, \ldots, k$. An alternative test is based on the Wald statistic:

$$T = \frac{n}{2} \sum_{i \neq j}^{k} \operatorname{tr} \left\{ \left(\frac{1}{n_i} \mathbf{S}_i - \frac{1}{n_j} \mathbf{S}_j \right) \left(\frac{1}{n} \mathbf{S} \right)^{-1} \right\}^2$$

$$= \frac{n}{2} \left\{ \sum_{i=1}^{k} \frac{n_i}{n} \operatorname{tr} \mathbf{S}_i \mathbf{S}^{-1} \mathbf{S}_i \mathbf{S}^{-1} - \sum_{i \neq j}^{k} \frac{n_i n_j}{n^2} \operatorname{tr} \mathbf{S}_i \mathbf{S}^{-1} \mathbf{S}_j \mathbf{S}^{-1} \right\}.$$

This statistic is valid as long as $\mathbf{S}$ is nonsingular (i.e., as long as $n \geq p$). It is known [Schott, 2001] that if p is fixed, the asymptotic null distribution of T, as $n_i \to \infty$ for $i = 1, \ldots, k$, is $\chi^2((k-1)p(p+1)/2)$.

As a test statistic for high-dimensional settings, Schott (2007) proposed

$$\begin{aligned} T_{np} = \sum_{i<j} \big(& \operatorname{tr} \{(\mathbf{S}_i - \mathbf{S}_j)^2\} \\ & - (n_i \eta_j)^{-1} [n_i(n_i - 2) \operatorname{tr} (\mathbf{S}_i^2) + n_i^2 \{ \operatorname{tr} (\mathbf{S}_i)\}^2] \\ & - (n_j \eta_j)^{-1} [n_j(n_j - 2) \operatorname{tr} (\mathbf{S}_j^2) + n_j^2 \{ \operatorname{tr} (\mathbf{S}_j)\}^2] \big), \end{aligned} \tag{8.5.11}$$

where $\eta_j = (n_i + 2)(n_i - 1)$. This test was proposed as an unbiased estimator for $\sum_{i \neq j} \operatorname{tr} (\Sigma_i - \Sigma_j)^2$.

As a framework for high-dimensional asymptotics, assume that:

$$\begin{aligned} & (1)\ p_i/n \to b_i \in [0, \infty) \\ & \quad \text{and at least one } b_i > 0. \\ & (2)\ \lim p^{-1} \operatorname{tr} \Sigma^i = \gamma_i \in (0, \infty) \quad \text{for } i = 1, \ldots, 8. \end{aligned} \tag{8.5.12}$$

The asymptotic normality of T_{np} was established [Schott, 2007] by using the martingale limit theorem in Section 5.9.

Theorem 8.5.5 *Suppose that sample covariance matrices $\mathbf{S}_1, \ldots, \mathbf{S}_k$ are to be obtained from a set of random samples taken independently from multivariate normal distributions with covariance matrices $\Sigma_1, \ldots, \Sigma_q$, respectively. If $\Sigma_1 = \cdots = \Sigma_q$ and if conditions* (8.5.12) *are satisfied, then T_{np} converges in distribution to* $\mathrm{N}(0, \theta^2)$, *where*

$$\theta^2 = 4 \left[\sum_{i<j} (b_i + b_j)^2 + (k-1)(k-2) \sum_{i=1}^{k} b_i^2 \right] \gamma_2^2.$$

Problems

8.1 Let $\mathbf{W}$ be distributed as $\mathrm{W}(n, \Sigma_I)$, where $\Sigma_I = \sigma^2\{(1-\rho)\mathrm{I}_p + \rho \mathbf{1}_p \mathbf{1}_p'\}$. $U = (1/p)\mathbf{1}_p' \mathbf{W} \mathbf{1}_p$ and $V = \operatorname{tr} \mathbf{W} - U$. Show that U and V are independently distributed and $U/\tau_1 \sim \chi^2(n)$ and $V/\tau_2 \sim \chi^2((p-1)n))$,where $\tau_1 = \sigma^2\{1+(p-1)\rho\}$, $\tau_2 = \sigma^2(1-\rho)$.

8.2 Consider a mixed-effect ANOVA and

$$Y_{ij} = \mu + \delta_i + \beta_j + \epsilon_{ij}, \quad i = 1, \ldots, n, \quad j = 1, \ldots, p,$$

where $\delta_1, \ldots, \delta_n$ are independent and identically distributed as $\mathrm{N}(0, \sigma_\delta^2)$, the ϵ_{ij}'s are independent and identically distributed as $\mathrm{N}(0, \sigma^2)$, and the δ_i and the ϵ_{ij} are independent. The μ and β_j are unknown parameters satisfying $\sum_{j=1}^p \beta_j = 0$. Let $\boldsymbol{Y}_i = (Y_{i1}, \ldots, Y_{ip})'$, $i = 1, \ldots, n$. Show that

$$\begin{aligned} \mathrm{E}(\boldsymbol{Y}_i) &= \mu \mathbf{1}_p + \boldsymbol{\beta} \equiv \boldsymbol{\eta}, \quad \boldsymbol{\beta} = (\beta_1, \ldots, \beta_p)', \\ \mathrm{Var}(\boldsymbol{Y}_i) &= \sigma_\delta^2 \mathrm{G}_p + \sigma^2 \mathrm{I}_p \\ &= \sigma^2 \left(\mathrm{I}_p - \frac{1}{p}\mathrm{G}_p \right) + \tau^2 \frac{1}{p} \mathrm{G}_p \equiv \Sigma_s, \end{aligned}$$

where $\tau^2 = \sigma^2 + p\sigma_\delta^2$ and G_p is a $p \times p$ matrix whose elements are all 1. Further, show that $\boldsymbol{Y}_1, \ldots, \boldsymbol{Y}_n$ are independent and identically distributed as $\mathrm{N}(\boldsymbol{\eta}, \Sigma_s)$.

8.3 (Continuation) Let $\mathrm{H} = (\boldsymbol{h}_1, \mathrm{H}_2)$ be an orthogonal matrix such that $\boldsymbol{h}_1 = (1/\sqrt{p})\mathbf{1}_p$. Consider the transformation

$$\boldsymbol{X}_i = \mathrm{H}_2' \boldsymbol{Y}_i, \quad i = 1, \ldots, n.$$

Then the random vectors $\boldsymbol{X}_1, \ldots, \boldsymbol{X}_n$ are independent and normally distributed $(\mathrm{N}(\boldsymbol{\xi}, \sigma^2 \mathrm{I}_{p-1})$, with $\boldsymbol{\xi} = \mathrm{H}_2'\boldsymbol{\eta})$. Using this propertym, construct a test for $\Sigma = \Sigma_s$.

Hint: Use a test for sphericity.

8.4 Let λ^* be the LR statistic for independence given by

$$\lambda^* = |\mathbf{R}\mathbf{D}_{\mathbf{R}}^{-1}| = |\mathbf{R}| \Big/ \prod_{i=1}^k |\mathbf{R}_{ii}|,$$

where $\mathbf{R}_{ii}$ $(i = 1, \ldots, k)$ is the submatrix of $\mathbf{R}$ corresponding to $\mathbf{V}_{ii}$. Examine the following special cases:

(a) $p_1 = p_2 = \cdots = p_k = 1$; $\quad \lambda^* = |\mathbf{V}|/(V_{11}V_{22}\cdots V_{pp}) = |\mathbf{R}|$.

(b) $k = 2$, $p_1 = 1$, $p_2 = p - 1$; $\quad \lambda^* = |\mathbf{V}|/(V_{11}|\mathbf{V}_{22}|) = 1 - r_{1(2,\ldots,p)}^2$,

where $r_{1(2,\ldots,p)}$ stands for the multiple correlation coefficient between X_1 and $(X_2, \ldots, X_p)$.

(c) $k = 2$, $p_1 + p_2 = p$;

$$\lambda^* = \frac{|\mathbf{V}|}{|\mathbf{V}_{11}||\mathbf{V}_{22}|} = \frac{|\mathbf{V}_{11} - \mathbf{V}_{12}\mathbf{V}_{22}^{-1}\mathbf{V}_{21}|}{|\mathbf{V}_{11}|} = \frac{|\mathbf{V}_{22} - \mathbf{V}_{21}\mathbf{V}_{11}^{-1}\mathbf{V}_{12}|}{|\mathbf{V}_{22}|}$$
$$= \prod_{j=1}^{s}(1 - \ell_j^2), \quad s = \min\{p_1, p_2\}.$$

Numbers ℓ_j, $j = 1, \ldots, s$ are called the *sample canonical correlation coefficients* between $(X_1, \ldots, X_{p_1})$ and $(X_{p_1+1}, \ldots, X_p)$. For details, see Chapter 11.

8.5 Show that the modified LR statistic W when $k = 2$ can be reduced to

$$W = \frac{|\mathbf{V}_1|^{n_1/2}|\mathbf{V}_2|^{n_2/2}}{|\mathbf{V}_1 + \mathbf{V}_2|^{(n_1+n_2)/2}} = \prod_{i=1}^{p} \ell_i^{n_1/2} \prod_{i=1}^{p}(1 + \ell_i)^{-(n_1+n_2)/2},$$

where $\ell_1 \geq \cdots \geq \ell_p > 0$, are ordered roots of the equation

$$|\mathbf{V}_1 - \ell\mathbf{V}_2| = 0.$$

Chapter 9

Discriminant Analysis

Discriminant analysis is used in situations where the populations or groups are known a priori. One of the aims of discriminant analysis is to classify an observation, or several observations, into these known populations. In this case we know that it or they have come from exactly one of those populations but we do not know from which. The other aim is to interpret the differences between the populations, in terms of a few linear functions of the original variables. In the following we use *classification* instead of *discrimination* although Kendall (1966) has extensive notes on the terminology.

Significance tests for discriminant functions and the evaluation of probabilities of misclassifications are discussed. Further, problems of selecting the canonical discriminant variables as well as the original variables are considered based on model selection criteria.

9.1 Classification Rules for Known Distributions

In discriminant analysis we have populations Π_i, $i = 1, \ldots, q$, and we need to allocate an observation $\boldsymbol{x}$ of $\boldsymbol{X} = (X_1, \ldots, X_p)'$ to one of these populations. We assume that for each i, the density $f_i(\boldsymbol{x})$ of $\boldsymbol{X}$ in population Π_i is known and we write $\boldsymbol{X} \in \Pi_i$ to mean that $\boldsymbol{X}$ has a distribution that characterizes the population Π_i. First, we give some concepts and notations. The mean vectors and the covariance matrices of $\boldsymbol{X}$ when $\boldsymbol{X} \in \Pi_i$ are

$$\mathrm{E}(\boldsymbol{X}|\Pi_i) = \boldsymbol{\mu}_i, \quad \mathrm{Var}(\boldsymbol{X}|\Pi_i) = \Sigma_i.$$

When $\boldsymbol{x} \in \Pi_i$, the distance D_i (≥ 0) between $\boldsymbol{x}$ and the center $\boldsymbol{\mu}_i$ of Π_i is defined by

$$D_i^2 = (\boldsymbol{x} - \boldsymbol{\mu}_i)'\Sigma_i^{-1}(\boldsymbol{x} - \boldsymbol{\mu}_i).$$

If two populations Π_i and Π_j have an identical variance, $\Sigma \equiv \Sigma_i = \Sigma_j$, say, the distance $\Delta(i,j)$ (≥ 0) between them is defined by

$$\Delta^2(i,j) = (\boldsymbol{\mu}_i - \boldsymbol{\mu}_j)'\Sigma^{-1}(\boldsymbol{\mu}_i - \boldsymbol{\mu}_j).$$

These distances are called *Mahalanobis distances.* Let ρ_i be the prior probability of populations Π_i, where *prior* means the a priori probability that an individual selected at random belongs to Π_i.

In general, a classification rule is expressed as a separation of the sample space (p-dimensional Euclidean space) R^p into sets R_j such that for $\boldsymbol{x} \in R_i$, it is identified as a member of Π_i; that is,

$$\boldsymbol{x} \in R_i \implies \boldsymbol{x} \in \Pi_i.$$

By classifying the observation into a certain population we may encounter a misclassification (or misdiscrimination). It is important to evaluate the probability of misclassification. For $q = 2$ populations, let R^p partition into R_1 and R_2. Then the probability of putting $\boldsymbol{X}$ into Π_2 although it is known from Π_1 is expressed as

$$\mathrm{P}(2|1) = \Pr\{\boldsymbol{X} \in R_2|\Pi_1\} = \int_{R_2} f_1(\boldsymbol{x})\,\mathrm{d}\boldsymbol{x}.$$

Similarly, the probability of classifying an object as belonging to Π_1 although it actually comes from Π_2 is

$$\mathrm{P}(1|2) = \Pr\{\boldsymbol{X} \in R_1|\Pi_2\} = \int_{A_1} f_2(\boldsymbol{x})\,\mathrm{d}\boldsymbol{x}.$$

The expected probability of misclassification is

$$\rho_1\mathrm{P}(2|1) + \rho_2\mathrm{P}(1|2).$$

Now we give several classification (discrimination) rules when some or all of the mean vectors $\boldsymbol{\mu}_i$, covariance matrices Σ_i, density functions f_i, or prior probabilities ρ_i are known. Discriminant analysis has been discussed in various text books on multivariate analysis. McLachlan (1991) discusses various topics in details.

Classification Rule Based on Mahalanobis Distance. Let D_i be the Mahalanobis distance between $\boldsymbol{x}$ and Π_i or the Mahalanobis distance from $\boldsymbol{x}$ to the center of Π_i. Then the classification rule is

$$\min_{j=1,\ldots,q} D_j^2 = D_i^2 \implies \boldsymbol{x} \in \Pi_i.$$

When $q = 2$, the rule is equivalent to

$$D_1^2 \leq D_2^2 \implies \boldsymbol{x} \in \Pi_1, \quad D_2^2 < D_1^2 \implies \boldsymbol{x} \in \Pi_2,$$

if $\Pr\{D_1^2 = D_2^2\} = 0$.

Maximum Likelihood Classification Rule. This rule is

$$\max_{j=1,\ldots,q} f_j(\boldsymbol{x}) = f_i(\boldsymbol{x}) \implies \boldsymbol{x} \in \Pi_i.$$

When $q = 2$, the rule becomes

$$f_1(\boldsymbol{x}) \geq f_2(\boldsymbol{x}) \implies \boldsymbol{x} \in \Pi_1; \quad f_1(\boldsymbol{x}) < f_2(\boldsymbol{x}) \implies \boldsymbol{x} \in \Pi_2,$$

if $\Pr\{f_1(\boldsymbol{X}) = f_2(\boldsymbol{X})\} = 0$.

Decision-Theoretic Approach. First we consider the case $q = 2$. Suppose that a discriminant rule is a separation of the entire p-dimensional vector space R^p into two regions, R_1 and R_2, in a certain way and make a rule such that if $\boldsymbol{x} \in R_i$, we assign $\boldsymbol{x}$ to $\Pi_i, i = 1, 2$. Assume that the misclassifications create a cost (loss) $c(i|j)$ when a Π_j observation is assigned to R_i, where $c(i|i) = 0$. The costs are summarized in Table 9.1.1. Then the expected cost

Table 9.1.1 Costs

		Classified population	
		Π_1	Π_2
True population	Π_1	0	$c(2\|1)$
	Π_2	$c(1\|2)$	0

of misclassification or risk is

$$EC = c(2|1)\rho_1 \mathrm{P}(2|1) + c(1|2)\rho_2 \mathrm{P}(1|2). \tag{9.1.1}$$

Theorem 9.1.1 *The classification rule that minimizes the risk* (9.1.1) *is given by*

$$\frac{f_1(\boldsymbol{x})}{f_2(\boldsymbol{x})} \geq k \implies \boldsymbol{x} \in \Pi_1, \quad \frac{f_1(\boldsymbol{x})}{f_2(\boldsymbol{x})} < k \implies \boldsymbol{x} \in \Pi_2, \tag{9.1.2}$$

or

$$R_1\colon \frac{f_1(\boldsymbol{x})}{f_2(\boldsymbol{x})} \geq k, \quad R_2\colon \frac{f_1(\boldsymbol{x})}{f_2(\boldsymbol{x})} < k, \tag{9.1.3}$$

where $k = [c(1|2)\rho_2]/[c(2|1)\rho_1]$. *If*

$$\Pr(c(2|1)\rho_1 f_1(\boldsymbol{x}) = c(1|2)\rho_2 f_2(\boldsymbol{x}) \mid \Pi_i) = 0, \quad i = 1, 2,$$

the rule is determined uniquely except for sets of probability zero.

Proof. The risk is expressed as

$$EC = \rho_1 c(2|1) \int_{R_2} f_1(\boldsymbol{x})\,\mathrm{d}\boldsymbol{x} + \rho_2 c(1|2) \int_{R_1} f_2(\boldsymbol{x})\,\mathrm{d}\boldsymbol{x}$$
$$= \rho_1 c(2|1) + \int_{R_1} \{\rho_2 c(1|2) f_2(\boldsymbol{x}) - \rho_1 c(2|1) f_1(\boldsymbol{x})\}\,\mathrm{d}\boldsymbol{x}.$$

Since the first term in this equation is constant, the minimum is obviously obtained for

$$R_1 = \{\boldsymbol{x} : \rho_2 c(1|2) f_2(\boldsymbol{x}) - \rho_1 c(2|1) f_1(\boldsymbol{x}) \le 0\},$$

which is the same as in (9.1.2) except for sets of probability zero. □

This rule is called a *Bayes rule* for given (ρ_1, ρ_2). If $\rho_1 = \rho_2$ and $c(2|1) = c(1|2)$, the Bayes rule is the same as the maximum likelihood classification method.

Example 9.1.1 Let Π_i be multivariate normal populations $\mathrm{N}_p(\boldsymbol{\mu}_i, \Sigma)$:

$$f_i(\boldsymbol{x}) = 2^{p/2}|\Sigma_i|^{-1/2} \exp\left\{-\frac{1}{2}(\boldsymbol{x} - \boldsymbol{\mu}_i)'\Sigma_i^{-1}(\boldsymbol{x} - \boldsymbol{\mu}_i)\right\}, \quad i = 1, 2.$$

Then

$$U(\boldsymbol{x}) = \log\{f_1(\boldsymbol{x})/f_2(\boldsymbol{x})\}$$
$$= \frac{1}{2}\boldsymbol{x}'(\Sigma_2^{-1} - \Sigma_1^{-1})\boldsymbol{x} + (\boldsymbol{\mu}_1'\Sigma_1^{-1} - \boldsymbol{\mu}_2'\Sigma_2^{-1})\boldsymbol{x}$$
$$+ \frac{1}{2}(\boldsymbol{\mu}_2'\Sigma_2^{-1}\boldsymbol{\mu}_2 - \boldsymbol{\mu}_1'\Sigma_1^{-1}\boldsymbol{\mu}_1) + \frac{1}{2}\log\{|\Sigma_2|/|\Sigma_1|\}.$$

By Theorem 9.1.1, Bayes' rule for a priori probability (ρ_1, ρ_2) is given by

$$R_1 : U(\boldsymbol{x}) \ge c, \quad R_2 : U(\boldsymbol{x}) < c, \tag{9.1.4}$$

where $c = \log k = \log[\rho_2 c(1|2)/\{\rho_1 c(2|1)\}]$.

When $\Sigma_1 = \Sigma_2$, $U(\boldsymbol{x})$ reduces to

$$U(\boldsymbol{x}) = (\boldsymbol{\mu}_1 - \boldsymbol{\mu}_2)'\Sigma^{-1}\left\{\boldsymbol{x} - \frac{1}{2}(\boldsymbol{\mu}_1 + \boldsymbol{\mu}_2)\right\}, \tag{9.1.5}$$

which is known as *Fisher's population discriminant function.*

Next we extend the decision-theoretic approach for two populations to that for q populations. Let $\Pi_1, \ldots, \Pi_q$ be q populations with density functions $f_1(\boldsymbol{x}), \ldots, f_q(\boldsymbol{x})$, respectively. We wish to divide the sample space R^p into

q mutually exclusive and exhaustive regions $R_1, \ldots, R_q$. If an observation falls into R_i, we say that it comes from Π_i. Let $c(i|j)$ be a loss when a Π_j observation is assigned to R_i, and $c(i|i) = 0$. Then the risk is

$$\sum_{i=1}^{q} \sum_{j \neq i} c(j|i) \rho_i p(j|i). \tag{9.1.6}$$

Theorem 9.1.2 *The classification rule that minimizes the risk* (9.1.6) *is given by*

$$\min_{j=1,\ldots,q} h_j(\boldsymbol{x}) = h_i(\boldsymbol{x}) \implies \boldsymbol{x} \in \Pi_i,$$

where

$$h_j(\boldsymbol{x}) = \sum_{k=1, k \neq j}^{q} \rho_k c(j|k) f_k(\boldsymbol{x}), \quad j = 1, \ldots, q.$$

Proof. Let the R and R^* classification rules be defined by partitions: $\mathrm{R}^p = R_1 + \cdots + R_q$ and $\mathrm{R}^p = R_1^* + \cdots + R_q^*$, respectively, where R^* is the classification rule described in the theorem. The risk (9.1.6) for a classification rule R is written as

$$\begin{aligned} EC_R &= \sum_{i=1}^{q} \sum_{j \neq i} c(j|i) \rho_i \int_{R_j} f_i(\boldsymbol{x}) \, \mathrm{d}\boldsymbol{x} = \sum_{i=1}^{q} \int_{R_i} h_i(\boldsymbol{x}) \, \mathrm{d}\boldsymbol{x} \\ &= \int_{R_i} h(\boldsymbol{x}|R) \, \mathrm{d}\boldsymbol{x}, \end{aligned}$$

where $h(\boldsymbol{x}|R) = h_i(\boldsymbol{x})$ for $\boldsymbol{x}$ in R_i. Similarly, the risk for R^* is written as

$$EC_{R^*} = \int h(\boldsymbol{x}|R^*) \, \mathrm{d}\boldsymbol{x},$$

where $h(\boldsymbol{x}|R^*) = \min_i h_i(\boldsymbol{x})$. Thus

$$\begin{aligned} EC_R - EC_{R^*} &= \int [h(\boldsymbol{x}|R) - h(\boldsymbol{x}|R^*)] \, \mathrm{d}\boldsymbol{x} \\ &= \sum_{i=1} \int_{R_i} [h_i(\boldsymbol{x}) - \min h_j(\boldsymbol{x})] \, \mathrm{d}\boldsymbol{x} \geq 0. \qquad \square \end{aligned}$$

When $\rho_1 = \cdots = \rho_q$ and $c(i|j) \equiv c$, for $i \neq j$, Bayes' rule is reduced to the maximum likelihood classification rule.

Example 9.1.2 As in Example 9.1.2, let Π_i be multivariate normal populations $\mathrm{N}_p(\boldsymbol{\mu}_i, \Sigma)$, $i = 1, \ldots, q$. Assume that the costs of misclassification are equal. Then

$$h_i(\boldsymbol{x}) \leq h_j(\boldsymbol{x}) \Leftrightarrow U_{ij}(\boldsymbol{x}) \equiv \log \frac{f_i(\boldsymbol{x})}{f_j(\boldsymbol{x})} > \log \frac{\rho_j}{\rho_i}$$

and

$$U_{ij}(\boldsymbol{x}) = (\boldsymbol{\mu}^{(i)} - \boldsymbol{\mu}^{(j)})'\Sigma^{-1}\Big[\boldsymbol{x} - \frac{1}{2}(\boldsymbol{\mu}^{(i)} + \boldsymbol{\mu}^{(j)})\Big]. \tag{9.1.7}$$

We can write the region R_j^* in the Bayes procedure as

$$\begin{aligned} R_i^* &= \{\boldsymbol{x}\colon h_i(\boldsymbol{x}) \geq h_j(\boldsymbol{x}), \quad j = 1, \ldots, q, \quad j \neq i\} \\ &= \{\boldsymbol{x}\colon U_{ij} \geq \log(\rho_j/\rho_i), \quad j = 1, \ldots, q, \quad j \neq i\}, \end{aligned}$$

and hence

$$R_i^* : U_{ij}(\boldsymbol{x}) > \log\frac{\rho_j}{\rho_i}, \quad j = 1, \ldots, q, \quad j \neq i.$$

Bayes' Method. We consider a classification of q groups. Let $Y \in \{1, 2, \ldots, q\}$ be a dependent (population label) variable associated with each object or observation. Suppose that the group conditional population densities $f_i(\boldsymbol{x}) = f(\boldsymbol{x}|Y = i)$ and the population prior ρ_i are unknown. Then Bayes' theorem may be used to express the posterior probability $f(i|\boldsymbol{x})$ of population i given observation vector $\boldsymbol{x}$. In fact, if $\boldsymbol{X}$ is discrete,

$$\begin{aligned} \Pr\{Y = i|\boldsymbol{X} = \boldsymbol{x}_0\} &= \frac{\Pr\{Y = i, \boldsymbol{X} = \boldsymbol{x}_0\}}{\Pr\{\boldsymbol{X} = \boldsymbol{x}_0\}} \\ &= \frac{\Pr\{Y = i\}\Pr\{\boldsymbol{X} = \boldsymbol{x}_0|Y = i\}}{\sum_{\ell=1}^{q} \Pr\{Y = \ell\}\Pr\{\boldsymbol{X} = \boldsymbol{x}_0|Y = \ell\}} \end{aligned}$$

and obtain

$$f(i|\mathbf{x}) = \frac{\rho_i f_i(\mathbf{x})}{\sum_{\ell=1}^{q} \rho_\ell f_\ell(\mathbf{x})}.$$

If the distribution of $\boldsymbol{X}$ is continuous, it is sufficient to replace $\boldsymbol{x}_0$ by a δ neighborhood $N_\delta(\boldsymbol{x}_0)$ of $\boldsymbol{x}_0$ and to tend $\delta \to 0$. Bayes' rule predicts the population of an observation vector $\boldsymbol{x}$ by that with the highest posterior probability:

$$\max_j \rho_j f_j(\boldsymbol{x}) = \rho_i f_i(\boldsymbol{x}) \implies \boldsymbol{x} \in G_i.$$

Fisher's Approach. When the Π_i have the same covariance matrix Σ, the (sample) linear discriminant functions $\boldsymbol{b}_i'\boldsymbol{X}$, $i = 1, \ldots, m = \min\{p, q-1\}$ that were introdued by Fisher (1936) achieve a good separation. Such functions were found by considering the linear combination $\boldsymbol{b}'\boldsymbol{X}$ that that maximizes the ratio of the between-group sum of squares to the within-group sum of squares. Statistical problems as well as this method are discussed in Section 9.4.

We consider Fisher's approach when the population mean vectors $\boldsymbol{\mu}_i$ and the population covariance matrix Σ and the prior probabilities ρ_i's are known. Let Z be a random variable such that

$$\mathrm{E}(Z|\Pi_i) = \xi_i, \quad \mathrm{Var}(Z|\Pi_i) = \sigma^2, \quad i = 1, \ldots, q.$$

Then we may define the population between-group and within-group variations as

$$\sigma_b^2 = \sum_{i=1}^{q} \rho_i(\xi_i - \bar{\xi})^2 \quad \text{and} \quad \sigma_w^2 = \sum_{i=1}^{q} \rho_i \sigma^2 = \sigma^2,$$

respectively, where $\bar{\xi} = \rho_1\xi_1 + \cdots + \rho_q\xi_q$. Consider a linear combination

$$\boldsymbol{\beta}'\boldsymbol{X} = \beta_1 X_1 + \cdots + \beta_p X_p.$$

Then the variations σ_b^2 and σ_w^2 of $\boldsymbol{\beta}'\boldsymbol{X}$ are expressed as $\sigma_b^2 = \boldsymbol{\beta}'\Sigma_b\boldsymbol{\beta}$ and $\sigma_w^2 = \boldsymbol{\beta}'\Sigma\boldsymbol{\beta}$, respectively, where

$$\Sigma_b = \sum_{i=1}^{q} \rho_i(\boldsymbol{\mu}_i - \bar{\boldsymbol{\mu}})'(\boldsymbol{\mu}_i - \bar{\boldsymbol{\mu}}) \equiv \Psi, \tag{9.1.8}$$

and $\bar{\boldsymbol{\mu}} = \rho_1\boldsymbol{\mu}_1 + \cdots + \rho_q\boldsymbol{\mu}_q$. Fisher's idea is to seek a coefficient vector $\boldsymbol{\beta}$ that maximizes the ratio

$$\frac{\sigma_b^2}{\sigma_w^2} = \frac{\boldsymbol{\beta}'\Psi\boldsymbol{\beta}}{\boldsymbol{\beta}'\Sigma\boldsymbol{\beta}}. \tag{9.1.9}$$

The solution is bound by applying [A.2.5]. Let the characteristic roots of $\Psi\Sigma^{-1}$ be denoted by $\lambda_1 \geq \cdots \geq \lambda_m \geq \lambda_{m+1} = \cdots = \lambda_p = 0$. The corresponding characteristic vectors are denoted by $\boldsymbol{\beta}_1, \ldots, \boldsymbol{\beta}_p$ with normalizations $\boldsymbol{\beta}_i'\Sigma\boldsymbol{\beta}_i = 1$, $i = 1, \ldots, q$. They are the solutions of

$$\Psi\boldsymbol{\beta}_i = \lambda_i\Sigma\boldsymbol{\beta}_i, \quad \boldsymbol{\beta}_i'\Sigma\boldsymbol{\beta}_i = \delta_{ij}. \tag{9.1.10}$$

The coefficient vector $\boldsymbol{\beta}$ that maximizes (9.1.9) is $\boldsymbol{\beta}_1$. In general, the functions $Z_i = \boldsymbol{\beta}_i'\boldsymbol{X}$, $i = 1, \ldots, m$ are called *Fisher's population discriminant functions.* The number m of nonzero characteristic roots λ_i is called the *dimensionality in discriminant analysis.*

When $q = 2$,

$$\Psi = \rho_1\rho_2(\boldsymbol{\mu}_1 - \boldsymbol{\mu}_2)(\boldsymbol{\mu}_1 - \boldsymbol{\mu}_2)'.$$

It is easily found that the coefficient vector $\boldsymbol{\beta}_1$ is proportional to $\Sigma^{-1}(\boldsymbol{\mu}_1 - \boldsymbol{\mu}_2)$. We use the linear combination $Z = (\boldsymbol{\mu}_1 - \boldsymbol{\mu}_2)'\Sigma^{-1}\boldsymbol{Z}$ as a discriminator. This is the same as in (9.1.5) except for the constant term. For discriminant procedures based on the Z_i, see Section 9.4.

9.2 Sample Classification Rules for Normal Populations

If the populations Π_i, $i = 1, \ldots, q$, are completely known, we can use the optimum classification rule described in Section 9.1. If, as is usually the case, the parameters are not known, the rules are devised on the basis of available samples from each population, mostly by the plug-in method or maximum likelihood method. We give some classification rules of p-variate normal populations $\Pi_i : \mathrm{N}_p(\boldsymbol{\mu}_q, \Sigma_i)$.

In this section we use the following notation. Let the n_i samples from Π_i be denoted by

$$\boldsymbol{X}_{i1}, \ldots, \boldsymbol{X}_{in_i}, \quad i = 1, \ldots, q. \tag{9.2.1}$$

The sample mean vectors and the sample covariance matrices are denoted by

$$\bar{\boldsymbol{X}}_i = \frac{1}{n_i}\sum_{j=1}^{n_i} \boldsymbol{X}_{ij}, \quad \mathbf{S}_i = \frac{1}{n_i - 1}\sum_{j=1}^{n_i}(\boldsymbol{X}_{ij} - \bar{\boldsymbol{X}}_i)(\boldsymbol{X}_{ij} - \bar{\boldsymbol{X}}_i)', \quad i = 1, \ldots, q.$$

Further, let $\bar{\boldsymbol{X}}$ and $\mathbf{S}$ be the overall mean vector and the pooled sample covariance matrix defined by

$$\bar{\boldsymbol{X}} = \frac{1}{n}\sum_{i=1}^{q} n_i \bar{\boldsymbol{X}}_i, \quad \mathbf{S} = \frac{1}{n-q}\sum_{i=1}^{q}(n_i - 1)\mathbf{S}_i,$$

where $n = n_1 + \cdots + n_q$.

9.2.1 Two Normal Populations with $\Sigma_1 = \Sigma_2$

Plug-in Rule. When the parameters are known, we have the rules based on the linear discriminant function

$$\begin{aligned} U(\boldsymbol{X}) &= (\boldsymbol{\mu}_1 - \boldsymbol{\mu}_2)\Sigma^{-1}\Big[\boldsymbol{X} - \frac{1}{2}(\boldsymbol{\mu}_1 + \boldsymbol{\mu}_2)\Big] \\ &= \beta_0 + \boldsymbol{\beta}'\boldsymbol{X}, \end{aligned} \tag{9.2.2}$$

where

$$\beta_0 = -\frac{1}{2}(\boldsymbol{\mu}_1 - \boldsymbol{\mu}_2)'\Sigma^{-1}(\boldsymbol{\mu}_1 + \boldsymbol{\mu}_2), \ \boldsymbol{\beta} = \Sigma^{-1}(\boldsymbol{\mu}_1 - \boldsymbol{\mu}_2).$$

Replacing the unknown parameters $(\boldsymbol{\mu}_1, \boldsymbol{\mu}_2, \Sigma)$ by $(\bar{\boldsymbol{X}}_1, \bar{\boldsymbol{X}}_1, \mathbf{S})$, we have the sample linear discriminant function [Wald, 1944: Anderson, 1951]

$$W(\boldsymbol{X}) = (\bar{\boldsymbol{X}}_1 - \bar{\boldsymbol{X}}_2)'\mathbf{S}^{-1}\left[\boldsymbol{X} - \frac{1}{2}(\bar{\boldsymbol{X}}_1 + \bar{\boldsymbol{X}}_2)\right]. \tag{9.2.3}$$

The discriminant rule based on $W(\boldsymbol{x})$ is defined by

$$\text{allocate } \boldsymbol{x} \text{ to } \Pi_1 \text{ if } W(\boldsymbol{x}) \geq c \text{ and to } \Pi_2 \text{ if } W(\boldsymbol{x}) < c, \tag{9.2.4}$$

where c is a cutoff point. A commonly used cutoff point is $c = 0$.

LR or ML Rule. Next we consider the LR criterion for testing the null hypothesis H: $\boldsymbol{X} \in \Pi_1$ vs. K: $\boldsymbol{X} \in \Pi_2$. It is easy to see that under H, the MLEs of $\boldsymbol{\mu}_1$, $\boldsymbol{\mu}_2$, and Σ are

$$\hat{\boldsymbol{\mu}}_1 = (n_1\bar{\boldsymbol{X}}_1 + \boldsymbol{X})/(n_1+1), \quad \hat{\boldsymbol{\mu}}_2 = \bar{\boldsymbol{X}}_2,$$

$$\begin{aligned}\hat{\Sigma} &= \frac{1}{n_1+n_2+1}\left\{\sum_{i=1}^{n_1}(\boldsymbol{X}_{1i} - \hat{\boldsymbol{\mu}}_1)(\boldsymbol{X}_{1i} - \hat{\boldsymbol{\mu}}_1)' + (\boldsymbol{X} - \hat{\boldsymbol{\mu}}_1)(\boldsymbol{X} - \hat{\boldsymbol{\mu}}_1)'\right.\\ &\qquad\qquad \left. + \sum_{i=1}^{n_2}(\boldsymbol{X}_{2i} - \hat{\boldsymbol{\mu}}_2)(\boldsymbol{X}_{2i} - \hat{\boldsymbol{\mu}}_2)'\right\}\\ &= \frac{1}{n_1+n_2+1}[mS + \{n_1/(n_1+1)\}(\boldsymbol{X} - \bar{\boldsymbol{x}}_1)(\boldsymbol{X} - \bar{\boldsymbol{X}}_1)'].\end{aligned}$$

Similarly, the MLEs of $\boldsymbol{\mu}_1$, $\boldsymbol{\mu}_2$, and Σ under K are, respectively,

$$\tilde{\boldsymbol{\mu}}_1 = \bar{\boldsymbol{X}}_1, \quad \tilde{\boldsymbol{\mu}}_2 = \{1/(n_2+1)\}(n_2\bar{\boldsymbol{X}}_2 + \boldsymbol{X}),$$

$$\tilde{\Sigma} = \frac{1}{n_1+n_2+1}[mS + \{n_2/(n_2+1)\}(\boldsymbol{X} - \bar{\boldsymbol{X}}_2)(\boldsymbol{X} - \bar{\boldsymbol{X}}_2)'].$$

Hence the LR criterion is the $[(n_1+n_2+1)/2]$th power of

$$\begin{aligned}\frac{|\tilde{\Sigma}|}{|\hat{\Sigma}|} &= \frac{|m\mathbf{S} + \{n_2/(n_2+1)\}(\boldsymbol{X} - \bar{\boldsymbol{X}}_2)(\boldsymbol{X} - \bar{\boldsymbol{X}}_2)'|}{|m\mathbf{S} + \{n_1/(n_1+1)\}(\boldsymbol{X} - \bar{\boldsymbol{X}}_1)(\boldsymbol{X} - \bar{\boldsymbol{X}}_1)'|}\\ &= \frac{m + d_2(\boldsymbol{X})}{m + d_1(\boldsymbol{X})},\end{aligned}$$

where $m = n_1 + n_2 - 2$,

$$d_i(\boldsymbol{X}) = \{n_i/(n_i+1)\}(\boldsymbol{X} - \bar{\boldsymbol{X}}_i)'\mathbf{S}^{-1}(\boldsymbol{X} - \bar{\boldsymbol{X}}_i), \quad i = 1, 2.$$

Thus the LR rule is given by

assign $\boldsymbol{x}$ to Π_1 if $(m + d_2(\boldsymbol{x}))/(m + d_1(\boldsymbol{x})) \geq c$, and to Π_2 otherwise.

If we take $c = 1$, the rule becomes

assign $\boldsymbol{x}$ to Π_1 if $d_2(\boldsymbol{x})/d_1(\boldsymbol{x}) \geq c$, and to Π_2 otherwise.

This is called the *maximum likelihood* (ML) *rule* in the case of two p-variate normal populations.

Let

$$\begin{aligned}Z(\boldsymbol{x}) &= d_1(\boldsymbol{x}) - d_2(\boldsymbol{x})\\ &= \frac{n_1}{n_1+1}(\boldsymbol{x} - \bar{\boldsymbol{x}}_1)'S^{-1}(\boldsymbol{x} - \bar{\boldsymbol{x}}_1) - \frac{n_2}{n_2+1}(\boldsymbol{x} - \bar{\boldsymbol{x}}_2)'S^{-1}(\boldsymbol{x} - \bar{\boldsymbol{x}}_2).\end{aligned}$$

Then the ML rule is written as

$$\text{assign } \boldsymbol{x} \text{ to } \Pi_1 \text{ if } Z \geq 0, \text{ and to } \Pi_2 \text{ if } Z > 0.$$

Thus we call the ML rule the Z-rule for short. The criterion Z was introduced by Kudo (1959, 1960) and John (1960) as a competitor to Wald-Anderson's W-rule. It is obvious that when $n_1 = n_2 \equiv n$, $Z = -\{2n/(n+1)\}W$, and the Z-rule is also a sort of minimum-distance rule.

9.2.2 Case of Several Normal Populations

Let

$$W_{ij}(\boldsymbol{x}) = (\bar{\boldsymbol{x}}_i - \bar{\boldsymbol{x}}_j)'\Sigma^{-1}\Big[\boldsymbol{x} - \frac{1}{2}(\bar{\boldsymbol{x}}_i - \bar{\boldsymbol{x}}_j)\Big].$$

Then the plug-in rule corresponding to the rule of (9.1.7) is defined by

$$\text{assign } \boldsymbol{x} \text{ to } \Pi_i \text{ if } W_{ij}(\boldsymbol{x}) > 0 \text{ for all } j = 1, \ldots, k,\ j \neq i.$$

For two populations the ML or Z-rule is extended to the present case as follows: Let

$$d_i(\boldsymbol{x}) = [n_i/(n_i + 1)](\boldsymbol{x} - \bar{\boldsymbol{x}}_j)'\Sigma^{-1}(\boldsymbol{x} - \bar{\boldsymbol{x}}_j), \quad i = 1, \ldots, k.$$

Then the rule is defined by

$$\text{assign } \boldsymbol{x} \text{ to } \Pi_i \text{ if } d_i(\boldsymbol{x}) = \min_j\{d_j(\boldsymbol{x})\},$$

which is also a minimum-distance rule taking account of different sample sizes.

9.3 Probability of Misclassifications

It is important to know the probability of misclassification (PM) of any proposed classification rule as a measure of goodness. In general, the probability of misclassification involves unknown parameters, and hence we also need to estimate them. In this section we consider the problems of the W and Z-rule given in Section 9.2. Siotani (1982) gave a review on large-sample approximations of these classification statistics.

The exact evaluation of these probabilities is very difficult, hence the evaluation here is based on asymptotic approximations under two asymptotic frameworks:

$$\text{A1}: \ n_1 \to \infty,\ n_2 \to \infty,\ n_1/n_2 \to r(> 0). \tag{9.3.1}$$

$$\text{A2}: \ n_i \to \infty,\ p/n_i \to r_i(0 < r_i < 1),\ i = 1, 2. \tag{9.3.2}$$

The first framework is called a *large-sample framework*. Under the second framework, we have

$$p \to \infty, \text{ and } n_i - p = n_i(1 - p/n_i) \to \infty.$$

The second framework is called a *high-dimensional framework*.

9.3.1 W-Rule

Let W be the linear discriminant function

$$W = (\bar{\boldsymbol{X}}_1 - \bar{\boldsymbol{X}}_2)'\mathbf{S}^{-1}\Big[\boldsymbol{X} - \frac{1}{2}(\bar{\boldsymbol{X}}_1 + \bar{\boldsymbol{X}}_2)\Big]$$

based on samples of size n_i from $\mathrm{N}(\boldsymbol{\mu}_i, \Sigma), i = 1, 2$. Let Δ be the Mahalanobis distance given by $\Delta^2 = (\boldsymbol{\mu}_1 - \boldsymbol{\mu}_2)'\Sigma^{-1}(\boldsymbol{\mu}_1 - \boldsymbol{\mu}_2)$. The W-rule is

$$W \geq c \implies \boldsymbol{x} \in \Pi_1, \quad W < c \implies \boldsymbol{x} \in \Pi_2,$$

where c is a cutoff point, and commonly c is set as $c = 0$. Then there are two types of PM. One is the probability of putting $\boldsymbol{X}$ into Π_2 although it is known from Π_1, which is expressed as

$$\mathrm{P}_W(2|1; c) = \Pr(W \leq c|\Pi_1). \tag{9.3.3}$$

The other is the probability of classifying an object as belonging to Π_1 although it actually comes from Π_2, which is expressed as

$$\mathrm{P}_W(1|2; c) = \Pr\{W \geq c|\Pi_1\}. \tag{9.3.4}$$

In particular, we write the PM of the W-rule with cutoff point 0 as

$$\mathrm{P}_W(2|1) = \mathrm{P}_W(2|1; 0), \quad \mathrm{P}_W(1|2) = \mathrm{P}_W(1|2; 0).$$

Large Sample Approximations. First we consider the limiting distribution of W under A1. It is easy to see that

$$\mathrm{plim}_{n_i \to \infty} \bar{\boldsymbol{X}}_i = \boldsymbol{\mu}_i, \ i = 1, 2, \quad \mathrm{plim}_{n_1, n_2 \to \infty} \mathbf{S} = \Sigma,$$

where plim means the limit in probability. Therefore, the limiting distribution of W is the same as that of

$$U = (\boldsymbol{\mu}_1 - \boldsymbol{\mu}_2)'\Sigma^{-1}\Big[\boldsymbol{X} - \frac{1}{2}(\boldsymbol{\mu}_1 + \boldsymbol{\mu}_2)\Big].$$

Therefore, the limiting distribution of W under A1 is $\mathrm{N}((1/2)\Delta^2, \Delta^2)$ when $\boldsymbol{x}$ comes from Π_1 and $\mathrm{N}(-(1/2)\Delta^2, \Delta^2)$ when $\boldsymbol{x}$ comes from Π_2. The result is extended to asymptotic expansion formulas. Okamoto (1963, 1969) gave asymptotic expansion formulas for the distribution of W up to terms of $\mathrm{O}(n^{-2})$. Siotani and Wang (1977) extended the formulas to terms of $\mathrm{O}(n^{-3})$. We discuss the validity of the asymptotic formulas and their error bounds in Section 16.1.

Theorem 9.3.1 *Under A1,*

$$\begin{aligned} \mathrm{P}_1(u; \Delta) &= \Pr\left\{(W - \Delta^2/2)/\Delta \leq u \mid \Pi_1\right\} \\ &= \Phi(u) + \varphi(u)\left(\frac{1}{n_1}a_1 + \frac{1}{n_2}a_2 + \frac{1}{m}a_3\right) + \mathrm{O}(n^{-2}), \end{aligned} \tag{9.3.5}$$

where Φ *and* φ *are the distribution and density functions of* $\mathrm{N}(0,1)$, *and* $m = n - 2$. *The coefficients* a_i *are given by*

$$\begin{aligned} a_1 &= -\frac{1}{2\Delta^2}\{u^3 + (p-3)u - p\Delta\}, \\ a_2 &= -\frac{1}{2\Delta^2}\{u^3 + 2\Delta u^2 + (p-3+\Delta^2)u + (p-2)\Delta\}, \\ a_3 &= -\frac{1}{4}\{4u^3 + 4\Delta u^2 + (6p-6+\Delta^2)u + (2p-1)\Delta\}. \end{aligned}$$

For the distribution of W *when* $\boldsymbol{X}$ *comes from* Π_2,

$$\mathrm{P}_2(u;\Delta) = \Pr\left\{(W + \Delta^2/2)/\Delta \le u \mid \Pi_2\right\} = 1 - \mathrm{P}_1^*(-u;\Delta), \qquad (9.3.6)$$

where $\mathrm{P}_1^*(v;\Delta)$ *is the* $\mathrm{P}_1(v;\Delta)$ *with the interchange of* n_1 *and* n_2.

The PMs of the W-rule with $c = 0$ are obtained by putting $u = -\Delta/2$ in (9.3.5) and interchanging n_1 and n_2 as follows:

Corollary 9.3.1 *The* PMs *of the W-rule with cutoff point* $c = 0$ *are given by*

$$\begin{aligned} \mathrm{P}_W(2|1) = \Phi\left(-\frac{1}{2}\Delta\right) + \varphi\left(-\frac{1}{2}\Delta\right)&\left[\frac{1}{16n_1\Delta}\left\{\Delta^2 + 12(p-1)\right\}\right. \\ &\left. + \frac{1}{16n_2\Delta}\left\{\Delta^2 - 4(p-1)\right\} + \frac{1}{4m}(p-1)\Delta\right] + \mathrm{O}(n^{-2}), \end{aligned}$$

$\mathrm{P}_W(1|2)$ = the expression obtained by interchanging n_1 and n_2 in $\mathrm{P}_W(2|1)$.

Estimation of PM. Now we consider estimating $\mathrm{P}_W(2|1)$ and $\mathrm{P}_W(1|2)$. One of the methods is to apply a resubstitution method which classifies the original (initial) samples by using a proposed classification rule. Let n_{ij} be the number of samples misclassified as a sample from Π_j for the n_i samples belonging to Π_i. Then the ratio n_{ij}/n_i is an estimator of $\mathrm{P}(j|i)$. Unfortunately, it tends to underestimate the true misclassification. The problem with the resubstitution method is that the same observations are used to define the classification rule and to judge its accuracy. As one way to avoid this problem, we may use a modified estimator called a *leave-one-out estimator*. One of its generalized versions is known as *cross-validation method* [see Stone, 1974]. In this method a sample $\boldsymbol{x}$ is judged on the basis of the classification rule constructed from $n-1$ samples by deleting the sample $\boldsymbol{x}$. This modification is an improvement of the resubstitution method. For further details, see Kshirsagar (1972).

The other method is to use asymptotic expansion formulas in Corollary 9.3.1. For this use, we need to replace Δ by an estimator. A natural estimator of Δ^2 is

$$D^2 = (\bar{\boldsymbol{X}}_1 - \bar{\boldsymbol{X}}_2)'\mathbf{S}^{-1}(\bar{\boldsymbol{X}}_1 - \bar{\boldsymbol{X}}_2),$$

which is a consistent estimator under A_1. The other estimator is

$$\tilde{\Delta}^2 = [(n-p-1)/n]D^2. \tag{9.3.7}$$

The estimator is based on an unbiased estimator given (see Theorem 3.1.2) by

$$\tilde{\Delta}^2 - np/(n_1 n_2)$$

Since this estimator might be negative, we merely ignore the constant term, which is $\mathrm{O}(n^{-1})$ under A1. A more reasonable estimator is to find an asymptotic estimator of $\mathrm{P}_W(i|j)$ whose bias is of $\mathrm{O}(n^{-1})$ or higher order. Based on this idea, McLachlan (1973) proposed some estimators. Such an estimator is constructed using the following result.

Lemma 9.3.1 *Let $\Phi(x)$ and $\varphi(x)$ be the distribution and density functions of $\mathrm{N}(0,1)$, and let D be the sample Mahalanobis distance. Then, under* A1,

$$\mathrm{E}\left[\Phi(-\frac{1}{2}D)\right] = \Phi\left(-\frac{1}{2}\Delta\right) + \varphi\left(-\frac{1}{2}\Delta\right)\left[\left(\frac{1}{n_1}+\frac{1}{n_2}\right)\left\{\Delta - \frac{4}{\Delta}(p-1)\right\}\right.$$
$$\left. + \frac{\Delta}{2m}\left\{\Delta^2 - 4(2p+1)\right\}\right] + \mathrm{O}(n^{-2}), \tag{9.3.8}$$

where $n = n_1 + n_2$ and $m = n-2$.

From Lemma 9.3.1 we have asymptotically unbiased estimators for PM of W-rule with cutoff point $c = 0$.

Theorem 9.3.2 *An estimator of $\mathrm{P}_W(2|1)$ whose bias is of $\mathrm{O}(n^{-2})$ is given by*

$$\hat{\mathrm{P}}_W(2|1) = \Phi\left(-\frac{1}{2}D\right) + \frac{1}{16}\varphi\left(-\frac{1}{2}D\right)$$
$$\times \left[\frac{p-1}{n_1 D} + \frac{D}{32m}\left\{4(4p-1) - D^2\right\}\right]. \tag{9.3.9}$$

9.3.2 Z-Rule

Let Z be the discriminant function

$$Z = \frac{n_1}{n_1+1}(\boldsymbol{X} - \bar{\boldsymbol{X}}_1)'\mathbf{S}^{-1}(\boldsymbol{X} - \bar{\boldsymbol{X}}_1) - \frac{n_2}{n_2+1}(\boldsymbol{X} - \bar{\boldsymbol{X}}_2)'\mathbf{S}^{-1}(\boldsymbol{X} - \bar{\boldsymbol{X}}_2)$$

based on samples of size n_i from $\mathrm{N}(\boldsymbol{\mu}_i, \Sigma)$, $i = 1, 2$. The Z-rule is

$$Z \le c \implies \boldsymbol{x} \in \Pi_1, \quad Z > c \implies \boldsymbol{x} \in \Pi_2,$$

where c is a cutoff point, and commonly c is set as $c = 0$. Similarly, as in the W-rule, we express two types of PM as

$$\mathrm{P}_Z(2|1;c) = \Pr\{(W \geq c|\Pi_1\}, \quad \mathrm{P}_Z(1|2;c) = \Pr\{W < c|\Pi_1\}. \tag{9.3.10}$$

Large Sample Approximations. Noting that $n_i/(n_i+1) = 1 + \mathrm{O}_p(n_i^{-1})$, we have

$$Z = -2W + \mathrm{O}_p(n^{-1})$$

under A1, where O_p is the order in probability. Therefore, the limiting distribution of Z is the same as that of $-2W$, and hence is $\mathrm{N}(-\Delta^2, 4\Delta^2)$ when $\boldsymbol{X}$ comes from Π_1 and $\mathrm{N}(\Delta^2, 4\Delta^2)$ when $\boldsymbol{X}$ comes from Π_2. The result is extended to asymptotic expansion formulas. Memon and Okamoto (1971) gave asymptotic expansion formulas for the distribution of Z up to terms of $\mathrm{O}(n^{-2})$. Siotani and Wang (1977) extended the formulas to terms of $\mathrm{O}(n^{-3})$. The results up to $\mathrm{O}(n^{-1})$ are given in the following theorem.

Theorem 9.3.3 *Under* A1,

$$\begin{aligned} Q_1(u;\Delta) &= \mathrm{P}\left\{(Z+\Delta^2)/(2\Delta) \leq u \mid \Pi_1\right\} \\ &= \Phi(u) + \varphi(u)\left(\frac{1}{n_1}\tilde{a}_1 + \frac{1}{n_2}\tilde{a}_2 + \frac{1}{m}\tilde{a}_3\right) + \mathrm{O}(n^{-2}), \end{aligned} \tag{9.3.11}$$

where Φ *and* φ *are the distribution and density functions of* $\mathrm{N}(0,1)$, *and* $m = n-2$. *The coefficients* $\tilde{a}_i$ *are given by*

$$\begin{aligned} \tilde{a}_1 &= -\frac{1}{2\Delta^2}\{u^3 - \Delta u^2 (p-3)u + \Delta\}, \\ \tilde{a}_2 &= -\frac{1}{2\Delta^2}\{u^3 - \Delta u^2 + (p-3-\Delta^2)u + \Delta(\Delta^2+1)\}, \\ \tilde{a}_3 &= -\frac{1}{4}\{4u^3 - 4\Delta u^2 + (6p-6+\Delta^2)u - (2p-1)\Delta\}. \end{aligned}$$

For the distribution of Z *when* $\boldsymbol{X}$ *comes from* Π_2,

$$Q_2(u;\Delta) = \mathrm{P}\left\{(Z-2\Delta^2)/(2\Delta) \leq u \mid \Pi_2\right\} = 1 - Q_1^*(-u;\Delta), \tag{9.3.12}$$

where $Q_1^*(v;\Delta)$ *is the* $Q_1(v;\Delta)$ *with the interchange of* n_1 *and* n_2.

The PMs of the Z-rule with $c = 0$ are obtained by putting $u = \Delta/2$ in (9.3.11) and interchanging n_1 and n_2 as follows:

Corollary 9.3.2 *The* PMs *of the Z-rule with cutoff point* $c = 0$ *are given by*

$$
\begin{aligned}
\mathrm{P}_Z(2|1) = \Phi\left(-\frac{1}{2}\Delta\right) + \varphi\left(-\frac{1}{2}\Delta\right)\left[\frac{1}{16n_1\Delta}\left\{-\Delta^2 + 4(p-1)\right\}\right.\\
\left. + \frac{1}{16n_2\Delta}\left\{3\Delta^2 + 4(p-1)\right\} + \frac{1}{4m}(p-1)\Delta\right] + \mathrm{O}(n^{-2}),
\end{aligned}
$$

$\mathrm{P}_Z(1|2) =$ the expression obtained by interchanging n_1 and n_2 in $\mathrm{P}_Z(2|1)$.

Estimation of PM. The estimation method for $\mathrm{P}_W(2|1)$ and $\mathrm{P}_W(1|2)$ is applicable to the estimation of $\mathrm{P}_Z(2|1)$ and $\mathrm{P}_Z(1|2)$. Here we give asymptotically unbiased estimators for PM of Z-rule with cut-off point $c = 0$.

Theorem 9.3.4 *An estimator of* $\mathrm{P}_Z(2|1)$ *whose bias is of* $\mathrm{O}(n^{-2})$ *is given by*

$$
\begin{aligned}
\hat{P}_Z(2|1) = \Phi\left(-\frac{1}{2}D\right) + \varphi\left(-\frac{1}{2}D\right)\\
\times\left[\frac{1}{8n_1}\left(-D + \frac{4}{D}(p-1)\right) + \frac{1}{8n_2}\left(D + \frac{4}{D}(p-1)\right)\right.\\
\left. + \frac{D}{32m}\left\{4(4p-1) - D^2\right\}\right]. \qquad (9.3.13)
\end{aligned}
$$

9.3.3 High-Dimensional Asymptotic Results

We consider similar problems under A2. For a study on high-dimensional approximations of $\mathrm{P}_W(j|i)$, it may be noted that there are some earlier works by Deev (1970), Raudys (1972), and others. For a survey on the former Soviet Union literature, see Raudys and Young (2004). Raudys (1972) proposed a high-dimensional approximation for $P_W(j|i)$ in the case $n_1 = n_2$. His result was extended to a class including W and Z by Fujikoshi and Seo (1998). In Section 16.1 we give a method of obtaining high-dimensional asymptotic approximations for the distribution of W and their error bound.

Theorem 9.3.5 *Suppose that* $\boldsymbol{X}$ *comes from* Π_i: $\mathrm{N}_p(\boldsymbol{\mu}_i, \Sigma)$, $i = 1, 2$. *Then under* A2, W *and* $\tilde{Z} = -(1/2)Z$ *are asymptotically distributed as* $\mathrm{N}(\zeta_W^{(i)}, \sigma^2)$ *and* $\mathrm{N}(\zeta_{\tilde{Z}}^{(i)}, \sigma^2)$, *respectively, where*

$$
\begin{aligned}
\zeta_W^{(i)} &= \frac{1}{2}q\left\{\Delta^2 + (-1)^i p(n_1^{-1} - n_2^{-1})\right\},\\
\zeta_{\tilde{Z}}^{(1)} &= \zeta_{\tilde{Z}}^{(2)} = \frac{1}{2}q\Delta^2,\ \sigma^2 = q^3\tilde{\Delta}^2,
\end{aligned}
$$

and
$$q = n/(n-p), \quad \tilde{\Delta} = \{\Delta^2 + pn/(n_1 n_2)\}^{1/2}.$$

Corollary 9.3.3 *Under A2, the PMs for W and Z are evaluated asymptotically as follows:*

$$\mathrm{P}_W(2|1) = \Phi(\gamma_W^{(1)}) + \mathrm{O}_1^*, \quad \mathrm{P}_W(1|2) = \Phi(\gamma_W^{(2)}) + \mathrm{O}_1^*,$$
$$\mathrm{P}_Z(2|1) = \Phi(\gamma_Z^{(1)}) + \mathrm{O}_1^*, \quad \mathrm{P}_Z(1|2) = \Phi(\gamma_Z^{(2)}) + \mathrm{O}_1^*,$$

where O_j^* *denotes the term of jth order with respect to* $(n_1^{-1}, n_2^{-1}, p^{-1})$,

$$\gamma_W^{(i)} = -\frac{1}{2}q^{1/2}\{\Delta^2 + (-1)^i p(n_1^{-1} - n_2^{-1})\}\tilde{\Delta}^{-1},$$
$$\gamma_Z^{(1)} = \gamma_Z^{(2)} = -\frac{1}{2}q^{1/2}\Delta^2\tilde{\Delta}^{-1}.$$

In a special case $n_1 = n_2$, we obtain

$$\gamma_W^{(1)} = \gamma_W^{(2)} = -\frac{1}{2}q^{1/2}\Delta^2\tilde{\Delta}^{-1},$$

which was proposed by Raudys (1972). Lachenbruch (1968) proposed similar approximations $\Phi(\gamma_L^{(i)})$ for $\mathrm{P}_W(j|i)$ without considering their asymptotic properties. However, from Corollary 9.3.3 we can state his approximations as

$$\mathrm{P}_W(2|1) = \Phi(\gamma_L^{(1)}) + \mathrm{O}_1^*, \quad \mathrm{P}_W(1|2) = \Phi(\gamma_L^{(2)}) + \mathrm{O}_1^*, \tag{9.3.14}$$

where

$$\gamma_L^{(i)} = \left\{\frac{n(n-p-2)(n-p-5)}{(n-3)(n-p)(n-p-3)}\right\}^{1/2}\gamma_W^{(i)}, \quad i = 1, 2. \tag{9.3.15}$$

We are interested in comparing the degrees of accuracy of large-sample approximations with high-dimensional approximations. Table 9.3.1 gives PMs of the W-rule for various values of (p, n_1, n_2, Δ) and their approximations: $\Phi(\gamma_W^{(1)})$, $\Phi(\gamma_L^{(1)})$, and $AE_W^{(1)}$, where $AE_W^{(1)}$ denotes the asymptotic expansion up to order n^{-1} in Corollary 9.3.1. The comparison shows that the high-dimensional approximations are also extremely accurate, and in particular, $\Phi(\gamma_L^{(1)})$ is more accurate than the other approximations in most cases. For more details, see Wyman et al. (1990) and Fujikoshi and Seo (1998). For a practical use of Corollary 9.3.3, we need to estimate Δ^2. It is recommended to use

$$D_u^2 = \frac{n-p-1}{n}D^2 - \frac{pn}{n_1 n_2},$$

which is unbiased and consistent under the high-dimensional framework. In this estimator, if D_u is negative, we need to replace it by 0.

Table 9.3.1 Values of approximations and simulation $p = 5$

(n_1, n_2)	Δ	$\Phi(\gamma_W^{(1)})$	$\Phi(\gamma_L^{(1)})$	$AE_W^{(1)}$	Simulation
(10,10)	1.05	0.3808	0.3710	0.3908	0.3711
	1.68	0.2822	0.2660	0.2659	0.2648
	2.56	0.1706	0.1509	0.1446	0.1504
	3.29	0.0105	0.0864	0.0793	0.0877
(10,20)	1.05	0.3914	0.3875	0.3990	0.3836
	1.68	0.2746	0.2672	0.2634	0.2631
	2.56	0.1524	0.1435	0.1377	0.1413
	3.29	0.0857	0.0780	0.0731	0.0779
(10,30)	1.05	0.3957	0.3934	0.4007	0.3871
	1.68	0.2712	0.2666	0.2613	0.2611
	2.56	0.1449	0.1395	0.1342	0.1367
	3.29	0.0785	0.0739	0.0701	0.0730
(20,10)	1.05	0.3310	0.3251	0.3328	0.3275
	1.68	0.2372	0.2291	0.2300	0.2292
	2.56	0.1340	0.1252	0.1240	0.1256
	3.29	0.0760	0.0687	0.0669	0.0693
(20,20)	1.05	0.3452	0.3419	0.3448	0.3398
	1.68	0.2395	0.2345	0.2325	0.2320
	2.56	0.1296	0.1242	0.1218	0.1230
	3.29	0.0707	0.0662	0.0642	0.0660

9.4 Canonical Discriminant Analysis

9.4.1 Canonical discriminant Method

In Section 9.1 we stated Fisher's approach in population, and this section a sample version is given. Suppose that q populations Π_i have the same covariance matrix Σ. Suppose that we have the n_i samples from Π_i given in (9.2.1). We use similar notation there. Further, let us denote the between-group and within-group SSP matrices by

$$\begin{aligned} \mathrm{S}_b &= n_1(\bar{\boldsymbol{x}}^{(1)} - \bar{\boldsymbol{x}})(\bar{\boldsymbol{x}}^{(1)} - \bar{\boldsymbol{x}})' + \cdots + n_q(\bar{\boldsymbol{x}}^{(q)} - \bar{\boldsymbol{x}})(\bar{\boldsymbol{x}}^{(q)} - \bar{\boldsymbol{x}})', \\ \mathrm{S}_w &= (n_1 - 1)\mathrm{S}^{(1)} + \cdots + (n_q - 1)\mathrm{S}^{(q)}. \end{aligned}$$

Now consider a linear combination of the x_i:

$$Z = b_1X_1 + \cdots + b_pX_p = \boldsymbol{b}'\boldsymbol{X},$$

where $\boldsymbol{b} = (b_1, \ldots, b_p)'$. The between-group and within-group sums of squares of z are expressed as $\boldsymbol{b}'\mathrm{S}_b\boldsymbol{b}$ and $\boldsymbol{b}'\mathrm{S}_w\boldsymbol{b}$. As a coefficient vector $\boldsymbol{b}$ with a good separation of q groups, Fisher proposed to maximizing the ratio

$$\ell = \frac{\boldsymbol{b}'\mathrm{S}_b\boldsymbol{b}}{\boldsymbol{b}'\mathrm{S}_w\boldsymbol{b}}. \tag{9.4.1}$$

Optimum solutions are obtained by considering the characteristic equation

$$\mathrm{S}_b\boldsymbol{b}_i = \ell_i\mathrm{S}_w\boldsymbol{b}_i, \quad \boldsymbol{b}_i'\mathrm{S}_w\boldsymbol{b}_j = n\delta_{ij}, \tag{9.4.2}$$

where $\ell_1 > \cdots > \ell_m \geq 0$ are possibly nonzero characteristic roots of $\mathrm{S}_b\mathrm{S}_w^{-1}$, $m = \min\{p, q-1\}$, δ_{ij} is Kronecker's delta, and $\delta_{ii} = 1, \delta_{ij} = 0$ for $i \neq j$. Now we have m (meaningful) functions

$$Z_i = \boldsymbol{b}_i'\boldsymbol{X}, \ i = 1, \ldots, m, \tag{9.4.3}$$

called *canonical disctiminant functions*, which have the following properties.

Theorem 9.4.1 *Let $Z = \boldsymbol{b}'\boldsymbol{X}$ be a linear combination of the X_i.*

(1) *Z_1 is a linear combination that maximizes the ratio ℓ in (9.4.3). The maximum value of ℓ is ℓ_1.*

(2) *For $k < m$, suppose that z is uncorrelated with $Z_1, \ldots, Z_k$ in the sense of $\boldsymbol{b}'\mathrm{S}_b\boldsymbol{b}_i = 0, i = 1, \ldots, k$. Then Z_{k+1} is a linear combination maximizing the ratio ℓ in (9.4.1). The maximum value of ℓ is ℓ_{k+1}.*

Proof. The result is a consequence of [A.2.6]. □

The canonical discriminant functions may also be defined by considering maximizing

$$d = \frac{\boldsymbol{b}'\mathrm{S}_b\boldsymbol{b}}{\boldsymbol{b}'\mathrm{S}_t\boldsymbol{b}}, \tag{9.4.4}$$

where $\mathrm{S}_t = \mathrm{S}_w + \mathrm{S}_b$. Optimum solutions are obtained by considering the characteristic equation

$$\mathrm{S}_b\boldsymbol{b}_i = d_i\mathrm{S}_t\boldsymbol{b}_i, \quad \boldsymbol{b}_i'\mathrm{S}_t\boldsymbol{b}_j = n\delta_{ij}, \tag{9.4.5}$$

where $d_1 > \cdots > d_m \geq 0$ are the possibly nonzero characteristic roots of $\mathrm{S}_b\mathrm{S}_t^{-1}$. There is a correspondence between d_i and ℓ_i such that $d_i = \ell_i/(\ell_i+1)$.

As measures of discriminability of the ith canonical discriminant function, we use

$$\ell_i/(\ell_1 + \cdots + \ell_m) \quad \text{or} \quad d_i/(d_1 + \cdots + d_m), \tag{9.4.6}$$

which are called the contribution ratio of the canonical discriminant function. Similarly, the cumulative contribution ratio of the first k canonical discriminant function is defined by

$$(\ell_1 + \cdots + \ell_k)/(\ell_1 + \cdots + \ell_m) \quad \text{or} \quad (d_1 + \cdots + d_k)/(d_1 + \cdots + d_m). \quad (9.4.7)$$

The discriminant scores of $\boldsymbol{x}_j^{(i)}$ based on the first k discriminant functions are

$$\boldsymbol{z}_j^{(i)} = \begin{pmatrix} z_{j1}^{(i)} \\ \vdots \\ z_{jk}^{(i)} \end{pmatrix} = \begin{pmatrix} \boldsymbol{b}_1' \\ \vdots \\ \boldsymbol{b}_k' \end{pmatrix} \boldsymbol{x}_j^{(i)}, \quad j = 1, \ldots, n_i; \quad i = 1, \ldots, q.$$

The scores are used as the coordinates in the k-dimensional Euclidean space. The coordinates for representing the $\boldsymbol{x}_j^{(i)}$ are as follows:

Unit	dim 1	$\cdots$	dim k
$\vdots$	$\vdots$		$\vdots$
$\boldsymbol{x}_j^{(i)}$	$z_{j1}^{(i)}$	$\cdots$	$z_{jk}^{(i)}$
$\vdots$	$\vdots$		$\vdots$

These discriminant functions are also used for a practical classification. The rule is defined as follows: Let $\boldsymbol{x}_0$ be a new observation vector. Let

$$\boldsymbol{z}_0 = \mathrm{B}\boldsymbol{x}_0, \quad \mathrm{B} = (\boldsymbol{b}_1, \ldots, \boldsymbol{b}_k)'$$

and let $\bar{\boldsymbol{z}}^{(i)} = \mathrm{B}'\bar{\boldsymbol{x}}^{(i)}$, $i = 1, \ldots, q$. Compute

$$d_i = \|\boldsymbol{z}_0 - \bar{\boldsymbol{z}}^{(i)}\|, \quad i = 1, \ldots, q.$$

Then the rule is to assign $\boldsymbol{x}_0$ to G_i if $\min\{d_1, \ldots, d_q\} = d_i$. When $q = 2$ the rule is

$$(\bar{\boldsymbol{x}}^{(1)} - \bar{\boldsymbol{x}}^{(2)})'\mathrm{S}^{-1}\left[\boldsymbol{x}_0 - \frac{1}{2}(\bar{\boldsymbol{x}}^{(1)} + \bar{\boldsymbol{x}}^{(2)})\right] > 0 \implies \boldsymbol{x}_0 \in G_1,$$

and otherwise, assigns $\boldsymbol{x}_0$ to G_2. This rule is the same as the rule based on the usual linear function.

9.4.2 Test for Additional Information

In canonical discriminant analysis, we are interested in which set of variables are important, or which set of variables are redundant. In two-group discriminant analysis, the coefficients of the population discriminant function are given by

$$\boldsymbol{\beta} = \Sigma^{-1}(\boldsymbol{\mu}_1 - \boldsymbol{\mu}_2) = (\beta_1, \ldots, \beta_p)'.$$

Partition $\boldsymbol{\beta}' = (\boldsymbol{\beta}_1', \boldsymbol{\beta}_2')$ and $\boldsymbol{X}' = (\boldsymbol{X}_1', \boldsymbol{X}_2')$, where $\boldsymbol{\beta}_1$ and $\boldsymbol{X}_1$ have k components. Suppose that $\boldsymbol{\beta}_2 = \boldsymbol{0}$. This means that the variables $X_{k+1}, \ldots, X_p$ have no discriminating power once the variables $X_1, \ldots, X_k$ have been taken into account. Then we say that $\boldsymbol{X}_2$ has no *additional information* in *discriminant analysis* in the presence of $\boldsymbol{X}_1$, or simply that $\boldsymbol{X}_1$ is sufficient and $\boldsymbol{X}_2$ is redundant. Partition $\boldsymbol{\mu}^{(i)}$ and Σ as

$$\boldsymbol{\mu}^{(i)} = \begin{pmatrix} \boldsymbol{\mu}_1^{(i)} \\ \boldsymbol{\mu}_2^{(i)} \end{pmatrix}, \quad \Sigma = \begin{pmatrix} \Sigma_{11} & \Sigma_{12} \\ \Sigma_{21} & \Sigma_{22} \end{pmatrix}$$

and let

$$\begin{aligned} \boldsymbol{\mu}_{2\cdot 1}^{(i)} &= \boldsymbol{\mu}_2^{(i)} - \Gamma\boldsymbol{\mu}_1^{(i)}, \quad i = 1,\ 2,\ \Gamma = \Sigma_{21}\Sigma_{11}^{-1}, \\ \Delta_1^2 &= (\boldsymbol{\mu}_1^{(1)} - \boldsymbol{\mu}_1^{(2)})'\Sigma_{11}^{-1}(\boldsymbol{\mu}_1^{(1)} - \boldsymbol{\mu}_1^{(2)}), \\ \boldsymbol{\beta} &= \Sigma^{-1}(\boldsymbol{\mu}^{(1)} - \boldsymbol{\mu}^{(2)}) = (\boldsymbol{\beta}_1'\ \boldsymbol{\beta}_2')'. \end{aligned}$$

For the Mahalanobis distance Δ based on $\boldsymbol{X}$ and the correspondence of Δ_1 based on $\boldsymbol{X}_1$, this is the relation

$$\Delta^2 = \Delta_1^2 + \Delta_{2\cdot 1}^2,$$

where

$$\Delta_{2\cdot 1}^2 = (\boldsymbol{\mu}_{2\cdot 1}^{(1)} - \boldsymbol{\mu}_{2\cdot 1}^{(2)})'\Sigma_{22\cdot 1}^{-1}(\boldsymbol{\mu}_{2\cdot 1}^{(1)} - \boldsymbol{\mu}_{2\cdot 1}^{(2)}), \quad \Sigma_{22\cdot 1} = \Sigma_{22} - \Sigma_{21}\Sigma_{11}^{-1}\Sigma_{21}.$$

This relation is obtained by using

$$\Sigma^{-1} = \begin{pmatrix} \Sigma_{11}^{-1} & \mathrm{O} \\ \mathrm{O} & \mathrm{O} \end{pmatrix} + \begin{pmatrix} -\Sigma_{12}\Sigma_{22}^{-1} \\ \mathrm{I}_{p-k} \end{pmatrix} \Sigma_{22\cdot 1}^{-1} \left(-\Sigma_{22}^{-1}\Sigma_{21} \quad \mathrm{I}_{p-k}\right).$$

On the other hand, the coefficient vector of the linear discriminant function is expressed as

$$\boldsymbol{\beta}_1 = \Sigma_{11}^{-1}(\boldsymbol{\mu}_1^{(1)} - \boldsymbol{\mu}_1^{(2)}) - \Sigma_{12}\Sigma_{22}^{-1}\boldsymbol{\beta}_2, \quad \boldsymbol{\beta}_2 = \Sigma_{22\cdot 1}^{-1}(\boldsymbol{\mu}_{2\cdot 1}^{(1)} - \boldsymbol{\mu}_{2\cdot 1}^{(2)}).$$

From these results we can see that the following three statements are equivalent, which was proved by Rao (1970):

$$(1)\ \ \Delta^2 = \Delta_1^2, \quad (2)\ \ \boldsymbol{\mu}_{2\cdot 1}^{(1)} = \boldsymbol{\mu}_{2\cdot 1}^{(2)}, \quad (3)\ \ \boldsymbol{\beta}_2 = 0.$$

Now we consider extending the results above to the q-group case. This extension has been given in Section 3.3, based on the extension of (1) and (2). Here we extend statement (3) to the q-group case. Let Ψ be the population between-group matrix in (9.1.8) with $\rho_i = n_i/n$:

$$\Psi = \sum_{i=1}^{q}(n_i/n)(\boldsymbol{\mu}^{(i)} - \bar{\boldsymbol{\mu}})(\boldsymbol{\mu}^{(i)} - \bar{\boldsymbol{\mu}})',$$

where $\bar{\boldsymbol{\mu}} = (1/n)\sum_{i=1}^{q} n_i \boldsymbol{\mu}^{(i)}$. Then the coefficient vectors $\boldsymbol{\beta}_i$ of the population discriminant functions are defined in (9.1.10) as the characteristic vectors of Ψ with respect to Σ. We partition

$$\boldsymbol{\beta}_j = \begin{pmatrix} \boldsymbol{\beta}_{1j} \\ \boldsymbol{\beta}_{2j} \end{pmatrix}, \quad \Psi = \begin{pmatrix} \Psi_{11} & \Psi_{12} \\ \Psi_{21} & \Psi_{22} \end{pmatrix}$$

in the same way as $\boldsymbol{\mu}^{(i)}$ and Σ. Then the statements on additional information are extended [see, e.g., Fujikoshi, 1982 & 1992] as follows:

Theorem 9.4.2 *The following statements are equivalent.*

(1) $\operatorname{tr}\Sigma^{-1}\Psi = \operatorname{tr}\Sigma_{11}^{-1}\Psi_{11}$.

(2) $\boldsymbol{\mu}_{2\cdot 1}^{(1)} = \cdots = \boldsymbol{\mu}_{2\cdot 1}^{(q)}$.

(3) $\boldsymbol{\beta}_{2j} = 0, \quad j = 1, \ldots, m, \quad m = \operatorname{rank}(\Psi)$.

If one of the statements in Theorem 9.4.2 is true, we say that $\boldsymbol{X}_1$ is sufficient, or $\boldsymbol{X}_2$ has no additional information in discriminant analysis in the presence of $\boldsymbol{X}_1$. Each statement is also called the redundancy hypothesis.

As we saw in Section 3.2, the LR test for additional information is closely related to the LR test for H_0: $\boldsymbol{\mu}^{(1)} = \cdots = \boldsymbol{\mu}^{(q)}$. Let $\mathbf{B}$, $\mathbf{W}$, and $\mathbf{T}$ be the SSP matrices due to the between-group, within-group and total variations, based on the sample $\boldsymbol{X}_1^{(i)}, \ldots, \boldsymbol{X}_{n_i}^{(i)}$ drawn from Π_i, $i = 1, \ldots, q$. In the test of H_0, we are interested in whether a subset of variables $X_1, \ldots, X_p$ is sufficient, or the set of remainder variables has no additional information or is redundant. Without loss of generality we consider the sufficiency of a subvector $\boldsymbol{X}_1 = (X_1, \ldots, X_k)'$ of $\boldsymbol{X}$, or redundancy of the remainder vector $\boldsymbol{X}_2 = (X_{k+1}, \ldots, X_p)'$. Let $\boldsymbol{\mu}^{(j)}$, $\mathbf{W}$, and $\mathbf{T}$ partition in the same manner as the partition of $\boldsymbol{X}$:

$$\boldsymbol{\mu}^{(j)} = \begin{pmatrix} \boldsymbol{\mu}_1^{(j)} \\ \boldsymbol{\mu}_2^{(j)} \end{pmatrix}, \quad \mathbf{W} = \begin{pmatrix} \mathbf{W}_{11} & \mathbf{W}_{12} \\ \mathbf{W}_{21} & \mathbf{W}_{22} \end{pmatrix}, \quad \mathbf{T} = \begin{pmatrix} \mathbf{T}_{11} & \mathbf{T}_{12} \\ \mathbf{T}_{21} & \mathbf{T}_{22} \end{pmatrix}.$$

Then the LR test for H_0 is based on

$$\Lambda = \frac{|\mathbf{W}|}{|\mathbf{W} + \mathbf{B}|} = \frac{|\mathbf{W}|}{|\mathbf{T}|}.$$

We consider testing

$$H_{2\cdot 1}\colon \boldsymbol{\mu}_{2\cdot 1}^{(1)} = \cdots = \boldsymbol{\mu}_{2\cdot 1}^{(q)}, \tag{9.4.8}$$

where $\boldsymbol{\mu}_{2\cdot 1}^{(j)} = \boldsymbol{\mu}_2^{(j)} - \Sigma_{21}\Sigma_{11}^{-1}\boldsymbol{\mu}_1^{(j)}$, $j = 1, \ldots, q$. The hypothesis $H_{2\cdot 1}$ was introduced by Rao (1964). The likelihood ratio criterion λ for the hypothesis $H_{2\cdot 1}$ is given (see Section 3.3) by

$$\lambda = \Lambda_{2\cdot 1}^{n/2},$$

where

$$\Lambda_{2\cdot 1} = \frac{|\mathbf{W}_{22\cdot 1}|}{|\mathbf{T}_{22\cdot 1}|} = \frac{|\mathbf{W}|}{|\mathbf{W}_{11}|} \cdot \frac{|\mathbf{T}|}{|\mathbf{T}_{11}|}$$
$$= \frac{|\mathbf{W}|}{|\mathbf{T}|} \cdot \left\{ \frac{|\mathbf{W}_{11}|}{|\mathbf{T}_{11}|} \right\}^{-1}.$$

From Theorem 3.3.2 the null distribution of $\Lambda_{2\cdot 1}$ is

$$\Lambda_{2\cdot 1} \sim \Lambda_{p-k}(q-1, n-q-k). \tag{9.4.9}$$

When $q = 2$, the hypothesis may be expressed as

$$M_k\text{: } \Delta^2 = \Delta_1^2.$$

The LR test is equivalent to the F test; that is,

$$F_k = \frac{n-p-1}{p-k} \frac{c(D_p^2 - D_k^2)}{n + cD_k^2}, \tag{9.4.10}$$

whose null distribution is $F(p-k, n-p-1)$. Here D_p or D_k are sample Maharanobis distances based on $\boldsymbol{X}$ and $\boldsymbol{X}_1$, respectively.

We use the notation,

$$\Lambda(X_1, \ldots, X_p) = \frac{|\mathbf{W}|}{|\mathbf{T}|}, \quad \Lambda(X_1, \ldots, X_k) = \frac{|\mathbf{W}_{11}|}{|\mathbf{T}_{11}|},$$

and

$$\Lambda_{2\cdot 1} = \frac{\Lambda(X_1, \ldots, X_p)}{\Lambda(X_1, \ldots, X_k)}$$
$$\equiv \Lambda(X_{k+1}, \ldots, X_p | X_1, \ldots, X_k).$$

9.4.3 Selection of Variables

First we consider the case $q = 2$. The coefficients in the linear discriminant function may be used as a measure of the importance of the variables $X_1, \ldots, X_p$. In this case we use those where the X_i are standardized as $\tilde{X}_i = X_i/\sqrt{s_{ii}}$, where s_{ii} is an estimate of $\mathrm{Var}(\boldsymbol{x}_i)$. Let the linear discriminant function be denoted by

$$\begin{aligned} W &= b_1 \boldsymbol{X}_1 + \cdots + b_p \boldsymbol{X}_p + b_0 \\ &= b_1\sqrt{s_{11}} \cdot \tilde{X}_i + \cdots + b_p\sqrt{s_{pp}} \cdot \tilde{X}_p + b_0. \end{aligned} \tag{9.4.11}$$

The coefficients $b_i\sqrt{s_{ii}}$ are called *scaled discriminant coefficients.*

In discriminant analysis, it is important to select a set of variables such that it minimizes the expected probability of misclassification. When we use the W-rule with cutoff point 0, its expected probability of misclassification is

$$R_W = \rho_1 \mathrm{P}_W(2|1) + \rho_2 \mathrm{P}_W(1|2).$$

An estimator is used as a measure of selection of variables. Usually, the prior probabilities ρ_i are unknown, so we use an estimator with $\rho_1 = \rho_2 = 1/2$ as a conventional approach. Then, for example, based on Theorem 9.3.2 we have a criterion,

$$\hat{R}_W = \Phi(Q), \tag{9.4.12}$$

where

$$\begin{aligned} Q = &-\frac{1}{2}D + \frac{1}{2}(p-1)\{n_1^{-1} + n_2^{-1}\}D^{-1} \\ &+ (32m)^{-1}D\{4(4p-1) - D^2\}, \end{aligned}$$

$m = n_1 + n_2 - 2$, and D is the sample Mahalanobis distance.

An alternative approach is to use the additional information hypothesis or model. The hypothesis $H_{2|1}$ in (9.4.8) means that $\boldsymbol{X}_2$ is redundant, or has no additional information, in the presence of $\boldsymbol{X}_1$. It also indicates a sufficiency of $\boldsymbol{X}_1$. Let the hypothesis $H_{2|1}$ be denoted by

$$M(X_{k+1}, \ldots, X_p | X_1, \ldots, X_k), \tag{9.4.13}$$

which may be regarded as a model. By considering various subvectors as $\boldsymbol{X}_1 = (X_1, \ldots, X_k)$, we have a family of redundancy (or additional information) models. Then we apply a model selection criterion for selecting a best redundancy model. Now we consider AIC for $M(X_{k+1}, \ldots, X_p | X_1, \ldots, X_k)$ which was given by Fujikoshi (1985). Let X be the observation matrix, and we denote its likelihood by $f(\mathrm{X}; \Theta)$, where $\Theta = (\boldsymbol{\mu}^{(1)}, \ldots, \boldsymbol{\mu}^{(k)}, \Sigma)$. Then, in the derivation of the LR test, it is shown that the MLEs under $M(X_{k+1}, \ldots, X_p | X_1, \ldots, X_k)$ are

$$\begin{aligned} &\bar{\boldsymbol{\mu}}_1^{(i)} = \bar{\boldsymbol{y}}_1^{(i)}, \ \bar{\boldsymbol{\mu}}_{2\cdot 1}^{(i)} = \bar{\boldsymbol{y}}_2 - \hat{\Gamma}\bar{\boldsymbol{y}}_1, \\ &\hat{\Gamma} = T_{21}T_{11}^{-1}, \ n\hat{\Sigma}_{11} = W_{11}, \\ &n\hat{\Sigma}_{22\cdot 1} = T_{22\cdot 1} = T_{22} - T_{21}T_{11}^{-1}T_{12}. \end{aligned} \tag{9.4.14}$$

Further,

$$\begin{aligned} -2\log f(\mathrm{X}; \hat{\Theta}) &= -n \log |n^{-1}\mathrm{W}_{11}| \\ &\quad + n \log |n^{-1}\mathrm{T}_{22\cdot 1}| + np(1 + \log 2\pi) \\ &= -n \log\{|\mathrm{W}_{22\cdot 1}|/|\mathrm{T}_{22\cdot 1}|\} \\ &\quad + n \log |n^{-1}\mathrm{W}| + np(1 + \log 2\pi). \end{aligned}$$

Therefore,

$$\begin{aligned} AIC &= -n \log\{|\mathrm{W}_{22\cdot 1}|/|\mathrm{T}_{22\cdot 1}|\} + n \log |n^{-1}\mathrm{W}| \\ &\quad + np(1 + \log 2\pi) + 2\left\{kq + p - k + \frac{1}{2}p(p+1)\right\}. \end{aligned} \tag{9.4.15}$$

When $q = 2$,

$$\begin{aligned} AIC &= n \log\left\{1 + \frac{c(D_p^2 - D_k^2)}{n - 2 + cD_k^2}\right\} + n \log |n^{-1}\mathrm{W}| \\ &\quad + np(1 + \log 2\pi) + 2\{2k + p - k + \frac{1}{2}p(p+1)\}, \end{aligned}$$

where D_p and D_k are the Mahalanobis distances based on $\boldsymbol{X}$ and $\boldsymbol{X}_1$, respectively, and $c = \sqrt{n_1 n_2/n}$. It is known [Fujikoshi, 1997a] that the variable selection methods based on $\hat{R}_W$ and AIC are asymptotically equivalent.

In Section 7.3 we have explained some stepwise procedures in the multivariate regression model based on the LR test for additional information in the model. The methods there are applicable for the selection of variables in discriminant analysis. It only has to change the Λ statistics.

Example 9.4.1 Fisher (1936) gathered data [see Mardia et al., 1979] on three species of iris: *setosa*, *versicolor*, and *virginica* based on the following variables: X_1, sepal length; X_2, sepal width; X_3, petal length; X_4: petal width. He exemplified the discriminant on *setosa* and *versicolor*. Kendall (1975) noted that petal width alone separated the two varieties with virtual certainty. The same note can be made for discrimination between *setosa* and *virginica*. Now we consider the problem of selecting a best subset of variables in the discrimination of *versicolor* and *virginica*, based on $n_1 = n_2 = 50$ observations.

The coefficient vector of the linear discriminant function is given by

$$\begin{aligned} \boldsymbol{b} &= (b_1, b_2, b_3, b_4)' \\ &= (3.05, 18.02, -21.77, -30.84)'. \end{aligned}$$

The scaled linear discriminant coefficients are

$$\begin{aligned} \tilde{\boldsymbol{b}} &= (b_1\sqrt{s_{11}}, b_2\sqrt{s_{22}}, b_3\sqrt{s_{33}}, b_4\sqrt{s_{44}})' \\ &= (1.35, 6.27, -7.71, -4.89)'. \end{aligned}$$

An order of importance based on the absolute values of the coefficients is

$$X_3 \to X_2 \to X_4 \to X_1.$$

Then we are interested in the following subsets:

$$\{X_3\}, \{X_3, X_2\}, \{X_3, X_2, X_4\}, \{X_3, X_2, X_4, X_1\}.$$

However, since dimension p is small, we consider the values of Q in (9.4.12 and AIC for all the subsets in Table 9.4.1., which are given as follows. Both methods select $\{X_2, X_3, X_4\}$ as the best subset.

Table 9.4.1 Selection criteria: Q, AIC

k	Variables	D_k	Q_k	AIC_k
1	X_1	2.10	−1.05	286.52
	X_2	1.89	−0.94	297.28
	X_3	7.90	−4.08	79.28
	X_4	6.82	−3.48	106.76
2	X_1, X_2	4.47	−2.22	183.49
	X_1, X_3	8.76	−4.51	61.70
	X_1, X_4	6.93	−3.51	105.64
	X_2, X_3	9.52	−4.95	45.75
	X_2, X_4	8.85	−4.56	59.83
	X_3, X_4	8.09	−4.14	76.73
3	X_1, X_2, X_3	9.61	−4.95	45.97
	X_1, X_2, X_4	9.01	−4.61	58.31
	X_1, X_3, X_4	8.97	−4.58	59.23
	X_2, X_3, X_4	10.12	−5.25	36.04
4	X_1, X_2, X_3, X_4	10.16	−5.21	37.29

9.4.4 Estimation of Dimensionality

The coefficient vectors $\boldsymbol{\beta}_i, i = 1, \ldots, m$ of population canonical discriminant functions were defined in (9.1.10) as solutions of

$$\Psi\boldsymbol{\beta}_i = \lambda_i \Sigma \boldsymbol{\beta}_i, \quad \boldsymbol{\beta}_i' \Sigma \boldsymbol{\beta}_i = \delta_{ij},$$

where

$$\Psi = \sum_{i=1}^{q} \rho_i (\boldsymbol{\mu}_i - \bar{\boldsymbol{\mu}})'(\boldsymbol{\mu}_i - \bar{\boldsymbol{\mu}})$$

and $\bar{\boldsymbol{\mu}} = \rho_1 \boldsymbol{\mu}_1 + \cdots + \rho_q \boldsymbol{\mu}_q$. The dimensionality m is defined by the number of nonzero characteristic roots λ_i [i.e. rank(Ψ)]. The statement rank(Ψ) $= m$ is equivalent to rank(Ξ) $= k$, where

$$\Xi = (\boldsymbol{\mu}_1 - \boldsymbol{\mu}_q, \ldots, \boldsymbol{\mu}_{q-1} - \boldsymbol{\mu}_q).$$

In this section we estimate the dimensionality by using a model selection criterion, AIC. In Section 6.6 we obtained the LR test for

$$M_k\text{: rank}(\Xi) = k. \tag{9.4.16}$$

In the process of its derivation, it is shown that $-2\log\max$ likelihood is

$$n\log(1+\ell_{k+1})\cdots(1+\ell_t) + n\log|n^{-1}\mathrm{W}| + np(\log 2\pi + 1),$$

where $t = \min\{p, q-1\}$, $\ell_1 > \cdots > \ell_t > 0$ are the characteristic roots of BW^{-1}. The number of independent parameters under M_k is

$$\begin{aligned} d(M_k) &= (q-1)k - \frac{1}{2}k(k+1) + k + pk - \frac{1}{2}k(k+1) \\ &\quad + p + \frac{1}{2}p(p+1) \\ &= k(p+q-1-k) + p + \frac{1}{2}p(p+1). \end{aligned}$$

Therefore, the AIC for M_k is

$$\begin{aligned} AIC_k &= n\log(1+\ell_{k+1})\cdots(1+\ell_t) + n\log|n^{-1}\mathrm{W}| + np(\log 2\pi + 1) \\ &\quad + 2\left\{k(p+q-1-k) + p + \frac{1}{2}p(p+1)\right\}. \end{aligned} \tag{9.4.17}$$

Based on AIC, if

$$\min\{AIC_0, AIC_1, \ldots, AIC_t\} = AIC_k,$$

we estimate the dimensionality as k. Instead of AIC, consider

$$\begin{aligned} A_k &= AIC_k - AIC_t \\ &= n\log\prod_{j=k+1}^{t}(1+\ell_j) - 2(p-k)(q-1-k), \quad k = 0, \ldots, t. \end{aligned} \tag{9.4.18}$$

Here $A_t = 0$. Then the estimation method is equivalent to that of

$$\min\{A_0, A_1, \ldots, A_t\} = A_k$$

and we estimate the dimensionality as k. This method was considered by Fujikoshi and Veitch (1979).

Example 9.4.2 Rao and Slater (1949) studied the problem of identifying the neurotic state of a person referred for psychiatric examination. The analysis is based on the scores of three measurements: X_1, X_2, and X_3 classified by five types of neurotics: G_1, anxiety state; G_2, hysteria; G_3, psychopathy; G_4, obsession; G_5, personal change; G_6, normal. The mean vectors and pooled sample covariance matrix are given as follows:

Means

Group	Sample Size	$\bar{X}_1$	$\bar{X}_2$	$\bar{X}_3$
G_1	114	2.930	1.167	0.728
G_2	33	3.030	1.242	0.546
G_3	32	3.813	1.844	0.813
G_4	17	4.705	1.588	1.118
G_5	5	1.400	0.200	0.000
G_6	55	0.600	0.146	0.218

$n = n_1 + \cdots + n_6 = 256$

Covariance Matrix S

	X_1	X_2	X_3
X_1	2.3009	0.2516	0.4742
X_2	0.2516	0.6075	0.0358
X_3	0.4742	0.0358	0.5951

d.f. $= 256 - 6 = 250$

The within-group, between-group and total SSP matrices are given by:

Within: W

	X_1	X_2	X_3
X_1	575.213	62.895	118.543
X_2	62.895	151.868	8.943
X_3	118.543	8.943	148.773

Between: B

	X_1	X_2	X_3
X_1	367.725	161.777	76.239
X_2	161.777	76.473	33.033
X_3	76.239	33.033	17.711

Total: T = B + W

	X_1	X_2	X_3
X_1	575.213	62.895	118.543
X_2	62.895	151.868	8.943
X_3	118.543	8.943	148.773

The characteristic roots of BW^{-1} (i.e. the roots of $|\mathrm{B} - \ell\mathrm{W}| = 0$) are

$$\ell_1 = 0.9281, \quad \ell_2 = 0.0257, \quad \ell_3 = 0.0145.$$

Some numerical values of A_ks are given as follow:

k	0	1	2	3
A_k	148.25	−5.82	−2.31	0.00

Therefore, we estimate the dimensionality as 1.

9.5 Regression Approach

In two-group discriminant analysis, it is known that if one sets an appropriate dummy variable Y, a formal regression coefficient vector Y to $\boldsymbol{X}$ is proportional to the coefficient vector of the linear discriminant function. In fact, let the observation vectors of $\boldsymbol{x}$ be denoted by

$$\Pi_1; \boldsymbol{X}_1^{(1)}, \ldots, \boldsymbol{X}_{n_1}^{(1)}, \quad \Pi_2; \boldsymbol{X}_1^{(2)}, \ldots, \boldsymbol{X}_{n_2}^{(2)}.$$

Suppose that the values of Y are

$$y_j^{(1)} = \sqrt{n_2/nn_1}, \ j = 1, \ldots, n_1, \quad y_j^{(2)} = -\sqrt{n_1/nn_2}, \ j = 1, \ldots, n_2.$$

Let $\bar{\boldsymbol{x}}^{(i)}, i = 1, 2$ and $\bar{\boldsymbol{x}}$ be the mean vectors of Π_i and all the observation vectors. Note that the mean of y is 0. Consider the regression of Y to $\boldsymbol{X}$, and minimize

$$\sum_{i=1}^{2} \sum_{j=1}^{n_i} [y_j^{(i)} - \boldsymbol{\beta}'(\boldsymbol{x}_j^{(i)} - \bar{\boldsymbol{x}})]^2.$$

Then the optimum solution of $\boldsymbol{\beta}$ is proportional to $\mathrm{S}^{-1}(\bar{\boldsymbol{x}}^{(1)} - \bar{\boldsymbol{x}}^{(2)})$, where S is the pooled sample covariance matrix.

Now we consider the case of q-group discriminant analysis. To this end, set $t = \min\{p, q-1\}$, and consider t-dimensional dummy variate $\boldsymbol{Y} = (Y_1, \ldots, Y_t)'$. We denote the observation matrices of $\boldsymbol{y}$ and $\boldsymbol{x}$ by

$$\begin{aligned} \mathrm{Y} &= (\boldsymbol{y}_1^{(1)}, \ldots, \boldsymbol{y}_{n_1}^{(1)}, \ldots, \boldsymbol{y}_1^{(q)}, \ldots, \boldsymbol{y}_{n_q}^{(q)})' \\ &= (\boldsymbol{y}_{(1)}, \ldots, \boldsymbol{y}_{(t)}), \\ \mathrm{X} &= (\boldsymbol{x}_1^{(1)}, \ldots, \boldsymbol{x}_{n_1}^{(1)}, \ldots, \boldsymbol{x}_1^{(q)}, \ldots, \boldsymbol{x}_{n_q}^{(q)})'. \end{aligned}$$

Suppose that a scoring of Y is of the form

$$\mathrm{Y} = \mathrm{Z}(\boldsymbol{\theta}_1, \ldots, \boldsymbol{\theta}_m) = \mathrm{Z}\Theta.$$

Here $\mathrm{Z} = (z_{ij})$, where $z_{ij} = 1$ if the group of the ith sample is Π_j, and 0 otherwise. In the following, without loss of information we assume that $\bar{\boldsymbol{x}} = \boldsymbol{0}$; that is, the observations of $\boldsymbol{X}$ have been standardized with respect to the mean. The regression of Y to X is $\mathrm{X}(\boldsymbol{\beta}_1, \ldots, \boldsymbol{\beta}_t) = \mathrm{X}\mathcal{B}$. The optimal scoring problem involves finding the coefficients $\boldsymbol{\beta}_j (j = 1, \ldots, t)$ and the parameter Θ that minimize the following average squared residual:

$$\begin{aligned} ASR &= \frac{1}{n} \sum_{j=1}^{t} \|\boldsymbol{y}_{(i)} - \mathrm{X}\boldsymbol{\beta}_j\|^2 \\ &= \frac{1}{n} \sum_{j=1}^{t} \|\mathrm{Z}\boldsymbol{\theta}_j - \mathrm{X}\boldsymbol{\beta}_j\|^2 \\ &= \frac{1}{n} \operatorname{tr} (\mathrm{Z}\Theta - \mathrm{X}\mathcal{B})'(\mathrm{Z}\Theta - \mathrm{X}\mathcal{B}) \equiv ASR(\Theta, \mathcal{B}), \end{aligned} \tag{9.5.1}$$

under the restriction $\Theta'Z'Z\Theta = nI_p$.

When Θ is fixed, ASR is minimized (see [A.2.11]) at

$$\hat{\mathcal{B}}_0 = (X'X)^{-1}X'Z\Theta \tag{9.5.2}$$

and

$$ASR(\Theta, \hat{\mathcal{B}}_0) = \frac{1}{n}\{\operatorname{tr}(Z\Theta)'Z\Theta - \operatorname{tr}(Z\Theta)'P_X Z\Theta\},$$

where $P_X = X(X'X)^{-1}X'$. The first term is t, since the restriction $\Theta'Z'Z\Theta = nI_p$ is satisfied. The optimal problem with respect to Θ is to find Θ such that $(1/n)\operatorname{tr}(Z\Theta)'P_X Z\Theta$ is maximized under $\Theta'Z'Z\Theta = nI_p$. The optimal Θ is the solution of the general characteristic root problem:

$$Z'P_X Z\Theta = Z'Z\Theta D, \quad \Theta'Z'Z\Theta = nI_t, \tag{9.5.3}$$

where $D = \operatorname{diag}(d_1, \ldots, d_t)$, and $\ell_1 > \cdots > \ell_t$ are the characteristic roots of $(Z'Z)^{-1}Y'P_X Y$. Further, let

$$S_t = X'X, \quad S_b = X'\ Z(Z'Z)^{-1}Z'X.$$

Then S_t and S_b are the total and between-variation matrices, respectively. The values $d_i (i = 1, \ldots, t)$ are also the characteristic roots $S_b S_t^{-1}$.

The optimum $\mathcal{B}$ is obtained by substituting the optimum Θ in (9.5.2). Multiplying $X'Z(Z'Z)^{-1}$ for both sides of the first expression of (9.5.3) from the left, and using (9.5.2), we can see that the optimum $\mathcal{B}$ is the solution of the general characteristic root problem:

$$S_b\mathcal{B} = S_t\mathcal{B}D, \quad \mathcal{B}'S_t\mathcal{B} = nD, \tag{9.5.4}$$

which is also expressed as

$$S_b\mathcal{B} = S_w\mathcal{B}L, \quad \mathcal{B}'S_w\mathcal{B} = nD(I_t - D), \tag{9.5.5}$$

where $S_w = S_t - S_b$ is the within-variation matrix, and $L = D(I_t - D)^{-1}$. We have considered the coefficient vectors of the canonical discriminant functions given by (9.4.2) and (9.4.5). Note that the solution $\tilde{\mathcal{B}}$ in (9.5.5) comprises the coefficient vectors in the canonical discriminant functions. In fact, let B^* and $\tilde{B}$ be the coefficient matrices which consist of the coefficient vectors given by (9.4.2) and (9.4.5), respectively. Then there are the following relationships:

$$B^* = \mathcal{B}\{D(I_t - D)\}^{-1/2}, \quad \tilde{B} = \mathcal{B}D^{-1/2}. \tag{9.5.6}$$

Based on the regression approach, we now have a discriminant rule for classifying future observations. A nearest centroid rule is given as follows: We assign a future observation $\boldsymbol{x}$ to the population Π_j that minimizes

$$\|\{D(I_t - D)\}^{-1/2}\mathcal{B}'(\boldsymbol{x} - \bar{\boldsymbol{x}}^{(j)})\|^2.$$

Using the regression approach, it is possible to consider the discriminant analysis based on a transformed variate:

$$\boldsymbol{T}\colon \boldsymbol{X} \to \boldsymbol{T}(\boldsymbol{X}) = (T_1(\boldsymbol{X}), \ldots, T_s(\boldsymbol{X}))'$$

instead of $\boldsymbol{X} = (X_1, \ldots, X_p)'$. Such a discriminant method is called *flexible discriminant analysis* [Hastie et al., 1994].

9.6 High-Dimensional Approach

9.6.1 Penalized Discriminant Analysis

The discriminant method based on the idea of ridge regression is to use

$$\hat{\mathcal{B}}_* = \{\mathrm{X}'\mathrm{X} + (1/n)\Omega\}^{-1}\mathrm{X}'\mathrm{Z}\Theta \tag{9.6.1}$$

instead of the coefficient vectors of the discriminant functions in (9.5.2). Here $\Omega = \lambda \mathrm{I}_p$, and λ is the ridge parameter. This method can be regarded as a penalized discriminant analysis [Friedman, 1989; Hastie et al., 1994, 1995; Ghosh, 2003]. For the ridge regression-based optimal scoring method, the following objective function is used:

$$ASR_*(\Theta, \mathcal{B}) = \frac{1}{n}\{\,\mathrm{tr}\,(\mathrm{Z}\Theta - \mathrm{X}\mathcal{B})'(\mathrm{Z}\Theta - \mathrm{X}\mathcal{B}) + \mathrm{tr}\,\mathcal{B}'\Omega\mathcal{B}]\,. \tag{9.6.2}$$

The optimal scoring is obtained under the restriction $\Theta'\mathrm{Z}'\mathrm{Z}\Theta = n\mathrm{I}_p$. Then, similarly, the optimum solution of $\mathcal{B}$ is given by (9.6.1) when Θ is fixed, and

$$ASR_*(\Theta, \hat{\mathcal{B}}_*) = \frac{1}{n}\{\,\mathrm{tr}\,(\mathrm{Z}\Theta)'\mathrm{Z}\Theta - \mathrm{tr}(\mathrm{Y}\Theta)'(\mathrm{X}'\mathrm{X} + \Omega)\mathrm{X}'\mathrm{Z}\Theta\}.$$

Therefore, the optimum Θ is given by the solution

$$\mathrm{Z}'\mathrm{X}\{\mathrm{S}_t + (1/n)\Omega\}\mathrm{X}'\mathrm{Z}\Theta = (1/n)\mathrm{Z}'\mathrm{Z}\Theta D_\alpha, \quad \Theta'\mathrm{Z}'\mathrm{Z}\Theta = n\mathrm{I}_t, \tag{9.6.3}$$

where $\mathrm{D}_\alpha = \mathrm{diag}\,(\alpha_1, \ldots, \alpha_m)$ and $\alpha_1 > \cdots > \alpha_t$ are the characteristic roots of $(\mathrm{Z}'\mathrm{Z})^{-1}\mathrm{Z}'\mathrm{X}\{\mathrm{S}_t + (1/n)\Omega\}\mathrm{X}'\mathrm{Z}$.

9.6.2 Other Approaches

When $p > n$, the classical discrimination rule becomes inapplicable since S is singular. In this section we explain some high-dimensional ($p > n$) discriminant methods.

Saranadasa (1993) proposed a method of looking at the classification problem in terms of the one-way MANOVA problem. The method is to classify a new observation by minimizing a suitable norm of the matrix of sums of squares and products due to within groups. Let W_i be the new matrix of sums of squares and products due to within groups when $\boldsymbol{X}$ is placed in Π_i. Then

$$\mathbf{W}_i = m\mathbf{S} + a_i(\boldsymbol{X} - \bar{\boldsymbol{X}}_1)(\boldsymbol{X} - \bar{\boldsymbol{X}}_1)',$$

where S is the pooled sample covariance matrix, $m = n_1 + n_2 - 2$, and $a_i = n_i/(n_i+1), i = 1, 2$. Using the trace criterion, we may classify $\boldsymbol{X}$ into Π_1 if

$$\operatorname{tr}\mathbf{W}_1 < \operatorname{tr}\mathbf{W}_2.$$

This is equivalent to the classification procedure based on the classification statistic defined by

$$T = a_1(\boldsymbol{X} - \bar{\boldsymbol{X}}_1)'(\boldsymbol{X} - \bar{\boldsymbol{X}}_1) - a_2(\boldsymbol{X} - \bar{\boldsymbol{X}}_2)'(\boldsymbol{X} - \bar{\boldsymbol{X}}_2). \tag{9.6.4}$$

If $T \geq 0 (< 0)$, we classify $\boldsymbol{x} \in \Pi_1(\Pi_2)$. In the procedure, the sample covariance matrix S is not used. Under a high-dimensional framework:

$$\text{A3: } p \to \infty, \ n_i (i = 1, 2) \quad \text{fixed.}$$

Saranadasa (1993) obtained asymptotic distribution of T.

The maximum likelihood ratio (MLR) procedure classifies $\boldsymbol{X}$ into group Π_1 if

$$(1 + n_1^{-1})^{-1} D_{\mathbf{S}}^2(\boldsymbol{X}, \bar{\boldsymbol{X}}_1) < (1 + n_2^{-1})^{-1} D_{\mathbf{S}}^2(\boldsymbol{X}, \bar{\boldsymbol{X}}_2);$$

otherwise, it is classified into population Π_2. Here

$$D_{\mathbf{S}}^2(\boldsymbol{X}, \bar{\boldsymbol{X}}_i) = (\boldsymbol{X} - \bar{\boldsymbol{X}}_i)'\mathbf{S}^{-1}(\boldsymbol{X} - \bar{\boldsymbol{X}}_i), i = 1, 2.$$

Srivastava (2007) proposed the MLR rule for $p > n$ by using Moore-Penrose inverse $\mathbf{S}^+$ instead of $\mathbf{S}^{-1}$. The Moore-Penrose generalized inverse of a matrix A is defined by A^+, which satisfies the following four conditions:

$$AA^+A = A, \ A^+AA^+ = A^+, \ (AA^+)' = AA^+, \ (A^+A)' = A^+A. \tag{9.6.5}$$

Furthermore, the author gives an expression for the error of misclassification for large p and $n = O(p^\delta)$, $0 < \delta < 1$. Let the spectral decomposition of S be denoted by

$$\mathbf{S} = \mathbf{HLH}', \quad \mathbf{H}'\mathbf{H} = I_k, \quad \mathbf{L} = \operatorname{diag}(\ell_1, \ldots, \ell_k), \quad \ell_1 \geq \cdots \geq \ell_k > 0.$$

Then the Moore-Penrose inverse matrix is defined by

$$\mathbf{S}^+ = \mathbf{HL}^{-1}\mathbf{H}'.$$

The discriminant function is

$$W^+ = (\bar{\boldsymbol{X}}_1 - \bar{\boldsymbol{X}}_2)'\mathbf{S}^+\left\{\boldsymbol{X} - \frac{1}{2}(\bar{\boldsymbol{X}}_1 + \bar{\boldsymbol{X}}_2)\right\}.$$

Similarly, for the quadratic discriminant, let

$$D_i^{+2} = (1 + N_i)^{-1}(\boldsymbol{X} - \bar{\boldsymbol{X}}_i)'\mathbf{S}^+(\boldsymbol{X} - \bar{\boldsymbol{X}}_i), \quad i = 1, 2.$$

We classify as follows:

$$D_1^{+2} < D_2^{+2} \implies \boldsymbol{x} \in \Pi_1, \quad D_1^{+2} > D_2^{+2} \implies \boldsymbol{x} \in \Pi_2.$$

Srivastava (2007) derived asymptotic approximations for the probability of misclassification under $\lim_{n_i \to \infty} \lim_{p \to \infty}$.

The modification of AIC has been examined [see Fujikoshi, 2002b] for the case when both the dimension p and the sample size n are large, but $p < n$.

Problems

9.1 Let $\hat{\boldsymbol{\beta}} = \mathbf{S}^{-1}(\bar{\boldsymbol{X}}^{(1)} - \bar{\boldsymbol{X}}^{(2)})$ be the coefficient vector in the linear discriminant function. Under normality we may assume that

$$\begin{aligned}
&\bar{\boldsymbol{X}}^{(1)} \sim \mathrm{N}_p(\boldsymbol{\mu}^{(1)}, n_1^{-1}\Sigma), \quad \bar{\boldsymbol{X}}^{(2)} \sim \mathrm{N}_p(\boldsymbol{\mu}^{(2)}, n_2^{-1}\Sigma), \\
&n\mathbf{S} \sim \mathrm{W}_p(n, \Sigma), \quad n = n_1 + n_2 - 2,
\end{aligned}$$

and $\bar{\boldsymbol{X}}^{(1)}$, $\bar{\boldsymbol{X}}^{(2)}$, and $\mathbf{S}$ are independent. Then show [see also Das Gupta, 1968] that

$$\begin{aligned}
\mathrm{E}[\hat{\boldsymbol{\beta}}] &= \frac{n}{n-p-1}\Sigma^{-1}(\boldsymbol{\mu}^{(1)} - \boldsymbol{\mu}^{(2)}), \\
\mathrm{Var}(\hat{\boldsymbol{\beta}}) &= \frac{n^2(n-1)}{(n-p)(n-p-1)(n-p-3)}\left[(n-1)\frac{n_1+n_2}{n_1 n_2} + \delta^2\right. \\
&\left.+\frac{n-p+1}{n-p-1}\Sigma^{-1}(\boldsymbol{\mu}^{(1)} - \boldsymbol{\mu}^{(2)})(\boldsymbol{\mu}^{(1)} - \boldsymbol{\mu}^{(2)})'\right]\Sigma^{-1},
\end{aligned}$$

where $\delta^2 = (\boldsymbol{\mu}^{(1)} - \boldsymbol{\mu}^{(2)})'\Sigma^{-1}(\boldsymbol{\mu}^{(1)} - \boldsymbol{\mu}^{(2)})$.

Hint: Use Theorems 2.2.7 and 2.2.8.

9.2 Let B and W be the between-group and within-group SSP matrices based on the samples of p-dimensional random vector $\boldsymbol{X}$ and $\mathrm{T} = \mathrm{B} + \mathrm{W}$. Let W_{11} and T_{11} be the first $k \times k$ submatrices of W and T, respectively. When $q = 2$, show that

$$\frac{|\mathrm{T}|/|\mathrm{W}|}{|\mathrm{T}_{11}|/|\mathrm{W}_{11}|} = 1 + \frac{c(D_p^2 - D_k^2)}{n - 2 + cD_k^2},$$

where D_p^2 and D_k^2 are the sample Mahalanobis distances based on $\boldsymbol{X}$ and $\boldsymbol{X}_1$.

Hint: When $q = 2$, we can write $\mathrm{B} = \boldsymbol{a}\boldsymbol{a}'$. Using

$$\begin{vmatrix} \mathrm{A}_{11} & \mathrm{A}_{12} \\ \mathrm{A}_{21} & \mathrm{A}_{22} \end{vmatrix} = |\mathrm{A}_{11}| \cdot |\mathrm{A}_{22\cdot 1}| = |\mathrm{A}_{22}| \cdot |\mathrm{A}_{11\cdot 2}|,$$

we have

$$|T| = |W + aa'| = |W|(1 + a'W^{-1}a).$$

9.3 Let $\boldsymbol{x}$ be a p-dimensional observation vector on an individual to be classified into one of the multivariate normal populations $\Pi_i : N_p(\boldsymbol{\mu}^{(i)}, \Sigma^{(i)})$, $i = 1, \ldots, q$. The random vector associated with $\boldsymbol{x}$ is denoted by $\boldsymbol{X} = (X_1, \ldots, X_p)'$. Let $\boldsymbol{X} = (\boldsymbol{X}'_1, \boldsymbol{X}'_2)'$, where $\boldsymbol{X}_1 = (X_1, \ldots, X_k)'$, and partition $\boldsymbol{\mu}^{(i)}$ and $\Sigma^{(i)}$ as

$$\boldsymbol{\mu}^{(i)} = \begin{pmatrix} \boldsymbol{\mu}_1^{(i)} \\ \boldsymbol{\mu}_2^{(i)} \end{pmatrix}, \quad \Sigma^{(i)} = \begin{pmatrix} \Sigma_{11}^{(i)} & \Sigma_{12}^{(i)} \\ \Sigma_{21}^{(i)} & \Sigma_{22}^{(i)} \end{pmatrix}, \quad i = 1, \ldots, q.$$

Following Rao (1970), Pynnönen (1987) found $\boldsymbol{X}_1$ to be sufficient for discrimination between the populations $\Pi_1, \ldots, \Pi_q$ if the conditional distributions of $\boldsymbol{X}$ given $\boldsymbol{X}_1$ are the same in each population. Then show that $\boldsymbol{X}_1$ is sufficient if and only if

$$\begin{aligned} &\boldsymbol{\mu}_{2\cdot1}^{(1)} = \cdots = \boldsymbol{\mu}_{2\cdot1}^{(q)}, \\ &\Sigma_{21}^{(1)}(\Sigma_{11}^{(1)})^{-1} = \cdots = \Sigma_{21}^{(q)}(\Sigma_{11}^{(q)})^{-1}, \quad \Sigma_{22\cdot1}^{(1)} = \cdots = \Sigma_{22\cdot1}^{(q)}. \end{aligned}$$

where $\boldsymbol{\mu}_{2\cdot1}^{(i)} = \boldsymbol{\mu}_2^{(i)} - \Sigma_{21}^{(i)}(\Sigma_{11}^{(i)})^{-1}\boldsymbol{\mu}_1^{(i)}$ and $\Sigma_{22\cdot1}^{(i)} = \Sigma_{22}^{(i)} - \Sigma_{21}^{(i)}(\Sigma_{11}^{(i)})^{-1}\Sigma_{12}^{(i)}$.

9.4 Let B and W be the between-group and within-group SSP matrices, respectively, both based on samples from the distribution of $\boldsymbol{X} = (X_1, \ldots, X_p)$. Let $Y = \boldsymbol{b}'\boldsymbol{X}$ be a linear combination of the X_i. The between-group and within-group sums of squares of Y are expressed as $\boldsymbol{b}'B\boldsymbol{b}$ and $\boldsymbol{b}'W\boldsymbol{b}$. Consider a transformed random vector $\boldsymbol{U} = C'\boldsymbol{X}$, where Cis a $p \times q$ constant matrix. Let $\boldsymbol{Z} = (\boldsymbol{U}', \boldsymbol{X}')'$, and consider a linear combination $\boldsymbol{a}'\boldsymbol{Z} = \boldsymbol{a}'_1\boldsymbol{U} + \boldsymbol{a}'_2\boldsymbol{X}$. Let us denote the between-group and within-group sums of squares of $\boldsymbol{a}'\boldsymbol{Z}$ by $\boldsymbol{a}'B^*\boldsymbol{a}$ and $\boldsymbol{a}'W^*\boldsymbol{a}$, respectively. Then, show that

$$\frac{\boldsymbol{a}'B^*\boldsymbol{a}}{\boldsymbol{a}'W^*\boldsymbol{a}} \le \ell_1$$

where ℓ_1 is the largest characteristic root of BW^{-1}. Examine the case that the equlity holds.

9.5 Let R_W be the approximation for $(1/2)P_W(2|1) + (1/2)P_W(1|2)$ up to the order $O(n^{-1})$ (see Corollary 9.3.1). Then, show that

$$\begin{aligned} R = \Phi\left(-\frac{1}{2}\Delta\right) + \varphi\left(-\frac{1}{2}\Delta\right)\Bigg[&\frac{1}{4m}(p-1)\Delta \\ &+\frac{1}{16\Delta}\left(\frac{1}{n_1} + \frac{1}{n_2}\right)\{\Delta^2 + 4(p-1)\}\Bigg]. \end{aligned}$$

Further, let the quantities R and Δ for a subset J of all the variables $X_1, \ldots,$ X_p denote by $R(J)$ and $\Delta(J)$, respectively. Then, if $\Delta(J) = \Delta$ for a subset J, then show that

$$R - R(J) = \frac{1}{4}\varphi\left(-\frac{1}{2}\Delta\right)(p - N(J))\left\{\left(\frac{1}{n_1} + \frac{1}{n_2}\right)\frac{1}{\Delta} + \frac{1}{m}\Delta\right\} > 0,$$

where $N(J)$ denotes the number of the elements in J.

9.6 (Continuation) Examine the same properties of R_Z: the approximation for $(1/2)\mathrm{P}_Z(2|1) + (1/2)\mathrm{P}_Z(1|2)$ up to the order $\mathrm{O}(n^{-1})$.

Chapter 10

Principal Component Analysis

The aim of principal component analysis is to describe the total variation of p-variate random vector $\boldsymbol{X} = (X_1, \ldots, X_p)'$ into variations of a few uncorrelated linear combinations of the X_i.

The notion and technique of principal component analysis were introduced by Pearson (1901) as a tool for fitting planes to a system of points in space and were later generalized by Hotelling (1933) for analyzing correlation structures.

The sampling distributions of principal components are discussed mainly under a large-sample framework. Related to the selection of principal components, we discuss a method based on model selection.

The problem is treated as one of selecting a covariance structure based on DIC which is similar to AIC. We also treat a test for redundancy of some original variables for a set of pricipal components. Models related to principal components are also discussed.

10.1 Definition of Principal Components

In this section we consider the principal components in a population. First we define the principal components of a random vector $\boldsymbol{X}$. Their concepts will be explained through two basic theorems.

Definition 10.1.1 Let $\boldsymbol{X}$ be a p-dimensional random vector with mean $\boldsymbol{\mu}$ and covariance matrix $\Sigma = (\sigma_{ij})$. Let $\lambda_1 \geq \cdots \geq \lambda_p$ be the ordered characteristic roots of Σ and $\boldsymbol{\gamma}_1, \ldots, \boldsymbol{\gamma}_p$ the corresponding characteristic vectors satisfying $\gamma_i'\gamma_j = \delta_{ij}$, $i, j = 1, \ldots, p$. Then the ith principal component of $\boldsymbol{X}$ is defined by

$$Y_i = \gamma_i'(\boldsymbol{X} - \boldsymbol{\mu}), \quad i = 1, \ldots, p. \tag{10.1.1}$$

The characteristic roots λ_i and vectors $\boldsymbol{\gamma}_i$ of Σ are defined by solutions of the equations

$$\Sigma\boldsymbol{\gamma}_i = \lambda_i\boldsymbol{\gamma}_i, \quad i = 1, \ldots, p, \tag{10.1.2}$$

where the $\boldsymbol{\gamma}_i$ are normalized as

$$\gamma_i'\gamma_i = 1, \quad \gamma_i'\gamma_j = 0, \quad i \neq j.$$

Let

$$\Lambda = \mathrm{diag}\,(\lambda_1, \ldots, \lambda_p), \quad \Gamma = (\gamma_1, \ldots, \gamma_p).$$

Then

$$\Sigma = \Gamma\Lambda\Gamma' = \lambda_1\gamma_1\gamma_1' + \cdots + \lambda_p\gamma_p\gamma_p'. \tag{10.1.3}$$

If λ_i is simple, the ith principal component is defined uniquely except for the multiplication by -1. If some of the λ_i are equal, the corresponding principal components are not defined uniquely but can be given by any orthonormal basis of the latent subspace belonging to that same root.

In the following theorems, some fundamental properties of principal components are given.

Theorem 10.1.1 *For $i = 1, \ldots, p$, let Y_i be the ith principal component and set $\boldsymbol{Y} = (Y_1, \ldots, Y_p)'$. Then:*

(1) $\mathrm{E}(\boldsymbol{Y}) = \boldsymbol{0}$, $\mathrm{Var}(\boldsymbol{Y}) = \Lambda$.

(2) $\boldsymbol{X} = \boldsymbol{\mu} + \Gamma\boldsymbol{Y} = \boldsymbol{\mu} + \sum_{i=1}^p \gamma_i Y_i$.

(3) $\sum_{i=1}^p \mathrm{Var}(Y_i) = \sigma_{11} + \cdots + \sigma_{pp}$.

Proof. We can write

$$\boldsymbol{Y} = \Gamma'(\boldsymbol{X} - \boldsymbol{\mu}). \tag{10.1.4}$$

Therefore,

$$\mathrm{E}(\boldsymbol{Y}) = \mathrm{E}(\boldsymbol{X} - \boldsymbol{\mu}) = \boldsymbol{0}.$$
$$\mathrm{Var}(\boldsymbol{Y}) = \mathrm{E}(\boldsymbol{Y}\boldsymbol{Y}') = \Gamma'\Sigma\Gamma = \Lambda.$$

Multiplying Γ from the left of (10.1.4), we obtain

$$\Gamma\boldsymbol{Y} = \Gamma\Gamma'(\boldsymbol{X} - \boldsymbol{\mu}) = \boldsymbol{X} - \boldsymbol{\mu},$$

which implies (2). To show (3), from (2) it is enough to see that

$$\mathrm{tr}\,\Sigma = \lambda_1 + \cdots + \lambda_p.$$

This is obtained by $\mathrm{tr}\,\Sigma = \mathrm{tr}\,\Gamma\Lambda\Gamma' = \mathrm{tr}\,\Lambda\Gamma'\Gamma = \mathrm{tr}\,\Lambda$. □

Theorem 10.1.2 *Let $Y = \boldsymbol{\gamma}'(\boldsymbol{X} - \boldsymbol{\mu})$ be a linear function of $\boldsymbol{X} - \boldsymbol{\mu}$ with $\boldsymbol{\gamma}'\boldsymbol{\gamma} = 1$. Then:*

(1) *The variance of Y is maximized when $Y = Y_1$, the first principal component.*

(2) *If Y is uncorrelated with the first k $(< p)$ principal components $Y_1, \ldots, Y_k$, the variance of Y is maximized when $Y = Y_{k+1}$.*

Proof. Note that $\mathrm{Var}(Y) = \boldsymbol{\gamma}'\Sigma\boldsymbol{\gamma}$. We can write

$$\boldsymbol{\gamma} = c_1\boldsymbol{\gamma}_1 + \cdots + c_p\boldsymbol{\gamma}_p = \Gamma\boldsymbol{c},$$

where $\boldsymbol{c}$ satisfies $\boldsymbol{c}'\boldsymbol{c} = 1$, since $\boldsymbol{\gamma}'\boldsymbol{\gamma} = \boldsymbol{c}'\Gamma'\Gamma\boldsymbol{c} = \boldsymbol{c}'\boldsymbol{c}$. Therefore,

$$\mathrm{Var}(\boldsymbol{Y}) = \boldsymbol{c}'\Gamma'\Sigma\Gamma\boldsymbol{c} = c_1^2\lambda_1 + \cdots + c_p^2\lambda_p. \tag{10.1.5}$$

This is maximized when $c_1 = 1$, $c_2 = \cdots = c_p = 0$ (i.e., $\boldsymbol{\gamma} = \boldsymbol{\gamma}_1$). For the second statement, note that

$$\begin{aligned}\mathrm{Cov}(\boldsymbol{Y}, \boldsymbol{Y}_i) &= \mathrm{E}[\boldsymbol{\gamma}'(\boldsymbol{X} - \boldsymbol{\mu})(\boldsymbol{X} - \boldsymbol{\mu})\boldsymbol{\gamma}_i] \\ &= \boldsymbol{c}'\Gamma'\Sigma\boldsymbol{\gamma}_i = \lambda_i\boldsymbol{c}'\Gamma'\boldsymbol{\gamma}_i = \lambda_i c_i.\end{aligned}$$

The second problem is to maximize (10.1.5) under $c_1 = \cdots = c_k = 0$. It is easy to see that the maximization occurs when $\boldsymbol{\gamma} = \boldsymbol{\gamma}_{k+1}$. □

From Theorem 10.1.1 (3) the contribution to the total variance $\mathrm{tr}\,\Sigma$ due to the ith principal component is measured by $\lambda_i/(\lambda_1 + \cdots + \lambda_p) = \lambda_i/\mathrm{tr}\,\Sigma$. The quantity $\lambda_i/\mathrm{tr}\,\Sigma$ is called a contribution ratio of the ith principal component.

An important use of principal components is to explain most variations of $\boldsymbol{X}$ by fewer variates. Usually, components will be computed until a certain large proportion of the total variance is explained. Thus the cumulative contribution ratios

$$\begin{aligned}\psi_k &= (\lambda_1 + \cdots + \lambda_k)/(\lambda_1 + \cdots + \lambda_p) \\ &= (\lambda_1 + \cdots + \lambda_k)/\mathrm{tr}\,\Sigma\end{aligned} \tag{10.1.6}$$

are computed successively until $\psi_k \geq c$; for example, $c = 0.8$. If ψ_k satisfies this criterion with a small k, a big reduction in dimensionality is achieved, and if k is large (≥ 4 or 5), the analysis may be fruitless, because of difficulty in the interpretation of principal components.

When the units of the X_i are different, we commonly consider standardized variates defined by

$$\boldsymbol{Z} = (Z_1, \ldots, Z_p)', \quad Z_i = (X_i - \mu_i)/\sqrt{\sigma_{ii}}.$$

The covariance matrix of $\boldsymbol{Z}$ is equal to the correlation matrix,

$$\mathrm{P} = \begin{pmatrix} \rho_{11} & \cdots & \rho_{1p} \\ \vdots & \ddots & \vdots \\ \rho_{p1} & \cdots & \rho_{pp} \end{pmatrix},$$

where

$$\rho_{ij} = \frac{\mathrm{Cov}(X_i, X_j)}{\sqrt{\mathrm{Var}(X_i)}\sqrt{\mathrm{Var}(X_j)}} = \frac{\sigma_{ij}}{\sigma_i \sigma_j}, \quad \rho_{ii} = 1.$$

In this case, the coefficient vector of the principal components of $\boldsymbol{Z}$ is defined by the characteristic vectors of the correlation matrix P. By the way, the ith principal component is expressed as

$$\begin{aligned} Y_i &= \gamma_{1i}(X_1 - \mu_1) + \cdots + \gamma_{pi}(X_p - \mu_p) \\ &= \gamma_{1i}\sigma_1 Z_1 + \cdots + \gamma_{pi}\sigma_p Z_p. \end{aligned}$$

However, the coefficient vector is not the same as the ith characteristic vector of P.

The correlation between variable X_i and the jth principal component Y_j is given by

$$\begin{aligned} \rho(X_i, Y_j) &= \frac{\mathrm{Cov}(X_i, Y_j)}{\sqrt{\mathrm{Var}(X_i)\mathrm{Var}(Y_j)}} \\ &= \frac{\lambda_j \gamma_{ij}}{\sqrt{\lambda_j \sigma_{ii}}} = \frac{\sqrt{\lambda_j}\gamma_{ij}}{\sqrt{\sigma_{ii}}}, \end{aligned}$$

since

$$\begin{aligned} \mathrm{Cov}(\boldsymbol{X}, \boldsymbol{Y}) &= \mathrm{E}[(\boldsymbol{X} - \boldsymbol{\mu}) \cdot \{\Gamma'(\boldsymbol{X} - \boldsymbol{\mu})\}'] \\ &= \Sigma\Gamma = \Gamma\Lambda. \end{aligned}$$

This correlation is called the *factor loading* of X_i on Y_j.

10.2 Optimality of Principal Components

One of important optimality properties of principal components has been given in Theorem 10.1.2. This is called *variation optimality*. By the relation $\boldsymbol{Y} = \Gamma'(\boldsymbol{X} - \boldsymbol{\mu})$, the original random vector $\boldsymbol{X}$ can be expressed in terms of the principal component vector $\boldsymbol{Y}$ as

$$\boldsymbol{X} = \boldsymbol{\mu} + \Gamma\boldsymbol{Y} = \boldsymbol{\mu} + Y_1\gamma_1 + \cdots + Y_p\gamma_p$$

and in the case when the first k principal components are used, $\boldsymbol{X}$ is approximated by

$$\hat{\boldsymbol{X}} = \boldsymbol{\mu} + Y_1\gamma_1 + \cdots + Y_k\gamma_k.$$

The approximation is writtem as

$$\hat{\boldsymbol{X}} = Y_1\gamma_1 + \cdots + Y_k\gamma_k + \boldsymbol{\mu} = (\gamma_1, \ldots, \gamma_k)\begin{pmatrix} Y_1 \\ \vdots \\ Y_k \end{pmatrix} + \boldsymbol{\mu}.$$

This can be regarded as an approximation of the form

$$\mathrm{A}\boldsymbol{Z} + \boldsymbol{b}, \quad \mathrm{A}\colon p \times k, \quad \boldsymbol{b}\colon p \times 1,$$

where $\boldsymbol{Z}$ is a k-variate random vector and A and $\boldsymbol{b}$ are a fixed matrix and vector. Suppose that we want to approximate $\boldsymbol{X}$ by $\mathrm{A}\boldsymbol{Z} + \boldsymbol{b}$. Then the error of approximation may be measured by the characteristic roots of

$$\mathrm{M} = \mathrm{E}[(\boldsymbol{X} - \mathrm{A}\boldsymbol{Z} - \boldsymbol{b})(\boldsymbol{X} - \mathrm{A}\boldsymbol{Z} - \boldsymbol{b})']. \tag{10.2.1}$$

Rao (1964) considered the problem for $\boldsymbol{Z} = \mathrm{B}'\boldsymbol{X}$, where B is a $p \times k$ matrix and the criterion of goodness of approximation is $\mathrm{tr}\,\mathrm{M}$ or $\|\mathrm{M}\| = \{\mathrm{tr}\,(\mathrm{MM}')\}^{1/2}$. Darroch (1965) extended Rao's result to any k-variate random vector $\boldsymbol{Z}$. Okamoto and Kanazawa (1968) and Okamoto (1969) solved the problem in the general case, which is given in the following theorem.

Theorem 10.2.1 *(Information loss optimality) Let* M *be the error matrix defined by* (10.2.1). *Then it holds that*

$$ch_i(\mathrm{M}) \geq \lambda_{i+k}, \quad i = 1, \ldots, p, \tag{10.2.2}$$

where $\lambda_1 \geq \cdots \geq \lambda_p$ *are the characteristic roots of* Σ *and* $\lambda_j = 0$ *for* $j > p$. *The equality in* (10.2.2) *is attained for all* i *when*

$$\mathrm{A}\boldsymbol{Z} + \boldsymbol{b} = \hat{\boldsymbol{X}} = Y_1\gamma_1 + \cdots + Y_k\gamma_k + \boldsymbol{\mu}. \tag{10.2.3}$$

Proof. Without loss of generality, we may assume that $\mathrm{E}(\boldsymbol{Z}) = \boldsymbol{0}$ and $\mathrm{Var}(\boldsymbol{Z}) = \mathrm{I}_k$. Then we have

$$\mathrm{M} = \mathrm{M}_0 + (\boldsymbol{\mu} - \boldsymbol{b})(\boldsymbol{\mu} - \boldsymbol{b})',$$

where $\mathrm{M}_0 = \mathrm{E}[(\boldsymbol{X} - \boldsymbol{\mu} - \mathrm{A}\boldsymbol{Z})(\boldsymbol{X} - \boldsymbol{\mu} - \mathrm{A}\boldsymbol{Z})']$. It is easy to see that M_0 can be written as

$$\mathrm{M}_0 = \Sigma - \mathrm{BB}' + (\mathrm{B} - \mathrm{A})(\mathrm{B} - \mathrm{A})',$$

where $\mathrm{B} = \mathrm{Cov}(\boldsymbol{X}, \boldsymbol{Z}) = \mathrm{E}[(\boldsymbol{X} - \boldsymbol{\mu})\boldsymbol{Z}']$. Therefore, we have, using [A.2.9],

$$\begin{aligned} ch_i(\mathrm{M}) &\geq ch_i(\mathrm{M}_0) \\ &\geq ch_i(\Sigma - \mathrm{BB}') \\ &\geq ch_{k+i}(\Sigma) = \lambda_{k+i} \end{aligned}$$

and the equality in (10.2.2) holds for all i when $\mathrm{A}\boldsymbol{Z} + \boldsymbol{b}$ is the right-hand side of (10.2.3). □

We observe from Theorem 10.2.1 that $\hat{\boldsymbol{X}}$ is best in the class of $\mathrm{A}\boldsymbol{Z}+\boldsymbol{b}$ for any criterion $g(ch_1(\mathrm{M}),\ldots,ch_p(\mathrm{M}))$ such that $g(\cdot)$ is an increasing function in each argument. For other optimal properties of principal components, see Okamoto (1969), Kshirsagar (1972, Chap. 11), Obenchain (1972), and Chen (1974).

10.3 Sample Principal Components

In practice, the covariance matrix Σ of $\boldsymbol{X}$ is rarely known, and hence the characteristic roots $\lambda_1,\ldots,\lambda_p$ of Σ and the corresponding vectors $\gamma_1,\ldots,\gamma_p$ must be estimated from a random sample. Let

$$\mathbf{X} = \begin{pmatrix} X_{11} & \cdots & X_{1p} \\ \vdots & \vdots & \vdots \\ X_{n1} & \cdots & X_{np} \end{pmatrix} = \begin{pmatrix} \boldsymbol{X}'_1 \\ \vdots \\ \boldsymbol{X}'_n \end{pmatrix}$$

be the observation matrix based on a random sample of size n ($> p$) drawn from the population of $\boldsymbol{X}$. We denote the centralized observation matrix by

$$\tilde{\mathbf{X}} = \begin{pmatrix} X_{11}-\bar{X}_1 & \cdots & X_{1p}-\bar{X}_p \\ \vdots & \vdots & \vdots \\ X_{n1}-\bar{X}_1 & \cdots & X_{np}-\bar{X}_p \end{pmatrix},$$

where $\bar{\boldsymbol{X}} = (1/n)\sum_{i=1}^{n}\boldsymbol{X}_i$. Then we compute the ordinary estimator $\mathbf{S}$ of Σ:

$$\mathbf{S} = \frac{1}{n-1}\sum_{i=1}^{n}(\boldsymbol{X}_i-\bar{\boldsymbol{X}})(\boldsymbol{X}_i-\bar{\boldsymbol{X}})' = \frac{1}{n-1}\tilde{\mathbf{X}}'\tilde{\mathbf{X}}.$$

Let $\ell_1 \geq \cdots \geq \ell_p$ be the characteristic root of $\mathbf{S}$, that is, the roots of the equation

$$|\mathbf{S}-\ell\mathrm{I}_p| = 0,$$

and $\boldsymbol{h}_1,\ldots,\boldsymbol{h}_p$ the corresponding characteristic vector satisfying

$$\mathbf{S}\boldsymbol{h}_i = \ell_i\boldsymbol{h}_i, \quad \boldsymbol{h}_i\boldsymbol{h}_j = \delta_{ij}.$$

We note that the sample roots ℓ_i's are positive and distinct with probability 1 under certain regularity conditions on the distribution of $\boldsymbol{X}$ [see Muirhead, 1982] and hence the $\boldsymbol{h}_i$ are defined uniquely except for multiplication by -1.

Definition 10.3.1 Let

$$Y_i = \boldsymbol{h}'_i(\boldsymbol{X}-\bar{\boldsymbol{X}}), \quad i = 1,\ldots,p.$$

Then Y_i is called the ith *sample principal component.*

We have defined the population principal components by maximizing the variances of normalized linear combinations of the original vector variate and have shown that they have some optimum properties. The arguments there can also be carried out in terms of the sample. For example, let $\boldsymbol{h}$ be the normalized vector. Then the sample variance of n observations $\boldsymbol{h}'\boldsymbol{X}_i$, $i = 1, \ldots, n$ on $\boldsymbol{h}'\boldsymbol{X}$ is $\boldsymbol{h}'\mathbf{S}\boldsymbol{h}$. Hence by maximizing $\boldsymbol{h}'\mathbf{S}\boldsymbol{h}$ under the condition $\boldsymbol{h}'\boldsymbol{h} = 1$, we obtain the first sample principal component, and so on.

The contributions to the total variance explained by the ith sample principal component and by the first k sample principal components are expressed, respectively, as

$$\ell_i / \operatorname{tr} \mathbf{S}, \quad (\ell_1 + \cdots + \ell_k) / \operatorname{tr} \mathbf{S},$$

where $\operatorname{tr} \mathbf{S} = \ell_1 + \cdots + \ell_p$ is the sample total variance. The quantities are called the contribution ratio of the ith principal component and the cumulative contribution ratio of the first k principal components, respectively.

Let Y_{ij} be the values or scores of the jth principal component for the ith observation $\boldsymbol{X}_i$, that is,

$$Y_{ij} = \boldsymbol{h}_j'(\boldsymbol{X}_i - \bar{\boldsymbol{X}}), \quad i = 1, \ldots, n, \quad j = 1, \ldots, p.$$

The principal scores are expressed as

$$\mathbf{Y} = \begin{pmatrix} Y_{11} & \cdots & Y_{1p} \\ \vdots & \vdots & \vdots \\ Y_{n1} & \cdots & Y_{np} \end{pmatrix} = \tilde{\mathbf{X}}(\boldsymbol{h}_1, \ldots, \boldsymbol{h}_p),$$

where $\mathbf{H} = (\boldsymbol{h}_1, \ldots, \boldsymbol{h}_p)$. The coordinates for presenting the n subjects in a k-dimensional Euclidian space are as follows:

Subject	dim 1	$\cdots$	dim k
1	Y_{11}	$\cdots$	Y_{1k}
2	Y_{21}	$\cdots$	Y_{2k}
$\vdots$	$\vdots$	$\vdots$	$\vdots$
n	Y_{n1}	$\cdots$	Y_{nk}

The validity of such coordinates is obtained by considering a singular value decomposition of $\tilde{\mathbf{X}}$:

$$\tilde{\mathbf{X}} = \mathbf{U}\mathbf{L}^{1/2}\mathbf{H},$$

where $\mathbf{H} = (\boldsymbol{h}_1, \ldots, \boldsymbol{h}_p)$ and $\mathbf{U}$ is $n \times n$ and $p \times p$ orthogonal matrices. This implies that

$$Y = \tilde{\mathbf{X}}\mathbf{H} = \mathbf{U}\mathbf{L}^{1/2}\mathbf{H}'\mathbf{H} = \mathbf{U}\mathbf{L}^{1/2}.$$

Therefore,

$$\tilde{\mathbf{X}} = \mathbf{Y}\mathbf{H}' \quad \text{or} \quad \tilde{\mathbf{X}}' = \mathbf{H}\mathbf{Y}'.$$

Example 10.3.1 The principal components were obtained for the data on the five mathematics subjects considered in Example 4.2.1. Table 10.3.1 are the principal components for $\boldsymbol{X} = (X_1, \ldots, X_5)'$. The principal components for its standardized variate are given in Table 10.3.2. The second half of Table 10.3.1 is not the same as the ones of Table 10.3.2. This shows that the principal components are not invariant under scale transformation. The first principal component denotes a weighted mean. The second principal component denotes the difference between $\{X_1, X_2,\}$ and $\{X_3, X_4, X_5\}$.

Table 10.3.1 Principal components for S

Ch. root ℓ_i	686.99	202.11	103.75	84.63	32.15
$\ell_i/\operatorname{tr} S$	0.62	0.18	0.09	0.08	0.03
$\sum \ell_i/\operatorname{tr} S$	0.62	0.80	0.89	0.97	1.00
X_1	0.51	0.75	−0.30	0.30	−0.08
X_2	0.37	0.21	0.42	−0.78	−0.19
X_3	0.35	−0.08	0.15	0.00	0.92
X_4	0.45	−0.30	0.60	0.52	−0.29
X_5	0.53	−0.55	−0.60	−0.18	−0.15
X_1/s_1	8.84	13.09	−5.24	5.18	−1.39
X_2/s_2	4.84	2.73	5.46	−10.29	−2.48
X_3/s_3	3.67	−0.81	1.54	−0.03	9.82
X_4/s_4	6.70	−4.47	8.86	7.69	−4.24
X_5/s_5	9.23	−9.45	−10.36	−3.03	−2.61

Table 10.3.2 Principal components for R

Ch. root ℓ_i	3.18	0.74	0.44	0.39	0.25
$\ell_i/\operatorname{tr} S$	0.64	0.15	0.09	0.08	0.05
$\sum \ell_i/\operatorname{tr} S$	0.64	0.78	0.87	0.95	1.00
X_1	0.40	0.65	−0.62	−0.15	−0.13
X_2	0.43	0.44	0.71	0.30	−0.18
X_3	0.50	−0.13	0.04	−0.11	0.85
X_4	0.46	−0.39	0.14	−0.67	−0.42
X_5	0.44	−0.47	−0.31	0.66	−0.23

10.4 MLEs of the Characteristic Roots and Vectors

We will study the MLEs of $\lambda_1, \ldots, \lambda_p$ and $\boldsymbol{\gamma}_1, \ldots, \boldsymbol{\gamma}_p$ when $\boldsymbol{X}$ has a p-variate normal distribution $\mathrm{N}_p(\boldsymbol{\mu}, \Sigma)$. We assume that the first nonzero component of $\boldsymbol{\gamma}_i$ $(i = 1, \ldots, p)$ is positive. Then we have

Theorem 10.4.1 *Let $\boldsymbol{X}_1, \ldots, \boldsymbol{X}_N$ be a random sample of size $N(> p)$ from $\mathrm{N}_p(\boldsymbol{\mu}, \Sigma)$. Let λ_i and $\boldsymbol{\gamma}_i$ be the ith largest characteristic root of Σ and its associated characteristic vector satisfying $\boldsymbol{\gamma}_i'\boldsymbol{\gamma}_i = 1$ and $\gamma_{1i} > 0$, respectively. If the p characteristic roots are distinct, the MLEs of λ_i and $\boldsymbol{\gamma}_i$ are given by*

$$\hat{\lambda}_i = (n/N)\ell_i, \quad \hat{\boldsymbol{\gamma}}_i = \boldsymbol{h}_i, \quad i = 1, \ldots, p, \tag{10.4.1}$$

where ℓ_i and $\boldsymbol{h}_i$ are the characteristic roots of $\mathbf{S}$ and the corresponding characteristic vectors satisfying $\mathbf{S}\boldsymbol{h}_i = \ell_i\boldsymbol{h}_i$, $\boldsymbol{h}_i'\boldsymbol{h}_j = \delta_{ij}$, and $h_{1i} > 0$.

Proof. Let Δ be the set of $p \times p$ positive definite matrices whose characteristic roots are all distinct. Then there exists one-to-one correspondence between $\Sigma \in \Delta$ and $\{\lambda_1, \ldots, \lambda_p, \boldsymbol{\gamma}_1, \ldots, \boldsymbol{\gamma}_p\}$ $(\gamma_{1i} \geq 0, i = 1, \ldots, p)$. Since $\bar{\boldsymbol{X}}$ and $(n/N)\mathbf{S}$ are the MLEs of $\boldsymbol{\mu}$ and Σ, it follows from Lemma 4.1.1 that the MLEs of the λ_i and the $\boldsymbol{\gamma}_i$ are given by the $\hat{\lambda}_i$ and the $\hat{\boldsymbol{\gamma}}_i$ given in (10.4.1).

Next we consider the case when the characteristic roots of Σ are $\theta_1, \ldots, \theta_r$ $(\theta_1 > \theta_2 > \cdots > \theta_r > 0)$ with multiplicities $q_1, \ldots, q_r$, respectively. In this case the covariance matrix Σ is expressed as

$$\Sigma = \Gamma\Lambda\Gamma' = (\Gamma_1, \ldots, \Gamma_r)\begin{pmatrix} \theta_1\mathrm{I}_{q_1} & & \mathrm{O} \\ & \ddots & \\ \mathrm{O} & & \theta_r\mathrm{I}_{q_r} \end{pmatrix}\begin{pmatrix} \Gamma_1' \\ \vdots \\ \Gamma_r' \end{pmatrix}, \tag{10.4.2}$$

where $\Gamma = (\boldsymbol{\gamma}_1, \ldots, \boldsymbol{\gamma}_p) = (\Gamma_1, \ldots, \Gamma_r)$, $\Gamma_i : p \times q_i$ and $\Lambda = \mathrm{diag}(\lambda_1, \ldots, \lambda_p)$. Similarly, the sample covariance matrix is expressed as

$$\mathbf{S} = \mathbf{H}\mathbf{D}_\ell\mathbf{H}'$$

where $\mathbf{H} = (\boldsymbol{h}_1, \ldots, \boldsymbol{h}_p)$ and $\mathbf{D}_\ell = \mathrm{diag}(\ell_1, \ldots, \ell_p)$. We partition $\mathbf{H}$ as

$$\mathbf{H} = (\mathbf{H}_1, \ldots, \mathbf{H}_r), \quad \mathbf{H}_i : p \times q_i. \tag{10.4.3}$$

Anderson (1963) has given the following theorem.

Theorem 10.4.2 *Let $\boldsymbol{X}_1, \ldots, \boldsymbol{X}_N$ be a random sample of size $N(> p)$ drawn from $\mathrm{N}_p(\boldsymbol{\mu}, \Sigma)$. If the roots of Σ are $\theta_1 > \theta_2 > \cdots > \theta_r > 0$ with multiplicities $q_1, q_2, \ldots, q_r$, respectively, the MLEs of the θ_i are*

$$\hat{\theta}_i = (1/q_i)(n/N)\sum_{j \in J_i} \ell_j, \quad i = 1, \ldots, r, \tag{10.4.4}$$

where J_i is the set of integers $q_1 + \cdots + q_{i-1} + 1, \ldots, q_1 + \cdots + q_i$, $q_0 = 0$, and MLE $\hat{\Gamma}_i$ of Γ_i is $\mathbf{H}_i$ multiplied on the right by any $q_i \times q_i$ orthogonal matrix with $\gamma_{1j} > 0$.

Proof. After maximizing with respect to mean $\boldsymbol{\mu}$, the logarithm of the likelihood function is, except for a constant term,

$$\begin{aligned}-\frac{1}{2}N \log|\Sigma| - \frac{n}{2}\operatorname{tr}(\Sigma^{-1}\mathbf{S}) &= -\frac{1}{2}N\log|\Gamma\Lambda\Gamma'| - \frac{n}{2}\operatorname{tr}(\Gamma\Lambda^{-1}\Gamma'\mathbf{H}\mathbf{D}_\ell\mathbf{H}') \\ &= -\frac{1}{2}N\log|\Lambda| - \frac{n}{2}\operatorname{tr}(\Lambda^{-1}\mathbf{Q}'\mathbf{D}_\ell\mathbf{Q}) \qquad (10.4.5)\end{aligned}$$

where $\mathbf{Q} = \mathbf{H}'\Gamma$. Since $\mathbf{Q}$ is an orthogonal matrix, it follows [see Anderson, 1963] that

$$\operatorname{tr}(\Lambda^{-1}\mathbf{Q}'\mathbf{D}_\ell\mathbf{Q}) \leq \sum_{i=1}^{r}\sum_{j\in J_i}(\ell_j/\theta_i) \qquad (10.4.6)$$

and equality is attained when $\Gamma_i = \mathbf{H}_i\mathrm{G}_i$, $\mathrm{G}_i \in O(q_i)$. Hence the function (10.4.5) is less than or equal to

$$-\frac{1}{2}N\sum_{i=1}^{r} q_i\log\theta_i - \frac{n}{2}\sum_{i=1}^{r}\sum_{j\in J_i}(\ell_j/\theta_i),$$

which is maximized by $\hat{\theta}_i$ of (10.4.4). This completes the proof. □

10.5 Distributions of the Characteristic Roots and Vectors

In this section we assume that the sample covariance matrix $\mathbf{S}$ is based on a sample of size $N = n + 1$, so that under normality, $n\mathbf{S}$ is distributed as a Wishart distribution $\mathrm{W}_p(n, \Sigma)$. The density function is

$$\frac{(\frac{1}{2}n)^{pn/2}}{\Gamma_p(\frac{1}{2}n)|\Sigma|^{n/2}}\operatorname{etr}\left(-\frac{1}{2}n\Sigma^{-1}\mathbf{S}\right)|\mathbf{S}|^{(n-p-1)/2}, \qquad (10.5.1)$$

where $\Gamma_p(a) = \pi^{p(p-1)/4}\prod_{i=1}^{p}\Gamma[a - \frac{1}{2}(i-1)]$. Let the characteristic roots of $\mathbf{S}$ be $\ell_1 > \cdots > \ell_p > 0$ and the corresponding orthonormal characteristic vectors $\boldsymbol{h}_1, \ldots, \boldsymbol{h}_p$ satisfy

$$\mathbf{S} = \mathbf{HLH}', \qquad (10.5.2)$$

where $\mathbf{L} = \operatorname{diag}(\ell_1, \ldots, \ell_p)$, and $\mathbf{H} = (\boldsymbol{h}_1, \ldots, \boldsymbol{h}_p)$.

10.5.1 Exact Distributions

When we consider the distribution of **L**, without loss of generalty we may assume that $\Sigma = \Lambda = \text{diag}(\lambda_1, \ldots, \lambda_p)$, since the characteristic roots of **S** are the same as those of $\Gamma' S \Gamma$ for any orthogonal matrix. The exact distribution of **L** in a special case $\Sigma = \lambda \mathrm{I}_p$ is given in the following theorem. However, the distribution in a general case was expressed in terms of zonal polynomials by James (1960). It is complicated. For details, see Muirhead (1982).

Theorem 10.5.1 *Suppose that $n\mathbf{S}$ is distributed as a Wishart distribution $\mathrm{W}_p(n, \Sigma)$. Let $\ell_1 > \cdots > \ell_p > 0$ and $\lambda_1 \geq \cdots \geq \lambda_p > 0$ be the characteristic roots of $\mathbf{S}$ and Σ, respectively. If $\lambda_1 = \cdots = \lambda_p = \lambda$, the joint distribution of $\ell_1, \ldots, \ell_p$ is given by*

$$\left(\frac{n}{2\lambda}\right)^{pn/2} \frac{\pi^{p^2/2}}{\Gamma_p(\frac{1}{2}n)\Gamma_p(\frac{1}{2}p)} \exp\left(-\frac{n}{2\lambda}\sum_{i=1}^{p} \ell_i\right) \times \prod_{i=1}^{p} \ell_i^{(n-p-1)/2} \prod_{i<j}(\ell_i - \ell_j), \quad \ell_1 > \cdots > \ell_p > 0.$$

Proof. Consider the transformation from S to its characteristic roots L and vectors H given by $\mathrm{S} = \mathrm{HLH}'$. The Jacobian is

$$J(\mathrm{S} \to \mathrm{L}, \mathrm{H}) = \frac{2^p \pi^{p^2/2}}{\Gamma(\frac{1}{2}p)} \prod_{i<j}(\ell_i - \ell_j) q_p(\mathrm{H}).$$

where $q_p(\mathrm{H})$ is the invariant measure on the orthogonal group $\mathrm{O}(p) = \{\mathrm{H};\ \mathrm{H}'\mathrm{H} = \mathrm{I}_p\}$ satisfying $\int_{\mathrm{O}(p)} q_p(\mathrm{H}) d\mathrm{H} = 1$. Therefore, the density function of $\ell_1, \ldots, \ell_p$ is

$$\left(\frac{n}{2}\right)^{pn/2} \frac{\pi^{p^2/2}|\Sigma|^{-n/2}}{\Gamma_p(\frac{1}{2}n)\Gamma_p(\frac{1}{2}p)} \prod_{i=1}^{p} \ell_i^{(n-p-1)/2} \prod_{i<j}(\ell_i - \ell_j) \times \int_{\mathrm{O}(p)} \text{etr}\left(-\frac{1}{2}\Sigma^{-1}\mathrm{HLH}'\right) q_p(\mathrm{H})\, d\mathrm{H}.$$

When $\Sigma = \lambda \mathrm{I}_p$,

$$\text{etr}\left(-\frac{1}{2}\Sigma^{-1}\mathrm{HLH}'\right) = \text{etr}\left(-\frac{1}{2}\lambda^{-1}\mathrm{I}_p\mathrm{HLH}'\right) = \text{etr}\left(-\frac{1}{2}\lambda^{-1}\mathrm{L}\right),$$

and hence we obtain the result. □

10.5.2 Large-sample Case

First we consider asymptotic joint distribution of the characteristic roots $\ell_1 > \cdots > \ell_p$ of the sample covariance matrix $\mathbf{S}$ and the corresponding characteristic vectors $\boldsymbol{h}_1, \ldots, \boldsymbol{h}_p$ satisfying $\boldsymbol{h}_i'\boldsymbol{h}_j = \delta_{ij}$, and $h_{1i} > 0$. For our present aim, it is convenient to deal with the characteristic roots and vectors of $\mathbf{W} = \Gamma'\mathbf{S}\Gamma$, since $n\mathbf{W} \sim \mathrm{W}_p(n, \Lambda)$ and the characteristic roots of $\mathbf{W}$ are the same as the ones of $\mathbf{S}$. The characteristic vectors $\boldsymbol{g}_1, \ldots, \boldsymbol{g}_p$ of $\mathbf{W}$ are expressed as

$$\boldsymbol{g}_i = (g_{1i}, \ldots, g_{pi})' = \Gamma'\boldsymbol{h}_i, \quad i = 1, \ldots, p. \tag{10.5.3}$$

Let

$$\mathbf{W} = \Lambda + \frac{1}{\sqrt{n}}\mathbf{V}. \tag{10.5.4}$$

Then it follows from Theorem 2.5.1 that the limiting distribution of $\mathbf{V} = (V_{ij})$ is normal with mean zero and covariance matrix with elements

$$\mathrm{Cov}(V_{ij}, V_{k\ell}) = \lambda_i\lambda_j(\delta_{ik}\delta_{j\ell} + \delta_{i\ell}\delta_{jk}). \tag{10.5.5}$$

Further, if λ_i is simple, using Theorem 5.5.1 we obtain

$$\ell_i = \lambda_i + \frac{1}{\sqrt{n}}V_{ii} + O_p(n^{-1}), \tag{10.5.6}$$

$$g_{ji} = \begin{cases} 1 + O_p(n^{-1}), \\ \frac{1}{\sqrt{n}}\lambda_{ij}V_{ji} + O_p(n^{-1}), \end{cases} \tag{10.5.7}$$

where $\lambda_{ij} = (\lambda_i - \lambda_j)^{-1}$. Using these results we obtain the following Theorem.

Theorem 10.5.2 *Let $n\mathbf{S} \sim \mathrm{W}_p(n, \Sigma)$, and let ℓ_i's and λ_i's be the ordered characteristic roots of $\mathbf{S}$ and Σ, respectively. Let $\mathbf{H} = (\boldsymbol{h}_i, \ldots, \boldsymbol{h}_p)$ and $\Gamma = (\boldsymbol{\gamma}_1, \ldots, \boldsymbol{\gamma}_p)$ be the matrices of the corresponding characteristic vectors of $\mathbf{S}$ and Σ with $\boldsymbol{h}_i'\boldsymbol{h}_j = \boldsymbol{\gamma}_i'\boldsymbol{\gamma}_j = \delta_{ij}$, $h_{1i} > 0$, $\gamma_{1i} > 0$. If λ_i's are distinct, then*

$$\sqrt{n}(\ell_1 - \lambda_1), \ldots, \sqrt{n}(\ell_p - \lambda_p), \text{ and } \sqrt{n}(\mathbf{H} - \Gamma)$$

are asymptotically independently distributed. Their limiting distributions are normal with mean zeros and covariance matrix with the following elements:

$$\mathrm{Var}(\sqrt{n}(\ell_i - \lambda_i)) = 2\lambda_i^2, \quad i = 1, \ldots, p, \tag{10.5.8}$$

$$\mathrm{Cov}(\sqrt{n}(\boldsymbol{h}_i - \boldsymbol{\gamma}_i), \sqrt{n}(\boldsymbol{h}_j - \boldsymbol{\gamma}_j)) = \begin{cases} \sum_{k=i}^{p} \lambda_i\lambda_k\lambda_{ik}^2\boldsymbol{\gamma}_k\boldsymbol{\gamma}_k', & i = j, \\ -\lambda_i\lambda_j\lambda_{ij}^2\boldsymbol{\gamma}_j\boldsymbol{\gamma}_i', & i \neq j. \end{cases} \tag{10.5.9}$$

Proof. From (10.5.6) and (10.5.7), we obtain

$$
\begin{aligned}
\sqrt{n}(\ell_i - \lambda_i) &= V_{ii} + O_p(n^{-1/2}), \quad i = 1, \ldots, p \\
\sqrt{n}(\boldsymbol{h}_i - \boldsymbol{\gamma}_i) &= \sqrt{n}(\Gamma \boldsymbol{g}_i - \boldsymbol{\gamma}_i) \\
&= \Gamma(\lambda_{i1}V_{1i}, \ldots, \lambda_{i,i-1}V_{i1,i}, 0, \lambda_{i,i+1}V_{i+1,i}, \ldots, \lambda_{ip}V_{pi}) \\
&\qquad + O_p(n^{-1/2}).
\end{aligned}
$$

Thus the limiting distributions of the left-hand sides are equal to those of the leading terms of the right-hand sides. The final results are obtained from the limiting distribution of $\mathbf{V}$. □

Now we consider asymptotic distributions of $\mathbf{L}$ when the underlying distribution is not necessarily normal. Let

$$
\boldsymbol{U} = (U_1, \ldots, U_p)' = \Gamma'(\boldsymbol{X} - \boldsymbol{\mu}).
$$

Then

$$
\mathrm{E}(\boldsymbol{U}) = 0, \quad \mathrm{Var}(\boldsymbol{U}) = \Lambda = \mathrm{diag}\,(\lambda_1, \ldots, \lambda_p).
$$

Further, letting

$$
\begin{aligned}
\boldsymbol{U}_j &= \Gamma'(\boldsymbol{X}_j - \boldsymbol{\mu}), \quad j = 1, \ldots, n, \\
\tilde{\mathbf{S}} &= \Gamma'\mathbf{S}\Gamma = \frac{1}{n-1}\sum_{j=1}^{n}(\boldsymbol{U}_j - \bar{\boldsymbol{U}})(\boldsymbol{U}_j - \bar{\boldsymbol{U}})',
\end{aligned}
$$

where $\bar{\boldsymbol{U}} = (1/n)\sum_{j=1}^{n} \boldsymbol{U}_j$.

Since Γ is an orthogonal matrix, the characteristic roots of $\mathbf{S}$ and $\tilde{\mathbf{S}}$ are the same. In the following we consider the characteristic roots of $\tilde{\mathbf{S}}$. Put

$$
\mathbf{V} = \sqrt{n}(\tilde{\mathbf{S}} - \Lambda) \quad \text{or} \quad \tilde{\mathbf{S}} = \Lambda + \frac{1}{\sqrt{n}}\mathbf{V}.
$$

If λ_α is simple, then we have from Theorem 5.5.1

$$
\begin{aligned}
Y_\alpha &= \sqrt{n}(\ell_\alpha - \lambda_\alpha) \\
&= V_{\alpha\alpha} + \frac{1}{\sqrt{n}}\sum_{j \neq \alpha} \lambda_{\alpha j} V_{\alpha j}^2 + \mathrm{O}_p(n^{-1}).
\end{aligned} \tag{10.5.10}
$$

The cumulant of Y_α is of the form

$$
\begin{aligned}
\kappa_1(Y_\alpha) &= \frac{1}{\sqrt{n}} a_1 + \mathrm{O}(n^{-3/2}), \\
\kappa_2(Y_\alpha) &= \sigma^2 + \mathrm{O}(n^{-1}), \\
\kappa_3(Y_\alpha) &= \frac{6}{\sqrt{n}} a_3 + \mathrm{O}(n^{-3/2}), \\
\kappa_j(Y_\alpha) &= \mathrm{O}(n^{-(j-2)/2}), \quad j \geq 2.
\end{aligned}
$$

The coefficients σ^2, a_1, and a_3 are computed by using the perturbation expansion (10.5.10) of Y_α. Thus, we obtain an asymptotic expansion of the characteristic function of Y_α as

$$\begin{aligned} C_{Y_\alpha}(t) &= \exp\left\{\sum_{j=1}^{3} \frac{(it)^j}{j!}\kappa_j(Y_\alpha) + \mathrm{O}(n^{-1})\right\} \\ &= \mathrm{e}^{-(1/2)\sigma^2}\left[1 + \frac{1}{\sqrt{n}}\left\{a_1(it) + a_3(it)^3\right\}\right] + \mathrm{O}(n^{-1}). \end{aligned}$$

Theorem 10.5.3 *Suppose that the α th root λ_α of Σ is simple. Then*

$$\begin{aligned} \mathrm{P}\left\{\frac{\sqrt{n}(\ell_\alpha - \lambda)}{\sigma} \le x\right\} &= \Phi(x) \\ &\quad -\varphi(x)\frac{1}{\sqrt{n}}\left\{a_1\sigma^{-1} + a_3\sigma^{-3}H_2(x)\right\} + \mathrm{o}\left(\frac{1}{\sqrt{n}}\right), \end{aligned}$$

where $\lambda_{\alpha j} = (\lambda_\alpha - \lambda_j)^{-1}$,

$$\begin{aligned} \sigma^2 &= 2\lambda_\alpha^2 + \kappa_{\alpha\alpha\alpha\alpha}, \\ a_1 &= \sum_{j\ne\alpha} \lambda_{\alpha j}\left(\lambda_\alpha\lambda_j + \kappa_{\alpha\alpha jj}\right), \\ a_3 &= \frac{1}{6}\left\{8\lambda_\alpha^3 + \kappa_{\alpha\alpha\alpha\alpha\alpha\alpha} + 12\lambda_\alpha\kappa_{\alpha\alpha\alpha\alpha} + 4\left(\kappa_{\alpha\alpha\alpha}\right)^2\right\} \\ &\quad + \sum_{j\ne\alpha} \lambda_{\alpha j}\left(\kappa_{\alpha\alpha\alpha j}\right)^2. \end{aligned}$$

Here $\kappa_{\alpha\alpha\alpha}$, $\kappa_{\alpha\alpha\alpha j}$, ..., are the cumulants of the population principal component $\boldsymbol{U} = (U_1, \ldots, U_p)'$.

When $\boldsymbol{X}$ is normal,

$$\sigma^2 = 2\lambda_\alpha^2, \quad a_1 = \sum_{j\ne\alpha} \lambda_{\alpha j}\lambda_\alpha\lambda_j, \quad a_3 = \frac{4}{3}\lambda_\alpha^3,$$

which was obtained by Sugiura (1973). The extension to the nonnormal case was made by Fujikoshi (1980). For asymptotic expansions of the characteristic vectors, see Sugiura (1973, 1976a).

Now we consider the case that the λ_i have multiplicities. Suppose that

$$\lambda_1 = \cdots = \lambda_{q_1} = \theta_1, \ldots, \lambda_{p-q_r+1} = \cdots = \lambda_p = \theta_r,$$

where $\theta_1 > \cdots > \theta_r$. Let J_α be the index set of the characteristic roots corresponding to θ_α (i.e., $J_\alpha = \{\sum_{i=1}^{\alpha-1} q_\alpha + 1, \ldots, \sum_{i=1}^{\alpha} q_\alpha\}$). Let

$$Y_i = \sqrt{\frac{n}{2}}\left(\frac{\ell_i - \theta_\alpha}{\theta_\alpha}\right), \quad i \in J_\alpha, \quad \alpha = 1, \ldots, r$$

when the underlying distribution is normal. Anderson (1963) obtained the limiting distribution of the Y_i. Their asymptotic expansion was obtained by Fujikoshi (1978) as

$$\prod_{\alpha=1}^{r}\left\{\frac{\pi^{q_\alpha(q_\alpha-1)/4}}{2^{q_\alpha/2}\Gamma_{q_\alpha}(q_\alpha/2)}\exp\left(-\frac{1}{2}\sum_{i\in J_\alpha}y_i^2\right)\prod_{i<j;i,j\in J_\alpha}(y_i-y_j)\right\}$$
$$\times\left\{1+\frac{1}{\sqrt{n}}Q_1+\mathrm{O}\left(\frac{1}{n}\right)\right\},$$

where

$$Q_1=\sqrt{2}\left\{\sum_{i=1}^{p}\left(\frac{1}{3}y_i^3-\frac{1}{2}(p+1)y_i\right)+\frac{1}{2}\sum_{\alpha<\beta}\sum_{i\in J_\alpha}\sum_{j\in J_\beta}\left(\frac{\theta_\alpha y_i-\theta_\beta y_j}{\theta_\alpha-\theta_\beta}\right)\right\}.$$

Now we consider the distribution of

$$T_\alpha=\sum_{i\in J_\alpha}Y_i=\sum_{i\in J_\alpha}\sqrt{\frac{n}{2}}\left(\frac{\ell_i-\theta_\alpha}{\theta_\alpha}\right).$$

Note that the limiting distribution of $\{Y_i, i\in J_\alpha\}$ is the same as the distribution of the characteristic roots of $\mathbf{W}$, where $\mathbf{W}=(W_{ij})$ is a $q_\alpha\times q_\alpha$ symmetric random matrix, $W_{ii}\sim\mathrm{N}(0,1)$, $W_{ij}\sim\mathrm{N}(0,1/2)$, and the W_{ij} are independent. Therefore,

$$T_\alpha\xrightarrow{d}\mathrm{N}(0,q_\alpha). \tag{10.5.11}$$

We consider a large-sample approximation to the distribution of the plug-in statistic for the constribution ratio Ψ_k,

$$\hat{\psi}_k=(\ell_1+\cdots+\ell_k)/\operatorname{tr}\mathbf{S}. \tag{10.5.12}$$

Theorem 10.5.4 *Assume that* $\lambda_1\geq\cdots\geq\lambda_k>\lambda_{k+1}\geq\cdots\geq\lambda_p\geq 0$. *Then the limiting distribution of* $T=\sqrt{n}(\hat{\psi}_k-\psi_k)$ *is normal with mean 0 and variance*

$$\tau^2=2[\operatorname{tr}(\Sigma^2)/\{\operatorname{tr}\Sigma\}^2](\psi_k^2-2\psi_k\xi_k+\xi_k),$$

where $\xi_k=(\lambda_1^2+\cdots+\lambda_k^2)/\{\operatorname{tr}(\Sigma^2)\}$.

Proof. We may assume without loss of generality that the characteristic roots of Σ are $\theta_1,\ldots,\theta_r$ $(\theta_1>\cdots>\theta_r>0)$ with multiplicities $q_1,\ldots,q_r$ and $q_1+\cdots+q_c=k$ $(c\leq r)$. Let

$$Z_j=\sqrt{n/2}(\ell_j-\theta_\alpha)/\theta_\alpha \quad \text{for } j\in J_\alpha,\ \alpha=1,\ldots,r,$$

where J_α denotes the set of integers $q_{\alpha-1}+1,\ldots,q_{(\alpha)}$ and $q_{(\alpha)}=\sum_{j=1}^{\alpha}q_j$, $q_{(0)}=0$. We have shown (see (10.5.11)) that the limiting distribution of $Y_\alpha=\sum_{j\in J_\alpha}Z_j$ is $\mathrm{N}(0,q_\alpha)$ and the $Y_1,\ldots,Y_r$ are mutually independent. We can write

$$\begin{aligned}
\hat{\psi}_k &= \sum_{\alpha=1}^{c}\sum_{j\in J_\alpha}\ell_j \Big/ \sum_{\alpha=1}^{r}\sum_{j\in J_\alpha}\ell_j \\
&= \left[\sum_{\alpha=1}^{c}\left\{\theta_\alpha q_\alpha+\left(\frac{2}{n}\right)^{1/2}\theta_\alpha Y_\alpha\right\}\right] \Big/ \left[\sum_{\alpha=1}^{r}\left\{\theta_\alpha q_\alpha+\left(\frac{2}{n}\right)^{1/2}\theta_\alpha Y_\alpha\right\}\right] \\
&= \left[\sum_{\alpha=1}^{c}\left\{\theta_\alpha q_\alpha+\left(\frac{2}{n}\right)^{1/2}\theta_\alpha Y_\alpha\right\}\right]\{\operatorname{tr}\Sigma\}^{-1} \\
&\qquad\times\left[1-\{\operatorname{tr}\Sigma\}^{-1}\left(\frac{2}{n}\right)^{1/2}\sum_{\alpha=1}^{r}\theta_\alpha Y_\alpha\right]+\mathrm{O}_p(n^{-1}).
\end{aligned} \tag{10.5.13}$$

From this expression it follows that T is asymptotically equivalent to

$$\sqrt{2}\{\operatorname{tr}\Sigma\}^{-1}\left(\sum_{\alpha=1}^{c}\theta_\alpha Y_\alpha-\sum_{\alpha=1}^{r}\theta_\alpha Y_\alpha\right),$$

whose limiting distribution is $\mathrm{N}(0,\tau^2)$. □

Sugiyama and Tong (1976) obtained an asymptotic expansion for the distribution of T when all the characteristic roots are distinct. Konishi (1977) has given an asymptotic expansion for the distribution of the general function of $\ell_1,\ldots,\ell_p$. Fujikoshi (1978) has extended Konishi's result to the case when multiplicities of characteristic roots are arbitrary.

Let $\hat{\tau}$ be an estimator of τ obtained by replacing λ_i in the expression of τ with ℓ_i. Then noting that the limiting distribution of $T/\hat{\tau}$ is $\mathrm{N}(0,1)$, we obtain a confidence interval for ψ_k with an asymptotic confidence coefficient equal to $1-\alpha$ as

$$\hat{\psi}_k-(\hat{\tau}/\sqrt{n})u_\alpha<\psi_k<\hat{\psi}_k+(\hat{\tau}/\sqrt{n})u_\alpha, \tag{10.5.14}$$

where u_α is the upper $100(\alpha/2)\%$ point of $\mathrm{N}(0,1)$.

Now we consider testing hypothesis that the smallest $q=p-k$ characteristic roots $\lambda_{k+1},\ldots,\lambda_p$ are equal, i.e.

$$H_k:\lambda_{k+1}=\cdots=\lambda_p(=\lambda). \tag{10.5.15}$$

If $k=0$, the testing problem reduces to the sphericity test in Section 8.2. If their common value in H_k is small compared with the first k roots, then the last q principal components can be ignored and a reduction in dimensionality is achieved. In this sense, we can call the number k dimensionality in principal component analysis. From Theorem 10.4.2 the LR criterion for testing H_k is found as

$$Q_k = \left[\prod_{j=k+1}^{p} \ell_i \Big/ \left\{ \frac{1}{q} \sum_{j=k+1}^{p} \ell_j \right\}^q \right]. \tag{10.5.16}$$

Theorem 10.5.5 *The LR criterion for testing the hypothesis H_k is given by Q_k in (10.5.11). Further, as n=N-1 tends to infinity, the null distribution of $-2\log Q_k$ converges to the chi-square distribution with $f = (q-1)(q+2)/2$ degrees of freedom.*

Proof. Put

$$Z_j = \sqrt{n}(\ell_j - \lambda)/(\sqrt{2}\lambda), \quad j = k+1, \ldots, p.$$

Then it follows from the results on the limiting distribution of Y_i in (10.5.10) that $Z_{k+1}, \ldots, Z_p$ are asymptotically distributed like the characteristic roots of the $q \times q$ symmetric matrix $\mathbf{U} = (U_{ij})$ whose nonduplicated elements are independent and normally distributed with mean 0 and $\mathrm{Var}(U_{ii}) = 1$, $\mathrm{Var}(U_{ij}) = 1/2$ $(i \neq j)$. On the other hand, under H_k we can reduce $-2(n/N)\log Q_k$ to an asymptotically equivalent form

$$\begin{aligned}
&- n\log\left[\prod_{j=k+1}^{p} \lambda\left\{1 + \sqrt{\frac{2}{n}} Z_j\right\}\right] + nq \cdot \log\left[\lambda\left\{1 + \sqrt{\frac{2}{n}\frac{1}{q}} \sum_{j=k+1}^{p} Z_j\right\}\right] \\
&= n\Bigg[- \sum_{j=k+1}^{p} \left\{\sqrt{\frac{2}{n}} Z_j - \frac{1}{n} Z_j^2 + \cdots\right\} \\
&\qquad + q\left\{\sqrt{\frac{2}{n}\frac{1}{q}} \sum_{j=k+1}^{p} Z_j - \frac{1}{n}\left(\frac{1}{q}\sum_{j=k+1}^{p} Z_j\right)^2 + \cdots\right\}\Bigg] \\
&= \mathrm{tr}\,\mathbf{U}^2 - \frac{1}{q}\{\mathrm{tr}\,\mathbf{U}\}^2 + O_p(n^{-1/2}).
\end{aligned}$$

Note that

$$\mathrm{tr}\,\mathbf{U}^2 - \frac{1}{2}\{\mathrm{tr}\,\mathbf{U}\}^2 = T_1 + T_2,$$

where

$$T_1 = \sum_{i<j}^{q} U_{ij}^2, \quad T_2 = \sum_{i=1}^{q} U_{ii}^2 - \frac{1}{q}\left(\sum_{i=1}^{q} U_{ii}\right)^2.$$

It is easy to see that $T_1 \sim \chi^2(q(q-1)/2)$, $T_2 \sim \chi^2(q)$, and T_1 and T_2 are independent, which imply the required result. □

Theorem 10.5.5 was given by Anderson (1963). Lawley (1956) proposed to use

$$-\frac{2}{N}\left\{n-k-\frac{1}{69}(2q^2+q+2)+\sum_{i=1}^{k}\{\bar{\ell}_q/(\ell_i-\bar{\ell}_q)\}^2\right\}\log Q_k,$$

where $\bar{\ell}_q = (1/q)\sum_{j=k+1}^{p}\ell_j$. This gives a better chi-square test than the use of the statistic $-2\log Q_k$. For a further study of this approximation, see Fujikoshi (1977b).

Finally we consider a test or a confidence region of a subvector of $\boldsymbol{\gamma}_i$. For a notational simplicity, let $\boldsymbol{\gamma}_i$ partition into q and $r = p - q$ components, i.e., $\boldsymbol{\gamma}_i = (\boldsymbol{\gamma}_{1i}', \boldsymbol{\gamma}_{2i}')'$, and consider testing the hypothesis

$$M_0 : \ \boldsymbol{\gamma}_{2i} = \boldsymbol{g} \text{ vs. } M_1 : \ \boldsymbol{\gamma}_{2i} \neq \boldsymbol{g}, \tag{10.5.17}$$

where $\boldsymbol{b}$ is a given vector such that $\boldsymbol{b}'\boldsymbol{b} \leq 1$. In Theorem 10.5.2, we showed that if all the λ_i's are simple, $\boldsymbol{U} = \sqrt{n}(\boldsymbol{h}_i - \boldsymbol{\gamma}_i)$ has an asymptotic normal distribution with mean vector $\mathbf{0}$ and covariance matrix $\Psi = \Gamma_i \Delta \Gamma_i'$, where

$$\begin{aligned}
\Gamma_i &= (\boldsymbol{\gamma}_1, \ldots, \boldsymbol{\gamma}_{i-1}, \boldsymbol{\gamma}_{i+1}, \ldots, \boldsymbol{\gamma}_p), \\
\Delta &= \lambda_i \operatorname{diag}(\lambda_1\lambda_{1i}^2, \ldots, \lambda_{i-1}\lambda_{i-1,i}^2, \lambda_{i+1}\lambda_{i+1,i}^2, \ldots, \lambda_p\lambda_{pi}^2).
\end{aligned}$$

The assumption that all the λ_i's are simple can be relaxed as the one that λ_i is simple. The covariance matrix Ψ is singular, but its Moore-Penrose inverse matrix is expressed as

$$\begin{aligned}
\Psi^+ &= \Gamma_i \Delta^{-1} \Gamma_i \\
&= \lambda_i \Sigma^{-1} - 2\mathrm{I}_p + \lambda_i^{-1}\Sigma.
\end{aligned}$$

The last reduction is obtained by using that $\Sigma^{-1} = \sum_{j=1}^{p}\lambda_j^{-1}\boldsymbol{\gamma}_j\boldsymbol{\gamma}_j'$ and $\mathrm{I}_p = \Gamma\Gamma' = \Gamma_i\Gamma_i' + \boldsymbol{\gamma}_i\boldsymbol{\gamma}_i'$. Along to the partition of $\boldsymbol{\gamma}_i$, we partition $\boldsymbol{h}_i, \mathbf{S}, \Sigma, \Psi$, and write, for example

$$\mathbf{S} = \begin{pmatrix} \mathbf{S}_{11} & \mathbf{S}_{12} \\ \mathbf{S}_{21} & \mathbf{S}_{22} \end{pmatrix}, \quad \Psi = \begin{pmatrix} \Psi_{11} & \Psi_{12} \\ \Psi_{21} & \Psi_{22} \end{pmatrix}.$$

Then, $\boldsymbol{U}_2 = \sqrt{n}(\boldsymbol{h}_2 - \boldsymbol{\gamma}_2)$ has an asymptotic normal distribution with mean vector $\mathbf{0}$ and covariance matrix Ψ_{22}. This suggests using

$$R = n(\boldsymbol{h}_{2i} - \boldsymbol{g})'\hat{\Psi}_{22}^{-1}(\boldsymbol{h}_{2i} - \boldsymbol{g}) \tag{10.5.18}$$

whose limiting null distribution is the chi-square distribution with r degrees of freedom, where $\hat{\Psi}_{22}$ is a sample analogue of Ψ_{22}. A confidence region is obtained by using that

$$n(\boldsymbol{h}_{2i} - \boldsymbol{\gamma}_2)'\hat{\Psi}_{22}^{-1}(\boldsymbol{h}_{2i} - \boldsymbol{\gamma}_2)$$

is asymptotically distributed as $\chi^2(r)$.

In the above setting we assume that γ_2 is a true subvector, $r > 0$. For a test of $\gamma_i = g$ with $g'g = 1$ or a confidence interval of γ, we may use

$$\begin{aligned} R &= n(\boldsymbol{h}_i - \boldsymbol{g})'\hat{\Psi}^+(\boldsymbol{h}_i - \boldsymbol{g}) \\ &= n(\boldsymbol{h}_i - \boldsymbol{g})'(\ell_i \mathbf{S}^{-1} - 2\mathrm{I}_p + \ell_i^{-1}\mathbf{S})(\boldsymbol{h}_i - \boldsymbol{g}) \\ &= n(\ell_i \boldsymbol{g}'\mathbf{S}^{-1}\boldsymbol{g} - 2 + \ell_i^{-1}\boldsymbol{g}'\mathbf{S}\boldsymbol{g}) \end{aligned} \tag{10.5.19}$$

whose limiting null distribution is the chi-square distribution with $p-1$ degrees of freedom. The limiting distribution may be concluded by using that $\hat{\Psi}^+$ converges to Ψ^+ in probability, and using Problem 1.16. Similarly, a confidence region for γ_i is obtained. These results were given by Anderson (1963).

10.5.3 High-Dimensional Case

When $n < p$, the LR test can not be used, since $\mathbf{S}$ becomes singular. For the test of sphericity, the test was generalized as the statistic U in (8.2.12). The natural generalization of U is

$$U_k = \frac{(1/q)\sum_{i=k+1}^{p} \ell_i^2}{\left[(1/q)\sum_{i=k+1}^{p} \ell_i\right]^2} - 1. \tag{10.5.20}$$

It is shown [see Schott, 2006a] or Problem 10.9) that under the large sample framework, the null distribution of $(nq/2)U_k$ is asymptotically distributed as the $\chi^2(f_k)$, where $f_k = q(q+1)/2 - 1$. This shows that the asymptotic null distribution of U_k is identical to that of the LR test statistic Q_k. For a high-dimensional case, the assumptions 1, 2 and 3 in Section 5.9.2 are modified as

$$\begin{aligned} &\text{Assumption 1}: \quad q/n \;\to\; c \in (0, \infty), \\ &\text{Assumption 2}: \quad \frac{1}{q}\sum_{i=k+1}^{p} \lambda_i^s = \mathrm{O}(1), \quad s = 1, 2, \\ &\text{Assumption 3}: \quad \frac{1}{q}\sum_{i=k+1}^{p} \lambda_i^s = \mathrm{O}(1), \quad s = 3, 4. \end{aligned}$$

Let

$$\frac{1}{q}\sum_{i=k+1}^{p} \lambda_i = \alpha, \quad \frac{1}{q}\sum_{i=k+1}^{p} (\lambda_i - \alpha)^2 = \delta^2.$$

Then, Schott (2006a) showed that if $\delta^2 = 0$,

$$(n-k)U_k - q \xrightarrow{d} \mathrm{N}(1, 4). \tag{10.5.21}$$

In general, the chi-square-type approximation to the LR test statistic Q_k becomes worse as p increase in value. A high-dimensional approximation of Q_k was studied in Fujikoshi et al. (2007) under the assumption

$$\begin{aligned}
&\text{A1}: \ k;\text{fix}, \ n \to \infty, \ q \to \infty, \ q/n \to c \in (0,1). \\
&\text{A2}: \ \lambda_j = O(q), \ j = 1, \ldots, k.
\end{aligned}$$

Then, they showed that

$$\frac{\log Q_k - \mu_{q,n}}{\sigma_{q,n}} \xrightarrow{d} N(0,1), \tag{10.5.22}$$

where

$$\begin{aligned}
\mu_{q,n} &= m \log q - q\psi\left(\frac{1}{2}qn\right) + \psi_q\left(\frac{1}{2}n\right), \\
\sigma_{q,n}^2 &= \psi_q'\left(\frac{1}{2}n\right) - q^2\psi'\left(\frac{1}{2}qn\right),
\end{aligned}$$

$\psi_q(a) = \sum_{j=1}^{q} \psi\left(a - \frac{1}{2}(j-1)\right)$, and $\psi(\cdot)$ is the digamma function (see Section 5.4.2). They also show by numerical simulations that the high-dimensional approximation is more accurate than the classical chi-square type approximation as p increases in value.

Recently there are considerable interests in high-dimensional situations such that $N = n + 1 < p$ or $N \ll r$. The principal components are defined for n non-zero characteristic roots of $\mathbf{S}$, even though $\mathbf{S}$ is singular. The consistency and spherisity for these principal components are studied under $p/n \to c \in (0, \infty)$, see, e.g., Johnstone and Lu (2009).

10.6 Model Selection Approach for Covariance Structures

10.6.1 General Approach

In this section we give a general approach for the selection of covariance structures based on ideas of Akaike (1973). Let $\boldsymbol{Y}$ be a p-dimensional random vector with mean vector $\boldsymbol{\mu}$ and covariance matrix Σ. Suppose that there is available a sample of size $N = n+1$ (i.e., $\boldsymbol{Y}_1, \ldots, \boldsymbol{Y}_N$ of $\boldsymbol{Y}$ from a population). Let $\mathbf{S}$ be the sample covariance matrix based on $\mathbf{Y} = (\boldsymbol{Y}_1, \ldots, \boldsymbol{Y}_N)'$. Consider a general covariance structure model denoted as

$$M: \ \Sigma \in \mathcal{S}.$$

A measure of closeness between $\mathbf{S}$ and an element Σ in $\mathcal{S}$ may be defined as

$$D(\mathbf{S}, \Sigma) = -n \log |\Sigma^{-1}\mathbf{S}| + n \operatorname{tr} \Sigma^{-1}(\mathbf{S} - \Sigma). \tag{10.6.1}$$

This measure has been used in problems of estimating and fitting a covariance structure. We set the minimum distance estimator $\hat{\Sigma}$ of Σ under M such that

$$\min_{\Sigma \in \mathcal{S}} D(\mathbf{S}, \Sigma) = D(\mathbf{S}, \hat{\Sigma}).$$

When $\boldsymbol{Y}$ is distributed as $\mathrm{N}(\boldsymbol{\mu}, \Sigma)$, using $(n/N)\mathbf{S}$ instead of $\mathbf{S}$ in the definition of D is equivalent to estimating by the maximum likelihood estimator. One of the merits in the use of the distance measure D is that it is not necessary to assume the normality of $\boldsymbol{Y}$.

Following Akaike (1973), we consider

$$R_D = \mathrm{E}^*_{\mathbf{Y}} \mathrm{E}^*_{\mathbf{Z}} [D(\mathbf{S}_{\mathbf{Z}}, \hat{\Sigma})] \tag{10.6.2}$$

as a measure of goodness or a risk for M_k. Here $\mathbf{S}_{\mathbf{Z}}$ is the covariance matrix based on a future observation matrix $\mathbf{Z} = (\boldsymbol{Z}_1, \ldots, \boldsymbol{Z}_N)'$. It is assumed that $\mathbf{Z}$ has the same distribution as $\mathbf{Y}$ and is independent of $\mathbf{Y}$. The expectations $\mathrm{E}^*_{\mathbf{Y}}$ and $\mathrm{E}^*_{\mathbf{Z}}$ denote those with respect to the true distributions of $\mathbf{Y}$ and $\mathbf{Z}$, respectively. A naive estimator of R_D is obtained by substituting $\mathbf{S}_{\mathbf{Y}}$ for $\mathbf{S}_{\mathbf{Z}}$ in $D(\mathbf{S}_{\mathbf{Z}}, \hat{\Sigma})$ (i.e., it is given by $D(\mathbf{S}, \hat{\Sigma})$). Then we can express R_D as

$$R_D = \mathrm{E}^*_{\mathbf{Y}} \mathrm{E}^*_{\mathbf{Z}} [D(\mathbf{S}, \hat{\Sigma})] + b_D,$$

where

$$b_D = \mathrm{E}^*_{\mathbf{Y}} \mathrm{E}^*_{\mathbf{Z}} [D(\mathbf{S}_{\mathbf{Z}}, \hat{\Sigma}) - D(\mathbf{S}, \hat{\Sigma})],$$

and it is called a *bias term* in the estimation of R_D by $D(\mathbf{S}, \hat{\Sigma})$. If we could construct an estimator $\hat{b}_D$ for b_D, we can propose a general criterion for selecting M as

$$-n \log |\hat{\Sigma}^{-1}\mathbf{S}| + n \operatorname{tr} \hat{\Sigma}^{-1}(\mathbf{S} - \hat{\Sigma}_k) + \hat{b}_D. \tag{10.6.3}$$

Here the bias term can be expressed as

$$\begin{aligned} b_D &= \mathrm{E}^*_{\mathbf{Y}} \mathrm{E}^*_{\mathbf{Z}} [-n \log |\mathbf{S}_{\mathbf{Z}}| + n \log |\mathbf{S}| + n \operatorname{tr} \hat{\Sigma}^{-1}\mathbf{S}_{\mathbf{Z}}| - n \operatorname{tr} \hat{\Sigma}^{-1}\mathbf{S}] \\ &= \mathrm{E}^*_{\mathbf{Y}} [n \operatorname{tr} \hat{\Sigma}^{-1}(\Sigma - \mathbf{S})] \\ &= \mathrm{E}^*_{\mathbf{Y}} [n \operatorname{tr} \hat{\Sigma}^{-1}\Sigma] - np, \quad \text{if} \quad \operatorname{tr} \hat{\Sigma}^{-1}\mathbf{S} = p. \end{aligned} \tag{10.6.4}$$

One typical approach for constructing $\hat{b}_D$ is to evaluate b_D under the assumption that M includes the true model. Then under appropriate regularity conditions it is shown that b_D is asymptotically equal to

$$b_D^{(0)} = 2 \times (\text{the number of independent parameters under } M). \tag{10.6.5}$$

This enables us to use a distance information criterion (DIC),

$$DIC = -n \log |\hat{\Sigma}^{-1}\mathbf{S}| + n \operatorname{tr} \hat{\Sigma}^{-1}(\mathbf{S} - \hat{\Sigma}) + b_D^{(0)}. \tag{10.6.6}$$

If one can evaluate b_D more precisely, we can use it instead of $b_D^{(0)}$ and have a modified DIC, which is denoted by CDIC.

For computation of the bias term $b_D^{(0)}$ in the case of Σ having a covariance structure $\Sigma(\boldsymbol{\theta})$, see Problem 10.13. On the theoretical basis of Konishi and Kitagawa (1996), Ichikawa and Konishi (1999) discussed the criteria in the situations where more general estimators are used.

Another approach to the evaluation of $\hat{b}_D$ is based on numerical technique: the bootstrap method due to Efron (1979) (see Section 5.8). This can be used without assuming the normality of $\boldsymbol{Y}$. Let

$$\mathbf{Y}^* = (\boldsymbol{Y}_1^*, \boldsymbol{Y}_2^*, \ldots, \boldsymbol{Y}_N^*)'$$

be a bootstrap sample from $\{\boldsymbol{Y}_1, \ldots, \boldsymbol{Y}_N\}$ (i.e., when $\mathbf{Y}$ is given). Then $\boldsymbol{Y}_1^*, \ldots, \boldsymbol{Y}_N^*$ are independently and identically distributed with

$$\mathrm{P}\{\boldsymbol{Y}_1^* = \boldsymbol{Y}_i\} = 1/N, \quad i = 1, \ldots, N.$$

Further, let $\mathbf{S}^*$ and $\hat{\Sigma}^*$ be $\mathbf{S}$ and $\hat{\Sigma}$, respectively, with a substitution for $\mathbf{Y}^*$ to $\mathbf{Y}$, and consider

$$\begin{aligned} b_D^* &= \mathrm{E}_{\mathbf{Y}^*}^*[n \operatorname{tr} \left(\hat{\Sigma}^*\right)^{-1} (\mathbf{S} - \mathbf{S}^*)] \\ &= \mathrm{E}_{\mathbf{Y}^*}^*[n \operatorname{tr} \left(\hat{\Sigma}^*\right)^{-1} \mathbf{S}] - np, \quad \text{if} \quad \operatorname{tr} \hat{\Sigma}^{-1}\mathbf{S} = p. \end{aligned}$$

Then it holds that for large N, b_D can be approximated by b_D^*. Using this result and simulation experiments on b_D^*, we can asymptotically approximate the value of b_D. The criterion is denoted by

$$DIC^* = -n \log |\hat{\Sigma}^{-1}\mathbf{S}| + n \operatorname{tr} \hat{\Sigma}^{-1}(\mathbf{S} - \hat{\Sigma}) + b_D^*, \tag{10.6.7}$$

which we call bootstrap DIC.

Note that DIC is just a modification of AIC (see Section 7.3.3) when the underlying distribution is normal. However, we use the term DIC or CDIC [see Fujikoshi and Kurata, 2008; Fujikoshi, 2009] since it can be used under a mild assumption of the underlying distribution. Now we will see a relationship between AIC and DIC under normality. Let $f(\mathrm{Y}; \Theta)$ be the density function of $\mathbf{Y}$ under M, where $\Theta = (\boldsymbol{\mu}, \Sigma)$. Then the AIC-type risk function is given as

$$\begin{aligned} R_A &= \mathrm{E}_{\mathbf{Y}}^* \mathrm{E}_{\mathbf{Z}}^*[-2 \log f(\mathrm{Z}; \tilde{\Theta})] \\ &= \mathrm{E}_{\mathbf{Y}}^*[-2 \log f(\mathrm{Y}; \tilde{\Theta})] + b_A, \end{aligned}$$

where $\tilde{\Theta}$ is the maximum likelihood estimator of Θ under M, and

$$b_A = \mathrm{E}_{\mathbf{Y}}^* \mathrm{E}_{\mathbf{Z}}^*[2 \log\{f(\mathbf{Y}; \tilde{\Theta}^{-1})/f(\mathbf{Z}; \tilde{\Theta}^{-1})\}]. \tag{10.6.8}$$

It is easy to see that

$$
\begin{aligned}
-2\log f(\mathbf{Y};\tilde{\boldsymbol{\mu}},\tilde{\Sigma}) &= N\log|\tilde{\Sigma}| + \operatorname{tr}\tilde{\Sigma}^{-1}n\mathbf{S} + Np\log 2\pi,\\
-2\log f(\mathbf{Z};\tilde{\boldsymbol{\mu}},\tilde{\Sigma}) &= N\log|\tilde{\Sigma}| + \operatorname{tr}\tilde{\Sigma}^{-1}n\mathbf{S}_{\mathbf{Z}}\\
&\quad + N\operatorname{tr}\tilde{\Sigma}^{-1}(\bar{\boldsymbol{Z}}-\tilde{\boldsymbol{\mu}})(\bar{\boldsymbol{Z}}-\tilde{\boldsymbol{\mu}})' + Np\log 2\pi,
\end{aligned}
$$

where $\tilde{\boldsymbol{\mu}} = \bar{\boldsymbol{Y}}$. Therefore,

$$
\begin{aligned}
b_A &= \mathrm{E}_{\mathbf{Y}}^{*}\mathrm{E}_{\mathbf{Z}}^{*}[-\operatorname{tr}\tilde{\Sigma}^{-1}n\mathbf{S} + \operatorname{tr}\tilde{\Sigma}^{-1}n\mathbf{S}_{\mathbf{Z}}\\
&\qquad + N\operatorname{tr}\tilde{\Sigma}^{-1}(\bar{\boldsymbol{Z}}-\tilde{\boldsymbol{\mu}})(\bar{\boldsymbol{Z}}-\tilde{\boldsymbol{\mu}})']\\
&= \mathrm{E}_{\mathbf{Y}}^{*}[n\operatorname{tr}\tilde{\Sigma}^{-1}(\Sigma-\mathbf{S}) + 2\operatorname{tr}\tilde{\Sigma}^{-1}\Sigma].
\end{aligned}
$$

Similarly, under an appropriate regularity condition it is shown that b_A is asymptotically equal to

$$b_A^{(0)} = b_D^{(0)} + 2p. \tag{10.6.9}$$

We can write AIC as

$$AIC = N\log|\tilde{\Sigma}| + n\operatorname{tr}\tilde{\Sigma}^{-1}\mathbf{S} + Np\log 2\pi + b_A^{(0)}.$$

A closer connection between DIC and AIC is obtained by noting that $\tilde{\Sigma} = (n/N)\hat{\Sigma}$.

10.6.2 Models for Equality of the Smaller Roots

In principal component analysis it is important to decide the number of useful or meaningful principal components. An approach to this problem is to consider a set of covariance structure models $M_k^{(1)}$, $k = 0, 1, \ldots, p-1$, on the equality of the smaller $p-k$ characteristic roots of Σ,

$$M_k\colon \lambda_k > \lambda_{k+1} = \cdots = \lambda_p (= \lambda), \tag{10.6.10}$$

where $\lambda_1 \geq \cdots \geq \lambda_p > 0$ are the characteristic roots of Σ.

The minimum distance estimator of Σ under M_k can be obtained by modifying the MLE slightly. Let $\ell_1 > \cdots > \ell_p$ be the characteristic roots of $\mathbf{S}$ and $\boldsymbol{h}_i$, the corresponding orthonormal characteristic vectors. These are expressed in matrix notation as $L = \operatorname{diag}(\ell_1, \ldots, \ell_p)$ and $\mathbf{H} = (\boldsymbol{h}_1, \ldots, \boldsymbol{h}_p)$. Then the minimum distance estimator is given (see Section 10.3) as

$$\hat{\Sigma}_k = \mathbf{H}\begin{pmatrix} \mathbf{L}_k & O \\ O & \bar{\ell}_{(q)}\mathrm{I}_q \end{pmatrix}\mathbf{H}', \tag{10.6.11}$$

where

$$\mathbf{L}_k = \operatorname{diag}(\ell_1, \ldots, \ell_k), \quad \bar{\ell}_{(q)} = \frac{1}{q}\sum_{j=k+1}^{p}\ell_j, \quad q = p-k.$$

Therefore, we have

$$DIC = -n\left\{ \sum_{j=k+1}^{p} \log \ell_j - \log\left(\frac{1}{q}\sum_{j=k+1}^{p} \ell_j\right)^q \right\} + 2f_k, \qquad (10.6.12)$$

where $f_k\big(\equiv pk - \frac{1}{2}k(k+1) + k + 1\big)$ is the number of independent parameters for Σ under M_k. The bias term is

$$b_D = -np + n\mathrm{E}^*[\operatorname{tr} \hat{\Sigma}^{-1}\Sigma]. \qquad (10.6.13)$$

Now we consider evaluating b_D under M_k, assuming that M_k includes the true model. For $0 \le k \le p-1$ it is shown (see Problem 10.8) that

$$b_D = -np + n\mathrm{E}^*[\operatorname{tr} \hat{\Sigma}_k^{-1}\Sigma] = 2f_k + \mathrm{O}(n^{-1}). \qquad (10.6.14)$$

For data on the five mathematics subjects considered in Example 4.2.1, we examine the model of equality of the smallest roots. The values of DIC are given in Table 10.6.1. The model $\lambda_1 > \cdots > \lambda_5$ is selected by DIC.

Table 10.6.1 Selection of equality of the smallest roots

k	0	1	2	3	4
S	232.54	80.76	50.05	45.85	30.00

10.6.3 Selecting a Subset of Original Variables

In the preceding subsection we discussed a method of selecting the number $p-k$ of the last principal components with the same variance. The method may be regarded as an approach for deciding how many principal components should be retained in order to account for most of the variation of $\boldsymbol{X}$, in particular if the same variance is small. Perhaps the most popular rule is use of what is called a cumulative contribution to the total variance, ψ_k in (10.1.6). It is usually desired that the principal components selected should contribute, say, 80% or 90%. The number of principal components required is then the smallest value of m for which this chosen percentage is exceeded. For other rules, mostly ad hoc, see Jollife (2002).

In principal component analysis the main objective is to replace the p variables by a smaller number m of principal components. This reduces the dimensionality of the problem. But the values of all p variables are still needed in order to calculate the principal components, since each principal component is a function of all p variables. This might cause difficulty in the interpretation of the principal components chosen. It is important to examine whether a set of original variables is taken without serious information loss. For some ad hoc rules for this problem, see Jollife (2002). In the following we give a formulation for redundancy of a subset of original variables, and consider its testing problem.

Let $\boldsymbol{X}$ be a p-dimensional random vector with mean $\boldsymbol{\mu}$ and covariance matrix Σ. Suppose that Σ is positive definite. Let $\lambda_1 \geq \cdots \geq \lambda_p$ be the ordered characteristic roots of Σ and $\boldsymbol{\gamma}_1, \ldots, \boldsymbol{\gamma}_p$ the corresponding characteristic vectors satisfying $\boldsymbol{\gamma}_i'\boldsymbol{\gamma}_j = \delta_{ij}$, $i, j = 1, \ldots, p$. Then the jth principal component of $\boldsymbol{X}$ is defined by $Y_j = \boldsymbol{\gamma}_j'(\boldsymbol{X} - \boldsymbol{\mu})$. For notational simplicity we consider formulating the redundancy hypothesis of the last $p - q$ variables $X_{q+1}, \ldots, X_p$ (i.e., $\boldsymbol{X}_2 = (X_{q+1}, \ldots, X_p)'$), in the presence of the first q variables $X_1, \ldots, X_q$ (i.e., $\boldsymbol{X}_1 = (X_1, \ldots, X_q)'$). Along the partition $\boldsymbol{X} = (\boldsymbol{X}_1', \boldsymbol{X}_2')'$, we partition $\boldsymbol{\gamma}_i$ and Σ as

$$\boldsymbol{\gamma}_j = \begin{pmatrix} \boldsymbol{\gamma}_{1j} \\ \boldsymbol{\gamma}_{2j} \end{pmatrix}, \quad \Sigma = \begin{pmatrix} \Sigma_{11} & \Sigma_{12} \\ \Sigma_{21} & \Sigma_{22} \end{pmatrix}.$$

The redundancy of $\boldsymbol{X}_2$ is defined only for a proper subset of the total set of all the principal components. Let J denote a subset of $m(\leq q)$ distinct integers between 1 and p. It is assumed that

$$\lambda_i \neq \lambda_j \text{ for all } i \in J, \ j \notin J. \tag{10.6.15}$$

Then we define the redundancy hypothesis of $\boldsymbol{X}_2$ for the m principal components $Y_i = \boldsymbol{\gamma}_i'(\boldsymbol{X} - \boldsymbol{\mu}), i \in J$ as

$$H_J(p, q): \ \boldsymbol{\gamma}_{2j} = \mathbf{0} \text{ for all } j \in J \tag{10.6.16}$$

In this case we also say that $\boldsymbol{X}_2$ has no additional information for the m principal components $Y_i, i \in J$, or $\boldsymbol{X}_1$ is sufficient for the m principal components $Y_i, i \in J$. Flury (1985, 1986) introduced the definition above for the characteristic vectors in the comparison of two covariance matrices. We note that such redundancy is defined only for a subset of principal components. In fact, if all the last $p - q$ components of $\boldsymbol{\gamma}_j$ are zero, then $\lambda_{q+1} = \cdots = \lambda_p = 0$ which contradicts the nonsingularity of Σ. More precisely, it is required that $m \leq q$. This is different from the redundancy hypothesis in discriminant analysis.

Now we consider testing the redundancy hypothesis $H_J(p, q)$ based on a random sample of size $N = n + 1 (> p)$ from a multivariate normal population $\mathrm{N}_p(\boldsymbol{\mu}, \Sigma)$. Taylor (1981) has proposed a test statistic for a general testing problem including the hypothesis $H_J(p, q)$. Flurly (1985, 1986) applied Taylor's approach to a test of the redundancy hypothesis in the comparison of two covariance matrices. The test statistic can be expressed as follows. Let $\ell_1 > \cdots > \ell_p > 0$ be the ordered characteristic roots of the sample covariance matrix $\mathbf{S}$ and $\boldsymbol{h}_1, \ldots, \boldsymbol{h}_p$, the corresponding characteristic vectors satisfying

$\boldsymbol{h}_i'\boldsymbol{h}_j = \delta_{ij}$, $i, j = 1, \ldots, p$. We partition $\boldsymbol{h}_j$ as $(\boldsymbol{h}_{1j}', \boldsymbol{h}_{2j}')'$, along with the partition of $\boldsymbol{X}$. Then we can express the test statistic as

$$T = n \sum_{j \in J} \boldsymbol{h}_{2j}' \left\{ \sum_{i \in K} \ell_i \ell_j (\ell_i - \ell_j)^{-2} \boldsymbol{h}_{2j} \boldsymbol{h}_{2j}' \right\}^{-1} \boldsymbol{h}_{2j}, \tag{10.6.17}$$

whose asymptotic null distribution is a chi-square distribution with $m(p-q)$ degrees of freedom.

We will see how the statistic T is proposed. Let

$$P_J = \sum_{i \in J} \boldsymbol{\gamma}_i \boldsymbol{\gamma}_i' \quad \text{and} \quad P_K = \sum_{i \in K} \boldsymbol{\gamma}_i \boldsymbol{\gamma}_i' \tag{10.6.18}$$

which are the projection matrices for the range spaces $\mathcal{S}_J = \mathcal{R}[\boldsymbol{\gamma}_i, i \in J]$ and $\mathcal{S}_K = \mathcal{R}[\boldsymbol{\gamma}_i, i \in K]$, respectively. Further, let

$$A = \begin{pmatrix} \mathrm{I}_q \\ \mathrm{O} \end{pmatrix}, \quad B = \begin{pmatrix} \mathrm{O} \\ \mathrm{I}_{p-q} \end{pmatrix}.$$

Then we can see that

$$H_J(p, q) \quad \Leftrightarrow \quad \mathcal{R}[A] \supset \mathcal{S}_J \quad \Leftrightarrow \quad \mathcal{R}[B] \supset \mathcal{S}_K \quad \Leftrightarrow \quad P_K B = B.$$

Let $\hat{P}_K$ be the sample analog of P_K. It is seen that $\sqrt{n}\,\mathrm{vec}\,[(\mathrm{I}_p - \hat{P}_K)B]$ converges in distribution to $\mathrm{N}(\boldsymbol{0}, \Psi(B))$. The test statistic proposed by Tylor (1981) is

$$T = n\ \mathrm{vec}\,[(\mathrm{I}_p - \hat{P}_K)B]'[\hat{\Psi}(B)]^+ \mathrm{vec}\,[(\mathrm{I}_p - \hat{P}_K)B],$$

where $\hat{\Psi}(B)$ is the sample analog of $\Psi(B)$, and $[\hat{\Psi}(B)]^+$ is the Moore-Penrose generalized inverse satisfying (9.6.5).

10.7 Methods Related to Principal Components

10.7.1 Fixed-Effect Principal Component Model

In this section we consider two methods related to principal components. First we consider a fixed-effect principal component model. Suppose that a p-dimensional variate $\boldsymbol{Y} = (Y_1, \ldots, Y_p)'$ is measured for N individuals, and let $\mathbf{Y} = (Y_{ij}) = (\boldsymbol{Y}_1, \ldots, \boldsymbol{Y}_N)'$ be an observation matrix. Using a singular value decomposition (see [A.1.12]) the model with dimension k is given as

$$M_k^{(1)}\text{:}\ Y_{ij} = \mu_j + \xi_{ij} + \varepsilon_{ij}, \quad i = 1, \ldots, N; j = 1, \ldots, p, \tag{10.7.1}$$

where ε_{ij} are independently distributed as $\mathrm{N}(0, \sigma^2)$, and

$$\Xi = (\xi_{ij}) = \sum_{j=1}^{k} \sqrt{\lambda_j}\, \boldsymbol{\beta}_j \boldsymbol{\gamma}_j'.$$

Here the parameters $\boldsymbol{\beta}_j$, $\boldsymbol{\gamma}_j$, and λ_j satisfy the restriction conditions

$$\boldsymbol{\beta}_i' \boldsymbol{\beta}_j = \delta_{ij}, \quad \boldsymbol{\gamma}_i' \boldsymbol{\gamma}_j = \delta_{ij}, \quad \lambda_1 \geq \cdots \geq \lambda_k > 0.$$

The model means that the rank of $\mathrm{E}(\mathbf{Y} - \mathbf{1}_N \boldsymbol{\mu}')$ is equal to k and the variables Y_{ij} are independenly normally distributed with identical variance σ^2.

Putting $\bar{\boldsymbol{Y}} = (1/N) \sum_{i=1}^{N} \boldsymbol{Y}_i$, consider the singular value decomposition of the centralized data matrix $\mathbf{Y} - \mathbf{1}_N \bar{\boldsymbol{Y}}'$ given as

$$\mathbf{Y} - \mathbf{1}_N \bar{\boldsymbol{Y}}' = \mathbf{U}\mathbf{L}^{1/2}\mathbf{H}' = \sum_{j=1}^{p} \sqrt{\ell_j}\, \boldsymbol{U}_j \boldsymbol{h}_j',$$

where

$$\mathbf{U} = (\boldsymbol{U}_1, \ldots, \boldsymbol{U}_p)', \ \mathbf{H} = (\boldsymbol{h}_1, \ldots, \boldsymbol{h}_k), \ \Lambda = \mathrm{diag}\,(\lambda_1, \ldots, \lambda_p).$$

These satisfy $\mathbf{U}'\mathbf{U} = \mathrm{I}_p$, $\mathbf{H}'\mathbf{H} = \mathrm{I}_p$, and $\ell_1 > \cdots > \ell_p > 0$. Then the maximum likelihood estimators of $\boldsymbol{\mu}$ and Ξ are given [see, e.g., Montel, 1971 and Siotani et al., 1985] as

$$\hat{\boldsymbol{\mu}} = \bar{\boldsymbol{X}}, \quad \hat{\Xi} = \sum_{j=1}^{k} \sqrt{\ell_j}\, \boldsymbol{U}_j \boldsymbol{h}_j'.$$

Further, let

$$\begin{aligned} \hat{\mathbf{Y}} &= \mathbf{1}_N \hat{\boldsymbol{\mu}}' + \hat{\Xi} \\ &= \mathbf{1}_N \bar{\boldsymbol{Y}} + \sum_{j=1}^{k} \sqrt{\ell_j}\, \boldsymbol{U}_j \boldsymbol{h}_j'. \end{aligned}$$

Then the maximum likelihood estimator of σ^2 is given as

$$\begin{aligned} Np\hat{\sigma}^2 &= \|\mathbf{Y} - \hat{\mathbf{Y}}\|^2 \\ &= \mathrm{tr}\,(\mathbf{Y} - \hat{\mathbf{Y}})'(\mathbf{Y} - \hat{\mathbf{Y}}) \\ &= \ell_{k+1} + \cdots + \ell_p. \end{aligned}$$

Here $\ell_1 > \cdots > \ell_p > 0$ are the characteristic roots of $n\mathbf{S}$, since $(\mathbf{Y} - \mathbf{1}_N \bar{\boldsymbol{Y}}')(\mathbf{Y} - \mathbf{1}_N \bar{\boldsymbol{Y}}') = n\mathbf{S}$. The dimension of Ξ under M_k is equal to

$$Nk - \frac{1}{2}k(k+1) + k + pk - \frac{1}{2}k(k+1) = k(N + p - k).$$

Therefore, AIC is formally given by

$$AIC = Np\log\hat{\sigma}^2 + Np(\log 2\pi + 1) + 2\left\{k(N + p - k) + p + 1\right\}.$$

However, it is left to examine whether or not the estimator of the bias term is appropriate. In fact, the bias term is expressed as

$$\begin{aligned} b_A &= -Np + \mathrm{E}^*_{\mathbf{Y}}\mathrm{E}^*_{\mathbf{Z}}\left[\frac{1}{\hat{\sigma}^2}\|\mathbf{Z} - \hat{\mathbf{Y}}\|^2\right] \\ &= -Np + \mathrm{E}^*_{\mathbf{Y}}\left[\frac{Np}{\ell_{k+1} + \cdots + \ell_p}\left\{Np\sigma^2 + \|\mathbf{1}_N + \Xi - \hat{\mathbf{Y}}\|^2\right\}\right], \end{aligned}$$

of which expectation is complicated. Some methods based on $\|\mathbf{Y} - \hat{\mathbf{Y}}\|^2$ have been proposed. For details, see, e.g. Jolliffe (2002) and Ferré (1995).

10.7.2 Random-Effect Principal Components Model

The model is defined as (10.7.1) with the following $\boldsymbol{\beta}_1, \ldots, \boldsymbol{\beta}_k$ or Ξ:

$$(\boldsymbol{\beta}_1, \ldots, \boldsymbol{\beta}_k) = \begin{pmatrix} \boldsymbol{b}'_1 \\ \vdots \\ \boldsymbol{b}'_n \end{pmatrix} \equiv B, \quad \text{or} \quad \Xi = B\Lambda_k^{1/2}\Gamma'_k,$$

where $\Lambda_k = \mathrm{diag}(\lambda_1, \cdots, \lambda_k)$ and $\Gamma_k = (\gamma_1, \ldots, \gamma_k)$. Here the terms $\boldsymbol{b}_i$ denote variation of the ith individual, and it is assumed that $\boldsymbol{b}_1, \ldots, \boldsymbol{b}_N$ are independently distributed as $\mathrm{N}_k(\mathbf{0}, \Delta)$ with positive definite covariance matrix Δ, and $\{\boldsymbol{b}_1, \ldots, \boldsymbol{b}_N\}$ and $\{\varepsilon_{ij}\}$ are independent. The model is equivalent to assuming that

$$M_k^{(2)}\colon \boldsymbol{Y}_i = \boldsymbol{\mu} + \Gamma_k\Lambda_k^{1/2}\boldsymbol{b}_i + \boldsymbol{\varepsilon}_i, \ i = 1, \ldots, N, \tag{10.7.2}$$

where $\boldsymbol{Y}_i = (Y_{i1}, \ldots, Y_{ip})'$ and $\boldsymbol{\varepsilon}_i = (\epsilon_{i1}, \ldots, \epsilon_{ip})'$. Further, we can write the model $M_k^{(2)}$ in terms of covariance structures as

$$\tilde{M}_k^{(2)}\colon \Sigma = \mathrm{CC}' + \sigma^2\mathrm{I}_p, \tag{10.7.3}$$

where C is a $p \times k$ matrix given by $\mathrm{C} = \Gamma_k\Lambda_k^{1/2}\Delta^{1/2}$. Note that covariance structure (10.7.3) is the one in which the unique covariance matrix in factor analysis is $\sigma^2\mathrm{I}_p$. Therefore, the maximum likelihood estimators of C and σ^2 can be obtained by modifying the algorithms for factor analysis. Therefore, for selection of $\tilde{M}_k^{(2)} (k = 0, 1, \ldots, p-1)$, we will be able to use DIC or AIC.

Problems

10.1 Consider predicting a random variable Y by a linear function of p-dimensional random vector $\boldsymbol{X}$ (i.e., $c + \boldsymbol{a}'\boldsymbol{X}$). Then show that the minimum of $\mathrm{E}[(Y - c - \boldsymbol{a}'\boldsymbol{X})^2]$ is attained at

$$\hat{Y} = \mu_Y + \boldsymbol{\sigma}'_{XY}\Sigma_{XX}^{-1}(\boldsymbol{X} - \boldsymbol{\mu}_X)$$

and

$$\begin{aligned}\mathrm{E}[(Y - \hat{Y})^2] &= \mathrm{Var}(Y - \hat{Y}) \\ &= \sigma_{YY} - \boldsymbol{\sigma}'_{XY}\Sigma_{XX}^{-1}\boldsymbol{\sigma}_{XY} = \sigma_{YY\cdot X},\end{aligned}$$

where

$$\begin{aligned}&\mathrm{E}(Y) = \mu_Y, \quad \mathrm{E}(\boldsymbol{X}) = \Sigma_{XX}, \quad \mathrm{Var}(Y) = \sigma_{YY} = \sigma_Y^2, \\ &\mathrm{Cov}(\boldsymbol{X}, Y) = \boldsymbol{\sigma}_{XY}, \quad \mathrm{Var}(\boldsymbol{X}) = \Sigma_{XX}.\end{aligned}$$

10.2 Consider a q-dimensional random vector $\boldsymbol{Y}$ by linear functions of p dimensional random vector $\boldsymbol{X}$:

$$\boldsymbol{c} + \mathrm{A}\boldsymbol{X} = (c_1 + \boldsymbol{a}_1'\boldsymbol{X}, \ldots, c_q + \boldsymbol{a}_q'\boldsymbol{X})'.$$

Then show that the minimum of $\mathrm{E}[(\boldsymbol{Y} - \boldsymbol{c} - \mathrm{A}\boldsymbol{X})'(\boldsymbol{Y} - \boldsymbol{c} - \mathrm{A}\boldsymbol{X})]$ is attained at

$$\hat{\boldsymbol{Y}} = \boldsymbol{\mu}_Y + \Sigma_{YX}\Sigma_{XX}^{-1}(\boldsymbol{X} - \boldsymbol{\mu}_X)$$

and

$$\begin{aligned}\mathrm{E}[(\boldsymbol{Y} - \hat{\boldsymbol{Y}})(\boldsymbol{Y} - \hat{\boldsymbol{Y}})'] &= \mathrm{Var}(\boldsymbol{Y} - \hat{\boldsymbol{Y}}) \\ &= \Sigma_{YY} - \Sigma_{YX}\Sigma_{XX}^{-1}\Sigma_{XY} = \Sigma_{YY\cdot X},\end{aligned}$$

where

$$\boldsymbol{\mu} = \begin{pmatrix} \mu_Y \\ \boldsymbol{\mu}_X \end{pmatrix}, \qquad \Sigma = \begin{pmatrix} \sigma_{YY} & \boldsymbol{\sigma}_{YX} \\ \boldsymbol{\sigma}_{XY} & \Sigma_{XX} \end{pmatrix}.$$

Hint: Note that

$$\mathrm{E}[(\boldsymbol{Y} - \boldsymbol{c} - \mathrm{A}\boldsymbol{X})'(\boldsymbol{Y} - \boldsymbol{c} - \mathrm{A}\boldsymbol{X})] = \sum_{i=1}^{p} \mathrm{E}[(Y_i - c_i - \boldsymbol{a}_i'\boldsymbol{X})^2].$$

Therefore, the result is obtained from Problem 10.1.

10.3 Let $\boldsymbol{Z} = \mathrm{B}'(\boldsymbol{X} - \boldsymbol{\mu}_X)$, where B is a $p \times k$ matrix. Consider predicting $\boldsymbol{X}$ by a linear function (i.e., $\boldsymbol{c} + \mathrm{A}\boldsymbol{Z}$). Then show that the minimum of

$$\mathrm{E}[(\boldsymbol{X} - \boldsymbol{c} - \mathrm{A}'\boldsymbol{Z})'(\boldsymbol{X} - \boldsymbol{c} - \mathrm{A}'\boldsymbol{Z})]$$

is attained at

$$\hat{\boldsymbol{X}} = \boldsymbol{\mu} + \Sigma\mathrm{B}(\mathrm{B}\Sigma\mathrm{B})^{-1}\mathrm{B}'(\boldsymbol{X} - \boldsymbol{\mu}_X)$$

and

$$\begin{aligned}\mathrm{E}[(\boldsymbol{X} - \hat{\boldsymbol{X}})(\boldsymbol{X} - \hat{\boldsymbol{X}})'] &= \mathrm{Var}(\boldsymbol{X} - \hat{\boldsymbol{X}}) \\ &= \Sigma - \Sigma\mathrm{B}(\mathrm{B}'\Sigma\mathrm{B})^{-1}\mathrm{B}'\Sigma.\end{aligned}$$

Hint: Note that

$$\mathrm{Var}(\boldsymbol{X}) = \Sigma, \quad \mathrm{Cov}(\boldsymbol{X}, \boldsymbol{Z}) = \Sigma\mathrm{B}, \quad \mathrm{Var}(\boldsymbol{Z}) = \mathrm{B}\Sigma\mathrm{B}.$$

The result is obtained from Problem 10.2.

10.4 Consider the residual covariance matrix appeared in Problem 10.3,

$$R(\mathrm{B}) = \Sigma - \Sigma\mathrm{B}(\mathrm{B}'\Sigma\mathrm{B})^{-1}\mathrm{B}'\Sigma,$$

where B is a $p \times k$ matrix with rank k. Show that the two measures of overall predictive efficiency,

(a) $\mathrm{tr}\,(\Sigma - \Sigma\mathrm{B}(\mathrm{B}'\Sigma\mathrm{B})^{-1}\mathrm{B}'\Sigma)$,

(b) $||\Sigma - \Sigma\mathrm{B}(\mathrm{B}'\Sigma\mathrm{B})^{-1}\mathrm{B}'\Sigma||$,

lead to the same optimum choice of $\mathrm{B} = (\boldsymbol{b}_1, \ldots, \boldsymbol{b}_k)$. Let $\lambda_1 \geq \cdots \geq \lambda_p$ be the characteristic roots of Σ, and $\boldsymbol{\gamma}_1, \ldots, \boldsymbol{\gamma}_p$ the corresponding characteristic vectors satisfying $\boldsymbol{\gamma}_i'\boldsymbol{\gamma}_i = 1$ and $\boldsymbol{\gamma}_i'\boldsymbol{\gamma}_j = 0$ $(i \neq j)$. Then the same optimum choice may be taken as $\boldsymbol{b}_i = \boldsymbol{\gamma}_i$, $i = 1, \ldots, k$ (Rao (1973)).

Hint: For (a), use

$$\mathrm{tr}\,\Sigma\mathrm{B}(\mathrm{B}'\Sigma\mathrm{B})^{-1}\mathrm{B}'\Sigma = \mathrm{tr}\,U'\Sigma U,$$

where $U = \Sigma^{1/2}\mathrm{B}(\mathrm{B}'\Sigma\mathrm{B})^{-1/2}$. Note that U satisfy $U'U = I_k$, and use [A.2.7]. For (b), use [A.2.8].

10.5 Let $\hat{\Sigma}_k$ be the estimator defined by (10.6.11). Then, show that $\mathrm{tr}\,\hat{\Sigma}_k\mathbf{S} = p$.

10.6 Let $\hat{\Sigma}_k$ be the minimum distance estimator of Σ under under the equality of the smaller $p-k$ characteristic roots of Σ, i.e., M_k: $\lambda_k > \lambda_{k+1} = \cdots = \lambda_p (= \lambda)$, based on the sample covariance matrix $\mathbf{S}$ with n degrees of reedom from $\mathrm{N}(\boldsymbol{\mu}, \Sigma)$. Then, show that under M_k with $\lambda_1 > \cdots > \lambda_k$

$$\mathrm{E}[\operatorname{tr} \hat{\Sigma}_k^{-1}\Sigma] = p + \frac{2}{n} f_k + \mathrm{O}(n^{-2}).$$

where $f_k = pk - \frac{1}{2}k(k+1) + k + 1$.

Hint: we have

$$\begin{aligned}\operatorname{tr} \hat{\Sigma}_k^{-1}\Sigma &= \operatorname{tr} L_k^{-1} G'(\Lambda_k - \lambda I_k)G - (1/\bar{\ell}_{(q)}) \operatorname{tr} G'(\Lambda_k - \lambda I_k)G \\ &\quad + \lambda \operatorname{tr} L_k^{-1} + (1/\bar{\ell}_{(q)}) \operatorname{tr}(\Lambda_k - \lambda I_k) + (1/\bar{\ell}_{(q)}) q\lambda,\end{aligned}$$

where $G = \Gamma_k' H_k$. Let $\tilde{S} = \Gamma' S \Gamma$, and $\tilde{S} = \Lambda + \frac{1}{\sqrt{n}} V$, $V = (v_{\alpha\beta})$. Then, letting $F = \Gamma' H$, we have

$$\tilde{S} F = F L.$$

We use perturbation expansions of the characteristic roots and vectors of $\tilde{S}$. Since the characteristic roots are the same as the ones of S, we have, for $\alpha = 1, \ldots, k$

$$\ell_\alpha = \lambda_\alpha + \frac{1}{\sqrt{n}} v_{\alpha\alpha} + \frac{1}{n} \sum_{\beta \neq \alpha}^{p} \lambda_{\alpha\beta} v_{\alpha\beta}^2 + \cdots,$$

where $\lambda_{\alpha\beta} = 1/(\lambda_\alpha - \lambda_\beta)$. Note that G is a submatrix of F, and the columns of F are the characteristic vectors of $\tilde{S}$. Therefore, we can expand G as

$$G = I_k + \frac{1}{\sqrt{n}} G^{(1)} + \frac{1}{n} G^{(2)} + \cdots,$$

where $G^{(1)} = (g_{\alpha\beta}^{1)})$, $G^{(2)} = (g_{\alpha\beta}^{2)})$ and

$$\begin{aligned}g_{\alpha\alpha}^{(1)} &= 0, \\ g_{j\alpha}^{(1)} &= -\lambda_{j\alpha} v_{j\alpha}, \ j \neq \alpha, \\ g_{\alpha\alpha}^{(2)} &= -\frac{1}{2} \sum_{\beta \neq \alpha} \lambda_{\alpha\beta}^2 v_{\alpha\beta}^2, \\ g_{j\alpha}^{(2)} &= -\lambda_{j\alpha} \left[\lambda_{j\alpha} v_{j\alpha} v_{\alpha\alpha} + \sum_{\beta \neq \alpha} \lambda_{\alpha\beta} v_{j\beta} v_{\beta\alpha} \right], \ j \neq \alpha.\end{aligned}$$

10.7 Let $\lambda_1 \geq \cdots \geq \lambda_p$ be the characteristic roots of $p \times p$ covariance matrix Σ. Consider two chasses of positive definite matrices Σ such that $\mathcal{S}_1 = \{\Sigma : \lambda_{k+1} = \cdots = \lambda_p\}$ and $\mathcal{S}_2 = \{\Sigma : \Sigma = \Sigma_0 + \sigma^2 I_p, \operatorname{rank}(\Sigma_0) = k\}$. Then, show that $\dim(\mathcal{S}_1) = pk - \frac{1}{2}k(k-1) + 1$, and $\dim(\mathcal{S}_2) = pk - \frac{1}{2}k(k-1) + 1$.

10.8 Let $\boldsymbol{X}_1, \ldots, \boldsymbol{X}_N$ be a random sample of size $N = n+1$ from $\mathrm{N}(\boldsymbol{\mu}, \Sigma)$. Assume that $\Sigma = \Psi + \sigma^2 \mathrm{I}_p$, where Ψ is a positive semi definite matrix of rank k. Then, show that the MLE of σ^2 is given by $\hat{\sigma}^2 = (n/N)(\ell_{k+1} + \cdots + \ell_p)/(p-k)$, where $\ell_1 > L \cdots > \ell_p > 0$ are the characyteristic roots of the sample covariance matrix $\mathbf{S}$.

10.9 Derive the limiting distribution of U_k in (10.5.20) by the method similar to that of Q_k in Theorem 10.5.5.

Hint: Without loss generality we may assume $\lambda = 1$. Use the following expressions:

$$\sum_{i=k+1}^{p} \ell_i = \sum_{i=k+1}^{p} \left(1 + \sqrt{\frac{2}{n}} Z_i\right) = q + \sqrt{\frac{2}{n}} \operatorname{tr} \mathbf{U},$$

$$\sum_{i=k+1}^{p} \ell_i^2 = \sum_{i=k+1}^{p} \left(1 + \sqrt{\frac{2}{n}} Z_i\right)^2 = q + 2\sqrt{\frac{2}{n}} \operatorname{tr} \mathbf{U} + \frac{2}{n} \operatorname{tr} \mathbf{U}^2.$$

10.10 Suppose $\Sigma = \sigma^2\{(1-\rho)\mathrm{I}_p + \rho \mathbf{1}_p \mathbf{1}_p'\}$, $-1/(p-1) < \rho < 1$, and let $\lambda_1 \leq \cdots \leq \lambda_p > 0$ be the characteristic roots of Σ, and $\boldsymbol{\gamma}_1, \ldots, \boldsymbol{\gamma}_p$ the corresponding characteristic vectors satisfying $\boldsymbol{\gamma}_i' \boldsymbol{\gamma}_j = \delta_{ij}$.

(1) Show that $\lambda_1 = \sigma^2\{1 + (p-1)\rho\}$, $\lambda_2 = \cdots = \lambda_p = \sigma^2(1-\rho)$, $\boldsymbol{\gamma}_1 = \pm(1/\sqrt{p})\mathbf{1}_p$, and $\Gamma_2 = (\boldsymbol{\gamma}_2, \ldots, \boldsymbol{\gamma}_p)$ is any matrix such that $\Gamma_2'\Gamma_2 = \mathrm{I}_{p-1}$ and $\Gamma_2' \mathbf{1}_p = \mathbf{0}$.

(2) Put $\mathrm{A}_1 = (1/p)\mathbf{1}_p \mathbf{1}_p'$, $\mathrm{A}_2 = \mathrm{I}_p - \mathrm{A}_1$. Then, examine that $\mathrm{A}_i^2 = \mathrm{A}_i$, $i = 1, 2$, $\mathrm{A}_1'\mathrm{A}_2 = \mathrm{O}$, and

$$\Sigma = \lambda_1 \mathrm{A}_1 + \lambda_2 \mathrm{A}_2, \quad \Sigma^{-1} = \lambda_1^{-1} \mathrm{A}_1 + \lambda_2^{-1} \mathrm{A}_2.$$

10.11 Consider the case that the covariance matrix Σ has a covariance structure such that $\Sigma = \sigma_1 \mathrm{G}_1 + \cdots + \sigma_k \mathrm{G}_k$, where G_i's are given $p \times p$ symmetric matrices and σ_i's are unknown. Further, G_i's are linearly independent, and σ_i's are parameters such that Σ is positive definite. Let $\hat{\sigma}_i$ be the MLE's of σ_i, based on a random sample of size $N = n+1$ from $\mathrm{N}(\boldsymbol{\mu}, \Sigma)$.

(1) Show that the MLE's $\hat{\sigma}_i$ of σ_i are the solutions of

$$\operatorname{tr}\left[\hat{\Sigma}^{-1}(\hat{\Sigma} - \mathbf{S}*)\hat{\Sigma}^{-1}\mathrm{G}_i\right] = 0, \ i = 1, \ldots, k,$$

where $\mathbf{S}^* = (n/N)\mathbf{S}$, $\mathbf{S}$ is the sample covariance matrix, and $\hat{\Sigma} = \sum_{i=1}^{k} \hat{\sigma}_i \mathrm{G}_i$.

(2) Show that $\operatorname{tr} \hat{\Sigma}^{-1}\mathbf{S}^* = p$.

10.12 Let $\boldsymbol{X}$ be a p-dimensional random vector with mean $\boldsymbol{\mu}$ and covariance matrix Σ. Let λ_1 be the largest characteristic root of Σ and γ_1 the corresponding characteristic vector satisfying $\gamma_1'\gamma_1 = 1$. Consider a transformed random vector $\boldsymbol{U} = \mathrm{C}'\boldsymbol{X}$, where C is a $p \times q$ constant matrix. Let $\boldsymbol{Z} = (\boldsymbol{U}', \boldsymbol{X}')'$, and consider a linear combination $\boldsymbol{a}'\boldsymbol{Z} = \boldsymbol{a}_1'\boldsymbol{U} + \boldsymbol{a}_2'\boldsymbol{X}$. Then, show that for any $\boldsymbol{a}$ satisfying $\boldsymbol{a}'\boldsymbol{a} = 1$, $\mathrm{Var}(\boldsymbol{a}'\boldsymbol{Z}) \leq \lambda_1$, and the equality is atained when $\mathrm{C}\boldsymbol{a}_1 + \boldsymbol{a}_2 = \pm\gamma_1$.

10.13 Assuming $\mathbf{W} = n\mathbf{S} \sim \mathrm{W}_p(n, \Sigma)$, let $f(\mathrm{S}; \Sigma)$ be the density function of $\mathbf{S}$, and prove the following:

$$-2 \log f(\mathrm{S}; \Sigma) = D(\mathrm{S}, \Sigma) + C_{p,n}(\mathrm{S}),$$

where

$$C_{p,n}(\mathrm{S}) = (p+1) \log |\mathrm{S}| + np\{1 + \log(2/n)\} + 2 \log \Gamma(\frac{1}{2}n).$$

Consider a covariance structure model $M: \Sigma = \Sigma(\boldsymbol{\theta})$. Use the relation above to show that the b_D in (10.6.4) is asymptotically equal to $b_D^{(0)}$ in (10.6.5) whenever M includes the true model.

Chapter 11

Canonical Correlation Analysis

The canonical correlation analysis is concerned with the correlation structure between two sets of variables, and aims at summarizing the relationships between them in terms of as few linear combinations of variables as possible.

The *first canonical variable* is the pair of linear combinations of variables such that have the maximum correlation, called the first canonical correlation. The second canonical correlation and variable are defined similarly by the pair of linear combinations of variables, uncorrelated with the first pair, that have the maximum correlation, and so on.

Large-sample and high-dimensional approximations are obtained for the distribution of canonical correlations, the former as a special case of the latter. Some inferential problems on dimensionality and selection of variables are treated as those of selecting special types of covariance structures.

11.1 Definition of Population Canonical Correlations and Variables

Suppose that we are given two random vectors $\boldsymbol{X} = (X_1, \ldots, X_p)'$ and $\boldsymbol{Y} = (Y_1, \ldots, Y_q)'$. The canonical correlations measure the associations between the random vectors $\boldsymbol{X}$ and $\boldsymbol{Y}$, which is a generalization of the multiple correlation. In this section we define them in a population.

Let Σ be the population covariance matrix of $(\boldsymbol{X}', \boldsymbol{Y}')'$, which includes partition into p and q rows and columns:

$$\Sigma = \begin{pmatrix} \Sigma_{11} & \Sigma_{12} \\ \Sigma_{21} & \Sigma_{22} \end{pmatrix}, \quad \Sigma_{12}: p \times q.$$

We may assume without loss of generality that $p \leq q$. Then the canonical correlation is based on linear combinations

$$\xi = \alpha_1 X_1 + \cdots + \alpha_p X_p = \boldsymbol{\alpha}' \boldsymbol{X},$$
$$\eta = \beta_1 Y_1 + \cdots + \beta_q Y_q = \boldsymbol{\beta}' \boldsymbol{Y}.$$

The correlation between ξ and η is expressed as

$$\rho(\xi, \eta) = \frac{\boldsymbol{\alpha}' \Sigma_{12} \boldsymbol{\beta}}{(\boldsymbol{\alpha}' \Sigma_{11} \boldsymbol{\alpha})^{1/2} (\boldsymbol{\beta}' \Sigma_{22} \boldsymbol{\beta})^{1/2}}. \tag{11.1.1}$$

In the problem of maximizing $\rho(\xi, \eta)$, we may assume without loss of generality that

$$\mathrm{Var}(\xi) = \boldsymbol{\alpha}' \Sigma_{11} \boldsymbol{\alpha} = 1, \quad \mathrm{Var}(\eta) = \boldsymbol{\beta}' \Sigma_{22} \boldsymbol{\beta} = 1, \tag{11.1.2}$$

since $\rho(c\xi, d\eta) = \rho(\xi, \eta)$ for $c > 0$ and $d > 0$. Then we may look for ξ and η, or $\boldsymbol{\alpha}$ and $\boldsymbol{\beta}$ maximizing

$$\rho(\xi, \eta) = \boldsymbol{\alpha}' \Sigma_{12} \boldsymbol{\beta} \tag{11.1.3}$$

under condition (11.1.2). The canonical correlations and variables are defined by finding vectors $\boldsymbol{\alpha}_1, \ldots, \boldsymbol{\alpha}_p$ in R^p and $\boldsymbol{\beta}_1, \ldots, \boldsymbol{\beta}_q$ in R^q, or

$$\xi_i = \boldsymbol{\alpha}_i' \boldsymbol{X}, \ i = 1, \ldots, p; \ \eta_j = \boldsymbol{\beta}_j' \boldsymbol{Y}, \ j = 1, \ldots, q, \tag{11.1.4}$$

such that:

(1) ξ_1 and η_1 are two linear combinations with variances 1 having the maximum correlation among the correlations between any two linear combinations $\boldsymbol{\alpha}' \boldsymbol{X}$ and $\boldsymbol{\beta}' \boldsymbol{Y}$.

(2) For $k \leq p$, ξ_k and η_k are two linear combinations with variances 1 having the maximum correlation among the correlations between any two linear combinations $\boldsymbol{\alpha}' \boldsymbol{X}$ and $\boldsymbol{\beta}' \boldsymbol{Y}$ under the conditions; $\rho(\xi, \xi_i) = 0, \ \rho(\eta, \eta_i) = 0, \ i = 1, \ldots, k-1$.

(3) For $k > p$, η_k is a linear combination with variance 1 satisfying $\rho(\eta, \eta_j) = 0, \ j = 1, \ldots, k-1$.

Let $\rho(\xi_i, \eta_i) = \rho_i, \ i = 1, \ldots, p$. We note that the variables in (11.1.4) satisfy

$$\rho(\xi_i, \eta_j) = 0 \quad \text{for } i \neq j \tag{11.1.5}$$

automatically. The variances and covariances of ξ_i and η_j are summarized as follows: Let $\boldsymbol{\xi} = (\xi_1, \ldots, \xi_p)'$ and $\boldsymbol{\eta} = (\eta_1, \ldots, \eta_q)'$. Then

$$\mathrm{Var}[\begin{pmatrix} \boldsymbol{\xi} \\ \boldsymbol{\eta} \end{pmatrix}] = \begin{pmatrix} \mathrm{I}_p & \mathrm{P} \\ \mathrm{P}' & \mathrm{I}_q \end{pmatrix},$$

where $\mathrm{P} = (\mathrm{D}_\rho, \mathrm{O})$, $\mathrm{D}_\rho = \mathrm{diag}\,(\rho_1, \cdots, \rho_p)$.

This problem was originated by Hotelling (1936). We assume here that $\Sigma > \mathrm{O}$ and $\Sigma_{11} > \mathrm{O}$, $\Sigma_{22} > \mathrm{O}$. Now we consider $\rho(\xi, \eta)$ in (11.1.3) under (11.1.2) by the Lagrange method. Let

$$\phi(\boldsymbol{\alpha}, \boldsymbol{\beta}) = (\boldsymbol{\alpha}'\Sigma_{12}\boldsymbol{\beta})^2 - \lambda(\boldsymbol{\alpha}'\Sigma_{11}\boldsymbol{\alpha} - 1) - \theta(\boldsymbol{\beta}'\Sigma_{22}\boldsymbol{\beta} - 1), \tag{11.1.6}$$

where λ and θ are the Lagrange multipliers. Differentiation of $\phi(\boldsymbol{\alpha}, \boldsymbol{\beta})$ with respect to $\boldsymbol{\alpha}$ and $\boldsymbol{\beta}$ yields the following two simultaneous equations:

$$\begin{aligned} -\lambda\Sigma_{11}\boldsymbol{\alpha} + (\boldsymbol{\alpha}'\Sigma_{12}\boldsymbol{\beta})\Sigma_{12}\boldsymbol{\beta} &= \mathbf{0}, \\ (\boldsymbol{\alpha}'\Sigma_{12}\boldsymbol{\beta})\Sigma_{21}\boldsymbol{\alpha} - \theta\Sigma_{22}\boldsymbol{\beta} &= \mathbf{0}. \end{aligned} \tag{11.1.7}$$

Premultiplication of the first equation by $\boldsymbol{\alpha}'$ and the second by $\boldsymbol{\beta}'$ shows that

$$\lambda = \theta = (\boldsymbol{\alpha}'\Sigma_{12}\boldsymbol{\beta})^2 = \rho^2(\boldsymbol{\alpha}, \boldsymbol{\beta}).$$

In other words, λ $(= \theta)$ is equal to the maximum of the squared correlation coefficient between $\xi = \boldsymbol{\alpha}'\boldsymbol{X}$ and $\eta = \boldsymbol{\beta}'\boldsymbol{Y}$. If $\Sigma_{12} \neq \mathrm{O}$, the maximum correlation coefficient is not zero. Thus (11.1.7) becomes

$$\begin{aligned} -\sqrt{\lambda}\Sigma_{11}\boldsymbol{\alpha} + \Sigma_{12}\boldsymbol{\beta} &= \mathbf{0}, \\ \Sigma_{21}\boldsymbol{\alpha} - \sqrt{\lambda}\Sigma_{22}\boldsymbol{\beta} &= \mathbf{0}. \end{aligned} \tag{11.1.8}$$

Eliminating $\boldsymbol{\beta}$ from the equations in (11.1.8), we obtain

$$\Sigma_{12}\Sigma_{22}^{-1}\Sigma_{21}\boldsymbol{\alpha} = \lambda\Sigma_{11}\boldsymbol{\alpha}, \tag{11.1.9}$$

and similarly, by eliminating $\boldsymbol{\alpha}$,

$$\Sigma_{21}\Sigma_{11}^{-1}\Sigma_{12}\boldsymbol{\beta} = \lambda\Sigma_{22}\boldsymbol{\beta}. \tag{11.1.10}$$

Thus the maximum value of $\lambda = \rho^2(\boldsymbol{\alpha}, \boldsymbol{\beta})$ is the largest root of the equation

$$|\Sigma_{12}\Sigma_{22}^{-1}\Sigma_{21} - \lambda\Sigma_{11}| = 0 \quad \text{or} \quad |\Sigma_{21}\Sigma_{11}^{-1}\Sigma_{12} - \lambda\Sigma_{22}| = 0. \tag{11.1.11}$$

In other words, it is the largest characteristic root of $\Sigma_{11}^{-1}\Sigma_{12}\Sigma_{22}^{-1}\Sigma_{21}$, and the corresponding vectors $\boldsymbol{\alpha}$ and $\boldsymbol{\beta}$ of coefficients of two linear combinations are the characteristic vectors, which are normalized solutions of equations (11.1.9) and (11.1.10) with the largest characteristic root. Let us denote these characteristic root and characteristic vectors by λ_1 and $(\boldsymbol{\alpha}_1, \boldsymbol{\beta}_1)$, respectively, where $\boldsymbol{\alpha}_1'\Sigma_{11}\boldsymbol{\alpha}_1 = \boldsymbol{\beta}_1'\Sigma_{22}\boldsymbol{\beta}_1 = 1$.

The same procedure can be continued for finding the second coefficient vector $\boldsymbol{\alpha}_2, \boldsymbol{\beta}_2$ and the second correlation ρ_2. For details, see Anderson (2003) or Siotani et al. (1985). In general, it is shown that the correlation coefficients $\rho_1^2 \geq \cdots \geq \rho_p^2 \geq 0$ are solutions of the characteristic equation

$$| \Sigma_{12}\Sigma_{22}^{-1}\Sigma_{21} - \rho^2\Sigma_{11} | = 0 \tag{11.1.12}$$

and the coefficient vectors $\boldsymbol{\alpha}_i$ and $\boldsymbol{\beta}_i$ are the solutions of the corresponding equations:

$$\begin{aligned} &\Sigma_{12}\Sigma_{22}^{-1}\Sigma_{21}\boldsymbol{\alpha}_i = \rho_i^2\Sigma_{11}\boldsymbol{\alpha}_i, \quad \boldsymbol{\alpha}_i'\Sigma_{11}\boldsymbol{\alpha}_j = \delta_{ij}, \quad i,j = 1,\ldots,p, \\ &\Sigma_{21}\Sigma_{11}^{-1}\Sigma_{12}\boldsymbol{\beta}_j = \rho_j^2\Sigma_{22}\boldsymbol{\beta}_j, \quad \boldsymbol{\beta}_i'\Sigma_{11}\boldsymbol{\beta}_j = \delta_{ij}, \quad i,j = 1,\ldots,q. \end{aligned} \tag{11.1.13}$$

Theorem 11.1.1 *The coefficient vectors $\boldsymbol{\alpha}_1, \ldots, \boldsymbol{\alpha}_p$ and $\boldsymbol{\beta}_1, \ldots, \boldsymbol{\beta}_q$ satisfying properties* (1), (2), *and* (3) *are given by the solutions of* (11.1.13). *Further, letting $\rho_i = \rho(\xi_i, \eta_i)$, $i = 1, \ldots, p$, the ρ_i^2 are the solutions of* (11.1.12).

Proof. For a complete proof based on the Lagrange method, see Anderson (2003). Here we give a proof based on a matrix-theoretic approach. Consider a singular value decomposition (see [A.1.12]) of the matrix $\Psi = \Sigma_{11}^{-1/2}\Sigma_{12}\Sigma_{22}^{-1/2}$:

$$\Psi = \mathrm{Q}_1\mathrm{P}\mathrm{Q}_2 \tag{11.1.14}$$

where $\mathrm{Q}_1 : p \times p$ and $\mathrm{Q}_2 : q \times q$ are orthogonal matrices whose columns are characteristic vectors of $\Psi\Psi'$ and $\Psi'\Psi$, respectively, and

$$\mathrm{P}: p \times q = (\mathrm{D}_\rho \ \ \mathrm{O}), \quad \mathrm{D}_\rho = \mathrm{diag}\,(\rho_1, \ldots, \rho_p), \tag{11.1.15}$$

where $\rho_i \geq 0$ and $\rho_1^2 \geq \rho_2^2 \geq \cdots \geq \rho_p^2$ are nonnegative characteristic roots of $\Psi\Psi'$ or $\Psi'\Psi$. Now we define $\mathcal{A} : p \times p = (\boldsymbol{\alpha}_1, \ldots, \boldsymbol{\alpha}_p)$ and $\mathcal{B} : q \times q = (\boldsymbol{\beta}_1, \ldots, \boldsymbol{\beta}_q)$ by

$$\mathcal{A} = \Sigma_{11}^{-1/2}\mathrm{Q}_1, \quad \mathcal{B} = \Sigma_{22}^{-1/2}\mathrm{Q}_2. \tag{11.1.16}$$

Then it follows from (11.1.14) that

$$\mathcal{A}'\Sigma_{12}\mathcal{B} = \mathrm{P}, \quad \mathcal{A}'\Sigma_{11}\mathcal{A} = \mathrm{I}_p, \quad \mathcal{B}'\Sigma_{22}\mathcal{B} = \mathrm{I}_q.$$

We also observe that $\mathcal{A}$ and $\mathcal{B}$ are defined as the solution of

$$\begin{aligned} &\Sigma_{11}^{-1}\Sigma_{12}\Sigma_{22}^{-1}\Sigma_{21}\mathcal{A} = \mathcal{A}\mathrm{D}_{1\rho}^2 \quad \text{with} \quad \mathcal{A}'\Sigma_{11}\mathcal{A} = \mathrm{I}_p, \\ &\Sigma_{22}^{-1}\Sigma_{21}\Sigma_{11}^{-1}\Sigma_{12}\mathcal{B} = \mathcal{B}\mathrm{D}_{2\rho}^2 \quad \text{with} \quad \mathcal{B}'\Sigma_{22}\mathcal{B} = \mathrm{I}_q, \end{aligned} \tag{11.1.17}$$

where

$$\begin{aligned} &\mathrm{D}_{1\rho}: p \times p = \mathrm{diag}\,(\rho_1, \ldots, \rho_p) \\ &\mathrm{D}_{2\rho}: q \times q = \mathrm{diag}\,(\rho_1, \ldots, \rho_p, 0, \ldots, 0). \end{aligned}$$

Now we apply [A.2.7] to the present maximization problem; that is, the maximum value of $\rho^2(\boldsymbol{\alpha}, \boldsymbol{\beta})$ is given by the largest root ρ_1^2 of (11.1.12), this maximum value being attained when $\boldsymbol{\alpha} = \boldsymbol{\alpha}_1$ and $\boldsymbol{\beta} = \boldsymbol{\beta}_1$. This means that taking $\rho_1 > 0$, the pair $\xi_1 = \boldsymbol{\alpha}_1'\boldsymbol{X}$ and $\eta_1 = \boldsymbol{\beta}_1'\boldsymbol{Y}$ give the maximum correlation ρ_1, the pair of $\xi_i = \boldsymbol{\alpha}_i'\boldsymbol{X}$ and $\eta_i = \boldsymbol{\beta}_i'\boldsymbol{Y}$, $i \leq p$ satisfy the requirements stated in (2) having the ith maximum correlation ρ_i, and $\eta_j = \boldsymbol{\beta}_j'\boldsymbol{Y}$ for $j > p$ satisfy requirement (3). □

Definition 11.1.1 The pair (ξ_i, η_i) $(i = 1, \ldots, p)$ is called the ith *pair of the canonical variables* or the canonical correlation variables, and the correlation ρ_i between ξ_i and η_i is called the ith *canonical correlation*, which is nonnegative.

The canonical correlations are unique since they are nonnegative square roots of p nonnegative characteristic roots of equation (11.1.14). However, canonical variables are not determined uniquely even if the nonzero characteristic roots of (11.1.14) are simple, since (ξ_i, η_i) and $(-\xi_i, -\eta_i)$ are both the ith pair of canonical variates satisfying requirements (1), (2), and (3).

It is often the case in practice that the first few correlations, $\rho_1, \ldots, \rho_k$, say, are moderately large, while the remaining correlations, $\rho_{k+1}, \ldots, \rho_p$, are small. In such cases, nearly all the correlation structure between $\boldsymbol{X}$ and $\boldsymbol{Y}$ can be explained by the first k pairs of canonical variables, $(\xi_1, \eta_1), \ldots, (\xi_k, \eta_k)$. As a measure of how much in the correlation structure has been explained by the first k pairs of canonical variables, we use

$$\phi_k = \sum_{i=1}^{k} \rho_i^2 \Big/ \sum_{i=1}^{p} \rho_i^2 = \sum_{i=1}^{k} \rho_i^2 \Big/ \operatorname{tr}(\Sigma_{11}^{-1}\Sigma_{12}\Sigma_{22}^{-1}\Sigma_{21}), \tag{11.1.18}$$

which is called the (cumulative) contribution ratio of the first k canonical variables. If rank$\Sigma_{12} = k$, then $\phi_k = 1$. This means that we can reduce the dimensionality from $(p+q)$ of $(\boldsymbol{X}', \boldsymbol{Y}')'$ to $2k$ of (ξ_i, η_i), $i = 1, \ldots, k$ without any loss of information about the correlation structure between $\boldsymbol{X}$ and $\boldsymbol{Y}$.

The canonical correlations and canonical variables have the following invariant property.

Theorem 11.1.2 *If* $\boldsymbol{U} = \mathrm{F}\boldsymbol{X} + \boldsymbol{f}$ *and* $\boldsymbol{V} = \mathrm{G}\boldsymbol{Y} + \boldsymbol{g}$, *where* $\mathrm{F}: p \times p$ *and* $\mathrm{G}: q \times q$ *are nonsingular matrices and* $\boldsymbol{f}$: $p \times 1$ *and* $\boldsymbol{g}$: $q \times 1$ *are fixed vectors, then:*

(1) *The canonical correlations between* $\boldsymbol{U}$ *and* $\boldsymbol{V}$ *are the same as those between* $\boldsymbol{X}$ *and* $\boldsymbol{Y}$.

(2) $(\mathrm{F}^{-1}\boldsymbol{\alpha}_i)'\boldsymbol{U}$ *and* $(\mathrm{G}^{-1}\boldsymbol{\beta}_i)'\boldsymbol{V}$ *are the ith pair of canonical variables between* $\boldsymbol{U}$ *and* $\boldsymbol{V}$, *where* $\boldsymbol{\alpha}_i'\boldsymbol{X}$ *and* $\boldsymbol{\beta}_i'\boldsymbol{Y}$ *are the ith pair of canonical variates between* $\boldsymbol{X}$ *and* $\boldsymbol{Y}$.

Proof. The covariance matrix of $(\boldsymbol{U}', \boldsymbol{V}')'$ is written as

$$\Psi = \begin{pmatrix} \Psi_{11} & \Psi_{12} \\ \Psi_{21} & \Psi_{22} \end{pmatrix} = \begin{pmatrix} \mathrm{F}\Sigma_{11}\mathrm{F}' & \mathrm{F}\Sigma_{12}\mathrm{G}' \\ \mathrm{G}\Sigma_{21}\mathrm{F}' & \mathrm{G}\Sigma_{22}\mathrm{G}' \end{pmatrix}.$$

Hence the squared ith canonical correlation between $\boldsymbol{U}$ and $\boldsymbol{V}$ is

$$\begin{aligned}\mathrm{ch}_i(\Psi_{11}^{-1}\Psi_{12}\Psi_{22}^{-1}\Psi_{21}) &= \mathrm{ch}_i(\mathrm{F}'^{-1}\Sigma_{11}^{-1}\Sigma_{12}\Sigma_{22}^{-1}\Sigma_{21}\mathrm{F}') \\ &= \mathrm{ch}_i(\Sigma_{11}^{-1}\Sigma_{12}\Sigma_{22}^{-1}\Sigma_{21}),\end{aligned}$$

which is the squared ith canonical correlation between $\boldsymbol{X}$ and $\boldsymbol{Y}$, where $\mathrm{ch}_i(\mathrm{A})$ denotes the ith largest characteristic root of a matrix A. This completes the proof of part (1). For (2) we observe that

$$\tilde{\boldsymbol{\alpha}}_i = \mathrm{F}^{-1}\boldsymbol{\alpha}_i, \quad \tilde{\boldsymbol{\beta}}_i = \mathrm{G}^{-1}\boldsymbol{\beta}_i, \quad \tilde{\boldsymbol{\alpha}}_i'\Psi_{11}\tilde{\boldsymbol{\alpha}}_j = \delta_{ij}, \quad \tilde{\boldsymbol{\beta}}_i'\Psi_{22}\tilde{\boldsymbol{\beta}}_j = \delta_{ij}$$

and

$$\Psi_{11}^{-1}\Psi_{12}\Psi_{22}^{-1}\Psi_{21}\tilde{\boldsymbol{\alpha}}_i = \rho_i^2\tilde{\boldsymbol{\alpha}}_i, \quad \Psi_{22}^{-1}\Psi_{21}\Psi_{11}^{-1}\Psi_{12}\tilde{\boldsymbol{\beta}}_i = \rho_i^2\tilde{\boldsymbol{\beta}}_i$$

and $\tilde{\boldsymbol{\alpha}}_i\Psi_{12}\tilde{\boldsymbol{\beta}}_i = \rho_i$. This proves part (2).

Remark 11.1.1 Let $\tilde{\boldsymbol{X}}$ and $\tilde{\boldsymbol{Y}}$ be the standardized vectors of $\boldsymbol{X} = (X_1, \ldots, X_p)'$ and $\boldsymbol{Y} = (Y_1, \ldots, Y_q)'$, given by

$$\tilde{\boldsymbol{X}} = \left(\frac{X_1}{\sqrt{\mathrm{Var}(X_1)}}, \ldots, \frac{X_p}{\sqrt{\mathrm{Var}(X_p)}}\right)', \quad \tilde{\boldsymbol{Y}} = \left(\frac{Y_1}{\sqrt{\mathrm{Var}(Y_1)}}, \ldots, \frac{Y_q}{\sqrt{\mathrm{Var}(Y_p)}}\right)'.$$

Then, from Theorem 11.1.2, the canonical correlations between $\boldsymbol{X}$ and $\boldsymbol{Y}$ are the same as those between $\tilde{\boldsymbol{X}}$ and $\tilde{\boldsymbol{Y}}$. Let $(\tilde{\boldsymbol{\alpha}}_i'\tilde{\boldsymbol{X}}, \tilde{\boldsymbol{\beta}}_i'\tilde{\boldsymbol{Y}})$ be the ith canonical variables between $\tilde{\boldsymbol{X}}$ and $\tilde{\boldsymbol{Y}}$. Further,

$$\boldsymbol{\alpha}_i'\boldsymbol{X} = \tilde{\boldsymbol{\alpha}}_i'\tilde{\boldsymbol{X}}, \quad \boldsymbol{\beta}_i'\boldsymbol{Y} = \tilde{\boldsymbol{\alpha}}_i'\tilde{\boldsymbol{Y}},$$

except for the sign.

11.2 Sample Canonical Correlations

In Section 11.1 we defined the canonical correlations ρ_i and canonical variates $\xi_i = \boldsymbol{\alpha}_i'\boldsymbol{X}$, $\eta_i = \boldsymbol{\beta}_i'\boldsymbol{Y}$ between $\boldsymbol{X}$ and $\boldsymbol{Y}$ in the population having the covariance matrix Σ. We now consider estimates of them on the basis of N observations:

$$(\boldsymbol{X}_i', \boldsymbol{Y}_i')', \quad i = 1, \ldots, N$$

from the population of $(\boldsymbol{X}', \boldsymbol{Y}')'$: $(p+q) \times 1$, where $N > p + q$.

Let

$$\mathbf{S} = \begin{pmatrix} \mathbf{S}_{11} & \mathbf{S}_{12} \\ \mathbf{S}_{21} & \mathbf{S}_{22} \end{pmatrix} = \frac{1}{n}\sum_{i=1}^{N} \begin{pmatrix} \boldsymbol{X}_i - \bar{\boldsymbol{X}} \\ \boldsymbol{Y}_i - \bar{\boldsymbol{Y}} \end{pmatrix} \begin{pmatrix} \boldsymbol{X}_i - \bar{\boldsymbol{X}} \\ \boldsymbol{Y}_i - \bar{\boldsymbol{Y}} \end{pmatrix}',$$

where $n = N - 1$, $\bar{\boldsymbol{X}} = (1/N)\sum_{i=1}^{N} \boldsymbol{X}_i$, and $\bar{\boldsymbol{Y}} = (1/N)\sum_{i=1}^{N} \boldsymbol{Y}_i$. Since $n \geq p + q$, $\mathbf{S} > \mathrm{O}$ with probability 1 [see, e.g., Muirhead, 1982]. The sample canonical correlations $r_1 \geq \cdots \geq r_p$ and sample canonical vectors $\mathbf{A} = (\boldsymbol{a}_1, \ldots, \boldsymbol{a}_p)$ and $\mathbf{B} = (\boldsymbol{b}_1, \ldots, \boldsymbol{b}_q)$ are defined by replacing Σ with S in the definition of the ρ_i, $\mathcal{A}$ and $\mathcal{B}$ in Section 11.1. That is, the r_i are defined by the positive square roots of the equation

$$|\mathbf{S}_{12}\mathbf{S}_{22}^{-1}\mathbf{S}_{21} - r^2\mathbf{S}_{11}| = 0 \quad \text{or} \quad |\mathbf{S}_{21}\mathbf{S}_{11}^{-1}\mathbf{S}_{12} - r^2\mathbf{S}_{22}| = 0. \tag{11.2.1}$$

The vectors $\boldsymbol{a}_i$ and $\boldsymbol{b}_i$ are defined by the solutions of

$$\begin{aligned} \mathbf{S}_{12}\mathbf{S}_{22}^{-1}\mathbf{S}_{21}\boldsymbol{a}_i &= r^2 S_{11}\boldsymbol{a}_i, \quad \boldsymbol{a}_i'\mathbf{S}_{11}\boldsymbol{a}_j = \delta_{ij}, \quad i, j = 1, \ldots, p, \\ \mathbf{S}_{21}\mathbf{S}_{11}^{-1}\mathbf{S}_{12}\boldsymbol{b}_i &= r^2\mathbf{S}_{22}\boldsymbol{b}_i, \quad \boldsymbol{b}_i'\mathbf{S}_{22}\boldsymbol{b}_j = \delta_{ij}, \quad i, j = 1, \ldots, q, \end{aligned} \tag{11.2.2}$$

where $r_{p+1} = \cdots = r_q = 0$ identically, and signs of $\boldsymbol{a}$ and $\boldsymbol{b}$ are chosen so that $\boldsymbol{a}_i' S_{12}\boldsymbol{b}_i$ are nonnegative. These may be summarized in the following form:

$$\begin{aligned} \mathbf{A}'\mathbf{S}_{11}^{-1}\mathbf{S}_{12}\mathbf{S}_{22}^{-1}\mathbf{S}_{21}\mathbf{A} &= \mathbf{R}_1^2, \quad \mathbf{A}'\mathbf{S}_{11}\mathbf{A} = \mathrm{I}_p, \\ \mathbf{B}'\mathbf{S}_{22}^{-1}\mathbf{S}_{21}\mathbf{S}_{11}^{-1}\mathbf{S}_{12}\mathbf{B} &= \mathbf{R}_2^2, \quad \mathbf{B}'\mathbf{S}_{22}\mathbf{B} = \mathrm{I}_q, \end{aligned} \tag{11.2.3}$$

where $\mathbf{R}_1 = \mathrm{diag}\,(r_1, \ldots, r_p) = \mathbf{D}_r$ and $\mathbf{R}_2 = (\mathbf{D}_r \ \mathrm{O}) \colon p \times q$.

Definition 11.2.1 Let

$$\begin{aligned} \boldsymbol{U} &= (U_1, \ldots, U_p)' = \mathbf{A}'\boldsymbol{X}, \\ \boldsymbol{V} &= (V_1, \ldots, V_p)' = \mathbf{B}'\boldsymbol{Y}. \end{aligned} \tag{11.2.4}$$

Then (U_i, V_i) $(i = 1, \ldots, p)$ is called the ith *pair of sample canonical variables* or sample canonical correlation variables and r_i is called the ith *sample canonical correlation.*

We have defined the population canonical correlations and variates by maximizing the correlations of linear combinations of two sets of variates, and have shown that they have some optimum properties. The arguments there can also be made in terms of a sample. For example, $\boldsymbol{a}_i'\boldsymbol{X}$ and $\boldsymbol{b}_i'\boldsymbol{Y}$ have the maximum sample correlation between any two linear combinations of $\boldsymbol{X}$ and $\boldsymbol{Y}$; similarly, $\boldsymbol{a}_i'\boldsymbol{X}$ and $\boldsymbol{b}_i'\boldsymbol{Y}$ have the second largest sample correlations; and so on.

When the sample is taken from a $(p + q)$-variate normal population $\mathrm{N}_{p+q}(\boldsymbol{\mu}, \Sigma)$, $\Sigma > \mathrm{O}$, $\mathbf{S}^* = \{(N - 1)/N\}\mathbf{S}$ is the MLE of Σ. Then it follows from Lemma 4.1.1 that the MLEs $\hat{\rho}_i$, $\hat{\mathcal{A}}$ and $\hat{\mathcal{B}}$ of the population quantities ρ_i, and $\mathcal{A}$ and $\mathcal{B}$ are given, respectively, by

$$\hat{\rho}_i = r_i, \quad \hat{\mathcal{A}} = \sqrt{N/(N-1)}\mathbf{A}, \quad \hat{\mathcal{B}} = \sqrt{N/(N-1)}\mathbf{B}. \tag{11.2.5}$$

11.3 Distributions of Canonical Correlations

Let $\mathbf{S}$ be the sample covariance matrix formed from a sample of size $N = n+1$ of $(\boldsymbol{X}', \ \boldsymbol{Y}')'$ from $(p+q)$-dimensional normal distribution $\mathrm{N}_{p+q}(\mathbf{0}, \Sigma)$, with $\boldsymbol{X} : p \times 1$ and $\boldsymbol{Y} : q \times 1$. In this section we assume that $q \geq p$, although in previous sections we that assume $p \leq q$. In this way we need only note that the number of possibly nonzero canonical correlations is q.

Let $\rho_1 \geq \cdots \geq \rho_q \geq 0$ and $r_1 > \cdots > r_q > 0$ be the population and sample canonical correlations between $\boldsymbol{X}$ and $\boldsymbol{Y}$. Corresponding to a partition $(\boldsymbol{X}', \boldsymbol{Y}')$, we partition Σ and $\mathbf{S}$ as

$$\Sigma = \begin{pmatrix} \Sigma_{11} & \Sigma_{12} \\ \Sigma_{21} & \Sigma_{22} \end{pmatrix}, \quad \mathbf{S} = \begin{pmatrix} \mathbf{S}_{11} & \mathbf{S}_{12} \\ \mathbf{S}_{21} & \mathbf{S}_{22} \end{pmatrix}.$$

Then $\rho_1^2 \geq \cdots \geq \rho_q^2 \geq 0$ and $r_1^2 > \cdots > r_q^2 > 0$ are the characteristic roots of $\Sigma_{22}^{-1}\Sigma_{21}\Sigma_{11}^{-1}\Sigma_{12}$ and $\mathbf{S}_{22}^{-1}\mathbf{S}_{21}\mathbf{S}_{11}^{-1}\mathbf{S}_{12}$, respectively.

11.3.1 Distributional Reduction

The transformation from the original variables to the population canonical variables is expressed as

$$\begin{pmatrix} \boldsymbol{X} \\ \boldsymbol{Y} \end{pmatrix} \to \begin{pmatrix} \boldsymbol{\xi} \\ \boldsymbol{\eta} \end{pmatrix} = \begin{pmatrix} \mathcal{A}' & \mathrm{O} \\ \mathrm{O} & \mathcal{B}' \end{pmatrix} \begin{pmatrix} \boldsymbol{X} \\ \boldsymbol{Y} \end{pmatrix}$$

such that

$$\begin{pmatrix} \mathcal{A}' & \mathrm{O} \\ \mathrm{O} & \mathcal{B}' \end{pmatrix} \begin{pmatrix} \Sigma_{11} & \Sigma_{12} \\ \Sigma_{21} & \Sigma_{22} \end{pmatrix} \begin{pmatrix} \mathcal{A} & \mathrm{O} \\ \mathrm{O} & \mathcal{B} \end{pmatrix} = \begin{pmatrix} \mathrm{I}_p & \mathcal{P}' \\ \mathcal{P} & \mathrm{I}_q \end{pmatrix},$$

where $\mathcal{P} = (\mathcal{P}_1 \ \mathrm{O})'$, $\mathcal{P}_1 = \mathrm{diag}\,(\rho_1, \ldots, \rho_q)$. Since the canonical correlations are invariant under the transformation above, without loss of generality we assume that

$$\Sigma = \begin{pmatrix} \mathrm{I}_q & \mathcal{P}' \\ \mathcal{P} & \mathrm{I}_p \end{pmatrix}$$

when we treat the distributions of the canonical correlations themselves or their function. Let $\mathbf{A} = n\mathbf{S}$, which is distributed as a Wishart distribution $\mathrm{W}_{q+p}(n, \Sigma)$, and partition $\mathbf{A}$ as

$$\mathbf{A} = \begin{pmatrix} \mathbf{A}_{11} & \mathbf{A}_{12} \\ \mathbf{A}_{21} & \mathbf{A}_{22} \end{pmatrix},$$

corresponding to a partition of $\mathbf{S}$. First we consider the distribution of ℓ_α^2 ($\ell_1^2 > \ldots > \ell_q^2$), defined by

$$\ell_\alpha^2 = r_\alpha^2/(1 - r_\alpha^2), \quad \alpha = 1, \ldots, q, \tag{11.3.1}$$

which are the characteristic roots of $\mathbf{A}_{22\cdot 1}^{-1}\mathbf{A}_{21}\mathbf{A}_{11}^{-1}\mathbf{A}_{12}$ instead of r_α^2 or r_α ($r_1 > \cdots > r_q$). Here $\mathbf{A}_{22\cdot 1} = \mathbf{A}_{22} - \mathbf{A}_{21}\mathbf{A}_{11}^{-1}\mathbf{A}_{12}$. The distribution of r_α^2 or r_α is treated as the distribution of a function $f(\ell_\alpha^2)$. For example, r_α^2 and r_α are expressed in terms of ℓ_α as

$$r_\alpha^2 = \ell_\alpha^2/(1 + \ell_\alpha^2) \text{ and } r_\alpha = \ell_\alpha/\sqrt{1 + \ell_\alpha^2},$$

respectively. The population quantity corresponding to ℓ_α^2 is denoted by

$$\gamma_\alpha^2 = \rho_\alpha^2/(1 - \rho_\alpha^2), \quad \alpha = 1, \ldots, q. \tag{11.3.2}$$

The following distributional reduction is useful in the derivation of the distribution of canonical correlations.

Lemma 11.3.1 *When we consider the distribution of a function of the canonical correlations $r_1 > \cdots > r_q$, without loss of generality we may assume that:*

(1) $\mathbf{A}_{22\cdot 1}$ *is distributed as a central Wishart distribution* $\mathrm{W}_q(m, \mathrm{I}_q)$, *where* $m = n - p$.

(2) *Let* $\mathbf{B}$ *be the first* $q \times q$ *submatrix of* $\mathbf{A}_{11}$. *Then, given* $\mathbf{B}$, $\mathbf{A}_{21}\mathbf{A}_{11}^{-1}\mathbf{A}_{12}$ *is distributed as a noncentral Wishart distribution* $\mathrm{W}_q(p, \mathrm{I}_p; \Gamma\mathbf{B}\Gamma)$, *where* $\Gamma = \mathrm{diag}\,(\gamma_1, \ldots, \gamma_q)$ *and* $\gamma_\alpha = \rho_\alpha/\sqrt{1 - \rho_\alpha^2}$.

(3) $\mathbf{A}_{21}\mathbf{A}_{11}^{-1}\mathbf{A}_{12}$ *and* $\mathbf{A}_{22\cdot 1}$ *are independent.*

(4) $\mathbf{B}$ *is distributed as a central Wishart distribution* $\mathrm{W}_q(n, \mathrm{I}_q)$.

Proof. The results were used by Sugiura and Fujikoshi (1969), except that (2) and (3) were given under a conditional setup. Noting that $\mathbf{A} \sim \mathrm{W}_{p+q}(n, \mathcal{P})$, we can write

$$\mathbf{A} = \mathbf{U}'\mathbf{U}, \quad \mathbf{U} = (\mathbf{U}_1 \ \mathbf{U}_2),$$

where $\mathbf{U}_1 : n \times p$, $\mathbf{U}_2 : n \times q$, and the rows of $\mathbf{U}$ are independently distributed as $\mathrm{N}_{p+q}(\mathbf{0}, \Sigma)$. The conditional distribution of $\mathbf{U}_2$ given $\mathbf{U}_1$ is normal with

$$\mathrm{E}(\mathbf{U}_2|\mathbf{U}_1) = \mathbf{U}_1(\mathcal{P}_1 \ \mathrm{O}),$$

and each row is distributed independently with the same covariance matrix $\mathrm{I} - \mathcal{P}_1^2$. Since

$$\mathbf{A}_{21}\mathbf{A}_{11}^{-1}\mathbf{A}_{12} = \mathbf{U}_2'\mathbf{U}_1(\mathbf{U}_1'\mathbf{U}_1)^{-1}\mathbf{U}_1'\mathbf{U}_2,$$

the conditional distribution of $\mathbf{A}_{21}\mathbf{A}_{11}^{-1}\mathbf{A}_{12}$ is a noncentral Wishart distribution $\mathrm{W}_q(p, \mathrm{I} - \mathcal{P}_1^2; \Omega)$, where

$$\Omega = (\mathcal{P}_1 \ \mathrm{O})'\mathbf{U}_1'\mathbf{U}_1(\mathcal{P}_1 \ \mathrm{O}) = \mathcal{P}_1'\mathbf{B}\mathcal{P}_1$$

and $\mathbf{B}$ is the first $q \times q$ matrix of $\mathbf{U}_1'\mathbf{U}_1$. Further, the canonical correlations are invariant under

$$\mathbf{A}_{22\cdot1} \to \mathcal{P}_1^{-1/2}\mathbf{A}_{22\cdot1}\mathcal{P}_1^{-1/2}, \quad \mathbf{A}_{21}\mathbf{A}_{11}^{-1}\mathbf{A}_{12} \to \mathcal{P}_1^{-1/2}\mathbf{A}_{21}\mathbf{A}_{11}^{-1}\mathbf{A}_{12}\mathcal{P}_1^{-1/2}.$$

Using these results, we can check (1) through (4).

11.3.2 Large-Sample Asymptotic Distribuitons

In this section we consider asymptotic distributions of the canonical correlations and their transformations under a large-sample framework:

$$\text{A1: } p \text{ and } q \text{ are fixed}, \quad n \to \infty.$$

Many asymptotic results have been obtained [see, e.g., Siotani et al., 1985; Anderson, 2003]. These results were obtained by a perturbation method. For that it is important to give a perturbation expansion for the matrix

$$\mathbf{Q} = \mathbf{A}_{22\cdot1}^{-1/2}\mathbf{A}_{21}\mathbf{A}_{11}^{-1}\mathbf{A}_{12}\mathbf{A}_{22\cdot1}^{-1/2}. \tag{11.3.3}$$

Let

$$\mathbf{A}_{21}\mathbf{A}_{11}^{-1}\mathbf{A}_{12} = (\mathbf{U}_1 - \Gamma\mathbf{B}^{1/2}\ \mathbf{U}_2)(\mathbf{U}_1 - \Gamma\mathbf{B}^{1/2}\ \mathbf{U}_2)', \tag{11.3.4}$$

$$\mathbf{V} = \sqrt{m}\left(\frac{1}{m}\mathbf{A}_{22\cdot1} - \mathrm{I}_q\right), \quad \mathbf{Z} = \sqrt{n}\left(\frac{1}{n}\mathbf{B} - \mathrm{I}_q\right), \tag{11.3.5}$$

where $\mathbf{U} = (\mathbf{U}_1\ \mathbf{U}_2)$, $\mathbf{U}_1 : q \times q$, $\mathbf{U}_2 : q \times (p-q)$, and the elements of $\mathbf{U}$ are independently distributed as $\mathrm{N}(0,1)$. Substituting (11.3.4) and (11.3.5), we can get a perturbation expansion for $\mathbf{Q}$:

$$\mathbf{Q} = \Theta^2 + \frac{1}{\sqrt{n}}\mathbf{Q}^{(1)} + \frac{1}{n}\mathbf{Q}^{(2)} + \mathrm{O}_{3/2}, \tag{11.3.6}$$

where $\Theta^2 = \mathrm{diag}\,(\theta_1^2, \ldots, \theta_q^2) = (1-c)^{-1}(c\mathrm{I}_q + \Gamma^2)$; that is, $\theta_\alpha^2 = (c+\gamma_\alpha^2)/(1-c)$, $\alpha = 1, \ldots, q$,

$$\begin{aligned}
\mathbf{Q}^{(1)} &= \frac{\sqrt{c}}{1-c}\mathbf{U} - \frac{1}{2}(1-c)^{-1/2}\left(\Theta^2\mathbf{V} + \mathbf{V}\Theta^2\right), \\
\mathbf{Q}^{(2)} &= -\frac{1}{2}(1-c)^{-3/2}\sqrt{c}\,(\mathbf{U}\mathbf{V} + \mathbf{V}\mathbf{U}) \\
&\quad + \frac{3}{8}(1-c)^{-1}\left(\Theta^2\mathbf{V}^2 + \mathbf{V}^2\Theta^2\right) + \frac{1}{4}(1-c)^{-1}\mathbf{V}\Theta^2\mathbf{V}.
\end{aligned}$$

We assume that $q \le p$ and the first $k(\le q)$ of the population canonical correlations ρ_i are distinct and positive and the remaining roots are zero:

$$1 > \rho_1 > \cdots > \rho_k > \rho_{k+1} = \cdots = \rho_q = 0. \tag{11.3.7}$$

Consider the standardized variables

$$\begin{aligned} X_\alpha &= \sqrt{n}(r_\alpha^2 - \rho_\alpha^2)/[2\rho_\alpha(1-\rho_\alpha^2)], \quad 1 \le \alpha \le k, \\ X_\alpha &= nr_\alpha^2, \quad k+1 \le \alpha \le q. \end{aligned} \tag{11.3.8}$$

Then the following result was obtained by Hsu (1941).

Theorem 11.3.1 *Let $X_1, \ldots, X_q$ be the standardized variables in* (11.3.7). *Then, under* (11.3.7), *the limiting joint pdf of $X_1, \ldots, X_q$ is*

$$\left\{ \prod_{\alpha=1}^{k} \varphi(x_\alpha) \right\} f_0(x_{k+1}, \ldots, x_q; s, t), \tag{11.3.9}$$

where $\varphi(\cdot)$ denotes the standard normal density function, $s = p - k$, $t = q - k$, and $f_0(x_{k+1}, \ldots, x_q; s, t)$ is the same as Theorem 6.5.2.

Sugiura (1976) extended Theorem 11.3.1 in the case of all the population canonical correlations being positive to obtain an asymptotic expansion. It may be noted that these results will not work well when dimensions q and/or p become large. Such examples can be seen in Srivastava (2002), pp. 521–527. One example is concerned with a canonical correlation analysis between the $p = 14$ explanatory variables and the $q = 8$ criteria variables based on $N = 128$ observations. In this example we are also interested in a canonical correlation analysis between the $p = 14$ explanatory variables and a subset of the $q = 8$ criteria variables.

11.3.3 High-Dimensional Asymptotic Distributions

To overcome the weakness of large-sample approximations, we consider the distributions of the canonical correlations and their functions under a high-dimensional framework such that

$$\begin{aligned} \text{A2:} \quad & q : \text{fixed}, \quad p \to \infty, \quad n \to \infty, \quad m = n - p \to \infty, \\ & c = p/n \to c_0 \in [0, 1). \end{aligned} \tag{11.3.10}$$

Here the constant c_0 is assumed to be $0 < c_0 < 1$.

In this case we need to modifiy the perturbation expansion of Q in (11.3.3), since the distribution of $\mathbf{A}_{21}\mathbf{A}_{11}^{-1}\mathbf{A}_{12}$ depends on p and p tends to infinity. To catch its behavior, let

$$\mathbf{U} = \sqrt{p}\left\{ \frac{1}{p}\mathbf{A}_{21}\mathbf{A}_{11}^{-1}\mathbf{A}_{12} - \left(\mathrm{I}_q + \frac{n}{p}\Gamma^2 \right) \right\}. \tag{11.3.11}$$

It is shown that the limiting distribution of $\mathbf{U}$ is normal under A2. Similarly, expanding $\mathbf{A}_{22}^{-1}$ under A2, we have the perturbation expansion of

$$\mathbf{Q} = \Theta^2 + \frac{1}{\sqrt{n}}\mathbf{Q}^{(1)} + \frac{1}{n}\mathbf{Q}^{(2)} + \mathrm{O}_{3/2}^*, \tag{11.3.12}$$

where $\Theta^2 = \mathrm{diag}\,(\theta_1^2, \ldots, \theta_q^2)$; that is, $\theta_\alpha^2 = (c + \gamma_\alpha^2)/(1-c), \quad \alpha = 1, \ldots, q$,

$$\begin{aligned}
\Theta^2 &= (1-c)^{-1}(c\mathrm{I}_q + \Gamma^2), \\
\mathbf{Q}^{(1)} &= \frac{\sqrt{c}}{1-c}\mathbf{U} - \frac{1}{2}(1-c)^{-1/2}\left(\Theta^2\mathbf{V} + \mathbf{V}\Theta^2\right), \\
\mathbf{Q}^{(2)} &= -\frac{1}{2}(1-c)^{-3/2}\sqrt{c}\,(\mathbf{UV} + \mathbf{VU}) \\
&\quad + \frac{3}{8}(1-c)^{-1}\left(\Theta^2\mathbf{V}^2 + \mathbf{V}^2\Theta^2\right) + \frac{1}{4}(1-c)^{-1}\mathbf{V}\Theta^2\mathbf{V}.
\end{aligned}$$

Here the notation O_i^* denotes a term of the ith order with respect to (n^{-1}, p^{-1}, m^{-1}). Now we can use a perturbation method on characteristic roots in Section 5.4. Suppose:

A3: The αth root ρ_α of $\Sigma_{11}^{-1}\Sigma_{12}\Sigma_{22}^{-1}\Sigma_{12}$ is simple and is not zero. (11.3.13)

Then ℓ_α^2 approaches

$$\theta_\alpha^2 = (1-c)^{-1}\left\{\rho_\alpha^2/(1-\rho_\alpha^2) + c\right\}, \tag{11.3.14}$$

not $\rho_\alpha^2/(1-\rho_\alpha^2)$, and its limiting distribution is given in the following theorem.

Theorem 11.3.2 *Let $\ell_\alpha^2 = r_\alpha^2/(1-r_\alpha^2)$, where r_α is the αth canonical correlation, and let $h(\ell_\alpha^2)$ be a function of ℓ_α^2 such that the first derivative is continuous in the neighborhood of $\ell_\alpha^2 = \theta_\alpha^2$ and $h'(\theta_\alpha^2) \neq 0$. Then, under A1 and A2, we have*

$$\begin{aligned}
Y_\alpha &= \sqrt{n}(\ell_\alpha^2 - \theta_\alpha^2) \ \xrightarrow{d}\ \mathrm{N}(0, \tau_\alpha^2), \\
Z_\alpha &= \sqrt{n}\left\{h(\ell_\alpha^2) - h(\theta_\alpha^2)\right\} \ \xrightarrow{d}\ \mathrm{N}(0, \left\{h'(\theta_\alpha^2)\right\}^2 \tau_\alpha^2),
\end{aligned}$$

where "$\xrightarrow{d}$" denotes convergence in the distribution, and the asymptotic variance τ_α^2 of Y_α is expressed in terms of θ_α^2 and ρ_α^2 as

$$\begin{aligned}
\tau_\alpha^2 &= \frac{2(2-c)}{1-c}(\theta_\alpha^2 + 1)\left(\theta_\alpha^2 - \frac{c}{2-c}\right) \\
&= \frac{2(2-c)}{(1-c)^3}(1-\rho_\alpha^2)^{-1}\left\{\frac{\rho_\alpha^2}{1-\rho_\alpha^2} + \frac{c}{2-c}\right\},
\end{aligned} \tag{11.3.15}$$

respectively.

Theorem 11.3.2 is shown by using a perturbation expansion of ℓ_α^2 in terms of $\mathbf{U}$ and $\mathbf{V}$. Therefore, we find that the limiting distribution of Y_α is the same as that of

$$q_{\alpha\alpha}^{(1)} = \sqrt{c}(1-c)^{-1}U_{\alpha\alpha} - \theta_\alpha^2 V_{\alpha\alpha},$$

where $\theta^{(1)} = (q_{\alpha\beta}^{(1)})$. The final result follows from the limiting distributions of $U_{\alpha\alpha}$ and $V_{\alpha\alpha}$.

In a rigorous terminology of convergence in distribution, the results of Theorem 11.3.2 for a high-dimensional framework should be considered $n = pc$ with a constant $0 < c < 1$ or c should be read as c_0. However, for actual use the statement in Theorem 11.3.2 will be more useful. Note that

$$r_\alpha^2 = \ell_\alpha^2/(1+\ell_\alpha^2) \quad \text{and} \quad r_\alpha = \sqrt{\ell_\alpha^2/(1+\ell_\alpha^2)}.$$

Therefore, r_α^2 and r_α can be expressed as functions of ℓ_α^2 given by

$$h(x) = x/(1+x) \quad \text{and} \quad h(x) = \sqrt{x/(1+x)},$$

respectively. From Theorem 11.3.2, we obtain the following results:

$$\sqrt{n}(r_\alpha^2 - \tilde{\rho}_\alpha^2) \xrightarrow{d} \mathrm{N}(0, \sigma_\alpha^2), \tag{11.3.16}$$

$$\sqrt{n}(r_\alpha - \tilde{\rho}_\alpha) \xrightarrow{d} \mathrm{N}\left(0, \frac{1}{4}\sigma_\alpha^2\tilde{\rho}_\alpha^{-2}\right), \tag{11.3.17}$$

where

$$\tilde{\rho}_\alpha = \{\rho_\alpha^2 + c(1-\rho_\alpha^2)\}^{1/2},$$
$$\sigma_\alpha^2 = 2(1-c)(1-\rho_\alpha^2)^2\{2\rho_\alpha^2 + c(1-2\rho_\alpha^2)\}.$$

In particular, letting $c = 0$ in (11.3.16) and (11.3.17), we have the following well-known large-sample results:

$$\sqrt{n}(r_\alpha^2 - \rho_\alpha^2) \xrightarrow{d} \mathrm{N}(0, 4\rho_\alpha^2(1-\rho_\alpha^2)^2), \tag{11.3.18}$$

$$\sqrt{n}(r_\alpha - \rho_\alpha) \xrightarrow{d} \mathrm{N}(0, (1-\rho_\alpha^2)^2). \tag{11.3.19}$$

Here, note that the high-dimensional asymptotic results (11.3.16) and (11.3.17) depend on p through $c = p/n$, but the large-sample results (11.3.18) and (11.3.19) do not depend on p and are the same for all p.

To make our asymptotic results more elaborate, we derive their asymptotic expansions. Let Y_α be the random variable in Theorem 11.3.2. Then we see (see Section 5.3) that the first four cumulants have the form

$$\begin{aligned}
\kappa_1(Y_\alpha) &= \mathrm{E}(Y_\alpha) = \frac{1}{\sqrt{n}}a_{1\alpha} + \mathrm{O}_{3/2}^*, \\
\kappa_2(Y_\alpha) &= \mathrm{Var}(Y_\alpha) = a_{2\alpha} + \mathrm{O}_1^*, \\
\kappa_3(Y_\alpha) &= \frac{1}{\sqrt{n}}a_{3\alpha} + \mathrm{O}_{3/2}^*, \\
\kappa_4(Y_\alpha) &= \mathrm{O}_1^*.
\end{aligned} \tag{11.3.20}$$

In fact, the coefficients can be expressed as follows:

$$
\begin{aligned}
a_{1\alpha} &= (1-c)^{-1}\left[-(q-1)-(q-3)\theta_\alpha^2+c(q-1)(1+\theta_\alpha^2)\right.\\
&\qquad \left.+(1+\theta_\alpha^2)\left\{2\theta_\alpha^2-c(1+\theta_\alpha^2)\right\}d_\alpha\right],\\
a_{2\alpha} &= 2(1-c)^{-1}(1+\theta_\alpha^2)\left\{2\theta_\alpha^2-c(1+\theta_\alpha^2)\right\},\\
a_{3\alpha} &= 8(1-c)^{-2}(1+\theta_\alpha^2)\left\{3\theta_\alpha^2(1+2\theta_\alpha^2)\right.\\
&\qquad \left.+c\left\{c-2+(2c-7)\theta_\alpha^2+(c-5)\theta_\alpha^4\right\}\right\},
\end{aligned}
\tag{11.3.21}
$$

where $d_\alpha = \sum_{\beta\neq\alpha}(\theta_\alpha^2-\theta_\beta^2)^{-1}$. Using the cumulant formulas, it is possible to give an asymptotic expansion of the distribution function of Y_α. In the following we give a general result for the distribution of r_α^2 and its function.

Consider the transformed variable defined by

$$
Z_\alpha = \sqrt{n}\left\{h(\ell_\alpha^2)-h(\theta_\alpha^2)\right\}
\tag{11.3.22}
$$

in Theorem 11.3.2. Here it is assumed that $h(x)$ is three times continuously differentiable in the neighborhood of $x=\theta_\alpha^2$. Then we can expand Z as

$$
\begin{aligned}
Z_\alpha &= \sqrt{n}\left\{h(\ell_\alpha^2)-h(\theta_\alpha^2)\right\}\\
&= \sqrt{n}\left\{h'(\theta_\alpha^2)(\ell_\alpha^2-\theta_\alpha^2)+\frac{1}{2}h''(\theta_\alpha^2)(\ell_\alpha^2-\theta_\alpha^2)^2+\cdots\right\}\\
&= h'(\theta_\alpha^2)Y_\alpha+\frac{1}{\sqrt{n}}\cdot\frac{1}{2}h''(\theta_\alpha^2)Y_\alpha^2+\cdots.
\end{aligned}
$$

This implies that

$$
\begin{aligned}
\kappa_1(Z_\alpha) &= \mathrm{E}(Z_\alpha) = \frac{1}{\sqrt{n}}b_{1\alpha}+\mathrm{O}_{3/2}^*,\\
\kappa_2(Z_\alpha) &= \mathrm{Var}(Z_\alpha) = b_{2\alpha}+\mathrm{O}_1^*,\\
\kappa_3(Z_\alpha) &= \frac{1}{\sqrt{n}}b_{3\alpha}+\mathrm{O}_{3/2}^*,
\end{aligned}
\tag{11.3.23}
$$

where

$$
\begin{aligned}
b_{1\alpha} &= h'(\theta_\alpha^2)a_{1\alpha}+\frac{1}{2}h''(\theta_\alpha^2)a_{2\alpha},\\
b_{2\alpha} &\equiv \tau_\alpha^2 = h'(\theta_\alpha^2)^2a_{2\alpha},\\
b_{3\alpha} &= h'(\theta_\alpha^2)^3a_{3\alpha}+3h'(\theta_\alpha^2)^2h''(\theta_\alpha^2)a_{2\alpha}^2.
\end{aligned}
\tag{11.3.24}
$$

We are especially interested in obtaining an asymptotic expansion of the distribution function of

$$
Z = \sqrt{n}[r_\alpha^2-\{\rho_\alpha^2+c(1-\rho_\alpha^2)\}],
$$

which can be expressed as $\sqrt{n}\{h(\ell_\alpha^2) - h(\theta_\alpha^2)\}$ with $h(x) = x/(1+x)$. Then

$$h'(\theta_\alpha^2) = (1+\theta_\alpha^2)^{-2} = (1-c)^2(1-\rho_\alpha^2)^2,$$
$$h''(\theta_\alpha^2) = -2(1+\theta_\alpha^2)^{-3} = -2(1-c)^3(1-\rho_\alpha^2)^3.$$

Using (11.3.21) the coefficients in (11.3.24) can be expressed as

$$\begin{aligned} b_{1\alpha} &= (1-\rho_\alpha^2)\Big[-(q-1) + 2(q-2)\rho_\alpha^2 + c\{2(q-1) - 2(q-2)\rho_\alpha^2\} \\ &\quad + \{2\rho_\alpha^2 + c(1-2\rho_\alpha^2)\}(1-\rho_\alpha^2)\sum_{\beta\neq\alpha}(\rho_\alpha^2 - \rho_\beta^2)^{-1}\Big], \\ b_{2\alpha} &\equiv \sigma_\alpha^2 = 2(1-c)(1-\rho_\alpha^2)^2\left[2\rho_\alpha^2 + c(1-2\rho_\alpha^2)\right], \qquad (11.3.25) \\ b_{3\alpha} &= 8(1-c)(1-\rho_\alpha^2)^3\{3\rho_\alpha^2 + c(1-3\rho_\alpha^2)\}\{1 - 3\rho_\alpha^2 - c(2-3\rho_\alpha^2)\}. \end{aligned}$$

Theorem 11.3.3 *Let r_α be the α th canonical correlation. Then under* A2 *and* A3 *we have*

$$\begin{aligned} \mathrm{P}\{\sqrt{n}(r_\alpha^2 - \tilde{\rho}_\alpha^2)/\sigma_\alpha \le x\} &= \Phi(x) \\ &\quad + \varphi(x)\frac{1}{\sqrt{n}}\{(b_{1\alpha}/\sigma_\alpha) + (b_{3\alpha}/\sigma_\alpha^3)(x^2-1)\} + \mathrm{O}_{3/2}^*, \qquad (11.3.26) \end{aligned}$$

where Φ and φ are the probability distribution function and probability density function of N(0, 1), *respectively,*

$$\tilde{\rho}_\alpha = \{\rho_\alpha^2 + c(1-\rho_\alpha^2)\}^{1/2}, \quad \sigma_\alpha = \sqrt{b_{2\alpha}}.$$

The coefficients $b_{1\alpha}$, $b_{2\alpha}$, and $b_{3\alpha}$ are given by (11.3.25). *Furthermore, let $h(r_\alpha^2)$ be a function of r_α^2 such that the third derivative is continuous in the neighborhood of $r_\alpha^2 = \tilde{\rho}_\alpha^2$ and $h'(\tilde{\rho}_\alpha^2) \neq 0$. Then the distribution function of*

$$\sqrt{n}\left\{h(r_\alpha^2) - h(\tilde{\rho}_\alpha^2)\right\}/\tau_\alpha$$

can be expanded as in (11.3.26) *with the coefficients $\tilde{b}_{i\alpha}$ instead of $b_{i\alpha}$, where $\tau_\alpha = \tilde{b}_{2\alpha}^{1/2}$,*

$$\begin{aligned} \tilde{b}_{1\alpha} &= h'(\tilde{\rho}_\alpha^2)b_{1\alpha} + \frac{1}{2}h''(\tilde{\rho}_\alpha^2)b_{2\alpha}, \\ \tilde{b}_{2\alpha} &= h'(\tilde{\rho}_\alpha^2)^2 b_{2\alpha}, \qquad (11.3.27) \\ \tilde{b}_{3\alpha} &= h'(\tilde{\rho}_\alpha^2)^3 b_{3\alpha} + 3h'(\tilde{\rho}_\alpha^2)^2 h''(\tilde{\rho}_\alpha^2) b_{2\alpha}^2. \end{aligned}$$

From Theorem 11.3.3, the upper percentage point of $\sqrt{n}(r_\alpha^2 - \tilde{\rho}_\alpha^2)/\sigma_\alpha$ is given by

$$u + \frac{1}{\sqrt{n}}\{(b_{1\alpha}/\sigma_\alpha) + (b_{3\alpha}/\sigma_\alpha^3)(u^2-1)\} \qquad (11.3.28)$$

in terms of the upper percentage point u of $\mathrm{N}(0,1)$. This expansion is called the *Cornish–Fisher expansion* (see Section 5.6).

The distribution of $\sqrt{n}(r_\alpha - \tilde{\rho}_\alpha)/\tau_\alpha$ can be obtained from Theorem 3.1 with $h(x) = \sqrt{x}$. In this case,

$$h'(\tilde{\rho}_\alpha^2) = \frac{1}{2}\tilde{\rho}_\alpha, \quad h''(\tilde{\rho}_\alpha^2) = -\frac{1}{4}\tilde{\rho}_\alpha^3.$$

Now we will see that the large-sample asymptotic expansion [see, e.g., Siotani et al., 1985] is obtained by considering a large-sample expansion for the high-dimensional asymptotic expansion (11.3.26). The statistic r_α^2 can be expanded for large n as

$$\sqrt{n}(r_\alpha^2 - \tilde{\rho}_\alpha^2) = \sqrt{n}(r_\alpha^2 - \rho_\alpha^2) + \frac{1}{\sqrt{n}}(1 - \rho_\alpha^2)p + \mathrm{O}(n^{-1}).$$

The coefficient $b_{i\alpha}$ in (11.3.25) can be expanded as $g_{i\alpha} + \mathrm{O}(n^{-1})$ when n is large, where

$$\begin{aligned} g_{1\alpha} &= (1 - \rho_\alpha^2)\left[-(q-1) + 2(q-2)\rho_\alpha^2 + 2\rho_\alpha^2(1 - \rho_\alpha^2)\sum_{\beta \neq \alpha}(\rho_\alpha^2 - \rho_\beta^2)^{-1}\right], \\ g_{2\alpha} &= 4(1 - \rho_\alpha^2)^2\rho_\alpha^2, \\ g_{3\alpha} &= 24(1 - \rho_\alpha^2)^3\rho_\alpha^2(1 - 3\rho_\alpha^2). \end{aligned} \tag{11.3.29}$$

These equations imply the well-known asymptotic expansion in a large-sample case given by

$$\begin{aligned} \mathrm{P}\{\sqrt{n}(r_\alpha^2 - \rho_\alpha^2)/\sqrt{g_{2\alpha}} \le x\} &= \Phi(x) + \phi(x)\frac{1}{\sqrt{n}}\{(g_{1\alpha}/\sqrt{g_{2\alpha}}) \\ &\quad + (g_{3\alpha}/(\sqrt{g_{2\alpha}})^3)(x^2 - 1)\} + \mathrm{O}(n^{-3/2}). \end{aligned} \tag{11.3.30}$$

Finally, we take up high-dimensional asymptotic distribution of the canonical correlations under (11.3.7). Consider the standardized variables

$$\begin{aligned} Z_\alpha &= \sqrt{n}(r_\alpha^2 - \tilde{\rho}_\alpha^2)/\sigma_\alpha, \quad \alpha = 1, \ldots, k \\ Z_\alpha &= \sqrt{n}(r_\alpha^2 - c)/\sqrt{2c(1-c)} \quad \alpha = k+1, \ldots, q \end{aligned} \tag{11.3.31}$$

Theorem 11.3.4 *Let $Z_1, \ldots, Z_q$ be the standardized variables in* (11.3.31). *Then, under* (11.3.7), *the limiting joint pdf of $Z_1, \ldots, Z_q$ is*

$$\left\{\prod_{\alpha=1}^{k} \varphi(z_\alpha)\right\} f_1(z_{k+1}, \ldots, z_q; t), \tag{11.3.32}$$

where $\varphi(\cdot)$ denotes the standard normal density function, $t = q - k$, and $f_1(z_{k+1}, \ldots, z_q; t)$ is the same as in (6.5.9).

11.3.4 Fisher's z-Transformation

It is of interest to find a transformation such that its asymptotic variance becomes parameter-free. In particular, a transformation whose asymptotic variance becomes 1 is generally used. This is equivalent to finding a function h such that the high-dimensional asymptotic variance of $z = \sqrt{n}(h(\ell_\alpha^2) - h(\theta_\alpha^2))$ is equal to 1. From Theorem 11.3.2, this is equivalent to finding a function h satisfying

$$h'(x)^2 \frac{2(2-c)}{1-c}(x+1)\left(x - \frac{c}{2-c}\right) = 1 \quad \text{for all } x > c/(2-c). \qquad (11.3.33)$$

It is easy to find a solution defined by

$$h(x) = \sqrt{\frac{1-c}{2(2-c)}} \log\left\{g(x) + \sqrt{g(x)^2 - 1}\right\}, \qquad (11.3.34)$$

where $g(x) = (2-c)x + 1 - c$. Fujikoshi and Sakurai (2008) found a solution given by

$$z = \frac{1}{2}\log\frac{1 + \sqrt{r_\alpha^2 - c(1-r_\alpha^2)/(2-c)}}{1 - \sqrt{r_\alpha^2 - c(1-r_\alpha^2)/(2-c)}}. \qquad (11.3.35)$$

Then, from Theorem 11.3.2 we have

$$\sqrt{\frac{n(1-c)}{(1-c/2)}}(z - \zeta) \ \xrightarrow{d}\ \mathrm{N}(0,1), \qquad (11.3.36)$$

where

$$\zeta = \frac{1}{2}\log\frac{1 + \sqrt{\tilde{\rho}_\alpha^2 - c(1-\tilde{\rho}_\alpha^2)/(2-c)}}{1 - \sqrt{\tilde{\rho}_\alpha^2 - c(1-\tilde{\rho}_\alpha^2)/(2-c)}}.$$

Similarly, letting $c = 0$ in (11.3.36), we have Fisher's z-transformation and its asymptotic normality in the large-sample case,

$$\sqrt{n}\left\{\frac{1}{2}\log\frac{1+r_\alpha}{1-r_\alpha} - \frac{1}{2}\log\frac{1+\rho_\alpha}{1-\rho_\alpha}\right\} \xrightarrow{d} \mathrm{N}(0,1). \qquad (11.3.37)$$

Results (11.3.36) and (11.3.37) can be used to construct a confidence interval for ρ. If A2 and A3 are satisfied, we have approximately

$$\Pr\{u_1 \le \zeta \le u_2\} \approx 1 - \delta,$$

where

$$u_1 = z - \frac{u(\delta/2)}{\sqrt{n(1-c)/(1-c/2)}}, \quad u_2 = z + \frac{u(\delta/2)}{\sqrt{n(1-c)/(1-c/2)}}$$

and $u(\delta)$ is the upper δ point of N(0, 1). The relationship $u_1 \leq \zeta \leq u_2$ is converted to the form of an interval as follows:

$$\sqrt{\frac{2-c}{2(1-c)}\left(\tanh^2(u_1)-\frac{c}{2-c}\right)} \leq \rho_\alpha \leq \sqrt{\frac{2-c}{2(1-c)}\left(\tanh^2(u_2)-\frac{c}{2-c}\right)}. \tag{11.3.38}$$

Letting $c = 0$ in (11.3.38), we obtain a confidence interval based on a large-sample framework as follows:

$$\tanh u_1 \leq \rho_\alpha \leq \tanh u_2. \tag{11.3.39}$$

Transformation (11.3.35) is defined for $r_\alpha^2 \geq c/2$. As a general transformation, replacing

$$\sqrt{r_\alpha^2 - c(1-r_\alpha^2)/(2-c)} \quad \text{by} \quad \sqrt{|r_\alpha^2 - c(1-r_\alpha^2)/(2-c)|}$$

is suggested. The other modification will be replaced by zero when $r_\alpha^2 < c/2$. These two modifications are almost the same, but it is noted [Fujikoshi and Sakurai, 2008] that the former is more useful. This is examined in the following numerical example.

The high-dimensional confidence interval is also applicable to a moderate sample size with $p = 2$ or 3. Consider two settings:

(1) $N = 25, p = q = 2, \rho_1 = 0.7886, \rho_2 = 0.054.$

(2) $N = 37, q = 2, p = 3, \rho_1 = 0.6889, \rho_2 = 0.1936.$

These are based on real data in Anderson (2003, p. 505) and Siotani (1990, p. 208), respectively. Then we have large-sample and high-dimensional confidence intervals with a 95% confidence coefficient. Using a simulation, the actual confidence coefficients are obtained as follows:

Case	Large	High
(1)	0.9122	0.9320
(2)	0.9148	0.9420

This result shows that the high-dimensional confidence interval is more useful than the large-sample confidence interval, even in a situation where the large-sample asymptotic is suitable.

11.4 Inference for Dimensionality

Let $\rho_1 \geq \cdots \geq \rho_p$ be the population canonical correlations between $\boldsymbol{X} : p \times 1$ and $\boldsymbol{Y} : q \times 1$ $(q \leq p)$. We are interested as the number of nonzero canonical correlations, which is called the *dimensionality* in canonical correlation analysis. Related to inference of the dimensionality we consider a dimensionality hypothesis or model:

$$
\begin{aligned}
M_k: \quad & \rho_k > \rho_{k+1} = \cdots = \rho_q = 0 \\
& \Leftrightarrow \ \mathrm{rank}(\Sigma_{12}) = k.
\end{aligned} \tag{11.4.1}
$$

If this is true, we can explain the correlation structure between $\boldsymbol{X}$ and $\boldsymbol{Y}$ by the first k canonical variables, since the remaining canonical variables have no power of prediction. We consider testing the hypothesis M_k as well as determining the dimensionality.

11.4.1 Test of Dimensionality

We consider an LR test for M_k against all the alternatives, based on N observations $(\boldsymbol{X}_i', \boldsymbol{Y}_i')', i = 1, \ldots, N = n + 1$ from $\mathrm{N}_{p+q}(\boldsymbol{\mu}, \Sigma)$. The LR criterion is closely related to $\min_{M_k} D(\mathbf{S}, \Sigma)$, where the quantity D is defined in Section 10.5 as

$$
D(\mathbf{S}, \Sigma) = -n \log |\Sigma^{-1}\mathbf{S}| + n \,\mathrm{tr}\, \Sigma^{-1}(\mathbf{S} - \Sigma). \tag{11.4.2}
$$

The concrete relation is given for the LR criterion for testing a general covariance structure in the following theorem.

Theorem 11.4.1 *Let λ be the LR criterion for testing a hypothesis H_ω: $\Sigma \in \mathcal{S}_\omega$ against all the alternatives H_Ω: $\Sigma \in \mathcal{S}_\Omega$, where $\mathcal{S}_\omega$ is a subset of $\mathcal{S}_\Omega = \{\Sigma;\ \Sigma > \mathrm{O}\}$. Then it holds that*

$$
-2 \log \lambda = (N/n) D(\mathbf{S}, \hat{\Sigma}_\omega), \tag{11.4.3}
$$

where $\hat{\Sigma}_\omega$ is the minimum distance estimator under M_ω and is defined by

$$
\min_{M_\omega} D(\mathbf{S}, \Sigma) = D(\mathbf{S}, \hat{\Sigma}_\omega).
$$

If $\mathrm{tr}\, \hat{\Sigma}_\omega^{-1}\mathbf{S} = q$,

$$
\lambda^{N/2} = |\mathbf{S}|/|\hat{\Sigma}_\omega|, \quad D(\mathbf{S}, \hat{\Sigma}_\omega) = -n \log |\mathbf{S}|/|\hat{\Sigma}_\omega| \tag{11.4.4}
$$

Proof. The $-2\log$ "likelihood function" after maximizing with respect to $\boldsymbol{\mu}$ is expressed as

$$
L(\Sigma) = N \log |\Sigma| + \mathrm{tr}\, \Sigma^{-1} n\mathbf{S} + N \log 2\pi.
$$

Let $\tilde{\Sigma}_\omega$ and $\tilde{\Sigma}_\Omega$ be the ML estimators of Σ under H_ω and H_Ω, respectively. Then

$$\tilde{\Sigma}_\omega = (n/N)\hat{\Sigma}_\omega, \quad \tilde{\Sigma}_\Omega = (n/N)\mathbf{S}.$$

Using the relations above, we have

$$\begin{aligned}
-2\log\lambda &= L(\tilde{\Sigma}_\omega) - L(\tilde{\Sigma}_\omega) \\
&= -N\log|\tilde{\Sigma}_\omega^{-1}(n/N)\mathbf{S}| + \operatorname{tr}\tilde{\Sigma}_\omega^{-1}(n\mathbf{S} - N\tilde{\Sigma}_\omega) \\
&= (N/n)D(\mathbf{S}, \hat{\Sigma}_\omega).
\end{aligned}$$

□

Theorem 11.4.1 shows that $D(\mathbf{S}, \hat{\Sigma}_\omega)$ is an LR statistic, and its rejection region is of the form $D(\mathbf{S}, \hat{\Sigma}_\omega) > c$.

Theorem 11.4.2 *An LR test for testing the hypothesis M_k in (11.4.1) is given by*

$$\min_{M_k} D(\mathbf{S}, \Sigma) = -\sum_{i=k+1}^{q} n\log(1 - r_i^2), \tag{11.4.5}$$

where $r_1 > \cdots > r_q > 0$ is the sample canonical correlation between $\boldsymbol{X}$ and $\boldsymbol{Y}$.

Proof. Using [A.1.1] and [A.1.2], that is,

$$\begin{aligned}
|\Sigma| &= |\Sigma_{11}| \cdot |\Sigma_{22\cdot 1}| \\
\operatorname{tr}\Sigma^{-1}\mathbf{S} &= \operatorname{tr}\Sigma_{11}^{-1}\mathbf{S}_{11} + \operatorname{tr}\Sigma_{22\cdot 1}^{-1}\mathbf{S}_{22\cdot 1} \\
&\quad + \operatorname{tr}\Sigma_{22\cdot 1}^{-1}(\mathrm{B}_{21} - \mathbf{S}_{21}\mathbf{S}_{11}^{-1})\mathbf{S}_{11}(\mathrm{B}_{21} - \mathbf{S}_{21}\mathbf{S}_{11}^{-1})',
\end{aligned}$$

we can write

$$\begin{aligned}
D(\mathbf{S}, \Sigma) &= n\left[\log|\Sigma_{11}| + \operatorname{tr}\Sigma_{11}^{-1}\mathbf{S}_{11} - p - \log|\mathbf{S}_{11}|\right] \\
&\quad + n\left[\log|\Sigma_{22\cdot 1}| - \log|\mathbf{S}_{22\cdot 1}| - q\right. \\
&\quad \left. + \operatorname{tr}\Sigma_{22\cdot 1}^{-1}\left\{\mathbf{S}_{22\cdot 1} + (\mathrm{B}_{21} - \mathbf{S}_{21}\mathbf{S}_{11}^{-1})\mathbf{S}_{11}(\mathrm{B}_{21} - \mathbf{S}_{21}\mathbf{S}_{11}^{-1})'\right\}\right],
\end{aligned}$$

where $\mathrm{B}_{21} = \Sigma_{21}\Sigma_{11}^{-1}$. Considering the minimization with respect to Σ_{11} and $\Sigma_{22\cdot 1}$:

$$\begin{aligned}
&\min_{\Sigma_{11}, \Sigma_{22\cdot 1}} D(\mathbf{S}, \Sigma) \\
&= n\left\{\log|\mathbf{S}_{22\cdot 1} + (\mathbf{S}_{21}\mathbf{S}_{11}^{-1} - \mathrm{B}_{21})\mathbf{S}_{11}(\mathbf{S}_{21}\mathbf{S}_{11}^{-1} - \mathrm{B}_{21})'| - \log|\mathbf{S}_{22\cdot 1}|\right\} \\
&= n\log|\mathrm{I}_q + (\mathbf{S}_{22\cdot 1}^{-1/2}\mathbf{S}_{21}\mathbf{S}_{11}^{-1/2} - \tilde{\mathrm{B}}_{21})(\mathbf{S}_{22\cdot 1}^{-1/2}\mathbf{S}_{21}\mathbf{S}_{11}^{-1/2}\tilde{\mathrm{B}}_{21})'|,
\end{aligned}$$

where $\tilde{\mathrm{B}}_{21} = \mathbf{S}_{22\cdot 1}^{-1/2}\mathrm{B}_{21}\mathbf{S}_{11}^{-1/2}$. The minimum is attained at

$$\begin{aligned}
\hat{\Sigma}_{11} &= \mathbf{S}_{11}, \\
\hat{\Sigma}_{22\cdot 1} &= \mathbf{S}_{22\cdot 1} + (\hat{\mathrm{B}}_{21} - \mathbf{S}_{21}\mathbf{S}_{11}^{-1})\mathbf{S}_{11}(\hat{\mathrm{B}}_{21} - \mathbf{S}_{21}\mathbf{S}_{11}^{-1})'.
\end{aligned}$$

Now we may consider minimizing

$$|\mathrm{I}_q + (\tilde{\mathrm{B}}_{21} - \mathbf{S}_{22\cdot 1}^{-1/2}\mathbf{S}_{21}\mathbf{S}_{11}^{-1/2})(\tilde{\mathrm{B}}_{21} - \mathbf{S}_{22\cdot 1}^{-1/2}\mathbf{S}_{21}\mathbf{S}_{11}^{-1/2})'|$$

under $\mathrm{rank}(\tilde{\mathrm{B}}_{21}) = k$. Since the ith singular value of $\mathbf{S}_{22}^{-1/2}\mathbf{S}_{21}\mathbf{S}_{11}^{-1/2}$ is r_i, the ith singular value of $\mathbf{S}_{22\cdot 1}^{-1/2}\mathbf{S}_{21}\mathbf{S}_{11}^{-1/2}$ is $r_i/\sqrt{1-r_i^2}$. Therefore, from [A.2.8], we have

$$\min_{\mathrm{rank}(\tilde{\mathrm{B}}_{21})=k} |\mathrm{I}_q + (\tilde{\mathrm{B}}_{21} - \mathbf{S}_{22\cdot 1}^{-1/2}\mathbf{S}_{21}\mathbf{S}_{11}^{-1/2})(\tilde{\mathrm{B}}_{21} - \mathbf{S}_{22\cdot 1}^{-1/2}\mathbf{S}_{21}\mathbf{S}_{11}^{-1/2})'|$$
$$= \prod_{j=k+1}^{q} (1-r_i^2)^{-1},$$

which is the result required. □

Using Theorem 11.3.1, the asymptotic null distribution of the test statistic $-\sum_{i=k+1}^{p} n\log(1-r_i^2)$ is known to be $\chi^2(st)$. On the other hand, using Theorem 11.3.4, the high-dimensional asymptotic null distribution is obtained. These results are summarized as follows.

Theorem 11.4.3 *Suppose that* (11.3.7) *is satisfied. Then, under* A1 *the asymptotic null distribution of the test statistic* $-\sum_{i=k+1}^{p} n\log(1-r_i^2)$ *is* $\chi^2(st)$. *On the other hand, under* A2 *the asymptotic null distribution of the standardized test statistic:*

$$\sqrt{n}\left[-\sum_{i=k+1}^{p}\log(1-r_i^2) + (q-k)\log(1-c)\right] \Big/ \sqrt{2(q-k)c/(1-c)} \quad (11.4.6)$$

is $\mathrm{N}(0,1)$.

11.4.2 Estimation of Dimensionality

A method to determine the dimensionality is to consider selection of the models $M_0, M_1, \ldots, M_p$ by using a model selection criterion. Here we use DIC in Section 10.5. If M_k is selected, we estimate the dimensionality as k. The dimension of $\mathcal{S}_k = \{\Sigma;\ \mathrm{rank}(\Sigma_{12}) = k\}$ is computed by considering a singular value decomposition of Σ_{12}. It is given by

$$\begin{aligned} b_D^{(0)} &= \frac{1}{2}p(p+1) + \frac{1}{2}q(q+1) \\ &\quad + pk - \frac{1}{2}k(k+1) + k + qk - \frac{1}{2}k(k+1) \\ &= \frac{1}{2}p(p+1) + \frac{1}{2}q(q+1) + k(p+q-k). \end{aligned}$$

Therefore, DIC for M_k is

$$DIC = -\sum_{i=k+1}^{q} n \log(1 - r_i^2) + 2\left\{\frac{1}{2}p(p+1) + \frac{1}{2}q(q+1) + k(p+q-k)\right\}. \tag{11.4.7}$$

11.5 Selection of Variables

In this section we consider a redundancy hypothesis or model of (1) a subvector of $\boldsymbol{X}$, and (2) subvectors of $\boldsymbol{X}$ and $\boldsymbol{Y}$. Related to these cases, the random vectors $\boldsymbol{X}$ and $\boldsymbol{Y}$ and their dimensions are expressed as follows:
For case (1):

$$\boldsymbol{X} = \begin{pmatrix} \boldsymbol{X}_1 \\ \boldsymbol{X}_2 \end{pmatrix}, \quad \boldsymbol{Y} = \boldsymbol{X}_3, \quad \boldsymbol{X}_i : p_i \times 1, \ i = 1, 2, 3.$$

For case (2):

$$\boldsymbol{X} = \begin{pmatrix} \boldsymbol{X}_1 \\ \boldsymbol{X}_2 \end{pmatrix}, \quad \boldsymbol{Y} = \begin{pmatrix} \boldsymbol{X}_3 \\ \boldsymbol{X}_4 \end{pmatrix}, \quad \boldsymbol{X}_i : p_i \times 1, \ i = 1, 2, 3, 4.$$

Similarly, the corresponding vectors of coefficients in the canonical variables are expressed, for case (1);

$$\boldsymbol{\alpha}_i = \begin{pmatrix} \boldsymbol{\alpha}_{1i} \\ \boldsymbol{\alpha}_{2i} \end{pmatrix}, \quad \boldsymbol{\beta}_j = \boldsymbol{\alpha}_{3j}, \quad \boldsymbol{\alpha}_{kj} : p_k \times 1, \ k = 1, 2, 3$$

and for case (2);

$$\boldsymbol{\alpha}_i = \begin{pmatrix} \boldsymbol{\alpha}_{1i} \\ \boldsymbol{\alpha}_{2i} \end{pmatrix}, \quad \boldsymbol{\beta}_j = \begin{pmatrix} \boldsymbol{\alpha}_{3j} \\ \boldsymbol{\alpha}_{4j} \end{pmatrix}, \quad \boldsymbol{\alpha}_{ki} : p_k \times 1, \ k = 1, 2, 3, 4.$$

We consider testing the redundancy hypothesis as well as selecting a best set of variables.

11.5.1 Test for Redundancy

Case (1). Along the partition $\boldsymbol{X} = (\boldsymbol{X}_1', \boldsymbol{X}_2', \boldsymbol{X}_3')'$, we partition Σ as

$$\Sigma = \begin{pmatrix} \Sigma_{11} & \Sigma_{12} & \Sigma_{13} \\ \Sigma_{21} & \Sigma_{22} & \Sigma_{23} \\ \Sigma_{31} & \Sigma_{32} & \Sigma_{33} \end{pmatrix},$$

where $\Sigma_{ij} : p_i \times p_j$. As a measure of the association between $(\boldsymbol{X}_1', \boldsymbol{X}_2')'$ and $\boldsymbol{X}_3$, it is natural to consider the sum of squares of canonical correlations between them:

$$\rho_1^2 + \cdots + \rho_m^2 = \operatorname{tr} \Sigma_{(12)(12)}^{-1} \Sigma_{(12)3} \Sigma_{33}^{-1} \Sigma_{3(12)}$$
$$\equiv \delta^2(\boldsymbol{X}_1, \boldsymbol{X}_2; \boldsymbol{X}_3), \tag{11.5.1}$$

where $m = \min\{p_1 + p_2, p_3\}$. Now we may formulate a redundancy hypothesis or no additional information on $\boldsymbol{X}_2$ as

$$\delta^2(\boldsymbol{X}_1, \boldsymbol{X}_2; \boldsymbol{X}_3) = \delta^2(\boldsymbol{X}_1; \boldsymbol{X}_3), \tag{11.5.2}$$

which is equivalent to

$$H_1\text{:}\ \ \Sigma_{23\cdot 1} \equiv \Sigma_{23} - \Sigma_{21}\Sigma_{11}^{-1}\Sigma_{13} = \mathrm{O}. \tag{11.5.3}$$

For the proof, see problem 11.4. Under normality it is also equivalent to

$$\boldsymbol{X}_2 \perp \boldsymbol{X}_3 \mid \boldsymbol{X}_1 \tag{11.5.4}$$

(i.e., $\boldsymbol{X}_2$ and $\boldsymbol{X}_3$ are independent given $\boldsymbol{X}_1$). The statement can be described in terms of canonical variables [see, e.g., Fujikoshi, 1982] as

$$\boldsymbol{\alpha}_{2j} = \mathbf{0}, \ \ i = 1, \ldots, m = \min\{p_1 + p_2, p_3\}.$$

Note that the hypothesis H_1 is equivalent to the covariance selection model (4.3.9). Therefore, the LR test and its null distribution have been treated in Section 4.3.3. Here, we give an alternative derivation.

Theorem 11.5.1 *The LR criterion λ_1 for testing H_1 in* (11.5.3) *is*

$$\lambda_1^{2/N} = \Lambda_1 = \frac{|\mathbf{S}_{(23)(23)\cdot 1}|}{|\mathbf{S}_{22\cdot 1}| \cdot |\mathbf{S}_{33\cdot 1}|},$$

whose null distribution is $\Lambda_{p_2}(p_3, n - p_1 - p_3))$.

Proof. We obtain the LR criterion λ_1 using (11.4.3) in Theorem 11.4.1. From [A.1.1] and Theorem 2.2.2 it follows that under H_1

$$|\Sigma| = |\Sigma_{11}| \cdot |\Sigma_{(23)(23)\cdot 1}| = |\Sigma_{11}| \cdot |\Sigma_{22\cdot 1}| \cdot |\Sigma_{33\cdot 1}|, \tag{11.5.5}$$

$$\begin{aligned}
\operatorname{tr} \Sigma^{-1}\mathbf{S} &= \operatorname{tr} \Sigma_{11}^{-1}\mathbf{S}_{11} + \operatorname{tr} \Sigma_{(23)(23)\cdot 1}^{-1} \{\mathbf{S}_{(23)(23)\cdot 1} \\
&\quad + (\mathrm{B}_{(23)1} - \mathbf{S}_{(23)1}\mathbf{S}_{11}^{-1})\mathbf{S}_{11}(\mathrm{B}_{(23)1} - \mathbf{S}_{(23)1}\mathbf{S}_{11}^{-1})'\} \\
&= \operatorname{tr} \Sigma_{11}^{-1}\mathbf{S}_{11} \\
&\quad + \operatorname{tr} \Sigma_{22\cdot 1}^{-1} \{\mathbf{S}_{22\cdot 1} + (\mathrm{B}_{21} - \mathbf{S}_{21}\mathbf{S}_{11}^{-1})\mathbf{S}_{11}(\mathrm{B}_{21} - \mathbf{S}_{21}\mathbf{S}_{11}^{-1})'\} \\
&\quad + \operatorname{tr} \Sigma_{33\cdot 1}^{-1} \{\mathbf{S}_{33\cdot 1} + (\mathrm{B}_{31} - \mathbf{S}_{31}\mathbf{S}_{11}^{-1})\mathbf{S}_{11}(\mathrm{B}_{31} - \mathbf{S}_{31}\mathbf{S}_{11}^{-1})'\},
\end{aligned} \tag{11.5.6}$$

where $\mathrm{B}_{(23)1} = \Sigma_{(23)1}\Sigma_{11}^{-1}, \mathrm{B}_{21} = \Sigma_{21}\Sigma_{11}^{-1}$, and $\mathrm{B}_{31} = \Sigma_{31}\Sigma_{11}^{-1}$. Using [A.2.8], we obtain the minimum distance estimator under H_1:

$$\hat{\Sigma}_{22\cdot 1} = \mathbf{S}_{22\cdot 1},\ \hat{\Sigma}_{33\cdot 1} = \mathbf{S}_{33\cdot 1},\ \hat{\Sigma}_{23\cdot 1} = \mathrm{O},$$
$$\hat{\mathrm{B}}_{21} = \mathbf{S}_{21}\mathbf{S}_{11}^{-1},\ \hat{\mathrm{B}}_{31} = \mathbf{S}_{31}\mathbf{S}_{11}^{-1},\ \hat{\Sigma}_{11} = \mathbf{S}_{11}.$$

Then we have

$$\begin{aligned}|\mathbf{S}|/|\hat{\Sigma}| &= |\mathbf{S}|/(|\mathbf{S}_{11}|\cdot|\mathbf{S}_{22\cdot 1}|\cdot|\mathbf{S}_{33\cdot 1}|)\\ &= |\mathbf{S}_{(23)(23)\cdot 1}|/(|\mathbf{S}_{22\cdot 1}|\cdot|\mathbf{S}_{33\cdot 1}|).\end{aligned}$$

The null distribution of Λ_1 follows by using the fact that Λ_1 is an LR test for independence of $\boldsymbol{X}_2$ and $\boldsymbol{X}_3$ given $\boldsymbol{X}_1$, and the result in Section 8.4. □

Case (2). Let $\boldsymbol{X} = (\boldsymbol{X}_1', \boldsymbol{X}_2', \boldsymbol{X}_3', \boldsymbol{X}_4')'$ with $\boldsymbol{X}_i : p_i \times 1$, and partition Σ along the partition of $\boldsymbol{X}$. The redundancy hypothesis is defined by

$$\delta^2(\boldsymbol{X}_1, \boldsymbol{X}_2; \boldsymbol{X}_3, \boldsymbol{X}_4) = \delta^2(\boldsymbol{X}_1; \boldsymbol{X}_3) \tag{11.5.7}$$

or

$$\mathrm{tr}\,\Sigma_{11}^{-1}\Sigma_{13}\Sigma_{33}^{-1}\Sigma_{31} = \mathrm{tr}\,\Sigma_{(12)(12)}^{-1}\Sigma_{(12)(34)}\Sigma_{(34)(34)}^{-1}\Sigma_{(34)(12)}.$$

Similarly, it is shown that (11.5.7) is equivalent (see problem 11.5) to

$$\begin{aligned}H_2\colon\ &\Sigma_{24\cdot 13} = \mathrm{O},\quad \Sigma_{23\cdot 1} = \mathrm{O},\quad \Sigma_{14\cdot 3} = \mathrm{O}.\\ \Leftrightarrow\ &\Sigma_{24\cdot 13} = \mathrm{O},\quad B_{23} = \mathrm{O},\quad B_{41} = \mathrm{O},\end{aligned} \tag{11.5.8}$$

where $\Sigma_{24\cdot 13} = \Sigma_{24} - \Sigma_{2(13)}\Sigma_{(13)(13)}^{-1}\Sigma_{(13)4}$ and

$$\mathrm{B}_{(24)(13)} = \begin{pmatrix} \mathrm{B}_{21} & \mathrm{B}_{23} \\ \mathrm{B}_{41} & \mathrm{B}_{43} \end{pmatrix} = \begin{pmatrix} \Sigma_{21} & \Sigma_{23} \\ \Sigma_{41} & \Sigma_{43} \end{pmatrix} \begin{pmatrix} \Sigma_{11} & \Sigma_{13} \\ \Sigma_{31} & \Sigma_{33} \end{pmatrix}^{-1}.$$

Theorem 11.5.2 *The LR criterion λ_2 for testing H_2 in* (11.5.3) *is*

$$\lambda_2^{2/N} = \Lambda_2 = \frac{|\mathbf{S}_{(24)(24)\cdot 13}|}{|\mathbf{S}_{22\cdot 1}|\cdot|\mathbf{S}_{44\cdot 3}|} = L_{(1)}\cdot L_{(2)}$$

where

$$L_{(1)} = \frac{|\mathbf{S}_{22\cdot 134}|}{|\mathbf{S}_{22\cdot 1}|},\quad L_{(2)} = \frac{|\mathbf{S}_{44\cdot 13}|}{|\mathbf{S}_{44\cdot 3}|}.$$

Further, under the hypothesis H_2 $L_{(1)}$ and $L_{(2)}$ are independently distributed as $\Lambda_{p_2}(p_3 + p_4, n - p_1 - p_3 - p_4)$ and $\Lambda_{p_4}(p_1, n - p_1 - p_3)$, respectively.

Proof. First we obtain the LR criterion λ_2 by using (11.4.3) in Theorem 11.4.1. It follows from [A.1.1] and Theorem 2.2.2 that, under the hypothesis H_2,

$$\begin{aligned}|\Sigma| &= |\Sigma_{(13)(13)}| \cdot |\Sigma_{(24)(24)\cdot 13}| \\ &= |\Sigma_{(13)(13)}| \cdot |\Sigma_{22\cdot 13}| \cdot |\Sigma_{44\cdot 13}|,\end{aligned} \tag{11.5.9}$$

and that, under the hypothesis H_2

$$\begin{aligned}\text{tr } \Sigma^{-1}S &= \text{tr } \Sigma_{(13)(13)}^{-1}\mathbf{S}_{(13)(13)} + \text{tr } \Sigma_{(24)(24)\cdot 13}^{-1}\mathbf{S}_{(24)(24)\cdot 13} \\ &\quad + \text{tr } \Sigma_{(24)(24)\cdot 13}^{-1}\big(\mathrm{B}_{(24)(13)} - \mathbf{S}_{(24)(13)}\mathbf{S}_{(13)(13)}^{-1}\big) \\ &\qquad \times \mathbf{S}_{(13)(13)}\big(\mathrm{B}_{(24)(13)} - \mathbf{S}_{(24)(13)}\mathbf{S}_{(13)(13)}^{-1}\big)' \\ &= \text{tr } \Sigma_{(13)(13)}^{-1}\mathbf{S}_{(13)(13)} \\ &\quad + \text{tr } \Sigma_{22\cdot 13}^{-1}\big[\mathbf{S}_{22\cdot 1} + (\mathrm{B}_{21} - \mathbf{S}_{21}\mathbf{S}_{11}^{-1})\mathbf{S}_{11}(\mathrm{B}_{21} - \mathbf{S}_{21}\mathbf{S}_{11}^{-1})'\big] \\ &\quad + \text{tr } \Sigma_{33\cdot 1}^{-1}\big[\mathbf{S}_{44\cdot 3} + (\mathrm{B}_{43} - \mathbf{S}_{43}\mathbf{S}_{33}^{-1})\mathbf{S}_{33}(\mathrm{B}_{43} - \mathbf{S}_{43}\mathbf{S}_{33}^{-1})'\big].\end{aligned} \tag{11.5.10}$$

Therefore, the distance estimators under H_2 are given by

$$\begin{aligned}&\hat{\Sigma}_{22\cdot 13} = \mathbf{S}_{22\cdot 1},\ \hat{\Sigma}_{44\cdot 13} = \mathbf{S}_{44\cdot 3},\ \hat{\Sigma}_{24\cdot 13} = \mathrm{O},\ \hat{\Sigma}_{(13)(13)} = \mathbf{S}_{(13)(13)}, \\ &\hat{\mathrm{B}}_{21} = \mathbf{S}_{21}\mathbf{S}_{11}^{-1},\ \hat{\mathrm{B}}_{24} = \mathrm{O},\ \hat{\mathrm{B}}_{41} = \mathrm{O},\ \hat{\mathrm{B}}_{43} = \mathbf{S}_{43}\mathbf{S}_{33}^{-1}.\end{aligned}$$

We can write $\lambda_2 = |\mathbf{S}|/|\hat{\Sigma}|$ as

$$\Lambda_2 = \frac{|\mathbf{S}|}{|\mathbf{S}_{(13)(13)}| \cdot |\mathbf{S}_{22\cdot 13}| \cdot \mathbf{S}_{44\cdot 13}|}.$$

The expressions required are obtained by using

$$|\mathbf{S}| = |\mathbf{S}_{(13)(13)}| \cdot |\mathbf{S}(24)(24) \cdot 13|, \quad |\mathbf{S}(24)(24) \cdot 13| = |\mathbf{S}_{44\cdot 13}| \cdot |\mathbf{S}_{22\cdot 134}|.$$

Next we consider the null distributions of $L_{(1)}$ and $L_{(2)}$ by using the redundancy condition (11.5.8). Note that this condition is effectively used in the term of the conditional distribution of $(\boldsymbol{X}_2', \boldsymbol{X}_4')'$ given $(\boldsymbol{X}_1', \boldsymbol{X}_4')'$. In fact, using the conditional set-up the LR test statistic Λ_2 is easily obtained. We can write

$$n\mathbf{S} = \mathbf{U}'\mathbf{U} = (\mathbf{U}_1, \mathbf{U}_2, \mathbf{U}_3, \mathbf{U}_4)'(\mathbf{U}_1, \mathbf{U}_2, \mathbf{U}_3, \mathbf{U}_4)',\ \mathbf{U} : n \times p_i,$$

where each row of $\mathbf{U}$ is independently distributed as $\mathrm{N}_p(\mathbf{0}, \Sigma)$, where $p = p_1 + p_2 + p_3 + p_4$. The statistics $L_{(1)}$ and $L_{(2)}$ can be expresses as

$$L_{(1)} = \frac{|\mathbf{U}_2'(\mathrm{I}_n - \mathrm{P}_{(134)})\mathbf{U}_2|}{|\mathbf{U}_2'(\mathrm{I}_n - \mathrm{P}_{(1)})\mathbf{U}_2|}, \quad L_{(2)} = \frac{|\mathbf{U}_4'(\mathrm{I}_n - \mathrm{P}_{(13)})\mathbf{U}_4|}{|\mathbf{U}_4'(\mathrm{I}_n - \mathrm{P}_{(3)})\mathbf{U}_4|},$$

where $P_{(i)}, P_{(ij)}$, etc. denote the projection matrices to $\mathcal{R}[\mathbf{U}_i], \mathcal{R}[(\mathbf{U}_i, \mathbf{U}_j)]$, respectively. The distribution of $L_{(1)}$ can be obtained by considering the conditional distribution of $\mathbf{U}_2$ given $(\mathbf{U}_1, \mathbf{U}_3, \mathbf{U}_4)$ and using Cochran's theorem. In the derivation, note that

$$\mathrm{E}[\mathbf{U}_2|(\mathbf{U}_1, \mathbf{U}_3, \mathbf{U}_4)] = \mathrm{E}[\mathbf{U}_2|(\mathbf{U}_1, \mathbf{U}_3)] = \mathbf{U}_1 B_{21}',$$

since $\mathbf{U}_2$ and $\mathbf{U}_4$ are conditionally independent when $\mathbf{U}_1$ and $\mathbf{U}_3$ are given. Then we can see that the distribution of $L_{(1)}$ is $\Lambda_{p_2}(p_3 + p_4, n - p_1 - p_3 - p_4)$ and is independent from $(\mathbf{U}_1, \mathbf{U}_3, \mathbf{U}_4)$. Similarly the distribution of $L_{(2)}$ is obtained by noting that

$$\mathrm{E}[\mathbf{U}_4|(\mathbf{U}_1, \mathbf{U}_3)] = \mathbf{U}_3 B_{43}'.$$

□

The statistic $L_{(1)}$ is decomposed as

$$L_{(1)} = \frac{|\mathbf{S}_{(24)(24)\cdot 13}|}{|\mathbf{S}_{22\cdot 13}| \cdot |\mathbf{S}_{44\cdot 13}|} \cdot \frac{|\mathbf{S}_{22\cdot 13}|}{|\mathbf{S}_{22\cdot 1}|} \; (= L_{(11)} \cdot L_{(12)}) \tag{11.5.11}$$

There are some interesting properties:

(1) The statistics $L_{(11)}$, $L_{(12)}$, and $l_{(2)}$ are likelihood ratio statistics for "$\Sigma_{24\cdot 13} = \mathrm{O}$", "$\mathrm{B}_{23} = \mathrm{O}$", and "$\mathrm{B}_{41} = \mathrm{O}$", respectively.

(2) The statistic $L_{(1)}$ is the likelihood ratio statistic for "$\Sigma_{24\cdot 13} = \mathrm{O}$ and $\mathrm{B}_{23} = \mathrm{O}$".

(3) The statistics $L_{(11)}$ and $L_{(12)}$ are independent, the statistics $L_{(1)}$ and $L_{(2)}$ are independent, but $L_{(11)}$, $L_{(12)}$, and $L_{(2)}$ are not independent.

The null distribution of Λ_2 can be approximated by using the method described in Problem 5.11. Sakurai (2009) has proposed various approximations in high-dimensional as well as large-sample situations.

11.5.2 Selection of Variables

Stepwise Methods. Based on the LR tests on additional information given in Theorems 11.5.1 and 11.5.2 it is possible to propose some stepwise methods. Suppose that in a stepwise method for canonical correlation analysis between $\boldsymbol{X} = (X_1, \ldots, X_p)'$ and $\boldsymbol{Y} = (Y_1, \ldots, Y_q)'$, the variables $(X_1, \ldots, X_j)$ and $(Y_1, \ldots, Y_k)$ have been entered. We wish to determine a significance of $X_a, a \neq 1, \ldots j$ or $Y_b, b \neq 1, \ldots k$, or both of X_a and Y_b. For a significance test of X_a, we can use the test statistic Λ_1 which is denoted by

$$\Lambda_1(X_a|X_1, \ldots, X_j; Y_1, \ldots, Y_k)$$

whose null distribution is the same as that of $\lambda_1(k, n-j-k)$. Similarly, for a significance test of Y_a, we can use the test statistic Λ_1 which is denoted by

$$\Lambda_1(Y_b|X_1, \ldots, X_j; Y_1, \ldots, Y_k)$$

whose null distribution is the same as that of $\lambda_1(j, n-j-k)$. For a significance test of X_a and Y_b, we can use the test statistic Λ_2 which is denoted by

$$\Lambda_1(X_a; Y_b|X_1, \ldots, X_j; Y_1, \ldots, Y_k)$$

whose null distribution is the same as that of a product of $\Lambda_1(k+1, n-j-k-1)$ and $\Lambda_1(j, n-j-k)$.

DIC and CDIC for H_1**.** From Theorem 11.5.1 it is easy to see that

$$DIC = -n \log \frac{|\mathbf{S}_{(23)(23)\cdot 1}|}{|\mathbf{S}_{22\cdot 1}| \cdot |\mathbf{S}_{33\cdot 1}|} + B_0, \tag{11.5.12}$$

where $B_0 = 2\{p(p+1)/2 - p_2 p_3\}$.

Next we derive CDIC. Using (11.5.6) with Σ and $\mathbf{S}$ replaced by $\hat{\Sigma}$ and Σ, respectively, we can see that

$$\begin{aligned}
\operatorname{tr} \hat{\Sigma}^{-1}\Sigma &= \operatorname{tr} \mathbf{S}_{11}^{-1}\Sigma_{11} + \left(\operatorname{tr} \mathbf{S}_{(12)(12)}^{-1}\Sigma_{(12)(12)} - \operatorname{tr} \mathbf{S}_{11}^{-1}\Sigma_{11} \right) \\
&\quad + \left(\operatorname{tr} \mathbf{S}_{(13)(13)}^{-1}\Sigma_{(13)(13)} - \operatorname{tr} \mathbf{S}_{11}^{-1}\Sigma_{11} \right) \\
&= \operatorname{tr} \mathbf{S}_{(12)(12)}^{-1}\Sigma_{(12)(12)} + \operatorname{tr} \mathbf{S}_{(13)(13)}^{-1}\Sigma_{(13)(13)} - \operatorname{tr} \mathbf{S}_{11}^{-1}\Sigma_{11}.
\end{aligned}$$

This enables us to compute the expectation of $\operatorname{tr} \hat{\Sigma}^{-1}\mathbf{S}$ by using Theorem 2.2.8. Therefore, CDIC is given by (11.5.12), with the bias term (B_0) replaced by

$$B_c = n \left[\frac{n(p_1 + p_2)}{n - p_1 - p_2 - 1} + \frac{n(p_1 + p_3)}{n - p_1 - p_3 - 1} - \frac{np_1}{n - p_1 - 1} - p \right].$$

DIC and CDIC for $\mathbf{H}_2$**.** From Theorem 11.5.2 we have

$$DIC = -n \log \frac{|\mathbf{S}_{(24)(24)\cdot 13}|}{|\mathbf{S}_{22\cdot 1}| \cdot |\mathbf{S}_{44\cdot 3}|} + B_0, \tag{11.5.13}$$

where $B_0 = 2\{p(p+1)/2 - p_2 p_4 - p_1 p_4 - p_2 p_3\}$.

Next we derive CDIC. Using (11.5.10) with Σ and $\mathbf{S}$ replaced by $\hat{\Sigma}$ and Σ, respectively, we have

$$\begin{aligned}
\operatorname{tr} \hat{\Sigma}^{-1}\Sigma &= \operatorname{tr} \mathbf{S}_{(13)(13)}^{-1}\Sigma_{(13)(13)} + \left(\operatorname{tr} \mathbf{S}_{(12)(12)}^{-1}\Sigma_{(12)(12)} - \operatorname{tr} \mathbf{S}_{11}^{-1}\Sigma_{11} \right) \\
&\quad + \left(\operatorname{tr} \mathbf{S}_{(34)(34)}^{-1}\Sigma_{(34)(34)} - \operatorname{tr} \mathbf{S}_{33}^{-1}\Sigma_{33} \right).
\end{aligned}$$

Therefore, CDIC is given by (11.5.13) with the bias term (B_0) replaced by

$$B_c = n\left[\frac{n(p_1+p_3)}{n-p_1-p_3-1} + \frac{n(p_1+p_2)}{n-p_1-p_2-1} + \frac{n(p_3+p_4)}{n-p_3-p_4-1} - \frac{np_1}{n-p_1-1} - \frac{np_3}{n-p_3-1} - p\right].$$

Example 11.5.1 For the data of five mathematics subjects considered in Example 4.2.1, we obtained the canonical correlations between $\boldsymbol{X}_1 = (X_1, X_2)'$ and $\boldsymbol{X}_2 = (X_3, X_4, X_5)'$. The canonical correlations are $r_1 = 0.66$ and $r_2 = 0.04$. Let the canonical correlation variates be denoted by $(\boldsymbol{\alpha}_i'\boldsymbol{X}_1, \boldsymbol{\beta}_i'\boldsymbol{X}_2), i = 1, 2$. The coefficient vectors and thier scaled vectors $\tilde{\boldsymbol{\alpha}}_i$ and $\tilde{\boldsymbol{\beta}}_i$ are given as follows:

$$\boldsymbol{\alpha}_1 = \begin{pmatrix} 0.03 \\ 0.05 \end{pmatrix}, \quad \boldsymbol{\alpha}_2 = \begin{pmatrix} 0.06 \\ -0.08 \end{pmatrix},$$

$$\tilde{\boldsymbol{\alpha}}_1 = \begin{pmatrix} 0.45 \\ 0.68 \end{pmatrix}, \quad \tilde{\boldsymbol{\alpha}}_2 = \begin{pmatrix} 1.11 \\ -0.99 \end{pmatrix},$$

$$\boldsymbol{\beta}_1 = \begin{pmatrix} 0.08 \\ 0.01 \\ 0.00 \end{pmatrix}, \quad \boldsymbol{\beta}_2 = \begin{pmatrix} 0.09 \\ -0.10 \\ 0.01 \end{pmatrix},$$

$$\tilde{\boldsymbol{\beta}}_1 = \begin{pmatrix} 0.87 \\ 0.12 \\ 0.06 \end{pmatrix}, \quad \tilde{\boldsymbol{\beta}}_2 = \begin{pmatrix} 0.96 \\ -1.46 \\ 0.25 \end{pmatrix},$$

The absolute values of the coefficients of X_4 and X_5 are small. To see whether or not the variables X_4 and X_5 are redundant, we examined DIC and CDIC for redundancy models, which are given as follows. The minimums are attained for both the pair $\{X_1, X_2\}$ and $\{X_3\}$. This means that X_4 and X_5 are redundant.

Table 11.5.1

DIC	$\{X_3\}$	$\{X_4\}$	$\{X_5\}$	$\{X_3,X_4\}$	$\{X_3,X_5\}$	$\{X_4,X_5\}$	$\{X_3,X_4,X_5\}$
$\{X_1\}$	39.64	54.57	56.25	41.53	41.49	53.24	43.43
$\{X_2\}$	30.11	47.19	52.15	31.35	31.86	45.49	33.28
$\{X_1,X_2\}$	22.89	45.87	50.71	26.13	26.59	44.96	30.00

Table 11.5.2

CDIC	$\{X_3\}$	$\{X_4\}$	$\{X_5\}$	$\{X_3,X_4\}$	$\{X_3,X_5\}$	$\{X_4,X_5\}$	$\{X_3,X_4,X_5\}$
$\{X_1\}$	40.55	55.49	57.16	42.64	42.60	54.35	44.82
$\{X_2\}$	31.02	48.11	53.06	32.46	32.97	46.60	34.66
$\{X_1, X_2\}$	24.00	46.98	51.82	27.71	28.17	46.54	32.22

Problems

11.1 Let S and R be the covarince matrix and the correlation matrix of $(\boldsymbol{X}', \boldsymbol{Y}')'$, which includes partition into p and q rows and columns:

$$\mathrm{S} = \begin{pmatrix} \mathrm{S}_{11} & \mathrm{S}_{12} \\ \mathrm{S}_{21} & \mathrm{S}_{22} \end{pmatrix}, \quad \mathrm{R} = \begin{pmatrix} \mathrm{R}_{11} & \mathrm{R}_{12} \\ \mathrm{R}_{21} & \mathrm{R}_{22} \end{pmatrix},$$

where $\mathrm{S}_{12} : p \times q$ and $\mathrm{R}_{12} : p \times q$. Suppose that $p \leq q$, and let $r_1 > \cdots > r_p > 0$ be the canonical correlations. Then prove the following formula:

$$\prod_{i=1}^{p} r_i^2 = |\mathrm{S}_{12}\mathrm{S}_{22}^{-1}\mathrm{S}_{21}|/|\mathrm{S}_{11}| = |\mathrm{R}_{12}\mathrm{R}_{22}^{-1}\mathrm{R}_{21}|/|\mathrm{R}_{11}|.$$

11.2 (Continuation) Prove the following formulas:

$$\begin{aligned} \prod_{i=1}^{p}(1 - r_i^2) &= |\mathrm{S}|/(|\mathrm{S}_{11}||\mathrm{S}_{22}|) = |\mathrm{S}_{11} - \mathrm{S}_{12}\mathrm{S}_{22}^{-1}\mathrm{S}_{21}|/|\mathrm{S}_{11}| \\ &= |\mathrm{R}|/(|\mathrm{R}_{11}||\mathrm{R}_{22}|) = |\mathrm{R}_{11} - \mathrm{R}_{12}\mathrm{R}_{22}^{-1}\mathrm{R}_{21}|/|\mathrm{R}_{11}|. \end{aligned}$$

11.3 Along the partition $\boldsymbol{X} = (\boldsymbol{X}_1', \boldsymbol{X}_2', \boldsymbol{X}_3')'$, we partition Σ and $\Lambda = \Sigma^{-1}$ as

$$\Sigma = \begin{pmatrix} \Sigma_{11} & \Sigma_{12} & \Sigma_{13} \\ \Sigma_{21} & \Sigma_{22} & \Sigma_{23} \\ \Sigma_{31} & \Sigma_{32} & \Sigma_{33} \end{pmatrix}, \quad \Lambda = \begin{pmatrix} \Lambda_{11} & \Lambda_{12} & \Lambda_{13} \\ \Lambda_{21} & \Lambda_{22} & \Lambda_{23} \\ \Lambda_{31} & \Lambda_{32} & \Lambda_{33} \end{pmatrix},$$

where $\Sigma_{ij}; p_i \times p_j$ and $\Lambda_{ij}; p_i \times p_j$. Then, show that $\Sigma_{13\cdot 2} = \mathrm{O} \Leftrightarrow \Lambda_{13} = \mathrm{O}$.

Hint : This equivalence follows from

$$\Lambda_{(13)(13)} \equiv \begin{pmatrix} \Lambda_{11} & \Lambda_{13} \\ \Lambda_{31} & \Lambda_{33} \end{pmatrix} = \Sigma_{(13)(13)\cdot 2}^{-1} \equiv \begin{pmatrix} \Sigma_{11\cdot 2} & \Sigma_{13\cdot 12} \\ \Sigma_{31\cdot 2} & \Sigma_{33\cdot 2} \end{pmatrix}^{-1}.$$

11.4 (Continuation) Relating to the redundancy of $\boldsymbol{X}_2$ or the sufficiency of $\boldsymbol{X}_1$ and $\boldsymbol{X}_3$ in canonical correlation analysis between $(\boldsymbol{X}_1', \boldsymbol{X}_2')'$ and $\boldsymbol{X}_3$, prove the followig results:

(1) $\delta^2(\boldsymbol{X}_1, \boldsymbol{X}_2; \boldsymbol{X}_3) = \delta^2(\boldsymbol{X}_1; \boldsymbol{X}_3) + \operatorname{tr} \Sigma_{22\cdot 1}^{-1}\Sigma_{23\cdot 1}\Sigma_{33}^{-1}\Sigma_{23\cdot 1}$, where

$$\delta^2(\boldsymbol{X}_1; \boldsymbol{X}_3) = \Sigma_{11}^{-1}\Sigma_{13}\Sigma_{33}^{-1}\Sigma_{31},$$
$$\delta^2(\boldsymbol{X}_1, \boldsymbol{X}_2; \boldsymbol{X}_3) = \operatorname{tr} \Sigma_{(12)(12)}^{-1}\Sigma_{(12)3}\Sigma_{33}^{-1}\Sigma_{3(12)}.$$

(2) $\delta^2(\boldsymbol{X}_1, \boldsymbol{X}_2; \boldsymbol{X}_3) = \delta^2(\boldsymbol{X}_1, \boldsymbol{X}_2; \boldsymbol{X}_3) \Leftrightarrow \Sigma_{23\cdot 1}$.

11.5 Let Σ be the covariance matrix of $\boldsymbol{X}$. Along to the partition $\boldsymbol{X} = (\boldsymbol{X}_1', \boldsymbol{X}_2', \boldsymbol{X}_3', \boldsymbol{X}_4')'$ with $\boldsymbol{X}_i : p_i \times 1$, we partition $\Sigma = (\Sigma_{ij}), \Sigma_{ij} : p_i \times p_j$. Relating to the redundancy of $\boldsymbol{X}_2$ and $\boldsymbol{X}_4$ or the sufficency of $\boldsymbol{X}_1$ and $\boldsymbol{X}_3$ in canonical correlation analysis between $(\boldsymbol{X}_1', \boldsymbol{X}_2')'$ and $(\boldsymbol{X}_3', \boldsymbol{X}_4')'$, show that the following conditions (a) and (b) are equivalent.

(a) $\operatorname{tr} \Sigma_{11}^{-1}\Sigma_{13}\Sigma_{33}^{-1}\Sigma_{31} = \operatorname{tr} \Sigma_{(12)(12)}^{-1}\Sigma_{(12)(34)}\Sigma_{(34)(34)}^{-1}\Sigma_{(34)(12)}$.

(b) $\Sigma_{24\cdot 13} = \mathrm{O}, \quad \Sigma_{23\cdot 1} = \mathrm{O}, \quad \Sigma_{14\cdot 3} = \mathrm{O}$.

Hint :

11.6 Assume that $\operatorname{Var}(\alpha'\boldsymbol{X}) = 1$ and $\operatorname{Var}(\beta'\boldsymbol{Y}) = 1$. Let ρ_1 be the first canonical correlation between $\boldsymbol{X}$ and $\boldsymbol{Y}$, and $(\boldsymbol{\alpha}_1'\boldsymbol{X}, \boldsymbol{\beta}_1'\boldsymbol{Y})$ is the first canonical variable. prove the the following inequality:

$$\mathrm{E}[(\alpha'\boldsymbol{X} - \boldsymbol{\beta}'\boldsymbol{Y})^2] \geq 2(1 - \rho_1^2).$$

11.7 Consider the problem of predicting the components of a $\boldsymbol{Y} = (Y_1, \ldots, Y_q)'$ by m linear combinations $\mathrm{C}'\boldsymbol{X}$ of $\boldsymbol{X} = (X_1, \ldots, X_p)'$, where C is a $p \times m$ matrix of rank $m (\leq p)$. Then the covariance matrix of $((\mathrm{C}'\boldsymbol{X})', \boldsymbol{Y}')'$ is given by

$$\begin{pmatrix} \mathrm{C}'\Sigma_{11}\mathrm{C} & \mathrm{C}'\Sigma_{12} \\ \Sigma_{21}\mathrm{C} & \Sigma_{22} \end{pmatrix}.$$

(1) Show that The best linear predictor of $\boldsymbol{Y}$ is given by $\hat{\boldsymbol{Y}} = \Sigma_{21}\Omega\Omega'\boldsymbol{X}$, where $\Omega = \mathrm{C}(\mathrm{C}'\Sigma_{11}\mathrm{C})^{-1/2}$.

(2) Show that the residual covariance matrix of $\boldsymbol{Y}$ when it is predicted by $\hat{\boldsymbol{Y}}$ is $R(\Omega) = \Sigma_{22} - \Sigma_{21}\Omega\Omega'\Sigma_{21}$, and Ω satsisfies the condition $\Omega'\Sigma_{11}\Omega = \mathrm{I}_m$.

(3) Obtain the matrix Ω such that $\operatorname{tr} R(\Omega)$ is minimum.

Hint: For a proof of (3), use [A.2.7].

11.8 Show that the LR criterion λ_2 in Theorem 11.5.2 can be also decomposed as

$$\lambda_2^{2/N} = \Lambda_2 = \frac{|\mathbf{S}_{22\cdot 13}|}{|\mathbf{S}_{22\cdot 1}|} \cdot \frac{|\mathbf{S}_{44\cdot 123}|}{|\mathbf{S}_{44\cdot 3}|}$$

Find that the distribution of a product of two independent $\Lambda_{p_2}(p_3 + p_4, n - p_1 - p_3 - p_4)$ and $\Lambda_{p_4}(p_1, n - (p_1 + p_3)$ is the same as that of two independent $\Lambda_{p_2}(p_3, n - p_1 - p_3)$ and $\Lambda_{p_4}(p_1 + p_2, n - p_1 - p_2 - p_3)$.

Chapter 12

Growth Curve Analysis

In this chapter we consider data obtained by measuring a response (characteristic, dependent) variable repeatedly. In particular, the data assume that a single variable is measured at p time points or different conditions for all subjects. The growth curve model is a typical model for such data. It is a generalization of MANOVA model. Various statistical methods are given, including for testing for adequacy, estimating the parameters, testing linear hypotheses, and constructing confidence intervals. The model selection criteria AIC and CAIC are also given.

12.1 Growth Curve Model

Suppose that a single variable Y is measured at p time points $t_1, \ldots, t_p$ (or different conditions) on n subjects chosen at random from a group. As an example of such data, consider the data in Table 12.1.1 on the ramus heights of 20 boys. We denote the variable Y at time point t_j by Y_j. One way to analyze such repeated measures is to specify a polynomial regression for Y on the time variable t, and to assume that the covariance matrix of $\boldsymbol{Y} = (Y_1, \ldots, Y_p)$ is unknown positive semi-definite.

For example, consider modeling for ramus height data. Let the observations of the ith subject be denoted by

$$\boldsymbol{Y}_i = (Y_{i1}, \ldots, Y_{i4})', \quad i = 1, \ldots, n \ (= 20).$$

Then $\boldsymbol{Y}_1, \ldots, \boldsymbol{Y}_n$ are random samples from a population with mean $\boldsymbol{\mu} = (\mu_1, \mu_2, \mu_3, \mu_4)'$ and covariance matrix Σ. If the growth or change does not depend on times, we consider a model of

$$\mu_1 = \mu_2 = \mu_3 = \mu_4 = \theta_1.$$

On the other hand, if the growth or change is linear as to time, we consider a model of

$$\mu_j = \theta_1 + t_j\theta_2, \quad j = 1, \dots, p\ (= 4),$$

which is expressed in matrix form as

$$\boldsymbol{\mu} = \begin{pmatrix} \mu_1 \\ \mu_2 \\ \mu_3 \\ \mu_4 \end{pmatrix} = \begin{pmatrix} 1 & t_1 \\ 1 & t_2 \\ 1 & t_3 \\ 1 & t_4 \end{pmatrix} \begin{pmatrix} \theta_1 \\ \theta_2 \end{pmatrix},$$

where $t_1 = 8.0$, $t_2 = 8.5$, $t_3 = 9.0$, and $t_4 = 9.5$.

Table 12.1.1 Ramus height of 20 boys

Subject	Age in years			
	8	8.5	9	9.5
1	47.8	48.8	49.0	49.7
2	46.4	47.3	47.7	48.4
3	46.3	46.8	47.8	48.5
4	45.1	45.3	46.1	47.2
5	47.6	48.5	48.9	49.3
6	52.5	53.2	53.3	53.7
7	51.2	53.0	54.3	54.5
8	49.8	50.0	50.3	52.7
9	48.1	50.8	52.3	54.4
10	45.0	47.0	47.3	48.3
11	51.2	51.4	51.6	51.9
12	48.5	49.2	53.0	55.5
13	52.1	52.8	53.7	55.0
14	48.2	48.9	49.3	49.8
15	49.6	50.4	51.2	51.8
16	50.7	51.7	52.7	53.3
17	47.2	47.7	48.4	49.5
18	53.3	54.6	55.1	55.3
19	46.2	47.5	48.1	48.4
20	46.3	47.6	51.3	51.8

(Elston and Grizzle, 1962) unit (mm)

Let the observations $Y_{i1}, \dots, Y_{ip}$ of the ith subject be denoted by

$$\boldsymbol{Y}_i = (Y_{i1}, \dots, Y_{ip})', \quad i = 1, \dots, n.$$

Then, in the growth curve model, it is assumed that for $i = 1, \dots, n$,

$$\mathrm{E}(\boldsymbol{Y}_i) = \boldsymbol{\mu} = \mathrm{X}\boldsymbol{\theta}, \tag{12.1.1}$$

and $\mathrm{Var}(\boldsymbol{Y}_i) = \Sigma$, where X is a given $p \times q$ matrix with rank q, $\boldsymbol{\theta} = (\theta_1, \ldots, \theta_q)'$ is an unknown parameter vector, and Σ is unknown positive definite. The matrix X is called a *within-design matrix.* If we consider a polynomial regression of degree $q-1$ of y on the time variable t, then

$$\mathrm{X} = \begin{pmatrix} 1 & t_1 & \cdots & t_1^{q-1} \\ \vdots & \vdots & \cdots & \vdots \\ 1 & t_p & \cdots & t_p^{q-1} \end{pmatrix}.$$

In the growth curve model, we wish to examine the following problems [Rao, 1959]:

(1) Is the specification (12.2.1) adequate?

(2) How can estimators of $\theta_1, \ldots, \theta_q$ be obtained and the precision of estimators be expressed?

(3) How can general linear hypotheses concerning $\theta_1, \ldots, \theta_q$ be tested?

(4) How can simultaneous confidence limits to a class of linear functions of $\theta_1, \ldots, \theta_q$ be obtained?

In addition, we wish to examine the following problem.

(5) If several other specifications are considered, how can we decide a best specification?

The first four problems are discussed in Section 12.2. In Section 12.3 we extend the model and the methods to those of several groups. The derivations of the methods are given in Section 12.4. Related to problem (5), the model selection criteria AIC and CAIC are discussed in Section 12.5.

In a polynomial growth curve model it is possible to use a within-design matrix X such that the columns of X are mutually orthogonal (i.e., X'X is a diagonal matrix). For example, a variable Y is measured at the four time points $t_1 = 8$, $t_2 = 8.5$, $t_3 = 9$, and $t_4 = 9.5$, and a growth curve model with a second-degree polynomial is considered. Then we have the two expressions,

$$\mathrm{E}(Y_t) = \theta_1 + \theta_2 t + \theta_3 t^2,$$
$$\mathrm{E}(Y_t) = \beta_1 + \beta_2 p_1(t) + \beta_3 p_2(t),$$

where $p_i(t)$ is an $(i-1)$th-degree polynomial. For the second expression, its within-design matrix is

$$\mathrm{X} = \begin{pmatrix} 1 & p_1(t_1) & p_2(t_1) \\ 1 & p_1(t_2) & p_2(t_2) \\ 1 & p_1(t_3) & p_2(t_3) \\ 1 & p_1(t_4) & p_2(t_4) \end{pmatrix}. \tag{12.1.2}$$

From the orthogonality condition of the columns of X, we have $p_1(t) = c_1(t - 8.75)$ and $p_2(t) = c_2(t^2 - 17.5t + 76.25)$, where c_1 and c_2 are any nonzero constants. Taking $c_1 = 4$ and $c_2 = 4$, then $p_1(t) = 4t - 35$ and $p_2(t) = 4t^2 - 70t + 305$. The within-design matrix is

$$\mathrm{X} = \begin{pmatrix} 1 & p_1(t_1) & p_2(t_1) \\ 1 & p_1(t_2) & p_2(t_2) \\ 1 & p_1(t_3) & p_2(t_3) \\ 1 & p_1(t_4) & p_2(t_4) \end{pmatrix} = \begin{pmatrix} 1 & -3 & 1 \\ 1 & -1 & -1 \\ 1 & 1 & -1 \\ 1 & 3 & 1 \end{pmatrix}.$$

The polynomials $p_1(t)$ and $p_1(t)$ are called *orthogonal polynomials.*

12.2 Statistical Inference: One Group

Suppose that $\boldsymbol{Y}_1, \ldots, \boldsymbol{Y}_n$ are random samples from a p-dimensional normal population $\mathrm{N}_p(\boldsymbol{\mu}, \Sigma)$ with

$$\boldsymbol{\mu} = \mathrm{X}\boldsymbol{\theta}, \tag{12.2.1}$$

where X is a $p\times q$ known matrix and $\boldsymbol{\theta} = (\theta_1, \ldots, \theta_q)$ is an unknown parameter vector. Further, we assume that Σ is unknown positive definite and that $\mathrm{rank}(\mathrm{X}) = q$. Statistical methods are based on the sample mean vector $\bar{\boldsymbol{Y}}$ and the sample covariance matrix $\mathbf{S}$, defined by

$$\bar{\boldsymbol{Y}} = \frac{1}{n}\sum_{i=1}^{n} \boldsymbol{Y}_i, \quad \mathbf{S} = \frac{1}{m}\sum_{i=1}^{n}(\boldsymbol{Y}_i - \bar{\boldsymbol{Y}})(\boldsymbol{Y}_i - \bar{\boldsymbol{Y}})',$$

where $m = n - 1$. More precisely,

(1)
$$\mathrm{E}(\bar{\boldsymbol{Y}}) = \mathrm{X}\boldsymbol{\theta}. \tag{12.2.2}$$

(2) The covariance matrix of $\bar{\boldsymbol{Y}}$ is $(1/n)\Sigma$ and the unbiased estimator $\mathbf{S}$ of Σ with m degrees of freedom are available.

For the distributions of $\bar{\boldsymbol{Y}}$ and $\mathbf{S}$, it holds that:

(a) $\bar{\boldsymbol{Y}} \sim \mathrm{N}_p(X\boldsymbol{\theta}, (1/n)\Sigma)$.

(b) $m\mathbf{S} \sim \mathrm{W}_p(m, \Sigma)$, and $\bar{\boldsymbol{Y}}$ and $\mathbf{S}$ are independent.

12.2.1 Test for Adequacy

To check the adequacy of (12.2.1) we need to test the hypothesis

$$H\colon \boldsymbol{\mu} = \mathrm{X}\boldsymbol{\theta} \quad \text{vs.} \quad K\colon \boldsymbol{\mu} \neq \mathrm{X}\boldsymbol{\theta}. \tag{12.2.3}$$

Let $\tilde{\mathrm{X}}$ be a $p \times (p-q)$ matrix such that $\tilde{\mathrm{X}}'\mathrm{X} = \mathrm{O}$ and $\tilde{\mathrm{X}}'\tilde{\mathrm{X}} = \mathrm{I}_{p-q}$, and consider the transformed random vector

$$\boldsymbol{Z} = \tilde{\mathrm{X}}'\bar{\boldsymbol{Y}}.$$

Since $\bar{\boldsymbol{Y}}$ is distributed as $\mathrm{N}_p(\boldsymbol{\mu}, (1/n)\Sigma)$,

$$\boldsymbol{Z} \sim \mathrm{N}_{p-q}\left(\boldsymbol{\zeta}, \frac{1}{n}\tilde{\mathrm{X}}'\Sigma\tilde{\mathrm{X}}\right),$$

where $\boldsymbol{\zeta} = \tilde{\mathrm{X}}'\boldsymbol{\mu}$. The adequacy hypothesis H is equivalent to $\boldsymbol{\zeta} = \mathbf{0}$. The covariance matrix $n^{-1}\tilde{\mathrm{X}}'\Sigma\tilde{\mathrm{X}}$ of $\boldsymbol{Z}$ is estimated by $n^{-1}\tilde{\mathrm{X}}'\mathbf{S}\tilde{\mathrm{X}}$. Then

$$\sqrt{n}\boldsymbol{Z} \sim \mathrm{N}_{p-q}(\sqrt{n}\boldsymbol{\zeta}, \tilde{\mathrm{X}}'\Sigma\tilde{\mathrm{X}}), \quad m\tilde{\mathrm{X}}'\mathbf{S}\tilde{\mathrm{X}} \sim \mathrm{W}_{p-q}(m, \tilde{\mathrm{X}}'\Sigma\tilde{\mathrm{X}}),$$

and they are independent. Therefore, we can test (12.2.3) by using Hotelling's statistic

$$\begin{aligned} T_{p-q}^2 &= m(\sqrt{n}\boldsymbol{Z})'(\tilde{\mathrm{X}}'\mathbf{S}\tilde{\mathrm{X}})^{-1}(\sqrt{n}\boldsymbol{Z}) \\ &= mn\boldsymbol{Z}'(\tilde{\mathrm{X}}'\mathbf{S}\tilde{\mathrm{X}})^{-1}\boldsymbol{Z}. \end{aligned}$$

Under the hypothesis H,

$$T_{p-q}^2 \cdot \frac{m-(p-q)+1}{p-q} \sim F(p-q, m-(p-q)+1).$$

The test statistic T_{p-q}^2 can be written as

$$\begin{aligned} T_{p-q}^2 &= mn(\tilde{\mathrm{X}}'\bar{\boldsymbol{Y}})'(\tilde{\mathrm{X}}'\mathbf{S}\tilde{\mathrm{X}})^{-1}\tilde{\mathrm{X}}'\bar{\boldsymbol{Y}} \\ &= mn\bar{\boldsymbol{Y}}'\left\{\mathbf{S}^{-1} - \mathbf{S}^{-1}\mathrm{X}(\mathrm{X}'\mathbf{S}^{-1}\mathrm{X})^{-1}\mathrm{X}'\mathbf{S}^{-1}\right\}\bar{\boldsymbol{Y}}. \end{aligned}$$

The reduction is obtained by using [A.1.4]. An alternative way of deriving the test criterion is to use the idea of regression or least squares. Consider the residuals under a general model and a hypothesis defined by

$$\begin{aligned} s_\Omega^2 &= \min_{\text{model}}(\bar{\boldsymbol{Y}} - \boldsymbol{\mu})'\mathbf{S}^{-1}(\bar{\boldsymbol{Y}} - \boldsymbol{\mu}), \\ s_\omega^2 &= \min_H(\bar{\boldsymbol{Y}} - \boldsymbol{\mu})'\mathbf{S}^{-1}(\bar{\boldsymbol{Y}} - \boldsymbol{\mu}), \end{aligned} \qquad (12.2.4)$$

and we derive a test based on $(s_\omega^2 - s_\Omega^2)$. In this case, $s_\Omega^2 = 0$. To obtain s_ω^2, let $\bar{\boldsymbol{Y}}$ and $\mathbf{S}$ transform as

$$\begin{aligned} \begin{pmatrix} \boldsymbol{U} \\ \boldsymbol{Z} \end{pmatrix} &= \begin{pmatrix} \mathrm{G}_1' \\ \mathrm{G}_2' \end{pmatrix} \bar{\boldsymbol{Y}}, \\ \mathbf{W} &= \begin{pmatrix} \mathrm{G}_1' \\ \mathrm{G}_2' \end{pmatrix} \mathbf{S}\,(\mathrm{G}_1\ \mathrm{G}_2) = \begin{pmatrix} \mathbf{W}_{11} & \mathbf{W}_{12} \\ \mathbf{W}_{21} & \mathbf{W}_{22} \end{pmatrix}, \end{aligned}$$

where $\mathrm{G}_1 = \mathrm{X}(\mathrm{X}'\mathrm{X})^{-1}$ and $\mathrm{G}_2 = \tilde{\mathrm{X}}$. Then, using [A.2.2], we have

$$\begin{aligned} s_\omega^2 &= \min_{\boldsymbol{\theta}} \begin{pmatrix} \boldsymbol{U}-\boldsymbol{\theta} \\ \boldsymbol{Z} \end{pmatrix}' \begin{pmatrix} \mathbf{W}_{11} & \mathbf{W}_{12} \\ \mathbf{W}_{21} & \mathbf{W}_{22} \end{pmatrix}^{-1} \begin{pmatrix} \boldsymbol{U}-\boldsymbol{\theta} \\ \boldsymbol{Z} \end{pmatrix} \\ &= \boldsymbol{Z}'\mathbf{W}_{22}^{-1}\boldsymbol{Z} = \boldsymbol{Z}'(\tilde{\mathrm{X}}'\mathbf{S}\tilde{\mathrm{X}})^{-1}\boldsymbol{Z}, \end{aligned}$$

which is the same as $T_{p-q}^2/(nm)$.

The hypothesis H is equivalent to $\tilde{\mathrm{X}}'\boldsymbol{\mu} = \mathbf{0}$, where $\mathrm{E}(\bar{\boldsymbol{Y}}) = \boldsymbol{\mu}$. The test statistic is also obtained as a likelihood ratio test, which is shown as a special case in Section 12.4.3. The SSP matrices due to the hypothesis and the errors are

$$\tilde{\mathbf{S}}_e = m\tilde{\mathrm{X}}'\mathbf{S}\tilde{\mathrm{X}}, \quad \tilde{\mathbf{S}}_h = n\tilde{\mathrm{X}}'\bar{\boldsymbol{Y}}\bar{\boldsymbol{Y}}'\tilde{\mathrm{X}}.$$

The likelihood ratio is $\Lambda = |\tilde{\mathbf{S}}_e|/|\tilde{\mathbf{S}}_e + \tilde{\mathbf{S}}_h|$, whose null distribution is $\Lambda_{p-q}(1, m)$. We can use

$$\frac{1-\Lambda}{\Lambda} \cdot \frac{m+1-(p-q)}{p-q} \sim F(p-q, m+1-(p-q)).$$

Furthermore,

$$\begin{aligned} \frac{1-\Lambda}{\Lambda} &= \frac{n}{m}\bar{\boldsymbol{Y}}'\left\{\mathbf{S}^{-1} - \mathbf{S}^{-1}\mathrm{X}(\mathrm{X}'\mathbf{S}^{-1}\mathrm{X})^{-1}\mathrm{X}'\mathbf{S}^{-1}\right\}\bar{\boldsymbol{Y}} \\ &= \frac{1}{m}T_{p-q}^2. \end{aligned}$$

12.2.2 Estimation and Test

Estimation. When Σ is known, a generalized least squares estimator of $\boldsymbol{\theta}$ is given by $(\mathrm{X}'\Sigma^{-1}\mathrm{X})^{-1}\mathrm{X}'\Sigma^{-1}\bar{\boldsymbol{Y}}$. Replacing Σ by $\mathbf{S}$ in this estimator, we have an estimator

$$\hat{\boldsymbol{\theta}} = (\mathrm{X}'\mathbf{S}^{-1}\mathrm{X})^{-1}\mathrm{X}'\mathbf{S}^{-1}\bar{\boldsymbol{Y}}.$$

This is an unbiased estimator. In fact, since $\bar{\boldsymbol{Y}}$ and $\mathbf{S}$ are independent, the conditional expectation of $\hat{\boldsymbol{\theta}}$ given $\mathbf{S}$ is given by

$$\mathrm{E}(\hat{\boldsymbol{\theta}} \mid \mathbf{S}) = (\mathrm{X}'\mathbf{S}^{-1}\mathrm{X})^{-1}\mathrm{X}'\mathbf{S}^{-1} \cdot \mathrm{X}\boldsymbol{\theta} = \boldsymbol{\theta},$$

and hence $\mathrm{E}(\hat{\boldsymbol{\theta}}) = \boldsymbol{\theta}$. Its conditional covariance matrix is

$$\mathrm{Var}(\hat{\boldsymbol{\theta}} \mid \mathbf{S}) = \frac{1}{n}(\mathrm{X}'\mathbf{S}^{-1}\mathrm{X})^{-1}\mathrm{X}'\mathbf{S}^{-1}\Sigma\mathrm{X}\mathbf{S}^{-1}(\mathrm{X}'\mathbf{S}^{-1}X)^{-1}.$$

Using Theorem 2.2.9, we have

$$\mathrm{Var}(\hat{\boldsymbol{\theta}}) = \frac{1}{n} \cdot \frac{n-2}{n-(p-q)-2}(\mathrm{X}'\Sigma^{-1}\mathrm{X})^{-1}.$$

To get an estimator of $\mathrm{Var}(\hat{\boldsymbol{\theta}})$, we need to estimate $(\mathrm{X}'\Sigma^{-1}\mathrm{X})^{-1}$, whose intuitive estimator is $(\mathrm{X}'\mathbf{S}\mathrm{X})^{-1}$, but it is not unbiased. For this, consider the transformation from $\mathbf{S}$ to $\mathbf{W}$ and put

$$\Psi = \begin{pmatrix} \mathrm{G}_1' \\ \mathrm{G}_2' \end{pmatrix} \Sigma \, (\mathrm{G}_1 \ \mathrm{G}_2) = \begin{pmatrix} \Psi_{11} & \Psi_{12} \\ \Psi_{21} & \Psi_{22} \end{pmatrix}.$$

Then

$$(\mathrm{X}'\Sigma^{-1}\mathrm{X})^{-1} = \Psi_{11\cdot 2}, \quad (\mathrm{X}'\mathbf{S}\mathrm{X})^{-1} = \mathbf{W}_{11\cdot 2} \tag{12.2.5}$$

In fact, using [A.1.4] the first is obtained as follows:

$$\begin{aligned}\Psi_{11\cdot 2} &= \mathrm{G}_1'\Sigma\mathrm{G}_1 - \mathrm{G}_1'\Sigma\{\Sigma^{-1} - \Sigma^{-1}\mathrm{G}_1(\mathrm{G}_1'\Sigma^{-1}\mathrm{G}_1)^{-1}\mathrm{G}_1'\Sigma^{-1}\}\Sigma\mathrm{G}_1 \\ &= \mathrm{G}_1'\mathrm{G}_1(\mathrm{G}_1'\Sigma^{-1}\mathrm{G}_1)^{-1}\mathrm{G}_1'\mathrm{G}_1 = (\mathrm{X}'\Sigma^{-1}\mathrm{X})^{-1}.\end{aligned}$$

Since $m\mathbf{W}_{11\cdot 2} \sim \mathrm{W}_q(m-(p-q), \Psi_{11\cdot 2})$, we have that $m/(m-(p-q)) \cdot \mathbf{W}_{11\cdot 2}$ is an biased estimator of $(\mathrm{X}'\Sigma^{-1}\mathrm{X})^{-1}$, and therefore that

$$\hat{\mathrm{Var}}(\hat{\boldsymbol{\theta}}) = \frac{m-1}{m-(p-q)-1} \cdot \frac{m}{m-(p-q)} \cdot \frac{1}{n}(\mathrm{X}'\mathbf{S}^{-1}\mathrm{X})^{-1}$$

is proposed as an unbiased estimator for $\mathrm{Var}(\hat{\boldsymbol{\theta}})$. **Tests.** Suppose that we wish to test the hypothesis

$$H\colon \boldsymbol{\theta}'\mathrm{D} = (0, \ldots, 0) \text{ or } \mathrm{D}'\boldsymbol{\theta} = \mathbf{0} \quad \text{vs.} \quad K\colon \mathrm{D}'\boldsymbol{\theta} \neq \mathbf{0},$$

where D is a given $q \times d$ matrix with rank d. If D is a q-dimensional vector $\boldsymbol{d} = (d_1, \ldots, d_q)'$, the hypothesis is

$$d_1\theta_1 + \cdots + d_q\theta_q = 0$$

and, moreover, the hypothesis in the case of

$$\boldsymbol{d}_1 = (1, 0, \ldots, 0)', \ldots, \boldsymbol{d}_q = (0, \ldots, 0, 1)'$$

corresponds to the significance tests of each component of $\boldsymbol{\theta}$. If $\mathrm{D} = \mathrm{I}_q$, the hypothesis is "$\boldsymbol{\theta} = \mathbf{0}$".

We derive a test statistic for $\mathrm{D}'\boldsymbol{\theta} = \mathbf{0}$ under the growth curve model, based on regression or least squares, although its LR test is obtained. In the same way as in a test of adequacy, we obtain s_Ω^2 and s_ω^2 in (12.2.4) and $s_\omega^2 - s_\Omega^2$. Consider the transformation matrix $\mathrm{H} = (\mathrm{H}_1 \ \mathrm{H}_2 \ \mathrm{H}_3)$ such that

$$\mathrm{H}_1 = \tilde{\mathrm{D}}'(\mathrm{X}'\mathrm{X})^{-1}\mathrm{X}', \ \mathrm{H}_2 = \mathrm{D}'(\mathrm{X}'\mathrm{X})^{-1}\mathrm{X}', \ \mathrm{H}_3 = \tilde{\mathrm{X}},$$

where $\tilde{\mathrm{D}}$ is a $q \times (q-d)$ matrix satisfying $\tilde{\mathrm{D}}\mathrm{D} = \mathrm{O}$. Using H, transform from $(\bar{\boldsymbol{Y}}, \mathbf{S})$ to $(\boldsymbol{U}, \mathbf{W})$:

$$\boldsymbol{U} = \begin{pmatrix} \boldsymbol{U}_1 \\ \boldsymbol{U}_2 \\ \boldsymbol{U}_3 \end{pmatrix} = \begin{pmatrix} \mathrm{H}_1' \\ \mathrm{H}_2' \\ \mathrm{H}_3' \end{pmatrix} \bar{\boldsymbol{Y}},$$

$$\mathbf{W} = \mathrm{H}'\mathbf{S}\mathrm{H} = \begin{pmatrix} \mathbf{W}_{11} & \mathbf{W}_{12} & \mathbf{W}_{13} \\ \mathbf{W}_{21} & \mathbf{W}_{22} & \mathbf{W}_{23} \\ \mathbf{W}_{31} & \mathbf{W}_{32} & \mathbf{W}_{33} \end{pmatrix}.$$

Then

$$\mathrm{E}(\boldsymbol{U}_1) = \tilde{\mathrm{D}}'\boldsymbol{\theta},\ \mathrm{E}(\boldsymbol{U}_2) = \mathrm{D}\boldsymbol{\theta},\ \mathrm{E}(\boldsymbol{U}_3) = \mathbf{0}$$

and

$$(\bar{\boldsymbol{Y}} - \mathrm{E}(\bar{\boldsymbol{Y}}))'\mathbf{S}^{-1}(\bar{\boldsymbol{Y}} - \mathrm{E}(\bar{\boldsymbol{Y}})) = (\boldsymbol{U} - \mathrm{E}(\boldsymbol{U}))'\mathbf{W}^{-1}(\boldsymbol{U} - \mathrm{E}(\boldsymbol{U})).$$

Therefore, using [A.2.2], we obtain

$$s_\Omega^2 = \boldsymbol{U}_3'\mathbf{W}_{33}^{-1}\boldsymbol{U}_3 = \boldsymbol{Z}'(\tilde{\mathrm{X}}'\mathbf{S}\tilde{\mathrm{X}})^{-1}\boldsymbol{Z},$$

$$s_\omega^2 = \begin{pmatrix} \boldsymbol{U}_2 \\ \boldsymbol{U}_3 \end{pmatrix}' \begin{pmatrix} \mathbf{W}_{22} & \mathbf{W}_{23} \\ \mathbf{W}_{32} & \mathbf{W}_{33} \end{pmatrix}^{-1} \begin{pmatrix} \boldsymbol{U}_2 \\ \boldsymbol{U}_3 \end{pmatrix},$$

and

$$s_\omega^2 = s_\Omega^2 + (\boldsymbol{U}_2 - \mathbf{W}_{23}\mathbf{W}_{33}^{-1}\boldsymbol{U}_3)'\mathbf{W}_{22\cdot 3}^{-1}(\boldsymbol{U}_2 - \mathbf{W}_{23}\mathbf{W}_{33}^{-1}\boldsymbol{U}_3),$$

where $\mathbf{W}_{22\cdot 3} = \mathbf{W}_{22} - \mathbf{W}_{23}\mathbf{W}_{33}^{-1}\mathbf{W}_{32}$. A derivation similar to (12.2.4), given by

$$\boldsymbol{U}_2 - \mathbf{W}_{23}\mathbf{W}_{33}^{-1}\boldsymbol{U}_3 = \mathrm{D}'\hat{\boldsymbol{\theta}},\ \mathbf{W}_{22\cdot 3} = \mathrm{D}'(\mathrm{X}'\mathbf{S}^{-1}\mathrm{X})^{-1}\mathrm{D}.$$

These suggest a test based on

$$\begin{aligned} L &= (\boldsymbol{U}_2 - \mathbf{W}_{23}\mathbf{W}_{33}^{-1}\boldsymbol{U}_3)'\mathbf{W}_{22\cdot 3}^{-1}(\boldsymbol{U}_2 - \mathbf{W}_{23}\mathbf{W}_{33}^{-1}\boldsymbol{U}_3) \\ &= (\mathrm{D}'\hat{\boldsymbol{\theta}})'\{\mathrm{D}'(\mathrm{X}'\mathbf{S}^{-1}\mathrm{X})^{-1}\mathrm{D}\}^{-1}\mathrm{D}'\hat{\boldsymbol{\theta}}. \end{aligned} \quad (12.2.6)$$

The statistic L is proportional to

$$(\mathrm{D}'\hat{\boldsymbol{\theta}})'\hat{\mathrm{Var}}(\mathrm{D}'\hat{\boldsymbol{\theta}})^{-1}(\mathrm{D}'\hat{\boldsymbol{\theta}}),$$

which provides a motivation for proposing a test statistic.

Although L is a test statistic, its distribution is complicated so we consider modifying L. When we consider the distribution of L, we may assume that $\Sigma = \mathrm{I}_p$ and $\Psi = \mathrm{I}_p$. Given $\boldsymbol{U}_3$ and $\mathbf{W}_{33}$:

(1) $\boldsymbol{U}_2 - \mathbf{W}_{23}\mathbf{W}_{33}^{-1}\boldsymbol{U}_3 \sim \mathrm{N}_d(\mathbf{0}, (1/n)(1 + n\boldsymbol{u}_3'\mathbf{W}_{33}^{-1}\boldsymbol{U}_3)\mathrm{I}_d)$.

(2) $m\mathbf{W}_{22\cdot 3} \sim \mathrm{W}_d(m - (p-q), \mathrm{I}_d)$.

(3) $\boldsymbol{U}_2 - \mathbf{W}_{23}\mathbf{W}_{33}^{-1}\boldsymbol{U}_3$ and $\mathbf{W}_{22\cdot 3}$ are independent.

Result (1) follows from Theorem 1.2.2, and results (2) and (3) follow from Theorem 2.2.2. Therefore, we can construct a Hotelling statistic

$$\begin{aligned} T^2_{d;m-(p-q)} &= \frac{n}{1 + n\boldsymbol{U}_3'\mathbf{W}_{33}^{-1}\boldsymbol{U}_3} mL \\ &= \frac{1}{R}(\mathrm{D}'\hat{\boldsymbol{\theta}})'\{m\mathrm{D}'(\mathrm{X}'\mathbf{S}^{-1}\mathrm{X})\mathrm{D}\}^{-1}(\mathrm{D}'\hat{\boldsymbol{\theta}}), \end{aligned}$$

where

$$R = \frac{1}{n}\left(1 + \frac{1}{m}T^2_{p-q}\right), \quad T^2_{p-q} = mn\boldsymbol{U}_3'\mathbf{W}_{33}^{-1}\boldsymbol{U}_3,$$

and

$$T^2_{p-q} = mn\bar{\boldsymbol{Y}}'\left\{\mathbf{S}^{-1} - \mathbf{S}^{-1}\mathrm{X}(\mathrm{X}'\mathbf{S}^{-1}\mathrm{X})^{-1}\mathrm{X}'\mathbf{S}^{-1}\right\}\bar{\boldsymbol{Y}}.$$

Finally, we can propose a Hotelling-type statistic whose null distribution is

$$T^2_{d;m-(p-q)} \cdot \frac{m-(p-q)-d+1}{d} \sim F(d, m-(p-q)-d+1).$$

This test is also given as the LR test. The SSP matrices due to the hypotheses and the error are

$$\mathbf{S}_h = \frac{1}{R}(\mathrm{D}'\hat{\theta})(\mathrm{D}'\hat{\theta})', \quad \mathbf{S}_e = m\mathrm{D}'(\mathrm{X}'\mathbf{S}^{-1}\mathrm{X})^{-1}\mathrm{D}.$$

The LR test is based on $\Lambda = |\mathbf{S}_e|/|\mathbf{S}_e + \mathbf{S}_h|$, whose null distribution is $\Lambda_d(1, m-(p-q))$. It holds that

$$\frac{1-\Lambda}{\Lambda} = \frac{1}{R}(\mathrm{D}'\hat{\theta})\mathbf{S}_e^{-1}(\mathrm{D}'\hat{\theta})' = T^2_{d;m-(p-q)}.$$

12.2.3 Confidence Intervals

We wish to obtain the $100(1-\alpha)\%$ simultaneous confidence intervals of $\boldsymbol{b}'\boldsymbol{\theta}$ for any q-dimensional vector $\boldsymbol{b}$. A method is based on the inequality

$$\begin{aligned} &\frac{\{\boldsymbol{b}'(\hat{\boldsymbol{\theta}} - \boldsymbol{\theta})\}^2}{m\boldsymbol{b}'(\mathrm{X}'\mathbf{S}^{-1}\mathrm{X})^{-1}\boldsymbol{b} \cdot R} \\ &\qquad \le \ell_1 = \frac{1}{mR}(\hat{\boldsymbol{\theta}} - \boldsymbol{\theta})'(\mathrm{X}'\mathbf{S}_e^{-1}\mathrm{X})(\hat{\boldsymbol{\theta}} - \boldsymbol{\theta}). \end{aligned}$$

The simultaneous confidence intervals are given by

$$\boldsymbol{b}'\boldsymbol{\theta} \in [\boldsymbol{b}'\hat{\boldsymbol{\theta}} \pm \{\ell_1(\alpha)R\boldsymbol{b}'(\mathrm{X}'\mathbf{S}_e^{-1}\mathrm{X})^{-1}\boldsymbol{b}\}^{1/2}],$$

where $\ell_1(\alpha)$ is the upper α-point of ℓ_1. Put $\mathbf{W} = m(\mathrm{X}'\mathbf{S}^{-1}\mathrm{X})^{-1}$ and $\boldsymbol{U} = (1/\sqrt{R})(\hat{\boldsymbol{\theta}} - \boldsymbol{\theta})$. Then we can express $\ell_1 = \boldsymbol{U}'W^{-1}\boldsymbol{U}$; therefore,

$$\ell_1(\alpha) = \frac{q}{m-(p-q)-q+1}F(q, m-(p-q)-q+1|\alpha).$$

If we choose the column vectors of the within-individual design matrix X as the coefficient vector $\boldsymbol{b}$, we can get the simultaneous confidence intervals for the means of Y at each time point.

Example 12.2.1 The data in Table 12.1.1 consist of the ramus height of 20 boys. The sample mean vector and the sample covariance matrix are given by

$$\bar{\boldsymbol{y}} = \begin{pmatrix} 48.66 \\ 49.63 \\ 50.57 \\ 51.45 \end{pmatrix}, \quad \mathrm{S} = \begin{pmatrix} 6.33 & 6.19 & 5.78 & 5.55 \\ 6.19 & 6.45 & 6.16 & 5.92 \\ 5.78 & 6.16 & 6.92 & 6.95 \\ 5.55 & 5.92 & 6.95 & 7.47 \end{pmatrix}.$$

Suppose that we hit a linear model of the form $\mathrm{E}(Y_t) = \theta_1 + \theta_2 t$, where $t = 4(\text{age} - 8.75)$. That is, take the value $t_1 = -3$, $t_2 = -1$, $t_3 = 1$, and $t_4 = 3$. The model is expressed as

$$\boldsymbol{\mu} = \mathrm{X}\begin{pmatrix} \theta_1 \\ \theta_2 \end{pmatrix}, \quad \mathrm{X} = \begin{pmatrix} 1 & 1 & 1 & 1 \\ -3 & -1 & 1 & 3 \end{pmatrix}'.$$

Estimates of the mean parameters are given by

$$\hat{\boldsymbol{\theta}} = \begin{pmatrix} \hat{\theta}_1 \\ \hat{\theta}_2 \end{pmatrix} = (\mathrm{X}'\mathrm{S}^{-1}\mathrm{X})^{-1}\mathrm{X}\bar{\boldsymbol{y}} = \begin{pmatrix} 50.05 \\ 0.465 \end{pmatrix}$$

and the estimate of its covariance matrix,

$$\begin{aligned} &\frac{1}{n} \cdot \frac{m-1}{m-(p-q)-1} \cdot \frac{m}{m-(p-q)}(\mathrm{X}'\mathrm{S}^{-1}\mathrm{X})^{-1} \\ &= \frac{1}{20} \cdot \frac{18}{16} \cdot \frac{19}{17}(\mathrm{X}'\mathrm{S}^{-1}\mathrm{X})^{-1} \\ &= \begin{pmatrix} 0.3891 & 0.0043 \\ 0.0043 & 0.0034 \end{pmatrix}. \end{aligned}$$

Consider

$$H_0\colon \theta_2 = 0 \quad \text{vs.} \quad H_1\colon \theta_2 \neq 0.$$

Then, taking $\boldsymbol{d} = (0, 1)'$, we have

$$\begin{aligned} \mathrm{S}_e &= m\boldsymbol{d}'(\mathrm{X}'\mathrm{S}^{-1}\mathrm{X})^{-1}\boldsymbol{d} = m(0,\ 1)(\mathrm{X}'\mathrm{S}^{-1}\mathrm{X})^{-1}(0,\ 1)' = 0.2839, \\ \mathrm{S}_h &= (\hat{\boldsymbol{\theta}}\boldsymbol{d})'R^{-1}(\hat{\boldsymbol{\theta}}\boldsymbol{d}) = \hat{\theta}_2^2/R = 3.6066, \end{aligned}$$

where

$$R = \frac{1}{n} + \frac{1}{m}\bar{\boldsymbol{y}}'\{\mathrm{S}^{-1} - \mathrm{S}^{-1}\mathrm{X}(\mathrm{X}'\mathrm{S}^{-1}\mathrm{X})^{-1}\mathrm{X}'\mathrm{S}^{-1}\} = 0.060.$$

Therefore, $\Lambda = S_e/(S_e + S_h) = 0.0073$, and the value of the corresponding F statistic is $[(\Lambda - 1)/\Lambda] \cdot [\{m - (p-q)\}/1] = 216$. The p-value of $F(1, 17)$ is very small, and hence the hypothesis is rejected.

Next, consider testing the adequacy of the model based on

$$\Lambda_0 = \frac{|\tilde{\mathrm{X}}'\mathrm{S}_e\tilde{\mathrm{X}}|}{|\tilde{\mathrm{X}}'(\mathrm{S}_e + \mathrm{S}_h)\tilde{\mathrm{X}}|},$$

where

$$\mathrm{S}_e = mS = \sum_{i=1}^{n}(\boldsymbol{y}_i - \bar{\boldsymbol{y}})(\boldsymbol{y}_i - \bar{\boldsymbol{y}})', \mathrm{S}_h = n\bar{\boldsymbol{y}}\bar{\boldsymbol{y}}',$$

$$\tilde{\mathrm{X}} = \begin{pmatrix} 1/2 & -1/2 & -1/2 & 1/2 \\ -1/\sqrt{20} & 3/\sqrt{20} & -3/\sqrt{20} & 1/\sqrt{20} \end{pmatrix}'.$$

Then $\Lambda_0 = 0.9895$ and the value of the corresponding F statistic is 0.0953. Therefore, the hypothesis is not rejected.

12.3 Statistical Methods: Several Groups

Suppose that there are k different groups or treatments and that a variable Y is measured at p time points $t_1, t_2, \ldots, t_p$ on n_i random samples drawn from the ith group $(i = 1, 2, \ldots, k)$. Let the n_i random samples be denoted by

$$G_i : \boldsymbol{Y}_{i1}, \ldots, \boldsymbol{Y}_{in_i}, \quad i = 1, \ldots, k,$$

and suppose that they are distributed independently as $\mathrm{N}_p(\boldsymbol{\mu}_i, \Sigma)$. Similar to the case of a single group, it is assumed that the mean vectors are expressed as

$$\boldsymbol{\mu}_1 = \mathrm{X}\boldsymbol{\theta}_1, \ldots, \boldsymbol{\mu}_k = \mathrm{X}\boldsymbol{\theta}_k,$$

where X is a known $p \times q$ matrix with rank(X) $= q$, and the components of unknown parameter vectors $\boldsymbol{\theta}_i$ are expressed as

$$\boldsymbol{\theta}_1 = \begin{pmatrix} \theta_{11} \\ \vdots \\ \theta_{1q} \end{pmatrix}, \ldots, \boldsymbol{\theta}_k = \begin{pmatrix} \theta_{k1} \\ \vdots \\ \theta_{kq} \end{pmatrix}.$$

The growth curve model for k groups is

$$\boldsymbol{Y}_{i1}, \ldots, \boldsymbol{Y}_{in_i} \sim \text{i.i.d. } \mathrm{N}_p(\mathrm{X}\boldsymbol{\theta}_i, \Sigma), \quad i = 1, \ldots, k. \tag{12.3.1}$$

In this section we state inferential methods along with those for a single group. Their derivations are discussed in the next section. Let $\bar{\boldsymbol{Y}}_i$ and S_i be the sample mean vector and the sample covariance matrix for the ith group. The pooled sample covariance matrix is expressed as

$$\mathbf{S} = \frac{1}{m}\{(n_1 - 1)\mathbf{S}_1 + \cdots + (n_k - 1)\mathbf{S}_k\},$$

where $m = n - k$ and $n = n_1 + \cdots + n_k$.

Estimation. The MLEs of $\boldsymbol{\theta}_i$ are

$$\hat{\boldsymbol{\theta}}_i = (\mathrm{X}'\mathbf{S}^{-1}\mathrm{X})^{-1}\mathrm{X}'\mathbf{S}^{-1}\bar{\boldsymbol{Y}}_i, \quad i = 1, \ldots, k,$$

and

$$\hat{\mathrm{Var}}(\hat{\boldsymbol{\theta}}_i) = \frac{m-1}{m-(p-q)-1} \cdot \frac{m}{m-(p-q)} \cdot \frac{1}{n_i}(\mathrm{X}'\mathbf{S}^{-1}\mathrm{X})^{-1}.$$

provides their covariance matrices with unbiased estimates. **Tests.** For a given k-dimensional vector $\boldsymbol{c} = (c_1, \ldots, c_k)'$ and a given $q \times d$ matrix D with rank d, consider testing the hypothesis $\boldsymbol{c}'\Theta\mathrm{D} = (0, \ldots, 0)$; that is,

$$(c_1, \ldots, c_k)\begin{pmatrix} \boldsymbol{\theta}_1' \\ \vdots \\ \boldsymbol{\theta}_k' \end{pmatrix}\mathrm{D} = (0, \ldots, 0).$$

If we choose $\boldsymbol{c}_1 = (1, 0, \ldots, 0)', \ldots, \boldsymbol{c}_k = (0, \ldots, 0, 1)'$ as $\boldsymbol{c}$, then $\boldsymbol{c}\Theta$ becomes $\theta_1, \ldots, \theta_k$, respectively. For these $\boldsymbol{c}$, the role of D is the same as that of a single group. That is, if we set

$$\boldsymbol{d}_1 = (1, 0, \ldots, 0)', \ \ldots, \boldsymbol{d}_q = (0, \ldots, 0, 1)'$$

as D, the hypotheses are those for significance tests of each component of $\boldsymbol{\theta}_i$. If we choose $\mathrm{D} = \mathrm{I}_q$, the hypotheses are those for significance tests of the $\boldsymbol{\theta}_i$ themselves.

For testing the hypothesis H: $\boldsymbol{c}'\Theta\mathrm{D} = (0, \ldots, 0)$ vs. K: $\boldsymbol{c}'\Theta\mathrm{D} \neq (0, \ldots, 0)$, let $\mathbf{S}_h$ and $\mathbf{S}_e$ be the SSP matrices due to the hypothesis and the error, respectively. Then they are

$$\mathbf{S}_h = \frac{1}{\boldsymbol{c}'\mathbf{R}\boldsymbol{c}}(\boldsymbol{c}'\hat{\Theta}\mathrm{D})'(\boldsymbol{c}'\hat{\Theta}\mathrm{D}), \quad \mathbf{S}_e = \mathrm{D}'(\mathrm{X}'\mathbf{S}_e^{-1}\mathrm{X})^{-1}\mathrm{D},$$

where

$$\mathbf{R} = \mathrm{diag}\left(\frac{1}{n_1}, \ldots, \frac{1}{n_k}\right) + \frac{1}{m}(\bar{\boldsymbol{Y}}_1, \ldots, \bar{\boldsymbol{Y}}_k)' \times \left\{\mathbf{S}^{-1} - \mathbf{S}^{-1}\mathrm{X}(\mathrm{X}'\mathbf{S}^{-1}\mathrm{X})^{-1}\mathrm{X}'\mathbf{S}^{-1}\right\}(\bar{\boldsymbol{Y}}_1, \ldots, \bar{\boldsymbol{Y}}_k).$$

The LR test is based on $\Lambda_0 = |\mathbf{S}_e|/|\mathbf{S}_e + \mathbf{S}_h|$, whose null distribution is $\Lambda_d(1, m-(p-q))$. The hypothesis H is rejected by

$$\frac{1-\Lambda_0}{\Lambda_0} \cdot \frac{m-(p-q)+1-d}{d} \geq F(d, m-(p-q)+1-d|\alpha).$$

The test statistic is expressed as

$$\frac{1-\Lambda_0}{\Lambda_0} = \operatorname{tr} \mathbf{S}_h \mathbf{S}_e^{-1}$$
$$= \frac{1}{\boldsymbol{c}'\mathbf{R}\boldsymbol{c}}(\boldsymbol{c}'\hat{\Theta}\mathrm{D})\{m\mathrm{D}'(\mathrm{X}'\mathbf{S}^{-1}\mathrm{X})^{-1}\mathrm{D}\}^{-1}(\boldsymbol{c}'\hat{\Theta}\mathrm{D}).$$

In the special case of $\boldsymbol{c} = \boldsymbol{c}_i$ and $\mathrm{D} = \boldsymbol{d}_j$, the hypothesis becomes H: $\theta_{ij} = 0$, and writing Λ_0 by Λ_1, the hypothesis is rejected if

$$\frac{1-\Lambda_1}{\Lambda_1} \cdot \frac{m-(p-q)}{1} \geq F(1, m-(p-q)|\alpha).$$

The test statistic is expressed as

$$\frac{1-\Lambda_1}{\Lambda_1} = \frac{(\boldsymbol{c}_i'\hat{\Theta}\boldsymbol{d}_\ell)^2}{m\boldsymbol{d}_\ell{}'(X'S^{-1}X)^{-1}\boldsymbol{d}_\ell\boldsymbol{c}_i'R\boldsymbol{c}_i}$$
$$= \frac{(\boldsymbol{c}_i'\hat{\Theta}\boldsymbol{d}_\ell)^2}{m\boldsymbol{d}_\ell{}'(\mathrm{X}'\mathbf{S}^{-1}\mathrm{X})^{-1}\boldsymbol{d}_\ell\boldsymbol{c}_i'R\boldsymbol{c}_i},$$

where

$$m\boldsymbol{d}_\ell{}'(\mathrm{X}'\mathbf{S}^{-1}\mathrm{X})^{-1}\boldsymbol{d}_\ell = \text{ the } (\ell,\ell) \text{ element of } m(\mathrm{X}'\mathbf{S}^{-1}\mathrm{X})^{-1} \quad \text{and}$$
$$\boldsymbol{c}_i'R\boldsymbol{c}_i = \frac{1}{n_i}\left\{1 + \frac{n_i}{m}\bar{\boldsymbol{Y}}_i'\left\{\mathbf{S}^{-1} - \mathbf{S}^{-1}\mathrm{X}(\mathrm{X}'\mathbf{S}^{-1}\mathrm{X})^{-1}\mathrm{X}'\mathbf{S}^{-1}\right\}\bar{\boldsymbol{Y}}_i\right\}.$$

For testing the hypothesis H: $\theta_i = \mathbf{0}$ vs. K: $\boldsymbol{\theta}_i \neq \mathbf{0}$, it sets only $\boldsymbol{c} = \boldsymbol{c}_i$ and $\mathrm{D} = \mathrm{I}_p$. Writing Λ_0 by Λ_2, the hypothesis is rejected if

$$\frac{1-\Lambda_2}{\Lambda_2} \cdot \frac{m-(p-q)}{1} \geq F(q, m-(p-q)+1-q|\alpha),$$

where

$$\frac{1-\Lambda_2}{\Lambda_2} = \frac{1}{m\boldsymbol{c}_i'\mathbf{R}\boldsymbol{c}_i}(\boldsymbol{c}_i'\hat{\Theta})(\mathrm{X}'\mathbf{S}^{-1}\mathrm{X})(\boldsymbol{c}_i'\hat{\Theta})'.$$

The hypothesis $\boldsymbol{\theta}_1 = \cdots = \boldsymbol{\theta}_k$ is expressed as

$$\mathrm{C}\Theta = \mathrm{O},$$

where $\mathrm{C} = (\mathrm{I}_{k-1}, -\mathbf{1}_{k-1})$. The SSP matrices due to the hypothesis and the error are given by

$$\mathbf{U} = (\mathrm{C}\hat{\Theta})'\{\mathbf{C}\mathbf{R}\mathbf{C}'\}^{-1}\mathrm{C}\hat{\Theta}, \quad \mathbf{V} = (\mathrm{X}'\mathbf{S}_e^{-1}\mathrm{X})^{-1}.$$

If q or $(k-1)$ is 2 or less, we can use the exact distribution for $\Lambda_q(k-1, m-(p-q))$. Otherwise, we can use large-sample and high-dimensional approximations. Using the large-sample approximation, the hypothesis is rejected if

$$-\left\{m-(p-q)+\frac{1}{2}(k-1-q-1)\right\}\log\frac{|\mathbf{V}|}{|\mathbf{V}+\mathbf{U}|}\geq\chi^2(q(k-1)|\alpha).$$

Test for Adequacy. To check the adequacy of the growth curve model (12.3.1), we need to test the hypothesis

$$H\colon \boldsymbol{\mu}_1=\mathrm{X}\boldsymbol{\theta}_1,\ldots,\boldsymbol{\mu}_k=\mathrm{X}\boldsymbol{\theta}_k,\quad \text{vs.}\quad K\colon \boldsymbol{\mu}_i\neq\mathrm{X}\boldsymbol{\theta}_i\quad\text{for some } i.$$

Let $\tilde{\mathrm{X}}$ be a $p\times(p-q)$ matrix satisfying $\tilde{\mathrm{X}}'\mathrm{X}=\mathrm{O}$ and $\tilde{\mathrm{X}}'\tilde{\mathrm{X}}=\mathrm{I}_{p-q}$. Then the hypothesis is equivalent to $\tilde{\mathrm{X}}'\boldsymbol{\mu}_i=\mathbf{0}$ for all $i=1,\ldots,k$. The SSP matrices due to the hypothesis and the error are given by

$$\tilde{\mathbf{S}}_h=\tilde{\mathrm{X}}'\{n_1\bar{\boldsymbol{Y}}_1\bar{\boldsymbol{Y}}_1'+\cdots+n_k\bar{\boldsymbol{Y}}_k\bar{\boldsymbol{Y}}_k'\}\tilde{\mathrm{X}},\quad \tilde{\mathbf{S}}_e=\tilde{\mathrm{X}}'\mathbf{S}_e\tilde{\mathrm{X}},$$

respectively. The LR test is based on $\Lambda_3=|\tilde{\mathbf{S}}_e|/|\tilde{\mathbf{S}}_e+\tilde{\mathbf{S}}_h|$, whose null distribution is $\Lambda_{p-q}(k,m)$. Using the large-sample approximation, the hypothesis is rejected if

$$-\left[m+\frac{1}{2}\{k-(p-q)-1\}\right]\log\Lambda_3\geq\chi^2((p-q)k|\alpha).$$

Confidence Intervals. Consider simultaneous confidence intervals for all the parameters of the form $\boldsymbol{b}'\Theta'\boldsymbol{c}$, where $\boldsymbol{c}=(c_1,\ldots,c_k)'$ and $\boldsymbol{b}=(b_1,\ldots,b_q)'$. A method is to use the inequality

$$\begin{aligned}&\frac{\{\boldsymbol{b}'(\hat{\Theta}'\boldsymbol{c}-\Theta'\boldsymbol{c})\}^2}{m\boldsymbol{b}'(\mathrm{X}'\mathbf{S}^{-1}\mathrm{X})^{-1}\boldsymbol{b}\cdot\boldsymbol{c}'\mathbf{R}\boldsymbol{c}}\\&\qquad\leq\ell_1=\frac{1}{m\boldsymbol{c}'\mathbf{R}\boldsymbol{c}}(\hat{\Theta}'\boldsymbol{c}-\Theta'\boldsymbol{c})'(\mathrm{X}'\mathbf{S}_e^{-1}\mathrm{X})(\hat{\Theta}'\boldsymbol{c}-\Theta'\boldsymbol{c}).\end{aligned}$$

The distribution of ℓ_1 is the same as that of the maximum root of $\mathbf{S}_h\mathbf{S}_e^{-1}$ with $\mathrm{D}=\mathrm{I}_q$ and $\Theta=\mathrm{O}$. Put $\mathbf{W}=m(\mathrm{X}'\mathbf{S}^{-1}\mathrm{X})^{-1}$ and $\boldsymbol{u}=(1/\sqrt{\boldsymbol{c}'\mathbf{R}\boldsymbol{c}})(\hat{\Theta}-\Theta)'\boldsymbol{c}$. Then $\ell_1=\boldsymbol{U}'\mathbf{W}^{-1}\boldsymbol{U}$ and

$$\ell\cdot[\{m-(p-q)-q+1\}/q]\sim F(q,m-(p-q)-q+1).$$

Therefore, a simultaneous $100(1-\alpha)\%$ confidence interval is

$$\boldsymbol{b}'\Theta'\boldsymbol{c}\in[\boldsymbol{b}'\hat{\Theta}'\boldsymbol{c}\pm\{\ell_1(\alpha)\boldsymbol{c}'\mathbf{R}\boldsymbol{c}\boldsymbol{b}'(\mathrm{X}'\mathbf{S}^{-1}\mathrm{X})^{-1}\boldsymbol{b}\}^{1/2}],$$

where

$$\ell_1(\alpha)=\frac{q}{m-(p-q)-q+1}F(q,m-(p-q)-q+1|\alpha).$$

As in one group, if we choose the column vectors of the within-individual design matrix X as the coefficient vector $\boldsymbol{b}$, we can get the simultaneous confidence intervals for the means of Y at each time point.

The data in Table 12.3.1 consist of dental measurements on $n_1 = 11$ girls and $n_2 = 14$ boys [Potthoff and Roy, 1964]. In the data set we delete two boys' data which may be outliers.

Table 12.3.1 Dental data on girls and boys (Potthoff and Roy, 1964)

	Age (Girls)				Age (Boys)			
Subject	8	10	12	14	8	10	12	14
1	21.0	20.0	21.5	23.0	26.0	25.0	29.0	31.0
2	21.0	21.5	24.0	25.5	21.5	22.5	23.0	26.5
3	20.5	24.0	24.5	26.0	23.0	22.5	24.0	27.5
4	23.5	24.5	25.0	26.5	25.5	27.5	26.5	27.0
5	21.5	23.0	22.5	23.5	20.0	23.5	22.5	26.0
6	20.0	21.0	21.0	22.5	24.5	25.5	27.0	28.5
7	21.5	22.5	23.0	25.0	22.0	22.0	24.5	26.5
8	23.0	23.0	23.5	24.0	24.0	21.5	24.5	25.5
9	20.0	21.0	22.0	21.5	27.5	28.0	31.0	31.5
10	16.5	19.0	19.0	19.5	23.0	23.0	23.5	25.0
11	–	–	–	–	21.5	23.5	24.0	28.0
12	–	–	–	–	22.5	25.5	25.5	26.0
13	–	–	–	–	23.0	24.5	26.0	30.0
14	–	–	–	–	22.0	21.5	23.5	25.0

(Potthoff and Roy, 1964) unit (mm)

Let Y be an observation matrix whose first 11 rows denote the data on girls. Assume that this is a growth curve model with a second-degree polynomial for Y. Using orthogonal polynomials the model is expressed as

$$\mathrm{E}(\mathbf{Y}) = \begin{pmatrix} 1 & 0 \\ \vdots & \vdots \\ 1 & 0 \\ 0 & 1 \\ \vdots & \vdots \\ 0 & 1 \end{pmatrix} \begin{pmatrix} \theta_{11} & \theta_{12} & \theta_{13} \\ \theta_{21} & \theta_{22} & \theta_{23} \end{pmatrix} \begin{pmatrix} 1 & 1 & 1 & 1 \\ -3 & -1 & 1 & 3 \\ 1 & -1 & -1 & 1 \end{pmatrix}$$

$$= \mathrm{A}\Theta\mathrm{X}'.$$

The sample mean vectors and pooled sample covariance matrix are given by

$$\bar{y}_1 = \begin{pmatrix} 21.18 \\ 22.23 \\ 23.09 \\ 24.09 \end{pmatrix}, \quad \bar{y}_2 = \begin{pmatrix} 23.29 \\ 24.00 \\ 25.32 \\ 27.43 \end{pmatrix}, \quad \mathrm{S} = \begin{pmatrix} 4.28 & 3.10 & 4.39 & 3.48 \\ 3.10 & 4.03 & 3.98 & 3.34 \\ 4.39 & 3.98 & 5.70 & 4.79 \\ 3.48 & 3.34 & 4.79 & 5.15 \end{pmatrix}.$$

The coefficient vectors $\boldsymbol{\theta}_1$ and $\boldsymbol{\theta}_2$ of girls and boys are estimated by

$$\left(\hat{\boldsymbol{\theta}}_1 \ \hat{\boldsymbol{\theta}}_2\right) = (\mathrm{X}'\mathrm{S}^{-1}\mathrm{X})^{-1}\mathrm{X}'\mathrm{S}^{-1}(\bar{y}_1 \ \bar{y}_2) = \begin{pmatrix} 22.70 & 25.04 \\ 0.47 & 0.69 \\ 0.002 & 0.36 \end{pmatrix}$$

Since

$$(\mathrm{A}'\mathrm{A})^{-1} = \begin{pmatrix} n_1^{-1} & 0 \\ 0 & n_2^{-1} \end{pmatrix},$$

$\mathrm{Cov}(\hat{\boldsymbol{\theta}}_1, \hat{\boldsymbol{\theta}}_2) = \mathrm{O}$. The covariance matrices of $\hat{\boldsymbol{\theta}}_1$ and $\hat{\boldsymbol{\theta}}_2$ are estimated by

$$\frac{23-1}{23-2} \cdot \frac{23}{23-1} \cdot \frac{1}{n_1}(\mathrm{X}'\mathrm{S}^{-1}\mathrm{X})^{-1} = \begin{pmatrix} 0.357 & 0.012 & -0.0121 \\ 0.012 & 0.006 & 0.002 \\ -0.021 & 0.002 & 0.010 \end{pmatrix}$$

and

$$\frac{23-1}{23-2} \cdot \frac{23}{23-1} \cdot \frac{1}{n_2}(\mathrm{X}'\mathrm{S}^{-1}\mathrm{X})^{-1} = \begin{pmatrix} 0.281 & 0.009 & -0.016 \\ 0.009 & 0.005 & 0.002 \\ -0.016 & 0.002 & 0.011 \end{pmatrix},$$

respectively.

We are interested in whether the girl's growth curve is linear, and consider testing the hypothesis $\theta_{13} = 0$, which is a general hypothesis $\mathrm{C\Theta D} = \mathrm{O}$ with $\mathrm{C} = (1,\ 0)$ and $\mathrm{D} = (0,\ 0,\ 1)'$. The sums of squares $\mathbf{S}_h$ and $\mathbf{S}_e$ due to the hypothesis and the error are given by

$$\mathrm{S}_h = \hat{\theta}_{13}^2 \times \text{the } (1,1) \text{ element of } \mathrm{R}^{-1} = 5.266 \times 10^{-5},$$
$$\mathrm{S}_e = \text{the } (3,3) \text{ element of } (\mathrm{X}'\mathrm{S}^{-1}\mathrm{X})^{-1} = 0.14547,$$

where

$$\mathrm{R} = \begin{pmatrix} n_1^{-1} & 0 \\ 0 & n_2^{-1} \end{pmatrix} + (\bar{y}_1 \ \bar{y}_2)' \left\{\mathrm{S}^{-1} - \mathrm{S}^{-1}\mathrm{X}(\mathrm{X}'\mathrm{S}^{-1}\mathrm{X})^{-1}\mathrm{S}^{-1}\right\}$$
$$\times (\bar{y}_1 \ \bar{y}_2) = \begin{pmatrix} 0.097 & 0.003 \\ 0.003 & 0.073 \end{pmatrix}.$$

The likelihood ratio test or F test is given by $F = \{m - (p-q)\}\mathrm{S}_h/\mathrm{S}_e = 3.67 \times 10^{-4}$. The p-value of $F(1, 22)$ is 0.98. The hypothesis $\theta_{13} = 0$ is not rejected.

12.4 Derivation of Statistical Inference

12.4.1 General Multivariate Linear Model

History of analysis of repeated measures goes back to the research by Wishart (1938). Statistical methods for the growth curve model (12.3.1) based on the likelihood were introduced by Rao (1959, 1965), Khatri (1966), Gleser and Olkin (1970), and others [For a review with applications see Grizzle and Allen (1969)]. Various statistical methods for analysis of repeated measures, see, e.g., Crowder and Hand (1990), Davis (2002). Let the observation vectors in (12.3.1) be denoted by

$$\mathbf{Y} = (\boldsymbol{Y}_{11}, \ldots, \boldsymbol{Y}_{1n_1}, \ldots, \boldsymbol{Y}_{k1}, \ldots, \boldsymbol{Y}_{kn_k})'.$$

Then the growth curve model (12.3.1) is expressed as

$$\mathrm{E}(\mathbf{Y}) = \mathrm{A}\Theta\mathrm{X}', \tag{12.4.1}$$

where A is the between-group design matrix,

$$\mathrm{A} = \begin{pmatrix} \mathbf{1}_{n_1} & \mathbf{0} & \cdots & \mathbf{0} \\ \mathbf{0} & \mathbf{1}_{n_2} & \cdots & \mathbf{0} \\ \vdots & \vdots & \ddots & \mathbf{0} \\ \mathbf{0} & \mathbf{0} & \cdots & \mathbf{1}_{n_k} \end{pmatrix} \tag{12.4.2}$$

and $\Theta = (\boldsymbol{\theta}_1, \ldots, \boldsymbol{\theta}_k)'$. The rows of $\mathbf{Y}$ are independently distributed as p-dimensional normal distributions with the same covariance matrix Σ. The growth curve model (12.4.1) is reduced to a MANOVA model when the within-individual design matrix X is I_p. In this sense the growth curve model is a generalized MANOVA model.

In order to relate the growth curve model to a multivariate linear model, consider the transformation from $\mathbf{Y}$ to $(\mathbf{U}\ \mathbf{V})$:

$$(\mathbf{U}\ \mathbf{V}) = \mathbf{Y}(\mathrm{G}_1\ \mathrm{G}_2), \tag{12.4.3}$$

where G_1 and G_2 are the same matrices as in the group $\mathrm{G}_1 = \mathrm{X}(\mathrm{X}'\mathrm{X})^{-1}$, $\mathrm{G}_2 = \tilde{\mathrm{X}}$, and $\tilde{\mathrm{X}}$ is a $p \times (p-q)$ matrix satisfying $\tilde{\mathrm{X}}'\mathrm{X} = \mathrm{O}$ and $\tilde{\mathrm{X}}'\tilde{\mathrm{X}} = \mathrm{I}_{p-q}$. Then, the rows of $(\mathbf{U}\ \mathbf{V})$ are independently distributed as p-dimensional normal distributions with means

$$\mathrm{E}[(\mathbf{U}\ \mathbf{V})] = (\mathrm{A}\Theta\ \mathrm{O})$$

and the common covariance matrix

$$\Psi = \mathrm{G}'\Sigma\mathrm{G} = \begin{pmatrix} \mathrm{G}_1'\Sigma\mathrm{G}_1 & \mathrm{G}_1'\Sigma\mathrm{G}_2 \\ \mathrm{G}_2'\Sigma\mathrm{G}_1 & \mathrm{G}_2'\Sigma\mathrm{G}_2 \end{pmatrix} = \begin{pmatrix} \Psi_{11} & \Psi_{12} \\ \Psi_{21} & \Psi_{22} \end{pmatrix},$$

where $G = (G_1\ G_2)$. This transformation can be regarded as one from $\boldsymbol{Y} = (Y_1, \ldots, Y_p)'$ to a q-dimensional and main variable $\boldsymbol{U} = (U_1, \ldots, U_q)'$ and a $(p-q)$-dimensional auxiliary variable $\boldsymbol{V} = (V_1, \ldots, V_{p-q})'$. The growth curve model is equivalent to the following two models:

(1) The conditional distribution of $\mathbf{U}$ given $\mathbf{V}$ is

$$\mathbf{U} \mid \mathbf{V} \sim \mathrm{N}_{n\times q}(\mathbf{A}^*\Xi,\ \Psi_{11\cdot 2}). \tag{12.4.4}$$

(2) The marginal distribution of $\mathbf{V}$ is

$$\mathbf{V} \sim \mathrm{N}_{n\times(p-q)}(\mathrm{O},\ \Psi_{22}), \tag{12.4.5}$$

where

$$\mathbf{A}^* = (\mathrm{A}\ \mathbf{V}), \quad \Xi = \begin{pmatrix} \Theta \\ \Gamma \end{pmatrix},$$
$$\Gamma = \Psi_{22}^{-1}\Psi_{21},\ \Psi_{11\cdot 2} = \Psi_{11} - \Psi_{12}\Psi_{22}^{-1}\Psi_{21}.$$

12.4.2 Estimation

The MLEs in the growth curve model are obtained by applying a general theory [A.2.11] of multivariate linear model to (12.4.4) and (12.4.5). The MLEs of $\hat{\Xi}$, $\hat{\Psi}_{11\cdot 2}$, and $\hat{\Psi}_{22}$ are given by

$$\hat{\Xi} = (\mathbf{A}^{*'}\mathbf{A}^*)^{-1}\mathbf{A}^{*'}\mathbf{U},\ n\hat{\Psi}_{11\cdot 2} = \mathbf{U}'(\mathrm{I}_n - \mathbf{P}_{\mathbf{A}^*})\mathrm{U},\ n\hat{\Psi}_{22} = \mathbf{V}'\mathbf{V}, \tag{12.4.6}$$

where $\mathrm{P_A} = \mathrm{A}(\mathrm{A}'\mathrm{A})^{-1}\mathrm{A}'$. Let $m = n - k$,

$$\mathbf{S} = \frac{1}{m}\mathbf{Y}'(\mathrm{I}_n - \mathrm{P_A})\mathbf{Y},\ \mathbf{W} = \mathrm{T}'\mathbf{S}\mathrm{T} = (\mathbf{U}\ \mathbf{V})'(\mathrm{I}_n - \mathrm{P_A})(\mathbf{U}\ \mathbf{V}),$$

and partition $\mathbf{W}$ as

$$\mathbf{W} = \begin{pmatrix} \mathbf{W}_{11} & \mathbf{W}_{12} \\ \mathbf{W}_{21} & \mathbf{W}_{22} \end{pmatrix}, \quad \mathbf{W}_{12} : q \times (p-q).$$

Theorem 12.4.1 *For an $n \times p$ observation matrix $\mathbf{Y}$, assume a growth curve model defined by* (12.3.1) *or* (12.4.1) *Then:*

(1) $\hat{\Theta} = \mathrm{A}(\mathrm{A}'\mathrm{A})^{-1}\mathrm{A}'\mathbf{Y}\mathbf{S}^{-1}\mathrm{X}(\mathrm{X}'\mathbf{S}^{-1}\mathrm{X})^{-1}$.

(2) $\mathrm{E}(\hat{\Theta}) = \Theta$.

(3) $\mathrm{Var}(\mathrm{vec}(\hat{\Theta}')) = \dfrac{m-1}{m-(p-q)-1}(\mathrm{A}'\mathrm{A})^{-1} \otimes \Psi_{11\cdot 2}$.

Proof. The MLE of Ξ is $\hat{\Xi} = (\mathbf{A}^{*'}\mathbf{A}^*)^{-1}\mathbf{A}^{*'}\mathbf{U}$. The inverse formula (see [A.1.2]) gives

$$\begin{aligned}\mathbf{Q} = (\mathbf{A}^{*'}\mathbf{A}^*)^{-1} &= \begin{pmatrix} (\mathbf{A}'\mathbf{A})^{-1} & \mathrm{O} \\ \mathrm{O} & \mathrm{O} \end{pmatrix} \\ &+ \begin{pmatrix} -(\mathrm{A}'\mathrm{A})^{-1}\mathrm{A}'\mathbf{V} \\ \mathrm{I}_{p-q} \end{pmatrix} [\mathbf{V}'(\mathrm{I}_n - \mathrm{P}_\mathrm{A})\mathbf{V}]^{-1} \begin{pmatrix} -(\mathrm{A}'\mathrm{A})^{-1}\mathrm{A}'\mathbf{V} \\ \mathrm{I}_{p-q} \end{pmatrix}' \\ &= \begin{pmatrix} \mathbf{Q}_{11} & \mathbf{Q}_{12} \\ \mathbf{Q}_{21} & \mathbf{Q}_{22} \end{pmatrix}.\end{aligned}$$

Therefore, we have

$$\begin{aligned}\hat{\Theta} &= (\mathbf{Q}_{11}\mathrm{A}' + \mathbf{Q}_{12}\mathbf{V}')\mathbf{U} \\ &= (\mathrm{A}'\mathrm{A})^{-1}\mathrm{A}'\mathbf{Y}\mathrm{G}_1 - (\mathrm{A}'\mathrm{A})^{-1}\mathrm{A}'\mathbf{Y}\mathrm{G}_2(\mathrm{G}_2'\mathbf{S}\mathrm{G}_2)^{-1}\mathrm{G}_2'SG_1.\end{aligned}$$

Using [A.1.4], we obtain (1). The conditional expectation of $\hat{\Xi}$ given $\mathbf{V}$ is

$$\mathrm{E}(\hat{\Xi}|\mathbf{V}) = (\mathbf{A}^{*'}\mathbf{A}^*)^{-1}\mathbf{A}^{*'} \cdot \mathbf{A}^*\Xi = \Xi,$$

which implies (2). Similarly,

$$\mathrm{Var}(\mathrm{vec}(\hat{\Xi}')|\mathbf{V}) = (\mathbf{A}^{*'}\mathbf{A}^*)^{-1} \otimes \Psi_{11\cdot 2},$$

and hence

$$\mathrm{Var}(\mathrm{vec}(\hat{\Theta}')) = \mathrm{E}(\mathbf{Q}_{11}) \otimes \Psi_{11\cdot 2},$$

where

$$\mathbf{Q}_{11} = (\mathrm{A}'\mathrm{A})^{-1} + (\mathrm{A}'\mathrm{A})^{-1}\mathrm{A}'\mathbf{V}\{\mathbf{V}'(\mathrm{I}_n - \mathrm{P}_\mathrm{A})\mathbf{V}\}^{-1}\mathbf{V}'\mathrm{A}(\mathrm{A}'\mathrm{A})^{-1}.$$

Note that $\mathrm{A}'\mathbf{V}$ and $\mathbf{V}'(\mathrm{I}_n - \mathrm{P}_\mathrm{A})\mathbf{V}$ are independent, and $\mathbf{V}'(\mathrm{I}_n - \mathrm{P}_\mathrm{A})\mathbf{V} \sim \mathrm{W}_{p-q}(m, \Psi_{22})$. Therefore, we have

$$\begin{aligned}\mathrm{E}(\mathbf{Q}_{11}) &= (\mathrm{A}'\mathrm{A})^{-1} + \frac{1}{m-(p-q)-1}\mathrm{E}\left\{(\mathrm{A}'\mathrm{A})^{-1}\mathrm{A}'\mathbf{V}\Psi_{22}^{-1}\mathbf{V}'\mathrm{A}(\mathrm{A}'\mathrm{A})^{-1}\right\} \\ &= \left\{1 + \frac{p-q}{m-(p-q)-1}\right\}(\mathrm{A}'\mathrm{A})^{-1},\end{aligned}$$

which implies (3). □

Theorem 12.4.2 *Under the growth curve model the followings hold:*

(1) *The MLE $\hat{\Psi}_{11\cdot 2}$ of $\Psi_{11\cdot 2}$ is given by*

$$n\hat{\Psi}_{11\cdot 2} = m\mathbf{W}_{11\cdot 2} = m(\mathrm{X}'\mathbf{S}^{-1}\mathrm{X})^{-1}.$$

(2) *Let*

$$\widehat{\mathrm{Var}}[\mathrm{vec}(\hat{\Theta}')] \equiv \frac{m-1}{m-(p-q)-1} \cdot \frac{1}{m-(p-q)} \times (\mathrm{A}'\mathrm{A})^{-1} \otimes (\mathrm{X}'\mathbf{S}^{-1}\mathrm{X})^{-1}.$$

Then $\widehat{\mathrm{Var}}[\mathrm{vec}(\hat{\Theta}')]$ *is an unbiased estimator of* $\mathrm{Var}[\mathrm{vec}(\hat{\Theta}')]$.

Proof. From (12.4.6) we have $n\hat{\Psi}_{11\cdot 2} = \mathbf{U}'(\mathrm{I}_n - \mathbf{P}_{\mathbf{A}^*})\mathbf{U}$. Let $\mathbf{B} = (\mathrm{I}_n - \mathrm{P}_\mathrm{A})\mathbf{V}$. Then, using $\mathbf{P}_{\mathbf{A}^*} = \mathrm{P}_\mathrm{A} + \mathbf{P}_\mathbf{B}$, the first expression of (1) is obtained. The second expression of (1) has been shown in (12.2.5). For a proof of (2), note that $m\mathbf{S} \sim \mathrm{W}_p(m, \Sigma)$ and $m\mathbf{W} \sim \mathrm{W}_p(m, \Psi)$. These imply that

$$n\hat{\Psi}_{11\cdot 2} \sim \mathrm{W}_q(m-(p-q),\ \Psi_{11\cdot 2}),$$

and hence

$$\mathrm{E}[(\mathrm{X}'\mathbf{S}^{-1}\mathrm{X})^{-1}] = \mathrm{E}[\mathbf{W}_{11\cdot 2}] = \{m-(p-q)\}\Psi_{11\cdot 2}.$$

This proves (2). □

12.4.3 LR Tests for General Linear Hypotheses

In this section we derive the LR criterion for testing a general linear hypothesis

$$H_0\colon \mathrm{C}\Theta\mathrm{D} = \mathrm{O} \text{ vs. } H_1\colon \mathrm{C}\Theta\mathrm{D} \neq \mathrm{O}, \qquad (12.4.7)$$

where $\mathrm{C}: c \times k$ and $\mathrm{D}: q \times d$ are known matrices with ranks c and d, respectively. For testing the hypothesis $\boldsymbol{\theta}_1 = \cdots = \boldsymbol{\theta}_k$, we may take

$$\mathrm{C} = (\mathrm{I}_{k-1} \ \ -\mathbf{1}_k), \quad \mathrm{D} = \mathrm{I}_q.$$

Theorem 12.4.3 *The LR test for* H_0: $\mathrm{C}\Theta\mathrm{D} = \mathrm{O}$ *vs.* H_1: $\mathrm{C}\Theta\mathrm{D} \neq \mathrm{O}$ *in the growth curve model* (12.3.1) *or* (12.4.1) *is based on*

$$\Lambda = |\mathbf{S}_e|/|\mathbf{S}_e + \mathbf{S}_h|,$$

where

$$\mathbf{S}_e = \mathrm{D}'(\mathrm{X}'\mathbf{S}^{-1}\mathrm{X})^{-1}\mathrm{D}, \quad \mathbf{S}_h = (\mathrm{C}\hat{\Theta}\mathrm{D})(\mathrm{C}\mathbf{R}\mathrm{C}')^{-1}\mathrm{C}\hat{\Theta}\mathrm{D}$$

and

$$\mathbf{R} = (\mathrm{A}'\mathrm{A})^{-1} + (\mathrm{A}'\mathrm{A})^{-1}\mathrm{A}'\mathbf{Y}\mathbf{S}^{-1}\{\mathbf{S} - \mathrm{X}'(\mathrm{X}\mathbf{S}^{-1}\mathrm{X}')^{-1}\mathrm{X}\} \times \mathbf{S}^{-1}\mathbf{Y}'\mathrm{A}(\mathrm{A}'\mathrm{A})^{-1}.$$

Further, the null distribution is $\Lambda_d(c, m-(p-q))$.

Proof. The testing problem (12.4.7) is equivalent to testing

$$H_0\text{: } \mathrm{C}^*\Xi\mathrm{D} = \mathrm{O} \text{ vs. } H_1\text{: } \mathrm{C}^*\Xi\mathrm{D} \neq \mathrm{O} \tag{12.4.8}$$

under the conditional model (12.4.4), where $\mathrm{C}^* = (\mathrm{C}\ \mathrm{O})$. Since the distribution of $\mathbf{V}$ does not depend on H_0, the LR test under the conditional model is the LR test under the unconditional model. Using Theorem 7.6.1, we have the results required with

$$\tilde{\mathbf{S}}_e = \mathrm{D}'\mathbf{U}(\mathrm{I}_n - \mathbf{A}^*(\mathbf{A}^{*'}\mathbf{A}^*)^{-1}\mathbf{A}^{*'})\mathbf{U}\mathrm{D},$$
$$\tilde{\mathbf{S}}_h = (\mathrm{C}\hat{\Xi}\mathrm{D})(\mathrm{C}^*(\mathbf{A}^{*'}\mathbf{A}^*)^{-1}\mathrm{C}^{*'})^{-1}\mathrm{C}\hat{\Xi}D$$

instead of $\mathbf{S}_e$ and $\mathbf{S}_h$. By reduction similar to those of MLEs, it is seen that $\tilde{\mathbf{S}}_e = S_e$ and $\tilde{\mathbf{S}}_h = \mathbf{S}_h$. This completes the proof. □

Next we consider testing the adequacy of growth curve model (12.3.1) or (12.4.1). This is to test

$$H_0\text{: } M \equiv (\boldsymbol{\mu}_1, \ldots, \boldsymbol{\mu}_k)' = \Theta\mathrm{X}'$$

in a MANOVA model $\mathrm{E}(\mathbf{Y}) = \mathrm{AM}$. Since the hypothesis is equivalent to "$\mathrm{MG}_2 = \mathrm{O}$", the testing problem is reduced to the special one in Theorem 7.6.1. The LR test is based on

$$\tilde{\Lambda} = |\mathrm{G}_2'\mathbf{S}\mathrm{G}_2| / |\mathrm{G}_2'(\mathbf{S} + \mathbf{B})\mathrm{G}_2|$$

whose null distribution is $\Lambda_{p-q}(k, m)$, where $\mathbf{B} = \mathbf{Y}'\mathrm{A}(\mathrm{A}'\mathrm{A})^{-1}\mathrm{A}'\mathbf{Y}$.

12.4.4 Confidence Intervals

We consider constructing a simultaneous confidence interval of $\boldsymbol{a}'\mathrm{C}\Theta\mathrm{D}\boldsymbol{b}$ for all constant vectors $\boldsymbol{a}$ and $\boldsymbol{b}$. From Theorem 12.4.1 we have

$$\mathrm{E}(\boldsymbol{a}'\mathrm{C}\hat{\Theta}\mathrm{D}\boldsymbol{b}) = \boldsymbol{a}'\mathrm{C}\Theta\mathrm{D}\boldsymbol{b},$$
$$\mathrm{Var}(\boldsymbol{a}'\mathrm{C}\hat{\Theta}\mathrm{D}\boldsymbol{b}) = \frac{m-1}{m-(p-q)-1}\boldsymbol{a}'\mathrm{C}(\mathrm{A}'\mathrm{A})^{-1}\mathrm{C}'\boldsymbol{a} \cdot \boldsymbol{b}'\mathrm{D}'\Psi_{11\cdot2}\mathrm{D}\boldsymbol{b}.$$

A standardized statistic of $\boldsymbol{a}'\mathrm{C}\Theta\mathrm{D}\boldsymbol{b}$ is

$$\frac{\boldsymbol{a}'\mathrm{C}(\widehat{\Theta} - \Theta)\mathrm{D}\boldsymbol{b}}{\boldsymbol{a}'\mathrm{C}\mathbf{R}\mathrm{C}'\boldsymbol{a} \cdot \boldsymbol{b}'\mathrm{D}'\mathbf{W}_{11\cdot2}\mathrm{D}\boldsymbol{b}}.$$

By estimating $\boldsymbol{b}'\mathrm{D}'\Psi_{11\cdot2}\mathrm{D}\boldsymbol{b}$ by $\{m - (p-q)\}^{-1}\boldsymbol{b}'\mathrm{D}'\mathbf{W}_{11\cdot2}\mathrm{D}\boldsymbol{b}$ and approximating $\boldsymbol{a}'\mathrm{C}(\mathrm{A}'\mathrm{A})^{-1}\mathrm{C}'\boldsymbol{a}$ by $\boldsymbol{a}'\mathrm{C}\mathbf{R}\mathrm{C}'\boldsymbol{a}$, it is suggested that the construction be based on

$$\frac{\{\boldsymbol{a}'\mathrm{C}(\widehat{\Theta} - \Theta)\mathrm{D}\boldsymbol{b}\}^2}{(\boldsymbol{a}'\mathrm{C}\mathbf{R}\mathrm{C}'\boldsymbol{a} \cdot \boldsymbol{b}'\mathrm{D}'\mathbf{W}_{11\cdot2}\mathrm{D}\boldsymbol{b})/\sqrt{m-(p-q)}}.$$

Note that

$$\begin{aligned} T(\hat{\Theta}, \mathbf{W}_{11\cdot 2}; \boldsymbol{a}, \boldsymbol{b}) &\equiv \frac{\{\boldsymbol{a}'\mathrm{C}(\widehat{\Theta} - \Theta)\mathrm{D}\boldsymbol{b}\}^2}{\boldsymbol{a}'\mathrm{C}\mathbf{R}\mathrm{C}'\boldsymbol{a} \cdot \boldsymbol{b}'\mathrm{D}'\mathbf{W}_{11\cdot 2}\mathrm{D}\boldsymbol{b}} \\ &\le \frac{\boldsymbol{b}'\{\mathrm{C}(\widehat{\Theta} - \Theta)\mathrm{D}\}'\{\mathrm{C}\mathbf{R}\mathrm{C}'\}^{-1}\{\mathrm{C}(\widehat{\Theta} - \Theta)\mathrm{D}\}\boldsymbol{b}}{\boldsymbol{b}'\mathrm{D}\mathbf{W}_{11\cdot 2}\mathrm{D}'\boldsymbol{b}} \\ &\le \ell_1, \end{aligned}$$

where ℓ_1 is the maximum characteristic root of $\mathbf{S}_h\mathbf{S}_e^{-1}$. Here

$$\begin{aligned} \mathbf{S}_e &= \mathrm{D}'\mathbf{W}_{11\cdot 2}\mathrm{D} \sim \mathrm{W}_d(m, \mathrm{D}\Sigma\mathrm{D}'), \\ \mathbf{S}_h &= \{\mathrm{C}(\widehat{\Theta} - \Theta)\mathrm{D}\}'\{\mathrm{C}\mathbf{R}\mathrm{C}'\}^{-1}\{\mathrm{C}(\widehat{\Theta} - \Theta)\mathrm{D}\} \\ &\sim \mathrm{W}_d(c, \mathrm{D}\Sigma\mathrm{D}), \end{aligned}$$

and $\mathbf{S}_e$ and $\mathbf{S}_h$ are independent. Let $\ell_1(\alpha)$ be the upper α-point of ℓ_1. Then

$$\mathrm{P}\{T(\hat{\Theta}, \mathbf{W}_{11\cdot 2}; \boldsymbol{a}, \boldsymbol{b}) \le \ell_1(\alpha)\} \le \mathrm{P}\{\ell_1 \le \ell_1(\alpha)\} = 1 - \alpha.$$

This implies the following simultaneous confidence interval with confidence coefficient $100(1-\alpha)\%$:

$$\boldsymbol{a}'\mathrm{C}\widehat{\Theta}\mathrm{D}\boldsymbol{b} \pm \{\ell_1(\alpha) \cdot \boldsymbol{a}'\mathrm{C}\mathbf{R}\mathrm{C}'\boldsymbol{a} \cdot \boldsymbol{b}'\mathbf{S}_e\boldsymbol{b}\}^{1/2}.$$

12.5 Model Selection

12.5.1 AIC and CAIC

It is important in a polynomial growth curve model to decide its degree. More generally, we need to select a best subset from the set of explanatory variables. Related to such problems, we derive model selection criteria AIC and CAIC for the growth curve model (12.3.1) or (12.3.1), which is denoted by

$$M_{k;q}\text{: } \boldsymbol{\mu} = \mathrm{X}\boldsymbol{\theta}_i, \quad i = 1, \ldots, k. \tag{12.5.1}$$

The model in the special case $q = p$, which is denoted by M_k, is essentially a one-way MANOVA model. Using (7.3.15) and (7.3.16), its AIC and $CAIC$ are given by

$$\begin{aligned} AIC_k &= n \log |\hat{\Sigma}_0| + np(\log 2\pi + 1) + 2\left\{kp + \frac{1}{2}p(p+1)\right\}, \\ CAIC_k &= n \log |\hat{\Sigma}_0| + np(\log 2\pi + 1) \\ &\quad + \frac{2n}{n-k-p-1}\left\{kp + \frac{1}{2}p(p+1)\right\}, \end{aligned}$$

where $\hat{\Sigma}_0$ is the MLE of Σ under the no-restriction condition of $\boldsymbol{\mu}_1, \ldots, \boldsymbol{\mu}_k$. Let S be the pooled sample covariance matrix. Then $\hat{\Sigma}_0 = (m/n)S, m = n-k$.

Under $M_{k;q}$, we have seen in Section 12.4 that the MLEs of $\hat{\boldsymbol{\theta}}_i$ and Σ are

$$\hat{\boldsymbol{\theta}}_i = (\mathrm{X}'\mathbf{S}^{-1}\mathrm{X})^{-1}\mathrm{X}'\mathbf{S}^{-1}\bar{\boldsymbol{Y}}_i, \quad i = 1, \ldots, k,$$

$$\hat{\Sigma} = \frac{1}{n}\sum_{i=1}^{k}\sum_{j=1}^{n_i}(\boldsymbol{Y}_{ij} - \mathrm{X}\hat{\boldsymbol{\theta}}_i)(\boldsymbol{Y}_{ij} - \mathrm{X}\hat{\boldsymbol{\theta}}_i)'.$$

Therefore, we have

$$AIC = n\log|\hat{\Sigma}| + np(\log 2\pi + 1) + 2\left\{kq + \frac{1}{2}p(p+1)\right\}. \tag{12.5.2}$$

A corrected version is

$$CAIC = n\log|\hat{\Sigma}| + np(\log 2\pi + 1) + b_c, \tag{12.5.3}$$

where

$$\begin{aligned} b_c = -np &+ \frac{(p-q)n^2}{n-p+q-1} \\ &+ \frac{qn(n-k-1)(n+k)}{(n-k-p-1)(n-k-p+q-1)}. \end{aligned}$$

A general bias term is defined by b_A in (7.3.14). The bias term above was obtained under the assumption that the true model is included in the present growth curve model.

It is also interesting in selection of the $(p-q)$-dimensional auxiliary variable $\boldsymbol{V} = (V_1, \ldots, V_{p-q})$ (see Section 12.4.1). For a discussion of this topic, see Grizzle and Allen (1969), and Fujikoshi and Rao (1991).

12.5.2 Derivation of CAIC

In this section we derive the bias term b_c, assuming that the present growth curve model includes the true model for the observations. Then, based on (7.3.14), we can write

$$b_c = \mathrm{E}_{\mathbf{Y}}\mathrm{E}_{\mathbf{Z}}[\operatorname{tr}\hat{\Sigma}^{-1}(\mathbf{Z} - \mathrm{A}\hat{\Theta}\mathrm{X}')(\mathbf{Z} - \mathrm{A}\hat{\Theta}\mathrm{X}')] - np, \tag{12.5.4}$$

where the distribution of $n \times p$ random matrix $\mathbf{Z}$ is the same as that of $\mathbf{Y}$ and is independent of $\mathbf{Y}$. Noting that

$$\mathbf{Z} - \mathrm{A}\hat{\Theta}\mathrm{X}' = \mathbf{Z} - \mathrm{A}\Theta\mathrm{X}' + \mathrm{A}(\hat{\Theta} - \Theta)\mathrm{X}'$$

and considering the expectation with respect to $\mathbf{Z}$,

$$b_c = n\mathrm{E}[\operatorname{tr}\hat{\Sigma}^{-1}\Sigma] + \mathrm{E}[\operatorname{tr}\hat{\Sigma}^{-1}\mathrm{X}(\Theta - \hat{\Theta})'\mathrm{A}'\mathrm{A}(\Theta - \hat{\Theta})\mathrm{X}'].$$

This quantity is shown to be the one in (12.5.3), based on the following lemma.

Lemma 12.5.1 *Under the growth curve model* (12.4.1) *it holds that*

$$(1)\ \mathrm{E}[\operatorname{tr}\hat{\Sigma}^{-1}\Sigma] = \frac{n(p-q)}{n-(p-q)-1} + c_0, \quad \text{and}$$
$$(2)\ \mathrm{E}[\operatorname{tr}\hat{\Sigma}^{-1}\mathrm{X}(\Theta-\hat{\Theta})'\mathrm{A}'\mathrm{A}(\Theta-\hat{\Theta})\mathrm{X}'] = kc_0,$$

where

$$c_0 = \frac{nq(n-k-1)}{(n-k-p-1)(n-k-p+q-1)}.$$

Proof. We use the transformation in (12.4.3) (i.e., $(\mathbf{U}\ \mathbf{V}) = \mathbf{Y}\mathrm{G} = \mathbf{Y}(\mathrm{G}_1\ \mathrm{G}_2)$) and the notations therein. Then

$$\begin{aligned} \operatorname{tr}\hat{\Sigma}^{-1}\Sigma &= \operatorname{tr}\hat{\Psi}^{-1}\Psi \\ &= \operatorname{tr}\hat{\Psi}_{11\cdot 2}^{-1}(\mathrm{I}_q\ -\hat{\Gamma}')\Psi(\mathrm{I}_q-\hat{\Gamma}')' + \operatorname{tr}\hat{\Psi}_{22}^{-1}\Psi_{22}. \end{aligned} \tag{12.5.5}$$

The last reduction is obtained by using

$$\hat{\Psi}^{-1} = \begin{pmatrix} \mathrm{O} & \mathrm{O} \\ \mathrm{O} & \hat{\Psi}_{22}^{-1} \end{pmatrix} + \begin{pmatrix} \mathrm{I}_q \\ -\hat{\Gamma} \end{pmatrix} \hat{\Psi}_{11\cdot 2}^{-1}(\mathrm{I}_q\ -\hat{\Gamma}').$$

Similarly,

$$\begin{aligned} &\operatorname{tr}\hat{\Sigma}^{-1}\mathrm{X}(\Theta-\hat{\Theta})'\mathrm{A}'\mathrm{A}(\Theta-\hat{\Theta})\mathrm{X}' \\ &\quad = \operatorname{tr}(\mathrm{T}'\hat{\Sigma}\mathrm{T})^{-1}\mathrm{T}'\mathrm{X}(\Theta-\hat{\Theta})'\mathrm{A}'\mathrm{A}(\Theta-\hat{\Theta})\mathrm{X}'\mathrm{T} \\ &\quad = \operatorname{tr}\hat{\Psi}_{11\cdot 2}^{-1}(\Theta-\hat{\Theta})'\mathrm{A}'\mathrm{A}(\Theta-\hat{\Theta}). \end{aligned} \tag{12.5.6}$$

Let

$$\begin{aligned} \mathbf{W} &= (\mathbf{U}\ \mathbf{V})'(\mathrm{I}_n-\mathrm{P_A})(\mathbf{U}\ \mathbf{V}) = (W_{ij}), \\ \mathbf{L} &= (\mathrm{A}'\mathrm{A})^{-1/2}\mathrm{A}'(\mathbf{U}-\mathrm{A}\Theta\ \mathbf{V}) = (\mathbf{L}_1\ \mathbf{L}_2). \end{aligned}$$

Then $\mathbf{W}$ and $\mathbf{L}$ are mutually independent and are distributed as $\mathrm{W}_{n-k}(n-k,\Psi)$ and $\mathrm{N}_{k\times p}(\mathrm{O},\Psi\otimes\mathrm{I}_n)$, respectively. We have seen (see Section 12.5.4) that the MLEs of Ψ and Θ are expressed in terms of $\mathbf{W}$ and $\mathbf{L}$ as

$$\begin{aligned} &n\hat{\Psi}_{11\cdot 2} = \mathbf{W}_{11\cdot 2}, \quad \hat{\Gamma} = \mathbf{W}_{22}^{-1}\mathbf{W}_{21}, \\ &n\hat{\Psi}_{22} = \mathbf{W}_{22} + \mathbf{L}_2'\mathbf{L}_2 = \tilde{\mathbf{W}}_{22}, \\ &\hat{\Theta} - \Theta = (\mathrm{A}'\mathrm{A})^{-1/2}(\mathbf{L}_1, -\mathbf{L}_2\mathbf{W}_{22}^{-1}\mathbf{W}_{21}). \end{aligned}$$

Therefore, from (12.5.5) and (12.5.6) we obtain

$$\begin{aligned} &\mathrm{E}[\operatorname{tr}\hat{\Sigma}^{-1}\Sigma] = \mathrm{E}[n\operatorname{tr}\tilde{\mathbf{W}}_{22}^{-1}\Psi_{22}] + c, \\ &\mathrm{E}[\operatorname{tr}\hat{\Sigma}^{-1}\mathrm{X}'(\Theta-\hat{\Theta})'\mathrm{A}'\mathrm{A}(\Theta-\hat{\Theta})\mathrm{X}] \\ &\quad = \mathrm{E}[n\operatorname{tr}\mathbf{W}_{11\cdot 2}^{-1}(\mathrm{I}_q, -\mathbf{W}_{12}\mathbf{W}_{22}^{-1})\mathbf{L}(\mathrm{I}_q, -\mathbf{W}_{12}\mathbf{W}_{22}^{-1})' = kc, \end{aligned}$$

where

$$c = n\mathrm{E}[\,\mathrm{tr}\,\mathbf{W}_{11\cdot 2}^{-1}(\mathrm{I}_q, -\mathbf{W}_{12}\mathbf{W}_{22}^{-1})\Psi(\mathrm{I}_q, -\mathbf{W}_{12}\mathbf{W}_{22}^{-1})'].$$

Since $\tilde{\mathbf{W}}_{22} \sim \mathrm{W}_{p-q}(n, \Psi_{22})$, $\mathrm{E}[n\,\mathrm{tr}\,\tilde{\mathbf{W}}_{22}^{-1}\Psi_{22}] = n(p-q)/[n-(p-q)-1]$. In the computation of c, we may assume that $\Psi = \mathrm{I}_p$, and hence

$$\begin{aligned}&\mathbf{W}_{11\cdot 2} \sim \mathrm{W}_q(n-k-(p-q), \mathrm{I}_q), \quad \mathbf{W}_{22} \sim \mathrm{W}_q(n-k, \mathrm{I}_q),\\ &\text{the elements of } \mathbf{H} = \mathbf{W}_{12}\mathbf{W}_{22}^{-1/2} \text{i.i.d. } \mathrm{N}(0,1).\end{aligned}$$

Therefore,

$$\begin{aligned}c &= n\mathrm{E}[\,\mathrm{tr}\,\mathbf{W}_{11\cdot 2}^{-1}\{\mathrm{I}_q + \mathbf{H}\mathbf{W}_{22}^{-1}\mathbf{H}'\}]\\ &= \frac{n}{n-k-(p-q)-q-1}\,\mathrm{tr}\left(\mathrm{I}_q + \frac{1}{n-k-(p-q)-1}\mathrm{E}(\mathbf{H}\mathbf{H}')\right)\\ &= \frac{n}{n-k-(p-q)-q-1}\left\{q + \frac{q(p-q)}{n-k-(p-q)-1}\right\}\end{aligned}$$

which is c. This proves (1) and (2). □

12.5.3 Extended Growth Curve Model

In Example 5.3.1 the dental data were treated as a second-degree growth curve model. However, the coefficient of the second degree was not significant for girls' data, but was significant for boys' data. This suggests using a linear model for girls' data and a quadratic model for boys' data.

In general, we consider a model such that the within-individual design matrix of the first k_1 groups is a $p \times q_1$ matrix X_1 and the within-individual design matrix of the remainder k_2 groups is a $p \times q_2$ matrix X_2, where X_1 is equal to the first q_1 columns of X_2. This model is not expressed as a growth curve model; such a model is an extended growth curve model. For various extended models, see Verbyla and Venables (1988). In the example, note that the within-individual design matrices X_1 and X_2 are hierarchical. The growth curve model with hierarchical within-individual design matrices has been studied by Fujikoshi et al. (1999, 2006).

Based on the motivation above, suppose that the k populations are classified into two types of populations: the first and second hierarchies. These populations are expressed as

$$\begin{aligned}&\Pi_{11} : \mathrm{N}_p(\boldsymbol{\mu}_{11}, \Sigma), \ \ldots, \ \Pi_{1k_1} : \mathrm{N}_p(\boldsymbol{\mu}_{1k_1}, \Sigma),\\ &\Pi_{21} : \mathrm{N}_p(\boldsymbol{\mu}_{21}, \Sigma), \ \ldots, \ \Pi_{2k_2} : \mathrm{N}_2(\boldsymbol{\mu}_{2k_2}, \Sigma),\end{aligned}$$

where $k_1 + k_2 = k$. Suppose that there are n_{ij} samples for Π_{ij}, and let $\bar{\boldsymbol{Y}}_{ij}$ and $\mathbf{S}_{ij}$ be the sample mean vectors and the sample covariance matrices. The sample covariance matrices of the first hierarchy, the second hierarchy, and all the populations are given by

$$\mathbf{S}_1 = \frac{1}{m_1}\{(n_{11}-1)\mathbf{S}_{11} + \cdots + (n_{1r_1}-1)\mathbf{S}_{1k_1}\},$$
$$\mathbf{S}_2 = \frac{1}{m_2}\{(n_{21}-1)\mathbf{S}_{21} + \cdots + (n_{2r_1}-1)\mathbf{S}_{2k_2}\},$$
$$\mathbf{S} = \frac{1}{m}\{(m_1-1)\mathbf{S}_1 + (m_2-1)\mathbf{S}_2\},$$

where

$$n_1 = n_{11} + \cdots + n_{1k_1}, \quad n_2 = n_{21} + \cdots + n_{2k_1}, \quad n = n_1 + n_2,$$
$$m_1 = n_1 - k_1, \quad m_2 = n_2 - k_2, \quad m = m_1 + m_2 = n - k.$$

Now we consider an extended growth curve model defined by $M_{k_1,k_2;q_1,q_2}$:

$$\begin{aligned}\boldsymbol{\mu}_{11} &= \mathrm{X}_1\boldsymbol{\theta}_{11}, \ldots, \boldsymbol{\mu}_{1k_1} = \mathrm{X}_1\boldsymbol{\theta}_{1k_1},\\ \boldsymbol{\mu}_{21} &= \mathrm{X}_2\boldsymbol{\theta}_{21}, \ldots, \boldsymbol{\mu}_{2k_2} = \mathrm{X}_2\boldsymbol{\theta}_{2k_2},\end{aligned} \tag{12.5.7}$$

where the regression coefficient vectors $\boldsymbol{\theta}_{11}, \ldots, \boldsymbol{\theta}_{1k_1}$ are $q_1 \times 1$, $\boldsymbol{\theta}_{21}, \ldots, \boldsymbol{\theta}_{2k_2}$ are $q_2 \times 1$, and these are unknown. Here, it is assumed that X_1 is equal to the first $p \times q_1$ submatrix of X_2.

Defining AIC for the model (12.5.7), some notations are introduced. Let B be a $p \times p$ matrix such that

$$\mathrm{B} = (\mathrm{B}_1\ \mathrm{B}_2\ \mathrm{B}_3),\ \mathrm{B}_1 : p \times p_1,\ \mathrm{B}_2 : p \times p_2,\ \mathrm{B}_3 : p \times p_3 \tag{12.5.8}$$

and $(\mathrm{B}_1\ \mathrm{B}_2) = \mathrm{X}_2(\mathrm{X}_2'\mathrm{X}_2)^{-1}$, $\mathrm{X}_2'\mathrm{B}_3 = O$, where $p_1 = q_1, p_2 = q_2 - q_1, p_3 = p - q_2$, and B_3 satisfies $\mathrm{B}_3'\mathrm{B}_3 = \mathrm{I}_{p_3}$. The pooled sample covariance matrix of $\mathrm{B}'\boldsymbol{Y}$ is

$$\mathbf{V}^{(0)} = (\mathrm{B}_1\ \mathrm{B}_2\ \mathrm{B}_3)'m\mathbf{S}(\mathrm{B}_1\ \mathrm{B}_2\ \mathrm{B}_3) = \begin{pmatrix} \mathbf{V}_{11}^{(0)} & \mathbf{V}_{12}^{(0)} & \mathbf{V}_{13}^{(0)} \\ \mathbf{V}_{21}^{(0)} & \mathbf{V}_{22}^{(0)} & \mathbf{V}_{23}^{(0)} \\ \mathbf{V}_{31}^{(0)} & \mathbf{V}_{32}^{(0)} & \mathbf{V}_{33}^{(0)} \end{pmatrix}.$$

Set

$$\begin{aligned}\mathbf{V}^{(1)} &= \mathbf{V}^{(0)} + (\mathrm{B}_1\ \mathrm{B}_2\ \mathrm{B}_3)'\sum_{i=1}^{k_1} n_{1i}\bar{\boldsymbol{Y}}_{1i}\bar{\boldsymbol{Y}}_{1i}'(\mathrm{B}_1\ \mathrm{B}_2\ \mathrm{B}_3)\\ &= \begin{pmatrix} \mathbf{V}_{11}^{(1)} & \mathbf{V}_{12}^{(1)} & \mathbf{V}_{13}^{(1)} \\ \mathbf{V}_{21}^{(1)} & \mathbf{V}_{22}^{(1)} & \mathbf{V}_{23}^{(1)} \\ \mathbf{V}_{31}^{(1)} & \mathbf{V}_{32}^{(1)} & \mathbf{V}_{33}^{(1)} \end{pmatrix}\end{aligned}$$

$$\mathbf{V}^{(2)} = \mathbf{V}^{(2)} + (\mathrm{B}_1\ \mathrm{B}_2\ \mathrm{B}_3)' \sum_{j=1}^{k_2} n_{2j} \bar{\boldsymbol{Y}}_{2j} \bar{\boldsymbol{Y}}'_{2j} (\mathrm{B}_1\ \mathrm{B}_2\ \mathrm{B}_3)$$
$$= \begin{pmatrix} \mathbf{V}_{11}^{(2)} & \mathbf{V}_{12}^{(2)} & \mathbf{V}_{13}^{(2)} \\ \mathbf{V}_{21}^{(2)} & \mathbf{V}_{22}^{(2)} & \mathbf{V}_{23}^{(2)} \\ \mathbf{V}_{31}^{(2)} & \mathbf{V}_{32}^{(2)} & \mathbf{V}_{33}^{(2)} \end{pmatrix}.$$

Or, in terms of matrix notation, we can rewrite them as

$$\mathbf{V}_{ij}^{(0)} = \mathrm{B}_i' m\mathbf{S}\mathrm{B}_j, \quad \mathbf{V}_{ij}^{(1)} = \mathbf{V}_{ij}^{(0)} + \mathrm{B}_i' \sum_{i=1}^{k_1} n_{1i} \bar{\boldsymbol{Y}}_{1i} \bar{\boldsymbol{Y}}'_{1i} \mathrm{B}_j,$$
$$\mathbf{V}_{ij}^{(2)} = \mathbf{V}_{ij}^{(1)} + \mathrm{B}_i' \sum_{j=1}^{k_2} n_{2j} \bar{\boldsymbol{Y}}_{2j} \bar{\boldsymbol{Y}}'_{2j} \mathrm{B}_j.$$

Then AIC is

$$AIC_{k_1,k_2;q_1\cdot q_2} = n \log \left\{ \left| \frac{1}{n} \mathbf{V}_{11\cdot 23}^{(0)} \right| \cdot \left| \frac{1}{n} \mathbf{V}_{22\cdot 3}^{(1)} \right| \cdot \left| \frac{1}{n} \mathbf{V}_{33}^{(2)} \right| \right\}$$
$$+ np(\log(2\pi) + 1) + n \log |\mathrm{X}_2'\mathrm{X}_2| + 2 \left\{ k_1 q_1 + k_2 q_2 + \frac{1}{2} p(p+1) \right\},$$

where

$$\mathbf{V}_{11\cdot 23}^{(0)} = \mathbf{V}_{11}^{(0)} - (\mathbf{V}_{12}^{(0)}\ \mathbf{V}_{13}^{(0)}) \begin{pmatrix} \mathbf{V}_{22}^{(0)} & \mathbf{V}_{23}^{(0)} \\ \mathbf{V}_{32}^{(0)} & \mathbf{V}_{33}^{(0)} \end{pmatrix}^{-1} \begin{pmatrix} \mathbf{V}_{21}^{(0)} \\ \mathbf{V}_{31}^{(0)} \end{pmatrix},$$
$$\mathbf{V}_{22\cdot 3}^{(1)} = \mathbf{V}_{22}^{(1)} - \mathbf{V}_{23}^{(1)} (\mathbf{V}_{33}^{(1)})^{-1} \mathbf{V}_{32}^{(1)}.$$

Its corrected version is given by [see Fujikoshi, 2003]

$$CAIC_{k_1,k_2;q_1\cdot q_2} = n \log \left\{ \left| \frac{1}{n} \mathbf{W}_{11\cdot 23}^{(0)} \right| \cdot \left| \frac{1}{n} \mathbf{W}_{22\cdot 3}^{(1)} \right| \cdot \left| \frac{1}{n} \mathbf{W}_{33}^{(2)} \right| \right\}$$
$$+ np(\log 2\pi + 1)) + n \log |\mathrm{X}_2'\mathrm{X}_2| + b_c,$$

where

$$b_c = -np + \frac{(p - q_2)n^2}{n - p + q_2 - 1}$$
$$+ \frac{q_1 n(n - k - 1)(n + k)}{(n - k - p - 1)(n - k - p + q_1 - 1)}$$
$$+ \frac{(q_2 - q_1)n(n - k_2 - 1)(n + k_2)}{(n - k_2 - p + q_1 - 1)(n - k_2 - p + q_2 - 1)}.$$

Example 12.5.1 We consider three polynomial growth curve models for the dental data in Table 12.3.1. Let girls' and boys' data be denoted by G and B, respectively. The models considered are

$M_{(1,1;2,2)}$: G and B are linear,
$M_{(1,1;3,3)}$: G and B are quadratic,
$M_{(1,1;2,3)}$: G is linear and B is quadratic.

For these models, AIC and $CAIC$ are given as follows:

	AIC	CAIC
$M_{(1,1;2,2)}$	362	373
$M_{(1,1;3,3)}$	356	363
$M_{(1,1;2,3)}$	353	336

Both criteria show that $M_{(1,1;2,3)}$ is better than $M_{(1,1;2,2)}$ and $M_{(1,1;3,3)}$. The differences among $CAIC$s are larger than those among AICs.

Problems

12.1 Let $\hat{\boldsymbol{\theta}}$ be the MLE of $\boldsymbol{\theta}$ based on a sample of size n from $\mathrm{N}_p(\mathrm{X}\boldsymbol{\theta}, \Sigma)$. Then it is given as $\hat{\boldsymbol{\theta}} = (\mathrm{X}'\mathbf{S}^{-1}\mathrm{X})^{-1}\mathrm{X}'\mathbf{S}^{-1}\bar{\boldsymbol{Y}}$. Show that

$$\sqrt{n}(\hat{\boldsymbol{\theta}} - \boldsymbol{\theta}) = \boldsymbol{Z} - \boldsymbol{U},$$

where

$$\boldsymbol{Z} = (\mathrm{X}'\Sigma^{-1}\mathrm{X})^{-1}\mathrm{X}'\Sigma^{-1}\boldsymbol{V}, \quad \boldsymbol{V} = \sqrt{n}(\bar{\mathrm{Y}} - \mathrm{X}\boldsymbol{\theta}),$$
$$\boldsymbol{U} = \mathbf{Q}\boldsymbol{V}, \quad \mathbf{Q} = (\mathrm{X}'\mathbf{S}^{-1}\mathrm{X})^{-1}\mathrm{X}'\mathbf{S}^{-1}\left[\mathrm{I}_p - \mathrm{X}(\mathrm{X}'\Sigma^{-1}\mathrm{X})^{-1}\mathrm{X}'\Sigma^{-1}\right]$$

Further, prove the following results: (1) $\boldsymbol{Z}$ and $\boldsymbol{U}$ are independent.
(2) $\boldsymbol{Z}$ is distributed as $\mathrm{N}_p(\mathbf{0}, (\mathrm{X}'\Sigma^{-1}\mathrm{X})^{-1})$.
(3) The conditional distribution of $\boldsymbol{U}$ given $\mathbf{S}$ is multivariate normal with mean vector $\mathbf{0}$ and covariance matrix $\mathbf{Q}\Sigma\mathbf{Q}'$.

12.2 Consider k growth curve populations Π_i : $\mathrm{N}_p(X\boldsymbol{\theta}_i, \Sigma)$ on a random vector $\boldsymbol{Y}$, and suppose that there are samples of size n_i from $\Pi_i, i = 1, \ldots, k$. Let $\boldsymbol{Y}$ transform as

$$\begin{pmatrix} \boldsymbol{U} \\ \boldsymbol{Z} \end{pmatrix} = \begin{pmatrix} \mathrm{G}_1' \\ \mathrm{G}_2' \end{pmatrix} \boldsymbol{Y}$$

where $G_1 = X(X'X)^{-1}$ and $G_2 = \tilde{X}$ and G_2 is a $p \times (p-q)$ matrix such that $G_2'X = O$ and $G_2'G_2 = I_{p-q}$. Show that the transformed populations are given as $N(\boldsymbol{\zeta}_i, \Psi), i = 1, \ldots, k$,where

$$\boldsymbol{\zeta}_i = \begin{pmatrix} \zeta_{1i} \\ \zeta_{2i} \end{pmatrix} = \begin{pmatrix} \boldsymbol{\theta}_i \\ \mathbf{0} \end{pmatrix}, \quad \Psi = \begin{pmatrix} G_1' \\ G_2' \end{pmatrix} \Sigma \ (G_1 \ G_2).$$

12.3 Let B be a $p \times p$ matrix expressed as $B = X'X$, where X is a $k \times p$ matrix. Let M be a $p \times p$ positive definite matrix, and put $T = B + W$. Along to a parttion of $X = (X_1 \ X_2)$, $X_1 : \ k \times q$, we partition B, W and T as

$$T = \begin{pmatrix} B_{11} & B_{12} \\ B_{21} & B_{22} \end{pmatrix} + \begin{pmatrix} W_{11} & W_{12} \\ W_{21} & W_{22} \end{pmatrix} = \begin{pmatrix} T_{11} & T_{12} \\ T_{21} & T_{22} \end{pmatrix}.$$

Show that

$$\begin{aligned} T_{22\cdot1} - W_{22\cdot1} &= (X_2 - X_1W_{11}^{-1}W_{12})'(I_k + X_1W_{11}^{-1}X_1')^{-1} \\ &\quad \times (X_2 - X_1W_{11}^{-1}W_{12}). \end{aligned}$$

Hint: Use $(I_k + X_1W_{11}^{-1}X_1')^{-1} = I_k - X_1(X_1'X_1 + W_{11})^{-1}X_1'$.

12.4 (Continuation) Let W and T be the SSP matrices due to within-group variation and the tatal variation based on the transformed observations, and decompose them as

$$W = \begin{pmatrix} W_{uu} & W_{uv} \\ W_{vu} & W_{vv} \end{pmatrix}, \quad T = \begin{pmatrix} T_{uu} & T_{uv} \\ T_{vu} & T_{vv} \end{pmatrix},$$

where $W_{uv}; q \times (p-q)$ and $T_{uv}; q \times (p-q)$. Then, show that the likelihood ratio criterion is a monoton function of

$$\Lambda_{u\cdot v} = \frac{|W_{uu\cdot v}|}{|T_{uu\cdot v}|} = \frac{|W|}{|W_{uu}|} \left(\frac{|T|}{|T_{uu}|} \right)^{-1},$$

where $W_{uu\cdot v} = W_{uu} - W_{uv}W_{vv}^{-1}W_{vu}$ and $T_{uu\cdot v} = T_{uu} - T_{uv}T_{vv}^{-1}T_{vu}$.

12.5 (Continuation)
(1) Show that the null distribution of $\Lambda_{u\cdot v}$ is $\Lambda_q(k-1, m-(p-q))$, where $m = n - (p-q)$ and n is the total sample size.
(2)Show that the likelihood ratio statistic $\Lambda_{u\cdot v}$ is the same as $\Lambda = |\mathbf{S}_e|/|\mathbf{S}_e + \mathbf{S}_h|$ in Theorem 12.4.3 with $C = (I_{k-1} \ \ -\mathbf{1}_k)$ and $D = I_q$.

Hint: For a proof of (2), use Problem 12.3.

12.6 Consider a test of adequancy for an extended growth curve model in Section 12.5.3 under a growth curve model. For two populations $\Pi_i : \mathrm{N}(\boldsymbol{\mu}_i, \Sigma)$, assume a growth curve model given by

$$M: \ \boldsymbol{\mu}_i = \mathrm{X}\boldsymbol{\theta}_i = (\mathrm{X}_1\ \mathrm{X}_2)\begin{pmatrix} \boldsymbol{\theta}_{i1} \\ \boldsymbol{\theta}_{i2} \end{pmatrix},$$

where $\mathrm{X}_1 : \ p \times q_1$ and $\mathrm{X}_2 : \ p \times q_2$. Based on the random samples $\Pi_i : \boldsymbol{Y}_{i1}, \cdots, \boldsymbol{Y}_{n_i}$, $i = 1, 2$, consider testing the hypothesis $M_0 : \ \boldsymbol{\theta}_{12} = \mathbf{0}$ against the alternative M.

(1) Let $\mathbf{Y} = (\boldsymbol{Y}_{11}, \ldots, \boldsymbol{Y}_{1n_1}, \boldsymbol{Y}_{21}, \ldots, \boldsymbol{Y}_{2n_2})'$. Then, confirm that

$$\begin{aligned} \mathrm{E}(\mathbf{Y}) &= \begin{pmatrix} \mathbf{1}_{n_1} & \mathbf{0} \\ \mathbf{0} & \mathbf{1}_{n_2} \end{pmatrix}\begin{pmatrix} \boldsymbol{\theta}'_{11} & \boldsymbol{\theta}'_{12} \\ \boldsymbol{\theta}'_{21} & \boldsymbol{\theta}'_{22} \end{pmatrix}\begin{pmatrix} \mathrm{X}'_1 \\ \mathrm{X}'_2 \end{pmatrix} \\ &= \mathrm{A}\Theta\mathrm{X}. \end{aligned}$$

(2) In order to obtain a canonical form for the above testing problem, consider a transformation $\mathbf{Z} = \mathrm{H}'\mathbf{Y}\mathrm{B}$, where $\mathrm{H} = (\boldsymbol{h}_1\ \boldsymbol{h}_2\ H_3)$ is an orthogonal matrix such that $(\boldsymbol{h}_1\ \boldsymbol{h}_2) = \mathrm{A}(\mathrm{A}'\mathrm{A})^{-1/2}$, and $\mathrm{B} = (\mathrm{B}_1\ \mathrm{B}_2\ \mathrm{B}_3)$ is a nonsigular matrix such that $(\mathrm{B}_1\ \mathrm{B}_2) = \mathrm{X}(\mathrm{X}'\mathrm{X})^{-1/2}$, $\mathrm{B}_i : p \times q_i$, $i = 1, 2$, and B_3 is a matrix satisfying $\mathrm{X}'\mathrm{B}_3 = \mathrm{O}$ and $\mathrm{B}'_3\mathrm{B}_3 = \mathrm{I}_{p-q}$, $q = q_1 + q_2$. Then, show that

$$\mathrm{E}(\mathbf{Z}) = \mathrm{E}[\begin{pmatrix} \boldsymbol{Z}'_{11} & \boldsymbol{Z}'_{12} & \boldsymbol{Z}'_{13} \\ \boldsymbol{Z}'_{21} & \boldsymbol{Z}'_{22} & \boldsymbol{Z}'_{23} \\ \boldsymbol{Z}_{31} & \mathbf{Z}_{32} & \mathbf{Z}_{33} \end{pmatrix}] = \begin{pmatrix} \boldsymbol{\xi}'_{11} & \boldsymbol{\xi}'_{12} & \mathbf{0}' \\ \boldsymbol{\xi}'_{21} & \boldsymbol{\xi}'_{22} & \mathbf{0}' \\ \mathrm{O} & \mathrm{O} & \mathrm{O} \end{pmatrix},$$

where $\boldsymbol{\xi}_{ij} = \sqrt{n_i}\boldsymbol{\theta}_{ij}$, $i, j = 1, 2$. The hypothesis M_0 is equivalent to $\boldsymbol{\xi}_{12} = \mathbf{0}$.

(3) Set

$$\begin{aligned} \mathbf{W} &= (\boldsymbol{Z}_{31}\ \mathbf{Z}_{32}\ \mathbf{Z}_{33})'(\boldsymbol{Z}_{31}\ \mathbf{Z}_{32}\ \mathbf{Z}_{33}) + (\boldsymbol{Z}'_{11}\ \boldsymbol{Z}'_{12}\ \boldsymbol{Z}'_{13})'(\boldsymbol{Z}'_{11}\ \boldsymbol{Z}'_{12}\ \boldsymbol{Z}'_{13}), \\ \mathbf{T} &= \mathbf{W} + (\boldsymbol{Z}'_{21}\ \boldsymbol{Z}'_{22}\ \boldsymbol{Z}'_{23})'(\boldsymbol{Z}'_{21}\ \boldsymbol{Z}'_{22}\ \boldsymbol{Z}'_{23}), \end{aligned}$$

and let them partition as

$$\mathbf{W} = \begin{pmatrix} \mathbf{W}_{11} & \mathbf{W}_{12} & \mathbf{W}_{13} \\ \mathbf{W}_{21} & \mathbf{W}'_{22} & \boldsymbol{W}_{23} \\ \mathbf{W}_{31} & \mathbf{W}_{32} & \mathbf{W}_{33} \end{pmatrix} \text{ and } \mathbf{T} = \begin{pmatrix} \mathbf{T}_{11} & \mathbf{T}_{12} & \mathbf{T}_{13} \\ \mathbf{T}_{21} & \mathbf{T}'_{22} & \boldsymbol{T}_{23} \\ \mathbf{T}_{31} & \mathbf{T}_{32} & \mathbf{T}_{33} \end{pmatrix}.$$

Prove that the likelihood ratio test is given by

$$\Lambda = |\mathbf{W}_{22\cdot 3}|/|\mathbf{T}_{22\cdot 3}|,$$

where $\mathbf{W}_{22\cdot 3} = \mathbf{W}_{22} - \mathbf{W}_{23}\mathbf{W}_{33}^{-1}\mathbf{W}_{32}$, and $\mathbf{T}_{22\cdot 3} = \mathbf{T}_{22} - \mathbf{T}_{23}\mathbf{T}_{33}^{-1}\mathbf{T}_{32}$.

(4) Show that the null distribution of Λ is $\Lambda_{q_3}(1, m + 1 - (p - q))$, where $m = n_1 + n_2 - 2$.

Chapter 13

Approximation to the Scale-Mixted Distributions

Chapters 13 through 16 are concerned with explicit and computable upper error bounds for asymptotic approximations. In this chapter we give a general theory of approximation to the distributions of scale mixtures, including simple examples of Student's t-distribition to start with and the F-distribution as a scale mixtures of the normal and t-distribution, respectively. Such scale mixtures appear as sampling distributions of various statistics such as the studentized version of some estimators. Errors of the approximation are evaluated in sup (uniform) and L_1-norms. The results are extended to some multivariate scale mixtures with error bounds evaluated in L_1-norm.

13.1 Introduction

13.1.1 Simple Example: Student's t-Distribution

We start with a simple and well-know example: Let $\{X_1, \ldots, X_{\mathrm{N}}\}$ be a sample of size N from the normal distribution $\mathrm{N}(0, \sigma^2)$. Then:

1. The sample mean and sample variance,

$$\bar{X} = \frac{\sum_{j=1}^{\mathrm{N}} X_j}{\mathrm{N}} \quad \text{and} \quad S^2 = \frac{1}{\mathrm{N}-1}\sum_{j=1}^{\mathrm{N}}(X_j - \bar{X})^2$$

   are mutually independent.

2. Normalized statistics

$$Z \equiv \sqrt{\mathrm{N}}\bar{X}/\sigma \quad \text{and} \quad \chi_n^2 \equiv (\mathrm{N}-1)S^2/\sigma^2$$

follow the standard normal distribution N(0, 1) and the chi-square distribution $\chi^2(n)$ with $n\,(= \mathrm{N}-1)$ degrees of freedom, respectively.

3. Unbiased estimate S^2 of σ^2 is consistent in the sense that

$$\Pr\{\lim_{n\to\infty} \chi_n^2/n = 1\} = \Pr\{\lim_{n\to\infty} S^2 = \sigma^2\} = 1. \tag{13.1.1}$$

4. The distribution $F(x;n)$ of the ratio

$$T \equiv \frac{Z}{\sqrt{\chi_n^2/n}} \left(= \frac{\sqrt{\mathrm{N}}\bar{X}}{S} \right) \tag{13.1.2}$$

is called Student's *t-distribution* with $n\,(= \mathrm{N}-1)$ degrees of freedom, which has the density

$$f(x;n) = \frac{\Gamma((n+1)/2)}{\sqrt{n\pi}\,\Gamma(n/2)} \left(1 + \frac{x^2}{n}\right)^{-(n+1)/2}. \tag{13.1.3}$$

The property (13.1.1) as well as the functional form of the density $f(x;n)$ suggests that as n increases, the distribution $F(x;n)$ will be approximated by the standard normal distribution $\Phi(x)$ with the density

$$\varphi(x) = \frac{1}{\sqrt{2\pi}} \mathrm{e}^{-x^2/2}.$$

This is true, first of all, in the following sense (in the sense of sup-norm):

$$\Delta(n) \equiv \sup |F(x;n) - \Phi(x)| \to 0 \qquad \text{as } n \to \infty. \tag{13.1.4}$$

We can prove an even stronger convergence statement (in the sense of the L_1-norm):

$$\begin{aligned} \vartriangle(n) &\equiv \frac{1}{2}\|f(x;n) - \varphi(x)\|_1 \\ &= \frac{1}{2}\int_{-\infty}^{\infty} |f(x;n) - \varphi(x)|\,\mathrm{d}x \to 0 \qquad \text{as } n \to \infty. \end{aligned} \tag{13.1.5}$$

To make the difference between (13.1.4) and (13.1.5) clear, we look at the problem in terms of the random variables Z and $Z/\sqrt{\chi_n^2/n}$. Note that condition (13.1.4) is tantamount to

$$\left|\Pr\{Z/\sqrt{\chi_n^2/n} \le x\} - \Pr\{Z \le x\}\right| \le \Delta(n) \to 0 \text{ for all } x,$$

while (13.1.5) implies the following convergence, which holds for all Borel subsets A of $\mathrm{R} = (-\infty, \infty)$:

$$\left|\Pr\{Z/\sqrt{\chi_n^2/n} \in \mathrm{A}\} - \Pr\{Z \in \mathrm{A}\}\right| \le \vartriangle(n) \to 0,$$

from which follows , in particular, that for any real Borel function $\xi(x)$ on R we have

$$\left|\Pr\{\xi(Z/\sqrt{\chi_n^2/n}) \le x\} - \Pr\{\xi(Z) \le x\}\right| \le \vartriangle(n) \to 0.$$

Despite this strong convergence result, we still have problems. Our concerns about the convergence is the accuracy of approximation [i.e., how quickly the errors $\Delta(n)$ and $\vartriangle(n)$ go to zero as n increases]. Table 13.1.1 provides some information about the question.

Table 13.1.1 Numerical values of $\Delta(n)$ and $\vartriangle(n)$

n	10	20	30	50	100
$\Delta(n)$	0.1555 (−1)	0.7844 (−2)	0.5245 (−2)	0.3154 (−2)	0.1580 (−2)
$\vartriangle(n)$	0.6204 (−1)	0.3137 (−1)	0.2098 (−1)	0.1262 (−1)	0.6319 (−2)

$(-p)$ stands for $\cdot 10^{-p}$.

13.1.2 Improving the Approximation

We can even improve the approximation: Assume without loss of generality that $\sigma^2 = 1$ and note that the variable is a ratio of two independent variables, i.e. $T = Z/S$. Then, given $S = s$, the conditional probability of the event $T \le x$ is

$$\Pr\{T \le x | S = s\} = \Pr\{X \le xs\} = \Phi(xs).$$

As we are interested in those values of s that are close to 1, we would approximate $\Phi(xs)$ by expanding it as a function of s^2 (not of s itself in this case) around $s^2 = 1$. To this end, we write

$$y = s^2 \quad \text{or} \quad s = y^{1/2}, \quad \text{and} \quad \eta(y;\, x) \equiv \Phi(xy^{1/2}).$$

Then we have

$$\eta'(y;\, x) = \frac{1}{2} x y^{-1/2} \varphi(xy^{1/2}) \tag{13.1.6}$$

$$\eta''(y;\, x) = -\frac{1}{4}\left(x^3 y^{-1/2} + x y^{-3/2}\right)\varphi(xy^{1/2}), \tag{13.1.7}$$

and the Taylor expansion of $\eta(y;\, x)$ around $y = 1$ up to the third term becomes

$$\begin{aligned}\eta_3(y;\, x) &\equiv \eta(1;\, x) + \eta'(1;\, x)(y-1) + \frac{1}{2}\eta''(1;\, x)(y-1)^2\\ &= \Phi(x) + \Big(\frac{1}{2}x(y-1) - \frac{1}{8}(x^3+x)(y-1)^2\Big)\varphi(x),\\ &= \Phi(x) + \Big(\frac{1}{2}x(s^2-1) - \frac{1}{8}(x^3+x)(s^2-1)^2\Big)\varphi(x).\end{aligned}$$

As

$$\mathrm{E}(S^2-1) = \mathrm{E}(\chi_n^2/n-1) = 0 \quad \text{and} \quad \mathrm{E}(S^2-1)^2 = \mathrm{Var}(\chi_n^2/n) = 2/n,$$

we can use

$$\Phi_3(x;n) = \Phi(x) - \frac{1}{4n}(x^3+x)\varphi(x), \tag{13.1.8}$$

for an approximation to $F(x;n)$ and the pdf

$$\varphi_3(x;n) = \varphi(x) + \frac{1}{4n}(x^4-2x^2-1)\varphi(x), \tag{13.1.9}$$

obtained by differentiating both sides of (13.1.8) for an approximation to the density $f(x;n)$. For the numerical values of the errors

$$\Delta_3(n) = \sup|F(x;n) - \Phi_3(x;n)|$$

and

$$\vartriangle_3(n) = \|f(x;n) - \varphi_3(x;n)\|_1$$

see Table 13.1.2.

Table 13.1.2 Numerical values of $\Delta_3(n)$ and $\vartriangle_3(n)$

n	10	20	30	50	100
$\Delta_3(n)$	0.1999 (−2)	0.5308 (−3)	0.2406 (−3)	0.8796 (−4)	0.2225 (−4)
$\vartriangle_3(n)$	0.9424 (−2)	0.2509 (−2)	0.1136 (−2)	0.4147 (−3)	0.1047 (−3)

$(-p)$ stands for $\cdot 10^{-p}$.

The purpose of the present chapter is to develop a theory of approximating the distribution of the scale mixture of some underlying distribution G, i.e. the distribution of the variable X of the form

$$X = S^\delta Z, \tag{13.1.10}$$

where Z follows G, while S is a positive random variable independent of X and $\delta = \pm 1$ is a parameter used to distinguish two types of mixtures: $X = SZ$ and $X = Z/S$. A relatively wide class of sampling distributions can be expressed as a mixture of the standard normal or chi-square distribution. For example, a studentized statistic Z/V of the form (13.1.10) with $\delta = -1$, $Z = U/\sigma$, and $S = V/\sigma$ (where σ is an unknown population parameter, Z has a known distribution G and V^2 is an estimator of σ^2 which is independent of Z). Naturally, the distribution of V^2 depends on the sample size n and often tends to σ^2 as $n \to \infty$. On the other hand, a scale mixture appears as a basic statistical distribution. Then we are interested in the distance of the mixture from its parent distribution.

Moreover, we will see in Chapter 15 that some multivariate test statistics can be expressed as a function of a multivariate scale mixture given by

$$\boldsymbol{X} = \mathbf{S}^{\delta}\boldsymbol{Z} = \mathrm{diag}\,(S_1^{\delta}, \ldots, S_p^{\delta})\boldsymbol{Z}, \tag{13.1.11}$$

where the S_is are positive random variables and $\mathbf{S}$ is independent of $\boldsymbol{Z} = (Z_1, \ldots, Z_p)'$. It is assumed that the $Z_1, \ldots, Z_p$ are independent and identically distributed random variables. We present a theory of approximation to the distribution of $\boldsymbol{X}$ and their error bounds. The approximation is based on knowledge of the underlying distribution G and information about some moments of the variable S. We tacitly assume that S is close to 1 in some sense. Not only do we propose some approximating formulas for the cumulative distribution function (cdf) $F(x)$ and its density $f(x)$ of the scale mixture X, but we give the upper bounds of their possible errors evaluated in sup and L_1-norms and in terms of the moment $\mathrm{E}(S^{\rho} - 1)^k$, where ρ is a positive integer to be chosen suitably.

A great number of research into asymptotic expansions have been carried out, most of which was concerned with either the methods of finding the first few terms of asymptotic expansion or the validity of the expansion in the sense of order evaluation. However, the works on explicit and computable error bounds are not much in comparison with those on the derivation and validity, except for the Berry-Esseen bounds [see Berry, 1941 and Esseen, 1942] for the normalized sums of independent and identically distributed random variables. For error bounds for its first order expansion, see Shimizu (1974), Dobrić and Ghosh (1996).

On the other hand, there are considerable results on explicit and computable error bounds for the distributions related to scale mixtures. For scale mixtures, the bounds of the distance of the mixture from its parent distribution have been studied in Kleisen and Steutel (1974), Heyde (1975), Heyde and Leslie (1976), Hall (1979). Kariya and Maekawa (1982) studied error bounds for the distributions of generalized least squares estimators with applications to the seemingly unrelated regression model. See also Kariya and Kurata (2004). The topics in this chapter were treated in the papers, Fujikoshi (1985b), Fujikoshi (1987), Shimizu (1987), Shimizu (1989), Fujikoshi and Shimizu (1990), Fujikoshi (1993b), Shimizu (1995), Shimizu and Fujikoshi (1997), etc. However, we note that the constants $\beta_{\delta,k}$ and $\tau_{\delta,k}$ appeared in the bounds have been improved in this book.

13.2 Error Bounds Evaluated in sup-Norm

13.2.1 General Theory

If a random variable Z follows the distribution G with the cdf $G(x)$, the cdf of the scale mixture $X = S^{\delta}Z$ is given by

$$F(x) \equiv \Pr\{X \leq x\} = \Pr\{Z \leq xS^{-\delta}\} = \mathrm{E}_s[\,G(xS^{-\delta})\,]. \tag{13.2.1}$$

Assuming that the scale factor S is close to 1 in some sense, we consider approximating the cdf $F(x)$. Our interest also lies in evaluating possible errors of approximations. Throughout this chapter we assume without further notice that $G(x)$ is $k+1$ times continuously differentiable on its support $D = \{x \in \mathrm{R} : g(x) > 0\}$ and that the scale factor S as well as its multivariate counterpart (random matrix) $\mathbf{S}$ have moments of required order. For technical reasons, in the following, we use a random variable Y defined by

$$Y = S^{\rho} \quad \text{or} \quad S = Y^{1/\rho}, \tag{13.2.2}$$

where ρ is a suitably chosen positive integer.

Let $g(x)$ be the probability density function (pdf) of Z. For any $s > 0$, the conditional distribution of $X = S^{\delta}Z$ given $S = s (= y^{1/\rho})$ has a cdf that is expressible as $G(xy^{-\delta/\rho})$, and mathematical induction shows that its derivatives with respect to y can be put in the form

$$\frac{\partial^j G(xy^{-\delta/\rho})}{\partial y^j} = y^{-j}c_{\delta,j}(xy^{-\delta/\rho})g(xy^{-\delta/\rho}), \quad j = 1, 2, \ldots, k, \tag{13.2.3}$$

where the $c_{\delta,j}$ are real functions obtained by

$$c_{\delta,j}(x)g(x) = \left.\frac{\partial^j G(xy^{-\delta/\rho})}{\partial y^j}\right|_{y=1}. \tag{13.2.4}$$

We can use

$$G_{\delta,k}(x,y) = G(x) + \sum_{j=1}^{k-1} \frac{1}{j!}c_{\delta,j}(x)g(x)(y-1)^j \tag{13.2.5}$$

as an approximation to $G(xy^{-\delta/\rho})$, which would in turn induce an approximation to $F(x)$ by means of

$$G_{\delta,k}(x) \equiv \mathrm{E}(G_{\delta,k}(x,Y)) = G(x) + \sum_{j=1}^{k-1} \frac{1}{j!} c_{\delta,j}(x) g(x) \mathrm{E}[(Y-1)^j]$$

$$= G(x) + \sum_{j=1}^{k-1} \frac{1}{j!} c_{\delta,j}(x) g(x) \mathrm{E}[(S^\rho - 1)^j]. \tag{13.2.6}$$

For $x \in D$ and $y > 0$, write

$$\Delta_{\delta,k}(x,y) \equiv G(xy^{-\delta/\rho}) - G_{\delta,k}(x,y) \tag{13.2.7}$$

and

$$\mathsf{J}_{\delta,k}(x,y) \equiv \begin{cases} \dfrac{|G(xy^{-\delta/\rho}) - G_{\delta,k}(x,y)|}{|y-1|^k}, & \text{for } y \neq 1, \\[2ex] \dfrac{1}{k!} |c_{\delta,k}(x)|\, g(x), & \text{for } y = 1. \end{cases} \tag{13.2.8}$$

By applying Taylor's theorem to $G(xy^{-\delta/\rho})$ as a function of y and using (13.2.5), the remainder term $\Delta_{\delta,k}(x,y)$ can be put in the form

$$\Delta_{\delta,k}(x,y) = \frac{1}{k!} c_{\delta,k}(u) g(u) y_0^{-k} (y-1)^k, \tag{13.2.9}$$

where $u = xy_0^{-\delta/\rho}$, and y_0 is a positive number lying between 1 and y.

We define the positive constants $\alpha_{\delta,k}$ and $\beta_{\delta,k}$ by

$$\begin{aligned} \alpha_{-1,0} &= \max\{1 - G(0),\ G(0)\}, \\ \alpha_{\delta,k} &= \frac{1}{k!} \sup_x \left|c_{\delta,k}(x) g(x)\right|, \\ \beta_{\delta,k} &= \sup_{x \in D, y \leq 1} \mathsf{J}_{\delta,k}(x,y). \end{aligned} \tag{13.2.10}$$

It follows from (13.2.9) that we have the inequalities $\alpha_{\delta,k} \leq \beta_{\delta,k}$ and

$$\sup_x \left|\Delta_{\delta,k}(x,y)\right| \leq \begin{cases} \alpha_{\delta,k}\ \left(y \vee y^{-1} - 1\right)^k \\[1ex] \beta_{\delta,k}\ \left|y - 1\right|^k. \end{cases} \tag{13.2.11}$$

Definition 13.2.1 For a set of positive numbers α, β, s (possibly random), a positive integer k and $\rho = 1$ or 2, such that $(s^\rho - 1)^k$ is nonnegative (a.s.), let the quantity $\mathbf{\Delta}_k(\alpha, \beta;\, s, \rho)$ be defined by

$$\mathbf{\Delta}_k(\alpha, \beta;\, s, \rho) \equiv \begin{cases} \beta \cdot (s^\rho - 1)^k, & \text{and} \\[1ex] \alpha \cdot (s^\rho - 1)^k, & \text{if } s \geq 1 \ \text{(a.s.)}. \end{cases} \tag{13.2.12}$$

The following theorem is a direct consequence of the definition of quantities involved and the above observation.

Theorem 13.2.1 *Suppose that X is the scale mixture defined by* (13.1.10). *Then we have*

$$\sup_x |F(x) - G_{\delta,k}(x)| \leq \mathrm{E}(\mathbf{\Delta}_k(\alpha_{\delta,k}, \beta_{\delta,k};\, S, \rho)).$$

The case of $\delta = -1$ is worth noticing in that a simple bound $\beta_{-1,k} = \alpha_0$ is available for all k, provided that an additional condition on G is fulfilled.

Theorem 13.2.2 *Consider the case $\delta = -1$ under the same setup as in Theorem 13.2.1. Suppose further that G is analytic on D and that $(-1)^{j+1}c_{-1,j}(x)$ has the same sign as x for all x and j. Then we have*

$$|F(x) - G_{-1,k}(x)| \leq \alpha_{-1,0}\mathrm{E}|S^{\rho} - 1|^k. \tag{13.2.13}$$

The inequality cannot be improved.

Proof. As (13.2.9) implies that

$$\sup_x \sup_{y \geq 1} \mathsf{J}_{-1,k}(x, y) \leq \alpha_{-1,k}, \tag{13.2.14}$$

we have only to prove

$$\sup_x \sup_{0<y\leq 1} \mathsf{J}_{-1,k}(x, y) = \alpha_{-1,0} \geq \alpha_{-1,k}. \tag{13.2.15}$$

We first note that the following inequality holds:

$$|\Delta_{-1,k}(x, y)| \leq \alpha_{-1,0} \quad \text{for all} \quad x \quad \text{and} \quad 0 < y < 1. \tag{13.2.16}$$

In fact, if $x > 0$, then since $(-1)^{j+1}c_{-1,j}(x) \geq 0$, we have from definition (13.2.7) of $\Delta_{-1,k}(x, y)$ together with (13.2.5) and its representation (13.2.9),

$$\begin{aligned}
0 \leq -\Delta_{-1,k}(x, y) &= \frac{1}{k!}(-1)^{k+1}c_{-1,k}(u)g(u)y_0^{-k}(1-y)^k \\
&= G_{-1,k}(x, y) - G(xy^{1/\rho}) \\
&= G(x) - \sum_{j=1}^{k-1} \frac{(1-y)^j}{j!}(-1)^{j+1}c_{-1,j}(x)g(x) - G(xy^{1/\rho}) \\
&\leq G(x) - G(xy^{1/\rho}) \leq \alpha_{-1,0}.
\end{aligned}$$

Similarly, if $x < 0$, we have the inequality $0 \geq -\Delta_{-1,k}(x, y) \geq G(x) - G(xy^{1/\rho}) \geq -\alpha_{-1,0}$.

To prove (13.2.15), let $0 < \epsilon < 1$ be an arbitrary number. If $0 < y \le \epsilon$, then using the result (13.2.16) we have

$$\mathsf{J}_{-1,k}(x,y) \le (1-\epsilon)^{-k}|\Delta_{-1,k}(x,y)| \le (1-\epsilon)^{-k}\alpha_{-1,0} \quad \text{for} \quad 0 < y < \epsilon.$$

Suppose next that $\epsilon < y \le 1$. For a given $x \neq 0$, $G(xy^{1/\rho})$, as a function of y, can be developed in a series around $y = 1$:

$$\begin{aligned} G(xy^{1/\rho}) &= G(x) - \sum_{j=1}^{\infty} \frac{(1-y)^j}{j!}(-1)^{j+1}c_{-1,j}(x)g(x) \\ &= G_{-1,k}(x,y) - \sum_{j=k}^{\infty} \frac{(1-y)^j}{j!}(-1)^{j+1}c_{-1,j}(x)\,g(x). \end{aligned} \tag{13.2.17}$$

If $x > 0$, each term of the summand is nonnegative and

$$\begin{aligned} \mathsf{J}_{-1,k}(x,y) &= (1-y)^{-k}\sum_{j=k}^{\infty} \frac{(1-y)^j}{j!}(-1)^{j+1}c_{-1,j}(x)g(x) \\ &\le \sum_{j=k}^{\infty} \frac{(1-\epsilon)^{j-k}}{j!}(-1)^{j+1}c_{-1,j}(x)g(x) \\ &= (1-\epsilon)^{-k}|\Delta_{-1,k}(x,\epsilon)| \le (1-\epsilon)^{-k}\alpha_{-1,0}. \end{aligned} \tag{13.2.18}$$

The same argument can apply to the case $x < 0$ to obtain the same inequality. The only difference in its inference is that the terms of the summand are all nonpositive in this case, so that $c_{-1,j}(x)$ should be replaced by $-c_{-1,j}(x)$. As ϵ is arbitrary, we have proved the inequality $\mathsf{J}_{-1,k}(x,y) \le \alpha_{-1,0}$ which is to hold for all x and $0 < y \le 1$. On the other hand, we have, for any fixed x,

$$\sup_{0<y\le 1} \mathsf{J}_{-1,k}(x,y) \ge \mathsf{J}_{-1,k}(x,1) \equiv \frac{1}{k!}|c_{-1,k}(x)|g(x).$$

In particular, the definition of $\alpha_{-1,k}$ leads to

$$\sup_{x}\sup_{0<y\le 1} \mathsf{J}_{-1,k}(x,y) \ge \sup_{x} \frac{1}{k!}|c_{-1,k}(x)|g(x) = \alpha_{-1,k}.$$

As we have already proved the inequality $\sup_x \sup_{0<y\le 1} \mathsf{J}_{-1,k}(x,y) \le \alpha_{-1,0}$, we obtain $\alpha_{-1,0} \ge \alpha_{-1,k}$. Finally, we will prove the equality in (13.2.15). In fact, we have, from (13.2.5) and (13.2.7),

$$\begin{aligned} &\lim_{x\to\infty}\lim_{y\to 0} \mathsf{J}_{-1,k}(x,y) \\ &= |G(0) - \lim_{x\to\infty}\Big(G(x) - \sum_{j=1}^{k-1}\frac{(-1)^{j+1}}{j!}c_{-1,j}(x)g(x)\Big)| \\ &= 1 - G(0), \end{aligned}$$

and similarly,

$$\lim_{x\to-\infty}\lim_{y\to 0} \mathsf{J}_{-1,k}(x,y) = G(0),$$

as was to be proved. □

13.2.2 Scale - Mixed Normal

By taking $\rho = 2$, the previous results applied to the normal case are summarized in the following theorem. We write Φ to denote the distribution function of the standard normal distribution $\mathrm{N}(0,1)$ and φ its density.

For $j = 1, 2, \ldots$, let $Q_{\delta,2j-1}(x)$ be the polynomials defined by

$$Q_{1,2j-1}(x) = H_{2j-1}(x), \tag{13.2.19}$$

$$Q_{-1,2j-1}(x) = K_{2j-1}(x) \equiv \sum_{m=0}^{j-1}(2m-1)!!\binom{j-1}{m}x^{2(j-m)-1}, \tag{13.2.20}$$

where we use the notation $(2m-1)!! = (2m-1)\cdot(2m-3)\cdots 3\cdot 1$ with the convention $(-1)!! = 1$, and $H_j(x)$ is the Hermite polynomial:

$$H_j(x)\varphi(x) = (-1)^j \frac{d^j}{dx^j}\varphi(x).$$

The polynomials H and K for $1 \le j \le 6$ are given in Tables 13.2.1 and 13.2.2, respectively, (more about the K in Section 13.3.2).

Now we have:

Theorem 13.2.3 *The cdf $F(x)$ of the scale mixture $X = S^\delta Z$ of the* $\mathrm{N}(0,1)$ *is approximated by*

$$\Phi_{\delta,k}(x) \equiv \Phi(x) - \sum_{j=1}^{k-1}\frac{\delta^j}{2^j j!}Q_{\delta,2j-1}(x)\mathrm{E}(S^2-1)^j\varphi(x). \tag{13.2.21}$$

The error estimate corresponding to (13.2.1) *is*

$$\sup_x |F(x) - \Phi_{\delta,k}(x)| \le \mathrm{E}(\mathbf{\Delta}_k(\alpha_{\delta,k},\beta_{\delta,k};\, s, 2)). \tag{13.2.22}$$

Numerical values of $\alpha_{\delta,k}$ are given in Table 13.2.3 for $k \le 6$. We can take $\beta_{\delta,k} = 1/2$ for $\delta = 1$, $k \le 6$ and $\beta_{-1,k} = 1/2$ for all $k = 1, 2, \ldots$.

Proof. We consider the transform $S \to Y = S^2$. Let the real function $c_{\delta,j}(x)$ be defined by

$$c_{\delta,j}(x)\varphi(x) = \left.\frac{\partial^j}{\partial y^j}\Phi(xy^{-\delta/2})\right|_{y=1}. \tag{13.2.23}$$

Then writing

$$Q_{\delta,2j-1}(x) = -(2\delta)^j c_{\delta,j}(x),$$

we get (13.2.20). The defining formula (13.2.20) for $K_{2j-1}(x)$ follows from

$$\frac{\partial^j}{\partial y^j}\Phi(xy^{1/2}) = \frac{(-1)^{j-1}}{2^j}\sum_{m=0}^{j-1}(2m-1)!!\binom{j-1}{m}x^{2(j-m)-1}y^{-\frac{1}{2}-m}\varphi(xy^{1/2}),$$

which will be proved by mathematical induction.

We can use Theorem 13.2.2 to obtain $\beta_{-1,k} = 1/2$. We have only to note that

$$(-1)^{j+1}c_{-1,j}(x) = \frac{1}{2^j}K_{2j-1}(x)\begin{cases} \geq 0 & \text{if } x \geq 0, \\ < 0 & \text{if } x < 0. \end{cases}$$

$\beta_{1,k} = 1/2$ was obtained by evaluating $J_{1,k}(y)$ of (13.2.8) numerically. □

Table 13.2.1 Polynomials $H_{2j-1}(x)$

j	$H_{2j-1}(x)$
1	x
2	$x^3 - 3x$
3	$x^5 - 10x^3 + 15x$
4	$x^7 - 21x^5 + 105x^3 - 105x$
5	$x^9 - 36x^7 + 378x^5 - 1260x^3 + 945x$
6	$x^{11} - 55x^9 + 990x^7 - 6930x^5 + 17325x^3 - 10395x$

Table 13.2.2 Polynomials $K_{2j-1}(x)$

j	$K_{2j-1}(x)$
1	x
2	$x^3 + x$
3	$x^5 + 2x^3 + 3x$
4	$x^7 + 3x^5 + 9x^3 + 15x$
5	$x^9 + 4x^7 + 18x^5 + 60x^3 + 105x$
6	$x^{11} + 5x^9 + 30x^7 + 150x^5 + 525x^3 + 945x$

Example 13.2.1 (t-Distribution) Let Z and χ^2_n be independent random variables following the standard normal $\mathrm{N}(0,1)$ and the chi-square distribution $\chi^2(n)$ with n degrees of freedom, respectively, and put $S = \sqrt{\chi^2_n/n}$. Then the scale mixture $S^{-1}Z$ follows the t-distribution $\mathrm{T}(n)$, with n degrees of free-

Table 13.2.3 Numerical values of $\alpha_{\delta,k}$ for $\mathrm{N}(0,1)$

k	1	2	3	4	5	6
$\alpha_{1,k}$	0.1210	0.0688	0.0481	0.0370	0.0300	0.0252
$\alpha_{-1,k}$	0.1210	0.0791	0.0608	0.0501	0.0431	0.0380

dom, which is approximated by (13.2.21) with $\delta = -1$. For $j \geq 1$ we have

$$\mathrm{E}(S^{2j}) = \frac{n(n+2)\cdots(n+2j-2)}{n^j}.$$

Taking $k = 4$, $q = 4$, and $\delta = -1$, and noting that

$$\begin{aligned} &\alpha_{-1,4} = 0.0501,\ \beta_{-1,4} = 0.5 \\ &K_3(x) = x^3 + x, \quad K_5(x) = x^5 + 2x^3 + 3x, \\ &\mathrm{E}[(S^2-1)] = 0, \quad \mathrm{E}[(S^2-1)^2] = 2/n, \\ &\mathrm{E}[(S^2-1)^3] = 8/n^2, \text{ and } \mathrm{E}[(S^2-1)^4] = 12(n+4)/n^3, \end{aligned}$$

we obtain

$$\Phi_{-1,4}(x) = \Phi(x) - \Big(\frac{x^3+x}{4n} - \frac{x^5+2x^3+3x}{6n^2}\Big)\varphi(x)$$

and

$$\sup|F_n(x) - \Phi_{-1,4}(x)| \leq \frac{n+4}{4n^3}.$$

13.2.3 Scale - Mixed Gamma

The class $\mathrm{G}(\sigma, \lambda)$ of gamma distributions is a two-parameter family consisting of all distributions with the density

$$g(x;\sigma,\lambda) = \frac{1}{\sigma\Gamma(\lambda)}\Big(\frac{x}{\sigma}\Big)^{\lambda-1} \mathrm{e}^{-x/\sigma}, \quad x > 0, \quad \sigma, \quad \lambda > 0. \tag{13.2.24}$$

When the underlying distribution is $\mathrm{G}(1,\lambda)$ with the pdf $g(x;\lambda) \equiv g(x;1,\lambda)$, for the approximation of the cdf $F(x;\lambda)$ of the scale mixture $X = S^\delta Z$, we take $\rho = 1$. The approximating function will be given in terms of the Laguerre polynomial defined by $L_{(\lambda,0)}(x) = 1$ and

$$\begin{aligned} L_{(\lambda,p)}(x) &\equiv (-1)^p x^{-\lambda} \mathrm{e}^x \frac{d^p}{dx^p}\left(x^{p+\lambda}\mathrm{e}^{-x}\right) \\ &= \sum_{\ell=0}^{p} (-1)^\ell (p+\lambda)^{(\ell)} \binom{p}{\ell} x^{p-\ell}. \end{aligned} \tag{13.2.25}$$

Setting

$$L_{1,j}(x;\lambda) = xL_{(\lambda,j-1)}(x) \quad \text{and} \quad L_{-1,j}(x;\lambda) = xL_{(\lambda-j,j-1)}(x),$$

Table 13.2.4 Polynomials $L_{1,k}(x;\lambda)$

k	$L_{1,k}(x;\lambda)$
1	x
2	$x\,(x - (\lambda+1))$
3	$x\,(x^2 - 2\,(\lambda+2)x + (\lambda+2)^{(2)})$
4	$x\,(x^3 - 3\,(\lambda+3)x^2 + 3\,(\lambda+3)^{(2)}x^1 - (\lambda+3)^{(3)})$

Table 13.2.5 Polynomials $L_{-1,k}(x;\lambda)$

k	$L_{-1,k}(x;\lambda)$
1	x
2	$x\,(x - (\lambda-1))$
3	$x\,(x^2 - 2\,(\lambda-1)x + (\lambda-1)^{(2)})$
4	$x\,(x^3 - 3\,(\lambda-1)x^2 + 3\,(\lambda-1)^{(2)}x - (\lambda-1)^{(3)})$

the approximating function for $G(xy^{-\delta};\lambda)$ corresponding to (13.2.5) is given by

$$G_{\delta,k}(x,y;\lambda) \equiv G(x;\lambda) - \sum_{j=1}^{k-1} \frac{\delta^j}{j!} L_{\delta,j}(x;\lambda) g(x;\lambda)(y-1)^j. \tag{13.2.26}$$

Let $\alpha_{\delta,k}(\lambda)$ and $\beta_{\delta,k}(\lambda)$ be positive numbers defined by

$$\begin{aligned} \alpha_{\delta,k}(\lambda) &= \sup_x \left| \frac{1}{k!} L_{\delta,k}(x;\lambda) g(x;\lambda) \right|, \\ \beta_{\delta,k}(\lambda) &= \sup_{x,0<y<1} \left| \frac{G(xy^{-\delta};\lambda) - G_{\delta,k}(x,y;\lambda)}{(y-1)^k} \right|. \end{aligned} \tag{13.2.27}$$

Then we have the following:

Theorem 13.2.4 *The cdf $F(x;\lambda)$ of the scale mixture $X = S^\delta Z$ of the gamma distribution* $\mathrm{G}(1;\lambda)$ *is approximated by*

$$G_{\delta,k}(x;\lambda) \equiv G(x;\lambda) - \sum_{j=1}^{k-1} \frac{\delta^j}{j!} L_{\delta,j}(x;\lambda) \mathrm{E}(S-1)^j g(x;\lambda), \tag{13.2.28}$$

The error estimate corresponding to (13.2.1) *is*

$$\sup_x |F(x;\lambda) - G_{\delta,k}(x;\lambda)| \le \mathrm{E}(\boldsymbol{\Delta}_k(\alpha_{\delta,k}(\lambda), \beta_{\delta,k}(\lambda); S, 1)). \tag{13.2.29}$$

Proof. Formula (13.2.28) follows from (13.2.3) and (13.2.5) by setting $\rho = 1$. □

The polynomials $L_{\delta,j}(x;\lambda)$ and some numerical values of $\alpha_{\delta,k}(\lambda)$ are given in Tables 13.2.4 and 13.2.5 and Tables 13.2.6 and 13.2.7, respectively.

Table 13.2.6 Numerical values of $\alpha_{1,k}(\lambda)$ for $\mathrm{G}(1,\lambda)$

λ	0.5	1.0	1.5	2.0	2.5	3.0
$\alpha_{1,1}(\lambda)$	0.2419	0.3678	0.4625	0.5413	0.6102	0.6721
$\alpha_{1,2}(\lambda)$	0.1376	0.2305	0.3134	0.3918	0.4674	0.5413
$\alpha_{1,3}(\lambda)$	0.0961	0.1681	0.2379	0.3085	0.3808	0.4551
$\alpha_{1,4}(\lambda)$	0.0738	0.1324	0.1919	0.2548	0.3557	0.4707

Table 13.2.7 Numerical Values of $\alpha_{-1,k}(\lambda)$ for $\mathrm{G}(1,\lambda)$

λ	0.5	1.0	1.5	2.0	2.5	3.0
$\alpha_{-1,1}(\lambda)$	0.2420	0.3679	0.4626	0.5414	0.6103	0.6722
$\alpha_{-1,2}(\lambda)$	0.1583	0.2707	0.3702	0.4630	0.5517	0.6375
$\alpha_{-1,3}(\lambda)$	0.1215	0.2241	0.3243	0.4250	0.5272	0.6310
$\alpha_{-1,4}(\lambda)$	0.1003	0.1954	0.2954	0.4016	0.5143	0.6334

Values of $\beta_{\delta,k}(\lambda)$ were obtained by maximizing the ratio $\mathrm{J}_{\delta,k}(x,y;\lambda)$ of (13.2.8) numerically. Tables 13.2.8 and 13.2.9 give some results. Note, however, that $\beta_{-1,k}(0.5) = \beta_{-1,k}(1.0) = 1$ follows from Theorem 13.2.2 and that $\beta_{\delta,1}(\lambda) = 1$ is a simple consequence of the fact that $G_{\delta,1}(x;\lambda)$ is a cdf for all $\lambda > 0$.

chi-square distribution. chi-square distribution $\chi^2(q)$ with q degrees of freedom is a special case $\mathrm{G}(2;q/2)$ of the gamma distribution, with the density

$$\begin{aligned} g_q(x) &= \frac{1}{2} g(x/2;\, q/2) \\ &= \frac{1}{2\Gamma(q/2)} \left(\frac{x}{2}\right)^{q/2-1} \mathrm{e}^{-x/2}, \quad x > 0,\ q = 1, 2, \ldots. \end{aligned} \tag{13.2.30}$$

Therefore, an approximating function for the cdf $F_{\delta,q}(x)$ of scale mixture $X = S^\delta Z$ of $\chi^2(q)$ is derived from (13.2.26). In fact, given the condition

Table 13.2.8 Numerical values of $\beta_{1,k}(\lambda)$ for $G(1,\lambda)$

λ	0.5	1.0	1.5	2.0	2.5	3.0	3.5	4.0
$\beta_{1,k}(k \leq 3)$	1.0	1.0	1.0	1.0	1.0	1.0	1.0	1.0
$\beta_{1,4}$	1.0	1.0	1.0	1.0	1.0	1.0	1.0	1.0565
$\beta_{1,5}$	1.0	1.0	1.0	1.0	1.0	1.0	1.0908	1.2719
$\beta_{1,6}$	1.0	1.0	1.0	1.0	1.0	1.0212	1.2290	1.4470

Table 13.2.9 Numerical values of $\beta_{-1,k}(\lambda)$ for $G(1,\lambda)$

λ	0.5	1.0	1.5	2.0	2.5	3.0	3.5	4.0
$\beta_{1,k}(k \leq 3)$	1.0	1.0	1.0	1.0	1.0	1.0	1.0	1.0
$\beta_{1,4}$	1.0	1.0	1.0	1.0	1.0	1.0	1.0	1.0963
$\beta_{1,5}$	1.0	1.0	1.0	1.0	1.0	1.0	1.1420	1.4217
$\beta_{1,6}$	1.0	1.0	1.0	1.0	1.0	1.0380	1.3991	1.7625

$S = s$, the conditional cdf of X is $G(xs^{-\delta}/2,\ q/2)$, and an approximation is given by

$$\begin{aligned} G_{\delta,k;q}(x) &\equiv G(x/2;\ q/2) - \sum_{j=1}^{k-1} \frac{\delta^j}{j!} 2\, L_{\delta,j}\left(\frac{x}{2};\ \frac{q}{2}\right) \cdot \frac{1}{2} g\left(\frac{x}{2};\ \frac{q}{2}\right) \cdot \mathrm{E}(S-1)^j \\ &= G_q(x) - \sum_{j=1}^{k-1} \frac{\delta^j}{2^{j-1} j!} U_{\delta,j;q}(x)\, \mathrm{E}(S-1)^j\, g_q(x), \end{aligned} \tag{13.2.31}$$

where $U_{\delta,j;q}(x)$ is the polynomial defined by

$$U_{\delta,j;q}(x) \equiv L_{\delta,j}(x/2;\ q/2) \tag{13.2.32}$$

(see Tables 13.2.10 and 13.2.11), and approximation errors are bounded by

$$\begin{aligned} \sup_x |F_{\delta,q}(x) - G_{\delta,k;\,q}(x)| &= \sup_x |F(x/2; q/2) - G_{\delta,k}(x/2;\ q/2)| \\ &\leq \mathrm{E}(\boldsymbol{\Delta}_k(\alpha_{\delta,k;q}, \beta_{\delta,k;q};\ S, 1)) \\ &\equiv \mathrm{E}(\boldsymbol{\Delta}_k(\alpha_{\delta,k}(q/2), \beta_{\delta,k}(q/2);\ S, 1)). \end{aligned} \tag{13.2.33}$$

This means that numerical values of α and β are available from Tables 13.2.6 to 13.2.9, by reading $\alpha_{\delta,k;q}$ for $\alpha_{\delta,k}(q/2)$ and $\beta_{\delta,k;q}$ for $\beta_{\delta,k}(q/2)$, respectively.

For example, if $k = 4$, the bound becomes

$$\sup_x |F_{-1,4}(x) - G_{-1,4;\,4}(x)| \le \beta_{-1,4;\,4} \cdot \mathrm{E}[(S-1)^4 = \mathrm{E}[(S-1)^4].$$

If $S > 1$ with probability 1, then $\beta_{-1,4;\,4}$ can be replaced by $\alpha_{-1,4;\,4} = 0.4016$.

Table 13.2.10 Polynomials $U_{1,k;q}(x)$

k	$U_{1,k;q}(x)$
1	x
2	$x\,(x - (q+2))$
3	$x\,(x^2 - 2\,(q+2)x + (q+2)\,(q+4)$
4	$x\,(x^3 - 3\,(q+2)x^2 + 3\,(q+2)\,(q+4)x - (q+2)\,(q+4)\,(q+6))$

Table 13.2.11 Polynomials $U_{-1,k;q}(x)$

k	$U_{-1,k;q}(x)$
1	x
2	$x\,(x - (q-2))$
3	$x\,(x^2 - 2\,(q-2)x + (q-2)\,(q-4)$
4	$x\,(x^3 - 3\,(q-2)x^2 + 3\,(q-2)\,(q-4)x - (q-2)\,(q-4)\,(q-6))$

Example 13.2.2 (F-distribution) Let χ_q^2 and χ_n^2 be mutually independent chi-squared variables with q and n degrees of freedom, respectively. Instead of the F variable $\chi_q^2/q/\chi_n^2/n$, we consider the scale mixture $T = \chi_q^2/\chi_n^2/n$ of the chi-square distribution $\chi^2(q)$, which appears in Chapter 15. Taking, for instance, $k = 4$, $q = 4$, and $\delta = -1$, and noting that

$$\begin{aligned}
&G_4(x) = 1 - (x/2+1)\mathrm{e}^{-x/2}, \quad g_4(x) = (x/4)\mathrm{e}^{-x/2}, \quad \beta_{-1,4}(2) = 1.0,\\
&U_{-1,1;\,4}(x) = x, \quad U_{-1,2;\,4}(x) = x(x-2), \quad U_{-1,3;\,4}(x) = x(x-2)(x+2),\\
&\mathrm{E}[(S-1)] = 0, \quad \mathrm{E}[(S-1)^2] = 2/n,\\
&\mathrm{E}[(S-1)^3] = 8/n^2, \quad \mathrm{E}[(S-1)^4] = 12(n+4)/n^3,
\end{aligned}$$

we obtain from (13.2.31) and (13.2.33) the fact that the cdf of T is approximated by

$$G_{-1,4;\,4}(x) = G_4(x) - \frac{x(x-2)}{n}\Big(\frac{1}{2} - \frac{1}{3\,n}(x+2)\Big)g_4(x)$$

with the approximation error bounded by

$$\Delta_4(n) = \sup \big|\Pr\{T \le x\} - G_{-1,4;\,4}(x)\big| \le \frac{12(n+4)}{n^3}.$$

13.3 Error Bounds Evaluated in L_1-Norm

13.3.1 Some Basic Results

In this section we study the asymptotic approximation of the pdf of the distribution of the scale mixture $X = S^\delta Z$ and evaluate its error in the L_1-norm, which will make it possible to approximate the probability $\Pr\{X \in \mathrm{A}\}$ for an arbitrary Borel set A. We extend the results to multivariate cases in the next section, which will be used to obtain error bounds of asymptotic expansions of some multivariate test statistics.

Suppose that the underlying distribution G of X has the density $g(x)$, and let the random variable Y be defined by $Y = S^\rho$. Then the conditional density function of X given the condition $S = s$ is

$$g(x|\, y) = y^{-\delta/\rho} g(xy^{-\delta/\rho}).$$

Therefore, the density $f(x)$ of X is given by

$$f(x) = \mathrm{E}_Y[g(x|\, Y)] = \mathrm{E}_Y[Y^{-\delta/\rho} g(xY^{-\delta/\rho})]. \tag{13.3.1}$$

Assuming that the density g is k-times continuously differentiable on its support $D \equiv \{x \in \mathrm{R} : g(x) > 0\}$, we introduce a set of functions $b_{\delta,j}(x)$, $j = 0, 1, \ldots, k-1$ by $b_{\delta,0}(x) = 1$ and

$$\frac{\partial^j}{\partial y^j}\left(y^{-\delta/\rho} g(xy^{-\delta/\rho})\right) = y^{-j} b_{\delta,j}(xy^{-\delta/\rho}) \cdot y^{-\delta/\rho} g(xy^{-\delta/\rho}),$$

or

$$b_{\delta,j}(x)g(x) = \frac{\partial^j}{\partial y^j}\left(y^{-\delta/\rho} g(xy^{-\delta/\rho})\right)\Big|_{y=1}. \tag{13.3.2}$$

For each $x \in \mathrm{R}$, we take the first k terms $g_{\delta,k}(x, y)$ of the Taylor expansion of $g(x|\, y)$ around $y = 1$ as its approximation:

$$g_{\delta,k}(x, y) = \sum_{j=0}^{k-1} \frac{1}{j!} b_{\delta,j}(x) g(x)(y-1)^j, \tag{13.3.3}$$

This suggests that the function

$$\begin{aligned} g_{\delta,k}(x) &\equiv \mathrm{E}[g_{\delta,k}(x, Y)] \\ &= \sum_{j=0}^{k-1} \frac{1}{j!} b_{\delta,j}(x) \mathrm{E}[(S^\rho - 1)^j] g(x) \end{aligned} \tag{13.3.4}$$

provides a reasonable approximation of the density $f(x)$. Comparing both sides of (13.2.6) and (13.3.4), we obtain

$$b_{\delta,j}(x)g(x) = \frac{\mathrm{d}}{\mathrm{d}x}\left(c_{\delta,j}(x)g(x)\right). \tag{13.3.5}$$

In fact, the approximating function $g_{\delta,k}(x)$ is obtained by differentiating $G_{\delta,k}(x)$ of (13.2.6). Note that $g_{\delta,k}(x)$ is not necessarily a pdf, but it satisfies the relation

$$\int_{\mathrm{R}} g_{\delta,k}(x)\,\mathrm{d}x = G_{\delta,k}(\infty) - G_{\delta,k}(-\infty) = 1. \tag{13.3.6}$$

Our error estimate is based on an evaluation of the quantity

$$\Delta_{\delta,k}(x,\, y) \equiv y^{-\delta/\rho} g(xy^{-\delta/\rho}) - g_{\delta,k}(x,y) \quad \Big(= \frac{\partial}{\partial x}\Delta_{\delta,k}(x,y)\Big). \tag{13.3.7}$$

Writing

$$\Delta_{\delta,k}(y) \equiv \int_{\mathrm{R}} \big|g(x|\, y) - g_{\delta,k}(x,y)\big|\,\mathrm{d}x \quad \Big(= \int_{\mathrm{R}} \big|\Delta_{\delta,k}(x,\, y)\big|\,\mathrm{d}x\Big) \tag{13.3.8}$$

put

$$\sigma_{\delta,k} = \frac{1}{k!}\left\|b_{\delta,k}(x)g(x)\right\|_1 \quad \text{and} \quad \tau_{\delta,k} = \sup_{0<y<1} \frac{\Delta_{\delta,k}(y)}{|y-1|^k}. \tag{13.3.9}$$

Compare these quantities with the α and β described in Section 13.2.1. We prove the following:

Lemma 13.3.1 *It holds that for $k \geq 1$ and $y > 0$*

(1) $\Delta_{\delta,k}(y) \leq \sigma_{\delta,k}\ (y \vee y^{-1} - 1)^k$,

(2) $\Delta_{\delta,k}(y) \leq \tau_{\delta,k}\ |y-1|^k$.

Proof. To prove (1), write the residual $\Delta_{\delta,k}(x,y)$ of the Taylor expansion in integral form:

$$\begin{aligned}\Delta_{\delta,k}(x,y) &= y^{-\delta/\rho} g(xy^{-\delta/\rho}) - \sum_{j=0}^{k-1}\frac{1}{j!} b_{\delta,j}(x)g(x)(y-1)^j \\ &= \frac{1}{(k-1)!}\int_1^y z^{-k} b_{\delta,k}(xz^{-\delta/\rho})\cdot z^{-\delta/\rho} g(xz^{-\delta/\rho})(y-z)^{k-1}\,\mathrm{d}z.\end{aligned}$$

Then we can use the Fubini theorem together with the definition of $\sigma_{\delta,k}$ to obtain

$$\begin{aligned}\Delta_{\delta,k}(y) &= \int_{\mathrm{R}} \big|\Delta_{\delta,k}(x,y)\big|\,\mathrm{d}y \leq \frac{k!}{(k-1)!}\ \sigma_{\delta,k}\Big|\int_1^y z^{-k}(y-z)^{k-1}\,\mathrm{d}z\Big| \\ &\leq \frac{k!}{(k-1)!}\ \sigma_{\delta,k}\ (y^{-k}\vee 1)\Big|\int_1^y (y-z)^{k-1}\,\mathrm{d}z\Big| = \sigma_{\delta,k}\ (y \vee y^{-1} - 1)^k,\end{aligned}$$

as was to be proved. Assertion (2) is clear from the definition of τ. □

Theorem 13.3.1 *Suppose that* $X = S^\delta Z$ *is a scale mixture of* Z *and has the pdf* $f(x)$*. Then we have for any Borel set* $\mathrm{A} \subset \mathrm{R}$,

$$\left|\Pr(X \in \mathrm{A}) - \int_{\mathrm{A}} g_{\delta,k}(x)\mathrm{d}x\right| \leq \frac{1}{2}\left\|f(x) - g_{\delta,k}(x)\right\|_1$$
$$\leq \frac{1}{2}\mathrm{E}\big[\boldsymbol{\Delta}_k(\sigma_{\delta,k}, \tau_{\delta,k};\ S, \rho)\big], \qquad (13.3.10)$$

where $\boldsymbol{\Delta}_k(\sigma, \tau;\ s, \rho)$ *is defined by* (13.2.12).

13.3.2 Scale - Mixed Normal Density

In this section we consider the normal case with $\rho = 2$, and write Φ and φ for G and g, respectively. Then the $b_{\delta,j}(x)$' become

$$b_{\delta,j}(x) = \left(\frac{\delta}{2}\right)^j Q_{\delta,2j}(x),$$

where

$$Q_{1,2j}(x) = H_{2j}(x) \qquad \text{(Hermite polynomial)}, \qquad (13.3.11)$$
$$Q_{-1,2j}(x) = K_{2j}(x) \equiv x^{2j} - \sum_{m=1}^{j}(2m-3)!!\binom{j}{m}x^{2(j-m)}. \qquad (13.3.12)$$

See Tables 13.3.1 and 13.3.2 for details. The approximation (13.3.5) to the pdf of X becomes

$$\varphi_{\delta,k}(x) = \varphi(x) + \sum_{j=1}^{k-1}\frac{\delta^j}{2^j j!}Q_{\delta,2j}(x)\varphi(x)\mathrm{E}[(S^2-1)^j]. \qquad (13.3.13)$$

Table 13.3.1 Polynomials $H_{2j}(x)$

j	$H_{2j}(x)$
1	$x^2 - 1$
2	$x^4 - 6x^2 + 3$
3	$x^6 - 15x^4 + 45x^2 - 15$
4	$x^8 - 28x^6 + 210x^4 - 420x^2 + 105$
5	$x^{10} - 45x^8 + 630x^6 - 3150x^4 + 4725x^2 - 945$
6	$x^{12} - 66x^{10} + 1485x^8 - 13860x^6 + 51975x^4 - 62370x^2 + 10395$

Table 13.3.2 Polynomials $K_{2j}(x)$)

j	$K_{2j}(x)$
1	$x^2 - 1$
2	$x^4 - 2x^2 - 1$
3	$x^6 - 3x^4 - 3x^2 - 3$
4	$x^8 - 4x^6 - 6x^4 - 12x^2 - 15$
5	$x^{10} - 5x^8 - 10x^6 - 30x^4 - 75x^2 - 105$
6	$x^{12} - 6x^{10} - 15x^8 - 60x^6 - 225x^4 - 630x^2 - 945$

The polynomials $K_j(x)$ with odd j's were given in (13.2.20). Note that the Q satisfy the recurrence relations

$$\begin{aligned} Q_{2j+1}(x) &= xQ_{2j}(x) + 2jQ_{2j-1}(x), \\ Q_{2j}(x) &= xQ_{2j-1}(x) - Q'_{2j-1}(x). \end{aligned} \tag{13.3.14}$$

The second of (13.3.14) is equivalent to

$$\int Q_{\delta,2j}(x)\varphi(x)\,\mathrm{d}x = -Q_{\delta,2j-1}(x)\varphi(x) \tag{13.3.15}$$

with $Q_{\delta,2j-1}(x)$ being defined by (13.2.20). It is also easy to connect K with the Hermite polynomials; for example,

$$\begin{aligned} K_1(x) &= H_1(x) \\ K_2(x) &= H_2(x) \\ K_3(x) &= H_3(x) + 4H_1(x) \\ K_4(x) &= H_4(x) + 4H_2(x) \end{aligned} \tag{13.3.16}$$

The next theorem follows from Theorems 13.3.1

Theorem 13.3.2 *The pdf $f(x)$ of a scale mixture $X = S^\delta Z$ of the standard normal distribution is approximated by* (13.3.13) *and we have, for any Borel set* A,

$$\begin{aligned} \left|\Pr\{X \in \mathrm{A}\} - \int_A \varphi_{\delta,k}\,\mathrm{d}x\right| &\le \frac{1}{2}\left\|f(x) - \varphi_{\delta,k}(x)\right\|_1 \\ &\le \frac{1}{2}\,\mathrm{E}\big[\mathbf{\Delta}_k(\sigma_{\delta,k}, \tau_{\delta,k};\ S, 2)\big]. \end{aligned} \tag{13.3.17}$$

Table 13.3.3 Numerical values of $\sigma_{\delta,k}$ fo $N(0,1)$

k	1	2	3	4	5	6
$\sigma_{1,k}$	0.4840	0.3501	0.2879	0.2503	0.2061	0.2054
$\sigma_{-1,k}$	0.4840	0.3165	0.2430	0.2005	0.1722	0.1519

Table 13.3.4 Numerical values of $\tau_{\delta,k}$ for $N(0,1)$

k	1	2	3	4	5	6
$\tau_{1,k}$	2.0	2.1419	2.2263	2.2864	2.3329	2.3708
$\tau_{-1,k}$	2.0	2.0	2.0	2.0	2.0	2.0

13.3.3 Scale - Mixed Gamma Density

In this section we consider the scale mixture of the gamma distribution with the density

$$g(x;\lambda) = \frac{1}{\Gamma(\lambda)} x^{\lambda-1} \mathrm{e}^{-x}, \quad x > 0. \tag{13.3.18}$$

The approximation (13.3.19) of the pdf of X is given by

$$g_{\delta,k}(x) \equiv \mathrm{E}[g_{\delta,k}(x,Y)], \tag{13.3.19}$$

$$g_{\delta,k}(x;\lambda) = g(x;\lambda) + \sum_{j=1}^{k-1} \frac{\delta^j}{j!} R_{\delta,j}(x;\lambda) \mathrm{E}[(s-1)^j]\, g(x;\lambda). \tag{13.3.20}$$

Here $R_{\delta,j}(x;\lambda)$ can be determined from $c_{\delta,k}(x;\lambda)$ of (13.2.6) through the relation

$$R_{\delta,j}(x;\lambda) g(x;\lambda) = \delta^j \frac{\mathrm{d}}{\mathrm{d}x} \left(L_{\delta,j}(x;\lambda) g(x;\lambda) \right), \tag{13.3.21}$$

which gives (see Tables 13.3.5 and 13.3.6)

$$\begin{aligned} R_{1,j}(x;\lambda) &= M_j(x;\lambda) \equiv \sum_{\ell=0}^{j} (-1)^\ell \binom{j}{\ell} (\lambda - 1 + j)^{(\ell)} x^{j-\ell}, \\ R_{-1,j}(x;\lambda) &= N_j(x;\lambda) \equiv \sum_{\ell=0}^{j} (-1)^\ell \binom{j}{\ell} \lambda^{(\ell)} x^{j-\ell}, \end{aligned} \tag{13.3.22}$$

or, in terms of the Laguerre polynomials (13.2.25), these expressions become

$$M_j(x;\lambda) = L_{(\lambda-1,j)}(x) \quad \text{and} \quad N_j(x;\lambda) = L_{(\lambda-j,j)}(x). \tag{13.3.23}$$

Table 13.3.5 Polynomials $M_j(x;\lambda)$ for $G(1,\lambda)$

j	$M_j(x;\lambda)$
1	$x-\lambda$
2	$x^2-2(\lambda+1)x+\lambda(\lambda+1)$
3	$x^3-3(\lambda+2)x^2+3(\lambda+1)(\lambda+2)x \quad -\lambda(\lambda+1)(\lambda+2)$
4	$x^4-4(\lambda+3)x^3+6(\lambda+2)(\lambda+3)x^2-4(\lambda+1)(\lambda+2)(\lambda+3)x$ $+\lambda(\lambda+1)(\lambda+2)(\lambda+3)$

Table 13.3.6 Polynomials $N_j(x;\lambda)$ for $G(1,\lambda)$

j	$N_j(x;\lambda)$
1	$x-\lambda$
2	$x^2-2\lambda x \quad +\lambda(\lambda-1)$
3	$x^3-3\lambda x^2+3\lambda(\lambda-1)x \quad -\lambda(\lambda-1)(\lambda-2)$
4	$x^4-4\lambda x^3-6\lambda(\lambda-1)x^2-4\lambda(\lambda-1)(\lambda-2)x-\lambda(\lambda-1)(\lambda-2)(\lambda-3)$

The following theorem is an immediate consequence of Theorem 13.3.1.

Theorem 13.3.3 *Suppose that* $X = S^\delta Z$ *is a scale mixture of the gamma distribution. Then we have the following two types* ($\delta = \pm 1$) *of inequalities, which hold for any* $k \geq 1$ *and any Borel set* $\mathrm{A} \subset \mathrm{R}$*:*

$$\left|\Pr\{X \in \mathrm{A}\} - \int_{\mathrm{A}} g_{\delta,k}(x;\lambda)\,\mathrm{d}x\right| \leq \frac{1}{2}\,\mathrm{E}[\mathbf{\Delta}_k(\sigma_{\delta,k}(\lambda), \tau_{\delta,k}(\lambda); S, 1)], \tag{13.3.24}$$

where the $\sigma_{\delta,k}$ *and* $\tau_{\delta,k}$ *are given by* (13.3.9) *with* $b_{\delta,k}(x)$ *and* $g(x)$ *being replaced by* $M_{\delta,k}(x;\lambda)$ *and* $g(x;\lambda)$*, respectively.*

Their numerical values are given in Tables 13.3.7 and 13.3.8 for some selected λ. We searched for the maximum values of $\mathrm{J}_{\delta,k}(y)$ on a computer and summarized them as $\tau_{\delta,k}$ (estimated maximum) in Tables 13.3.9 and 13.3.10.

Table 13.3.7 Numerical values of $\sigma_{1,k}(\lambda)$ for $G(1,\lambda)$

λ	0.5	1.0	1.5	2.0	2.5	3.0
$\sigma_{1,1}(\lambda)$	0.4840	0.7358	0.9251	1.0827	1.2205	1.3444
$\sigma_{1,2}(\lambda)$	0.3501	0.6201	0.8754	1.1253	1.3725	1.6181
$\sigma_{1,3}(\lambda)$	0.2879	0.5598	0.8541	1.1739	1.5185	1.8870
$\sigma_{1,4}(\lambda)$	0.2503	0.5210	0.8434	1.2217	1.6572	2.1504

Table 13.3.8 Numerical values of $\sigma_{-1,k}(\lambda)$ for $G(1,\lambda)$

λ	0.5	1.0	1.5	2.0	2.5	3.0
$\sigma_{-1,1}(\lambda)$	0.4840	0.7358	0.9251	1.0827	1.2205	1.3443
$\sigma_{-1,2}(\lambda)$	0.3165	0.5414	0.7681	1.0051	1.2446	1.4850
$\sigma_{-1,3}(\lambda)$	0.2430	0.4481	0.6991	0.9900	1.3091	1.6544
$\sigma_{-1,4}(\lambda)$	0.2005	0.3908	0.6550	0.9837	1.3688	1.8128

Table 13.3.9 Estimated values of $\tau_{1,k}(\lambda)$ for $G(1,\lambda)$

λ	0.5	1.0	1.5	2.0	2.5	3.0
$\tau_{1,1}(\lambda)$	2.0	2.0	2.0	2.0	2.0	2.0
$\tau_{1,2}(\lambda)$	2.142	2.271	2.390	2.517	2.672	2.852
$\tau_{1,3}(\lambda)$	2.227	2.461	2.698	2.959	3.271	3.631
$\tau_{1,4}(\lambda)$	2.287	2.610	2.964	3.368	3.858	4.378

Table 13.3.10 Estimated values of $\tau_{-1,k}(\lambda)$ for $G(1,\lambda)$

λ	0.5	1.0	1.5	2.0	2.5	3.0
$\tau_{-1,1}(\lambda)$	2.0	2.0	2.0	2.0	2.0	2.0
$\tau_{-1,2}(\lambda)$	2.0	2.0	2.088	2.238	2.419	2.627
$\tau_{-1,3}(\lambda)$	2.0	2.0	2.197	2.483	2.820	3.204
$\tau_{-1,4}(\lambda)$	2.0	2.0	2.291	2.699	3.195	3.772

13.3.4 Scale - Mixed Chi-square Density

The chi-square distributions belong to the class of two-parameter gamma distributions with the density

$$g(x;\sigma,\lambda)=\frac{1}{\sigma\Gamma(\lambda)}\left(\frac{x}{\sigma}\right)^{\lambda-1}\mathrm{e}^{-x/\sigma}, \tag{13.3.25}$$

and as we saw in Example 13.2.2, the problem of the scale mixture

$$X=S^{-1}\chi_q^2 \tag{13.3.26}$$

of $\chi^2(q)$ reduces to that of the gamma distribution with the density $g(x;1,\lambda)$. The process of reduction is quite simple, but it creates further notational complications. This subsection is devoted to the scale mixture (13.3.26), rearranging the results of the previous subsection in a way that will be useful in Chapter 15.

For a positive integer q, we write $G_q(x)$ and $g_q(x)$ to denote the cdf and density of the chi-square distribution with q degrees of freedom:

$$g_q(x)=\frac{1}{2\Gamma(q/2)}\left(\frac{x}{2}\right)^{q/2-1}\mathrm{e}^{-x/2}. \tag{13.3.27}$$

It is connected with $g(x;\lambda)$ of (13.3.18) by the relation

$$g_q(x)=\frac{1}{2}g\left(\frac{x}{2};\frac{q}{2}\right).$$

The conditional density $g_q(x|y)$ given $Y=y$ is approximated by the first k terms $g_{(k;q)}(x,y)$ of the Taylor expansion of $yg_q(xy)$:

$$g_{(k;q)}(x,y)=g_q(x)+\sum_{j=1}^{k-1}\frac{(-1)^j}{2^j j!}V_{j;q}(x)g_q(x)(y-1)^j, \tag{13.3.28}$$

where the $V_{j;q}$ are polynomials defined by

$$V_{j;q}(x)=\sum_{\ell=0}^{j}(-2)^\ell\binom{j}{\ell}\left(\frac{q}{2}\right)^{(\ell)}x^{j-\ell},\quad j=1,2,\ldots, \tag{13.3.29}$$

and where $a^{(\ell)}=a(a-1)\cdots(a-\ell+1)$ for any real number a and positive integer ℓ.

Now we have the following:

Theorem 13.3.4 *The pdf $f_q(x)$ of the scale mixture $X=S^{-1}Z$ of the chi-square distribution with q degrees of freedom is approximated by*

$$g_{(k;q)}(x)=g_q(x)+\sum_{j=1}^{k-1}\frac{(-1)^j}{2^{j-1}j!}V_{j;q}(x)\mathrm{E}(S-1)^j g_q(x), \tag{13.3.30}$$

with an error estimate corresponding to (13.3.10) [*see also* (13.2.12)]:

$$\left|\Pr\{X \in \mathrm{A}\} - \int_A g_{q,k}(x)\,\mathrm{d}x\right| \le \frac{1}{2}\left\|f(x) - g_{q,k}(x)\right\|_1$$

$$\le \frac{1}{2}\,\mathrm{E}\big[\boldsymbol{\Delta}_k(\sigma_{q,k}, \tau_{q,k};\ s, 1)\big]. \qquad (13.3.31)$$

Some numerical values of $\sigma_{q,k}$ and $\tau_{q,k}$, which are identical to $\sigma_{-1,k}(q/2)$ and $\tau_{-1,k}(q/2)$, respectively, are reproduced in Tables 13.3.12 and 13.3.13.

Table 13.3.11 Polynomials $V_{j;q}(x)$

j	$V_{j;q}(x)$
1	$x - q$
2	$x^2 - 2qx + q(q-2)$
3	$x^3 - 3qx^2 + 3q(q-2)x - q(q-2)(q-4)$
4	$x^4 - 4qx^3 + 6q(q-2)x^2 - 4q(q-2)(q-4)x + q(q-2)(q-4)(q-6)$

Table 13.3.12 Numerical values of $\sigma_{q,k}$ for $\chi^2(q)$

q	1	2	3	4	5	6
$\sigma_{q,1}$	0.4840	0.7358	0.9251	1.0827	1.2205	1.3443
$\sigma_{q,2}$	0.3165	0.5414	0.7681	1.0051	1.2446	1.4850
$\sigma_{q,3}$	0.2430	0.4481	0.6991	0.9900	1.3091	1.6544
$\sigma_{q,4}$	0.2005	0.3908	0.6550	0.9837	1.3688	1.8128

Table 13.3.13 Estimated values of $\tau_{q,k}$ for $\chi^2(q)$

q	1	2	3	4	5	6
$\tau_{q,1}$	2.0	2.0	2.0	2.0	2.0	2.0
$\tau_{q,2}$	2.0	2.0	2.088	2.238	2.419	2.627
$\tau_{q,3}$	2.0	2.0	2.197	2.483	2.820	3.204
$\tau_{q,4}$	2.0	2.0	2.291	2.699	3.195	3.772

Example 13.3.1 (F-distribution: continued from Example 13.2.2) If k is an even integer, then for $X = \chi_q^2/(\chi_n^2/n)$ we have

$$\left|\Pr\{X \in \mathrm{A}\} - \int_{\mathrm{A}} g_{(k;q)}(x)\mathrm{d}x\right| \leq \frac{1}{2}\sigma_{q,k}(\mathrm{E}[(S-1)^k] + \mathrm{E}[(S^{-1}-1)^k]).$$

For $k = 2$ and 4, the $g_{(k;q)}(x)$ are

$$\begin{aligned} g_{(2;q)}(x) &= g_q(x). \\ g_{(4;q)}(x) &= \left(1 - \frac{1}{4n}V_{2;q}(x) + \frac{1}{12n^2}V_{3;q}(x)\right)g_q(x). \end{aligned}$$

13.4 Multivariate Scale Mixtures

13.4.1 General Theory

In this section we extend the results obtained in previous sections to multivariate cases, which will have some important distributional applications to MANOVA test statistics (see Chapter 15).

Let $\boldsymbol{Z} = (Z_1, \cdots, Z_p)$ be a random vector and let $\mathbf{S}$ be a positive definite random matrix independent of $\boldsymbol{Z}$. Then for $\delta = \pm 1$,

$$\boldsymbol{X} = \mathbf{S}^{\delta}\boldsymbol{Z} \tag{13.4.1}$$

is called a *scale mixture* of $\boldsymbol{Z}$ with the *scale-factor matrix* $\mathbf{S}$. In this chapter we assume that $Z_1, \ldots, Z_p$ are *i.i.d.* random variables with the common distribution function G and density g. We tacitly assume that the scale-factor matrix $\mathbf{S}$ is close to the unit matrix I in some sense. We introduce the following transformation:

$$Y = S^{\rho} \quad \text{or} \quad S = Y^{1/\rho}, \tag{13.4.2}$$

where ρ is a positive integer. We start with the case where the scale factor $\mathbf{S}^{\delta}$ is a diagonal matrix with probability 1: $\mathbf{S} = \mathrm{diag}\,(S_1, \ldots, S_p)$. Then the transformation (13.4.2) becomes

$$Y_r = S_r^{\rho}, \quad r = 1, \ldots, p. \tag{13.4.3}$$

In the following, we often identify the diagonal matrix $\mathbf{Y} = \mathrm{diag}\,(Y_1, \ldots, Y_p)$ with the vector $\boldsymbol{Y} = (Y_1, \ldots, Y_p)'$.

The cdf of the random vector

$$\boldsymbol{X} = \mathrm{diag}\,(S_1^{\delta}, \ldots, S_p^{\delta})\boldsymbol{Z} \tag{13.4.4}$$

is given by

$$F_p(\boldsymbol{x}) \equiv \Pr\{X_1 \le x_1, \ldots, X_p \le x_p\} = \mathrm{E}\Big[\prod_{r=1}^{p} G(x_r S_r^{-\delta})\Big]$$
$$= \mathrm{E}\Big[\prod_{r=1}^{p} G(x_r Y_r^{-\delta/\rho})\Big],$$

where $\boldsymbol{x} = (x_1, \ldots, x_p)'$. We assume that the density g is k-times continuously differentiable on its support $D \equiv \{x \in \mathrm{R} : \; g(x) > 0\}$.

Let $\boldsymbol{G}_p(\boldsymbol{x}) = \prod_{j=1}^{p} G(x_j)$ be the joint cdf of the random vector $\boldsymbol{Z}$. As an approximation of the conditional distribution function

$$\boldsymbol{G}_p(\boldsymbol{x}|\, \boldsymbol{y}) = \prod_{r=1}^{p} G(x_r y_r^{-\delta/\rho})$$

of $\boldsymbol{X}$ given the condition $\boldsymbol{Y} = \boldsymbol{y}$, we take the first k terms of its Taylor expansion around $y_r = 1, r = 1, \ldots, p$:

$$\begin{aligned}
\boldsymbol{G}_{\delta,k,p}(\boldsymbol{x}, \boldsymbol{y}) &\equiv \sum_{j=0}^{k-1} \frac{1}{j!}\Big(\sum_{r=1}^{p}(y_r - 1)\frac{\partial}{\partial y_r}\Big)^j \prod_{r=1}^{p} G(x_r y_r^{-\delta/\rho})\Big|_{\boldsymbol{y}=\boldsymbol{1}} \\
&= \sum_{j=0}^{k-1}\sum_{(j)}\prod_{r=1}^{p}\frac{1}{j_r!}\Big((y_r - 1)^{j_r}\frac{d^{j_r}}{dy_r^{j_r}} G(x_r y_r^{-\delta/\rho})\Big|_{\boldsymbol{y}=\boldsymbol{1}}\Big) \qquad (13.4.5) \\
&= \boldsymbol{G}_p(\boldsymbol{x}) + \sum_{j=1}^{k-1}\frac{1}{j!}\sum_{(j)}\binom{j}{j_1, \ldots, j_p}\prod_{r=1}^{p} c_{\delta,j_r}(x_r)(y_r - 1)^{j_r}\, \boldsymbol{g}_p(\boldsymbol{x}), \\
&= \boldsymbol{G}_p(\boldsymbol{x}) + \sum_{j=1}^{k-1}\frac{1}{j!}\Big(\sum_{r=1}^{p} c^*_{\delta,1}(x_r)(y_r - 1)\Big)^j \cdot \boldsymbol{g}_p(\boldsymbol{x}), \qquad (13.4.6)
\end{aligned}$$

where the summation $\sum_{(j)}$ is taken over all non-negative integers such that $j_1 + \cdots + j_p = j$. The function $c_{\delta,j}(x)$ is given by (13.2.4) and we use:

Convention 1 We use the asterisk * of $c^*_{\delta,1}(x_r)$ in the following sense:

$$\begin{aligned}
c^*_{\delta,u}(x) &= c^{*1}_{\delta,u}(x) = c_{\delta,u}(x) \\
(c^*_{\delta,u})^q &= c^{*q}_{\delta,u} = c_{\delta,u,q} \quad \text{for } u,\; q = 0, 1, \ldots.
\end{aligned}$$

For instance,

$$\begin{aligned}
&(c^*_{\delta,1}(x_1)(y_1 - 1) + c^*_{\delta,1}(x_2)(y_2 - 1))^j \\
&\qquad = \sum_{t=0}^{j}\binom{j}{t} c_{\delta,t}(x_1) c_{\delta,j-t}(x_2)(y_1 - 1)^t (y_2 - 1)^{j-t}.
\end{aligned}$$

Our target function $F_p(\boldsymbol{x})$ will be approximated by

$$\boldsymbol{G}_{\delta,k,p}(\boldsymbol{x}) = \boldsymbol{G}_p(\boldsymbol{x}) + \sum_{j=1}^{k-1} \frac{1}{j!} \mathrm{E}\Big[\Big(\sum_{r=1}^{p} c^*_{\delta,1}(x_r)(S^\rho_r - 1)\Big)^j\Big] \cdot \boldsymbol{g}_p(\boldsymbol{x}) \qquad (13.4.7)$$

and an upper bound of its possible error is available in the form

$$\Delta_{\delta,k,p} \equiv \sup_x \big| F_p(\boldsymbol{x}) - \boldsymbol{G}_{\delta,k,p}(\boldsymbol{x}) \big| \leq \mathrm{B}_{\delta,k} \cdot \frac{1}{p} \mathrm{E}\big[\,\mathrm{tr}\,(\mathbf{S}^\rho - \mathrm{I})^k\big]. \qquad (13.4.8)$$

However, this gives only an approximation of the probability of the event $\boldsymbol{X} \in (-\infty, x_1] \times (-\infty, x_2] \times \cdots \times (-\infty, x_p]$. What we need, in many applications, is an approximation of the events related to a linear form $\boldsymbol{c}'\boldsymbol{X}$, a quadratic form $\boldsymbol{X}'C\boldsymbol{X}$, or more generally, a random q-vector $a(\boldsymbol{X})$, where a is a measurable function $\mathrm{R}^p \to \mathrm{R}^q$. This is tantamount to approximating the probability $\Pr\{\boldsymbol{X} \in \mathrm{A}\}$, where A is a suitably chosen Borel set in R^p. Therefore, our main focus in the rest of this chapter will be the approximation of the density $f_p(\boldsymbol{x})$ of $\boldsymbol{X}$, which is obtained by expanding it around the density $\boldsymbol{g}_p(\boldsymbol{x}) = \prod_{r=1}^{p} g(x_r)$ of $\boldsymbol{Z}$. For this purpose we start with the conditional density of $\boldsymbol{X}$ given $Y_r = y_r$, $r = 1, \ldots, p$:

$$\boldsymbol{g}_p(\boldsymbol{x} \,|\, \boldsymbol{y}) \equiv \prod_{r=1}^{p} g(x_r \,|\, y_r) \quad \text{with} \quad g(x \,|\, y) \equiv y^{-\delta/\rho} g(x y^{-\delta/\rho}). \qquad (13.4.9)$$

A reasonable approximation to $\boldsymbol{g}_p(\boldsymbol{x} \,|\, \boldsymbol{y})$ will be obtained by differentiating $\boldsymbol{G}_{\delta,k,p}(\boldsymbol{x}, \boldsymbol{y})$ of (13.4.5):

$$\boldsymbol{g}_{\delta,k,p}(\boldsymbol{x}, \boldsymbol{y}) \equiv \boldsymbol{g}_p(\boldsymbol{x}) + \sum_{j=1}^{k-1} \frac{1}{j!} \Big(\sum_{r=1}^{p} b^*_{\delta,1}(x_r)(y_r - 1)\Big)^j \cdot \boldsymbol{g}_p(\boldsymbol{x}) \qquad (13.4.10)$$

$$= \sum_{j=0}^{k-1} \sum_{(j)} \prod_{r=1}^{p} (y_r - 1)^{j_r} \prod_{r=1}^{p} \frac{1}{j_r!} b_{\delta,j_r}(x_r) \cdot \boldsymbol{g}_p(\boldsymbol{x}), \qquad (13.4.11)$$

where the function $b_{\delta,j}(x)$ is defined by (13.3.2) and we use Convention 1. We take the function

$$\boldsymbol{g}_{\delta,k,p}(\boldsymbol{x}) \equiv \mathrm{E}(\boldsymbol{g}_{\delta,k,p}(\boldsymbol{x}, \mathbf{S}^\rho))$$

as an approximation to $f_p(\boldsymbol{x})$ and will derive an upper bound for its error in the L_1-norm:

$$\Delta_{\delta,k,p} \equiv \big\| f_p(\boldsymbol{x}) - \boldsymbol{g}_{\delta,k,p}(\boldsymbol{x}) \big\|_1 = \int_{\mathrm{R}^p} | \, f_p(\boldsymbol{x}) - \boldsymbol{g}_{\delta,k,p}(\boldsymbol{x}) \, | \, \mathrm{d}\boldsymbol{x}.$$

For $\delta = \pm 1$ and $k \geq 1$, let the σ's and τ's be the positive constants given in Lemma 13.3.1. In the following lemma, we simplify the notation by suppressing the subscript δ.

Lemma 13.4.1 *For $k \geq 1$ and $p \geq 1$, let $\mathrm{S}_{k,p}$ and $\mathrm{T}_{k,p}$ be the positive constants determined by the recurrence relations*

$$\mathrm{S}_{k,p} = \sigma_k + \sum_{m=0}^{k-1} \mathrm{S}_{k-m,p-1} \cdot \sigma_m, \tag{13.4.12}$$

$$\mathrm{T}_{k,p} = \tau_k + \sum_{m=0}^{k-1} \mathrm{T}_{k-m,p-1} \cdot \sigma_m, \tag{13.4.13}$$

respectively, with the initial conditions $\sigma_0 = 1$, $\mathrm{S}_{k,0} = \mathrm{T}_{k,0} = 0$, $\mathrm{S}_{k,1} = \sigma_k$, and $\mathrm{T}_{k,1} = \tau_k$ for all $k \geq 1$. Then we have

$$\Delta_{k,p}(\boldsymbol{y}) \equiv \int_R \left|\boldsymbol{g}_p(\boldsymbol{x}\,|\,\boldsymbol{y}) - \boldsymbol{g}_{k,p}(\boldsymbol{x}, \boldsymbol{y})\right| \mathrm{d}\boldsymbol{x},$$

$$\leq \begin{cases} \mathrm{S}_{k,p} \cdot \dfrac{1}{p} \displaystyle\sum_{r=1}^{p} (y_r \vee y_r^{-1} - 1)^k \\ \mathrm{T}_{k,p} \cdot \dfrac{1}{p} \displaystyle\sum_{r=1}^{p} |y_r - 1|^k. \end{cases} \tag{13.4.14}$$

Proof. For $j, = 0, 1, \ldots, k-1$, write

$$u_j(x, y) = \frac{1}{j!} b_{\delta,j}(x)(y-1)^j g(x)$$

and

$$\ell_{j,p}(\boldsymbol{x}, \boldsymbol{y}) = \sum_{j_1+j_2+\cdots+j_p=j} \prod_{r=1}^{p} u_{j_r}(x_r, y_r). \tag{13.4.15}$$

Then

$$\boldsymbol{g}_1(x\,|\,y) = \sum_{j=0}^{k-1} u_j(x, y) + \Delta_{k,1}(x, y) \tag{13.4.16}$$

and

$$\boldsymbol{g}_{m,p}(\boldsymbol{x}, \boldsymbol{y}) = \sum_{j=0}^{m-1} \ell_{j,p}(\boldsymbol{x}, \boldsymbol{y}), \quad m = 1, 2, \ldots.$$

We prove Lemma 13.4.1 by mathematical induction on p. If $p = 1$, the inequalities of (13.4.14) reduce to Lemma 13.3.1. Assuming that the assertion of the lemma is valid for $q \equiv p-1$, write $\boldsymbol{x}_{(s)} = (x_1, x_2, \ldots, x_{s-1}, x_{s+1}, \ldots, x_p)'$ and $\boldsymbol{y}_{(s)} = (y_1, y_2, \ldots, y_{s-1}, y_{s+1}, \ldots, y_p)$, where s is an arbitrary positive integer between 1 and p. We can use (13.4.9) and (13.4.16) to obtain

$$\begin{aligned} \boldsymbol{g}_p(\boldsymbol{x}\,|\,\boldsymbol{y}) &= \boldsymbol{g}_q(\boldsymbol{x}_{(s)}\,|\,\boldsymbol{y}_{(s)}) \cdot \boldsymbol{g}_1(x_s\,|\,y_s) \\ &= \sum_{m=0}^{k-1} \boldsymbol{g}_q(\boldsymbol{x}_{(s)}\,|\,\boldsymbol{y}_{(s)}) u_m(x_s, y_s) \\ &\qquad + \boldsymbol{g}_q(\boldsymbol{x}_{(s)}\,|\,\boldsymbol{y}_{(s)}) \Delta_{k,1}(x_s, y_s). \end{aligned} \tag{13.4.17}$$

The hypothesis of induction asserts that $\boldsymbol{g}_q(\boldsymbol{x}_{(s)} \mid \boldsymbol{y}_{(s)})$ can be put in the form

$$\boldsymbol{g}_q(\boldsymbol{x}_{(s)} \mid \boldsymbol{y}_{(s)}) = g_{k-m,q}(\boldsymbol{x}_{(s)}, \boldsymbol{y}_{(s)}) + \Delta_{k-m,q}(\boldsymbol{x}_{(s)}, \boldsymbol{y}_{(s)}), \tag{13.4.18}$$

$$\|\Delta_{k-m,q}(\boldsymbol{x}_{(s)}, \boldsymbol{y}_{(s)})\|_1 \leq \mathrm{C}_{k-m,q} \cdot \frac{1}{q} \sum_{r \neq s} \lambda_r^{k-m}, \tag{13.4.19}$$

where and in what follows we write either $\gamma_k = \sigma_k$, $\mathrm{C}_{k-m,q} = \mathrm{S}_{k-m,q}$, and $\lambda_r = (y_r \vee y_r^{-1} - 1)$ or $\gamma_k = \tau_k$, $\mathrm{C}_{k-m,q} = \mathrm{T}_{k-m,q}$ and $\lambda_r = |y_r - 1|$, as the case may be.

For each m, substitute (13.4.18) for $\boldsymbol{g}_q(\boldsymbol{x}_{(s)} \mid \boldsymbol{y}_{(s)})$ in (13.4.17) to obtain

$$\boldsymbol{g}_p(\boldsymbol{x} \mid \boldsymbol{y}) = \sum_{m=0}^{k-1} g_{k-m,q}(\boldsymbol{x}_{(s)}, \boldsymbol{y}_{(s)}) u_m(x_s, y_s) + \Delta_{k,p;s}(\boldsymbol{x}_{(s)}, \boldsymbol{y}_{(s)}), \tag{13.4.20}$$

where

$$\begin{aligned}\Delta_{k,p;s}(\boldsymbol{x}_{(s)}, \boldsymbol{y}_{(s)}) &= \boldsymbol{g}_q(\boldsymbol{x}_{(s)} \mid \boldsymbol{y}_{(s)}) \Delta_{k,1}(x_s, y_s) \\ &\quad + \sum_{m=0}^{k-1} \Delta_{k-m,q}(\boldsymbol{x}_{(s)}, \boldsymbol{y}_{(s)}) u_m(x_s, y_s).\end{aligned}$$

The first term in the right-hand side of (13.4.20) is independent of the choice of s and is equal to $g_{k,p}(\boldsymbol{x}, \boldsymbol{y})$. In fact, for $j = 0, 1, \ldots, k-1$, we can rewrite $\ell_{j,p}(\boldsymbol{x}, \boldsymbol{y})$ of (13.4.15) as follows:

$$\begin{aligned}\ell_{j,p}(\boldsymbol{x}, \boldsymbol{y}) &= \sum_{m=0}^{j} \left(\sum{}' \prod_{r \neq s} u_{j_r}(x_r, y_r) \right) u_m(x_s, y_s) \\ &= \sum_{m=0}^{j} \ell_{j-m,q}(\boldsymbol{x}_{(s)}\ \boldsymbol{y}_{(s)}) u_m(x_s, y_s),\end{aligned} \tag{13.4.21}$$

where the summation $\sum'$ extends over the set of j's such that $j_1 + \cdots + j_{s-1} + j_{s+1} + \cdots + j_p = j - m$. The desired result follows by adding both sides of (13.4.21) for $j = 0, 1, \ldots, k-1$ and changing the order of summation on the right-hand side. Thus we have proved

$$\boldsymbol{g}_p(\boldsymbol{x} \mid \boldsymbol{y}) = g_{k,p}(\boldsymbol{x}, \boldsymbol{y}) + \Delta_{k,p;s}(\boldsymbol{x}_{(s)}, \boldsymbol{y}_{(s)}). \tag{13.4.22}$$

As s is arbitrary, adding both sides of (13.4.22) for $s = 1, 2, \ldots, p$ and dividing by p, we obtain

$$\boldsymbol{g}_p(\boldsymbol{x} \mid \boldsymbol{y}) = g_{k,p}(\boldsymbol{x}, \boldsymbol{y}) + \Delta_{k,p}(\boldsymbol{x}, \boldsymbol{y}),$$

where

$$\Delta_{k,p}(\boldsymbol{x}, \boldsymbol{y}) = \frac{1}{p} \sum_{s=1}^{p} \Delta_{k,p;s}(\boldsymbol{x}_{(s)}, \boldsymbol{y}_{(s)}).$$

To prove (13.4.14), use Lemma 13.3.1 and assumptions (13.4.18) and (13.4.19) to obtain

$$\begin{aligned} \left|\Delta_{k,p,s}(\boldsymbol{x},\boldsymbol{y})\right| &\le \left|\Delta_{k,1}(x_s,y_s)\right| \\ &\quad + \sum_{m=1}^{k-1} \left|\Delta_{k-m,q}(\boldsymbol{x}_{(s)},\boldsymbol{y}_{(s)})\right| \cdot \left|u_m(x_s,y_s)\right| \\ &\le \gamma_k \cdot \lambda_s^k + \frac{1}{q}\sum_{m=1}^{k-1} \mathrm{C}_{k-m,q}\cdot\sigma_m \sum_{r\neq s} \lambda_r^{k-m}\cdot\lambda_s^m. \end{aligned} \tag{13.4.23}$$

As s is arbitrary, we have arrived at

$$p\cdot\left|\Delta_{k,p}(\boldsymbol{x},\boldsymbol{y})\right| \le \gamma_k\cdot\sum_{s=1}^{p}\lambda_s^k + \frac{1}{q}\sum_{m=1}^{k-1}\mathrm{C}_{k-m,q}\cdot\sigma_m\sum_{s=1}^{p}\sum_{r\neq s}\lambda_r^{k-m}\cdot\lambda_s^m. \tag{13.4.24}$$

But as the function $\lambda_r^{k-x}\cdot\lambda_s^x + \lambda_r^x\cdot\lambda_s^{k-x}$ of x, defined on the finite interval $[0,k]$, is convex and is equal to $\lambda_r^k+\lambda_s^k$ at both ends of the interval, it follows that $\lambda_r^{k-m}\cdot\lambda_s^m + \lambda_r^m\cdot\lambda_s^{k-m} \le \lambda_r^k+\lambda_s^k$ for all $m\le k$, and in particular,

$$\begin{aligned} \sum_{s=1}^{p}\sum_{r\neq s}\lambda_r^{k-m}\cdot\lambda_s^m &= \sum_{r,s=1}^{p}\lambda_r^{k-m}\cdot\lambda_s^m - \sum_{s=1}^{p}\lambda_s^k \\ &= (1/2)\sum_{r,s=1}^{p}(\lambda_r^{k-m}\cdot\lambda_s^m + \lambda_r^m\cdot\lambda_s^{k-m}) - \sum_{s=1}^{p}\lambda_s^k \le q\sum_{r=1}^{p}\lambda_r^k. \end{aligned}$$

From this and (13.4.24), follow inequalities (13.4.14) with the constants S and T satisfying the relations (13.4.12) and (13.4.13), respectively, as was to be proved. □

Remark 13.4.1 For small values of k, the S and T are:

$$\mathrm{S}_{1,p} = p\sigma_1,\quad \mathrm{S}_{2,p} = p\sigma_2 + p(p-1)\sigma_1^2/2,\ldots$$
$$\mathrm{T}_{1,p} = p\tau_1,\quad \mathrm{T}_{2,p} = p\tau_2 + p(p-1)\sigma_1\tau_1/2,\ldots.$$

To state our main theorem, we need:

Definition 13.4.1 For a set of positive numbers S and T, symmetric matrix S (possibly random), integers $k\,(\ge 1)$ and $\rho\,(=1$ or $2)$, such that the matrix $(\mathrm{S}^\rho - \mathrm{I})^k$ is positive semidefinite (a.s.), let the quantity $\Delta_{k;p}(\mathrm{S},\mathrm{T};\mathrm{S},\rho)$ be defined by

$$\Delta_{k;p}(\mathrm{S},\mathrm{T};\mathrm{S},\rho) \equiv \begin{cases} \dfrac{\mathrm{T}}{p}\cdot\mathrm{tr}\,(\mathrm{S}^\rho-\mathrm{I})^k, & \text{and} \\ \dfrac{\mathrm{S}}{p}\cdot\mathrm{tr}\,(\mathrm{S}^\rho-\mathrm{I})^k, & \text{if } \mathrm{S}\ge\mathrm{I}\ \text{(a.s.)}. \end{cases} \tag{13.4.25}$$

The following theorem is a direct consequence of Lemma 13.4.1.

Theorem 13.4.1 *Let $Z_1, \ldots, Z_p$ be a set of i.i.d. random variables, and let $\mathbf{S} = \mathrm{diag}(S_1, \ldots, S_p)$ be a positive definite random matrix that is independent of $\boldsymbol{Z} = (Z_1, \ldots, Z_p)'$. Then, the pdf $f_{\delta,p}(\boldsymbol{x})$ of the joint distribution of the scale mixture $\boldsymbol{X} = \mathbf{S}^{\delta}\boldsymbol{Z}$ can be approximated by*

$$\boldsymbol{g}_{\delta,k,p}(\boldsymbol{x}) \equiv \boldsymbol{g}_p(\boldsymbol{x}) + \sum_{j=1}^{k-1} \frac{1}{j!} \mathrm{E}\Big(\sum_{r=1}^{p} b^*_{\delta,1}(x_r)(S_r^{\rho} - 1) \Big)^j \cdot \boldsymbol{g}_p(\boldsymbol{x}), \tag{13.4.26}$$

with the error being bounded by

$$\begin{aligned} \left|\Pr\{\boldsymbol{X} \in \mathrm{A}\} - \int_{\mathrm{A}} \boldsymbol{g}_{\delta,k,p}(\boldsymbol{x})\,\mathrm{d}\boldsymbol{x}\right| &\le \frac{1}{2} \left\| f_{\delta,p}(\boldsymbol{x}) - \boldsymbol{g}_{\delta,k,p}(\boldsymbol{x}) \right\|_1 \\ &\le \frac{1}{2} \mathrm{E}[\boldsymbol{\Delta}_{k;p}(\mathrm{s}_{\delta,k,p}, \mathrm{T}_{\delta,k,p};\, \mathbf{S}, \rho)]. \end{aligned} \tag{13.4.27}$$

Remark 13.4.2 For $k = 2$ and 3 the approximation (13.4.26) becomes

$$\boldsymbol{g}_{\delta,2,p}(\boldsymbol{x}) = \boldsymbol{g}_p(\boldsymbol{x}) + \sum_{u=1}^{p} b_{\delta,1}(x_r)\mathrm{E}(S_u^{\rho} - 1) \cdot \boldsymbol{g}_p(\boldsymbol{x}), \tag{13.4.28}$$

$$\begin{aligned} \boldsymbol{g}_{\delta,3,p}(\boldsymbol{x}) &= \boldsymbol{g}_{2,p}(\boldsymbol{x}) + \frac{1}{2} \sum_{u=1}^{p} b_{\delta,2}(x_u)\mathrm{E}(S_u^{\rho} - 1)^2 \cdot \boldsymbol{g}_p(\boldsymbol{x}) \\ &\quad + \sum_{u \ne v} b_{\delta,1}(x_u) b_{\delta,1}(x_v) \mathrm{E}[(S_u^{\rho} - 1)(S_v^{\rho} - 1)] \cdot \boldsymbol{g}_p(\boldsymbol{x}), \end{aligned} \tag{13.4.29}$$

respectively.

13.4.2 Normal Case

The normal case with $\rho = 2$ and $Z_j \sim \mathrm{N}(0,1)$ deserves special attention because of its characteristic property. Whenever the scale factor $\mathbf{S}_0$ is a diagonal matrix $\mathbf{S}_0 = \mathrm{diag}(S_1, \ldots, S_p)$, the approximation (13.4.11) to the conditional density

$$\varphi_{k,p}(\boldsymbol{x} \,|\, \mathbf{S}_0^2) = |\mathbf{S}_0|^{-\delta} \varphi(\mathbf{S}_0^{-\delta}\boldsymbol{x}) = \prod_{u=1}^{p} s_u^{-\delta} \varphi(x_u s_u^{-\delta})$$

reduces to

$$\varphi_{\delta,k,p}(\boldsymbol{x}) \equiv \varphi(\boldsymbol{x}) + \sum_{j=1}^{k-1} \frac{\delta^j}{2^j j!} \left(\sum_{r=1}^{p} Q^*_{\delta,2}(x_r)(s_r^2 - 1) \right)^j \varphi(\boldsymbol{x}), \tag{13.4.30}$$

where $Q_{1,2}(x) = H_2(x)$ and $Q_{-1,2}(x) = K_2(x)$. For details about them; see (13.3.12), see also Convention 1. Note, in particular, that we have

$$\Big(\sum_{r=1}^{p} Q^*_{1,2}(x_r)(s_r^2-1)\Big)^j \varphi(\boldsymbol{x}) = \Big(\sum_{r=1}^{p} H^*_2(x_r)(s_r^2-1)\Big)^j \varphi(\boldsymbol{x})$$
$$= \Big(\sum_{r=1}^{p}(s_r^2-1)\frac{\partial^2}{\partial x_r^2}\Big)^j \varphi(\boldsymbol{x})$$
$$= \Big(\boldsymbol{\partial}'_{\boldsymbol{x}}(\mathrm{S}_0^2-\mathrm{I})\boldsymbol{\partial}_{\boldsymbol{x}}\Big)^j \varphi(\boldsymbol{x}). \qquad (13.4.31)$$

In this way, using Theorem 13.4.1, we first arrive at

Theorem 13.4.2 *Assume that* $\boldsymbol{Z}$ *follows the standard normal distribution* $\mathrm{N}(\mathbf{0},\mathrm{I}_p)$ *and that the random matrix* $\mathbf{S}_0 = \mathrm{diag}(S_1, S_2, \dots, S_p)$ *is positive definite. Then the pdf* $f_{\delta,p}(\boldsymbol{x})$ *of the scale mixture* $\boldsymbol{X} = \mathbf{S}^{\delta}\boldsymbol{Z}$ *is approximated by*

$$\varphi_{\delta,k,p}(\boldsymbol{x}) \equiv \varphi(\boldsymbol{x}) + \sum_{j=1}^{k-1}\frac{\delta^j}{2^j j!}\mathrm{E}\left[\left(\sum_{r=1}^{p} Q^*_{\delta,2}(x_r)(s_r^2-1)\right)^j\right]\varphi(\boldsymbol{x}) \qquad (13.4.32)$$

with the error bounded by

$$\big\|f_0(x) - \mathrm{E}[\varphi_{1,k,p}(\boldsymbol{x},\mathbf{S}_0)]\big\|_1 \le \Delta_{k;p}(\mathrm{S}_{1,k,p}, \mathrm{T}_{1,k,p}; \mathbf{S}_0, 2). \qquad (13.4.33)$$

The relation (13.4.31) makes it possible to generalize Theorem 13.4.2 to the case where the scale-factor matrix is not necessarily diagonal.

Theorem 13.4.3 *Assume that* $\boldsymbol{Z}$ *follows the standard normal distribution* $\mathrm{N}(\mathbf{0},\mathrm{I}_p)$ *and that the random matrix* $\mathbf{S}$ *is positive definite. Then the pdf* $f_p(\boldsymbol{x})$ *of the scale mixture* $\boldsymbol{X} = \mathbf{S}\boldsymbol{Z}$ *is approximated by*

$$\varphi_{1,k,p}(\boldsymbol{x}) = \varphi(\boldsymbol{x}) - \sum_{j=1}^{k-1}\frac{1}{2^j j!}\mathrm{E}\boldsymbol{\partial}_{\boldsymbol{x}}\Big[\Big(\boldsymbol{\partial}'_{\boldsymbol{x}}(\mathrm{S}^2-\mathrm{I})\boldsymbol{\partial}_{\boldsymbol{x}}\Big)^j\Big]\varphi(\boldsymbol{x}). \qquad (13.4.34)$$

Its error in the L_1*-norm is bounded by*

$$\big\|f_p(\boldsymbol{x}) - \varphi_{1,k,p}(\boldsymbol{x})\big\|_1 \le \mathrm{E}[\Delta_{k;p}(\mathrm{S}_{1,k,p}, \mathrm{T}_{1,k,p}; \mathbf{S}, 2)]. \qquad (13.4.35)$$

Proof. If $\mathbf{S}$ is given, we can find an orthogonal matrix T that diagonalizes it: $T'\mathbf{S}T = \mathbf{S}_0 \equiv (s_1, \dots, s_p)$. Moreover, putting $\boldsymbol{v} = T'\boldsymbol{x}$, we obtain

$$\boldsymbol{\partial}'_{\boldsymbol{x}}(\mathrm{S}^2-\mathrm{I})\boldsymbol{\partial}_{\boldsymbol{x}} = \boldsymbol{\partial}'_{\boldsymbol{v}}(\mathrm{S}_0^2-\mathrm{I})\boldsymbol{\partial}_{\boldsymbol{v}}.$$

It then follows from Theorem 13.4.2 that

$$\begin{aligned}\Delta_{1,k,p} &= \left\|f_p(\boldsymbol{x}) - \varphi_{1,k,p}(\boldsymbol{x})\right\|_1 = \left\|\mathrm{E}[\varphi(\boldsymbol{x}|\,\mathbf{S})] - \mathrm{E}[\varphi_{1,k,p}(\boldsymbol{x},\mathbf{S})]\right\|_1 \\ &= \left\|\mathrm{E}[\varphi(\boldsymbol{v}|\,\mathbf{S}_0)] - \mathrm{E}[\varphi_{1,k,p}(\boldsymbol{v},\mathbf{S}_0)]\right\|_1 = \left\|f_0(\boldsymbol{v}) - \varphi_{1,k,p}(\boldsymbol{v})\right\|_1 \\ &\le \mathrm{E}[\boldsymbol{\Delta}_{k;p}(\mathrm{s}_{1,k,p},\mathrm{T}_{1,k,p};\,\mathbf{S}_0,2)] = \mathrm{E}[\boldsymbol{\Delta}_{k;p}(\mathrm{s}_{1,k,p},\mathrm{T}_{1,k,p};\,\mathbf{S},2)],\end{aligned}$$

as was to be proved. □

Table 13.4.1 Numerical values of $\mathrm{s}_{1,k,p}$ for N(0, 1)

k	$\mathrm{s}_{1,1,p}$	$\mathrm{s}_{1,2,p}$	$\mathrm{s}_{1,3,p}$	$\mathrm{s}_{1,4,p}$	$\mathrm{s}_{1,5,p}$	$\mathrm{s}_{1,6,p}$
$p = 1$	0.484	0.351	0.288	0.251	0.207	0.206
2	0.968	0.935	0.915	0.902	0.856	0.869
3	1.452	1.754	1.994	2.201	2.331	2.501
4	1.936	2.806	3.639	4.448	5.168	5.917

Table 13.4.2 Numerical values of $\mathrm{T}_{1,k,p}$ for N(0, 1)

k	$\mathrm{T}_{1,1,p}$	$\mathrm{T}_{1,2,p}$	$\mathrm{T}_{1,3,p}$	$\mathrm{T}_{1,4,p}$	$\mathrm{T}_{1,5,p}$	$\mathrm{T}_{1,6,p}$
$p = 1$	2.000	2.142	2.227	2.287	2.333	2.371
2	4.000	5.252	6.190	6.976	7.670	8.261
3	6.000	9.330	12.36	15.25	18.06	20.71
4	8.000	14.38	21.20	28.51	36.29	44.29

Numerical values of $\mathrm{s}_{1,k,p}$ and $\mathrm{T}_{1,k,p}$ for $\le k,\ p \le 6$ are given in Tables 13.4.1 and 13.4.2.

Approximation $\varphi_{\text{-}1,k,p}(\boldsymbol{x})$ to the density $\mathrm{E}[|\mathbf{S}|\varphi(\mathbf{S}\boldsymbol{x})]$ is obtained by differentiating $\Phi_{\text{-}1,k,p}(\boldsymbol{x})$, the approximation to the cdf $F_p(\boldsymbol{x}) = \mathrm{E}[\Phi(\mathbf{S}\boldsymbol{x})]$. The following theorem provides some examples without proof. Note that the relation $T'(\mathbf{S}-\mathrm{I})T = \mathbf{S}_0 - \mathrm{I}$ implies that $T'(\mathbf{S}-\mathrm{I})^2T = (\mathbf{S}_0-\mathrm{I})^2$.

Theorem 13.4.4 *Assuming the same conditions as in Theorem* 13.4.3, *the pdf* $f_p(\boldsymbol{x})$ *of the random vector* $\boldsymbol{X} = \mathbf{S}^{-1}\boldsymbol{Z}$ *is approximated by*

$$\varphi_{\text{-}1,2,p}(\boldsymbol{x}) = \varphi(\boldsymbol{x}) + \frac{1}{2}\mathrm{E}\big[\partial'_{\boldsymbol{x}}(\mathbf{S}^2-\mathrm{I})\partial_{\boldsymbol{x}}\big]\varphi(\boldsymbol{x}), \tag{13.4.36}$$

$$\begin{aligned}\varphi_{\text{-}1,3,p}(\boldsymbol{x}) &= \varphi(\boldsymbol{x}) - \sum_{j=1}^{2}\frac{(-1)^j}{2^j j!}\mathrm{E}\big[(\partial'_{\boldsymbol{x}}(\mathbf{S}^2-\mathrm{I})\partial_{\boldsymbol{x}})^j\big]\varphi(\boldsymbol{x}) \\ &\quad -\frac{1}{2}\mathrm{E}\Big[\partial'_{\boldsymbol{x}}(\mathbf{S}^2-\mathrm{I})^2\partial_{\boldsymbol{x}}\Big]\varphi(\boldsymbol{x}),\end{aligned} \tag{13.4.37}$$

$$\begin{aligned}\varphi_{-1,4,p}(\boldsymbol{x}) &= \varphi(\boldsymbol{x}) + \sum_{j=1}^{3} \frac{(-1)^j}{2^j j!} \mathrm{E}\Big[(\partial'_{\boldsymbol{x}}(\mathbf{S}^2 - \mathrm{I})\partial_{\boldsymbol{x}})^j\Big]\varphi(\boldsymbol{x}) \\ &\quad + \frac{1}{4}\mathrm{E}\Big[\partial'_{\boldsymbol{x}}(\mathbf{S}^2 - \mathrm{I})^2\partial_{\boldsymbol{x}} - \partial'_{\boldsymbol{x}}(\mathbf{S}^2 - \mathrm{I})^2\partial_{\boldsymbol{x}} \cdot \partial'_{\boldsymbol{x}}(\mathbf{S}^2 - \mathrm{I})\partial_{\boldsymbol{x}} \\ &\qquad -2\partial'_{\boldsymbol{x}}(\mathbf{S}^2 - \mathrm{I})^3\partial_{\boldsymbol{x}}\Big]\varphi(\boldsymbol{x}). \end{aligned} \tag{13.4.38}$$

with

$$\left\| f_p(\boldsymbol{x}) - \varphi_{-1,k,p}(\boldsymbol{x}) \right\|_1 \le \mathrm{E}[\mathbf{\Delta}_{k;p}(\mathrm{S}_{-1,k,p}, \mathrm{T}_{-1,k,p}; \mathbf{S}, 2)]$$

Some numerical values of $\mathrm{S}_{-1,k,p}$ and $\mathrm{T}_{-1,k,p}$ are given in Tables 13.4.3 and 13.4.4.

Table 13.4.3 Numerical values of $\mathrm{S}_{-1,k,p}$ for $\mathrm{N}(0, 1)$

k	$\mathrm{S}_{-1,1,p}$	$\mathrm{S}_{-1,2,p}$	$\mathrm{S}_{-1,3,p}$	$\mathrm{S}_{-1,4,p}$	$\mathrm{S}_{-1,5,p}$	$\mathrm{S}_{-1,6,p}$
$p = 1$	0.484	0.317	0.243	0.201	0.173	0.152
2	0.968	0.868	0.793	0.737	0.693	0.657
3	1.452	1.653	1.762	1.831	1.877	1.910
4	1.936	2.672	3.264	3.759	4.185	4.559

Table 13.4.4 Numerical values of $\mathrm{T}_{-1,k,p}$ for $\mathrm{N}(0, 1)$

k	$\mathrm{T}_{-1,1,p}$	$\mathrm{T}_{-1,2,p}$	$\mathrm{T}_{-1,3,p}$	$\mathrm{T}_{-1,4,p}$	$\mathrm{T}_{-1,5,p}$	$\mathrm{T}_{-1,6,p}$
$p = 1$	2.000	2.000	2.000	2.000	2.000	2.000
2	4.000	4.968	5.601	6.087	6.488	6.833
3	6.000	8.904	11.28	13.35	15.22	16.95
4	8.000	13.81	19.48	25.08	30.61	36.09

Example 13.4.1 Let $\boldsymbol{Y}_1$, $\boldsymbol{Y}_2$, ..., vY_{L} be a random sample from the m-variate normal distribution $\mathrm{N}(B'\boldsymbol{\beta}, \Sigma)$, where B is a given $p \times m$ matrix of rank p ($\le m$), $\boldsymbol{\beta}$ is a p-vector of unknown parameters, and $\mathbf{\Sigma}$ is an unknown positive definite matrix. The exact distribution of the maximum likelihood estimate

$$\hat{\boldsymbol{\beta}} = (\mathrm{B}\hat{\mathbf{\Sigma}}^{-1}\mathrm{B}')^{-1}\mathrm{B}\hat{\mathbf{\Sigma}}^{-1}\bar{\boldsymbol{Y}},$$

where $\bar{\boldsymbol{Y}} = N^{-1}\sum_{j=1}^{N} \boldsymbol{Y}_j$ and $\hat{\boldsymbol{\Sigma}} = N^{-1}\sum_{j=1}^{N}(\boldsymbol{Y}_j - \bar{\boldsymbol{Y}})(\boldsymbol{Y}_j - \bar{\boldsymbol{Y}})'$, was derived and studied by Gleser and Olkin (1972). The random vector

$$\boldsymbol{X}_0 \equiv \boldsymbol{Z} - \mathbf{L}\mathbf{W}^{-1/2}\boldsymbol{U}$$

where $\boldsymbol{Z}$, $\boldsymbol{U}$, $\mathbf{L}$ and $\mathbf{W}$ are mutually independent and such that $\boldsymbol{Z} \sim \mathrm{N}(\mathbf{0}, \mathrm{I})$, $\boldsymbol{U} \sim \mathrm{N}(\mathbf{0}, \mathrm{I}_r)$, and $\mathbf{W} \sim \mathrm{W}_r(n, \mathrm{I}_r)$, while the elements of the $p \times r$ random matrix $\boldsymbol{L}$ constitute a set of i.i.d. standard normal random variables. But the characteristic property of the multivariate normal distribution implies that the distribution of $\boldsymbol{X}_0$ is identical with that of $\boldsymbol{X} = \mathbf{S}\boldsymbol{Z}$, where $\mathbf{S}^2 = \mathrm{I}_p + \mathbf{L}\mathbf{W}^{-1}\mathbf{L}'$. As $\mathbf{L}\mathbf{L}' \sim \mathrm{W}_p(r, \mathrm{I})$ and $\mathbf{L}'\mathbf{L} \sim \mathbf{W}_r(p, \mathrm{I})$, we have [see, e.g., Muirhead, 1982]

$$\mathrm{E}(\mathbf{S}^2 - \mathrm{I}) = \mathrm{E}[\mathbf{L}\mathbf{W}^{-1}\mathbf{L}'] = \frac{r}{n-r-1}\mathrm{I}.$$

For the derivation of $\mathrm{E}[\,\mathrm{tr}\,(\mathbf{S}^2 - \mathrm{I})^k] = \mathrm{E}[\,\mathrm{tr}\,(\mathbf{L}'\mathbf{L}\mathbf{W}^{-1})^k]$, we can use Watamori (1990). For example, if $p = r = 2$, then,

$$\mathrm{E}[(\mathbf{S}^2 - \mathrm{I})] = \frac{2}{n-3}\mathrm{I} \quad \text{and} \quad \mathrm{E}[\,\mathrm{tr}\,(\mathbf{S}^2 - \mathrm{I})^2] = \frac{4\,(5n-9)}{(n-2)(n-3)(n-5)}.$$

Noting that $\mathbf{S}^2 - \mathrm{I} > 0$ with probability 1, and that

$$\begin{aligned}
\mathrm{E}\big[\partial'_{\boldsymbol{x}}(\mathbf{S}^2 - \mathrm{I})\partial_{\boldsymbol{x}})\varphi(\boldsymbol{x})\big] &= \partial'_{\boldsymbol{x}}\mathrm{E}[(\mathbf{S}^2 - \mathrm{I})]\partial_{\boldsymbol{x}}\varphi(\boldsymbol{x}) \\
&= \partial'_{\boldsymbol{x}}\,\mathrm{diag}\left(\frac{2}{n-3}, \frac{2}{n-3}\right)\partial_{\boldsymbol{x}}\varphi(\boldsymbol{x}) \\
&= \frac{2}{n-3}(H_2(x_1) + H_2(x_2))\varphi(\boldsymbol{x}), \\
&= \frac{2}{n-3}(x_1^2 + x_2^2 - 2)\varphi(\boldsymbol{x}),
\end{aligned}$$

we obtain from Theorem 13.4.3 and Table 13.4.1 that the pdf $f(\boldsymbol{x})$ of the variable $\boldsymbol{X} = (X_1, X_2)$ is approximated by

$$\varphi_{1,2,2}(\boldsymbol{x}) = \left(1 + \frac{x_1^2 + x_2^2 - 2}{n-3}\right)\varphi_2(\boldsymbol{x}),$$

with its error being bounded by

$$\Delta(n) \equiv \|f(\boldsymbol{x}) - \varphi_{1,2,2}(\boldsymbol{x})\|_1 \leq \mathbf{\Delta}_{2;2}(n) \equiv \frac{0.935}{2} \cdot \frac{4\,(5n-9)}{(n-2)(n-3)(n-5)}$$

In particular, for any Borel set $\mathrm{A} \subset \mathrm{R}^2$, we have

$$\left|\Pr\{\boldsymbol{X} \in \mathrm{A}\} - \int_{\mathrm{A}} \varphi_{1,2,2}(\boldsymbol{x})\,\mathrm{d}\boldsymbol{x}\right| \leq \frac{1}{2}\,\|f(\boldsymbol{x}) - \varphi_{1,2,2}(\boldsymbol{x})\|_1 \leq \frac{1}{2}\,\mathbf{\Delta}_{2;2}(n).$$

For example, by letting A be a circle with the radius x centered at the origin, the probability of the event $|\boldsymbol{X}| \le x$ can be put in the following form:

$$\mathrm{P}\{|\boldsymbol{X}| \le x\} = 1 - \left(1 + \frac{x^2}{n-3}\right) \mathrm{e}^{-x^2/2} + \theta \cdot \Delta_{2;2}(n),$$

where and in what follows, θ denotes a quantity such that $|\theta| \le 1$.

Similarly, we have

$$\begin{aligned}\mathrm{P}\{X_1 \le x_1,\ X_2 \le x_2\} &= \Phi(x_1)\Phi(x_2) \\ &\quad - \frac{1}{n-3}\big(x_1\varphi(x_1)\Phi(x_2) + x_2\varphi(x_2)\Phi(x_1)\big) + \theta \cdot \Delta_{2;2}(n),\end{aligned}$$

and in particular,

$$\Pr\{\max\{X_1, X_2\} \le x\} = \left(\Phi(x) - \frac{2x}{n-3}\varphi(x)\right)\Phi(x) + \theta \cdot \Delta_{2;2}(n).$$

13.4.3 Gamma Case

In this subsection we consider the multivariate extension of the scale mixture of the gamma distribution with the scale factor matrix $\mathbf{S} = \operatorname{diag}(S_1, \dots, S_p)$. Let

$$\boldsymbol{X} \equiv \mathbf{S}^{-1}\boldsymbol{Z} = (Z_1/S_1, \dots, Z_p/S_p)'$$

be the scale mixture of the multivariate gamma distribution with the joint pdf

$$\boldsymbol{g}(\boldsymbol{z}; \lambda) \equiv \prod_{u=1}^{p} g(z_u; \lambda) = \prod_{u=1}^{p} \frac{1}{\Gamma(\lambda)} z_u^{\lambda-1} \mathrm{e}^{-z_u}. \tag{13.4.39}$$

Taking $\rho = 1$, the approximation (13.4.11) to the conditional density

$$\begin{aligned}\boldsymbol{g}(\boldsymbol{x} \mid \mathbf{S}; \lambda) &= \prod_{u=1}^{p} s_u g(s_u x_u; \lambda) \\ &= \prod_{u=1}^{p} \frac{s_u}{\Gamma(\lambda)} (s_u x_u)^{\lambda-1} \mathrm{e}^{-s_u x_u}.\end{aligned} \tag{13.4.40}$$

becomes

$$g_k(\boldsymbol{x}, \mathbf{S}) \equiv \sum_{j=0}^{k-1} (-1)^j \sum_{(j)} \prod_{u=1}^{p} (S_u - 1)^{j_u} \prod_{u=1}^{p} \frac{1}{j_u!} N_{j_u}(x_u; \lambda)\ \boldsymbol{g}(\boldsymbol{x}; \lambda). \tag{13.4.41}$$

where $N_j(x; \lambda)$ is the polynomial given in (13.3.23).

Chi-square Distribution. If each Z_i follows the chi-square distribution $\chi^2(q)$ with q degrees of freedom, the joint density of $\boldsymbol{Z} = (Z_1, \ldots, Z_p)'$ is

$$\boldsymbol{g}_q(\boldsymbol{z}) \equiv \prod_{u=1}^{p} g_q(z_j) = \prod_{u=1}^{p} \frac{1}{2\,\Gamma(q/2)}\left(\frac{z_u}{2}\right)^{q/2-1} \mathrm{e}^{-z_u/2}, \tag{13.4.42}$$

and the conditional density

$$\begin{aligned} \boldsymbol{g}_q(\boldsymbol{x}|\,\mathbf{S}) &= \prod_{u=1}^{p} s_u \boldsymbol{g}_q(s_u x_u) \\ &= \prod_{u=1}^{p} \frac{s_u}{2\,\Gamma(q/2)}\left(\frac{s_u x_u}{2}\right)^{q/2-1} \mathrm{e}^{-s_u x_u/2} \end{aligned} \tag{13.4.43}$$

is approximated by

$$g_k(\boldsymbol{x}, \mathbf{S}) \equiv \sum_{j=0}^{k-1}(-1)^j \sum_{(j)} \prod_{u=1}^{p} (s_u - 1)^{j_u} \prod_{u=1}^{p} \frac{1}{2^{j_u} j_u!} V_{j_u;q}(x_u)\ \boldsymbol{g}_q(\boldsymbol{x}). \tag{13.4.44}$$

Approximations to the densities $f(\boldsymbol{x}; \lambda)$ and $f_q(\boldsymbol{x})$ of the scale mixtures of multivariate gamma and chi-square distributions are summarized in

Theorem 13.4.5 *The pdf* $f(\boldsymbol{x}; \lambda) = \mathrm{E}_{\mathbf{S}}[\boldsymbol{g}(\boldsymbol{x}|\,\mathbf{S}; \lambda)]$ *is approximated by*

$$g_{k,p}(\boldsymbol{x}; \lambda) \equiv \sum_{j=0}^{k-1}(-1)^j \sum_{(j)} \mathrm{E}\left[\prod_{u=1}^{p} (s_u - 1)^{j_u}\right] \prod_{u=1}^{p} \frac{1}{j_u!} N_{j_u}(x_u; \lambda)\ \boldsymbol{g}(\boldsymbol{x}; \lambda). \tag{13.4.45}$$

The approximation error is bounded by

$$\Delta_{k,p}(\lambda) \equiv \|f(\boldsymbol{x}; \lambda) - g_{k,p}(\boldsymbol{x}; \lambda)\|_1 \le \mathrm{E}[\Delta_k(\mathrm{S}_{k,p}(\lambda), \mathrm{T}_{k,p}(\lambda); \mathbf{S}, 1)]. \tag{13.4.46}$$

Similarly, the pdf $f_q(\boldsymbol{x}) = \mathrm{E}_{\mathbf{S}}[\boldsymbol{g}_q(\boldsymbol{x}|\mathbf{S})]$ *is approximated by*

$$\begin{aligned} g_{k,p;q}(\boldsymbol{x}) &\equiv \sum_{j=0}^{k-1}(-1)^j \sum_{(j)} \mathrm{E}\left[\prod_{u=1}^{p} (s_u - 1)^{j_u}\right] \prod_{u=1}^{p} \frac{1}{2^{j_u} j_u!} V_{j_u;q}(x_u) \cdot \boldsymbol{g}_q(\boldsymbol{x}). \\ &= \sum_{j=0}^{k-1} \sum_{(j)} \mathrm{E}\left[\prod_{u=1}^{p} (s_u - 1)^{j_u}\right] \prod_{u=1}^{p} \frac{(-1)^{j_u}}{2^{j_u} j_u!} V_{j_u;q}(x_u) \cdot \boldsymbol{g}_q(\boldsymbol{x}). \end{aligned} \tag{13.4.47}$$

The approximation error is bounded by

$$\Delta_{k,p;q} \equiv \|f_q(\boldsymbol{x}) - g_{k,p;q}(\boldsymbol{x})\|_1 \le \mathrm{E}[\Delta_k(\mathrm{S}_{k,p;q}, \mathrm{T}_{k,p;q}; \mathbf{S}, 1)]. \tag{13.4.48}$$

The upper bounds Δ_k on the right-hand side of (13.4.46) and (13.4.47) are given in (13.4.25), and S and T will be obtained from σ and τ (see Tables 13.3.7 - 13.3.10) using the recurrence relations of Lemma 13.4.1. Some of their numerical values are only given for the chi-distribution in Tables 13.4.5 and 13.4.6, respectively.

Table 13.4.5 Numerical values of $\mathrm{s}_{k,p;q}$ for $\chi^2(q)$

p	$\mathrm{s}_{2,p;2}$	$\mathrm{s}_{3,p;2}$	$\mathrm{s}_{4,p;2}$	$\mathrm{s}_{2,p;3}$	$\mathrm{s}_{3,p;3}$	$\mathrm{s}_{4,p;3}$	$\mathrm{s}_{2,p;4}$	$\mathrm{s}_{3,p;4}$	$\mathrm{s}_{4,p;4}$
1	0.55	0.45	0.40	0.77	0.70	0.66	1.01	0.99	0.99
2	1.63	1.70	1.74	2.40	2.83	3.20	3.19	4.16	5.13
3	3.25	4.14	4.91	4.88	7.16	9.59	6.54	10.77	15.95
4	5.42	8.17	11.09	8.21	14.50	22.55	11.06	22.10	38.38

Table 13.4.6 Numerical values of $\mathrm{T}_{k,p;q}$ for $\chi^2(q)$

p	$\mathrm{T}_{2,p;2}$	$\mathrm{T}_{3,p;2}$	$\mathrm{T}_{4,p;2}$	$\mathrm{T}_{2,p;3}$	$\mathrm{T}_{3,p;3}$	$\mathrm{T}_{4,p;3}$	$\mathrm{T}_{2,p;4}$	$\mathrm{T}_{3,p;4}$	$\mathrm{T}_{4,p;4}$
1	2.0	2.0	2.0	2.1	2.2	2.3	2.3	2.5	2.7
2	5.5	6.6	6.5	6.1	7.9	8.9	6.7	9.4	12.6
3	10.5	14.8	29.0	11.9	18.8	59.5	13.3	23.1	112.7
4	16.9	27.7	152.7	19.5	36.5	380.0	22.0	46.0	790.6

For some small values of k, the $\boldsymbol{g}_{k,p;q}(\boldsymbol{x})$ are

$$\begin{aligned}
\boldsymbol{g}_{0,p;q}(\boldsymbol{x}) &= \boldsymbol{g}_q(\boldsymbol{x}),\\
\boldsymbol{g}_{1,p;q}(\boldsymbol{x}) &= \boldsymbol{g}_{1,p;q'}(\boldsymbol{x}) - \frac{1}{2}\sum_{u=1}^{p} \mathrm{E}\big[(S_u-1)\big]\, V_{1;q}(x_u)\cdot \boldsymbol{g}_q(\boldsymbol{x}),\\
\boldsymbol{g}_{2,p;q}(\boldsymbol{x}) &= \boldsymbol{g}_{2,p;q}(\boldsymbol{x}) + \Bigg(\frac{1}{2^2 2!}\sum_{u=1}^{p} \mathrm{E}\big[(S_u-1)^2\big]\, V_{2;q}(x_u)\\
&\quad + \frac{1}{2\cdot 2}\sum_{u\neq v} \mathrm{E}[(S_u-1)(S_v-1)]\, V_{1;q}(x_u)V_{1;q}(x_v)\Bigg)\boldsymbol{g}_q(\boldsymbol{x}),\\
\boldsymbol{g}_{3,p;q}(\boldsymbol{x}) &= \boldsymbol{g}_{3,p;q}(\boldsymbol{x}) - \Bigg(\frac{1}{2^3 3!}\sum_{u=1}^{p} \mathrm{E}\big[(S_u-1)^3\big]\, V_{3;q}(x_u)\\
&\quad + \frac{1}{2^2 2!\cdot 2}\sum_{u\neq v} \mathrm{E}\big[(S_u-1)^2(S_u-1)\big]\, V_{2;q}(x_u)V_{1;q}(x_v)\\
&\quad + \frac{1}{2\cdot 2\cdot 2}\sum_{u\neq v\neq u} \mathrm{E}[(S_u-1)(S_u-1)(S_u-1)]\\
&\qquad \times V_{1;q}(x_u)V_{1;q}(x_v)V_{1;q}(x_w)\Bigg)\boldsymbol{g}_q(\boldsymbol{x}).
\end{aligned}$$

Problems

13.1 For a distribution function G, prove the inequality

$$\sup_{x,0<y<1} |G(xy^{-\delta/\rho}) - G(x)| \le \max\{1 - G(0), G(0)\},$$

where $\delta = 1$ or -1, and ρ is a positive integer.

13.2 Let $X_1 = S_1 Z_1$ and $X_2 = S_2 Z_2$ be scale mixtures of the standard normal distribution $\mathrm{N}(0,1)$. Assuming that all the variables involved are mutually independent, show that the variables $Y_1 = X_1 + X_2$ and $Y_2 = X_1 - X_2$ are independent and have an identical distribution, which again is a scale mixture of $\mathrm{N}(0,1)$. What are their respective scale factors?

13.3 Let $X = sZ$ be a scale mixture of the standard normal distribution with the scale factor $s \ge 1$. Then prove (13.2.13) in the following approach. (a) Show that we can express X in (13.1.10) as

$$X = Z - U,$$

where $U = \sqrt{S^2 - 1}V, V \sim \mathrm{N}(0,1)$, and V is independent of Z and S. Therefore, we have

$$F(x) \equiv \mathrm{P}\{SZ \le x\} = \mathrm{P}\{Z - U \le x\} = E_{S,V}\big(\Phi(x+U)\big).$$

(b) Consider the Taylor expansion given by

$$\Phi(x+u) = \sum_{j=0}^{2k-1} \frac{1}{j!}\Phi^{(j)}(x)u^j + R_{2k},$$

where $R_{2k} = \big((2k)!\big)^{-1}\Phi^{(2k)}(x+\theta u)u^{2k}$ and $0 < \theta < 1$.
(c) Show that

$$\mathrm{E}(V^j) = \begin{cases} \dfrac{(2m)!}{2^m m!} & \text{if } j = 2m \ \ \text{(even)}, \\ 0 & \text{if } j = 2m-1 \ \ \text{(odd)}. \end{cases}$$

(d) After checking that

$$\mathrm{E}_{s,V}\left[\sum_{j=0}^{2k-1} \frac{1}{j!}\Phi^{(j)}(x)U^j\right] = \Phi_{1,k}(x),$$

$$|R_{2k}| \le \alpha_{1,k}\mathrm{E}(\mathbf{S}^2 - 1)^k,$$

obtain (13.2.13).

13.4 Let X be a random variable with the pdf f. Let g be an integrable real function such tat $\int g(x)\mathrm{d}x = 1$, but that is not necessarily a pdf (i.e., g(x) may take negative values). Prove that the inequality

$$\left|\Pr\{X \in A\} - \int_A g(x)\mathrm{d}x\right| \leq \frac{1}{2}\int_{-\infty}^{\infty} |f(x) - g(x)|\mathrm{d}x$$

holds for any Borel subset A ⊂ R.

Hint: Use the following relations:

$$\left|\Pr\{X \in A\} - \int_A g(x)\mathrm{d}x\right| \leq \int_A |f(x) - g(x)|\mathrm{d}x,$$

$$\left|\Pr\{X \in A^c\} - \int_{A^c} g(x)\mathrm{d}x\right| \leq \int_{A^c} |f(x) - g(x)|\mathrm{d}x,$$

$$\left|\Pr\{X \in A\} - \int_A g(x)\mathrm{d}x\right| = \left|\Pr\{X \in A^c\} - \int_{A^c} g(x)\mathrm{d}x\right|.$$

13.5 For a distribution function G, let $\Delta_{\delta,k}(x, y)$, and $\alpha_{\delta,k}$ and $\beta_{\delta,k}$ be the quantities defined by (13.2.7) and (13.2.11), respectively. For $\delta = \pm 1$ and $\rho = 1 or 2$, prove the following:

(1) $\sup_{x,0<y<1} |G(xy^{-\delta/\rho}) - G(x)| \leq \max\{1 - G(0), G(0)\}.$

(2) $\sup_x \left|\Delta_{\delta,k}(x, y)\right| \leq \alpha_{\delta,k}\, \left(y \vee y^{-1} - 1\right)^k.$

(3) $\beta_{\delta,k} = \sup_{x\in D,y>0} \mathrm{J}_{\delta,k}(x, y).$

(4) $\alpha_{\delta,k} \leq \beta_{\delta,k}.$

Hint: For the proof of (2), use (13.2.9).

13.6 Let T be the random variable having t-distribution with n degrees of freedom whose density $f(x; n)$ is given by (13.1.3).

(1) Using that $-\log(1 - x) = x + (1/2)x^2 + \mathrm{O}(x^3)$ and $\Gamma(z + h)/\Gamma(z) = z^h[1 + (1/2)h(h - 1)z^{-1} + \mathrm{O}(z^{-2})]$, show that

$$f(x; n) = \varphi(x)\left[1 + \frac{1}{4n}(x^4 - 2x^2 - 1) + \mathrm{O}(n^{-2})\right]$$

and obtain an approximation [Fisher, 1925)] for $\Pr\{T \leq x\}$ by formally integrating the expansion

$$\Phi_a(x) \equiv \Phi(x) - \frac{1}{4n}(x^3 + x)\varphi(x).$$

(2) Show that

$$\begin{aligned}\sup|F_n(x)-\Phi_a(x)| &\le \frac{n+4}{4n^3}+\frac{1}{6n^2}\sup\varphi(x)|x^5+2x^3+3x| \\ &= \frac{n+4}{4n^3}+\frac{8}{n^2}\times 0.0608.\end{aligned}$$

Hint: For the proof of (2), use Example 13.2.1 and Table 13.2.3.

13.7 Consider a multivariate extension of Problem 13.3 [Fujikoshi and Shimizu, 1989] . Let $\boldsymbol{X} = \mathbf{S}\boldsymbol{Z}$ be a scale mixture of $\mathrm{N}_p(\mathbf{0}, \mathrm{I}_p$ with $\mathbf{S} - \mathrm{I}_p \ge \mathrm{O}$.

(1) prove the equivalence

$$\mathbf{S}^2 - \mathrm{I}_p > \mathrm{O} \iff \mathbf{S}^2 = \mathrm{I}_p + \mathbf{B}\mathbf{B}',$$

where $\mathbf{B}$ is a $p \times q$ matrix with $q \le p$. Prove also that we can express $\boldsymbol{X}$ as

$$\boldsymbol{X} = \boldsymbol{Z} - \mathbf{B}\boldsymbol{U},$$

where $\boldsymbol{U} \sim \mathrm{N}_q(\mathbf{0}, \mathrm{I}_q)$, and $\boldsymbol{U}$ is independent of $\boldsymbol{Z}$ and $\mathbf{B}$.

(2) Show that

$$F(\boldsymbol{x}) \equiv \mathrm{P}\{\mathbf{S}\boldsymbol{Z} \le \boldsymbol{x}\} = \mathrm{E}_{\mathbf{B},\boldsymbol{U}}\left[\Phi_p(\boldsymbol{x} + \mathbf{B}\boldsymbol{U})\right],$$

where $\Phi_p(\boldsymbol{x}) = \prod_{i=1}^{p} \Phi(x_i)$.

(3) Show that a Taylor expansion is given by

$$\Phi_p(\boldsymbol{x} + \mathbf{B}\boldsymbol{U}) = \Phi_p(\boldsymbol{x}) + \sum_{j=0}^{2k-1} \frac{1}{j!}(\boldsymbol{U}'\mathbf{B}\boldsymbol{\partial})^j \Phi_p(\boldsymbol{x}) + R_{2k},$$

where $R_{2k} = ((2k)!)^{-1}(\boldsymbol{U}'\mathbf{B}\boldsymbol{\partial})^{2k}\Phi_p(\boldsymbol{x} + \theta\mathbf{B}\boldsymbol{U}), 0 < \theta < 1$ and $\boldsymbol{\partial} = (\partial/\partial x_1, \ldots, \partial/\partial x_p)'$.

(4) Noting that for a $p \times 1$ constant vector, the odd moments of $\boldsymbol{a}'\boldsymbol{U}$ are zero, and that $\mathrm{E}[(\boldsymbol{a}'\boldsymbol{U})^{2j}] = (\boldsymbol{a}'\boldsymbol{a})^j j!/\{2^{j/2}(j/2)!\}$, prove the inequality

$$\sup_{\boldsymbol{x}} |F(\boldsymbol{x}) - \Phi_{p;k}(\boldsymbol{x})| \le \frac{m_{p;k}}{(2k)!}\mathrm{E}\left[(\boldsymbol{U}'\mathbf{B}'\mathbf{B}\boldsymbol{U})^k\right],$$

where

$$\Phi_{p;k}(\boldsymbol{x}) = \Phi_p(\boldsymbol{x}) + \sum_{j=1}^{k-1} \frac{1}{j!2^j}(\boldsymbol{\partial}'\mathbf{B}\mathbf{B}'\boldsymbol{\partial})^j \Phi_p(\boldsymbol{x}), \quad \text{and}$$

$$m_{p;k} = \sum_{\alpha_1+\cdots+\alpha_p=2k} \sup_{\boldsymbol{x}} \left| \frac{\partial^{2k}}{(\partial x_1)^{\alpha_1}\cdots(\partial x_p)^{\alpha_p}} \right|.$$

(5) Show that

$$
\begin{aligned}
m_{p,1} &= \sup_{i,j} \sup_{\boldsymbol{x}} \left| \frac{\partial^2}{\partial x_i \partial x_j} \Phi_p(\boldsymbol{x}) \right| = \sup_{x} \left| \Phi(x)^{(2)} \right| = \frac{1}{\sqrt{2\pi e}}, \\
m_{p,2} &= \sup_{i,j,k,\ell} \sup_{\boldsymbol{x}} \left| \frac{\partial^4}{\partial x_i \partial x_j \partial x_k \partial x_\ell} \Phi_p(\boldsymbol{x}) \right| = \sup_{x} \left| \Phi(x)^{(4)} \right| \\
&= \frac{1.38\cdots}{\sqrt{2\pi}}.
\end{aligned}
$$

(6) Using the result (4) with $k = 1$ and $k = 2$, and noting that for a $q \times q$ symmtric matrix Q, $\mathrm{E}[\boldsymbol{U}'\mathrm{Q}\boldsymbol{U}] = \operatorname{tr}\mathrm{Q}$ and $\mathrm{E}[(\boldsymbol{U}'\mathrm{Q}\boldsymbol{U})^2] = (\operatorname{tr}\mathrm{Q})^2 + 2\operatorname{tr}\mathrm{Q}^2$, obtain the follwing results:

$$
\begin{aligned}
\sup_{\boldsymbol{x}} |F(\boldsymbol{x}) - \Phi_p(\boldsymbol{x})| &\le \frac{p}{2\sqrt{2\pi e}} \mathrm{E}[\operatorname{tr}(\Sigma - \mathrm{I}_p)], \\
\sup_{\boldsymbol{x}} |F(\boldsymbol{x}) - \Phi_{p;2}(\boldsymbol{x})| &\le \frac{1.39p^2}{4!\sqrt{2\pi}} \mathrm{E}\left[\{\operatorname{tr}(\Sigma - \mathrm{I}_p)\}^2 + \operatorname{tr}(\Sigma - \mathrm{I}_p)^2 \right],
\end{aligned}
$$

where

$$
\Phi_{p;2}(\boldsymbol{x}) = \Phi_p(\boldsymbol{x}) + \sum_{i,j} \mathrm{E}[(\mathbf{B}\mathbf{B}')_{ij}] \Phi_p(\boldsymbol{x}) \frac{1}{\Phi(x_i)\Phi(x_j)} \frac{\partial^2 \Phi(x_i)\Phi(x_j)}{\partial x_i \partial x_j}.
$$

13.8 Let $\mathrm{S}_{k,p}$ and $\mathrm{T}_{k,p}$ be the positive constants determined by the recurrence relations in (13.4.12) and (13.4.13), respectively, with the initial conditions $\sigma_0 = 1$, $\mathrm{S}_{k,0} = \mathrm{T}_{k,0} = 0$, $\mathrm{S}_{k,1} = \sigma_k$, and $\mathrm{T}_{k,1} = \tau_k$ for all $k \ge 1$. Then show that for $k = 1, 2, 3$ and 4, they are expressed as follws:

$$
\begin{aligned}
\mathrm{S}_{1,p} &= p\sigma_1, \quad \mathrm{S}_{2,p} = p\sigma_2 + \frac{1}{2}p(p-1)\sigma_1^2, \\
\mathrm{S}_{3,p} &= p\sigma_3 + p(p-1)\sigma_1\sigma_2 + \frac{1}{6}p(p-1)(p-2)\sigma_1^3, \\
\mathrm{S}_{4,p} &= p\sigma_4 + \frac{1}{2}p(p-1)\sigma_2^2 + p(p-1)\sigma_1\sigma_3 \\
&\quad + \frac{1}{2}p(p-1)(p-2)\sigma_1^2\sigma_2 + \frac{1}{24}p(p-1)(p-2)(p-3)\sigma_1^4, \\
\mathrm{T}_{1,p} &= p\tau_1, \quad \mathrm{T}_{2,p} = p\tau_2 + \frac{1}{2}p(p-1)\tau_1^2, \\
\mathrm{T}_{3,p} &= p\tau_3 + p(p-1)\tau_1\tau_2 + \frac{1}{6}p(p-1)(p-2)\tau_1^3, \\
\mathrm{T}_{4,p} &= p\tau_4 + \frac{1}{2}p(p-1)\tau_2^2 + p(p-1)\tau_1\tau_3 \\
&\quad + \frac{1}{2}p(p-1)(p-2)\tau_1^2\tau_2 + \frac{1}{24}p(p-1)(p-2)(p-3)\tau_1^4.
\end{aligned}
$$

Chapter 14

Approximation to Some Related Distributions

In this chapter we extend the basic theory of approximation to the distributions of scale mixtures to some related distributions. First we study a location and scale mixture of the standard normal distribution. Some approximations and their error bounds are given. We also consider approximating the distributions of the maximum component of the multivariate t- and F-variables and that of scale mixture of F-distribution.

This chapter also includes a subsection on nonuniform error bounds, in which bounds are provided that are improvements to the uniform bounds at the tails of distribution. An alternative method of obtaining approximation to a scale mixture is the use of characterisitc function. It is shown that some previous results are obtained by inverting corresponding characteristic function.

14.1 Location and Scale Mixtures

Suppose that a random variable X is put in the form

$$X = SZ - U, \tag{14.1.1}$$

where S is a positive random variable and (U, S) are jointly independent of Z. Then X is called a *location and scale mixture* of Z with location factor U and scale factor S. The purpose of the present section is to propose some approximations to the d.f. $F(x)$ of X together with the upper bounds for their errors. Here, we confine ourselves to the case when Z follows the standard normal distribution, and we assume without loss of generality that $\mathrm{E}[U] = 0$

and that the variable S is close to 1. The results will be used in asymptotic expansions of the distribution function of the linear discriminant function.

Introducing the variable

$$Y = S^2 \quad \text{or} \quad S = Y^{1/2},$$

the distribution function $F(x)$ of X is given by

$$F(x) = \Pr\{X \le x\} = \mathrm{E}_{(U,Y)}[\Phi(Y^{-1/2}(x+U))]. \tag{14.1.2}$$

For every fixed u and y, we expand the conditional distribution function $\Phi\big(y^{-1/2}(x+u)\big)$ given $U = u$ and $Y = y$ to obtain

$$\Phi(y^{-1/2}(x+u)) = \Phi(x) + \sum_{j=1}^{k-1} \frac{1}{j!}(y-1)^j c_j(x+u)\varphi(x+u) + R_{0,k}(x;\, u, y), \tag{14.1.3}$$

where

$$c_j(x) = -\frac{1}{2^j} H_{2j-1}(x) \quad \text{for } j = 1, \ldots, k-1. \tag{14.1.4}$$

Theorem 13.2.3 asserts that the residual $R_{0,k}(x;\, u, y)$ is bounded by

$$|R_{0,k}(x;\, u, y)| \le \mathbf{\Delta}_k(\alpha_{1,k}, \beta_{1,k}, y^{1/2}, 2). \tag{14.1.5}$$

Now, for each $j\ (= 1, \ldots, k-1)$, expand $c_j(x+u)\varphi(x+u)$ around $u = 0$ up to the $(k-j)$th term and we get

$$c_j(x+u)\varphi(x+u) = \sum_{i=0}^{k-j-1} \frac{1}{i!} u^i c_{(i,j)}(x)\varphi(x) + R_j(x;\, u), \tag{14.1.6}$$

where

$$c_{(i,j)}(x) = -\frac{(-1)^i}{2^{i+j}} H_{i+2j-1}(x) \tag{14.1.7}$$

and

$$R_j(x;\, u) = \frac{1}{(k-j)!} u^{k-j} c_{(k-j,\ j)}(x+\theta_j u)\varphi(x+\theta_j u). \tag{14.1.8}$$

Substituting (14.1.6) into (14.1.3), we obtain

$$\begin{aligned}
\Phi(y^{-1/2}(x+u)) &- R_{0,k}(x;\, u, y) \\
&= \Phi(x) + \sum_{j=1}^{k-1} \sum_{i=0}^{k-j-1} \frac{1}{i!j!} u^i (y-1)^j c_{(i,j)}(x)\varphi(x) + R_{1,k}(x;\, u, y) \\
&= \Phi(x) + \sum_{j=1}^{k-1} \sum_{i=0}^{j} \frac{1}{i!(j-i)!} u^i (y-1)^{j-i} c_{(i,j-i)}(x)\varphi(x) \\
&\quad + R_{1,k}(x;\, u, y)
\end{aligned} \tag{14.1.9}$$

where

$$R_{1,k}(x;\, u, y) = \sum_{j=1}^{k-1} \frac{1}{j!(k-j)!} u^j (y-1)^{k-j} c_{(j,k-j)}(x_u)\varphi(x_u), \tag{14.1.10}$$

with $x_u = x + \theta u$.

To sum up, the conditional distribution function $\Phi((x+u)/s)$ can be approximated by

$$\Phi(x;\, u, s) = \Phi(x) - \sum_{j=1}^{k-1}\sum_{i=0}^{j} \frac{(-1)^i}{2^j i!(j-i)!} u^i (s^2-1)^{j-i} H_{2j-i-1}(x)\varphi(x),$$

with the error

$$\begin{aligned}\Delta_k(x, u, s) &\equiv \Phi((x+u)/s) - \Phi(x; u, s)\\ &= R_{0,k}(x;\, u, s^2) + R_{1,k}(x;\, u, s^2).\end{aligned} \tag{14.1.11}$$

We already know that the first term in the right-hand side is bounded by (14.1.5). To evaluate the second term, we need the following: For $0 \le j < k$,

$$\alpha_{(j,k-j)} \equiv \frac{1}{2^k j!(k-j)!} \sup_x |H_{2k-j-1}(x)| \tag{14.1.12}$$

and

$$\beta_{(k)} = \sup_{x, 0<y<1} \frac{|\Phi(y^{-1/2}x) - \Phi(x; 0, y)|}{|y-1|^k}. \tag{14.1.13}$$

Compare $c_{(i,j)}(x)$, $\alpha_{(j,k-j)}$, and $\beta_{(k)}$ just defined with $c_{1,j}(x)$, $\alpha_{1,k}$, and $\beta_{1,k}$ defined in Sections 13.2.1 and 13.2.2. Note, in particular, that $c_{(0,j)}(x)$ coincides with $c_{1,j}(x)$ used in (13.2.23), that $\alpha_{(0,k)} = \alpha_{1,k}$, and that we can take

$$\beta_{(k)} = \beta_{1,k} = 1/2 \quad \text{for } k \le 6. \tag{14.1.14}$$

Numerical values of α's are given in Table 14.1.1 for the pairs of j and $k-j$ such that $j < k (= i+j) \le 6$. The definitions of $R_{1,k}(u, y, x)$ (see (14.1.10) and $\alpha_{(i,j)}$ (see (14.1.12)) lead us immediately to the following:

$$\mathrm{E}\big[\sup_x |R_{1,k}(x;\, U, S)|\big] \le \sum_{j=1}^{k-1} \alpha_{(j,k-j)} \mathrm{E}\big[|U|^j |S^2 - 1|^{k-j}\big]. \tag{14.1.15}$$

Thus we have arrived at:

Table 14.1.1 Numerical values of $\alpha_{(j,k)}$ $(j < k \le 6)$

k	1	2	3	4	5	6
$\alpha_{(0,k-0)}$	0.1210	0.06883	0.04807	0.03693	0.029972	0.02523
$\alpha_{(1,k-1)}$	–	0.09974	0.07481	0.06234	0.054543	0.04909
$\alpha_{(2,k-2)}$	–	–	0.03442	0.03605	0.036922	0.03747
$\alpha_{(3,k-3)}$	–	–	–	0.01247	0.015584	0.018181
$\alpha_{(4,k-4)}$	–	–	–	–	0.003005	0.004616
$\alpha_{(5,k-5)}$	–	–	–	–	–	0.0007792

Theorem 14.1.1 *The distribution function $F(x)$ of the location scale mixture X of the form* (14.1.1) *can be approximated by*

$$\Phi_k \equiv \Phi(x) - \sum_{j=1}^{k-1}\sum_{i=0}^{j} \frac{(-1)^i}{2^{j-i} i!(j-i)!} \mathrm{E}[U^i(S^2-1)^{j-i}]\, H_{2j-i-1}(x)\varphi(x),$$

with the error being bounded by

$$|F(x) - \Phi_k(x)| \le \mathrm{E}\big[\boldsymbol{\Delta}_k(\alpha_{(0,k)}, 1/2, S, 2)\big] + \sum_{j=1}^{k-1} \alpha_{(j,k-j)} \mathrm{E}\big[|U|^{k-j}|S^2-1|^j\big].$$

14.2 Maximum of Multivariate Variables

14.2.1 Distribution of the Maximum Component of a Multivariate Variable

Let $Z_1, \ldots, Z_p$ be independent and identically distributed random variables following a univariate distribution G with the cdf $G(x)$ and pdf $g(x)$ and let S be a positive random variable independent of Z's. Put

$$X_i = Z_i/S, \quad i = 1, \ldots, p. \tag{14.2.1}$$

Then, the random vector

$$\boldsymbol{X} = (X_1, \ldots, X_p)' \tag{14.2.2}$$

follows the p-variate distribution with the cdf

$$F(\boldsymbol{x}) = \mathrm{E}\prod_{i=1}^{p}[G(x_i S)]. \tag{14.2.3}$$

We are interested in the univariate random variable X that is given simply by

$$X \equiv \max\{X_1, \ldots, X_p\} = S^{-1} \max\{Z_1, \ldots, Z_p\}. \tag{14.2.4}$$

The distribution of X is a scale mixture of the univariate distribution of the variable $Z_{[p]} = \max\{Z_1, \ldots, Z_p\}$ with the cdf

$$G_{[p]}(x) \equiv \Pr\{Z_1 \leq x, \ldots, Z_p \leq x\} = \prod_{i=1}^{p} \Pr\{Z_i \leq x\} = (G(x))^p. \tag{14.2.5}$$

Assuming that the variable S is close to 1, we can use the results of Chapter 13 to obtain an approximation to the cdf $F(x)$ in terms of moments of S together with the cdf $G_{[p]}(x)$ and its density

$$g_{[p]}(x) = p\, g(x)(G(x))^{p-1}. \tag{14.2.6}$$

14.2.2 Multivariate t-Distribution

When the underlying distribution G is the standard normal $\mathrm{N}(0,1)$ and nS^2 follows the chi-square distribution $\chi^2(n)$ with n degrees of freedom, the p-vector $\boldsymbol{X}$ of (14.2.2) follows, by definition, the p-variate t-distribution with the correlation matrix I_p [see, e.g., Kotz and Nadarajah (2004) for its use in relation to the procedures of multiple comparison]. Using the standard notation for the normal distribution (i.e. Φ for the cdf and φ for its pdf), (14.2.5) and pdf (14.2.6) become

$$\Phi_{[p]}(x) = \big(\Phi(x)\big)^p \quad \text{and} \quad \varphi_{[p]}(x) = p\,\varphi(x)\big(\Phi(x)\big)^{p-1}, \tag{14.2.7}$$

respectively. As S tends to 1 with probability 1 as n tends to infinity, we can use the results of Section 13.2.1 to obtain an asymptotic approximation to the distribution function $F_p(x)$ of X, with $\rho = 1/2$ and $\delta = -1$.

Note, in particular, that the conditional distribution function of X given the condition $Y(\equiv S^2) = y$ is

$$F_p(x|\, y) = \Phi_{[p]}(xy^{1/2}) = \big(\Phi(xy^{1/2})\big)^p, \tag{14.2.8}$$

and its approximation corresponding to $G_{\delta,k}(x,y)$ is given by, suppressing the suffix $\delta(= -1)$,

$$\Phi_{k;p}(x,y) = \Phi_{[p]}(x) - \sum_{j=1}^{k-1} \frac{(-1)^j}{j!} c_{j;p}(x)\varphi_{[p]}(x)(y-1)^j, \tag{14.2.9}$$

and Theorem 13.2.1 gives for each $y\,(=s^2)>0$,

$$\begin{aligned}\sup_x |F_p(x|\,y) \;-\; \Phi_{k;p}(x,y)| &\le \mathbf{\Delta}_k(\alpha_k,\beta_k;\, y^{1/2},2)\\ &= \alpha_{k;\,p}(s^2\vee s^{-2}-1)^k,\end{aligned} \tag{14.2.10}$$

where the function $c_{j;p}(x)$ is defined by

$$\frac{\partial^j}{\partial y^j}\Phi_{[p]}(xy^{1/2})\Big|_{y=1} = c_{j;p}(x)\varphi_{[p]}(x), \tag{14.2.11}$$

while the constants α_k and β_k are given by

$$\alpha_{k;p} = \frac{1}{k!}\sup_x \left|c_{k;p}(x)\varphi_{[p]}(x)\right| \tag{14.2.12}$$

and

$$\beta_{k;p} = \sup_x\sup_{y\le 1}\frac{\left|\Phi_p(xy^{1/2})-\Phi_{k,p}(x,y)\right|}{|1-y|^k}. \tag{14.2.13}$$

In the special case of $p=1$, we have $\Phi_{[1]}(x)=\Phi(x)$, and therefore, it is clear from Section 13.2.2, that

$$c_{j;1}(x) = K_{2j-1}(x)/2^j. \tag{14.2.14}$$

For the polynomial $K_{2j-1}(x)$, see Table 13.2.2. If $p>1$, the functions $c_{j;p}(x)$ for $j=1$ to 4 are given as follows:

$$\begin{aligned}
c_{1;p}(x) &= K_1(x)/2,\\
c_{2;p}(x) &= \left(K_3(x)+(p-1)K_1^2(x)Q(x)\right)/2^2,\\
c_{3;p}(x) &= \left(K_5(x)+3(p-1)K_1(x)K_3(x)Q(x)\right.\\
&\qquad \left.+\,(p-1)(p-2)K_1^3(x)Q^2(x)\right)/2^3, \qquad (14.2.15)\\
c_{4;p}(x) &= \Big(K_7(x)+(p-1)\big(3K_3^2(x)+4K_1(x)K_5(x)\big)Q(x)\\
&\qquad +6(p-1)(p-2)K_1^2(x)K_3(x)Q^2(x)\\
&\qquad\quad +(p-1)(p-2)(p-3)K_1^4(x)Q^3(x)\Big)\Big/2^4.
\end{aligned}$$

where $Q(x)=\varphi(x)/\Phi(x)$. Taking the expectation of both sides of (14.2.10) with respect to $Y(=S^2)$, we obtain,

$$\sup_x |F_p(x)-\mathrm{E}[\Phi_{k;p}(x,S^2)]| \le \mathrm{E}[\mathbf{\Delta}_k(\alpha_{k;p},\beta_{k;p};\,S,2)]. \tag{14.2.16}$$

Some numerical values of the $\alpha_{k;p}$ are given in Table 14.2.1. Note that the distribution of S depends on n, and the left-hand side of (14.2.16) goes to zero as n tends to infinity.

Table 14.2.1 Numerical values of $\alpha_{k;p}$

p	1	2	3	4	5	6
$\alpha_{1;\,p}$	0.1210	0.2074	0.2741	0.3284	0.3743	0.4141
$\alpha_{2;\,p}$	0.0792	0.1600	0.2415	0.3224	0.4016	0.4783
$\alpha_{3;\,p}$	0.0608	0.1264	0.1979	0.2740	0.3552	0.4415
$\alpha_{4;\,p}$	0.0502	0.1053	0.1671	0.2230	0.2930	0.3726

14.2.3 Multivariate F-Distribution

Similar arguments can be applied to the distribution of the maximum component of a p-variate scaled F random vector $\boldsymbol{X}$ of (14.2.2), in which $X_i = Z_i/S$, $Z_i \sim \chi^2(q)$, and $nS \sim \chi^2(n)$. Note that for each i, X_i/q follows the F distribution in the usual sense. In other words, the underlying distribution is the chi-square distribution $\chi^2(q)$ with q degrees of freedom and nS^2 follows the chi-square distribution $\chi^2(n)$ with n degrees of freedom.

Thus the maximum component X of the vector $\boldsymbol{X}$ is a scale mixture of the distribution with the cdf and pdf

$$\boldsymbol{G}_{q;\,p}(x) = \big(G_q(x)\big)^p \quad \text{and} \quad \boldsymbol{g}_{q;\,p}(x) = p\, g_q(x)\big(G_q(x)\big)^{p-1},$$

respectively. Taking $\rho = 1$ and $\delta = -1$, we can follow the same line of argument as that for the case of t-distribution to derive an approximation to the distribution function of X

$$F_{q;\,p}(x) = \mathrm{E}[\boldsymbol{G}_{q;\,p}(xS)].$$

The conditional distribution function of X given $Y = y$ (i.e., $S = s$),

$$F_{q;\,p}(x|\, s) = \boldsymbol{G}_{q;\,p}(xs),$$

is approximated by

$$\boldsymbol{G}_{k,q;\,p}(x, s) = \boldsymbol{G}_{q;\,p}(x) - \sum_{j=1}^{k-1} \frac{(-1)^j}{j!} c_{j,q;\,p}(x)\boldsymbol{g}_{q;\,p}(x)(s-1)^j,$$

where the functions $c_{j,q;\,p}(x)$ are defined by

$$\frac{\partial^j}{\partial y^j} vG_{q;\,p}(xy)\Big|_{y=1} = c_{j,q;\,p}(x)\boldsymbol{g}_{q;\,p}(x), \tag{14.2.17}$$

given by

$$
\begin{aligned}
c_{1,q;\,p}(x) &= U_{1;\,q}(x), \\
c_{2,q;\,p}(x) &= (U_{2;\,q}(x) + (p-1)U_{1;\,q}^2(x)R(x))/2, \\
c_{3,q;\,p}(x) &= (U_{3;\,q}(x) + 3(p-1)U_{1;\,q}(x)U_{2;\,q}(x)R(x) \\
&\qquad + (p-1)(p-2)U_{1;\,q}^3(x)R^2(x))/2^2, \qquad (14.2.18) \\
c_{4,q;\,p}(x) &= \Big(U_{4;\,q}(x) + (p-1)(3U_{2;\,q}^2(x) + 4U_{1;\,q}(x)U_{3;\,q}(x))R(x) \\
&\qquad + 6(p-1)(p-2)U_{1;\,q}^2(x)U_{2;\,q}(x)R^2(x) \\
&\qquad + (p-1)(p-2)(p-3)U_{1;\,q}^4(x)R^3(x)\Big)/2^3,
\end{aligned}
$$

where $R(x) = g_q(x)/G_q(x)$. Corresponding to (14.2.16), we have in this case

$$
\sup_x |F_p(x) - \mathrm{E}[\boldsymbol{G}_{k,q;\,p}(x, S)]| \le \mathrm{E}[\boldsymbol{\Delta}_k(\alpha_{k,q;p}, \beta_{k,q;p};\, S, 1)]. \qquad (14.2.19)
$$

Some numerical values of the $\alpha_{k;p}$ are given in Table 14.2.2.

Table 14.2.2 Numerical values of $\alpha_{k;q,p}$

p	2	3	4	5	6
$\alpha_{1;\,2,p}$	0.5207	0.6229	0.7004	0.7003	0.8161
$\alpha_{2;\,2,p}$	0.4905	0.6756	0.8362	0.8362	1.1075
$\alpha_{3;\,2,p}$	0.5003	0.8321	1.2011	1.2010	1.9837
$\alpha_{4;\,2,p}$	0.5024	0.9826	1.8453	1.8453	6.2579
$\alpha_{1;\,4,p}$	0.7365	0.8635	0.9585	1.0347	1.0984
$\alpha_{2;\,4,p}$	0.8233	1.1197	1.3730	1.5952	1.7939
$\alpha_{3;\,4,p}$	0.9756	1.6731	2.4637	3.2919	4.1296
$\alpha_{4;\,4,p}$	0.9928	1.9280	4.2710	9.5657	18.285

14.3 Scale Mixtures of the F-Distribution

With $\delta = \pm 1$, consider a sequence of scale mixtures

$$
X_n = Z_n^\delta S_n, \quad n = 1, 2, \ldots
$$

of the scaled F-distribution $F_q(n)$ with the distribution function $F_q(x;\, n)$ and the density

$$
f_q(x;\, n) = \mathrm{B}_0(q, n)\frac{1}{n}\left(\frac{x}{n}\right)^{q/2-1}\left(1 + \frac{x}{n}\right)^{-(q+n)/2}
$$

where

$$\mathrm{B}_0(q,n) \equiv \frac{\Gamma((q+n)/2)}{\Gamma(q/2)\,\Gamma(n/2)}.$$

This means that Z_n/q follows the F-distribution $\mathrm{F}(q,n)$ with (q,n) degrees of freedom. For an example of this situation, see Section 16.2.3.

Assuming that the scale factor S_n tends to 1 with probability 1, we use Theorem 13.3.1 to derive an approximation to the cdf $G_q(x;\, n)$ of the distribution of X_n for large n. To this end, write $Y = S_n$ and let

$$F_q(xy^{-\delta};\, n) \equiv \int_0^{xy^{-\delta}} f_q(x;\, n)\,\mathrm{d}x$$

be the conditional cdf given the condition $Y = y$. This function is formally approximated by

$$G_{k;\,q}(x; y, n) = F_q(x;\, n) - \delta \sum_{j=1}^{k-1} \frac{1}{j!} C_{j,q}(x; n)(y-1)^j f_q(x;\, n),$$

where $C_{j,q}(x; n)$'s are defined by

$$C_{j,q}(x;n) f_q(x;n) = \left.\frac{\partial^j F_q(xy^{-\delta}; n)}{\partial y^j}\right|_{y=1}. \qquad (14.3.1)$$

Naturally, the functions $C_{j,q}$'s and th approximating function $G_{k;\,q}$ depend on the choice of $\delta = 1$ and $\delta = -1$, In the following, however, we suppress the suffix indicating δ, to make the notation simpler. We also use the notation $a^{(\ell)}$ for a nonzero a and an integer ℓ to mean

$$a^{(\ell)} = a \cdot (a-1) \cdot \cdots \cdot (a-\ell-1), \quad \text{if} \quad \ell > 0, \quad \text{and} \quad a^{(0)} = 1.$$

Assume first that $\delta = 1$. Then, the defining equation (14.3.1) of $C_{j,q}(x; n)$ is equivalent to

$$\begin{aligned} C_{j,q}(x;n) &= x\Big(1+\frac{x}{n}\Big)^{(q+n)/2} \left.\frac{\partial^{j-1}}{\partial y^{j-1}} y^{n/2-1}\Big(y+\frac{x}{n}\Big)^{-(q+n)/2}\right|_{y=1} \\ &= x \sum_{i=0}^{j-1} \binom{j-1}{i} \Big(\frac{n-2}{2}\Big)^{(j-i-1)} \Big(\frac{-(q+n)}{2}\Big)^{(i)} \Big(1+\frac{x}{n}\Big)^i, \quad (14.3.2) \end{aligned}$$

which gives, for example,

$$\begin{aligned} C_{1,q}(x;n) &= x, \\ C_{2,q}(x;n) &= \frac{x}{2(x+n)}[(n-2)x - n(q+2)], \end{aligned}$$

$$\begin{aligned} C_{3,q}(x;n) &= \frac{x}{4(x+n)^2}\big[(n-2)(n-4)x^2 - 2(n-2)n(q+4)x \\ &\quad + n^2(q+2)(q+4))\big]. \end{aligned} \tag{14.3.3}$$

$$\begin{aligned} C_{4,q}(x;n) &= \frac{x}{8(x+n)^3}\big[(n-2)(n-4)(n-6)x^3 \\ &\quad -3n(n-2)(n-4)(6+q)x^2 \\ &\quad\quad +3(-2+n)n^2(q+4)(q+6)x \\ &\quad\quad\quad -n^3(q+2)(q+4)(q+6))\big]. \end{aligned}$$

If $\delta = -1$, equation (14.3.2) leads to

$$\begin{aligned} C_{j,q}(x;n) &= x\Big(1+\frac{x}{n}\Big)^{(q+n)/2} \frac{\partial^{j-1}}{\partial y^{j-1}} y^{n/2-1}\Big(y+\frac{x}{n}\Big)^{-(q+n)/2}\Bigg|_{y=1} \\ &= x\sum_{i=0}^{j-1}\binom{j-1}{i}\Big(\frac{q-2}{2}\Big)^{(j-i-1)}\Big(\frac{-(q+n)}{2}\Big)^{(i)}\Big(1+\frac{n}{x}\Big)^{i}, \end{aligned} \tag{14.3.4}$$

which gives, for example,

$$\begin{aligned} C_{1,q}(x;n) &= x, \\ C_{2,q}(x;n) &= \frac{x}{2(x+n)}[(q-2)(x+n)-(q+n)x], \\ C_{3,q}(x;n) &= \frac{x}{4(x+n)^2}\big[(q-2)(q-4)(x+n)^2 - 2\,(q-2)(q+n)x(x+n) \\ &\quad +(q+n)(q+n+2)x^2\big]. \end{aligned} \tag{14.3.5}$$

$$\begin{aligned} C_{4,q}(x;n) &= \frac{x}{8(x+n)^3}\big[(q-2)(q-4)(q-6)(x+n)^3 \\ &\quad -3(q-2)(q-4)(q+n)x(x+n)^2 \\ &\quad\quad +3(q-2)(q+n)(q+n+2)x^2(x+n) \\ &\quad\quad\quad -(q+n)(q+n+2)(q+n+4)x^3\big]. \end{aligned}$$

Compare the two expressions (14.3.2) and (14.3.4), and two sets of functions (14.3.3) and (14.3.3). Note, in particular that functions $C_{k,q}(x;n)$ are bounded as $n \to \infty$ and $O(x^k)$ as $x \to \infty$ for each fixed n. Also, note the following:

$$\lim_{n\to\infty} C_{k,q}(x;n) = \begin{cases} 2^{-k+1}U_{1,k;q}(x), & \text{if } \delta = 1 \\ 2^{-k+1}U_{-1,k;q}(x), & \text{if } \delta = -1 \end{cases} \tag{14.3.6}$$

where $U_{1,k;q}(x)$ and $U_{-1,k;q}(x)$ are polynomials defined by (13.2.32).

Setting

$$\alpha_{k;\,q}(n) = \sup_x \left| \frac{1}{k!} C_{k,\,q}(x;\, n) f_q(x;\, n) \right|$$

we obtain from the general theory (see Theorem 13.2.1) the following upper bound for the approximation error:

$$\left| F_q(x;\, n) - G_{k;\,q}(x; S, n) \right| \le \alpha_{k;\,q}(n) \mathrm{E}[(S_n \vee S_n^{-1} - 1)^k].$$

For numerical values of α's see Tables 14.3.1 and 14.3.2.

Table 14.3.1 Numerical values of $\alpha_{k;\,q}(n)$ $(\delta = 1)$

n	10	20	50	100	200	300
$\alpha_{2;\,2}(n)$	0.2067	0.2180	0.2254	0.2280	0.2293	0.2301
$\alpha_{2;\,4}(n)$	0.3154	0.3487	0.3731	0.3822	0.3870	0.3899
$\alpha_{2;\,6}(n)$	0.3942	0.4540	0.5019	0.5207	0.5308	0.5371
$\alpha_{4;\,2}(n)$	0.1175	0.1245	0.1291	0.1308	0.1316	0.1321
$\alpha_{4;\,4}(n)$	0.1980	0.2224	0.2407	0.2476	0.2512	0.2534
$\alpha_{4;\,6}(n)$	0.2674	0.3170	0.3630	0.4121	0.4401	0.4582

Table 14.3.2 Numerical values of $\alpha_{k;\,q}(n)$ $(\delta = -1)$

n	10	20	50	100	200	300
$\alpha_{2;\,2}(n)$	0.2278	0.2468	0.2604	0.2654	0.2680	0.2696
$\alpha_{2;\,4}(n)$	0.3371	0.3881	0.4291	0.4452	0.4539	0.4593
$\alpha_{2;\,6}(n)$	0.4102	0.4952	0.5703	0.6018	0.6190	0.6300
$\alpha_{4;\,2}(n)$	0.1446	0.1648	0.1814	0.1880	0.1916	0.1939
$\alpha_{4;\,4}(n)$	0.2317	0.2907	0.3471	0.3721	0.3862	0.3953
$\alpha_{4;\,6}(n)$	0.2950	0.3987	0.5114	0.5657	0.5976	0.6185

14.4 Nonuniform Error Bounds

It is natural for us to expect that the difference

$$R_{\delta,k}(x) \equiv F(x) - G_{\delta,k}(x)$$

between the d.f. $F(x)$ of the scale mixture $X = S^{\delta\rho} Z$ and its approximation

$G_{\delta,k}(x) \equiv \mathrm{E}_s[G_{\delta,k}(x,S)]$ should go to zero quickly as $|x|$ increases. The purpose of this section is to give a refined upper bound (nonuniform bound) for the approximation errors, reflecting its dependence on x as well as the moments of S.

For a continuous positive function $h(x)$ such that monotonically tends to infinity as $|x| \to \infty$, the nonuniform error bounds in the classical form are obtained by estimating

$$\sup_x \left[h(x)|R_{\delta,k}(x)|\right].$$

In general, the nonuniform bounds are improvements on the uniform bounds at the tails of the distribution of X. For $\delta = \pm 1$ and the given function $h(x)$, put

$$\alpha_{\delta,k,h} = \frac{1}{k!}\sup_x \left|c_{\delta,k}(x)g(x)h(x)\right|. \tag{14.4.1}$$

We first prove the following basic lemma:

Lemma 14.4.1 *For the weight function*

$$h(x) = 1 + |x|^r, \quad r > 0, \tag{14.4.2}$$

the error $\Delta_{\delta,k}(x,y)$ satisfies the inequality

$$|\Delta_{\delta,k}(x,y)| \le \frac{\alpha_{\delta,k,h}}{h(x)}\,(y^{r\delta/\rho} \vee 1)(y \vee y^{-1} - 1)^k). \tag{14.4.3}$$

Proof. We use two expressions for $\Delta_{\delta,k}(x,y)$ given in (13.2.7) and (13.2.9):

$$\begin{aligned}\Delta_{\delta,k}(x,y) &= G(xy^{-\delta/\rho}) - G_{\delta,k}(x,y) &(14.4.4)\\ &= \frac{1}{k!}c_{\delta,k}(u)g(u)y_0^{-k}(y-1)^k, &(14.4.5)\end{aligned}$$

where $u = xy_0^{-\delta/\rho}$ and y_0 is a positive number lying between 1 and y. The definition of $\alpha_{\delta,k,h}$ gives

$$|\Delta_{\delta,k}(x,y)| \le \frac{\alpha_{\delta,k,h}}{h(u)}y_0^{-k}|y-1|^k \le \frac{\alpha_{\delta,k,h}}{h(u)}(y \vee y^{-1} - 1)^k \tag{14.4.6}$$

and

$$h(u) \ge \begin{cases} h(x) & \text{if } 0 < y^\delta < 1, \\ y^{-\delta r/\rho}h(x) & \text{if } 1 \le y^\delta. \end{cases} \tag{14.4.7}$$

The claim (14.4.3) easily follows from (14.4.6) and (14.4.7). □

The proof of the following Theorem is straightforward.

Theorem 14.4.1 *For the weight function* (14.4.2), *we have*

$$|R_{\delta,k}(x)| \le \frac{\alpha_{\delta,k,h}}{h(x)} \left[\mathrm{E}\big(S^{r(\delta+1)/2}|S^{\rho}-1|^{k}\big)+\mathrm{E}\big(S^{r(\delta-1)/2}|S^{-\rho}-1|^{k}\big)\right]. \quad (14.4.8)$$

If, in particular, $S \ge 1$ *with probability 1, then*

$$|R_{\delta,k}(x)| \le \frac{\alpha_{\delta,k,h}}{h(x)}\ \mathrm{E}[S^{r(\delta+1)/2}(S^{\rho}-1)^{k}]. \quad (14.4.9)$$

Some numerical values of the α's are given in Tables 14.4.1 – 14.4.6.

Table 14.4.1 Values of $\alpha_{1,k,h}$ for $\mathrm{N}(0,1)$

k	1	2	3	4	5	6
$\alpha_{1,k,1}$	0.1210	0.0688	0.0484	0.0370	0.0300	0.0252
$\alpha_{1,k,1+x^2}$	0.3165	0.1295	0.1016	0.07720	0.06040	0.04876
$\alpha_{1,k,1+x^4}$	0.9533	0.8400	0.4141	0.3748	0.2906	0.2207

Table 14.4.2 Values of $\alpha_{-1,k,h}$ for $\mathrm{N}(0,1)$

k	1	2	3	4	5	6
$\alpha_{-1,k,1}$	0.1210	0.0791	0.0608	0.0501	0.0431	0.0380
$\alpha_{-1,k,1+x^2}$	0.3165	0.3385	0.3619	0.3837	0.4036	0.4220
$\alpha_{-1,k,1+x^4}$	0.9533	1.6004	2.3359	3.1452	4.0184	4.9481

Table 14.4.3 Values of $\alpha_{1,k,h}(\lambda)$ for $G(1,\lambda)$

λ	0.5	1.0	1.5	2.0	2.5	3.0
$\alpha_{1,1,1+x}(\lambda)$	0.4152	0.8400	1.3056	1.8130	2.3606	2.9461
$\alpha_{1,2,1+x}(\lambda)$	0.1843	0.3883	0.6584	1.0173	1.4434	1.9355
$\alpha_{1,3,1+x}(\lambda)$	0.1216	0.3296	0.6337	1.0427	1.5656	2.2100
$\alpha_{1,4,1+x}(\lambda)$	0.09637	0.2727	0.5460	0.9336	1.4535	2.1230

Table 14.4.4 Values of $\alpha_{1,k,h}(\lambda)$ for $G(1,\lambda)$

λ	0.5	1.0	1.5	2.0	2.5	3.0
$\alpha_{1,1,1+x^2}(\lambda)$	0.5419	1.5054	2.967	4.992	7.637	10.96
$\alpha_{1,2,1+x^2}(\lambda)$	0.4392	1.3328	2.799	4.954	7.911	11.78
$\alpha_{1,3,1+x^2}(\lambda)$	0.2225	0.7486	1.727	3.324	5.724	9.127
$\alpha_{1,4,1+x^2}(\lambda)$	0.1943	0.6953	1.683	3.375	6.028	9.920

Table 14.4.5 Values of $\alpha_{-1,k,h}(\lambda)$ for $G(1,\lambda)$

λ	0.5	1.0	1.5	2.0	2.5	3.0
$\alpha_{-1,1,1+x}(\lambda)$	0.4152	0.8400	1.3056	1.8131	2.3606	2.9461
$\alpha_{-1,2,1+x}(\lambda)$	0.4051	0.9066	1.5067	2.2025	2.9908	3.8693
$\alpha_{-1,3,1+x}(\lambda)$	0.4145	0.9821	1.7037	2.5795	3.6105	4.7977
$\alpha_{-1,4,1+x}(\lambda)$	0.4280	1.0559	1.8919	2.9446	4.2228	5.7359

Table 14.4.6 Values of $\alpha_{-1,k,h}(\lambda)$ for $G(1,\lambda)$

λ	0.5	1.0	1.5	2.0	2.5	3.0
$\alpha_{-1,1,1+x^2}(\lambda)$	0.5419	1.5054	2.967	4.992	7.637	10.96
$\alpha_{-1,2,1+x^2}(\lambda)$	0.8555	2.4956	5.087	8.785	13.75	20.11
$\alpha_{-1,3,1+x^2}(\lambda)$	1.2174	3.6520	7.612	13.41	21.34	31.71
$\alpha_{-1,4,1+x^2}(\lambda)$	1.6178	4.9538	10.510	18.81	30.38	45.77

14.5 Method of Characteristic Functions

In Chapter 13 we obtained asymptotic expansions of the distribution of a scale mixture and the error bounds by expanding its distribution function directly. In this section some of the results are obtained by inverting its characteristic function and using the following theorems.

Theorem 14.5.1 *Let F be a distribution function with the density f. Consider an approximation Q to F such that*

$$Q(x) = \int_{-\infty}^{x} q(u)\,\mathrm{d}u \quad \text{and} \quad Q(\infty) = 1.$$

We assume that q is bounded and that

$$\int_{-\infty}^{\infty} |x| f(x)\,\mathrm{d}x < \infty \quad \text{and} \quad \int_{-\infty}^{\infty} |x| \cdot |q(x)|\,\mathrm{d}x < \infty.$$

Further, if f and q have integral Fourier transforms C_f and C_q given by

$$C_f(t) = \int \mathrm{e}^{-itx} f(x) dx, \quad C_q(t) = \int \mathrm{e}^{-itx} q(x) dx,$$

then

$$\int_{-\infty}^{\infty} |t|^{-1} |C_f(t) - C_q(t)|\,\mathrm{d}t < \infty,$$

$$\sup_{x \in R^1} |F(x) - Q(x)| \le \frac{1}{2\pi} \int_{-\infty}^{\infty} |t|^{-1} \big| C_f(t) - C_q(t) \big|\,\mathrm{d}t. \tag{14.5.1}$$

For a proof of Theorem 14.5.1, see Bhattacharya and Ranga Rao (1976, p. 108), in which they treat the case when F is a finite measure and G is a finite signed measure on R^1. We consider the variable

$$X = SZ, \tag{14.5.2}$$

where Z is a random variable with the distribution function G and S is a positive random variable independent of Z. In this case we can use the following theorem instead of Theorem 14.5.1.

Theorem 14.5.2 *Let $\gamma(t)$ be the characteristic function of Z. Assume that G is absolutely continuous and*

$$\int_{-\infty}^{\infty} |t|^{-1} \big|\gamma(t)\big| dt < \infty.$$

Then

$$G(xs^{-1}) - G(x) = \frac{1}{2\pi} \int_{-\infty}^{\infty} \mathrm{e}^{-itx} \big(\gamma(ts) - \gamma(t)\big)(-it)^{-1} dt. \tag{14.5.3}$$

Proof. For a fixed s, consider the distribuion of $X = sZ$. Then the distribution function of X is $G(s^{-1}x)$, and hence its density function is $s^{-1}g(s^{-1}x)$. The characteristic function of X is $\gamma(st)$. Using the inversion theorem (see Theorem 1.1.5, p. 6) we have

$$s^{-1} g(xs^{-1}) - g(x) = \frac{1}{2\pi} \int_{-\infty}^{\infty} \mathrm{e}^{-itx} \{\gamma(ts) - \gamma(t)\}\,\mathrm{d}t.$$

The required result is obtained by integrating the both sides on the range $(0, x)$.

Consider the transformation

$$s^{-1} = y^{\alpha}, \quad (\text{i.e.,} \quad y = s^{-1/\alpha}), \tag{14.5.4}$$

where α is a constant. The values of α may be set as $\alpha = -1, -1/2, 1/2$, or 1 according to the easiness of calculating the first few moments of $s^{-1/\alpha}$ and $s^{1/\alpha}$. Let $G'(x) = g(x)$, and let for $j = 1, \ldots, k$,

$$\frac{\mathrm{d}^j}{\mathrm{d}y^j} G(xy^{\alpha}) = G_{(j)}(x, y),$$

$$G_{(j)}(x, 1) = G_{(j)}(x) = b_j(x) g(x).$$

We used an approximation for $G(xy^{\alpha})$:

$$G_k(x, y) = G(x) + \sum_{j=1}^{k-1} \frac{1}{j!} b_j(x) g(x) (y-1)^j.$$

Suppose that the support of G is R^+ or R, and G satisfies the following assumptions:

A1. G is k times continuously differentiable.

A2. $\sup |b_j(x) g(x)| = \bar{b}_k < \infty$.

Then

$$|G(xy^{\alpha}) - G_k(x, y)| \leq \frac{1}{k!} \bar{b}_k (y \vee y^{-1} - 1)^k,$$

where $a \vee b = \max\{a, b\}$.

On the other hand, consider expanding the characteristic function $\gamma(s) = \gamma(y^{-\alpha})$ in (14.5.3).

$$(\mathrm{d}^j/\mathrm{d}y^j)\gamma(ty^{-\alpha}) = \gamma_{(j)}(t, y),$$

$$\gamma_{(j)}(t, 1) = \gamma_{(j)}(t) = \beta_j(t)\gamma(t).$$

Substituting a Taylor's series expansion for $\gamma(ts) = \gamma(ty^{-\alpha})$ into (14.5.3), we obtain the following relations:

$$\frac{1}{2\pi} \int_{-\infty}^{\infty} \mathrm{e}^{-itx} \gamma(t) \beta_j(t) (-\mathrm{i}t)^{-1} \, \mathrm{d}t = b_j(x) g(x), \quad j = 1, \ldots, k-1,$$

$$\frac{1}{2\pi} \int_{-\infty}^{\infty} \mathrm{e}^{itx} \gamma_{(k)}(t, \tilde{y}) (-\mathrm{i}t)^{-1} \, \mathrm{d}t = G_{(k)}(x) g(x).$$

These relations will be useful in finding the functions $b_j(x)$'s. The remainder term in the method of characteristic functions can be expressed as

$$G(xy^{\alpha}) - G_k(x) = \left[\frac{1}{2\pi} \int_{-\infty}^{\infty} \mathrm{e}^{itx} \gamma_{(k)}(t, \tilde{y}) (-\mathrm{i}t)^{-1} \mathrm{d}t\right] \frac{1}{k!} (y-1)^k. \tag{14.5.5}$$

Here we note that $\gamma_{(k)}(t,\tilde{y}) = \tilde{y}^{-k}\gamma_{(k)}(t\tilde{y}^{-\alpha},1)$. By considering the transformation $t \to v = t\tilde{y}^{-\alpha}$, we can write the right-hand side of (14.5.5) as

$$\left[\frac{1}{2\pi}\int_{-\infty}^{\infty} \mathrm{e}^{\mathrm{i}v\tilde{x}}\gamma(v)\beta_k(v)(-itv)^{-1}\mathrm{d}t\right]\frac{1}{k!}\tilde{y}^{-k}(y-1)^k$$
$$= \frac{1}{k!}b_k(\tilde{x})g(\tilde{x})\tilde{y}^{-k}(1-y)^k,$$

where $\tilde{x} = x\tilde{y}^{\alpha}$. This shows that two methods give the same results.

Next we consider the standardized statistic Z defined by (5.4.28). Then the characteristic function $C_Z(t)$ of Z can be approximated as $C_{Z,s}(t)$ in (5.4.30). The inversion of $C_{Z,s}(t)$ gives an asymptotic approximation $Q_s(x)$ in (5.4.31). Using the inverse Fourier transformation, it is expected to get a uniform bound for the error of the asymptotic expansion above as

$$\begin{aligned}\sup_x |\Pr\{T \le x\} - Q_s(x)| &\le \frac{1}{2\pi}\int_{-\infty}^{\infty}\frac{1}{|t|}\,|C_Z(t) - C_{Z,s}(t)|\,\mathrm{d}t \\ &= \frac{1}{2\pi}(I_1[v] + I_2[v] + I_3[v]),\end{aligned} \tag{14.5.6}$$

where

$$I_1[v] = \int_{-mv}^{mv}\frac{1}{|t|}\,|C_Z(t) - C_{Z,s}(t)|\,dt,$$
$$I_2[v] = \int_{|t|>mv}\frac{1}{|t|}\,|C_{Z,s}(t)|\,dt \quad \text{and} \quad I_3[v] = \int_{|t|>mv}\frac{1}{|t|}\,|C_Z(t)|\,dt$$

with some positive constant $v < 1$. To find a bound for each integral I_1, I_2, and I_3, we heed to evaluate $I_i[v]$, $i = 1,2,3$. For some results, see Wakaki (2009), Kato et al. (2009).

Problems

14.1 Let $X = SZ - U$ be a location and scale mixture of the standard normal distribution. Theorem 14.1.1 gives an asymptotic approximation around $(S,U) = (1,0)$ and its error bound. Extend the result to an asymptotic approximation around $(S,U) = (1,u_0)$ and its error bound. For an asymptotic approximation around $(S,U) = (s_0,u_0)$ and its error bound, see Section 15.1

14.2 Let $X = SZ$ be a scal miture of the standard normal distribution. Then, from Theorem 14.5.2 we have

$$G(xs^{-1}) - G(x) = \frac{1}{2\pi}\int_{-\infty}^{\infty}\mathrm{e}^{-\mathrm{i}tx}\mathrm{e}^{-(1/2)t^2}\left[\mathrm{e}^{(1/2)(\mathrm{i}t^2)(s^2-1)} - 1\right](-it)^{-1}dt.$$

Expanding $\mathrm{e}^{(1/2)(\mathrm{i}t^2)(s^2-1)}$, obtain an asymptotic expansion $\Phi_{1,k}$ (see Theorem 13.2.3) of the distribution function of X.

14.3 (Continuation) For a scale mixture of the gamma distribution $G(1 : \lambda)$, obtain an asymptotic expansion by the method as in Problem 14.2

Chapter 15

Error Bounds for Approximations of Multivariate Tests

The main purpose of this chapter is to give error bounds for asymptotic expansions of the null distributions of Hotelling's T^2 and the lambda statistics when the sample size is large. The results are obtained by expressing these statistics in terms of a multivariate scale mixture, and using error bounds evaluated in L_1-norm. It may be noted that these results also establish the validity of asymptotic expansions in a strong sense.

15.1 Multivariate Scale Mixture and MANOVA Tests

Let $\mathbf{S}_h$ and $\mathbf{S}_e$ be independent $p \times p$ random matrices having the Wishart distributions $\mathrm{W}_p(q, \Sigma)$ and $\mathrm{W}_p(n, \Sigma)$, respectively. Consider two statistics

$$\begin{aligned} T_0^2 &= n \operatorname{tr} \mathbf{S}_h \mathbf{S}_e^{-1}, \\ T_{\mathrm{LR}} &= -n \log\{|\mathbf{S}_e| / |\mathbf{S}_e + \mathbf{S}_h|\}. \end{aligned} \tag{15.1.1}$$

These statistics appear as null distributions of the Lawley–Hotelling criterion and likelihood ratio statistic, respectively, for testing a linear hypothesis in the MANOVA model.

When we consider the distributions of T_0^2 and T_{LR}, we may assume that $\Sigma = I_p$, and it is assumed in the following.

We saw in Chapter 13 that they have identical limiting distributions $\chi^2(pq)$ when n tends to infinity while p and q are kept fixed. We also obtained their asymptotic expansions. In this chapter we apply the results of Chapter 13 to give error bounds for these asymptotic expansions as approximations to respective distributions.

The results in this chapter are mainly based on Fujikoshi et al. (2005) and Fujikoshi and Ulyanov (2006).

In this section we will see that each of the statistics T_0^2 and $T_{\rm LR}$ is expressible as a function $T = T(\boldsymbol{X})$ of multivariate scale mixture with $\delta = -1$ and $\rho = 1$:

$$\boldsymbol{X} = (X_1, \ldots, X_p)', \quad \mathbf{S} = \mathrm{diag}\,(S_1, \ldots, S_p) \quad \text{and}$$

$$\boldsymbol{X} = \mathbf{S}^{-1}\boldsymbol{Z} = (Z_1/S_1, \ldots, Z_p/S_p), \tag{15.1.2}$$

where $Z_1, \ldots, Z_p \sim i.i.d.\ \chi^2(q)$.

Lemma 15.1.1 *The statistic T_0^2 is a simple function of a multivariate scale mixture $\boldsymbol{X} = \mathbf{S}^{-1}\boldsymbol{Z}$:*

$$T_0^2 = X_1 + \cdots + X_p,$$

where $\boldsymbol{Z} = (Z_1, \ldots, Z_p)$ and $Z_1, \ldots, Z_p$ are i.i.d. random variables with the common distribution $\chi^2(q)$, and $S_1 > \cdots > S_p > 0$ are the characteristic roots of $\mathbf{W}$ such that $n\mathbf{W} \sim \mathrm{W}_p(n, I_p)$.

Proof. From the definition of Wishart distribution we can write $\mathbf{S}_h = \mathbf{U}'\mathbf{U}$, where $\mathbf{U}$ is a $q \times p$ random matrix whose elements are independent and identically distributed as $\mathrm{N}(0, 1)$. Therefore, the statistic T_0^2 can be put in the form

$$\begin{aligned} T_0^2 &= n\,\mathrm{tr}\,(\mathbf{U}'\mathbf{U})\mathbf{S}_e^{-1} \\ &= n\,\mathrm{tr}\,(\mathbf{H}'\mathbf{U}'\mathbf{U}\mathbf{H})(\mathbf{H}'\mathbf{S}_e\mathbf{H})^{-1}, \end{aligned}$$

where $\mathbf{H}$ is the orthogonal matrix that diagonalizes the matrix $\mathbf{S}_e$: $\mathbf{H}'\mathbf{S}_e\mathbf{H} = \mathrm{diag}\,(S_1, \ldots, S_p)$. Note that $\mathbf{H}'\mathbf{U}'\mathbf{U}\mathbf{H}$ is independent of $\mathbf{H}$ and that its distribution is identical with that of $\mathbf{U}'\mathbf{U}$. This is obtained by considering the conditional distribution of $\mathbf{H}'\mathbf{U}'\mathbf{U}\mathbf{H}$ given $\mathbf{H}$ and noting that $\mathbf{H}$ and $\mathbf{U}$ are independent. The independence of the variables involved is a consequence of the fact that the diagonal elements of $\mathbf{U}'\mathbf{U}$ are independent and identically distributed as $\chi^2(q)$. □

Lemma 15.1.2 *We can write $T_{\rm LR}$ in terms of a multivariate scale mixture $\boldsymbol{X} = (X_1, \ldots, X_p)' = \mathrm{diag}\,(S_1, \ldots, S_p)\boldsymbol{Z}$:*

$$T_{\rm LR} = n \log\left(1 + \frac{1}{n}X_1\right) + \cdots + n \log\left(1 + \frac{1}{n}X_p\right),$$

where
$$\boldsymbol{Z} = (Z_1, \ldots, Z_p) \quad \textit{with} \quad Z_r \sim \chi^2(q), \quad nS_r \sim \chi^2(m_r)$$
and
$$m_r = n - (r-1), \quad r = 1, \ldots, p.$$
All the variables $\{S_r, Z_r, r = 1, \ldots, p\}$ *are mutually independent.*

Proof. Note that $\Lambda = |\mathbf{S}_e|/|\mathbf{S}_e + \mathbf{S}_h|$ is distributed as a Lambda distribution $\Lambda_p(q,n)$, which, in turn, as it is well known (see Theorem 3.3.3), has identical distribution as a product of p independent beta variables:

$$\Lambda_p(q,n) = \prod_{r=1}^{p} V_r \quad \text{with} \quad V_r \sim \text{Be}\Big(\frac{n-r+1}{2}, \frac{q}{2}\Big).$$

Therefore, we can write

$$-n \log |\mathbf{S}_e|/|\mathbf{S}_e + \mathbf{S}_h| = -\sum_{r=1}^{p} \log V_r,$$

and V_r can be put in the form

$$V_r = \frac{\chi^2_{m_r}}{\chi^2_{m_r} + \chi^2_q} = \left(1 + \frac{1}{n} \cdot \frac{\chi^2_q}{\chi^2_{m_r}/n}\right)^{-1},$$

from which follows the assertion with $S_r = \chi^2_{m_r}/n, r = 1, \ldots, p.$ □

15.2 Function of a Multivariate Scale Mixture

Let $f(\boldsymbol{x})$ be the density function of the random vector $\boldsymbol{X}$ given in (15.1.2), and let T be a statistic defined as a real function of $\boldsymbol{X}$. It follows from Theorem 13.4.5 that the probability of the event $\{T \le t\}$ is given by

$$\Pr\{T \le t\} = \int_{\mathrm{A}} f(\boldsymbol{x})\mathrm{d}\boldsymbol{x} = \int_{\mathrm{A}} g_{k,p;q}(\boldsymbol{x})\mathrm{d}\boldsymbol{x} + R_k, \tag{15.2.1}$$

where $g_{k,p;q}(\boldsymbol{x})$ is defined by (13.4.47), and $\|R_k\|_1 \le \Delta_{k,p;q}$ (see Section 13.4.3 for its definition), while the set A is given, in general, by

$$\mathrm{A} = \{\boldsymbol{x} \in \mathrm{R}^p : \; T(\boldsymbol{x}) \le t\},$$

and in particular,

$$\mathrm{A} = \begin{cases} \Big\{\boldsymbol{x} \in \mathrm{R}^p : \; \sum_{u=1}^{p} x_u \le t\Big\}, & \text{if } T = T_0^2, \\ \Big\{\boldsymbol{x} \in \mathrm{R}^p : \; n\sum_{u=1}^{p} \log(1 + x_u/n) \le t\Big\}, & \text{if } T = T_{\mathrm{LR}}. \end{cases} \tag{15.2.2}$$

The main purpose of the rest of this chapter is to evaluate the integral

$$\mathrm{P}_{k,p}(\mathrm{A}) \equiv \int_{\mathrm{A}} \boldsymbol{g}_{k,p;q}(\boldsymbol{x})\,\mathrm{d}\boldsymbol{x} \tag{15.2.3}$$

for these cases. For explicit forms of $\boldsymbol{g}_{k,p;q}(\boldsymbol{x})$ for $k \leq 3$, see Section 13.4.3. Also take notice of the fact that the A_t for the statistics T_0^2 and T_{LR} are symmetric in the sense that they are invariant under any permutation of coordinates $x_1, \ldots, x_p$. If the set A is symmetric in this sense, we can carry out the integral (15.2.2) by splitting the integral into several parts.

First, we have

$$\mathrm{P}_{k,p}(\mathrm{A}) = Q_0(\mathrm{A}) + Q_1(\mathrm{A}) + \cdots + Q_{k-1}(\mathrm{A}), \tag{15.2.4}$$

where for $j = 0, 1, \ldots, k-1$,

$$Q_j(\mathrm{A}) = \sum_{(j)} \mathrm{E}\Big[\prod_{u=1}^{p}(s_u - 1)^{j_u}\Big] \prod_{u=1}^{p} \frac{(-1)^{j_u}}{2^{j_u} j_u!} \int_{\mathrm{A}} V_{j_u;q}(x_u) \cdot \boldsymbol{g}_q(\boldsymbol{x})\,\mathrm{d}\boldsymbol{x}.$$

In the following, we assume the symmetry of the set A and, for simplicity's sake, we suppress the suffixes $\delta\,(= -1)$, $k\,(\leq 4)$ and p. Note, however, that $\boldsymbol{g}_q(\boldsymbol{x})$ denotes the joint density of the underlying distribution:

$$\boldsymbol{g}_q(\boldsymbol{x}) = \prod_{u=1}^{p} g_q(x_u),$$

where

$$g_q(x_u) = \frac{1}{2\Gamma(q/2)}\left(\frac{x_u}{2}\right)^{q/2-1} \mathrm{e}^{-x_u/2}, \tag{15.2.5}$$

and that $V_{j;q}(x)$'s are the polynomials defined by (13.3.29). We will also use the relation

$$g_{q+2\ell}(x) = \frac{x^\ell}{q(q+2)\cdots(q+2\ell-2)} g_q(x), \tag{15.2.6}$$

which holds for all positive integers ℓ. For $i, j, k = 1, 2, \ldots, p$, write

$$\begin{aligned} I_i(\mathrm{A}) &= (-1)^i \int_{\mathrm{A}} V_{i;q}(x_1) \cdot g_q(\boldsymbol{x})\,\mathrm{d}\boldsymbol{x}, \\ I_{ij}(\mathrm{A}) &= (-1)^{i+j} \int_{\mathrm{A}} V_{i;q}(x_1) V_{j;q}(x_2) \cdot \boldsymbol{g}_q(\boldsymbol{x})\,\mathrm{d}\boldsymbol{x}, \\ I_{ijk}(\mathrm{A}) &= (-1)^{i+j+k} \int_{\mathrm{A}} V_{i;q}(x_1) V_{j;q}(x_2) V_{k;q}(x_3) \cdot \boldsymbol{g}_q(\boldsymbol{x})\,\mathrm{d}\boldsymbol{x}. \end{aligned} \tag{15.2.7}$$

and

$$E_i = \mathrm{E}\big[\sum_u (S_u - 1)^i\big],$$
$$E_{ij} = \mathrm{E}\big[\sum_{u \neq v} (S_u - 1)^i (S_v - 1)^j\big], \tag{15.2.8}$$
$$E_{ijk} = \mathrm{E}\big[\sum_{u \neq v \neq w \neq u} (S_u - 1)^i (S_v - 1)^j (S_w - 1)^k\big].$$

Then the $Q_j(\mathrm{A})$'s of (15.2.4) become

$$Q_1(\mathrm{A}) = \frac{1}{2} E_1 \cdot I_1(\mathrm{A}),$$
$$Q_2(\mathrm{A}) = \frac{1}{8} E_2 \cdot I_2(\mathrm{A}) + \frac{1}{4} E_{11} \cdot I_{11}(\mathrm{A}), \tag{15.2.9}$$
$$Q_3(\mathrm{A}) = \frac{1}{48} E_3 \cdot I_3(\mathrm{A}) + \frac{1}{16} E_{12} \cdot I_{12}(\mathrm{A}) + \frac{1}{8} E_{111} \cdot I_{111}(\mathrm{A}).$$

On the other hand, it follows from the definition of V's (see Table 13.3.11) and the relation (15.2.6) that we have

$$\begin{aligned} V_{1;q}(x) g_q(x) &= -q\big[g_q(x) - g_{q+2}(x)\big], \\ V_{2;q}(x) g_q(x) &= +q\big[(q-2) g_q(x) - 2q g_{q+2}(x) + (q+2) g_{q+4}(x)\big], \\ V_{3;q}(x) g_q(x) &= -q\big[(q-2)(q-4) g_q(x) - 3q(q-2) g_{q+2}(x) \\ &\quad + 3q(q+2) g_{q+4}(x) - (q+2)(q+4) g_{q+6}(x)\big]. \end{aligned} \tag{15.2.10}$$

15.3 Hotelling's T_0^2 Statistic

For the distribution of Hotelling's T_0^2 Statistic we only need to consider the special case when

(1) $\mathrm{A} = \mathrm{A}_x = \{(x_1, \ldots, x_p) \in \mathbf{R}^p : \sum_{r=1}^{p} x_r \leq x\}$,

(2) $S_1 \geq \cdots \geq S_p$ are the characteristic roots of a random matrix W,

where nW follows the Wishart distribution $\mathrm{W}_p(n, \mathrm{I}_p)$. Then, noting that

$$\int_{\mathrm{A}_x} \boldsymbol{g}_q(\boldsymbol{x}) \mathrm{d}\boldsymbol{x} = \int_0^x g_q^{*p}(x) \mathrm{d}x = \int_0^x g_{p \cdot q}(x) dx = G_{pq}(x),$$

where $g_q^{*p}(x)$ denotes the p-fold convolution of $g_q(x)$, the density of the random variable $Z_1 + Z_2 + \cdots + Z_p$, and writing $r = p \times q$, we obtain from (15.2.8) and (15.2.10),

$$\begin{aligned}
I_1(\mathrm{A}_x) &= q\left[G_r(x) - G_{r+2}(x)\right],\\
I_2(\mathrm{A}_x) &= q\left[(q-2)\,G_r(x) - 2\,q\,G_{r+2}(x) + (q+2)\,G_{r+4}(x)\right],\\
I_{11}(\mathrm{A}_x) &= q^2\left[G_r(x) - 2\,G_{r+2}(x) + G_{r+4}(x)\right],\\
I_3(\mathrm{A}_x) &= q\left[(q-2)(q-4)\,G_r(x) - 3\,q\,(q-2)\,G_{r+2}(x)\right.\\
&\qquad \left. +3\,q\,(q+2)\,G_{r+4}(x) - (q+2)(q+4)\,G_{r+6}(x)\right],\\
I_{12}(\mathrm{A}_x) &= q^2\left[(q-2)\,G_r(x) - (3q-2)\,G_{r+2}(x)\right.\\
&\qquad \left. +(3q+2)\,G_{r+4}(x) - (q+2)\,G_{r+6}(x)\right],\\
I_{111}(\mathrm{A}_x) &= q^3\left[G_r(x) - 3\,G_{r+2}(x) + 3\,G_{r+4}(x) - G_{r+6}(x)\right].
\end{aligned} \tag{15.3.1}$$

Write $\mathbf{V} = \mathbf{W} - \mathrm{I}$. Then it is easily seen that we have

$$\begin{aligned}
E_i &= \mathrm{E}\,[\operatorname{tr}\mathbf{V}^i], \qquad i = 1, 2, \ldots,\\
E_{11} &= \frac{1}{2}(\mathrm{E}\,[(\operatorname{tr}\mathbf{V})^2 - \operatorname{tr}\mathbf{V}^2]),\\
E_{12} &= \mathrm{E}\,[\operatorname{tr}\mathbf{V}\cdot\operatorname{tr}\mathbf{V}^2 - \operatorname{tr}\mathbf{V}^3],\\
E_{111} &= \frac{1}{6}(\mathrm{E}\,[(\operatorname{tr}\mathbf{V})^3 - 3\operatorname{tr}\mathbf{V}\cdot\operatorname{tr}\mathbf{V}^2 + 2\operatorname{tr}\mathbf{V}^3]),
\end{aligned} \tag{15.3.2}$$

which lead to the following expressions for $Q_j(\mathrm{A})$:

$$\begin{aligned}
Q_0(\mathrm{A}) &= G_r(x),\\
Q_1(\mathrm{A}) &= \frac{1}{2}I_1(\mathrm{A})\mathrm{E}[\operatorname{tr}\mathbf{V}],\\
Q_2(\mathrm{A}) &= \frac{1}{8}I_2(\mathrm{A})\mathrm{E}[\operatorname{tr}\mathbf{V}^2] + \frac{1}{8}I_{11}(\mathrm{A})(\mathrm{E}[(\operatorname{tr}\mathbf{V})^2 - \operatorname{tr}\mathbf{V}^2]),\\
Q_3(\mathrm{A}) &= \frac{1}{48}I_3(\mathrm{A})\mathrm{E}[\operatorname{tr}\mathbf{V}^3] + \frac{1}{16}I_{12}(\mathrm{A})\mathrm{E}[\operatorname{tr}\mathbf{V}\cdot\operatorname{tr}\mathbf{V}^2 - \operatorname{tr}\mathbf{V}^3]\\
&\qquad + \frac{1}{48}I_{111}(\mathrm{A})\mathrm{E}[(\operatorname{tr}\mathbf{V})^3 - 3\operatorname{tr}\mathbf{V}\cdot\operatorname{tr}\mathbf{V}^2 + 2\operatorname{tr}\mathbf{V}^3].
\end{aligned} \tag{15.3.3}$$

Further, we can use the following facts [see, e.g., Theorem 2.2.6; Fujikoshi, 1970; Mathai et al., 1995]:

$$\begin{aligned}
&\mathrm{E}[\operatorname{tr}\mathbf{V}] = 0, \qquad \mathrm{E}[\operatorname{tr}\mathbf{V}^2] = p\,(p+1)/n, \qquad \mathrm{E}[(\operatorname{tr}\mathbf{V})^2] = 2\,p/n,\\
&\mathrm{E}[\operatorname{tr}\mathbf{V}^3] = p\,(p^2+3p+4)/n^2, \quad \mathrm{E}[\operatorname{tr}\mathbf{V}\cdot\operatorname{tr}\mathbf{V}^2] = 4\,p\,(p+1)/n^2,\\
&\mathrm{E}[(\operatorname{tr}\mathbf{V})^3] = 8\,p/n^2, \qquad \mathrm{E}[\operatorname{tr}\mathbf{V}^4] = p\,(2\,p^2+5\,p+5)/n^2.
\end{aligned}$$

In this way we obtain from (15.2.10) the following expressions for $Q_{k,p}(\mathrm{A}_x)$:

$$
\begin{aligned}
Q_0(\mathrm{A}_x) &= G_r(x), \\
Q_1(\mathrm{A}_x) &= 0, \\
Q_2(\mathrm{A}_x) &= \frac{pq}{4n}\big(a_{21}(p,q)G_r(x) + a_{22}(p,q)\,G_{r+2}(x) + a_{23}(p,q)\,G_{r+4}(x)\big), \\
Q_3(\mathrm{A}_x) &= \frac{pq}{3n^2}\big(a_{31}(p,q)G_r(x) + a_{32}(p,q)\,G_{r+2}(x) + a_{33}(p,q)G_{r+4}(x) \\
&\qquad\qquad + a_{34}(p,q)G_{r+6}(x)\big),
\end{aligned}
\tag{15.3.4}
$$

where $r = p \times q$ and

$$
\begin{aligned}
a_{21}(p,q) &= q - p - 1, \\
a_{22}(p,q) &= -2q, \\
a_{23}(p,q) &= q + p + 1, \\
a_{31}(p,q) &= q^2 - 3q - 3pq + p^2 + 3p + 4. \\
a_{32}(p,q) &= -3q(q - p - 1). \\
a_{33}(p,q) &= 3q(q + p + 1). \\
a_{34}(p,q) &= -(q^2 + 3pq + 3q + p^2 + 3p + 4).
\end{aligned}
\tag{15.3.5}
$$

Using (15.3.5) and the inequality (13.4.48) for the definition (13.4.1) with $\delta = -1$, we have the following Theorem:

Theorem 15.3.1 *Let T_0^2 be the statistic defined by* (15.1.1) *or Lemma* 15.1.2. *Then:*

(1) $$\left|\Pr\{T_0^2 \le x\} - G_r(x)\right| \le \frac{1}{2n}\frac{\mathrm{T}_{2,p;q}}{p} \cdot p(p+1).$$

(2) $$\left|\Pr\{T_0 \le x\} - G_q(x) - Q_2(x) - Q_3(x)\right|$$
$$\le \frac{1}{2n^2}\frac{\mathrm{T}_{4,p;q}}{p} \cdot p(2p^2 + 5p + 5).$$

Here $Q_2(x)$ and $Q_3(x)$ are given by (15.3.5).

Example 15.3.1 If, in particular, $p = q = 2$, then

$$
\begin{aligned}
P_{1,2}(\mathrm{A}_x) &= G_4(x), \\
P_{2,2}(\mathrm{A}_x) &= G_4(x), \\
P_{3,2}(\mathrm{A}_x) &= G_4(x) - \frac{1}{n}\big(G_4(x) + 4\,G_6(x) - 5\,G_8(x)\big), \\
P_{4,2}(\mathrm{A}_x) &= P_{3,2}(\mathrm{A}_x) + \frac{4}{3n^2}\big(6\,G_6(x) + 30\,G_8(x) - 36\,G_{10}(x)\big).
\end{aligned}
\tag{15.3.6}
$$

Theorem 15.3.1, (13.4.48), and Table 13.4.6 give

$$\left|\Pr\{T_0^2 \le x\} - P_{k,2}(\mathrm{A}_x)\right| \le \frac{1}{2}\frac{\mathrm{T}_{k,2;2}}{2} \cdot \mathrm{E}[\ \mathrm{tr}\,(\mathbf{W}-\mathrm{I})^k]$$

$$= \begin{cases} \dfrac{5.472}{4}\ \mathrm{E}[\ \mathrm{tr}\,\mathbf{V}^2] \le \dfrac{8.21}{n} & \text{if } k=2 \\ \dfrac{6.442}{4}\ \mathrm{E}[\ \mathrm{tr}\,\mathbf{V}^4] \le \dfrac{75}{n^2} & \text{if } k=4. \end{cases}$$

15.4 Wilk's Lambda Distribution

15.4.1 Univariate Case

In this section we derive computable error bounds for asymptotic expansions of the distribution of

$$T = n\log\left(1+\frac{1}{n}X\right), \tag{15.4.1}$$

where $X = Z/S$, $nS \sim \chi^2(n)$, $Z \sim \chi^2(q)$, and Z and S are independent. The result will be used in the next section for deriving computable error bounds for asymptotic expansions of a likelihood ratio statistic in a MANOVA model. From Theorem 13.3.4 we have

$$\left|\Pr\left\{n\log\left(1+\frac{1}{n}X\right) \le u\right\} - \tilde{G}_{(k)}(u)\right| \le \frac{1}{2}\mathrm{E}\left[\boldsymbol{\Delta}_k(\sigma_{q,k}, \tau_{q,k};\ S, 1)\right], \tag{15.4.2}$$

where

$$\begin{aligned}
\tilde{G}_{(k)}(u) &= \int_{\tilde{\mathrm{A}}_u} g_{(k;q)}(x)\,\mathrm{d}x = \tilde{Q}_0(u) + \sum_{j=1}^{k-1}\frac{1}{j!}\tilde{Q}_j(u)M_j, \\
\tilde{Q}_0(u) &= \int_{\tilde{\mathrm{A}}_u} g_q(y)\,\mathrm{d}y, \\
\tilde{Q}_j(u) &= \int_{\tilde{\mathrm{A}}_u} \frac{(-1)^j}{2^j} V_{j;q}(y) g_q(y)\,\mathrm{d}y, \quad M_j = \mathrm{E}[(S-1)^j],
\end{aligned}$$

$\tilde{\mathrm{A}}_u = \{x : n\log(1+x/n) \le u\}$ and $V_{j;q}(y)$ are defined in (13.3.29).

We consider two approximations for $k = 2, 4$:

$$\begin{aligned}
\tilde{G}_{(2)}(u) &= \tilde{Q}_0(u), \\
\tilde{G}_{(4)}(u) &= \tilde{Q}_0(u) + \frac{1}{n}\tilde{Q}_2(u) + \frac{4}{3n^2}\tilde{Q}_3(u),
\end{aligned}$$

since

$$M_1 = 0, \quad M_2 = \frac{2}{n}, \quad M_3 = \frac{8}{n^2}.$$

The set $\tilde{\mathrm{A}}_u$ for large n is close to $\mathrm{A}_u = \{x : x \le u\}$, and hence it is easy to get

$$\lim_{n\to\infty} \tilde{Q}_0(u) = Q_0(u) = \int_{\mathrm{A}_u} g_q(x)\,\mathrm{d}x = G_q(u).$$

Thus, we can express $\tilde{G}_{(2)}(u)$ and therefore the distribution function of T in terms of a chi-square distribution up to some errors. A little more details follow: Let h_q be the density function of

$$T_0 = n \log\Big(1 + \frac{1}{n}\chi_q^2\Big). \tag{15.4.3}$$

Then we have

$$\begin{aligned} |\tilde{Q}_0(u) - Q_0(u)| &= \Big|\int_{\mathrm{A}_u} (h_q(x) - g_q(x))dx\Big| \\ &\le \frac{1}{2}\int_0^\infty |\Delta_1(x)|dx, \end{aligned} \tag{15.4.4}$$

where

$$\Delta_1(x) = h_q(x) - g_q(x). \tag{15.4.5}$$

After that we have to estimate how close the distributions of T and T_0 are using the scale mixture approach, (see (15.4.2)).

To get a more precise result we can construct asymptotic expansion for $\tilde{Q}_0(u)$:

$$\tilde{Q}_0(u) = G_q(u) + \frac{1}{4n}q(q+2)\big(G_{q+2}(u) - G_{q+4}(u)\big) + \mathrm{O}(n^{-2}). \tag{15.4.6}$$

It can be done by expanding the characteristic function of T_0:

$$\begin{aligned} \hat{C}_{T_0}(t) &= \mathrm{E}\Big[\exp\Big(\mathrm{i}tn\log\big(1 + \frac{1}{n}\chi_q^2\big)\Big)\Big] \\ &= \mathrm{E}\Big[\exp\Big(\mathrm{i}t\big(\chi_q^2 - \frac{1}{2n}(\chi_q^2)^2 + \dots\big)\Big)\Big] = \mathrm{E}\Big[\mathrm{e}^{\mathrm{i}t\chi_q^2}\Big(1 - \frac{it}{2n}\left(\chi_q^2\right)^2\Big)\Big] \\ &= (1-2\mathrm{i}t)^{-q/2}\Big[1 + \frac{1}{4n}q(q+2)\cdot\big((1-2\mathrm{i}t)^{-1} - (1-2\mathrm{i}t)^{-2}\big)\Big]. \end{aligned}$$

However, since we are interested in getting the computable error bounds, we shall proceed in another way. Let

$$\Delta_2(x) = h_q(x) - g_q(x) - \frac{1}{4n}q(q+2)\big(g_{q+2}(x) - g_{q+4}(x)\big) \tag{15.4.7}$$

be the remainder term of the expansion for the density corresponding to (15.4.6). Similar to (15.4.4), we have

$$\begin{aligned} &\Big|\tilde{Q}_0(u) - G_q(u) - \frac{1}{4n}q(q+2)\big(G_{q+2}(u) - G_{q+4}(u)\big)\Big| \\ &\qquad \le \frac{1}{2}\int_0^\infty |\Delta_2(x)|\,\mathrm{d}x. \end{aligned} \tag{15.4.8}$$

Now our aim is to obtain bounds of $\Delta_1(x)$ and $\Delta_2(x)$ in terms of the L_1-norm. Related to this problem, we introduce a continuous function $a_q(y,z)$ defined on the set $(0,\infty)\times[0,\infty)$:

$$a_q(y,z) \equiv \begin{cases} \left(\frac{y}{z}(e^z-1)\right)^{q/2-1} \exp\left[z - \frac{y}{2z}(e^z-1)\right], & \text{if } z>0, \\ y^{q/2-1}e^{-y/2}, & \text{if } z=0. \end{cases} \tag{15.4.9}$$

The density functions of T_0 and Z can be written in terms of a_q as

$$h_q(x) = \frac{1}{2^{q/2}\Gamma(q/2)} a_q(x, x/n) \quad \text{and} \quad g_q(x) = \frac{1}{2^{q/2}\Gamma(q/2)} a_q(x,0),$$

respectively. It is easy to check the following equalities:

$$\begin{aligned} \frac{\partial a_q(y,z)}{\partial z} &= \left(1 - \frac{y}{2}\frac{ze^z - e^z + 1}{z^2}\right) a_q(y,z) \\ &\quad + \left(\frac{q}{2}-1\right) y \frac{ze^z - e^z + 1}{z^2} \cdot a_{q-2}(y,z), \end{aligned} \tag{15.4.10}$$

$$\begin{aligned} \frac{\partial^2 a_q(y,z)}{\partial z^2} &= \left[\frac{y}{2}\cdot\frac{2-2e^z+2ze^z-z^2e^z}{z^3} + \left(1 - \frac{y}{2}\frac{ze^z-e^z+1}{z^2}\right)^2\right] a_q(y,z) \\ &+ \left(\frac{q}{2}-1\right) y \Big[2\left(1-\frac{y}{2}\cdot\frac{ze^z-e^z+1}{z^2}\right)\frac{ze^z-e^z+1}{z^2} \\ &\qquad + \frac{2-2e^z+2ze^z-z^2e^z}{z^3}\Big] a_{q-2}(y,z) \\ &\qquad + \left(\frac{q}{2}-1\right)\left(\frac{q}{2}-2\right)\left(\frac{ze^z-e^z+1}{z^2}\right)^2 a_{q-4}(y,z). \end{aligned} \tag{15.4.11}$$

We can define the value of $\partial a_q(y,z)/\partial z$ at $z=0$ similarly as

$$\begin{aligned} \left.\frac{\partial a_q(y,z)}{\partial z}\right|_{z=0} &= \lim_{z\downarrow 0} \frac{\partial a_q(y,z)}{\partial z} \\ &= \frac{1}{4}(q+2-y)y^{q/2-1}\exp(-y/2). \end{aligned} \tag{15.4.12}$$

Lemma 15.4.1 *The following inequalities hold for all $z>0$:*

(1) $\quad 1 < \dfrac{e^z-1}{z} < e^z,$

(2) $\quad \dfrac{|ze^z-e^z+1|}{z^2} < \dfrac{e^z}{2}$ *and*

(3) $\quad \dfrac{|z^2e^z-2ze^z+2e^z-2|}{z^3} < \dfrac{e^z}{3}.$

Proof. The first inequality is well known. All the inequalities can be obtained by dividing both sides by e^z and then considering Taylor's expansion for e^{-z} at 0 up to orders z, z^2 and z^3, respectively. □

Lemma 15.4.2 *Let χ_q^2 be a random variable having a chi-square distribution with q degrees of freedom. Let h_q and g_q be the density functions of $T_0 = n\log(1+\chi_q^2/n)$ and χ_q^2, respectively. Put*

$$\begin{aligned}\Delta_1(y) &= h_q(y) - g_q(y),\\ \Delta_2(y) &= h_q(y) - g_q(y) - \frac{1}{4n}q(q+2)\big(g_{q+2}(y) - g_{q+4}(y)\big).\end{aligned}$$

Then for $y > 0$ we have

$$|\Delta_i(y)| \le \bar{\Delta}_i(y), \quad i = 1, 2, \tag{15.4.13}$$

where

$$\begin{aligned}\bar{\Delta}_1(y) &= \frac{1}{4n}y\big[4 + |q-2| + ye^{y/n}\big]g_q(y)e^{qy/2n},\\ \bar{\Delta}_2(y) &= \frac{1}{96n^2}\Big[48 + 32|q-2| + \big|(q-2)(q-4)\big|\\ &\qquad + 2\big(16 + 3|q-2|\big)ye^{y/n} + 3y^2e^{2y/n}\Big]g_q(y)e^{qy/2n}.\end{aligned} \tag{15.4.14}$$

Proof. Write $\Delta_1(y)$ in terms of the function $a(y,z)$ defined in (15.4.9):

$$\Delta_1(y) = \frac{a_q(y, n^{-1}y) - a_q(y,0)}{2^{q/2}\Gamma(q/2)}$$

We expand $a(y,z)$ in Taylor's series with respect to z and use (15.4.9), (15.4.10), and (15.4.12) to obtain

$$\Big|a_q(y,z) - a_q(y,0)\Big| \le z\bigg|\frac{\partial a_q(y,\bar{z})}{\partial \bar{z}}\bigg|_{\bar{z}=\theta z}\bigg|,$$

which is to hold for all $y > 0, z > 0$, where $\theta = \theta(y,z)$ and $0 < \theta < 1$. Then we have

$$\begin{aligned}|a_q(y,z) - a_q(y,0)| &\le zy^{q/2-1}e^{-y/2}\cdot\Big[\Big(1+\frac{y}{4}e^z\Big)e^{qz/2} + \frac{1}{2}\Big|q/2-1\Big|e^{qz/2}\Big]\\ &= \frac{z}{4}\Big(ye^z + 4 + |q-2|\Big)\cdot y^{q/2-1}\exp\Big(-\frac{1}{2}(y-qz)\Big).\end{aligned}$$

Taking $z = y/n$ in the above we get the first result. Using (15.4.9), (15.4.12), Taylor's expansion of $a(y,z)$ at $z=0$, and the form of density function $g_q(x)$, we have

$$\Big|\Delta_2(y)\Big| \le \frac{1}{\Gamma(q/2)2^{q/2}}\frac{y^2}{2n^2}\sup_{\theta\in(0,1)}\bigg|\frac{\partial^2 a_q(y,z)}{\partial z^2}\bigg|_{z=\theta y/n}\bigg|.$$

Similarly, by using (15.4.11) and Lemma 15.4.1, we can get the second result.

Lemma 15.4.3 *Let $\bar{\Delta}_1(y)$ and $\bar{\Delta}_2(y)$ be the quantities defined by* (15.4.13). *Then*

(1) $$\int_0^\infty \bar{\Delta}_1(y)\,\mathrm{d}y = \frac{1}{n} d_1(q,n) \qquad \text{for } n > q+2,$$

(2) $$\int_0^\infty \bar{\Delta}_2(y)\,\mathrm{d}y = \frac{1}{n^2} d_2(q,n) \qquad \text{for } n > q+4,$$

where

$$\begin{aligned} d_1(q,n) &= \frac{1}{4} q(q+2)\Big[\Big(\frac{n}{n-q}\Big)^{q/2+1} + \Big(\frac{n}{n-q-2}\Big)^{q/2+2}\Big], \\ d_2(q,n) &= \frac{1}{96} q(q+2)(q+4)\Big[(3q+2)\Big(\frac{n}{n-q}\Big)^{q/2+2} \\ &\qquad +2(3q+10)\Big(\frac{n}{n-q-2}\Big)^{q/2+3} \\ &\qquad +3(q+6)\Big(\frac{n}{n-q-4}\Big)^{q/2+4}\Big]. \end{aligned}$$

Proof. The results are obtained by a direct computation. The conditions $n > q+2$ and $n > q+4$ are required for integrability. □

From (15.4.4), (15.4.8). and Lemma 15.4.2, we have:

Theorem 15.4.1 *For the random variable $T_0 \equiv n \log(1 + \chi_q^2/n)$, we have*

(1) $$\big|\Pr\{T_0 \le x\} - G_q(x)\big| \le \frac{1}{2n} d_1(q,n) \qquad \text{when } n > q+2.$$

(2) $$\Big|\Pr\{T_0 \le x\} - G_q(x) - \frac{1}{4n} q(q+2)\Big(G_{q+2}(x) - G_{q+4}(x)\Big)\Big| \le \frac{1}{2n^2} d_2(q,n) \qquad \text{when } n > q+4.$$

Where $d_1(q,n)$ and $d_2(q,n)$ are as given in Lemma 15.4.3.

The values of $\tilde{B}_i(q,n) = d_i(q,n)/2n^i$ are given in Table 15.4.1 for $q = 2, 3, 4$ and $n = 50, 75, 120$. We can see that the bounds $\tilde{B}_2(q,n)$ for the approximations up to $\mathrm{O}(n^{-2})$ are considerably smaller than the bounds $\tilde{B}_1(q,n)$.

Lemma 15.4.4 *Write*

$$\begin{aligned} Q_i(u) &\equiv \int_{\mathrm{A}_u} \frac{(-1)^j}{2^j} V_{j;q}(x) g_q(x)\,\mathrm{d}x \quad \text{with } \mathrm{A}_u = (-\infty, u] \\ \tilde{Q}_i(u) &\equiv \int_{\tilde{\mathrm{A}}_u} \frac{(-1)^j}{2^j} V_{j;q}(x) g_q(x)\,\mathrm{d}x \quad \text{with } \tilde{\mathrm{A}}_u = \{x; n\log(1+x/n) \le u\}. \end{aligned}$$

Table 15.4.1 Values of $\tilde{B}_i(q,n) = d_i(q,n)/2n^i$

q	n	$\tilde{B}_1$	$\tilde{B}_2$
2	50	0.0474	0.0099
	75	0.0298	0.0038
	120	0.0179	0.0013
3	50	0.0980	0.0299
	75	0.0595	0.0108
	120	0.0348	0.0037
4	50	0.1771	0.0754
	75	0.1030	0.0255
	120	0.0584	0.0084

Then it holds that

$$\begin{aligned} Q_1(u) &= \frac{1}{2}q\Big(G_q(u) - G_{q+2}(u)\Big), \\ Q_2(u) &= \frac{1}{4}q\Big((q-2)G_q(u) - 2qG_{q+2}(u) + (q+2)G_{q+4}(u)\Big), \end{aligned} \tag{15.4.15}$$

and for $i = 1, 2$,

$$|\tilde{Q}_i(u) - Q_i(u)| \le \frac{1}{2\,n}h_i(q,n), \tag{15.4.16}$$

where

$$\begin{aligned} h_1(q,n) &= \frac{q}{2}\left\{d_1(q,n) + d_1(q+2,n)\right\} \\ h_2(q,n) &= \frac{q}{4}\left\{|q-2|d_1(q,n) + 2qd_1(q+2,n) + (q+2)d_1(q+4,n)\right\}, \end{aligned}$$

while $d_1(q,n)$ *is as given in Lemma* 15.4.3.

Proof. Equalities (15.4.15) follow from the first two relations of (15.2.10)

$$\begin{aligned} V_{1;q}(x)g_q(x) &= -q\big[g_q(x) - g_{q+2}(x)\big], \\ V_{2;q}(x)g_q(x) &= +q\big[(q-2)g_q(x) - 2qg_{q+2}(x) + (q+2)g_{q+4}(x)\big], \end{aligned} \tag{15.4.17}$$

To prove (15.4.16), let $h_q(x)$ be the probability density of $T_0 = n\log(1+\chi_q^2/n)$.

Let us consider the case $i = 2$. By using the relations above we have

$$\begin{aligned}\left|\tilde{Q}_2(u) - Q_2(u)\right| &\leq \frac{q}{4}\Big|\int_0^u \big((q-2)(h_q(x) - g_q(x)) - 2q(h_{q+2}(x) - g_{q+2}(x)) \\ &\qquad + (q+2)(h_{q+4}(x) - g_{q+4}(x))\big)\,\mathrm{d}x\Big| \\ &\leq \frac{q}{8}\int_0^\infty \big(|q-2|\cdot|h_q(x) - g_q(x)| + 2q|h_{q+2}(x) - g_{q+2}(x)| \\ &\qquad + (q+2)|h_{q+4}(x) - g_{q+4}(x)|\big)\,\mathrm{d}x.\end{aligned}$$

Therefore, the desired result follows from Lemmas 15.4.2 and 15.4.3. □

Lemma 15.4.5 *For $\tilde{Q}_i(u)(i = 1,2,3)$ we have the following two types of upper bounds:*

(1) $|\tilde{Q}_i(u)| \leq i!\,\alpha_i(q)$.

(2) $|\tilde{Q}_i(u)| \leq \delta_i(q)$.

Where

$$\begin{aligned}\alpha_i(q) &= \frac{1}{i!}\sup_x \left|c_i(x)g_q(x)\right|; \\ c_1(x) &= x, \\ c_2(x) &= -\frac{1}{4}\left(x^2 - (q-2)x\right), \\ c_3(x) &= \frac{1}{4}\left(x^3 - 2(q-2)x^2 + (q-2)(q-4)x\right), \\ \delta_1(q) &= \frac{1}{2}q, \\ \delta_2(q) &= \frac{1}{4}q\left(\left|q-2\right| + q + 2\right), \\ \delta_3(q) &= \frac{1}{8}q\left(\left|(q-2)(q-4)\right| + 2|q-2|(q+4) + (q+2)(q+4)\right).\end{aligned}$$

Proof. Note that the relations

$$\tilde{Q}_j(u) = \int_{\tilde{A}_u} g_q(x)\,\mathrm{d}x = \int_0^{u^*} b_j(x)g_q(x)\,\mathrm{d}x,$$

and $\left(c_j(x)g_q(x)\right)' = (-1)^j/2^j \cdot V_{j;q}(x)g_q(x)$ lead to

$$|\tilde{Q}_j(u)| \leq \sup_x |c_j(x)g_q(x)| = j!\alpha_j(q).$$

Note also that we can write

$$\begin{aligned} V_{1;q}(x)g_q(x) &= -\big(g_q(x) - g_{q+2}(x)\big), \\ V_{2;q}(x)g_q(x) &= \quad (q-2)\big(g_q(x) - g_{q+2}(x)\big) - (q+2)\big(g_{q+2}(x) - g_{q+4}(x)\big) \\ V_{3;q}(x)g_q(x) &= -q\Big((q-2)(q-4)\big(g_q(x) - g_{q+2}(x)\big) \\ &\qquad +3(q-2)(q+2)\big(g_{q+2}(x) - g_{q+4}(x)\big) \\ &\qquad\qquad -(q+2)(q+4)\big(g_{q+4}(x) - g_{q+6}(x)\big)\Big). \end{aligned}$$

Therefore, the second assertion follows from

$$\sup_{\mathrm{A}} \left| \int_{\mathrm{A}} (f(x) - g(x))\,\mathrm{d}x \right| \le 1,$$

for any density functions $f(x)$ and $g(x)$.

Theorem 15.4.2 *Let* $T = n\log(1 + X/n)$, *where*

$$X = S^{-1}Z, \quad nS \sim \chi_n^2, \quad Z \sim \chi^2(q),$$

and S *is independent of* Z. *Then it holds that:*

(1) $\big|\Pr\{T \le x\} - G_q(x)\big| \le B_1(q,n)$, *for* $n > q + 2$.

(2) $\left|\Pr\{T \le x\} - G_q(x) - \dfrac{1}{4n}q(q-2)\big(G_q(x) - G_{q+2}(x)\big)\right| \le B_2(q,n)$,

for $n > q + 4$.

Where

$$\begin{aligned} B_1(q,n) &= \frac{1}{2\,n}\big(n\,\mathrm{E}\big[\boldsymbol{\Delta}_2(\sigma_{q,2}, \tau_{q,2};\ S, 1)\big] + d_1(q,n)\big), \\ B_2(q,n) &= \frac{1}{2\,n^2}\big(n^2\,\mathrm{E}\big[\boldsymbol{\Delta}_4(\sigma_{q,4}, \tau_{q,4};\ S, 1)\big] + d_2(q,n) + h_2(q,n) + 16\alpha_3(q)\big), \end{aligned}$$

and the $\boldsymbol{\Delta}$*'s,* d*'s and* $(h_2(q,n), \alpha_3)$ *are given in* (15.4.2), *Lemma* 15.4.3, *and Lemmas* 15.4.4–15.4.5, *respectively.*

Proof. Note that

$$\begin{aligned} \big|\Pr\{T \le u\} - G_q(u)\big| &\le \left|\Pr\{T \le u\} - \tilde{Q}_0(u)\right| + \left|\tilde{Q}_0(u) - G_q(u)\right| \\ &\le \left|\Pr\{T \le u\} - \tilde{G}_{(2)}(u)\right| + \left|\tilde{Q}_0(u) - G_q(u)\right|. \end{aligned}$$

Therefore, the first inequality follows from Lemma 3.4 (i) and (15.4.2) with $k = 2$. For a proof of the second assertion we use the following inequalities:

$$\begin{aligned}
&\left|\Pr\{T \le x\} - G_q(x) - \frac{1}{4n}q(q-2)\big(G_q(x) - G_{q+2}(x)\big)\right| \\
&\quad = \left|\Pr\{T \le u\} - G_q(u) - \frac{1}{4n}q(q+2)\big(G_{q+2}(u) - G_{q+4}(u)\big)\right. \\
&\qquad \left. - \frac{1}{4n}q\big((q-2)G_q(u) - 2qG_{q+2}(u) + (q+2)G_{q+4}(u)\big)\right| \\
&\quad \le \left|\Pr\{T \le x\} - \tilde{Q}_0(u) - \frac{1}{n}\tilde{Q}_2(u) - \frac{4}{3n^2}\tilde{Q}_3(u)\right| \\
&\qquad + \left|\tilde{Q}_0(u) - G_q(u) - \frac{1}{4n}q(q+2)\big(G_{q+2}(u) - G_{q+4}(u)\big)\right| \\
&\qquad + \left|\frac{1}{n}\tilde{Q}_2(u) - \frac{q}{4n}\big((q-2)G_q(u) - 2qG_{q+2}(u) + (q+2)G_{q+4}(u)\big)\right| \\
&\qquad + \frac{4}{3n^2}|\tilde{Q}_3(u)|.
\end{aligned}$$

The desired result follows by applying (15.4.2) with $k = 4$, and Lemmas 15.4.4 and 15.4.5 to the terms on the right-hand side of the inequality above.

The values of $B_i(q, n)$ in Theorem 15.4.2 are given in Table 15.4.2 for $q = 2, 3, 4$ and $n = 50, 75, 120$. Useful bounds are obtained if q/n is small.

Table 15.4.2 Values of $B_i(q, n)$ in Theorem 15.4.2

q	n	$B_1(q,n)$	$B_2(q,n)$
2	50	0.2090	0.1322
	75	0.1357	0.0522
	120	0.0832	0.0189
3	50	0.3298	0.3131
	75	0.2082	0.1146
	120	0.1253	0.0394
4	50	0.5064	0.7207
	75	0.3074	0.2424
	120	0.1801	0.0788

15.4.2 Multivariate Case

For the statistic

$$T_{\rm LR} = \sum_{r=1}^{p} n \log\left(1 + \frac{1}{n}X_r\right), \tag{15.4.18}$$

with p independent variables $X_1, \ldots, X_p$ such that for $i = 1, \ldots, p$,

$$X_i = Z_i/S_i, \quad Z_i \sim \chi^2(q), \quad Y_i = S_i\,(\delta = -1,\, \rho = 1), \quad nS_i \sim \chi^2(m_i),$$

where $m_i = n - (i-1)$, and all variables S_i and Z_i are independent, we can write

$$\Pr\{T_{\mathrm{LR}} \le x\} = \int_{\tilde{\mathrm{A}}_x} f_q(\boldsymbol{x})\,\mathrm{d}\boldsymbol{x},$$

where $f_q(\boldsymbol{x})$ is the density function of $\boldsymbol{X} = (X_1, \ldots, X_p)$ and

$$\tilde{\mathrm{A}}_x = \Big\{(x_1, \ldots, x_p) \in \mathrm{R}^p : \sum_{r=1}^{p} n \log(1 + x_r/n) \le x\Big\}.$$

Then by Theorem 13.4.5 (13.4.48), we get

$$|\Pr\{T_{\mathrm{LR}} \le x\} - \tilde{G}_{(k,p)}(x)| \le \frac{1}{2}\mathrm{E}[\boldsymbol{\Delta}_k(\mathrm{s}_{k,p;q}, \mathrm{T}_{k,p;q}; \mathbf{S}, 1)], \tag{15.4.19}$$

where $\mathbf{S} = \mathrm{diag}\,(S_1, \ldots, S_p)$ $\big[$i.e.. $\mathbf{Y} = \mathrm{diag}\,(Y_1, \ldots, Y_p)\big]$ and

$$\tilde{G}_{(k,p)}(x) = \int_{\tilde{\mathrm{A}}_x} \boldsymbol{g}_{k,p;q}(\boldsymbol{x})\,\mathrm{d}\boldsymbol{x} = \sum_{i=0}^{k-1} \tilde{Q}_{i,p}(x),$$

where $\boldsymbol{g}_{k,p;q}$ are defined in (13.4.47) and

$$\tilde{Q}_{i,p}(x) = \sum_{(j)} \mathrm{E}\left[\prod_{u=1}^{p}(S_u - 1)^{j_u}\right] \prod_{u=1}^{p} \frac{(-1)^{j_u}}{2^{j_u} j_u!} \int_{\tilde{\mathrm{A}}_x} V_{j_u;q}(x_u)\boldsymbol{g}_q(\boldsymbol{x}).$$

Similar to the one-dimensional case in Section 15.4 we shall use approximation (15.4.19) for $k = 2$ and $k = 4$. In order to make the notation shorter, we omit the index p below. We have

$$\begin{aligned} \tilde{G}_{(2)}(x) &= \tilde{I}_0(x) + \tilde{I}_1(x)\,M_1, \\ \tilde{G}_{(4)}(x) &= \tilde{G}_{(2)}(x) + \tilde{I}_2(x)\,M_2 + \frac{1}{2}\,\tilde{I}_{11}(x)\,M_{11} \\ &\qquad + \tilde{I}_3(x)\,M_3 + \frac{1}{2}\,\tilde{I}_{12}(x)\,M_{12} + \frac{1}{6}\,\tilde{I}_{111}(x)\,M_{111}, \end{aligned} \tag{15.4.20}$$

where for $i, j, k = 1, 2, 3$, we write

$$\begin{aligned} \tilde{I}_0(x) &= \int_{\tilde{\mathrm{A}}_x} \boldsymbol{g}_q(\boldsymbol{x})\,\mathrm{d}\boldsymbol{x}, \\ \tilde{I}_i(x) &= \frac{(-1)^i}{2^i\, i!} \int_{\tilde{\mathrm{A}}_x} V_{i;q}(x_1)\boldsymbol{g}_q(\boldsymbol{x})\,\mathrm{d}\boldsymbol{x}, \\ \tilde{I}_{ij}(x) &= \frac{(-1)^{i+j}}{2^{i+j}\, i!\, j!} \int_{\tilde{\mathrm{A}}_x} V_{i;q}(x_1)V_{j;q}(x_2)\cdot \boldsymbol{g}_q(\boldsymbol{x})\,\mathrm{d}\boldsymbol{x}, \\ \tilde{I}_{ijk}(x) &= \frac{(-1)^{i+j+k}}{2^{i+j+k}\, i!\, j!\, k!} \int_{\tilde{\mathrm{A}}_x} V_{i;q}(x_1)V_{j;q}(x_2)V_{k;q}(x_3)\cdot \boldsymbol{g}_q(\boldsymbol{x})\,\mathrm{d}\boldsymbol{x}. \end{aligned} \tag{15.4.21}$$

and the moments M's can be calculated as follows

$$\begin{aligned}
M_1 &= \mathrm{E}\Big[\sum_{i=1}^{p}(S_i-1)\Big] = -\frac{1}{2n}p(p-1),\\
M_2 &= \mathrm{E}\Big[\sum_{i=1}^{p}(S_i-1)^2\Big] = \frac{2}{n}p\big(1+\frac{1}{12n}(p-1)(2p-7)\big),\\
M_3 &= \mathrm{E}\Big[\sum_{i=1}^{p}(S_i-1)^3\Big] = -\frac{1}{n^2}p\big(3p-11+\frac{1}{4n}(p-1)(p-4)(p-5)\big),\\
M_{11} &= \mathrm{E}\Big[\sum_{i\neq j}^{p}(S_i-1)\,(S_j-1)\Big] = \frac{1}{12n^2}p(p-1)(p-2)(3p-1),\\
M_{12} &= \mathrm{E}\Big[\sum_{i\neq j}^{p}(S_i-1)\,(S_j-1)^2\Big] = -\frac{1}{n^2}p(p-1)^2\big(1+\frac{1}{6n}(p-1)(p-4)\big),\\
M_{111} &= \mathrm{E}\Big[\sum_{i\neq j\neq k}^{p}(S_i-1)(S_j-1)(S_j-1)\Big] = -\frac{1}{8n^3}p^2(p-1)^2(p-2)(p-3).
\end{aligned} \tag{15.4.22}$$

We can write

$$\tilde{I}_0(x) = \Pr\{T_0 \le x\},$$

where

$$T_0 = \sum_{r=1}^{p} n\log\left(1+\frac{1}{n}Z_r\right), \quad Z_1,\dots,Z_p \sim \text{i.i.d. } \chi^2(q).$$

We can extend Theorem 15.4.1 to the multivariate case as follows.

Theorem 15.4.3 *For a set of i.i.d. random variables $Z_1,\dots,Z_p$ following $\chi^2(q)$, let T_0 be defined by $T_0=\sum_{r=1}^{p} n\log(1+Z_r/n)$. Then it holds that*

(1) $$\big|\Pr\{T_0\le x\}-G_{pq}(x)\big| \le \frac{1}{2\,n}\,d_{1,p}(q,n),$$

(2) $$\big|\Pr\{T_0\le x\}-G_{pq}(x)-\frac{1}{4n}pq(q+2)\big(G_{r+2}(x)-G_{r+4}(x)\big)\big| \le \frac{1}{2\,n^2}\,d_{2,p}(q,n).$$

Where

$$\begin{aligned}
d_{1,p}(q,n) &= p\,d_1(q,n),\\
d_{2,p}(q,n) &= \big(p\,d_2(q,n)+\frac{1}{4}p(p-1)q(q+2)d_1(q,n)\big),
\end{aligned}$$

and $d_1(q,n)$ and $d_2(q,n)$ are given in Lemma 15.4.3

Proof. Note that

$$G_{pq}(x) = \Pr\{Z_1 + \cdots + Z_p \le x\} = \int_{A_x} \prod_{r=1}^{p} g_q(y_r)\,\mathrm{d}\boldsymbol{y},$$

where $A_x = \{(x_1, \cdots, x_p) \in \mathrm{R}^p : x_1 + \cdots + x_p \le x\}$. Let $h_q(x)$ be the density function of $n \log(1 + \chi_q^2/n)$. Then we have

$$\begin{aligned}\Big|\Pr\{T_0 \le x\} - G_{pq}(x)\Big| &= \Big|I_{0,p}(\tilde{A}_x) - I_{0,p}(A_x)\Big| \\ &= \Big|\int_{A_x} \Big(\prod_{r=1}^{p} h_q(y_r) - \prod_{r=1}^{p} g_q(y_r)\Big)\,\mathrm{d}\boldsymbol{y}\Big| \\ &\le \frac{1}{2}\int_0^\infty \cdots \int_0^\infty \Big|\prod_{r=1}^{p} h_q(y_r) - \prod_{r=1}^{p} g_q(y_r)\Big|\,\mathrm{d}\boldsymbol{y}. \qquad (15.4.23)\end{aligned}$$

Note also that

$$\prod_{r=1}^{p} h_q(y_r) - \prod_{r=1}^{p} g_q(y_r) = \sum_{r=1}^{p} \prod_{j=1}^{r-1} g_q(y_j)(h_q(y_i) - g_q(y_i)) \prod_{k=i+1}^{p} h_q(y_k). \qquad (15.4.24)$$

This will be obtained by decomposing $h_q(x_1)h_q(x_2)\cdots h_q(x_p)$ as follows. First we decompose $h_q(x_1)$ as $(h_q(x_1) - g_q(x_1)) + g_q(x_1)$. We can write

$$\begin{aligned} h_q(x_1)h_q(x_2)\cdots h_q(x_p) &= (h_q(x_1) - g_q(x_1))h_q(x_2)\cdots h_q(x_p) \\ &\quad + g_q(x_1)h_q(x_2)\cdots h_q(x_p). \end{aligned}$$

Next, we decompose $h_q(x_2)$ in the last term. Continuing this procedure, we get (15.4.24). By substituting (15.4.24) into (15.4.23) and using Lemmas 15.4.2 and 15.4.3, we obtain part (1) of the theorem.

In order to prove (2), we use two expressions (see Lemma 15.4.2):

$$\begin{aligned} h_q(x) &= g_q(x) + \Delta_1(x) \\ &= g_q(x) + g_q^{(1)}(x) + \Delta_2(x), \end{aligned}$$

where $g_q^{(1)}(x) = (4n)^{-1}q(q+2)\{g_{q+2}(x) - g_{q+4}(x)\}$. Constructing short asymptotic expansions for the distribution of T_0, we argue similar to the proof of (1): We first consider the expansion for $h_q(x_1)$, then for $h_q(x_2)$, and so on. In the beginning we get

$$\begin{aligned}\prod_{r=1}^{p} h_q(x_r) &= \left(g_q(x_1) + g_q^{(1)}(x_1) + \Delta_2(x_1)\right) \prod_{r=2}^{p} h_q(x_r) \\ &= g_q(x_1) \prod_{r=2}^{p} h_q(x_r) + g_q^{(1)}(x_1) \prod_{r=2}^{p} h_q(x_r) \\ &\quad + \Delta_2(x_1) \prod_{r=2}^{p} h_q(x_r).\end{aligned}$$

Then

$$\begin{aligned}h_q(x_1)\cdots h_q(x_p) = {}& g_q(x_1)\left(g_q(x_2) + g_q^{(1)}(x_2) + \Delta_2(x_2)\right) h_q(x_3)\cdots h_q(x_p) \\ &+ g_q^{(1)}(x_1)\left(g_q(x_2) + \Delta_1(x_2)\right) h_q(x_3)\cdots h_q(x_p) \\ &+ \Delta_2(x_1) g_q(x_2) h_q(x_3)\cdots h_q(x_p).\end{aligned}$$

and so on. The second assertion (2) follows from Lemmas 15.4.2 and 15.4.3.

□

Theorem 15.4.4 *Let*

$$T_{0r} = n\log(1 + Z_r/n), \quad Z_r \sim \chi^2(q_r),$$

where the Z_r *are independent. Then*

$$\left|\Pr\left\{\sum_{r=1}^{p} T_{0r} \le x\right\} - G_r(x)\right| \le \frac{1}{2n}\sum_{r=1}^{p} d_1(q_r, n),$$

where $r = \sum_{r=1}^{p} q_r$ *and* $d_1(q, n)$ *is as given in Lemma* 15.4.3.

Proof. The result can be proved by the same method as in the proof of Theorem 15.4.3 (1). In fact, we have

$$\begin{aligned}\left|\Pr\{\sum_{r=1}^{p} T_{0r} \le x\} - G_r(x)\right| &= \left|\int_{snA_x}\left(\prod_{r=1}^{p} f_{q_r}(y_r) - \prod_{r=1}^{p} g_{q_r}(y_r)\right)\mathrm{d}\boldsymbol{y}\right| \\ &\le \frac{1}{2}\int_0^{\infty}\cdots\int_0^{\infty}\left|\prod_{r=1}^{p} h_{q_r}(y_r) - \prod_{r=1}^{p} g_{q_r}(y_r)\right|\mathrm{d}y_1\cdots\mathrm{d}y_p.\end{aligned}$$

The final result can be obtained by using (15.4.24) and Lemmas 15.4.2 and 15.4.3. □

Lemma 15.4.6 *Let $I_i(x)$ be defined in the same way as $\tilde{I}_i(x)$ in (15.4.21) but*

$$\tilde{\mathrm{A}}_x = \left\{ \boldsymbol{x} = (x_1, \ldots, x_p) : \ \sum_{r=1}^{p} n \log(1 + x_r/n) \le x \right\}$$

is replaced by

$$A_x = \left\{ \boldsymbol{x} = (x_1, \ldots, x_p) : \ \sum_{r=1}^{p} x_r \le x \right\}.$$

Then it holds that

$$I_1(x) = \frac{1}{2} q\{G_r(x) - G_{r+2}(x)\},$$
$$I_2(x) = \frac{1}{8} q \left\{(q-2)G_r(x) - 2qG_{r+2}(x) + (q+2)G_{r+4}(x)\right\},$$

and for $i = 1, 2$,

$$|\tilde{I}_i(x) - I_i(x)| \le \frac{1}{2in} h_{i,p}(q, n),$$

where

$$h_{1,p}(q,n) = \frac{q}{2} \left\{(2p-1)d_1(q,n) + d_1(q+2,n)\right\},$$
$$h_{2,p}(q,n) = \frac{q}{4} \Big[|q-2| p d_1(q,n) + 2q\big\{(p-1)d_1(q,n) + d_1(q+2,n)\big\} + (q+2)\big\{(p-1)d_1(q,n) + d_1(q+4,n)\big\} \Big],$$

and $d_1(q, n)$ are given in Lemma 15.4.3.

Proof. The statement can be proved in the same way as in the proof of Lemma 15.4.4 with the help of Theorem 15.4.3.

Lemma 15.4.7 *Let A_x and $\tilde{\mathrm{A}}_x$ be the same as in Lemma* 15.4.6. *Let $I_{ij}(x)$ be defined in the same way as $\tilde{I}_{ij}(x)$ in (15.4.21) with replacement of $\tilde{\mathrm{A}}_x$ by A_x. Then*

(1) $|I_i(x)| \le \alpha_i(q)$.

(2) $|I_{ij}(x)| \le \dfrac{1}{j!} \alpha_i(q)\delta_j(q)$.

(3) $|I_{ijk}(x)| \le \dfrac{1}{j!k!} \alpha_i(q)\delta_j(q)$.

Where $\alpha_i(q)$ and $\delta_j(q)$ are as given in Lemma 15.4.5. These results hold for $\tilde{I}_i(x), \tilde{I}_{ij}$, and $\tilde{I}_{ijk}$ as well.

Proof. We derive (2) only, since (1) and (3) are obtained similarly. It is easy to see that

$$\begin{aligned}
|\tilde{I}_{ij}(u)| &= \frac{1}{i!\,j!}\left|\int_{\tilde{A}_u} \frac{(-1)^i}{2^i} V_{i;q}(x_1) g_q(x_1) \frac{(-1)^j}{2^j} V_{j;q}(x_2) \prod_{i=2}^{p} g_q(x_i)\,\mathrm{d}\boldsymbol{x}\right| \\
&= \frac{1}{i!\,j!}\left|\int_{\mathrm{R}^{r-1}} \int_0^{a(x_2,\ldots,x_p)\vee 0} \frac{(-1)^i}{2^i} V_{i;q}(x_1) g_q(x_1)\,\mathrm{d}x_1\right. \\
&\qquad \left.\times \frac{(-1)^j}{2^j} V_{j;q}(x_2) \prod_{r=2}^{p} g_q(x_r) dx_r\right| \\
&\le (j!)^{-1}\alpha_i(q) \int_0^\infty \left|\frac{1}{2^j} V_{j;q}\right| g_q(x_2)\,\mathrm{d}x_2 \le \alpha_i(q)(j!)^{-1}\delta_j(q),
\end{aligned}$$

where

$$a(x_2,\ldots,x_p) = n\Big[\exp\Big(\frac{1}{n}\big(u - \sum_{r=2}^{p} n\log(1 + x_r/n)\big)\Big) - 1\Big].$$

Combining this result, Lemmas 15.4.6 and 15.4.7, and Theorems 15.4.3 and 15.4.4, we get the following main theorem.

Theorem 15.4.5 *Let T_{LR} be the statistic defined by* (15.4.18). *Then:*

(1) *For* $n > \max(p-1, q+2)$,

$$|\Pr\{T_{\mathrm{LR}} \le x\} - G_r(x)| \le B_{1,p}(q,n),$$

(2) *For* $n > \max(p-1, q+4)$,

$$\begin{aligned}
|\Pr\{T_{\mathrm{LR}} \le x\} - G_r(x) - \frac{r(q-p-1)}{4n}\{G_r(x) - G_{r+2}(x)\}| \\
\le B_{2,p}(q,n),
\end{aligned}$$

where

$$\begin{aligned}
B_{1,p}(q,n) &= \frac{1}{2}\mathrm{E}\big[\boldsymbol{\Delta}_k(\mathrm{S}_{k,p;q}, \mathrm{T}_{k,p;q}; \mathbf{S}, 1)\big] + \frac{1}{2n} d_{1,p}(q,n) + \alpha_1(q)|M_1|, \\
B_{2,p}(q,n) &= \frac{1}{2}\mathrm{E}\big[\boldsymbol{\Delta}_k(\mathrm{S}_{k,p;q}, \mathrm{T}_{k,p;q}; \mathbf{S}, 1)\big] + \frac{1}{2n^2} d_{2,p}(q,n) \\
&+ \frac{1}{2n} h_{1,p}(q,n)|M_1| + \frac{1}{4n} h_{2,p}(q,n) M_2 \\
&\quad + \frac{1}{2}\alpha_1(q)\delta_1(q)|M_{11}| + \alpha_3(q)|M_3| \\
&\qquad + \alpha_2(q)\delta_1(q)|M_{21}| + \frac{1}{6}\alpha_1(q)\{\delta_1(q)\}^2|M_{111}| \\
&\qquad\quad + \frac{1}{6n^2} p(p-1)|2p-7|\alpha_2(q),
\end{aligned}$$

and, suppressing suffixes, the quantities $\mathrm{E}[\boldsymbol{\Delta}_k(\mathrm{S},\mathrm{T};\mathbf{S},1)], d(q,n), h(q,n), \alpha(q)$, $\delta_i(q)$, M, *and so on, are defined in Theorems 13.4.5 and* 15.4.3, *Lemmas* 15.4.6 *and* 15.4.7, and (15.4.22*), respectively.*

Proof. To prove (1) it is sufficient to consider the following decomposition:

$$\begin{aligned}&\Pr\{T_{\mathrm{LR}} \le x\} - G_r(x)\\ &\quad = \Big(\Pr\{T_{\mathrm{LR}} \le x\} - \tilde{G}_{(2)}(x)\Big) + \Big\{\tilde{I}_0(x) - G_r(x)\Big\} + \tilde{I}_1(x)\, M_1,\end{aligned}$$

which follows from (15.4.20). Therefore, we have

$$\begin{aligned}&|\Pr\{T_{\mathrm{LR}} \le x\} - G_r(x)|\\ &\quad \le |\Pr\{T_{\mathrm{LR}} \le x\} - \tilde{G}_{(2)}(x)| + |\tilde{I}_0(x) - G_r(x)| + |\tilde{I}_1(x)\, M_1|.\end{aligned}$$

The required result is obtained by using (15.4.19) with $k = 2$ and Lemmas 15.4.6 and 15.4.7. To prove (2), consider decomposing a main term and its remainder term defined by

$$\tilde{G}_{(4)}(x) = H_1(x) + R_1(x),$$

where

$$\begin{aligned}H_1(x) &= \tilde{I}_0(x) + \tilde{I}_1(x)\, M_1 + \tilde{I}_2(x)\, M_2,\\ R_1(x) &= \frac{1}{2}\, \tilde{I}_{11}(x)\, M_{11} + \tilde{I}_3(x)\, M_3 + \frac{1}{2}\, \tilde{I}_{12}(x)\, M_{12} + \frac{1}{6}\, \tilde{I}_{111}(x)\, M_{111}.\end{aligned}$$

Further, let $H_1(x)$ be decomposed as

$$H_1(x) = H_2(x) + R_2(x), \quad \text{with}$$

$$\begin{aligned}H_2(x) &= \bar{I}_0(x) + I_1(x)\, M_1 + I_2(x)\, M_2,\\ R_2(x) &= \big(\tilde{I}_1(x) - \bar{I}_0(x)\big) + \big(\tilde{I}_1(x) - I_1(x)\big)M_1 + \big(\tilde{I}_2(x) - I_2(x)\big)M_2,\\ \bar{I}_0(x) &= G_r(x) + \frac{1}{4n} pq\big(G_{r+2}(x) - G_{r+4}(x)\big).\end{aligned}$$

Noting that $M_2 = 2pn^{-1} + p(p-1)(2p-7)(6n^2)^{-1}$, put

$$H_2(x) = H_3(x) + R_3(x),$$

where

$$\begin{aligned}H_3(x) &= \bar{I}_{0,}(x) + I_1(x)M_1 + \frac{2p}{n} I_2,\\ R_3(x) &= \frac{1}{12n^2} p(p-1)(2p-7) I_2(A_x).\end{aligned}$$

It is easy to see that

$$H_3(x) = G_r(x) + \frac{r}{4n}(q-p-1)\{G_r(x) - G_{r+2}(x)\}, \text{ and}$$

$$\tilde{G}_{(4)}(x) = H_3(x) + R_1(x) + R_2(x) + R_3(x).$$

Therefore, we have

$$\begin{aligned}
\left|\Pr\{T_{\mathrm{LR}} \le x\} \right. &- \left. H_3(x)\right| \\
&= \left|\Pr\{T_{\mathrm{LR}} \le x\} - \tilde{G}_{(4)}(x) + R_1(x) + R_2(x) + R_3(x)\right| \\
&\le \left|\Pr\{T_{\mathrm{LR}} \le x\} - \tilde{G}_{(4)}(x)\right| + |R_1(x)| + |R_2(x)| + |R_3(x)|.
\end{aligned}$$

Result (2) is obtained by using (15.4.19) with $k = 4$ and Lemmas 15.4.6 and 15.4.7. □

Some numerical results of $B_{i,p}(q, n)$ are given in Tables 15.4.3 and 15.4.4. From the tables we can see that to get useful bounds, it will be required for us to have a large n in comparison with values of p and q as p or q is increasing. To overcome this weak point, it would be interesting to obtain bounds for some other approximations, especially under a high dimensional asymptotic framework, such as both p and n tending to infinity.

Table 15.4.3 Values of $B_{i,2}(q, n)$

q	n	$B_{1,2}(q, n)$	$B_{2,2}(q, n)$
2	120	0.1862	0.0475
	200	0.1106	0.0164
3	120	0.2768	0.1044
	200	0.1626	0.0352
4	120	0.3923	0.2165
	200	0.2269	0.0709

Table 15.4.4 Values of $B_{i,3}(q, n)$

q	n	$B_{1,3}(q, n)$	$B_{2,3}(q, n)$
2	120	0.3091	0.0903
	200	0.1838	0.0314
3	120	0.4548	0.2049
	200	0.2677	0.0697
4	120	0.6370	0.4323
	200	0.3695	0.1432

Problems

15.1 Let $T_0^2 = n \operatorname{tr} \mathbf{S}_h \mathbf{S}_e^{-1} = T_0^2(p,q,n)$, where $\mathbf{S}_h$ and $\mathbf{S}_e$ are mutually independent $p \times p$ random matrices following the Wishart distributions $\mathrm{W}_p(q,\Sigma)$ and $\mathrm{W}_p(n,\Sigma)$. Then, show that

$$T_0^2(1,q,n) = n\mathrm{F}(q,n),$$
$$T_0^2(p,1,n) = \frac{np}{n-p+1}\mathrm{F}(p,n-p+1).$$

15.2 Let

$$\varepsilon(x;p,q,n) = \Pr\{T_0^2 \le x\} - G_{pq}(x) - (pq/4n)F(x),$$

where $G_f(x) = \Pr\{\chi_f^2 \le x\}$,

$$F(x) = (q-p+1)G_{pq}(x) - 2qG_{pq+2}(x); (p+q+1)G_{pq+4}(x).$$

Obtain an error bound $B(p,q,n)$ such that

(1) $|\varepsilon(x;p,q,n)| \le B(p,q,n)$,
(2) $B(p,q,n) = \mathrm{O}(n^{-2})$,

and (3) $B(p,q,n)$ is numerically computable, does not involve an unknown constant.

15.3 Let $T_{LR} = n\log(|\mathbf{S}_e|/|\mathbf{S}_e+\mathbf{S}_h|) = T_{LR}(p,q,n)$, where $\mathbf{S}_h$ and $\mathbf{S}_e$ are mutually independent $p\times p$ random matrices following the Wishart distributions $\mathrm{W}_p(q,\Sigma)$ and $\mathrm{W}_p(n,\Sigma)$. Then, show the following results:

(1) $T_{LR}(1,q,n) = n\log\left(1+\frac{1}{n}\left(\frac{\chi_n^2}{n}\right)^{-1}\chi_q^2\right)$,

(2) $T_{LR}(p,1,n) = n\log\left(1+\frac{1}{n}T_0^2(p,1,n)\right)$,

(3) The distribution of $n\log\left(1+\frac{1}{n}\chi_q^2\right)$ tends to $\chi^2(q)$ as $n\to\infty$.

If, in particular, $q=2$, then $n\log\left(1+\frac{1}{n}\chi_2^2\right) \sim \chi^2(2)$. for all n.

15.4 Let T_0^2 be the Hotelling's statistic defined by (15.1.1). Show that T_0^2 can be expressed as a function of a multivariate normal mixed variable as follows: Let

$$\boldsymbol{Z} = (Z_{11},\ldots,Z_{1q},\ldots,Z_{p1},\ldots,Z_{pq}),$$

whose $m = pq$ components are independently distributed as $\mathrm{N}(0,1)$. Define

$$\boldsymbol{X} = [\mathbf{S}^{-1/2}\otimes \mathrm{I}_q]\boldsymbol{Z},$$

where $\mathbf{S} = \operatorname{diag}(S_1, \ldots, S_p)$, and $S_1 > \cdots > S_p > 0$ are the characteristic roots of $\mathbf{W}$ such that $n\mathbf{W} \sim \mathrm{W}_p(n, I_p)$, and $\boldsymbol{Z}$ and $\mathbf{S}$ are matually independent. Then, $T_0^2 = \boldsymbol{X}'\boldsymbol{X}$.

Chapter 16

Error Bounds for Approximations to Some Other Statistics

In this chapter, we first use the results of approximation to location and scale mixtures obtained in Section 14.1 to construct error bounds for the large-sample and high-dimensional approximations to the linear discriminant function. Then, we discuss the estimators in the profile analysis and growth curve analysis. Simultaneous confidence intervals are related to the distribution of a scale mixture of chi-square distribution. We will see that the distributions of these estimators are also regarded as a scale mixture of F-distribution. Finally some generalized least squares estimators are discussed.

16.1 Linear Discriminant Function

16.1.1 Representation as a Location and Scale Mixture

One of the most important subjects in discriminant analysis is the classification of a $p \times 1$ observed vector $\boldsymbol{X}$ into one of the two populations Π_1 and Π_2 (i.e., deciding which populations they have come from). For $i = 1, 2$, let Π_i be the p-variate normal populations $\mathrm{N}_p(\boldsymbol{\mu}_i, \Sigma)$ with $\boldsymbol{\mu}_1 \neq \boldsymbol{\mu}_2$ and the positive definite Σ. If all parameters are known, the observation $\boldsymbol{X}$ may be classified into Π_1 or Π_2 according to the conditions

$$(\boldsymbol{\mu}_1 - \boldsymbol{\mu}_2)'\Sigma^{-1}\Big\{\boldsymbol{X} - \frac{1}{2}(\boldsymbol{\mu}_1 + \boldsymbol{\mu}_2)\Big\} \gtrless 0. \qquad (16.1.1)$$

For the optimum properties of the rule, see Section 9.1. The probability of misclassification (PM) resulting from the rule is the same for observations from either population, and is $\Phi(-\frac{1}{2}\Delta)$, where $\Phi(\cdot)$ denotes the standard normal distribution function and $\Delta^2 = (\boldsymbol{\mu}_1 - \boldsymbol{\mu}_2)'\Sigma^{-1}(\boldsymbol{\mu}_1 - \boldsymbol{\mu}_2)$ is the Mahalanobis squared distance between the two populations.

Suppose that all the parameters are unknown. Then the observation $\boldsymbol{X}$ may be classified by the rules corresponding to (16.1.1), with parameters being replaced by their appropriate estimators. Assume that random samples of sizes N_1 and N_2 are available from Π_1 and Π_2, respectively. Let $\bar{\boldsymbol{X}}_1$, $\bar{\boldsymbol{X}}_2$, and $\mathbf{S}$ be the sample mean vectors and the sample covariance matrix. Then the observation $\boldsymbol{X}$ is usually classified into Π_1 or Π_2 according to the following rule:

$$W \equiv (\bar{\boldsymbol{X}}_1 - \bar{\boldsymbol{X}}_2)'\mathbf{S}^{-1}\left(\boldsymbol{X} - \frac{1}{2}(\bar{\boldsymbol{X}}_1 + \bar{\boldsymbol{X}}_2)\right) \gtrless 0. \tag{16.1.2}$$

The probabilities of misclassification by the linear discriminant rule (16.1.2) is expressed as

$$\mathrm{P}(2|1) = \Pr\{W \leq 0 \mid \Pi_1\}, \ \mathrm{P}(1|2) = \Pr\{W \geq 0 \mid \Pi_2\}. \tag{16.1.3}$$

Here $\mathrm{P}(2|1)$ is the probability of putting $\boldsymbol{X}$ in Π_2, although it is known from Π_1, and so on. While, in general, the exact PM's are difficult to obtain, there has been considerable work on their asymptotic approximations, including asymptotic expansions. For the results, see Section 9.3. The main purpose of this section is their derivations with validity and error bounds. They are obtained by expressing the linear discriminant function W as a location and scale mixture of the standard normal distribution and applying the general results of Section 14.1. These results were given by Fujikoshi (1994, 2000), and Fujikoshi and Ulyanov (2006a).

When the observation $\boldsymbol{X}$ is from Π_1, the statistic W in (16.1.2) is expressed as

$$W = V^{1/2}Z - U, \tag{16.1.4}$$

where

$$\begin{aligned}
V &= (\bar{\boldsymbol{X}}_1 - \bar{\boldsymbol{X}}_2)'\mathbf{S}^{-1}\Sigma\mathbf{S}^{-1}(\bar{\boldsymbol{X}}_1 - \bar{\boldsymbol{X}}_2) \ (= \tilde{D}^2), \\
Z &= V^{-1/2}(\bar{\boldsymbol{X}}_1 - \bar{\boldsymbol{X}}_2)'\mathbf{S}^{-1}(\boldsymbol{X} - \boldsymbol{\mu}_1), \\
U &= (\bar{\boldsymbol{X}}_1 - \bar{\boldsymbol{X}}_2)'\mathbf{S}^{-1}(\bar{\boldsymbol{X}}_1 - \boldsymbol{\mu}_1) - \frac{1}{2}D^2
\end{aligned}$$

and $D^2 = (\bar{\boldsymbol{X}}_1 - \bar{\boldsymbol{X}}_2)'\mathbf{S}^{-1}(\bar{\boldsymbol{X}}_1 - \bar{\boldsymbol{X}}_2)$ is the sample Mahalanobis squared distance between two populations. The result is obtained by noting that the conditional distribution of $(\bar{\boldsymbol{X}}_1 - \bar{\boldsymbol{X}}_2)'\mathbf{S}^{-1}(\boldsymbol{X} - \boldsymbol{\mu}_1)$ given $\{\bar{\boldsymbol{X}}_1, \bar{\boldsymbol{X}}_2, \mathbf{S}\}$ is $\mathrm{N}(\mathbf{0}, \tilde{D}^2)$ when $\boldsymbol{X}$ comes from Π_1. Thus, W is a location and scale mixture of the standard normal distribution and we have

$$\mathrm{P}(2|1) = \mathrm{E}_{(U,V)}\left[\Phi(V^{-1/2}U)\right]. \tag{16.1.5}$$

Note that P(1|2) is obtained from P(2|1) by interchanging N_1 and N_2, and we are interested in the asymptotic approximation of P(2|1) and their error bounds. More specifically, we will study asymptotic approximation of the distribution of W when $\boldsymbol{X}$ comes from Π_1. For this purpose, first we extend Theorem 14.1.1 on a general location and scale mixture

$$X = V^{1/2}Z - U \tag{16.1.6}$$

of the standard normal distribution, where $Z \sim \mathrm{N}(0,1)$, Z and (U,V) are independent, and $V > 0$. The approximation in Theorem 14.5.1 is the one around $(U,V) = (0,1)$. We generalize it to the one around $(U,V) = (u_0, v_0)$, where (u_0, v_0) is any given point in the range space of (U,V). A generalized approximation is defined by

$$\Phi_{k(u_0,v_0)}(x) = \Phi(x) - \sum_{i+j=1}^{k-1} c_{(i,j)}(x) v_0^{-j-i/2}\mathrm{E}\Big[(U-u_0)^i(V-v_0)^j\Big], \tag{16.1.7}$$

where

$$c_{(i,j)}(x) = \frac{(-1)^i}{2^j i! j!} H_{2j+i-1}(x)\varphi(x), \tag{16.1.8}$$

and the summation is taken over the pairs $\{i,j\}$ of nonnegative integers such that $j > 0$ and $i + j < k$.

Theorem 16.1.1 *Let X be a location and scale mixture of Z in* (16.1.6). *Suppose that* $\mathrm{E}(V^k) + \mathrm{E}(|U|^k) < \infty$. *Then, for the approximation $\Phi_{k(u_0,v_0)}(x)$ in* (16.1.7),

$$\begin{aligned} |\Pr\{X \le x\} - \Phi_{k(u_0,v_0)}(y)| &\le \beta_{1,k} v_0^{-k}\mathrm{E}(|V-v_0|^k) \\ &+ \sum_{j=1}^{k-1} \bar{\alpha}_{(k-j,j)} v_0^{-j-(k-j)/2}\mathrm{E}(|U-u_0|^{k-j}|V-v_0|^j), \end{aligned} \tag{16.1.9}$$

where $y = v_0^{-1/2}(x + u_0)$, *and*

$$\alpha_{(i,j-i)} = \sup |c_{(i,j-i)}(x)|, \quad 0 \le i < j \le k.$$

Proof. We can write

$$\begin{aligned} X &= V^{1/2}Z - U \\ &= v_0^{1/2}[(V/v_0)^{1/2}Z - v_0^{-1/2}(U-u_0)] - u_0 \\ &= v_0^{1/2}\tilde{X} - u_0, \end{aligned}$$

where

$$\tilde{X} = \tilde{V}^{1/2}Z - \tilde{U}, \quad \tilde{U} = v_0^{-1/2}(U - u_0), \quad \tilde{V} = V/v_0.$$

Note that

$$\Pr\{(X \le x\} = \Pr\{\tilde{X} \le y\}$$

and $\tilde{X}$ is a scale and location mixture with a location $\tilde{U}$ and a scale $\tilde{V}$. The required result is obtained by applying Theorem 14.5.1 to the right side of the equality above. □

For $k \le 6$, the values of $\alpha_{(k-j,j)}$ are given in Table 14.1.1, and $\beta_{1,k} = \frac{1}{2}$. When we approximate the distribution of X by $\Phi(y)$, its error bound is given in the following theorem.

Theorem 16.1.2 *Let X be a location and scale mixture of Z in* (16.1.6). *Assume that* $\mathrm{E}(U^2) < \infty$ *and* $\mathrm{E}(V^2) < \infty$. *Then*

$$\left|\Pr\{X \le x\} - \Phi(y)\right| \le B_{20} + B_{21}, \tag{16.1.10}$$

where $y = v_0^{-1/2}(x + u_0)$, *and*

$$\begin{aligned} B_{20} &= \frac{1}{2} v_0^{-2} \mathrm{E}[(V - v_0)^2]) + \frac{1}{2\sqrt{2\pi \mathrm{e}}} v_0^{-1} \mathrm{E}[(U - u_0)^2] \\ &\quad + \frac{1}{2\sqrt{2\pi}} v_0^{-3/2} \left\{ \mathrm{E}[(U - u_0)^2] \mathrm{E}[(V - v_0)^2] \right\}^{1/2}, \\ B_{21} &= \frac{1}{\sqrt{2\pi}} v_0^{-1/2} |\mathrm{E}(U - u_0)| + \frac{1}{2\sqrt{2\pi \mathrm{e}}} v_0^{-1} |\mathrm{E}(V - v_0)|. \end{aligned}$$

Proof. From Theorem 16.1.1 we have

$$\left|\Pr\{X \le x\} - \Phi(y) - \varphi(y) A_1(y)\right| \le \tilde{B}_{20}$$

and hence

$$\left|\Pr\{X \le x\} - \Phi(y)\right| \le \tilde{B}_{20} + \tilde{B}_{21},$$

where

$$\begin{aligned} A_1(y) &= v_0^{-1/2} \mathrm{E}(U - u_0) - \frac{1}{2} H_1(y) v_0^{-1} \mathrm{E}(V - v_0), \\ \tilde{B}_{20} &= \bar{\alpha}_{(0,2)} v_0^{-1} \mathrm{E}[(U - u_0)^2] + \beta_{1,2} v_0^{-2} \mathrm{E}[(V - v_0)^2] \\ &\quad + \bar{\alpha}_{(1,1)} v_0^{-3/2} \mathrm{E}[|(U - u_0)(V - v_0)|], \\ \tilde{B}_{21} &= |\varphi(y_0) A_1(y)|. \end{aligned}$$

Note that we have

$$\begin{aligned} \mathrm{E}\big[\big|(U - u_0)(V - v_0)\big|\big] &\le \left\{ \mathrm{E}[(U - u_0)^2] \mathrm{E}[(V - v_0)^2] \right\}^{1/2}, \\ \bar{\alpha}_{(0,2)} = \frac{1}{2\sqrt{2\pi \mathrm{e}}}, \quad \bar{\alpha}_{(1,1)} &= \frac{1}{2\sqrt{2\pi}}, \quad \beta_{1,2} = \frac{1}{2}. \end{aligned}$$

which imply the required result.

Corollary 16.1.1 *Under Theorem 16.1.2, assume that* $\mathrm{E}(U) = u_0$, *and* $\mathrm{E}(V) = v_0$. *Then*

$$\left|\Pr\{X \le x\} - \Phi(y)\right| \le B_{20}, \tag{16.1.11}$$

where

$$\begin{aligned} B_{20} &= \frac{1}{2\sqrt{2\pi e}} v_0^{-1}\mathrm{Var}(U) + \frac{1}{2} v_0^{-2}\mathrm{Var}(V) \\ &\quad + \frac{1}{2\sqrt{2\pi}} v_0^{-3/2}\mathrm{E}(|(U-u_0)(V-v_0)|. \end{aligned}$$

It may be noted that the inequality (16.1.10) is equivalent to

$$\left|\Pr\{v_0^{-1/2}(X+u_0) \le x\} - \Phi(x)\right| \le B_{20} + B_{21}. \tag{16.1.12}$$

In practice, in addition to the approximation $\Phi_{1(u_0,v_0)}(x) = \Phi(y)$, we can use

$$\begin{aligned} \Phi_{2(u_0,v_0)}(x) &= \Phi(y) + \varphi(y)A_1, \\ \Phi_{3(u_0,v_0)}(x) &= \Phi(y) + \varphi(y)(A_1 + A_2), \\ \Phi_{4(u_0,v_0)}(x) &= \Phi(y) + \varphi(y)(A_1 + A_2 + A_3), \end{aligned} \tag{16.1.13}$$

where $y = v_0^{-1/2}(x+u_0)$, and

$$\begin{aligned} A_1 &= v_0^{-1/2}\mathrm{E}(U-u_0) - \frac{1}{2}H_1(y)v_0^{-1}\mathrm{E}(V-v_0), \\ A_2 &= -\frac{1}{2}H_1(y)v_0^{-1}\mathrm{E}[(U-u_0)^2] + \frac{1}{2}H_1(y)v_0^{-3/2}\mathrm{E}[(U-u_0)(V-v_0] \\ &\quad - \frac{1}{8}H_3(y)v_0^{-2}\mathrm{E}[(V-v_0)^2], \\ A_3 &= \frac{1}{6}H_2(y)v_0^{-3/2}\mathrm{E}[(U-u_0)^3] - \frac{1}{4}H_3(y)v_0^{-2}\mathrm{E}[(U-u_0)^2(V-v_0)] \\ &\quad + \frac{1}{8}H_4(y)\mathrm{E}[(U-u_0)(V-v_0)^2] - \frac{1}{48}H_5(y)v_0^{-3}\mathrm{E}[(V-v_0)^3]. \end{aligned}$$

Approximation errors of $\Phi_{2(u_0,v_0)}(x)$ and $\Phi_{4(u_0,v_0)}(x)$ are bounded by B_2 and B_4, respectively, that are given below:

$$\begin{aligned} B_2 &= \frac{1}{2}v_0^{-2}\mathrm{E}[(V-v_0)^2] + \alpha_{(2,0)}v_0^{-1}\mathrm{E}[(U-u_0)^2] \\ &\quad + v_0^{-3/2}\bar{\alpha}_{(1,1)}\mathrm{E}[|(U-u_0)(V-v_0)|], \qquad (16.1.14) \\ B_4 &= \frac{1}{2}v_0^{-4}\mathrm{E}[(V-v_0)^4] + \alpha_{(4,0)}v_0^{-2}\mathrm{E}[(U-u_0)^4] \\ &\quad + \alpha_{(3,1)}v_0^{-5/2}\mathrm{E}[|(U-u_0)^3(V-v_0)|] \\ &\quad + \alpha_{(2,2)}v_0^{-3/2}\mathrm{E}[(U-u_0)^2(V-v_0)^2] \\ &\quad + \alpha_{(1,3)}v_0^{-7/2}\mathrm{E}[|(U-u_0)(V-v_0)^3|]. \qquad (16.1.15) \end{aligned}$$

16.1.2 Large-Sample Approximations

In this section we assume a large-sample framework:

$$N_1 \to \infty, \quad N_2 \to \infty, \text{ and } p \text{ is fixed.} \tag{16.1.16}$$

Then it is easily seen that

$$U \to -\frac{1}{2}\Delta^2 \text{ and } V \to \Delta^2.$$

Therefore, we define

$$(u_0,\ v_0) := \big(-\frac{1}{2}\Delta^2,\ \Delta^2\big). \tag{16.1.17}$$

Now we consider the moments of (U, V) in (16.1.4). Set

$$\begin{aligned} \boldsymbol{Z}_1 &\equiv N^{-1/2}\Sigma^{-1/2}(N_1\bar{\boldsymbol{X}}_1 + N_2\bar{\boldsymbol{X}}_2 - N_1\boldsymbol{\mu}_1 - N_2\boldsymbol{\mu}_2), \\ \boldsymbol{Z}_2 &\equiv \Big(\frac{N}{N_1N_2}\Big)^{-1/2}\Sigma^{-1/2}(\bar{\boldsymbol{X}}_1 - \bar{\boldsymbol{X}}_2 - \boldsymbol{\mu}_1 + \boldsymbol{\mu}_2) \quad \text{and} \\ \mathbf{B} &\equiv \Sigma^{-1/2}\mathbf{S}\Sigma^{-1/2}. \end{aligned} \tag{16.1.18}$$

Note that they are mutually independent and $\boldsymbol{Z}_1 \sim \mathrm{N}_p(0, \mathrm{I}_p)$, $\boldsymbol{Z}_2 \sim \mathrm{N}_p(0, \mathrm{I}_p)$, and $\mathbf{B} \sim \mathrm{W}_p(n, \mathrm{I}_p)$. We can express (U, V) in terms of $\boldsymbol{Z}_1$, $\boldsymbol{Z}_2$ and $\mathbf{B}$ as follows:

$$\begin{aligned} U &= -\frac{1}{2}\boldsymbol{\delta}'\mathbf{B}^{-1}\boldsymbol{\delta} + \frac{1}{N^{1/2}}\boldsymbol{\delta}'\mathbf{B}^{-1}z_1 - \Big(\frac{N_1}{NN_2}\Big)^{1/2}\boldsymbol{\delta}'\mathbf{B}^{-1}\boldsymbol{Z}_2 \\ &\quad + \frac{1}{(N_1N_2)^{1/2}}\boldsymbol{Z}_1'\mathbf{B}^{-1}z_2 - \frac{N_1 - N_2}{2N_1N_2}\boldsymbol{Z}_2'\mathbf{B}^{-1}\boldsymbol{Z}_2, \\ V &= \boldsymbol{\delta}'\mathbf{B}^{-2}\boldsymbol{\delta} + 2\Big(\frac{N}{N_1N_2}\Big)^{1/2}\boldsymbol{\delta}'\mathbf{B}^{-2}\boldsymbol{Z}_2 + \frac{N}{N_1N_2}\boldsymbol{Z}_2'\mathbf{B}^{-2}\boldsymbol{Z}_2, \end{aligned} \tag{16.1.19}$$

where $\boldsymbol{\delta} = \Sigma^{-1/2}(\boldsymbol{\mu}_1 - \boldsymbol{\mu}_2)$. The first and second moments of U and V can be computed with the help of Theorems 1.2.4 and 2.2.7. As an alternative approach is to use the following Lemmas.

Lemma 16.1.1 *Let* $\mathbf{W}$ *be a Wishart matrix having* $\mathrm{W}_p(n, \Sigma)$, *and* B *a* $p \times q$ *constant matrix with rank* q. *Put*

$$\mathbf{Q}_1 = (\mathrm{B}'\mathbf{W}^{-1}\mathrm{B})^{-1}, \quad \mathbf{Q}_2 = \mathrm{B}'\mathbf{W}^{-1}\Sigma\mathbf{W}^{-1}\mathrm{B}.$$

Then, $\mathbf{Q}_1$ *and* $\mathbf{Q}_2$ *are expressed as*

$$\begin{aligned} \mathbf{Q}_1 &= (\mathrm{B}\Sigma^{-1}\mathrm{B})^{1/2}\mathbf{V}_{11\cdot 2}(\mathrm{B}\Sigma^{-1}\mathrm{B})^{1/2}, \\ \mathbf{Q}_2 &= (\mathrm{B}'\Sigma^{-1}\mathrm{B})^{1/2}V_{11\cdot 2}^{-1}\{\mathrm{I}_q + \mathbf{V}_{12}\mathbf{V}_{22}^{-2}\mathbf{V}_{21}\}V_{11\cdot 2}^{-1}(\mathrm{B}\Sigma^{-1}\mathrm{B})^{1/2}, \end{aligned}$$

where $\mathbf{V} \sim \mathrm{W}_p(m, \mathrm{I}_p)$, *and*

$$\mathbf{V} = \begin{pmatrix} \mathbf{V}_{11} & \mathbf{V}_{12} \\ \mathbf{V}_{21} & \mathbf{V}_{22} \end{pmatrix}, \quad \mathbf{V}_{12} : q \times p.$$

Proof. The expression for $\mathbf{Q}_2$ has been obtained in the proof of Theorem 2.2.9. Similarly, we can get the expression of $\mathbf{Q}_1$. □

Lemma 16.1.2 *Let Y_1, Y_2, Y_3, Y_4 be a set of independent random variables with $\tau^2 = (N_1 N_2/N)\Delta^2$, and such that*

$$Y_1 \sim \chi^2(p; \tau^2),\ Y_2 \sim \chi^2(n-p+1),\ Y_3 \sim \chi^2(p-1),\ Y_4 \sim \chi^2(n-p+2).$$

Then the distributions of $D^2 = (\bar{\boldsymbol{X}}_1 - \bar{\boldsymbol{X}}_2)'\mathbf{S}^{-1}(\bar{\boldsymbol{X}}_1 - \bar{\boldsymbol{X}}_2)'$ (the sample Mahalanobis squared distance) and $\tilde{D}^2 = (\bar{\boldsymbol{X}}_1 - \bar{\boldsymbol{X}}_2)'\mathbf{S}^{-1}\Sigma\mathbf{S}^{-1}(\bar{\boldsymbol{X}}_1 - \bar{\boldsymbol{X}}_2)$ are identical with those of

$$\frac{Nn}{N_1 N_2}\frac{Y_1}{Y_2} \quad \textit{and} \quad \frac{Nn^2}{N_1 N_2}\frac{Y_1}{Y_2^2}\left(1 + \frac{Y_3}{Y_4}\right)$$

respectively. In particular, their ratio $\tilde{D}^2/D^2$ is distributed as

$$\frac{n}{Y_1}\left(1 + \frac{Y_3}{Y_4}\right).$$

Proof. The first assertion is essentially the same as that of Hotelling's T^2 statistic. (For its proof, see, e.g., Theorem 3.3.1). The second assertion is proved by taking $\boldsymbol{B} = \bar{\mathbf{X}}_1 - \bar{\mathbf{X}}_2$ in the second expression, $\mathbf{Q}_2$ in Lemma 16.1.1, and using Theorems 2.2.3 and 3.1.1. The final assertion is a direct consequence of the other two. □

Write

$$n = N - 2, \quad N = N_1 + N_2, \quad \text{and} \quad m = n - p.$$

Using Lemma 16.1.2 we get

$$\begin{aligned} \mathrm{E}(U) &= -\frac{n}{2(m-1)}\left(\Delta^2 + \frac{(N_1 - N_2)p}{N_1 N_2}\right), \quad m > 1, \\ \mathrm{E}(V) &= \frac{n^2(n-1)}{m(m-1)(m-3)}\left(\Delta^2 + \frac{Np}{N_1 N_2}\right), \quad m > 3. \end{aligned} \tag{16.1.20}$$

These imply that

$$\begin{aligned} M_U &= \mathrm{E}(U) - u_0 \\ &= -\frac{p+1}{2(m-1)}\Delta^2 + \frac{pn}{2(m-1)}\left(\frac{1}{N_1} - \frac{1}{N_2}\right) \end{aligned} \tag{16.1.21}$$

and

$$
\begin{aligned}
M_V &= \mathrm{E}(V) - v_0 \\
&= \left[\frac{n^2(n-1)}{m(m-1)(m-3)} - 1\right]\Delta^2 + \frac{n^2(n-1)}{m(m-1)(m-3)}\left(\frac{Np}{N_1N_2}\right). \qquad (16.1.22)
\end{aligned}
$$

Similarly we can express the variance of U as

$$
\begin{aligned}
\mathrm{Var}(U) &= \frac{n^2}{2m(m-1)(m-3)}\Big[\frac{1}{m-1}\Delta^4 \\
&+ \frac{2(n-1)}{mN_2}\Delta^2\Big(1 + \frac{N_1 - N_2}{(m-1)N_1}\Big) \qquad (16.1.23) \\
&\quad + \frac{2(n-1)p}{N_1N_2}\Big(\frac{1}{m} + \frac{(N_1-N_2)^2}{2(m-1)N_1N_2}\Big)\Big], \quad m > 3.
\end{aligned}
$$

Note that V is the same as $\tilde{D}^2$ in Lemma 16.1.2. It follows from Lemma 16.1.2

$$
\begin{aligned}
\mathrm{Var}(V) &= \frac{2(n-1)n^4}{m(m-1)^2(m-3)^2}\Big[\frac{1}{m}\Big(1 + \frac{8(m-4)}{(m-5)(m-7)}\Big) \\
&\quad \times \Big(\frac{p-1}{m}\Big(\Delta^2 + \frac{pN}{N_1N_2}\Big)^2 + \frac{N(n-3)}{N_1N_2}\Big(2\Delta^2 + \frac{pN}{N_1N_2}\Big)\Big) \\
&\quad + \frac{4(n-1)(m-4)}{m(m-5)(m-7)}\Big(\Delta^2 + \frac{pN}{N_1N_2}\Big)^2\Big], \quad m > 7. \qquad (16.1.24)
\end{aligned}
$$

Theorem 16.1.3 *Let W be the discriminant function defined by* (16.1.2). *Assume that $m > 7$. Then we have*

$$
\Big|\Pr\{W \le x \mid \Pi_1\} - \Phi\Big(\frac{x}{\Delta} - \frac{1}{2}\Delta\Big)\Big| \le B_{20}^{(1)}(L) + B_{21}^{(1)}(L), \qquad (16.1.25)
$$

where

$$
\begin{aligned}
B_{20}^{(1)}(L) &= \frac{1}{2\sqrt{2\pi \mathrm{e}}}\frac{1}{\Delta^2}\big(\mathrm{Var}(U) + M_U^2\big) + \frac{1}{2}\frac{1}{\Delta^4}\big(\mathrm{Var}(V) + M_V^2\big) \\
&\quad + \frac{1}{2\sqrt{2\pi}}\frac{1}{\Delta^3}\Big[\big(\mathrm{Var}(U) + M_U^2\big)\big(\mathrm{Var}(V) + M_V^2\big)\Big]^{1/2}, \\
B_{21}^{(1)}(L) &= \frac{1}{\sqrt{2\pi}}\frac{1}{\Delta}|M_U| + \frac{1}{2\sqrt{2\pi \mathrm{e}}}\frac{1}{\Delta^2}|M_V|.
\end{aligned}
$$

16.1.3 High - Dimensional Approximations

In this section we consider the distributions of W in (16.1.1) when $\boldsymbol{X}$ comes from Π_1 under a high-dimensional framework:

$$
N_1 \to \infty, \quad N_2 \to \infty, \quad p \to \infty, \quad m = n - p \to \infty. \qquad (16.1.26)
$$

In order to apply Theorem 16.1.1 we need to define (u_0, v_0). Here we define

$$(u_0, v_0) = (\mathrm{E}(U), \mathrm{E}(V)). \tag{16.1.27}$$

Then we have the quantity

$$\gamma = \mathrm{E}(V)^{-1/2}\mathrm{E}(U) = -\left(\frac{m(m-3)}{(n-1)(m-1)}\right)^{1/2} \gamma_0, \tag{16.1.28}$$

where

$$\gamma_0 = \left(-\frac{1}{2}\Delta^2 + p\Big(\frac{1}{N_1} - \frac{1}{N_2}\Big)\right)\left(\Delta^2 + \frac{Np}{N_1 N_2}\right)^{-1/2}. \tag{16.1.29}$$

Lachenbruch (1968) proposed an approximation

$$\mathrm{P}(2|1) \simeq \Phi(\gamma). \tag{16.1.30}$$

Fujikoshi and Seo (1998) have pointed out that the approximation $\Phi(\gamma)$ has the same property as in Theorem 16.1.2. As a more simple and accurate approximation it has been proposed to use

$$\tilde{\gamma} = -\Big(\frac{N-p}{N}\Big)^{1/2} \gamma_0 \tag{16.1.31}$$

and the approximation

$$\mathrm{P}(2|1) \simeq \Phi(\tilde{\gamma}). \tag{16.1.32}$$

This approximation was proposed by Raudys (1972) for $N_1 = N_2$ and by Fujikoshi and Seo (1998) for $N_1 \neq N_2$. Wyman et al. (1990) compared the accuracy of several approximations for P(2|1) in the case $N_1 = N_2$ and pointed that the approximation $\Phi(\tilde{\gamma})$ in (16.1.32) has overall the best accuracy for the combinations of parameters considered in the study. Similar properties hold for the case $N_1 \neq N_2$.

Theorem 16.1.4 *Let W be the discriminant function defined by* (16.1.2). *Let γ be the one defined by* (16.1.28) *and $v_0 = \mathrm{E}(V)$ in* (16.1.20). *Then we have*

$$\left|\Pr\{W \leq x \mid \boldsymbol{x} \in \Pi_1\} - \Phi(x v_0^{-1} + \gamma)\right| \leq B_{20}^{(1)}(H), \tag{16.1.33}$$

where

$$\begin{aligned} B_{20}^{(1)}(H) &= \frac{1}{2\sqrt{2\pi e}} v_0^{-1}\mathrm{Var}(U) + \frac{1}{2} v_0^{-2}\mathrm{Var}(V) \\ &\quad + \frac{1}{2\sqrt{2\pi}} v_0^{-3/2}\big(\mathrm{Var}(U)\mathrm{Var}(V)\big)^{1/2}. \end{aligned}$$

Here Var(U) *and* Var(V) *are given in* (16.1.24).

Table 16.1.1 Upper bounds for $B_{20}^{(1)}(H)$ in (16.1.33)

p	N_1	N_2	$B_{20}^{(1)}(H); \Delta = 1.68$	$B_{20}^{(1)}(H); \Delta = 2.56$
5	20	20	0.2854	0.2625
	75	75	0.0605	0.0522
10	20	20	0.3255	0.2625
	75	75	0.0606	0.0540
30	30	30	0.2922	0.2936
	60	60	0.0842	0.0813
	100	100	0.0457	0.0425

In Table 16.1.1 we give the values of $B_{20}^{(1)}(H)$ in some cases. The table shows that to get a reasonable bound. we need larger sample sizes and dimension.

Finally we consider a bound for the approximation $\Phi(\tilde{\gamma})$. Let

$$\tilde{u}_0 = cu_0, \quad c = \left(\frac{(n-1)(m-1)(m+2)}{(n+2)m(m-3)} \right)^{1/2}. \tag{16.1.34}$$

Then $\tilde{\gamma} = v_0^{-1/2}\tilde{u}_0$. Considering an approximation with $\tilde{u_0}$ instead of u_0 we have the following result from Theorems 16.1.2 and 16.1.4.

Theorem 16.1.5 *Under the same condition as in Theorem 16.1.4, it holds that*

$$|e(2|1) - \Phi(\tilde{\gamma})| \leq \tilde{B}^{(2)} + h_0 v_0^{-1/2}|1-c|, \tag{16.1.35}$$

where $\tilde{B}_{20}^{(1)}(H)$ *is defined from* $B_{20}^{(1)}(H)$ *by substituting into*

$$\mathrm{E}\{(U - \tilde{u_0})^2\} = \mathrm{Var}(U) + (1-c)^2 u_0^2 \tag{16.1.36}$$

for Var(U).

Note that $1-c = O_1^*$, and hence the bounds in Theorems 16.1.4 and 16.1.5 are O_1^*. In general, $\tilde{B}_{20}^{(1)}(H) \geq B_{20}^{(1)}(H)$. However, Fujikoshi and Seo (1998) have pointed out that $\Phi(\tilde{\gamma})$ is more accurate than $\Phi(\gamma)$. So, it is expected that the bound in Theorem 16.1.5 can be improved by a more elaborate method.

16.1.4 Some Related Topics

In this section we take up some related topics, focusing on the linear discriminant function W in (16.1.2) under a large-sample framework. **Studentized**

Classification Statistic $\boldsymbol{W}$. First we note that the result in Theorem 16.1.3 can be stated as

$$\left|\Pr\left\{\frac{W-\frac{1}{2}\Delta^2}{\Delta}\le x\,\middle|\,\Pi_1\right\}-\Phi(x)\right|\le B_{20}^{(1)}(L)+B_{21}^{(1)}(L). \tag{16.1.37}$$

Anderson (1973) derived asymptotic expansions for the distributions of the Studentized W, [i.e., $(W-\frac{1}{2}D^2)/D$ and $(W+\frac{1}{2}D^2)/D$ under Π_1 and Π_2, respectively]. As a special case, the limiting distributions are the standard normal distribution. So we are interested in obtaining an error bound for

$$\left|\Pr\left\{\frac{W-\frac{1}{2}D^2}{D}\le x\,\middle|\,\Pi_1\right\}-\Phi(x)\right|.$$

Using (16.1.4) we can write the Studentized statistic W as a location and scale mixture of the standard normal distribution defined by

$$\frac{W-\frac{1}{2}D^2}{D}=V^{1/2}Z-U, \tag{16.1.38}$$

where

$$U=\frac{1}{D}(\bar{\boldsymbol{x}}^{(1)}-\bar{\boldsymbol{x}}^{(2)})'S^{-1}(\bar{\boldsymbol{x}}^{(1)}-\boldsymbol{\mu}^{(1)}),\quad V=\frac{\tilde{D}}{D}. \tag{16.1.39}$$

Therefore we can apply Theorem 16.1.1 with an appropriate (u_0,v_0). In this case we take $(u_0,v_0)=(0,1)$.

It is not simple to obtain the first two exact moments of U. However we can get bounds for the moments by using

$$|U|\le\{(\bar{\boldsymbol{x}}^{(1)}-\boldsymbol{\mu}^{(1)})'S^{-1}(\bar{\boldsymbol{x}}^{(1)}-\boldsymbol{\mu}^{(1)})\}^{1/2}=T.$$

In fact, it holds that

$$\begin{aligned}\mathrm{E}(U^2)&\le\mathrm{E}(T^2)=\frac{pn}{N_1(n-p-1)}=M_1,\\ |\mathrm{E}(U)|&\le\{\mathrm{E}(U^2)\}^{1/2}\le M_1^{1/2}.\end{aligned}$$

The first two moments of V can be obtained by using

$$V=\frac{\tilde{D}^2}{D^2}=\frac{n}{Y_2}\left(1+\frac{Y_3}{Y_4}\right),$$

in Lemma 16.1.2, which does not depend on Δ. The computations of moments of V in this case is much simpler than that in the case of the W statistic. The moments are expressed as

$$\begin{aligned}M_2&\le\mathrm{E}(V-1)=\frac{p}{m^{(2)}}\big(2n-(p+1)\big),\\ M_3&\le\mathrm{E}(V-1)^2=\frac{1}{m^{(4)}}\big(2n^3+p^{[4]}\big),\end{aligned}$$

where for a pair of positive integers n and k we write $n^{(k)} = n(n-1)\dots(n-k+1)$ and $n^{[k]} = n(n+1)\cdots(n+k-1)$. Using the results just obtained, we have

$$\left|\Pr\left\{\frac{W-\frac{1}{2}D^2}{D} \le x \,\middle|\, \boldsymbol{x}\in\Pi_1\right\} - \Phi(x)\right|$$
$$\le \frac{1}{2\sqrt{2\pi \mathrm{e}}}M_1 + \frac{1}{2}M_3 + \frac{1}{2\sqrt{2\pi}}(M_1M_3)^{1/2} + \frac{1}{\sqrt{2\pi}}M_1^{1/2} + \frac{1}{2\sqrt{2\pi\mathrm{e}}}|M_2|. \qquad (16.1.40)$$

The bound in (16.1.40) depends on the sample sizes (N_1, N_2) only, but its order is $\mathrm{O}_{1/2}$. It is interesting to find a sharp bound for $|E(U)|$ whose order is O_1. We note that for the Studentized classification statistic, the inequality (16.1.40) is valid under the conditions $n-p-3>0$. On the other hand, the corresponding condition for the classification statistic W is $n-p-7>0$.

Expectation of $Q_D = \Phi(-D/2)$. The distribution of $Q_D = \Phi(-D/2)$ plays an important role in estimating the probabilities of misclassification. McLachlan (1973) gave an asymptotic expansion for the expectation of Q_D whose leading term is $\Phi(-\Delta/2)$. Now we will obtain an upper bound for $|\mathrm{E}(Q_D) - \Phi(-\Delta/2)|$. Let

$$X = V^{1/2}Z,$$

where Z is the standard normal variate, $V = \Delta^2/D^2$, and Z and V are independent. Then we have

$$\mathrm{E}(Q_D) = \Pr\left\{X \le -\frac{1}{2}\Delta\right\}.$$

Note that X is a scale mixture of the standard normal distribution. Therefore, using (4.29) with $\delta = -1$ and $k = 2$ we have

$$\left|\mathrm{E}(Q_D) - \Phi(-\frac{1}{2}\Delta)\right| \le \alpha_{-1,1}\left|\mathrm{E}\left(\frac{D^2}{\Delta^2}-1\right)\right| + \beta_{-1,2}\mathrm{E}\left(\frac{D^2}{\Delta^2}-1\right)^2$$
$$\le 0.08\cdot\left|\mathrm{E}\left(\frac{D^2}{\Delta^2}-1\right)\right| + \frac{1}{2}\mathrm{E}\left(\frac{D^2}{\Delta^2}-1\right)^2. \qquad (16.1.41)$$

The expectations in (16.1.41) can be computed by Lemma 16.1.2 (i) as follows.

$$\mathrm{E}\left(\frac{D^2}{\Delta^2}-1\right) = \frac{n}{m-1}\left(\frac{1}{n}p+1) + p\tau^{-2}\right),$$
$$\mathrm{E}\left(\frac{D^2}{\Delta^2}-1\right)^2 = \frac{n^2}{(m-1)(m-3)}\left[\frac{2}{n} + \frac{1}{n^2}(p+1)(p+3) + 2\left(2+\frac{1}{n}p(p+3)\right)\tau^{-2} + p(p+2)\tau^{-4}\right],$$

where $m = n - p$ and $\tau^2 = N_1N_2/N)\Delta^2$. We note that the inequality (16.1.41) is valid under the condition $n - p - 3 > 0$. Further, the order of the resulting upper bound is O_1 with respect to (N_1^{-1}, N_2^{-1}).

16.2 Profile Analysis

16.2.1 Parallelism Model and MLE

Suppose that we wish to compare the performance of students at school A_1 with those of school A_2, based on three tests Y_1, Y_2, Y_3. Let the mean vectors of (Y_1, Y_2, Y_3) at schools A_1 and A_2 be denoted by

$$\boldsymbol{\theta}_1 = (\theta_{11}, \theta_{12}, \theta_{13})' \quad \text{and} \quad \boldsymbol{\theta}_2 = (\theta_{21}, \theta_{22}, \theta_{23})',$$

respectively. The population mean vectors $\boldsymbol{\theta}_1$ and $\boldsymbol{\theta}_2$ are called *profiles* of A_1 and A_2. One of our interests in profile analysis is to examine whether or not the two profiles are parallel, (see Figure 16.2.1).

Figure 16.2.1 Profiles of schools A_1 and A_2

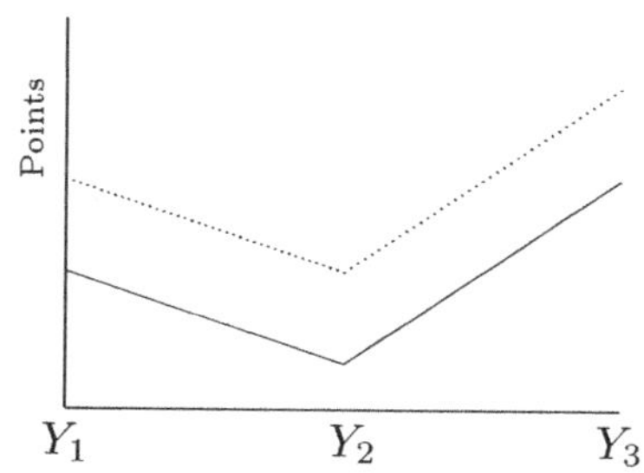

The parallelism hypothesis is expressed as

$$\mathrm{H}\colon \theta_{11} - \theta_{21} = \theta_{12} - \theta_{22} = \theta_{13} - \theta_{23}, \text{ or } \boldsymbol{\theta}_1 - \boldsymbol{\theta}_2 = \begin{pmatrix} 1 \\ 1 \\ 1 \end{pmatrix} \gamma,$$

where γ is an unknown parameter. Our interest is to estimate or find a confidence interval for the difference γ in the levels of profiles.

The two groups of parallelism problems can be extended to k groups. Consider k p-dimensional normal populations $\mathrm{N}_p(\boldsymbol{\theta}_i, \Sigma), i = 1, \ldots, k$. When the p responses are measured in commensurable units, the mean vectors $\boldsymbol{\theta}_1, \ldots, \boldsymbol{\theta}_k$ are called the *mean profiles* of the k groups. The parallelism hypothesis or model is expressed as

$$\mathrm{H}\colon \boldsymbol{\theta}_i - \boldsymbol{\theta}_k = \gamma_i \mathbf{1}_p, \quad i = 1, \ldots, k - 1, \tag{16.2.1}$$

where the γ_i are unknown parameters and $\mathbf{1}_p$ is a p-dimensional vector with all the elements 1. The γ_i denotes the level difference between the ith and kth groups. Let $\boldsymbol{Y}_{i1}, \ldots, \boldsymbol{Y}_{in_i}$ be a sample of size n_i from the ith population $\mathrm{N}_p(\boldsymbol{\theta}_i, \Sigma)$ and

$$\bar{\boldsymbol{Y}}_i = \frac{1}{n_i}\sum_{j=1}^{n_i} \boldsymbol{Y}_{ij}, \ i = 1, \ldots, k, \ n = n_1 + \cdots + n_k$$

$$\bar{\boldsymbol{Y}} = \frac{1}{n}\sum_{i=1}^{k}\sum_{j=1}^{n_i} \boldsymbol{Y}_{ij}, \ \mathbf{S}_e = \sum_{i=1}^{k}\sum_{j=1}^{n_i}(\boldsymbol{Y}_{ij} - \bar{\boldsymbol{Y}}_i)(\boldsymbol{Y}_{ij} - \bar{\boldsymbol{Y}}_i)'.$$

Let

$$\mathrm{Q} = \mathrm{diag}\left(\frac{1}{n_1}, \ldots, \frac{1}{n_{k-1}}\right) + \frac{1}{n_k}\mathbf{1}_{k-1}\mathbf{1}_{k-1}'.$$

Then

$$\mathrm{Q}^{-1} = \mathrm{diag}\,(n_1, \ldots, n_{k-1}) - \frac{1}{n}\boldsymbol{n}_{k-1}\boldsymbol{n}_{k-1}',$$

where $\boldsymbol{n}_{k-1} = (n_1, \ldots, n_{k-1})'$.

Some statistical problems have been discussed by Srivastava (1987, 2002) and others.

Theorem 16.2.1 *The MLE of* $\boldsymbol{\gamma} = (\gamma_1, \ldots, \gamma_{k-1})'$ *in the parallelism model* (16.2.1) *is given by*

$$\hat{\boldsymbol{\gamma}} = (\mathbf{1}_p'\mathbf{S}_e^{-1}\mathbf{1}_p)^{-1}\mathbf{W}'\mathbf{S}_e^{-1}\mathbf{1}_p, \tag{16.2.2}$$

where

$$\mathbf{W} = (\bar{\boldsymbol{Y}}_1 - \bar{\boldsymbol{Y}}_k, \ldots, \bar{\boldsymbol{Y}}_{k-1} - \bar{\boldsymbol{Y}}_k).$$

Proof. Srivastava (1987) obtained the MLE by solving the likelihood equations; here we obtain it by maximizing the likelihood function. Let L be the likelihood of all the observations $\boldsymbol{y}_{ij}$s. Then

$$\begin{aligned}-2\log L &= n\log|\Sigma| + np\log 2\pi \\ &\quad + \mathrm{tr}\,\Sigma^{-1}\left\{\sum_{i=1}^{k} n_i(\bar{\boldsymbol{y}}_i - \boldsymbol{\theta}_i)(\bar{\boldsymbol{y}}_i - \boldsymbol{\theta}_i)' + S_e\right\}.\end{aligned}$$

We can see that

$$\begin{aligned}&\sum_{i=1}^{k} n_i(\bar{\boldsymbol{y}}_i - \boldsymbol{\theta}_i)(\bar{\boldsymbol{y}}_i - \boldsymbol{\theta}_i)' \\ &\qquad = (W - \Xi)Q^{-1}(W - \Xi)' + n(\bar{\boldsymbol{y}} - \bar{\boldsymbol{\theta}})(\bar{\boldsymbol{y}} - \bar{\boldsymbol{\theta}})',\end{aligned}$$

where $\Xi = (\boldsymbol{\theta}_1 - \boldsymbol{\theta}_k, \ldots, \boldsymbol{\theta}_{k-1} - \boldsymbol{\theta}_k)$, and $\bar{\boldsymbol{\theta}} = (1/n)\sum_{i=1}^{k} n_i \boldsymbol{\theta}_i$. Then, the likelihood L_1 after maximizing L with respect to $\bar{\boldsymbol{\theta}}$ is

$$\begin{aligned} -2\log L_1 &= n\log|\Sigma| + np\log 2\pi \\ &\quad + \operatorname{tr}\Sigma^{-1}\left\{(W - \mathbf{1}_p\gamma')Q^{-1}(W - \mathbf{1}_p\gamma')' + S_e\right\}. \end{aligned}$$

Now we consider the transformation from Σ to $\Psi = B'\Sigma B$, where

$$B = (\boldsymbol{b}_1, B_2),\ \boldsymbol{b}_1 = (1/\sqrt{p})\mathbf{1}_p,\ B_2 = D = \begin{pmatrix} I_{p-1} \\ -\mathbf{1}_{p-1}' \end{pmatrix}.$$

Let

$$\begin{aligned} &\Psi = B'\Sigma B = \begin{pmatrix} \psi_{11} & \boldsymbol{\psi}_{12}' \\ \boldsymbol{\psi}_{21} & \Psi_{22} \end{pmatrix},\ V = B'S_eB = \begin{pmatrix} v_{11} & \boldsymbol{v}_{12}' \\ \boldsymbol{v}_{21} & V_{22} \end{pmatrix}, \\ &\boldsymbol{\delta} = \sqrt{p}Q^{-1/2}\gamma,\ U = B'ZQ^{-1/2} = (\boldsymbol{u}_1, U_2)'. \end{aligned}$$

There exists a one-to-one transformation between $\{\gamma, \Sigma\}$ and $\{\boldsymbol{\delta}, \Psi\}$. The MLEs of γ and Ψ are obtained from those of $\boldsymbol{\delta}$ and Ψ. The latter MLEs are obtained by minimizing

$$g(\boldsymbol{\delta}, \Psi) = n\log|\Psi| + \operatorname{tr}\Psi^{-1}\{(\boldsymbol{u}_1 - \boldsymbol{\delta},\ U_2)'(\boldsymbol{u}_1 - \boldsymbol{\delta},\ U_2) + V\}. \qquad (16.2.3)$$

For a detailed computation, see Problem. □

16.2.2 Distributions of $\hat{\gamma}$

In this section we consider the distribution of the MLE of $\hat{\gamma}$ of the level differences between the first $k-1$ groups and the last group. The MLE is written as

$$\hat{\gamma} = (\mathbf{1}_p'\mathbf{S}^{-1}\mathbf{1}_p)^{-1}\mathbf{W}'\mathbf{S}^{-1}\mathbf{1}_p,$$

where $\mathbf{S} = (1/m)\mathbf{S}_e$ is an unbiased estimator of Σ and $m = n - k$. When $k = 2$,

$$\hat{\gamma} = (\mathbf{1}_p'\mathbf{S}^{-1}\mathbf{1}_p)^{-1}\mathbf{1}_p'\mathbf{S}_e^{-1}(\bar{\boldsymbol{Y}}_1 - \bar{\boldsymbol{Y}}_2).$$

Theorem 16.2.2 *Let $\hat{\gamma}$ be the MLE of γ given by* (16.2.2). *Then*

$$(\mathbf{1}_p'\Sigma^{-1}\mathbf{1}_p)^{1/2}Q^{-1/2}(\hat{\gamma} - \gamma) = U\boldsymbol{Z}, \qquad (16.2.4)$$

where $\boldsymbol{Z}$ is distributed as $\mathrm{N}_{k-1}(\mathbf{0}, \mathrm{I}_{k-1})$,

$$U = \frac{(\mathbf{1}_p'\Sigma^{-1}\mathbf{1}_p)^{1/2}(\mathbf{1}_p'\mathbf{S}^{-1}\Sigma\mathbf{S}^{-1}\mathbf{1}_p)^{1/2}}{\mathbf{1}_p'\mathbf{S}^{-1}\mathbf{1}_p}, \qquad (16.2.5)$$

and $\boldsymbol{Z}$ *and* $\mathbf{S}$ *are independent. Further, for a fixed vector* $\boldsymbol{a} = (a_1, \ldots, a_{k-1})'$,

$$X_a \equiv \frac{(\mathbf{1}_p'\Sigma^{-1}\mathbf{1})^{1/2}}{(\boldsymbol{a}'Q\boldsymbol{a})^{1/2}}\boldsymbol{a}'(\hat{\gamma} - \gamma) = UZ, \tag{16.2.6}$$

where Z *is distributed as* $\mathrm{N}(0, 1)$ *and* Z *and* $\mathbf{S}$ *are independent.*

Proof. It is easy to see that

$$\hat{\gamma} - \gamma = \frac{1}{\boldsymbol{\xi}'\mathbf{1}_p}\begin{pmatrix} R_1 \\ \vdots \\ R_{k-1} \end{pmatrix} = \frac{1}{\boldsymbol{\xi}'\mathbf{1}_p}\boldsymbol{R},$$

where $\boldsymbol{\xi} = \mathbf{S}^{-1}\mathbf{1}_p$ and the ith element of $\boldsymbol{R}$ is

$$R_i = (\bar{\boldsymbol{Y}}_1 - \bar{\boldsymbol{Y}}_{k-1} - \gamma_i\mathbf{1}_p)'\boldsymbol{\xi}, \quad i = 1, \ldots, k-1.$$

The conditional distribution of $\boldsymbol{R}$ given $\mathbf{S}$ is normal with mean $\mathbf{0}$ and covariance matrix given by

$$\boldsymbol{\xi}'\Sigma\boldsymbol{\xi}Q = (\mathbf{1}_p'\mathbf{S}^{-1}\Sigma\mathbf{S}^{-1}\mathbf{1}_p)Q,$$

which yields the first result, and then we also obtain

$$\begin{aligned}(\mathbf{1}_p'\Sigma^{-1}\mathbf{1}_p)^{1/2}\boldsymbol{a}'(\hat{\gamma} - \gamma) &= U\boldsymbol{a}'Q^{1/2}\boldsymbol{Z} \\ &= U(\boldsymbol{a}'Q\boldsymbol{a})^{1/2} \cdot \frac{\boldsymbol{a}'Q^{1/2}\boldsymbol{Z}}{(\boldsymbol{a}'Q\boldsymbol{a})^{1/2}}.\end{aligned}$$

Here $\boldsymbol{a}'Q^{1/2}\boldsymbol{Z}/(\boldsymbol{a}'Q\boldsymbol{a})^{1/2}$ is distributed as $\mathrm{N}(0, 1)$. This proves the second result. □

From Theorem 16.2.2, X_a is a scale mixture of $\mathrm{N}(0, 1)$ with the scale factor U. For constructing a confidence interval for $\boldsymbol{a}'\boldsymbol{\gamma}$, it is important to consider the distribution of $\hat{X}_a$, which is defined from X_a by substituting $\mathbf{S}$ for Σ:

$$\begin{aligned}\hat{X}_a &= \frac{(\mathbf{1}_p'\mathbf{S}^{-1}\mathbf{1})^{1/2}}{(\boldsymbol{a}'Q\boldsymbol{a})^{1/2}}\boldsymbol{a}'(\hat{\gamma} - \gamma) = \frac{(\mathbf{1}_p'\mathbf{S}^{-1}\mathbf{1})^{1/2}}{(\mathbf{1}_p'\Sigma^{-1}\mathbf{1}_p)^{1/2}} \cdot UZ \\ &= VZ,\end{aligned} \tag{16.2.7}$$

where

$$V = \frac{(\mathbf{1}_p'\mathbf{S}^{-1}\Sigma\mathbf{S}^{-1}\mathbf{1}_p)^{1/2}}{(\mathbf{1}_p'\mathbf{S}^{-1}\mathbf{1}_p)^{1/2}}. \tag{16.2.8}$$

A simultaneous confidence interval for $\boldsymbol{a}'\gamma$ can be constructed based on

$$\hat{X}_a^2 \le \max_{\boldsymbol{a}} \hat{X}_a^2 \equiv T$$

and the distribution of T. We can write

$$\begin{aligned} T &= (\mathbf{1}_p'\mathbf{S}^{-1}\mathbf{1}_p) \max_{\boldsymbol{a}} \frac{(\boldsymbol{a}'(\hat{\gamma}-\gamma))^2}{\boldsymbol{a}'Q\boldsymbol{a}} \\ &= (\mathbf{1}_p'\mathbf{S}^{-1}\mathbf{1})^2(\hat{\gamma}-\gamma)'Q^{-1}(\hat{\gamma}-\gamma) \\ &= V^2\chi_{k-1}^2. \end{aligned} \tag{16.2.9}$$

The statistics X_a and $\hat{X}_a$ are scale mixtures of the standard normal distribution with scale factors U and V, respectively. On the other hand, T is a scale mixture of a $\chi^2(k-1)$ variable χ_{k-1}^2. First we derive some properties of the scale factors. The scale factor U of X_a satisfies

$$U \geq 1,$$

since

$$\begin{aligned} (\mathbf{1}_p'\mathbf{S}^{-1}\mathbf{1}_p)^2 &= (\mathbf{1}_p'\Sigma^{-1/2}\cdot\Sigma^{1/2}\mathbf{S}^{-1}\mathbf{1}_p)^2 \\ &\leq (\mathbf{1}_p'\Sigma^{-1}\mathbf{1}_p)(\mathbf{1}_p'\mathbf{S}^{-1}\Sigma\mathbf{S}^{-1}\mathbf{1}_p). \end{aligned}$$

Further, the scale factors U and V have simple structures described in the following lemma.

Lemma 16.2.1 *Let U and V be the random variables or scale factors defined by* (16.2.5) *and* (16.2.8), *respectively. Then U^2 and V^2 are distributed as*

$$U^2 = 1 + \frac{\chi_{p-1}^2}{\chi_{m-p+2}^2},$$

$$V^2 = \frac{m}{\chi_{m-p+1}^2}\left[1 + \frac{\chi_{p-1}^2}{\chi_{m-p+2}^2}\right],$$

where $m = n - k$, and $\chi_{p-1}^2, \chi_{m-p+1}^2$ and χ_{m-p+2}^2 are independent $\chi^2(\cdot)$ variables with $p-1, m-p+1$, and $m-p+2$ degrees of freedom, respectively.

Proof. The results are obtained by using Lemma 16.1.1 along the lines of the proof of Lemma 16.1.2. For details, see Problem 2. □

Lemma 16.2.2 *Let U and V be the random variables or scale factors defined by* (16.2.5) *and* (16.2.8). *respectively. The moments of $(U^{2\delta}-1)$ and $(V^{2\delta}-1)$ are given as follows with $r = p-1$, where δ is 1 or -1.*

(1) *For $m - r + 1 - 2j > 0$,*

$$\mathrm{E}[(U^2-1)^j] = \frac{r(r+2)\cdot(r+2(j-1))}{(m-r-1)(m-r-1)\cdots(m-r+1-2j)}.$$

(2) *For any j,*

$$\mathrm{E}[(U^{-2}-1)^j] = \frac{r(r+2)\cdot(r+2(j-1))}{(m+1)(m+3)\cdots(m+2j-1)}.$$

The moments of $(V^{2\delta}-1)$ *can be obtained through the relation,*

$$\mathrm{E}[(V^{2\delta}-1)^j] = \sum_{i=0}^{j}\binom{j}{i}(-1)^{j-i}\mathrm{E}[(V^{2\delta})^i].$$

and the moments of $V^{2\delta}$ *are given by* (3) *and* (4) *below.*

(3) *For* $m-r+1-2j>0$,

$$\mathrm{E}[V^{2j}] = \prod_{i=1}^{j}\frac{m(m-2i+1)}{(m-r-2i)(m-r-2i+1)}.$$

(4) *For any j,*

$$\mathrm{E}[V^{-2j}] = \prod_{i=1}^{j}\frac{(m-r+2(i-1))(m-r+2i-1))}{m(m+2i-1)}.$$

Proof. From Lemma 16.2.1 we can write

$$\begin{aligned} U^2-1 &= \chi_r^2/\chi_{m-r+1}^2, \\ U^{-2}-1 &= -\chi_r^2/(\chi_r^2+\chi_{m-r+1}^2) \\ &= (-1)\beta(r/2,(m-r+1)/2), \end{aligned}$$

which imply results (1) and (2). The results (3) and (4) follow from

$$V^2 = m\{\beta((m-r+1)/2, r/2)\}^{-1}. \qquad \square$$

Two types of asymptotic approximations can be used (see Chapter 13) for a scale mixture of N(0, 1). Let $\Phi(x)$ and $\varphi(x)$ be the distribution function and the density function of N(0, 1). The approximations in the case of the scale factor U are

$$\Phi_{\delta,k,U}(x) = \Phi(x) - \sum_{j=1}^{k-1}\frac{\delta^j}{2^j j!}Q_{\delta,2j-1}(x)\mathrm{E}[(U^{2\delta}-1)^j]\varphi(x). \qquad (16.2.10)$$

where $\delta=1$ or $\delta=-1$ denote two types of the approximations, and

$$\begin{aligned} Q_{1,2j-1}(x) &= H_{2j-1}(x) \quad \text{(Hermite polynomial)}, \\ Q_{-1,2j-1}(x) &= \sum_{m=0}^{j-1}(2m-1)!!\binom{j-1}{m}x^{2(j-m)-1}. \end{aligned} \qquad (16.2.11)$$

Some $Q_{\delta,2j-1}(x)$ are given as follows:

$$Q_{1,1}(x) = x, \; Q_{1,3}(x) = x^3 - 3x, \; Q_{1,5}(x) = x^5 - 10x^3 + 15x,$$
$$Q_{-1,1}(x) = x, \; Q_{-1,3}(x) = x^3 + x, \; Q_{-1,5}(x) = x^5 + 2x^3 + 3x.$$

For two types of the approximations, we have the following error bounds. Noting that $U > 1$,

$$|\Pr\{X_a \le x\} - \Phi_{-1,k,U}(x)| \le \alpha_{\delta,k}\mathrm{E}[|U^{2\delta} - 1|^k], \tag{16.2.12}$$

where the values of $\alpha_{\delta,k}$ for $k = 1$ to 4 are given in Table 16.2.1.

Table 16.2.1 Numerical values of $\alpha_{\delta,k}$

k	1	2	3	4
$\alpha_{1,k}$	0.1210	0.0688	0.0481	0.0370
$\alpha_{-1,k}$	0.1210	0.0791	0.0608	0.0501

Next we consider the distribution of $\hat{X}_a$ which is a scale mixture of $N(0,1)$ with scale factor V. In this case we can use the approximations obtained from $\Phi_{\delta,k,U}$ by using $V^{2\delta}$ instead of $U^{2\delta}$:

$$\Phi_{\delta,k,V}(x) = \Phi(x) - \sum_{j=1}^{k-1} \frac{\delta^j}{2^j j!} Q_{\delta,2j-1}(x)\mathrm{E}[(V^{2\delta} - 1)^j]\varphi(x). \tag{16.2.13}$$

However, the error bounds should be modified as follows. Compare this with (16.2.10). Unlike the U-case, however, the scale factor V does not satisfy the restrictions $V \ge 1$. So we need to replace the $\alpha_{\delta,k}$ in $\beta_{1,k}$. For the case of $\delta = -1$,

$$|\Pr\{\hat{X}_a \le x\} - \Phi_{-1,k,V}(x)| \le \frac{1}{2}\mathrm{E}[|V^{-2} - 1|^k] \;\text{ for }\; k = 1, 2, \ldots. \tag{16.2.14}$$

For the case of $\delta = 1$,

$$|\Pr\{\hat{X}_a \le x\} - \Phi_{1,k,V}(x)| \le \frac{1}{2}\mathrm{E}[|V^2 - 1|^k] \;\text{ for }\; k \le 6. \tag{16.2.15}$$

As a practical use, we consider the approximation $\Phi_{-1,k,V}(x)$ with $k = 3$, which is expressed as

$$\Phi_{-1,3,V}(x) = \Phi(x) + \varphi(x)\Big(\frac{1}{2}xd_1 - \frac{1}{8}(x^3 + 3x)d_2\Big), \tag{16.2.16}$$

where

$$\begin{aligned} d_1 &= \frac{1}{m(m+1)}\{-2(p-1)m+(p-1)(p-2)\}, \\ d_2 &= \frac{1}{m^2(m+1)(m+2)}\{3m^3+(4p^2-24p+29)m^2 \\ &\quad -2(2p^3-13p^2+29p-21)m+(p-1)(p-2)(p-3)(p-4)\}. \end{aligned}$$

It is easy to see that

$$\begin{aligned} &\mathrm{E}[(V^{-2}-1)] = \mathrm{O}(n^{-1}), &&\mathrm{E}[(V^{-2}-1)^2] = \mathrm{O}(n^{-1}), \\ &\mathrm{E}[(V^{-2}-1)^3] = \mathrm{O}(n^{-2}), &&\mathrm{E}[(V^{-2}-1)^4] = \mathrm{O}(n^{-2}). \end{aligned}$$

Therefore,

$$\Pr\{\hat{X}_a \le x\} = \Phi_{-1,3,V}(x) + \mathrm{O}(n^{-2}). \tag{16.2.17}$$

Let z_α be the upper α-point of $\mathrm{N}(0,1)$. Then, using the Cornish and Fisher expansion formula, we can asymptotically approximate the upper α-point x_α of $\hat{X}_a$ by z_α, or

$$\begin{aligned} \hat{x}_\alpha &= z_\alpha - \frac{1}{2}z_\alpha d_1 + \frac{1}{8}(z_\alpha^3 + z_\alpha)d_2 \\ &= z_\alpha\Big[1 - \frac{1}{2}d_1 + \frac{1}{8}(z_\alpha^2+1)d_2\Big]. \end{aligned} \tag{16.2.18}$$

16.2.3 Confidence Interval for γ

Based on the distributional result of $\hat{X}_a$, we can propose a confidence interval with confidence coefficient $1-\alpha$ for a given constant vector $\boldsymbol{a}$,

$$\boldsymbol{a}'\gamma \in \big[\boldsymbol{a}'\hat{\gamma} \pm \hat{x}_{\alpha/2}(\mathbf{1}_p' S^{-1}\mathbf{1}_p)^{-1/2}(\boldsymbol{a}'Q\boldsymbol{a})^{1/2}\big]. \tag{16.2.19}$$

For a simultaneous confidence interval, we need the distribution of $T = V^2\chi^2_{k-1}$, which is a mixture of $\chi^2(k-1)$ with scale factor V^2. We can use two similar types of approximations for the distribution of T by changing the polynomials $Q_{\delta,2j-1}$ in $\Phi_{\delta,k,V}$. For example, we have

$$\Pr\{T \le x\} = G_f(x) + g_f(x)\big(\frac{1}{2}xd_1 - \frac{1}{8}x(x+2-f)d_2\big) + \mathrm{O}(n^{-2}), \tag{16.2.20}$$

where $f = k-1$, $G_f(x)$, and $g_f(x)$ are the distribution function and the density function of $\chi^2(f)$. Let u be the upper α-point of $\chi^2(k-1)$. Then, using the Cornish and Fisher expansion formula, we can asymptotically approximate the upper α point t_α of T by u or

$$\begin{aligned} \hat{t}_\alpha &= u - \frac{1}{2}ud_1 + \frac{1}{8}u(u+2-f)d_2 \\ &= u\Big[1 - \frac{1}{2}d_1 + \frac{1}{8}(u+2-f)d_2\Big]. \end{aligned} \tag{16.2.21}$$

Based on the distributional result of T, we can propose a simultaneous confidence interval with confidence coefficient $1-\alpha$ for any $\boldsymbol{a}$,

$$\boldsymbol{a}'\gamma \in \left[\boldsymbol{a}'\hat{\gamma} \ \pm \ \hat{t}_\alpha^{1/2}(\mathbf{1}_p'\mathbf{S}^{-1}\mathbf{1}_p)^{-1/2}(\boldsymbol{a}'Q\boldsymbol{a})^{1/2}\right]. \tag{16.2.22}$$

On the other hand, Srivastava (1987) proposed the following simultaneous confidence interval based on an LR test:

$$\boldsymbol{a}'\gamma \in \left[\boldsymbol{a}'\hat{\gamma} \ \pm \ T_\alpha \frac{\{\boldsymbol{a}'(Q+\mathbf{W}'D(D'\mathbf{S}_eD)^{-1}D'\mathbf{W})\boldsymbol{a}\}^{1/2}}{(\mathbf{1}_p'\mathbf{S}_e^{-1}\mathbf{1}_p)^{1/2}}\right], \tag{16.2.23}$$

where $T_\alpha^2 = (k-1)(m-p+1)^{-1}\mathrm{F}(k-1, m-p+1|\alpha)$.

Now we compare two confidence intervals for an example given in Srivastava (2002, Example 7.2.1). The data are the times required to solve four different mathematics problems for each of two groups of n_1 and n_2 subjects, respectively. The level difference γ of two groups was estimated as $\hat{\gamma} = 12.79$. He gave a confidence interval with confidence coefficient 0.95 based on (16.2.23) as $(2.47, 23.12)$, which, actually should have read as

$$\gamma \in (1.24,\ 24.33).$$

Our confidence intervals $\left[\text{see (16.2.19) with } a = 1\right]$, based on the limiting distribution and its refinement through asymptotic expansion, give

$$\gamma \in (3.85,\ 21.72) \quad \textit{and} \quad \gamma \in (1.79,\ 23.78),$$

respectively. These are slightly better than the one obtained directly from (16.2.23).

Finally we note that not only can we regard the distribution of $T = V^2\chi_{k-1}^2$ as a scale mixture of the chi-square distribution $\chi^2(k-1)$, but it is a scale mixture of a scaled F distribution with the scale factor $S_m = 1 + \chi_{p-1}^2/\chi_{m-p+2}^2 (> 1)$. In fact, T can be put in the form

$$T = V^2\chi_{k-1}^2 = \left(1 + \frac{\chi_{p-1}^2}{\chi_{m-p+2}^2}\right) \cdot \frac{\chi_{k-1}^2}{\chi_{m-p+1}^2/m}.$$

In particular, we will be able to use the method of approximation to the distribution T of the scale mixtures of the chi-square and Fdistributions [see Sections 13.3.4 and 14.3].

16.3 Estimators in the Growth Curve Model

16.3.1 Error Bounds

We now reconsider the growth curve model introduced by Rao (1959) and Potthoff and Roy (1964), which was treated in Chapter 12. Let $\mathbf{Y}$ be an

$N \times p$ observation matrix based on p repeated measurements on N subjects. The model is given by

$$\mathrm{E}(\mathbf{Y}) = \mathrm{A}\Theta\mathrm{B}, \tag{16.3.1}$$

where A is a known $N \times k$ matrix of rank k, B is a known $q \times p$ matrix of rank $q(\leq p)$ and Θ is a $k \times q$ matrix of unknown parameters. Further, it is assumed that the rows of $\mathbf{Y}$ have independent p-variate normal distribution with unknown covariance matrix Σ. If, in particular, $\mathrm{B} = \mathrm{I}_p$, the model is a MANOVA model. Here we use somewhat different notations from those in Chapter 12.

Based on the results in Chapter 12, the MLE of Θ is given by

$$\hat{\Theta} = (\mathrm{A}'\mathrm{A})^{-1}\mathrm{A}'\mathbf{Y}\mathbf{S}^{-1}\mathrm{B}'(\mathrm{B}\mathbf{S}^{-1}\mathrm{B})^{-1}, \tag{16.3.2}$$

where $\mathbf{S} = (1/n)\mathbf{Y}'(\mathrm{A} - \mathrm{A}(\mathrm{A}'\mathrm{A})^{-1}\mathrm{A}')\mathbf{Y}$ and $n = N - k$. It is known that $\hat{\Theta}$ is an unbiased estimator,

$$\mathrm{E}(\hat{\Theta}) = \Theta, \tag{16.3.3}$$

and that its variance is given by

$$\mathrm{Var}[\mathrm{vec}(\hat{\Theta})] = \frac{n-1}{n-r-1}(\mathrm{B}\Sigma^{-1}\mathrm{B}')^{-1} \otimes (\mathrm{A}'\mathrm{A})^{-1}, \tag{16.3.4}$$

where $r = p - q$, and for a matrix C, vec(C) denotes the vector formed by stacking the columns of C under each other. The distribution of $\hat{\Theta}$ was studied by several authors. Gleser and Olkin (1970) and Kenward (1986) expressed the density function of $\hat{\Theta}$ with the help of hypergeometric functions. However, the expressions were extremely complicated, and all these results are difficult to apply. On the other hand, Fujikoshi (1987) derived an asymptotic expansion with an error bound for the density function of bilinear form of $\hat{\Theta}$. The result was extended to the density and the distribution function of $\hat{\Theta}$ itself by Fujikoshi (1993a), Fujikoshi and Shimizu (1989), and Shimizu (1995).

16.3.2 Distribution of the Bilinear Form

We consider the distribution of the bi-linear form $\boldsymbol{a}'\hat{\Theta}\boldsymbol{b}$. From (16.3.3) and (16.3.4) it follows that its mean and variance are given by

$$\mathrm{E}(\boldsymbol{a}'\hat{\Theta}\boldsymbol{b}) = \boldsymbol{a}'\Theta\boldsymbol{b}$$

and

$$\mathrm{Var}(\boldsymbol{a}'\hat{\Theta}\boldsymbol{b}) = \frac{n-1}{n-r-1}\tau^2,$$

respectively, where $r = p - q$ and $\tau^2 = \boldsymbol{a}'(A'A)^{-1}\boldsymbol{a} \cdot \boldsymbol{b}'(B\Sigma^{-1}B')^{-1}\boldsymbol{b}$. Noting that the ratio $(n-1)/(n-r-1)$ tends to 1 as n tends to infinity. Let us begin with the standardized statistic defined by

$$X_{a,b} = \boldsymbol{a}'(\hat{\Theta} - \Theta)\boldsymbol{b}/\tau. \tag{16.3.5}$$

Using

$$\hat{\Theta} - \Theta = (A'A)^{-1}A'(Y - A\Theta B)S^{-1}B'(BS^{-1}B')^{-1},$$

we can write $X_{a,b}$ as

$$X_{a,b} = UZ, \tag{16.3.6}$$

where

$$\begin{aligned} Z &= \frac{\boldsymbol{a}'(\hat{\Theta} - \Theta)\boldsymbol{b}}{\{\boldsymbol{a}'(A'A)^{-1}\boldsymbol{a}\}^{1/2}} \\ &\quad \times \frac{1}{\{\boldsymbol{b}'(BS^{-1}B')^{-1}BS^{-1}\Sigma S^{-1}B'(BS^{-1}B')^{-1}\boldsymbol{b}\}^{1/2}}, \end{aligned} \tag{16.3.7}$$

$$U = \left[\frac{\boldsymbol{b}'(BS^{-1}B')^{-1}BS^{-1}\Sigma S^{-1}B'(BS^{-1}B')^{-1}\boldsymbol{b}}{\boldsymbol{b}'(B\Sigma^{-1}B')^{-1}\boldsymbol{b}}\right]^{1/2}. \tag{16.3.8}$$

Then, considering the conditional distribution of Z given $\mathbf{S}$, we can see that Z is distributed as the standard normal distribution and that Z is independent of U. This means that $X_{a,b}$ is a scale mixture of the normal distribution.

Another statistic of interest is the Studentized version of $X_{a,b}$, defined by

$$\hat{X}_{a,b} = \boldsymbol{a}'(\hat{\Theta} - \Theta)\boldsymbol{b}/\hat{\tau}, \tag{16.3.9}$$

where $\hat{\tau}^2 = \boldsymbol{a}'(A'A)^{-1}\boldsymbol{a}\cdot\boldsymbol{b}'(BS^{-1}B')^{-1}\boldsymbol{b}$. Compare this with (16.3.5) and note that the population covariance matrix Σ is replaced by its sample counterpart S. Like $X_{a,b}$, the statistic $\hat{X}_{a,b}$ is expressible as a scale mixture of the standard normal distribution:

$$\hat{X}_{a,b} = VZ, \tag{16.3.10}$$

where Z is the same as in (16.3.7), and

$$V = \left[\frac{\boldsymbol{b}'(BS^{-1}B')^{-1}BS^{-1}\Sigma S^{-1}B'(BS^{-1}B')^{-1}\boldsymbol{b}}{\boldsymbol{b}'(BS^{-1}B')^{-1}\boldsymbol{b}}\right]^{1/2}. \tag{16.3.11}$$

It may be noted that the scale factor U in (16.3.8) and V in (16.3.11) have essentially the same distributional structures as those in (16.2.5) and (16.2.8). More precisely, their distributional structures are given in the following lemma.

Lemma 16.3.1 *Let U and V be the statistic in* (16.3.8) *and* (16.3.11). *Then we can express U and V as*

$$U^2 = 1 + \frac{\chi_r^2}{\chi_{n-r+1}^2}, \tag{16.3.12}$$

$$V^2 = \frac{n}{\chi_{n-r}^2}\left(1 + \frac{\chi_r^2}{\chi_{n-r+1}^2}\right), \tag{16.3.13}$$

where χ_r^2 *and* χ_{n-r+1}^2 *are independent* $\chi^2(\cdot)$*-variables with* r *and* $n-r+1$ *degrees of freedom, respectively.*

Proof. The results are obtained in the same way as those of Lemma 16.2.1 with the help of Lemma 16.1.1 □

The distributions of $X_{a,b}$ and $\hat{X}_{a,b}$ are obtained from those of X_a and $\hat{X}_a$ by changing

$$p-1 \to r, \quad m \to n.$$

Table 16.3.1 Numerical values of $\mathrm{E}[(U^2-1)^j]$

n	20	30	50	100
$r=1$				
$j=2$	$5.208 \cdot 10^{-2}$	$2.060 \cdot 10^{-2}$	$6.793 \cdot 10^{-3}$	$1.594 \cdot 10^{-3}$
3	$2.604 \cdot 10^{-2}$	$6.010 \cdot 10^{-3}$	$1.081 \cdot 10^{-3}$	$1.187 \cdot 10^{-4}$
4	$1.953 \cdot 10^{-2}$	$2.458 \cdot 10^{-3}$	$2.316 \cdot 10^{-4}$	$1.161 \cdot 10^{-5}$
$r=2$				
$j=2$	$1.882 \cdot 10^{-1}$	$7.111 \cdot 10^{-2}$	$2.270 \cdot 10^{-2}$	$5.209 \cdot 10^{-3}$
3	$1.158 \cdot 10^{-1}$	$2.473 \cdot 10^{-2}$	$4.222 \cdot 10^{-3}$	$4.481 \cdot 10^{-4}$
4	$1.053 \cdot 10^{-1}$	$1.178 \cdot 10^{-2}$	$1.030 \cdot 10^{-3}$	$4.924 \cdot 10^{-5}$

16.4 Generalized Least Squares Estimators

We consider the distribution of the linear combination of an estimator of $\boldsymbol{\beta}$ in a generalized linear regression model

$$\boldsymbol{y} = A\boldsymbol{\beta} + \boldsymbol{\varepsilon}, \tag{16.4.1}$$

where A is an $n \times g$ fixed matrix of rank g and $\boldsymbol{\varepsilon}$ is a random vector following the normal distribution $\mathrm{N}(\mathbf{0}, \Omega)$. Here $\Omega = \Omega(\boldsymbol{\xi})$ depends on an unknown parameter vector $\boldsymbol{\xi}$, but its functional form is known. A generalized least squares estimator (GLSE) is given by

$$\hat{\boldsymbol{\beta}} = (A'\hat{\Omega}^{-1}A)^{-1}A'\hat{\Omega}^{-1}\boldsymbol{y}, \tag{16.4.2}$$

where $\hat{\Omega} = \Omega(\hat{\boldsymbol{\xi}})$ and $\hat{\boldsymbol{\xi}}$ is an estimator.

We assume that

$$\mathrm{A}_0\text{: } \hat{\Omega} = \Omega(\hat{\boldsymbol{\xi}}) \textit{ and } \hat{\boldsymbol{\beta}} \text{ are independent.} \tag{16.4.3}$$

Then

$$\mathrm{A}_1\text{: } \hat{\boldsymbol{\beta}}| \text{ given } \hat{\Omega} \sim N(\boldsymbol{\beta}, H), \tag{16.4.4}$$

where $H = (A'\hat{\Omega}^{-1}A)^{-1}A'\hat{\Omega}^{-1}\Omega\hat{\Omega}^{-1}A(A'\hat{\Omega}^{-1}A)^{-1}$. The assumption A_0 or A_1 is satisfied automatically whenever the parameter $\boldsymbol{\xi}$ is estimated from a sample independent of $\boldsymbol{y}$. Furthermore, the GLSEs with the growth curve model (16.3.1), the usual estimators in the SUR (seemingly unrelated) model and a heteroscedastic model satisfy the assumption.

We consider the distribution of a linear combination $\boldsymbol{a}'\hat{\boldsymbol{\beta}}$, where $\boldsymbol{a} = (a_1, \ldots, a_g)'$ is a fixed nonnull vector. More precisely, we consider the distributions of its standardized version

$$X = \frac{\boldsymbol{a}'(\hat{\boldsymbol{\beta}} - \boldsymbol{\beta})}{\lambda} = \frac{\tilde{\lambda}}{\lambda} \cdot \frac{\boldsymbol{a}'(\hat{\boldsymbol{\beta}} - \boldsymbol{\beta})}{\tilde{\lambda}} = s \cdot Z, \tag{16.4.5}$$

and its Studentized version

$$\tilde{X} = \frac{\boldsymbol{a}'(\hat{\boldsymbol{\beta}} - \boldsymbol{\beta})}{\hat{\lambda}} = \frac{\tilde{\lambda}}{\hat{\lambda}} \cdot \frac{\boldsymbol{a}'(\hat{\boldsymbol{\beta}} - \boldsymbol{\beta})}{\tilde{\lambda}} = \tilde{s} \cdot Z, \tag{16.4.6}$$

where

$$\lambda = \{\boldsymbol{a}'(A'\Omega^{-1}A)^{-1}\boldsymbol{a}\}^{1/2}, \quad \hat{\lambda} = \{\boldsymbol{a}'(A'\hat{\Omega}^{-1}A)^{-1}\boldsymbol{a}\}^{1/2},$$
$$\tilde{\lambda} = \{\boldsymbol{a}'(A'\hat{\Omega}^{-1}A)^{-1}A'\hat{\Omega}^{-1}\Omega\hat{\Omega}^{-1}A(A'\hat{\Omega}^{-1}A)^{-1}\boldsymbol{a}\}^{1/2}.$$

From the assumption A_0 or A_1 the conditional distribution of $\boldsymbol{a}'(\hat{\boldsymbol{\beta}} - \boldsymbol{\beta})$ given $\hat{\Omega}$ is $N(0, \tilde{\lambda})$. Therefore, Z is distributed as $\mathrm{N}(0, 1)$, and Z is independent of $\hat{\Omega}$. This implies that X and $\tilde{X}$ are scale mixtures of the standardized normal distribution with scale factors s and $\tilde{s}$, respectively. So it is possible to apply a general theory to our problem as in Sections 4.3 and 9.1. However, in this case it is difficult to obtain the first few exact moments of $s^{\pm 2}$ or $\tilde{s}^{\pm 2}$. On the other hand, it is possible to give their approximations with appropriate error bounds. In fact, for certain estimators in a SUR model and a heteroscedastic model, Kariya and Maekawa (1982), Mukaihata and Fujikoshi (1993), Kurata and Kariya (1996), and others have obtained such results.

In the following we consider a SUR model which is expressed as a model (16.4.1) with

$$\begin{aligned} A &= \mathrm{diag}\,\{A_1, \ldots, A_k\}, \\ \boldsymbol{y} &= (\boldsymbol{y}_1, \ldots, \boldsymbol{y}_k)', \quad \boldsymbol{\varepsilon} = (\boldsymbol{\varepsilon}_1, \ldots, \boldsymbol{\varepsilon}_k)', \\ \boldsymbol{\beta} &= (\boldsymbol{\beta}_1, \ldots, \boldsymbol{\beta}_k)', \quad \Omega = \Sigma \otimes I_m. \end{aligned} \tag{16.4.7}$$

Here A_i is $m \times g_i, \boldsymbol{\varepsilon}_i$ is $m \times 1$, $n = km$, $g = \sum_{j=1}^k g_j$, and $\mathrm{diag}\,\{A_1, \ldots, A_k\}$ denotes the block diagonal matrix. The Zellner estimator in this model is the GLSE $\hat{\boldsymbol{\beta}} = \hat{\boldsymbol{\beta}}(\hat{\Omega})$ with $\hat{\Omega} = S \otimes I_m$, where

$$S = Y'[I_m - A_0(A_0'A_0)^+A_0']Y.$$

Here $Y = (\boldsymbol{y}_1, \ldots, \boldsymbol{y}_k)$ is $m \times k$, $A_0 = (A_1, \ldots, A_k)$ is $m \times k$ and B^+ is the Moore–Penrose inverse of a matrix B. The model is called the k-equation SUR model. For the case $k = 2$, some error bounds were obtained [see, e.g., Kariya and Maekawa, 1982; Mukaihata and Fujikoshi, 1993] for

$$\mathrm{E}[(s^{\pm 2} - 1)^j] \quad \text{and} \quad \mathrm{E}[(\tilde{s}^{\pm 2} - 1)^j],$$

with $j = 1$ and $j = 2$.

Problems

16.1 When the covariance matrix Σ is known, the linear discriminant fuction W may be modified as

$$W_0 = (\bar{\boldsymbol{X}}_1 - \bar{\boldsymbol{X}}_2)'\Sigma^{-1}\Big(\boldsymbol{X} - \frac{1}{2}(\bar{\boldsymbol{X}}_1 + \bar{\boldsymbol{X}}_2)\Big).$$

When the observation $\boldsymbol{X}$ is from Π_1, the statistic W_0 is expressed as a location and scale mixture of the standard normal distribution such that

$$W_0 = V_0^{1/2} Z - U_0,$$

where

$$\begin{aligned} V_0 &= (\bar{\boldsymbol{X}}_1 - \bar{\boldsymbol{X}}_2)'\Sigma^{-1}(\bar{\boldsymbol{X}}_1 - \bar{\boldsymbol{X}}_2) \\ Z_0 &= V_0^{-1/2}(\bar{\boldsymbol{X}}_1 - \bar{\boldsymbol{X}}_2)'\Sigma^{-1}(\boldsymbol{X} - \boldsymbol{\mu}_1), \\ U_0 &= (\bar{\boldsymbol{X}}_1 - \bar{\boldsymbol{X}}_2)'\Sigma^{-1}(\bar{\boldsymbol{X}}_1 - \boldsymbol{\mu}_1) - \frac{1}{2}D_0^2 \end{aligned}$$

and $D_0^2 = (\bar{\boldsymbol{X}}_1 - \bar{\boldsymbol{X}}_2)'\Sigma^{-1}(\bar{\boldsymbol{X}}_1 - \bar{\boldsymbol{X}}_2)$.

16.2 (Continuation) Show that the means and the variances of U_0 and V_0 are given as follows:

$$\begin{aligned} \mathrm{E}(U_0) &= -\frac{1}{2}\Delta^2 + \frac{1}{2}p\Big(\frac{1}{N_1} - \frac{1}{N_2}\Big), \\ \mathrm{E}(V_0) &= \Delta^2 + \frac{Np}{N_1 N_2}, \\ \mathrm{Var}(U_0) &= \frac{1}{N_2}\Delta^2 + \frac{1}{2}p\Big(\frac{1}{N_1^2} + \frac{1}{N_2^2}\Big), \\ \mathrm{Var}(V_0) &= 4\Delta^2\frac{N}{N_1 N_2} + 2p\Big(\frac{N}{N_1 N_2}\Big)^2. \end{aligned}$$

Using the above results, we can approximate $\mathrm{P}(2|1)$ by $\Phi(\gamma_0)$ with $\gamma_0 = \mathrm{E}(V_0)^{-1/2}\mathrm{E}(U_0)$. Obtain its error bound.

16.3 Consider an extended linear discriminant function

$$W_e \equiv (\bar{\boldsymbol{X}}_1 - \bar{\boldsymbol{X}}_2)'\mathbf{S}_e^{-1}\left(\boldsymbol{X} - \frac{1}{2}(\bar{\boldsymbol{X}}_1 + \bar{\boldsymbol{X}}_2)\right)$$

which is obtained by substituting an extended estimator $\mathbf{S}_e$ to $\mathbf{S}$. Suppose that $\mathbf{S}_e$ is independent of $\bar{\boldsymbol{X}}_1$ and $\bar{\boldsymbol{X}}_2$. Then, check that a general theory of asymptotic approximations and error bounds for W can be also used for W_e and only the differences are with respect to comuputations of thier moments.

16.4 Complete the proof in Theorem 16.2.1, or obtain the MLEs by minimizing

$$g(\boldsymbol{\delta}, \Psi) = n \log|\Psi| + \operatorname{tr}\Psi^{-1}\{(\boldsymbol{u}_1 - \boldsymbol{\delta},\ U_2)'(\boldsymbol{u}_1 - \boldsymbol{\delta},\ U_2) + V\},$$

along the following steps.
(1) Prove that

$$\begin{aligned}
&|\Psi| = \psi_{11\cdot 2}|\Psi_{22}|, \quad \psi_{11\cdot 2} = \psi_{11} - \boldsymbol{\beta}'\Psi_{22}^{-1}\boldsymbol{\beta}, \quad \boldsymbol{\beta} = \Psi_{22}^{-1}\boldsymbol{\psi}_{21},\\
&\operatorname{tr}\Psi^{-1}(\boldsymbol{u}_1 - \boldsymbol{\delta},\ U_2)'(\boldsymbol{u}_1 - \boldsymbol{\delta},\ U_2) = \operatorname{tr}\Psi_{22}^{-1}U_2'U_2\\
&\qquad + \frac{1}{\psi_{11\cdot 2}}(\boldsymbol{u}_1 - \boldsymbol{\delta} - U_2\boldsymbol{\beta})'(\boldsymbol{u}_1 - \boldsymbol{\delta} - U_2\boldsymbol{\beta}),\\
&\operatorname{tr}\Psi^{-1}V = \operatorname{tr}\Psi_{22}^{-1}V_{22}\\
&\qquad + \frac{1}{\psi_{11\cdot 2}}(\boldsymbol{v}_{21} - V_{22}\boldsymbol{\beta})'(\boldsymbol{v}_{21} - V_{22}\boldsymbol{\beta}).
\end{aligned}$$

(2) Prove that MLE's of $\boldsymbol{\delta}$ and Ψ are

$$\begin{aligned}
&\hat{\boldsymbol{\delta}} = \boldsymbol{u}_1 - U_2\hat{\boldsymbol{\beta}},\ \hat{\boldsymbol{\beta}} = (V_{22})^{-1}\boldsymbol{v}_{21},\\
&n\hat{\Psi}_{22} = U_2'U_2 + V_{22},\\
&n\hat{\psi}_{11\cdot 2} = v_{11} - \boldsymbol{v}_{21}'(V_{22})^{-1}\boldsymbol{v}_{21}.
\end{aligned}$$

(3) Prove that the MLE's $\hat{\psi}_{11\cdot 2}$ and $\hat{\Psi}_{22}$ are expressed in terms of the original observation matrix as

$$\begin{aligned}
n\hat{\psi}_{11\cdot 2} &= p^{-1}\left\{\mathbf{1}_p'S_e\mathbf{1}_p - \mathbf{1}_p'S_eD\{D'S_eD\}^{-1}D'S_e\mathbf{1}_p\right\}\\
&= p(\mathbf{1}_p'S_e^{-1}\mathbf{1}_p)^{-1},\\
n\hat{\Psi}_{22} &= D'(S_e + S_h)D,
\end{aligned}$$

by using

$$B_2(B_2'S_eB_2)^{-1}B_2' = S_e^{-1} - S_e^{-1}\boldsymbol{b}_1(\boldsymbol{b}_1'S_e^{-1}\boldsymbol{b}_1)^{-1}\boldsymbol{b}_1'S_e^{-1}.$$

16.5 Prove Lemma 16.2.2 along the following steps.
(1) Let $\boldsymbol{c} = \Sigma^{-1/2}\mathbf{1}_p$ and $\mathbf{W} = m\Sigma^{-1/2}\mathbf{S}\Sigma^{-1/2}$. Then we can write

$$U^2 = \boldsymbol{c}'\boldsymbol{c}\frac{\boldsymbol{c}'W^{-2}\boldsymbol{c}}{(\boldsymbol{c}'W^{-1}\boldsymbol{c})^2}, \quad V^2 = \frac{m\boldsymbol{c}'W^{-2}\boldsymbol{c}}{\boldsymbol{c}'W^{-1}\boldsymbol{c}},$$

where $\mathbf{W} \sim \mathrm{W}_p(m, I_p)$.
(2) Consider an orthogonal matrix H whose first column is $(\boldsymbol{c}'\boldsymbol{c})^{-1/2}\boldsymbol{c}$. Since $\mathrm{H}'\mathbf{W}\mathrm{H}$ has the same distribution with $\mathbf{W}$, we write $\mathrm{H}'\mathbf{W}\mathrm{H}$ as $\mathbf{W}$ simply to save the notation. Then, we have

$$\boldsymbol{c}'\mathbf{W}^{-1}\boldsymbol{c} = \boldsymbol{c}'\boldsymbol{c}(\mathbf{W}^{-1})_{11}, \quad \boldsymbol{c}'\mathbf{W}^{-2}\boldsymbol{c} = \boldsymbol{c}'\boldsymbol{c}(\mathbf{W}^{-2})_{11},$$

where $(\mathrm{A})_{ij}$ denotes the (i, j) element of a matrix A.
(3) We partition $\mathbf{W}$ as

$$\mathbf{W} = \begin{pmatrix} W_{11} & \boldsymbol{W}'_{21} \\ \boldsymbol{W}_{21} & \mathbf{W}_{22} \end{pmatrix}.$$

Then, show that

$$\boldsymbol{c}'\mathbf{W}^{-1}\boldsymbol{c} = \boldsymbol{c}'\boldsymbol{c}\boldsymbol{W}_{11\cdot 2}^{-1}, \quad \boldsymbol{c}'\mathbf{W}^{-2}\boldsymbol{c} = \boldsymbol{c}'\boldsymbol{c}\boldsymbol{W}_{11\cdot 2}^{-2}(1 + \boldsymbol{W}'_{21}\mathbf{W}_{22}^{-2}\boldsymbol{W}_{21}),$$

and hence

$$U^2 = 1 + \boldsymbol{W}'_{21}\mathbf{W}_{22}^{-2}\boldsymbol{W}_{21}, \quad V^2 = W_{11\cdot 2}^{-1}(1 + \boldsymbol{W}'_{21}\mathbf{W}_{22}^{-2}\boldsymbol{W}_{21}),$$

where $W_{11\cdot 2} = W_{11} - \boldsymbol{W}'_{21}\mathbf{W}_{22}^{-1}\boldsymbol{W}_{21}$.
(4) Use that $W_{11\cdot 2}^{-1} \sim \chi^2(m - (p-1))$, and

$$\begin{aligned} \boldsymbol{W}'_{21}\mathbf{W}_{22}^{-2}\boldsymbol{W}_{21} &= (\mathbf{W}_{22}^{-1/2}\boldsymbol{W}_{21})'\mathbf{W}_{22}^{-1}(\mathbf{W}_{22}^{-1/2}\boldsymbol{W}_{21}) \\ &\sim \chi^2_{p-1}/\chi^2_{n-p+2}. \end{aligned}$$

Examine tht these prove the required results.

Appendix

A.1 Some Results on Matrices

A.1.1 Determinants and Inverse Matrices

Let A be a $p \times p$ matrix that is partitioned as

$$\mathrm{A} = \begin{pmatrix} \mathrm{A}_{11} & \mathrm{A}_{12} \\ \mathrm{A}_{21} & \mathrm{A}_{22} \end{pmatrix},$$

where $\mathrm{A}_{ij} : p_i \times p_j$, and $p = p_1 + p_2$.

[A.1.1] Assuming that A_{22} is nonsingular, set

$$\mathrm{M} = \begin{pmatrix} \mathrm{I}_{p_1} & -\mathrm{A}_{12}\mathrm{A}_{22}^{-1} \\ \mathrm{O} & \mathrm{I}_{p_2} \end{pmatrix}.$$

Then we have

(1) $\mathrm{MAM}' = \begin{pmatrix} \mathrm{A}_{11} - \mathrm{A}_{12}\mathrm{A}_{22}^{-1}\mathrm{A}_{21} & \mathrm{O} \\ \mathrm{O} & \mathrm{A}_{22} \end{pmatrix}$, and

(2) $|\mathrm{A}| = |\mathrm{A}_{22}| \cdot |\mathrm{A}_{11} - \mathrm{A}_{12}\mathrm{A}_{22}^{-1}\mathrm{A}_{21}|$.

Similarly, $|\mathrm{A}| = |\mathrm{A}_{11}| \cdot |\mathrm{A}_{22} - \mathrm{A}_{21}\mathrm{A}_{11}^{-1}\mathrm{A}_{12}|$ holds provided that A_{11} is nonsingular.

[A.1.2] Assume that A, A_{11}, and A_{22} are all nonsingular, then $\mathrm{A}_{11\cdot 2} \equiv \mathrm{A}_{11} - \mathrm{A}_{12}\mathrm{A}_{22}^{-1}\mathrm{A}_{21}$ and $\mathrm{A}_{22\cdot 1} \equiv \mathrm{A}_{22} - \mathrm{A}_{21}\mathrm{A}_{11}^{-1}\mathrm{A}_{12}$ are also nonsingular and the inverse matrix of A is given by

$$\begin{aligned} \mathrm{A}^{-1} &= \begin{pmatrix} \mathrm{A}_{11}^{-1} & \mathrm{O} \\ \mathrm{O} & \mathrm{O} \end{pmatrix} + \begin{pmatrix} -\mathrm{A}_{11}^{-1}\mathrm{A}_{12} \\ \mathrm{I}_{p_2} \end{pmatrix} \mathrm{A}_{22\cdot 1}^{-1} \begin{pmatrix} -\mathrm{A}_{21}\mathrm{A}_{11}^{-1} & \mathrm{I}_{p_2} \end{pmatrix} \\ &= \begin{pmatrix} \mathrm{O} & \mathrm{O} \\ \mathrm{O} & \mathrm{A}_{22}^{-1} \end{pmatrix} + \begin{pmatrix} \mathrm{I}_{p_1} \\ -\mathrm{A}_{22}^{-1}\mathrm{A}_{21} \end{pmatrix} \mathrm{A}_{11\cdot 2}^{-1} \begin{pmatrix} \mathrm{I}_{p_1} & -\mathrm{A}_{12}\mathrm{A}_{22}^{-1} \end{pmatrix}. \end{aligned}$$

[A.1.3] Let A and B be $p \times p$ and $q \times q$ nonsingular matrices, and $\mathrm{U}: p \times q$, $\mathrm{V}: p \times q$. Then

$$(\mathrm{A} + \mathrm{UBV})^{-1} = \mathrm{A}^{-1} - \mathrm{A}^{-1}\mathrm{U}(\mathrm{B}^{-1} + \mathrm{VA}^{-1}\mathrm{U})^{-1}\mathrm{VA}^{-1}.$$

If, in particular, $\boldsymbol{u}$ and $\boldsymbol{v}$ are p dimensional vectors, then

$$(\mathrm{A} + \boldsymbol{u}\boldsymbol{v}')^{-1} = \mathrm{A}^{-1} - \mathrm{A}^{-1}\frac{\boldsymbol{u}\boldsymbol{v}'}{1 + \boldsymbol{v}'\mathrm{A}^{-1}\boldsymbol{u}}\mathrm{A}^{-1}.$$

[A.1.4] Let $\mathrm{G} = (\mathrm{G}_1\ \mathrm{G}_2)$ be a $p \times p$ nosigular matrix such that $\mathrm{G}_1'\mathrm{G}_2 = \mathrm{O}$. For a $p \times p$ positive definite matrix Q,

$$\mathrm{G}_2(\mathrm{G}_2'\mathrm{QG}_2)^{-1}\mathrm{G}_2' = \mathrm{Q}^{-1} - \mathrm{Q}^{-1}\mathrm{G}_1(\mathrm{G}_1'\mathrm{Q}^{-1}\mathrm{G}_1)^{-1}\mathrm{G}_1'\mathrm{Q}^{-1}.$$

Proof. Note that the column vectors of $\mathrm{Q}^{-1/2}\mathrm{G}_1$ are orthogonal to the column vector $\mathrm{Q}^{1/2}\mathrm{G}_2$. The column vector of each matrix are orthonormalized by

$$\mathrm{H}_1 = \mathrm{Q}^{-1/2}\mathrm{G}_1(\mathrm{G}_1'\mathrm{Q}^{-1}\mathrm{G}_1)^{-1/2}, \quad \mathrm{H}_2 = \mathrm{Q}^{1/2}\mathrm{G}_2(\mathrm{G}_2'\mathrm{QG}_2)^{-1/2}.$$

The result is obtained that $\mathrm{H} = (\mathrm{H}_1\ \mathrm{H}_2)$ is an orthogonal matrix and $\mathrm{I}_p = \mathrm{H}_1\mathrm{H}_1' + \mathrm{H}_2\mathrm{H}_2'$. □

A.1.2 Characteristic Roots and Vectors

Definition A.1.1 Let A be a $p \times p$ matrix. The characteristic root ℓ and characteristic vector $\boldsymbol{h}$ corresponding to ℓ are defined as solutions of

$$\mathrm{A}\boldsymbol{h} = \ell\boldsymbol{h}, \quad \boldsymbol{h} \neq \mathbf{0}.$$

A characteristic root of the matrix A is a solution to the characteristic equation defined by

$$|\mathrm{A} - \ell\mathrm{I}_p| = 0.$$

[A.1.5] Let A be a $p \times p$ symmetric matrix. Then

(1) The characteristic roots are all real.

(2) If ℓ_i and ℓ_j are two distinct characteristic roots of A, the corresponding characteristic vectors $\boldsymbol{h}_i$ and $\boldsymbol{h}_j$ are orthogonal.

(3) If ℓ is an m-multiple characteristic root of A, there exist m orthogonal characteristic vectors corresponding to ℓ.

[A.1.6] Let A and B be $n \times p$ matrices. Then the nonzero characteristic roots of AB' and $\mathrm{B}'\mathrm{A}$ are the same.

[A.1.7] Let $\ell_1, \ldots, \ell_p$ be the characteristic roots of a $p \times p$ symmetric matrix A. Then

(1) $\text{tr A} = \ell_1 + \cdots + \ell_p$.

(2) $|\text{A}| = \ell_1 \cdots \ell_p$.

Definition A.1.2 A $p \times p$ symmetric matrix A is called positive (negative) definite if

$$\boldsymbol{x}'\text{A}\boldsymbol{x} > 0 \quad (< 0)$$

for all vectors $\boldsymbol{x} \neq \mathbf{0}$; this is commonly expressed as $\text{A} > 0$. It is called positive (negative) semidefinite if $\boldsymbol{x}'\text{A}\boldsymbol{x} \geq 0$ (≤ 0) for all $\boldsymbol{x} \neq \mathbf{0}$, written as $\text{A} \geq 0$ (≤ 0).

Definition A.1.3 For $p \times p$ symmetric matrices A and B, if $\text{A} - \text{B} > 0$, it is expressed as $\text{A} > \text{B}$. Similarly, $\text{A} - \text{B} \geq 0$, it is expressed as $\text{A} \geq \text{B}$.

[A.1.8] Let $\ell_1 \geq \cdots \geq \ell_p$ be the characteristic roots of a $p \times p$ symmetric matrix A. Then

(1) $\text{A} > 0 \iff \ell_i > 0\ ,\ i = 1, \cdots, p.$

(2) $\text{A} \geq 0 \iff \ell_i \geq 0\ ,\ i = 1, \cdots, p\ .$

[A.1.9] Let A, B and C be $p \times p$ symmetric matrices. Then

(1) $\text{A} \geq \text{B}\ ,\ \text{B} \geq \text{C} \implies \text{A} \geq \text{C}.$

(2) if $\text{A} \geq \text{B}$,

(a) $\text{ch}_i(\text{A}) \geq \text{ch}_i(\text{B}), \quad i = 1, \ldots, p.$

(b) $\text{ch}_i(\text{A}) = \text{ch}_i(\text{B}), \quad i = 1, \ldots, p \iff \text{A} = \text{B},$
where $\text{ch}_i(\text{A})$ denotes the i-th characteristic root of a symmetric matrix A.

[A.1.10] Let A and B be $n \times p$ matrices such that $\text{A}'\text{A} = \text{B}'\text{B}$. Then there exist an orthogonal matrix H such that $\text{A} = \text{HB}$.

A.1.3 Matrix Factorizations

Let A be a $p \times p$ symmetric matrix. Then from [A.1.5] we may write the characteristic roots as $\ell_1 \geq \cdots \geq \ell_p$, and there exist the corresponding characteristic vectors $\boldsymbol{h}_1, \ldots, \boldsymbol{h}_p$ such that they are orthonormal, i.e.,

$$\boldsymbol{h}_i'\boldsymbol{h}_j = \delta_{ij} \quad i, j = 1, \cdots, p.$$

[A.1.11] (Spectral Decomposition) Let A be a $p \times p$ symmetric matrix with the characteristic roots $\ell_1 \leq \cdots \leq \ell_p$ and the corresponding orthonormal characteristic vectors $\boldsymbol{h}_i$, $i = 1, \ldots, p$. Then

(1) $\mathrm{AH} = \mathrm{HL}$,

(2) $\mathrm{A} = \mathrm{HLH}' = \ell_1 \boldsymbol{h}_1 \boldsymbol{h}_1' + \cdots + \ell_p \boldsymbol{h}_p \boldsymbol{h}_p'$,

where

$$\mathrm{L} = \mathrm{diag}\,(\ell_1, \cdots, \ell_p)\ ,\ \mathrm{H} = (\boldsymbol{h}_1, \cdots, \boldsymbol{h}_p).$$

[A.1.12] (Singular Values Decomposition) Let A be a $n \times p$ matrix of rank k. Then A can be expressed as

$$\mathrm{A} = \mathrm{UD}_\ell \mathrm{V}' = \ell_1 \boldsymbol{u}_1 \boldsymbol{v}_1' + \cdots + \ell_k \boldsymbol{u}_k \boldsymbol{v}_k',$$

where

$$\begin{aligned} \mathrm{U} &= (\boldsymbol{u}_1, \cdots, \boldsymbol{u}_k) \in \mathrm{O}(n \times p), \\ \mathrm{V} &= (\boldsymbol{v}_1, \cdots, \boldsymbol{v}_k) \in \mathrm{O}(p \times k), \\ \mathrm{D}_\ell &= \mathrm{diag}\,(\ell_1, \ldots, \ell_k)\ ,\ \ell_1 \geq \cdots \geq \ell_k > 0. \end{aligned}$$

Here $\mathrm{O}(n \times k)$ is the set of $n \times k$ matrix H satisfying $\mathrm{H}'\mathrm{H} = \mathrm{I}_k$. In this expression, $\ell_1^2, \ldots, \ell_k^2$ are non-zero characteristic roots of $\mathrm{A}'\mathrm{A}$ or AA', $\boldsymbol{u}_i$ is the characteristic vector of AA' with length 1 corresponding to ℓ_i^2, and $\boldsymbol{v}_i$ is the characteristic vector of $\mathrm{A}'\mathrm{A}$ with length 1 corresponding to ℓ_i^2.

[A.1.13] (Cholesky Decomposition) Let A be a $p \times p$ positive definite matrix. Then there exists a nonsingular lower triangular matrix T such that $\mathrm{A} = \mathrm{TT}'$. If all the diagonal elements are positive, then T is uniquely determined.

Proof. For $p = 3$, the equation $\mathrm{A} = \mathrm{TT}'$ is expressed as

$$\begin{pmatrix} a_{11} & a_{12} & a_{13} \\ a_{21} & a_{22} & a_{23} \\ a_{31} & a_{32} & a_{33} \end{pmatrix} = \begin{pmatrix} t_{11} & 0 & 0 \\ t_{21} & t_{22} & 0 \\ t_{31} & t_{32} & t_{33} \end{pmatrix} \begin{pmatrix} t_{11} & t_{21} & t_{31} \\ 0 & t_{22} & t_{32} \\ 0 & 0 & t_{33} \end{pmatrix},$$

and in element-wise expressions

$$\begin{aligned} &(1)\ a_{11} = t_{11}^2, \qquad (2)\ a_{21} = t_{21}t_{11}, \qquad (3)\ a_{21} = t_{31}t_{11}, \\ &(4)\ a_{22} = t_{21}^2 + t_{22}^2, \qquad (5)\ a_{32} = t_{31}t_{21} + t_{32}t_{22}, \\ &(6)\ a_{33} = t_{31}^2 + t_{32}^2 + t_{33}^2. \end{aligned}$$

Solving the simultaneous equations in the order of (1), (2), ..., (6), the solutions are sequentially determined as

$$t_{11} = \sqrt{a_{11}}, \qquad t_{21} = a_{21}/\sqrt{a_{11}}, \qquad t_{31} = a_{31}/\sqrt{a_{11}},$$
$$t_{22} = \sqrt{a_{22} - a_{21}^2/a_{11}}, \qquad t_{32} = a_{32} - a_{21}a_{11}^{-1}a_{31},$$
$$t_{33} = \sqrt{a_{33} - a_{31}^2/a_{11} - a_{32}^2/a_{11}}.$$

[A.1.14] Let A be a $p \times p$ positive definite matrix with rank $r(\leq p)$ and B be a $p \times p$ positive definite matrix. Then there exists a nonsingular matrix P such that

$$\mathrm{A} = \mathrm{PDP}', \quad \mathrm{B} = \mathrm{PP}'$$

where $\mathrm{D} = \mathrm{diag}\,(d_1, \ldots, d_p)$, $d_1 \geq \cdots \geq d_p \geq 0$, and d_i's are the roots of the equation $|\mathrm{A} - d\mathrm{B}| = 0$. If d_i's are distinct, A is positive definite, and the elements of the first row of p are positive, then the correspondence between (A, B) and (P, D) is one-to-one.

Proof. Consider a spectral decomposition of $\mathrm{B}^{-1/2}\mathrm{AB}^{-1/2}$ giving

$$\mathrm{B}^{-1/2}\mathrm{AB}^{-1/2} = \mathrm{HDH}',$$

where H is an orthogonal matrix and D is a diagonal matrix. Put $\mathrm{P} = \mathrm{B}^{1/2}\mathrm{H}$. Then P and D satisfy the required condition. For the proof of uniqueness, let P_1 and P_2 be two such matrices, i.e.

$$\mathrm{A} = \mathrm{P}_i\mathrm{DP}_i', \quad \mathrm{B} = \mathrm{P}_i\mathrm{P}_i', \quad i = 1, 2$$

which lead to

$$\mathrm{P}_1\mathrm{DP}_1' = \mathrm{P}_2\mathrm{DP}_2', \quad \mathrm{P}_1\mathrm{P}_1' = \mathrm{P}_2\mathrm{P}_2'$$

or

$$\mathrm{P}_2^{-1}\mathrm{P}_1\mathrm{D} = \mathrm{DP}_2'\mathrm{P}_1'^{-1}, \quad \mathrm{P}_2^{-1}\mathrm{P}_1 = \mathrm{P}_2'\mathrm{P}_1'^{-1}.$$

Putting $\mathrm{G} = (g_{ij}) = \mathrm{P}_2^{-1}P_1$, we have $\mathrm{GD} = \mathrm{DG}$. Therefore, G is a diagonal matrix, and let $\mathrm{G} = \mathrm{D}_g = \mathrm{diag}\,(g_1, \ldots, g_p)$. However, since

$$\mathrm{D}_g\mathrm{D}_g' = \mathrm{D}_g^2 = \mathrm{P}_2^{-1}\mathrm{P}_1 \cdot \mathrm{P}_1^{-1}\mathrm{P}_2 = \mathrm{I}_p.$$

We have $g_i^2 = 1$ and hence $g_i = \pm 1$ for all $i = 1, \ldots, p$. Thus if the elements of the first row of P are taken as positive number, then $\mathrm{P}_1 = \mathrm{P}_2$, which proves of the uniqueness of the decomposition. □

[A.1.15] If A is an $n \times p$ matrix of rank p $(n \geq p)$, then A can be uniquely represented as $\mathrm{A} = \mathrm{HT}$, where H is $n \times p$ with $\mathrm{H}'\mathrm{H} = \mathrm{I}_p$ and T is $p \times p$ upper-triangular matrix with positive diagonal elements.

Proof. Since A′A is $m \times m$ positive definite, we have from [A.1.13] that there exists a unique upper-triangular matrix with positive diagonal elements such that $A'A = T'T$. Now we show that there exists a unique $n \times p$ matrix H with $H'H = I_p$ such that $A = HT$. Let $H = AT'^{-1}$. Then

$$H'H = T^{-1}A'AT'^{-1} = T^{-1}(T'T)^{-1}T' = I_p.$$

To see the uniqueness of the decomposition, suppose

$$A = HT = H_1T_1,$$

where $H_1'H_1 = I_p$ and T_1 is an upper triangular matrix. Then, $A'A = T'T = T_1'T_1$. From the uniqueness of T, we have $T = T_1$ and hence $H = H_1$.

A.1.4 Idempotent Matrices

A $p \times p$ matrix A is called idempotent if $A^2 = A$. A symmetric and idempotent matrix is called projection matrix. Let R^n be the n dimensional Euclid space, and Ω be a subspace in R^n. Then, any $n \times 1$ vector $\boldsymbol{y}$ can be uniquely decomposed into direct sum, i.e.,

$$\boldsymbol{y} = \boldsymbol{u} + \boldsymbol{v}, \quad \boldsymbol{u} \in \Omega, \ \boldsymbol{v} \in \Omega^{\perp} \tag{A.1.1}$$

where $\Omega^{\perp}$ is the orthocomplement space. Using decomposition (A.1.1), consider a mapping

$$P_{\Omega} : \boldsymbol{y} \longrightarrow \boldsymbol{u}, \quad \text{i.e. } P_{\Omega}\boldsymbol{y} = \boldsymbol{u}. \tag{A.1.2}$$

The mapping is linear, and hence it is expressed as a matrix. In this case, $\boldsymbol{u}$ is called the orthogonal projection of $\boldsymbol{y}$ into Ω, and P_{Ω} is also called the orthogonal projection (operator) to Ω.

[A.1.16] P_{Ω} is uniquely defined.

Proof. If P_1 and P_2 satisfy (A.1.1), then $P_1\boldsymbol{y} = P_2\boldsymbol{y} = \boldsymbol{u}$. Therefore, $(P_1 - P_2)\boldsymbol{y} = \boldsymbol{0}$ for any $\boldsymbol{y} \in R^n$. By choosing $\boldsymbol{y} = (1, 0, \cdots, 0)', \cdots, \boldsymbol{y} = (0, 0, \cdots, 1)'$, we have $P_1 = P_2$. □

[A.1.17] $I_n - P_{\Omega}$ is the orhogonal projection to $\Omega^{\perp}$.

Proof. Let $P\boldsymbol{y} = \boldsymbol{v}$, then $P_{\Omega}\boldsymbol{y} = \boldsymbol{u}$ and hence $\boldsymbol{y} = P_{\Omega}\boldsymbol{y} + P\boldsymbol{y}$ for any $\boldsymbol{y} \in R^n$. This implies $(I_n - P_{\Omega})\boldsymbol{y} = P\boldsymbol{y}$. Therefore $P = I_n - P_{\Omega}$. □

[A.1.18] P_{Ω} is a symmetric idempotent matrix.

Proof. For any $\boldsymbol{a} \in R^n$ and $\boldsymbol{b} \in R^n$, $P_{\Omega}\boldsymbol{a} \in \Omega$, $(I_n - P_{\Omega})\boldsymbol{b} \in \Omega^{\perp}$. Therefore, we have $\boldsymbol{0} = (P_{\Omega}\boldsymbol{a})'(I_n - P_{\Omega})\boldsymbol{b}$, and hence $P_{\Omega}' = P_{\Omega}'P_{\Omega}$. This implies that $P_{\Omega}' = P_{\Omega}$ and $P_{\Omega} = P_{\Omega}^2$. □

[A.1.19] $\mathcal{R}[\mathrm{P}_\Omega] = \Omega$.

Proof. Since $\mathrm{P}_\Omega \boldsymbol{y} = \boldsymbol{u} \in \Omega$, $\mathcal{R}[\mathrm{P}_\Omega] \subseteq \Omega$. For $\boldsymbol{x} \in \Omega$, $\boldsymbol{x} = \boldsymbol{x} + \mathbf{0}$, $\boldsymbol{x} \in \Omega$, $\mathbf{0} \in \Omega^\perp$. This implies that $\mathrm{P}_\Omega \boldsymbol{y} = \boldsymbol{x}$. □

[A.1.20] Let X be an $n \times k$ matrix of rank k, $\Omega = \mathcal{R}[\mathrm{X}]$. Then $\mathrm{P}_\Omega = \mathrm{X}(\mathrm{X}'\mathrm{X})^{-1}\mathrm{X}'$.

Proof. Let $\hat{\boldsymbol{y}} = \mathrm{X}(\mathrm{X}'\mathrm{X})^{-1}\mathrm{X}'\boldsymbol{y}$ and consider a decomposion $\boldsymbol{y} = \hat{\boldsymbol{y}} + (\boldsymbol{y} - \hat{\boldsymbol{y}})$. Then, $\hat{\boldsymbol{y}} \in \mathcal{R}[\mathrm{X}]$. On the other hand, since $\hat{\boldsymbol{y}}'(\boldsymbol{y} - \hat{\boldsymbol{y}}) = \mathbf{0}$, $\boldsymbol{y} - \hat{\boldsymbol{y}} \in \mathcal{R}[\mathrm{X}]^\perp$. Therefore, $\mathrm{P}_\Omega \boldsymbol{y} = \hat{\boldsymbol{y}}$ and hence $\mathrm{P}_\Omega = \mathrm{X}(\mathrm{X}'\mathrm{X})^{-1}\mathrm{X}'$. □

Remark A.1.1 For any $\boldsymbol{\theta}$, $\|\boldsymbol{y} - \mathrm{X}\boldsymbol{\theta}\|^2 = \|\boldsymbol{y} - \hat{\boldsymbol{y}}\|^2 + \|\hat{\boldsymbol{y}} - \mathrm{X}\boldsymbol{\theta}\|^2 \geq \|\boldsymbol{y} - \hat{\boldsymbol{y}}\|^2$. The equality holds if and only if $\boldsymbol{\theta} = (\mathrm{X}'\mathrm{X})^{-1}\mathrm{X}'\boldsymbol{y}$.

[A.1.21] Let ω be a subspace of Ω. Then $\mathrm{P}_\Omega \mathrm{P}_\omega = \mathrm{P}_\omega \mathrm{P}_\Omega = \mathrm{P}_\omega$.

Proof. For $\boldsymbol{y} \in \mathrm{R}^n$, $\mathrm{P}_\omega \boldsymbol{y} \in \omega \subset \Omega$. This imply $\mathrm{P}_\Omega \mathrm{P}_\omega \boldsymbol{y} = \mathrm{P}_\omega \boldsymbol{y}$ and hence $\mathrm{P}_\Omega \mathrm{P}_\omega = \mathrm{P}_\omega$. Further, since P_ω, P_Ω, $\mathrm{P}_\omega = \mathrm{P}_\omega' = \mathrm{P}_\omega' \mathrm{P}_\Omega' = \mathrm{P}_\omega \mathrm{P}_\Omega$. □

[A.1.22] Let ω be a subspace of Ω. Then $\mathrm{P}_\Omega - \mathrm{P}_\omega = \mathrm{P}_{\omega^\perp \cap \Omega}$.

Proof. Consider a decomposition $\mathrm{P}_\Omega \boldsymbol{y} = \mathrm{P}_\omega \boldsymbol{y} + (\mathrm{P}_\Omega - \mathrm{P}_\omega)\boldsymbol{y}$. Then, $(\mathrm{P}_\Omega - \mathrm{P}_\omega)\boldsymbol{y} = \mathrm{P}_\Omega \boldsymbol{y} - \mathrm{P}_\omega \boldsymbol{y} \in \Omega$. Since $(\mathrm{P}_\omega \boldsymbol{y})'(\mathrm{P}_\Omega - \mathrm{P}_\omega)\boldsymbol{y} = \boldsymbol{y}'\mathrm{P}_\omega(\mathrm{P}_\Omega - \mathrm{P}_\omega)\boldsymbol{y} = \mathbf{0}$, $(\mathrm{P}_\Omega - \mathrm{P}_\omega)\boldsymbol{y} \in \omega^\perp \cap \Omega$. Then it means that the above decomposition is a direct decomposition of Ω to ω and $\omega^\perp \cap \Omega$.

Let B be a $q \times n$ matrix, and let $\mathrm{N}[\mathrm{B}] = \{\boldsymbol{y}; \mathrm{B}\boldsymbol{y} = \mathbf{0}\} = \mathcal{R}[\mathrm{B}']^\perp$.

[A.1.23] Let Ω_1 and Ω_2 be subspace in R^n. Then $(\Omega_1 \cap \Omega_2)^\perp = \Omega_1^\perp + \Omega_2^\perp$.

Proof. Let $\Omega_i = \mathrm{N}[\mathrm{B}_i] = \mathcal{R}[\mathrm{B}_i']^\perp$, $i = 1, 2$. Then

$$\begin{aligned} (\Omega_1 \cap \Omega_2)^\perp &= \left\{ \mathrm{N}\left[\begin{bmatrix} \mathrm{B}_1 \\ \mathrm{B}_2 \end{bmatrix} \right] \right\}^\perp \\ &= \mathcal{R}[[\mathrm{B}_1', \mathrm{B}_2']] \\ &= \mathcal{R}[\mathrm{B}_1'] + \mathcal{R}[\mathrm{B}_2'] = \Omega_1^\perp + \Omega_2^\perp \end{aligned}$$ □

[A.1.24] If $\omega = \mathrm{N}[\mathrm{B}] \cap \Omega$, then $\omega^\perp \cap \Omega = \mathcal{R}[\mathrm{P}_\Omega \mathrm{B}']$.

Proof. The assumption implies

$$\omega^\perp \cap \Omega = \{\Omega \cap \mathrm{N}[\mathrm{B}]\}^\perp \cap \Omega = \{\Omega^\perp + \mathcal{R}[\mathrm{B}']\} \cap \Omega.$$

Let $\boldsymbol{x} \in \omega^{\perp} \cap \Omega$. Then we have

$$\boldsymbol{x} = \mathrm{P}_{\Omega}\boldsymbol{x} = \mathrm{P}_{\Omega}\{(\mathrm{I}_n - \mathrm{P}_{\Omega})\boldsymbol{\alpha} + \mathrm{B}'\boldsymbol{\beta}\} = \mathrm{P}_{\Omega}\mathrm{B}'\boldsymbol{\beta} \in \mathcal{R}[\mathrm{P}_{\Omega}\mathrm{B}'].$$

Conversely, let $\boldsymbol{x} \in \mathcal{R}[\mathrm{P}_{\Omega}\mathrm{B}']$. Then $\boldsymbol{x} \in \mathcal{R}[\mathrm{P}_{\Omega}] = \Omega$. Further, for $\boldsymbol{z} \in \omega$, $\mathrm{B}\boldsymbol{z} = \mathbf{0}$, and hence $\boldsymbol{x}'\boldsymbol{z} = \boldsymbol{\beta}'\mathrm{BP}_{\Omega}\boldsymbol{z} = \boldsymbol{\beta}'\mathrm{B}\boldsymbol{z} = \mathbf{0}$. This implies $\boldsymbol{x} \in \omega^{\perp} \cap \Omega$. □

[A.1.25] Let X be an $n \times k$ matrix of rank k, and let $\Omega = \mathcal{R}[\mathrm{X}]$ which is defined also by the set $\{\boldsymbol{y} : \ \boldsymbol{y} = X\boldsymbol{\theta}\}$, where $\boldsymbol{\theta}$ is a $k \times 1$ unknown parameter vector. Let C be a $c \times k$ matrix of rank c, and define ω by the set $\{\boldsymbol{y} : \ \boldsymbol{y} = X\boldsymbol{\theta}, \ \mathrm{C}\boldsymbol{\theta} = \mathbf{0}\}$. Then,

$$\mathrm{P}_{\Omega} - \mathrm{P}_{\omega} = \mathrm{X}(\mathrm{X}'\mathrm{X})^{-1}\mathrm{C}'\{\mathrm{C}(\mathrm{X}'\mathrm{X})^{-1}\mathrm{C}\}^{-1}\mathrm{C}(\mathrm{X}'\mathrm{X})^{-1}\mathrm{X}.$$

Proof. Since $\mathrm{C}\boldsymbol{\theta} = \mathrm{C}(\mathrm{X}'\mathrm{X})^{-1}\mathrm{X}' \cdot \mathrm{X}\boldsymbol{\theta}$, we can write $\omega = \mathrm{N}[\mathrm{B}] \cap \Omega$, where $\mathrm{B} = \mathrm{C}(\mathrm{X}'\mathrm{X})^{-1}\mathrm{X}'$. Using [A.1.24],

$$\omega^{\perp} \cap \Omega = \mathcal{R}[\mathrm{P}_{\Omega}\mathrm{B}'] = \mathcal{R}[\mathrm{X}(\mathrm{X}'\mathrm{X})^{-}\mathrm{C}'].$$

The final result is obtained by using [A.1.20] and [A.1.22]. □

A.2 Inequalities and Max–Min Problems

[A.2.1] (Schwartz's inequality) Let $\boldsymbol{a}$ be a $p \times 1$ vector, and S a $p \times p$ positive definite matrix. Then

$$\frac{(\boldsymbol{a}'\boldsymbol{x})^2}{\boldsymbol{x}'\mathrm{S}\boldsymbol{x}} \leq \boldsymbol{a}'\mathrm{S}^{-1}\boldsymbol{a}, \quad \boldsymbol{a} \neq \mathbf{0}.$$

The equality holds if and only if $\boldsymbol{x}$ is proportional to $\mathrm{S}^{-1}\boldsymbol{a}$.

[A.2.2] Let $\boldsymbol{y}$ be a $p \times 1$ vector and S a $p \times p$ positive definite matrix that are decomposed as

$$\boldsymbol{y} = \begin{pmatrix} \boldsymbol{y}_1 \\ \boldsymbol{y}_2 \end{pmatrix}, \quad \mathrm{S} = \begin{pmatrix} \mathrm{S}_{11} & \mathrm{S}_{12} \\ \mathrm{S}_{21} & \mathrm{S}_{22} \end{pmatrix},$$

where $\boldsymbol{a}_i : p_i \times 1$, $\mathrm{S}_{ij} : p_i \times p_j$ and $p = p_1 + p_2$. Then

(1) $\boldsymbol{y}'\mathrm{S}^{-1}\boldsymbol{y} = \boldsymbol{y}_1'\mathrm{S}_{11}\boldsymbol{y}_1 + (\boldsymbol{y}_2 - \mathrm{S}_{21}\mathrm{S}_{11}^{-1}\boldsymbol{y}_1)'\mathrm{S}_{22\cdot 1}^{-1}(\boldsymbol{y}_2 - \mathrm{S}_{21}\mathrm{S}_{11}^{-1}\boldsymbol{y}_1)$,

(2) $\displaystyle \min_{\boldsymbol{\theta}} \begin{pmatrix} \boldsymbol{y}_1 \\ \boldsymbol{y}_2 - \boldsymbol{\theta} \end{pmatrix}' \begin{pmatrix} \mathrm{S}_{11} & \mathrm{S}_{12} \\ \mathrm{S}_{21} & \mathrm{S}_{22} \end{pmatrix}^{-1} \begin{pmatrix} \boldsymbol{y}_1 \\ \boldsymbol{y}_2 - \boldsymbol{\theta} \end{pmatrix} = \boldsymbol{y}_1'\mathrm{S}_{11}^{-1}\boldsymbol{y}_1$,

where θ is a $p_2 \times 1$ vector, and $\mathrm{S}_{22\cdot 1} = \mathrm{S}_{22} - \mathrm{S}_{21}\mathrm{S}_{11}^{-1}\mathrm{S}_{12}$.

Proof. (1) is consequence of [A.1.1]. The second result follows from the first result. □

[A.2.3] Let A be a $p \times p$ positive definite matrix, and $\boldsymbol{b}$ a $p \times 1$ vector. Then for any $p \times 1$ vector $\boldsymbol{x}$,

$$\boldsymbol{x}'\mathrm{A}\boldsymbol{x} - 2\,\boldsymbol{b}'\boldsymbol{x} \geq -\boldsymbol{b}'\mathrm{A}^{-1}\boldsymbol{b}.$$

The equality holds if and only if $\boldsymbol{x} = \mathrm{A}^{-1}\boldsymbol{b}$.

[A.2.4] Let $\boldsymbol{a}$ be a $p \times 1$ vector, and B a $p \times p$ positive definite matrix. Then

$$\max_{\boldsymbol{x}'\mathrm{B}\boldsymbol{x}=1} \boldsymbol{a}'\boldsymbol{x} = \sqrt{\boldsymbol{a}'\mathrm{B}^{-1}\boldsymbol{a}}.$$

The equality holds if and only if $\boldsymbol{x} = (\boldsymbol{a}'\mathrm{B}^{-1}\boldsymbol{a})^{-\frac{1}{2}}\mathrm{B}\boldsymbol{a}$.

[A.2.5] Suppose that a $p \times p$ symmetric matrix A has a spectral decomposition as in [A.1.11]. Then we have

(1)
$$\sup_{\boldsymbol{x}} \frac{\boldsymbol{x}'\mathrm{A}\boldsymbol{x}}{\boldsymbol{x}'\boldsymbol{x}} = \ell_1 \quad \text{and} \quad \inf_{\boldsymbol{x}} \frac{\boldsymbol{x}'\mathrm{A}\boldsymbol{x}}{\boldsymbol{x}'\boldsymbol{x}} = \ell_p,$$

where the supremum and infimum are attained at $\boldsymbol{x} = \boldsymbol{h}_1$ and $\boldsymbol{x} = \boldsymbol{h}_p$, respectively.

(2) For $1 \leq q < p$, put $\mathrm{H}_1 = (\boldsymbol{h}_1, \ldots, \boldsymbol{h}_q)$ and $\mathrm{H}_2 = (\boldsymbol{h}_{p-q+1}, \ldots, \boldsymbol{h}_p)$. Then

$$\sup_{\mathrm{H}_1'\boldsymbol{x}=\boldsymbol{0}} \frac{\boldsymbol{x}'\mathrm{A}\boldsymbol{x}}{\boldsymbol{x}'\boldsymbol{x}} = \ell_{q+1} \quad \text{and} \quad \inf_{\mathrm{H}_2'\boldsymbol{x}=\boldsymbol{0}} \frac{\boldsymbol{x}'\mathrm{A}\boldsymbol{x}}{\boldsymbol{x}'\boldsymbol{x}} = \ell_{p-q},$$

where the supremum and infimum are attained at $\boldsymbol{x} = \boldsymbol{h}_{q+1}$ and $\boldsymbol{x} = \boldsymbol{h}_{p-q}$, respectively.

(3) (Courant-Fisher min-max theorem) For $1 \leq q < p$, let $\mathrm{F} = (\boldsymbol{f}_1, \ldots, \boldsymbol{f}_q)$ be a $p \times q$ matrix, where $\boldsymbol{f}_i'\boldsymbol{f}_i = 1, i = 1, \ldots, q$. Then

$$\inf_{\mathrm{F}} \sup_{\mathrm{F}'\boldsymbol{x}=\boldsymbol{0}} \frac{\boldsymbol{x}'\mathrm{A}\boldsymbol{x}}{\boldsymbol{x}'\boldsymbol{x}} = \ell_{q+1} \quad \text{and} \quad \sup_{\mathrm{F}} \inf_{\mathrm{B}'\boldsymbol{x}=\boldsymbol{0}} \frac{\boldsymbol{x}'\mathrm{A}\boldsymbol{x}}{\boldsymbol{x}'\boldsymbol{x}} = \ell_{p-q},$$

where the values ℓ_{q+1} and ℓ_{p-q} are attained at $\mathrm{F} = (\boldsymbol{h}_1, \ldots, \boldsymbol{h}_q), \boldsymbol{x} = \boldsymbol{h}_{q+1}$, and $\mathrm{F} = (\boldsymbol{h}_{p-q+1}, \ldots, \boldsymbol{h}_p), \boldsymbol{x} = \boldsymbol{h}_{p-q}$, respectively.

Proof. The results on sup in (1) and (2) have been essentially proved in Theorem 10.1.2. The results on inf are similarly proved. For the Courant-Fisher min-max theorem, see Bellman (1960) or Rao (1973). □

[A.2.6] Let A and B be $p \times p$ symmetric matrices and assume that B is positive definite. Let $d_1 \geq \cdots \geq d_p$ be the roots of $|A - dB| = 0$, and $\boldsymbol{g}_i$, $i = 1, \ldots, p$ be the characteristic vectors associated with d_i's and normalized by $\boldsymbol{g}_i' B \boldsymbol{g}_i = \delta_{ij}$. Then

(1)
$$\sup_{\boldsymbol{x}} \frac{\boldsymbol{x}' A \boldsymbol{x}}{\boldsymbol{x}' B \boldsymbol{x}} = \ell_1 \quad \text{and} \quad \inf_{\boldsymbol{x}} \frac{\boldsymbol{x}' A \boldsymbol{x}}{\boldsymbol{x}' B \boldsymbol{x}} = \ell_p,$$

where the supremum and infimum are attained at $\boldsymbol{x} = \boldsymbol{g}_1$ and $\boldsymbol{x} = \boldsymbol{g}_p$, respectively.

(2) For $1 \leq q < p$, put $G_1 = (\boldsymbol{g}_1, \ldots, \boldsymbol{g}_q)$ and $H_2 = (\boldsymbol{g}_{p-q+1}, \ldots, \boldsymbol{g}_p)$. Then

$$\sup_{G_1' \boldsymbol{x} = \boldsymbol{0}} \frac{\boldsymbol{x}' A \boldsymbol{x}}{\boldsymbol{x}' B \boldsymbol{x}} = \ell_{q+1} \quad \text{and} \quad \inf_{G_2' \boldsymbol{x} = \boldsymbol{0}} \frac{\boldsymbol{x}' A \boldsymbol{x}}{\boldsymbol{x}' B \boldsymbol{x}} = \ell_{p-q},$$

where the supremum and infimum are attained at $\boldsymbol{x} = \boldsymbol{g}_{q+1}$ and $\boldsymbol{x} = \boldsymbol{g}_{p-q}$, respectively.

(3) For $1 \leq q < p$, let $F = (\boldsymbol{f}_1, \ldots, \boldsymbol{f}_q)$ be a $p \times q$ matrix, where $\boldsymbol{f}_i' \boldsymbol{f}_i = 1, i = 1, \ldots, q$. Then

$$\inf_{F} \sup_{F' \boldsymbol{x} = \boldsymbol{0}} \frac{\boldsymbol{x}' A \boldsymbol{x}}{\boldsymbol{x}' F \boldsymbol{x}} = \ell_{q+1} \quad \text{and} \quad \sup_{F} \inf_{F' \boldsymbol{x} = \boldsymbol{0}} \frac{\boldsymbol{x}' A \boldsymbol{x}}{\boldsymbol{x}' B \boldsymbol{x}} = \ell_{p-q},$$

where the values ℓ_{q+1} and ℓ_{p-q} are attained at $F = (\boldsymbol{g}_1, \ldots, \boldsymbol{g}_q), \boldsymbol{x} = \boldsymbol{F}_{q+1}$, and $F = (\boldsymbol{g}_{p-q+1}, \ldots, \boldsymbol{g}_p), \boldsymbol{x} = \boldsymbol{g}_{p-q}$, respectively.

Proof. Let $\boldsymbol{y} = B^{1/2} \boldsymbol{x}$ and $C = B^{-1/2} A B^{-1/2}$. Then

$$\frac{\boldsymbol{x}' A \boldsymbol{x}}{\boldsymbol{x}' B \boldsymbol{x}} = \frac{\boldsymbol{y}' C \boldsymbol{y}}{\boldsymbol{y}' \boldsymbol{y}},$$

and the characteristic vectors of matrix C are $\boldsymbol{h}_i = B^{1/2} \boldsymbol{g}_i$, $i = 1, \ldots, p$, and $\boldsymbol{g}_i' \boldsymbol{x} = \boldsymbol{h}_i' \boldsymbol{y}$, $i = 1, \ldots, p$. From these results, (1) and (2) follows from [A.2.5] immediately. □

[A.2.7] Let a $p \times p$ symmetric matrix A have a spectral decomposition as in [A.1.12], and let X be a $p \times (p - k)$ matrix. Then

$$\inf_{X'X = I_{n-k}} \operatorname{tr} X'AX = \ell_{k+1} + \cdots + \ell_p.$$

The equality holds when $X = (\boldsymbol{h}_{k+1}, \cdots, \boldsymbol{h}_p)$.

[A.2.8] Let a $n \times p$ matrix A has a singular value decomposition as in [A.1.16], and let X be an $n \times p$ matrix. Then

$$\inf_{\text{rank}(\mathrm{X})=k} \|\mathrm{A} - \mathrm{X}\|^2 = \ell_{k+1}^2 + \cdots + \ell_p^2.$$

The equality holds when $\mathrm{X} = \ell_1 \boldsymbol{u} \boldsymbol{v}_1' + \cdots + \ell_k \boldsymbol{u}_k \boldsymbol{v}_k'$.

[A.2.9] Let a $p \times p$ symmetric matrix A have a spectral decomposition as in [A.1.12], and let B be a $p \times p$ symmetric positive semi-definite matrix with rank k. Then

$$\mathrm{ch}_i(\mathrm{A} - \mathrm{B}) \geq \mathrm{ch}_{k+i}(\mathrm{A}), \ \text{for } i = 1, \ldots, p - k.$$

The equality is attained for all $i (1 \leq i \leq p-k)$ when $\mathrm{B} = \ell_1 \boldsymbol{h}_1 \boldsymbol{h}_1' + \cdots + \ell_k \boldsymbol{h}_k \boldsymbol{h}_k$.

Proof. We can write B as CC$'$, where C is a $p \times k$ matrix of rank k. For $1 \leq q < p - k$, let L be a $p \times q$ matrix. Then, using [A.2.5] (3), we have

$$\begin{aligned} \mathrm{ch}_q(\mathrm{A} - \mathrm{B}) &= \mathrm{ch}_q(\mathrm{A} - \mathrm{CC}') \\ &= \inf_{\mathrm{L}} \sup_{\mathrm{L}'\boldsymbol{x}=\boldsymbol{0}} \frac{\boldsymbol{x}'(\mathrm{A} - \mathrm{CC}')\boldsymbol{x}}{\boldsymbol{x}'\boldsymbol{x}} \\ &\geq \inf_{\mathrm{L}} \sup_{[\mathrm{L},\mathrm{C}]'\boldsymbol{x}=\boldsymbol{0}} \frac{\boldsymbol{x}'\mathrm{A}\boldsymbol{x}}{\boldsymbol{x}'\boldsymbol{x}} \geq \mathrm{ch}_{k+q}(\mathrm{A}). \end{aligned}$$

When $\mathrm{B} = \ell_1 \boldsymbol{h}_1 \boldsymbol{h}_1' + \cdots + \ell_k \boldsymbol{h}_k \boldsymbol{h}_k$,

$$\mathrm{A} - \mathrm{B} = \ell_{k+1} \boldsymbol{h}_{k+1} \boldsymbol{h}_{k+1}' + \cdots + \ell_p \boldsymbol{h}_p \boldsymbol{h}_{p\cdot},$$

and hence A $-$ B has the characteristic roots $\ell_{k+1}, \ldots, \ell_p$ and k zeroes. This completes the proof. □

For the proofs of [A.2.7] and [A.2.8], see Rao (1973). [A.2.8] was essentially proved by Okamoto and Kanazawa (1968). They also noted that, assuming $\mathrm{A} \geq \mathrm{B} \geq \mathrm{O}$, the equality in [A.2.9] holds if and only if $\mathrm{B} = \ell_1 \boldsymbol{h}_1 \boldsymbol{h}_1' + \cdots + \ell_k \boldsymbol{h}_k \boldsymbol{h}_k$.

[A.2.10] Consider a function $f(\Sigma) = \log|\Sigma| + \mathrm{tr}\,(\Sigma^{-1}\mathrm{S})$ of positive definite matrix S. Then, $f(\Sigma)$ takes uniquely the minimum at $\Sigma = \mathrm{S}$, and the minimum value is given by

$$\min_{\Sigma > 0} f(\Sigma) = f(\mathrm{S}) = \log|\mathrm{S}| + p.$$

Proof. Let $\ell_1, \ldots, \ell_p$ be the characteristic roots of $\Sigma^{-1}S$. Note that the characteristic roots of $\Sigma^{-1}S$ and $\Sigma^{-1/2}S\Sigma^{-1/2}$ are the same. The latter matrix is positive definite, and hence we may assume $\ell_1 \geq \cdots \geq \ell_p > 0$. Then

$$\begin{aligned} f(\Sigma) - f(S) &= \log|\Sigma S^{-1}| + \mathrm{tr}\,(\Sigma^{-1}S) - p \\ &= -\log|\Sigma^{-1}S| + \mathrm{tr}\,(\Sigma^{-1}S) - p \\ &= \sum_{i=1}^{p}(-\log \ell_i + \ell_i - 1) \geq 0. \end{aligned}$$

The last inequality follows from $x - 1 \geq \log x$ $(x > 0)$. The equality holds if and only if $\ell_1 = \cdots = \ell_p = 1 \Leftrightarrow \Sigma = S$. □

[A.2.11] Let Y be an $n \times p$ known matrix and A an $n \times k$ known matrix of rank k. Consider a function of $p \times p$ positive definite matrix Σ and $k \times p$ matrix $\Theta = (\theta_{ij})$ given by

$$g(\hat{\Theta}, \hat{\Sigma}) = m \log|\Sigma| + \mathrm{tr}\,\Sigma^{-1}(Y - A\Theta)'(Y - A\Theta),$$

where $m > 0$, $-\infty < \theta_{ij} < \infty$. Then, $g(\hat{\Theta}, \hat{\Sigma})$ takes the minimum at

$$\Theta = \hat{\Theta} = (A'A)^{-1}A'Y, \quad \Sigma = \hat{\Sigma} = \frac{1}{m}Y'(I_n - P_A)Y,$$

and the minimum value is given by $m \log|\hat{\Sigma}| + mp$.

Proof. For any positive definite matrix Σ, we have

$$\begin{aligned} &\mathrm{tr}\,\Sigma^{-1}(Y - A\Theta)'(Y - A\Theta) \\ &\quad = \mathrm{tr}\,\Sigma^{-1}(Y - A\hat{\Theta})'(Y - A\hat{\Theta}) + \mathrm{tr}\,\Sigma^{-1}\{A(\hat{\Theta} - \Theta)\}'A(\hat{\Theta} - \Theta) \\ &\quad \geq \mathrm{tr}\,\Sigma^{-1}Y'(I_n - P_A)Y. \end{aligned}$$

In the last step, the equality holds when $\Theta = \hat{\Theta}$. The required result is obtained by combining this result with [A.2.10]. □

[A.2.12] Let Y be a $k \times p$ matrix and S a $p \times p$ positive definite matrix that are decomposed as

$$Y = (Y_1\ Y_2), \quad S = \begin{pmatrix} S_{11} & S_{12} \\ S_{21} & S_{22} \end{pmatrix},$$

where $Y_i; k \times p_i$, and $S_{ij}; p_i \times p_j$. Let Θ be a $k \times p_1$ matrix. Then

$$\begin{aligned} g(\Theta) &= \mathrm{tr}\,S^{-1}(Y_1 - \Theta,\ Y_2)'(Y_1 - \Theta,\ Y_2) \\ &\geq \mathrm{tr}\,S_{22}^{-1}Y_2'Y_2. \end{aligned}$$

The equality holds when $\Theta = Y_1 - Y_2S_{22}^{-1}S_{21}$.

Proof. The inverse matrix of S is expressed (see [A.1.2]) as

$$\mathrm{S}^{-1} = \begin{pmatrix} \mathrm{O} & \mathrm{O} \\ \mathrm{O} & \mathrm{S}_{22}^{-1} \end{pmatrix} + \begin{pmatrix} \mathrm{I}_{p_1} \\ -\mathrm{S}_{22}^{-1}\mathrm{S}_{21} \end{pmatrix} \mathrm{S}_{11\cdot 2}^{-1} \begin{pmatrix} \mathrm{I}_{p_1} & -\mathrm{S}_{12}\mathrm{S}_{22}^{-1} \end{pmatrix}.$$

Then

$$\begin{aligned} g(\Theta) &= \mathrm{tr}\,\mathrm{S}_{11\cdot 2}^{-1}(\mathrm{Y}_1 - \Theta - \mathrm{Y}_2\mathrm{S}_{22}^{-1}\mathrm{S}_{21})'(\mathrm{Y}_1 - \Theta - \mathrm{Y}_2\mathrm{S}_{22}^{-1}\mathrm{S}_{21}) \\ &\quad + \mathrm{tr}\,\mathrm{S}_{22}^{-1}\mathrm{Y}_2'\mathrm{Y}_2. \end{aligned}$$

This implies the required result. □

[A.2.13] Let Σ and $\mathrm{W}^{(1)}$ be $p \times p$ positive definite matrices, and $\mathrm{W}^{(2)}$ a $p \times p$ positive semi-definite matrix. They are decomposed as

$$\Sigma^{(i)} = \begin{pmatrix} \Sigma_{11}^{(i)} & \Sigma_{12}^{(i)} \\ \Sigma_{21}^{(i)} & \Sigma_{22}^{(i)} \end{pmatrix}, \quad \mathrm{W}^{(i)} = \begin{pmatrix} \mathrm{W}_{11}^{(i)} & \mathrm{W}_{12}^{(i)} \\ \mathrm{W}_{21}^{(i)} & \mathrm{W}_{22}^{(i)} \end{pmatrix}, \quad i = 1, 2,$$

where $\Sigma_{ij} : p_i \times p_j$, $\mathrm{W}_{ij}^{(1)} : p_i \times p_j$ and $\mathrm{W}_{ij}^{(2)} : p_i \times p_j$. Consider a function of Σ given by

$$g(\Sigma) = n \log|\Sigma| + \mathrm{tr}\,\Sigma^{-1}W^{(1)} + \mathrm{tr}\,\Sigma_{22}^{-1}\mathrm{W}_{22}^{(2)}.$$

Then

$$g(\Sigma) \geq n \log(|\frac{1}{n}\mathrm{W}_{11\cdot 2}^{(1)}| \cdot |\frac{1}{n}(\mathrm{W}_{22}^{(1)} + \mathrm{W}_{22}^{(2)})| + np.$$

The equality holds when $\Sigma = \hat{\Sigma}$,

$$n\hat{\Sigma}_{11\cdot 2} = \mathrm{W}_{11\cdot 2}^{(1)}, \quad n\hat{\Sigma}_{22} = \mathrm{W}_{22}^{(1)} + \mathrm{W}_{22}^{(2)}, \quad \hat{\mathrm{B}} = (\mathrm{W}_{22}^{(1)})^{-1}\mathrm{W}_{21}^{(1)},$$

where $\mathrm{B} = (\Sigma_{22}^{(1)})^{-1}\Sigma_{21}^{(1)}$.

Proof. Using $|\Sigma| = |\Sigma_{11\cdot 2}| \cdot |\Sigma_{22}|$ ([A.1.1]) and an expression for Σ^{-1} (see, [A.1.2]), we have

$$\begin{aligned} g(\Sigma) &= n \log|\Sigma_{11\cdot 2}| + n \log|\Sigma_{22}| \\ &\quad + \mathrm{tr}\,\Sigma_{11\cdot 2}^{-1}(\mathrm{W}_{11}^{(1)} - \mathrm{B}'\mathrm{S}_{21}^{(1)} - \mathrm{W}_{12}^{(1)}\mathrm{B} + \mathrm{B}'\mathrm{W}_{22}^{(1)}) \\ &\quad + \mathrm{tr}\,\Sigma_{22}^{-1}(\mathrm{W}_{22}^{(1)} + \mathrm{W}_{22}^{(1)}). \end{aligned}$$

Note that

$$\begin{aligned} &\mathrm{W}_{11}^{(1)} - \mathrm{B}'\mathrm{S}_{21}^{(1)} - \mathrm{W}_{12}^{(1)}\mathrm{B} + \mathrm{B}'\mathrm{W}_{22}^{(1)} \\ &\quad = \mathrm{W}_{11\cdot 2}^{(1)} + \{(\mathrm{W}_{22}^{(1)})^{-1}\mathrm{W}_{21}^{(1)} - \mathrm{B}\}'\mathrm{W}_{22}^{(1)}\{(\mathrm{W}_{22}^{(1)})^{-1}\mathrm{W}_{21}^{(1)} - \mathrm{B}\}. \end{aligned}$$

From here we can get the required result. □

A.3 Jacobians of Transformations

Suppose that the transformation $\boldsymbol{y} = \boldsymbol{y}(\boldsymbol{x}) = (y_1(\boldsymbol{x}), \ldots, y_p(\boldsymbol{x}))'$ is such that the partial derivatives $\partial y_i / \partial x_j$ exist. Then the determinant of the square matrix

$$\begin{pmatrix} \frac{\partial y_1}{\partial x_1} & \cdots & \frac{\partial y_1}{\partial x_p} \\ \vdots & & \vdots \\ \frac{\partial y_p}{\partial x_1} & \cdots & \frac{\partial y_p}{\partial x_p} \end{pmatrix} = \frac{\partial(y_1, \ldots, y_p)}{\partial(x_1, \ldots, x_p)} = \frac{\partial \boldsymbol{y}}{\partial \boldsymbol{x}} \tag{A.3.1}$$

is called the Jacobian determinant (or simply Jacobian) of the transformation of $\boldsymbol{x}$ to $\boldsymbol{y}$ and is denoted by

$$J(\boldsymbol{x} \to \boldsymbol{y}).$$

In the following the notation $J(\boldsymbol{x} \to \boldsymbol{y})$ is also used for the absolute value of determinant of matrix (A.3.1).

[A.3.1] Let $\boldsymbol{y} : p \times 1,\ \boldsymbol{x} : p \times 1,\ \mathrm{X} : p \times q,\ \mathrm{Y} : p \times q.\ \mathrm{A} : p \times p$, and $\mathrm{B} : q \times q$. Then

(1) for $\boldsymbol{y} = \mathrm{A}\boldsymbol{x}$, $J(\boldsymbol{x} \to \boldsymbol{y}) = |\mathrm{A}|$.

(2) for $\mathrm{Y} = \mathrm{AX}$, $J(\mathrm{X} \to \mathrm{Y}) = |\mathrm{A}|^q$.

(3) for $\mathrm{Y} = \mathrm{XB}$, $J(\mathrm{X} \to \mathrm{Y}) = |\mathrm{B}|^p$.

(4) for $\mathrm{Y} = \mathrm{AXB}$, $J(\mathrm{X} \to \mathrm{Y}) = |\mathrm{A}|^q |\mathrm{B}|^p$.

Proof. (1) follows from that

$$\frac{\partial \boldsymbol{y}}{\partial \boldsymbol{x}} = \mathrm{A}.$$

For the second case, set $\mathrm{Y} = (\boldsymbol{y}_1, \ldots, \boldsymbol{y}_q)$ and $\mathrm{X} = (\boldsymbol{x}_1, \ldots, \boldsymbol{x}_q)$. Then, the transformation is expressed as

$$\begin{pmatrix} \boldsymbol{y}_1 \\ \vdots \\ \boldsymbol{y}_q \end{pmatrix} = \begin{pmatrix} \mathrm{A} & & \mathrm{O} \\ & \ddots & \\ \mathrm{O} & & \mathrm{A} \end{pmatrix} \begin{pmatrix} \boldsymbol{x}_1 \\ \vdots \\ \boldsymbol{x}_q \end{pmatrix}.$$

Therefore the result (2) follows from (1). The result (3) is obtained by considering $\mathrm{Y}' = \mathrm{B}'\mathrm{X}'$ and using (2). The result (4) is just a combination of (2) and (3): That is

$$\mathrm{Y} = \mathrm{AXB} \Leftrightarrow \mathrm{Z} = \mathrm{XB}, \quad \mathrm{Y} = \mathrm{AZ}.$$

Therefore

$$J(\mathrm{X} \to \mathrm{Y}) = J(\mathrm{Z} \to \mathrm{Y}) J(\mathrm{X} \to \mathrm{Z}) = |\mathrm{A}|^q |\mathrm{B}|^p.$$

□

[A.3.2] Consider a transformation $Y = AXA'$, where X and Y are $p \times p$ symmetric matrices. Then

$$J(X \to Y) = |A|^{-(p+1)}.$$

Proof. Let the Jacobian denote by $h(A)$. Then we have the functional equation

$$h(AB) = h(A)h(B) \quad \text{for all nonsingular } p \times p \text{ matrices A and B.}$$

Since $Y = ABX(AB)'$ may be defined by $Y = AZA'$ and $Z = BXB'$. The solution of this equation has the form $h(A) = |A|^c$ [see Olkin and Sampson, 1972]. The constant c can be determined by choosing $A = \text{diag}(a, 1, \ldots, 1)$.

[A.3.3] Let X, Y, A and B be $p \times p$ nonsingular lower triangular matrices. Then

(1) for $Y = XA$, $J(X \to Y) = \prod_{i=1}^{p} a_{ii}^{p-i+1}$.

(2) for $Y = BX$, $J(X \to Y) = \prod_{i=1}^{p} b_{ii}^{i}$.

(4) for $Y = BXA$, $J(X \to Y) = \prod_{i=1}^{p} a_{ii}^{p-i+1} b_{ii}^{i}$.

Proof. Let the Jacobian for $Y = XA$ be denoted by $h(A)$. Then, as in the proof of [A.3.2], we obtain the functional equation

$$h(AB) = h(A)h(B)$$

to hold for all $p \times p$ nonsingular lower triangular matrices A and B. According to Theorem 3 of Olkin and Sampson (1972), the solution is of the form $h(A) = \prod_{i=1}^{p} a_{ii}^{c_i}$. Letting $A = \text{diag}(a_1, \ldots, a_p)$, we obtain $c_i = p - i + 1$, which completes the proof of (1). The results (2) and (3) are obtained by arguments similar to ones in the proof of [A.3.1] (3) and (4). □

Let $f(\boldsymbol{x})$ be a scalar-valued function of $\boldsymbol{x} = (x_1, \ldots, x_p)'$. Then the total differential of $f(\boldsymbol{x})$ is defined by

$$df = df(\boldsymbol{x}) = \sum_{i=1}^{p} \frac{\partial f(\boldsymbol{x})}{\partial x_i} dx_i = \frac{\partial f(\boldsymbol{x})}{\partial \boldsymbol{x}'} d\boldsymbol{x},$$

where $d\boldsymbol{x} = (dx_1, \ldots, dx_p)$. For a matrix-valued function $F(\boldsymbol{x}) = (f_{ij}(\boldsymbol{x}))$, the total differential is defined by

$$dF = dF(\boldsymbol{x}) = (df_{ij}(\boldsymbol{x})) = \left(\frac{\partial f_{ij}(\boldsymbol{x})}{\partial \boldsymbol{x}'} d\boldsymbol{x} \right).$$

The Jacobian of a nonlinear transformation may be obtained from the Jacobian of the linear transformation in the differentials as in the following theorem.

[A.3.4] Consider a matrix valued transformation defined by $\mathrm{Y} = \mathrm{Y}(\mathrm{X}) = (y_{ij}(\mathrm{X}))$. Then

$$J(\mathrm{X} \to \mathrm{Y}) = J(d\mathrm{X} \to d\mathrm{Y}).$$

[A.3.5] If X is a nonsingular symmetric matrix of order p, then the Jacobian of $\mathrm{Y} = \mathrm{X}^{-1}$ is given by

$$J(\mathrm{X} \to \mathrm{Y}) = |\mathrm{X}|^{-(p+1)}.$$

If X is a general nonsingular matrix, then

$$J(\mathrm{X} \to \mathrm{Y}) = |\mathrm{X}|^{-2p}.$$

Proof. Considering the differentials of the both sides of $\mathrm{X} \cdot \mathrm{X}^{-1} = \mathrm{I}_p$, we have $d\mathrm{X} \cdot \mathrm{X}^{-1} + \mathrm{X} \cdot d\mathrm{X}^{-1} = \mathrm{O}$, and hence $d\mathrm{Y} = -\mathrm{X}^{-1}(d\mathrm{X})\mathrm{X}^{-1}$. The results are obtained by using [A.3.3] with the help of [A.3.1] (5) and [A.3.2] (2). □

[A.3.6] Let Y be a $p \times p$ positive definite matrix. If $\mathrm{Y} = \mathrm{TT}'$, where T is a $p \times p$ lower triangular matrix, then

$$J(\mathrm{T} \to \mathrm{Y}) = 2^p \prod_{i=1}^{p} t_{ii}^{p-i+1}.$$

Proof. Since $d\mathrm{Y} = \mathrm{T}(d\mathrm{T}') + (d\mathrm{T})\mathrm{T}'$, it is sufficient to show the following result: Let Y be a $p \times p$ symmetric matrix. If $\mathrm{Y} = \mathrm{AX}' + \mathrm{XA}$, where X and A are $p \times p$ symmetric matrices, then

$$J(\mathrm{X} \to \mathrm{Y}) = 2^p \prod_{i=1}^{p} a_{ii}^{p-i+1}.$$

Let $\mathrm{Y} = \mathrm{AZA}'$ and $\mathrm{W} = \mathrm{A}^{-1}X$. Then $\mathrm{Z} = \mathrm{W} + \mathrm{W}'$, and

$$J(\mathrm{X} \to \mathrm{Y}) = J(\mathrm{Z} \to \mathrm{Y})J(\mathrm{W} \to \mathrm{Z})J(\mathrm{X} \to \mathrm{W}).$$

Each factor on the right-hand side is evaluated as

$$\begin{aligned}
J(\mathrm{Z} \to \mathrm{Y}) &= |\mathrm{A}|^{p+1} = \prod_{i=1}^{p} a_{ii}^{p+1} \quad \text{(by [A.3.4])}, \\
J(\mathrm{W} \to \mathrm{Z}) &= 2^p, \\
J(\mathrm{X} \to \mathrm{W}) &= \prod_{i=1}^{p} a_{ii}^{-i} \quad \text{(by [A.3.3] (2))}.
\end{aligned}$$

For the last factor, we use that A^{-1} is also lower triangular matrix and its (i,i) element is a_{ii}^{-1}. Combining the above results completes the proof. □

[A.3.7] Let U and V be $p \times p$ positive definite matrices and consider the transformation $U = WLW'$, $V = WW'$, where $W : p \times p$ is nonsingular and $L = \text{diag}(\ell_1, \ldots, \ell_p)$, $\ell_1 > \cdots > \ell_p$. Then

$$J(W, L \to U, V) = 2^p |W|^{p+2} \prod_{i<j} (\ell_i - \ell_j).$$

Proof. Taking differentials of the transformation

$$\begin{aligned} dU &= (dW)LW' + W(dL)W' + WL(dW)', \\ dV &= (dW)W' + W(dW)'. \end{aligned}$$

Put

$$X = W^{-1}(dU)W'^{-1}, \quad Y = W^{-1}(dV)W'^{-1}, \quad Z = W^{-1}(dW).$$

Then

$$X = ZL + dL + LZ', \quad Y = Z + Z'.$$

Using [A.3.3] and the chain rule,

$$\begin{aligned} J(W, L \to U, V) &= J(dW, dL \to dU, dV) \\ &= J(X, Y \to dU, dV) J(Z, dL \to X, Y) \\ &\quad \times J(dW, dL \to Z, dL). \end{aligned}$$

We have

$$\begin{aligned} J(X, Y \to dU, dV) &= J(X \to dU) J(Y \to dV) = |W|^{2(p+1)}, \\ &\qquad \text{(by [A.3.3])} \\ J(dW, dL \to Z, dL) &= J(dW \to Z) = |W|^{-p}. \\ &\qquad \text{(by [A.3.3])} \end{aligned}$$

For finding $J(Z, dL \to X, Y)$, we may construct the linear transformation matrix in the order of (x_{ii} ($i = 1, \ldots, p$), y_{ii} ($i = 1, \ldots, p$), x_{ij} ($i < j$), y_{ij} ($i < j$)) and ($d\ell_i$ ($i = 1, \ldots, p$), z_{ii} ($i = 1, \ldots, p$), z_{ij} ($i < j$), z_{ij} ($i > j$)). For its detail, see Anderson (2003) or Siotani et al. (1985). □

[A.3.8] Let $S = HLH'$, where all matrices are $p \times p$, S is positive definite, H is orthogonal, and $L = \text{diag}(\ell_1, \ldots, \ell_p)$, $\ell_1 > \cdots > \ell_p$. Then

$$J(H, L \to S) = \frac{2^p \pi^{p^2/2}}{\Gamma_p(p/2)} \prod_{i<j} (\ell_i - \ell_j) g_p(H),$$

where $g_p(H)$ is a function of H and is the density function of the invariant measure over $O(p)$.

Proof. For a direct derivation, see Muirhead (1982). On the other side, Anderson (2003) has given an indirect proof.

Bibliography

Akaike, H. (1973), Information theory and an extension of the maximum likelihood principle, In *2nd International Symposium on Information Theory* (B.N. Petrov and F. Csáki, eds.). Akadémia Kiado, Budapest, Hungary, pp. 267–281, (194, 200, 302, 303)

Anderson, T.W. (1951), Estimating linear restrictions on regression coefficients for multivariate normal distributions, *Ann. Math. Statist.*, **22**, 327–351. (178)

Anderson, T.W. (1951a), Classification by multivariate analysis, *Psychometrika*, **16**, 31–50. (256)

Anderson, T.W. (1963), Asymptotic theory for principal components, *Ann. Math. Statist.*, **34**, 122–148. (291, 292, 297, 300, 301)

Anderson, T.W. (1973), An asymptotic expansion of the distribution of the Studentized classification statistic W, *Ann. Statist.*, **1**, 964–972. (147, 477)

Anderson, T.W. (2003), *Introduction to Multivariate Statistical Analysis*, 3rd ed., Wiley, Hoboken, NJ. (41, 161, 320, 326, 334, 511)

Arnold, S.F. (1981), *The Theory of Linear Models and Multivariate Analysis*, Wiley, Hoboken, N.J. (154)

Babu, G.J. (1989), Applications of Edgeworth expansions to bootstrap: a review., *Statistical Data Analysis and Inference*, (Y. Dodge, ed.), Elsevier, Amsterdam, pp. 223–237. (135)

Babu, G.J. and Bai, Z.D. (1993), Edgeworth expansions of a function of sample means under minimal moment conditions and partial Cramér's condition, *Sankhyā A*, **55**, 244–258. (108)

Babu, G.J. and Singh, K.(1984), On one term Edgeworth correction by Efron's bootstrap, *Sankkyā, A*, **46**, 219–232. (137)

Bai, Z.D. (1999), Methodologies in spectoral analysis of large dimensional random matrices: a review, *Statist. Sinica*, **9**, 611–677. (139, 140)

Bai, Z.D. and Rao, C.R. (1991), Edgeworth expansion of a function of sample means, *Ann. Statist.*, **19**, 1295–1315. (96)

Bai, Z.D. and Saranadasa, H. (1996), Effect of high dimension: by an example of a two sample problem, *Statistica Sinica*, **6**, 311–329. (181, 183)

Bai, Z.D., Yin, Y.Q. and Krishnaiah, P.R. (1987), On the limiting empirical distribution function of the eigenvalues of a multivariate F matrix, *Theory Probab. Appl.*, **32**, 490–500. (139)

Bai, Z.W., Yin, Y. and Krishnaiah, P.R. (1987), On the limiting empirical distribution of the eigenvalues of a multivariate F matrix, *Theory Probab. Appl.*, **32**, 490–500. (139)

Barnes, E.W. (1899), The theory of the gamma function, *Messeng. Math.*, **53**, 260–129. (113)

Bartlett, M.S. (1933), On the theory of statistical regression, *Proc. Roy. Soc. Edinb.*, **53**, 260–283. (38, 88)

Bedrick, E.J. and Tsai, C.L. (1994), Model selection for multivariate regession in small samples, *Biometrics*, **50**, 226–231. (201)

Bellman, R. (1960), *Introduction to Matrix Analysis*, MacGrow-Hill, New York (122, 503)

Berndt, E.R. and Savin, E. (1977), Conflict among criteria for testing hypotheses in the multivariate regression model, *Econometrica*, **45**, 1263-1277. (161)

Berry, A.C.(1941), The accuracy of the Gaussian approximation to the sum of independent variates, *Trans. Amer Math. Soc.*, **49**, 122–136. (383)

Bhattacharya, R.N. (1985), Some recent results on Cramér-Edgeworth expansions with applications, In *Multivariate Analysis*, Vol. VI (P.R. Krishnaiah, ed.). Elsevier, Amsterdam, pp. 57–75. (96)

Bhattacharya, R.N. (1987), Some aspects of Edgeworth expansions in statistics and probability, In *New Perspectives in Theoretical and Applied Statistics*, (M.L., Pure, J.P., Vilaplana, and W. Wertz eds), Wiley, New York. (137)

Bhattacharya, R.N. and Ghosh, J.K. (1978), On the validity of the formal Edgeworth expansion, *Ann. Statist.*, **6**, 434–451 (corrigendum ibid., **8**, 1980). (92, 96, 107)

Bhattacharya, R.N. and Ghosh, J.K. (1988), On moment conditions for valid formal Edgeworth expansions, *J. Multivariate Anal.*, **27**, 68–79. (108)

Bhattacharya, R.N. and Ranga Rao, R. (1976), *Normal Approximation and Asymptotic Expansions*, Wiley, New York. (437)

Bhattacharya, R.N.(1971), Rates of weak convergence and asymptotic expansions for classical central limit theorem, *Proc. Roy. Soc.Edinb.*, **53**, 260–283. (93)

Box, G.E.P. (1949), A general distribution theory for a class of likelihood criteria, *Ann. Math. Statist.*, **42**, 241–259. (114, 158)

Box, G.E.P. and Youle, P.V. (1995), The exploration of response surfaces: an example of the link between the fitted surface and the basic mechanism of the system, *Biometrics*, **11**, 287–323. (xv, 202)

Bulter, R.W. and Huzurbazar, S. (1992), Saddlepoint approximations for the generalalized variance and Wilks' statistic, *Biometrika*, **79**, 157–169. (117)

Chamber, J.M. (1967), On methods of asymptotic expansion for multivariate distributions, *Biometrika*, **36**, 317–346. (96, 125)

Chandra, T.K. and Ghosh, G.K. (1979), Valid asymptotic expansions for the likelihood ratio statistic and other perturbed chi-square variables, *Sankhyā A*, **41**, 22–47. (98, 106, 115)

Chase, G.R. and Bulgren, W.G. (1971), A Monte Carlo investigation of the robustness of T^2, *J. Amer. Statist. Assoc.*, **66**, 499–502. (167)

Chen, C.W. (1974), An optimal property of principal components, *Comm. Statist.*, **3**, 979–983. (288)

Chibisov, D.M.(1972), An asymptotic expansion for the distribution of a statistic admitting an asymptotic expasion, *Theory Probab. Appl.*, **17**, 620–630. (96)

Cochran, W.G. (1934), The distribution of quadratic forms in a normal system, with applications to the analysis of covariance, *Proceedings of the Cambridge Philosophical Society*, **30**, 178–191. (39)

Constantine, A.G. (1963), Some noncentral distribution problem in some multivariate analysis, *Ann. Math. Statist.*, **34**, 1270–1285. (158)

Cordeiro, G.M. and Ferrari, S.L.de P. (1991), A modified score test statistic having chi-squared distribution to order n^{-1}, *Biometrika*, **78**, 573–582. (134)

Cordeiro, G.M. and Ferrari, S.L.de P. (1998), A formula to improve score test statistics, *J. Statist. Comput. Simul.*, **62**, 123–136. (134)

Cornish, E.A. and Fisher, R.A. (1937), Moments and cumulants in the specification of distributions, *Rev. Inst. Internat. Statist.*, **4**, 307–320. (126)

Cramér, H. (1928), On the composition of elementary errors, *Skand. Aktuarietidskr. J.*, **11**, 13–74. (93)

Crowder, M.J. and Hand, D.J. (1990), *Analysis of Repeated Measures*, Chapman & Hall, London. (365)

Darroch, J.N. (1965), An optimum property of principal components, *Ann. Math. Statist.*, **36**, 1579–1582. (287)

Das Gupta, S. (1968), Some aspects of discrimination function coefficients, *Sankhyā A*, **30**, 387–400. (280)

Das Gupta, S. (1969), Properties of power functions of some tests concerning dispersion matrices of multivariate normal distributions, *Ann. Math. Statist.*, **40**, 697–701. (220)

Davis, C.S. (2002), *Statistical Methods for the Analysis of Repeated Measurements*, Springer-Verlag, New York. (365)

Deev, A.D. (1970), Representation of statistics of discriminant analysis and asymptotic expansions when space dimensions are comparable with sample size, *Soviet Math. Dokl.*, **11**, 1547–1550. (263)

Dempster, A.P. (1958), A high dimensional two sample significance test, *Ann. Math. Statist.*, **29**, 995–1010. (181)

Dempster, A.P. (1960), A significance test for the separation of two highly multivariate small samples, *Biometrics*, **16**, 41–50. (181)

Dempster, A.P. (1972), Covariance selection, *Biometrika*, **28**, 157–175. (83)

Diciccio, T.J. and Romano, J.P. (1988), A review of bootstrap confidence intervals, *J. R. Statist. Soc. B*, **50**, 338–354. (135)

Dobrić, V. and Ghosh, B.K.(1996), Some analogs of the Berry-Esseen bound for first-order Chebyshev-Edgeworth expansions, *Statistics & Decision*, **14**, 383–404. (383)

Edgeworth, E.Y. (1905), The law of errors, *Trans. Cambridge Philos. Soc.*, **20**, 63-65, 113–141. (93)

Efron, B. (1979), Bootstrap methods: another look at the jackknife, *Ann. Statist.*, **7**, 1–26. (135, 304)

Elston, R.C. and Grizzle, J.E. (1962), Estimation of time–response curves and their confidence bands, *Biometrics*, **18**, 148–159. (xv, 350)

Esseen, C.G.(1942), On the Liapounoff limit of error in the theory of probability, *Arkiv Mat., Astr. och Fysik*, **28A**, 1–19. (383)

Everitt, B.S. (1979), A Monte Carlo investigation of the robustness of Hotelling's one- and two-sample T^2 tests, *J. Amer. Statist. Assoc.*, **74**, 48–51. (167)

Ferré, L. (1995), Selection of components in principal component analysis: a comparison of methods, *Computat. Statist. & Data Anal.*, **19**, 669–682. (310)

Fisher, N.I. and Hall, P. (1990), On bootstrap hypothesis testing, *Austral. J. Statist.*, **32**, 177–190. (167, 168)

Fisher, R.A. (1921), On the "probable error" of a coefficient of correlation coefficient in samples from infinitely large population, *Metron*, **1**, 1–32. (74)

Fisher, R.A. (1925), Expansion of Student's integral in powers of n^{-1}, *Metron*, **5**, 109–112. (419)

Fisher, R.A. (1936), The use of multiple measurements in taxonomic problems, *Ann. Eugen*, **9**, 238–249. (51, 254, 272)

Fisher, R.A. and Cornish, E.A. (1960), The percentile points of distributions having known cumulants, *Technometrics*, **2**, 209–225. (126)

Flurly, B.N. (1985), Anlysis of linear combinations with extreme ratios of variance, *J. Amer. Statist. Assoc.* (307)

Flurly, B.N. (1986), An asymptotic test for redundancy of variables in the comparison of two covariance matrices, *Statist. Probab. Lett.*, **4**, 123–126. (307)

Friedman, J.H. (1989), Regularized discriminant analysis., *J. Amer. Statist. Assoc.*, **84**, 165–175. (278)

Fujikoshi, Y. (1970), Asymptotic expansions of the distributions of test statistics in multivariate analysis, *J. Sci. Hiroshima Univ. A-I*, **34**, 73–144. (158, 446)

Fujikoshi, Y. (1974), The likelihood ratio tests for the dimensionality of regression coefficients, *J. Multivariate Anal.*, **4**, 327–340. (178)

Fujikoshi, Y. (1977), Asymptotic expansion for the distributions of the latent roots in MANOVA and the canonical correlations, *J. Multivariate Anal.*, **7**, 386–396. (179)

Fujikoshi, Y. (1977a), An asymptotic expansion for the distributions of the latent roots of the Wishart matrix with multiple population roots, *Ann. Inst. Statist. Math.*, **29**, 379–387. (123)

Fujikoshi, Y. (1977b), Asymptotic expansions for the distributions of some multivariate tests, In *Multivariate Analysis, Vol. IV*, (P.R. Krishnaiah, ed.), North-Holland, Amsterdam, pp. 55–71. (123, 300)

Fujikoshi, Y. (1978), Asymptotic expansions for the distributions of some functions of the latent roots of matrices in three situations, *J. Multivariate Anal.*, **8**, 63–72. (297, 298)

Fujikoshi, Y. (1980), Asymptotic expansions for the distributions of the sample roots under nonnormality, *Biometrika*, **67**, 45–51. (296)

Fujikoshi, Y. (1982), A test for additional information in canonical correlation analysis, *Ann. Inst. Statist. Math.*, **34**, 137–144. (269, 339)

Fujikoshi, Y. (1985), Selection of variables in discriminant analysis and canonical correlation analysis, In *Multivariate Analysis, Vol. VI* (P. R. Krishnaiah, ed.), North-Holland, pp. 219–236. (271)

Fujikoshi, Y. (1985a), Selection of variables in two-group discriminant analysis by error rate and Akaike's information criteria, *J. Multivariate Anal.*, **17**, 27–37. (272)

Fujikoshi, Y. (1985b), An error bound for an asymptotic expansion of the distribution function of an estimate in a multivariate linear model, *Ann. Statist.*, **13**, 827–831. (383)

Fujikoshi, Y. (1987), Error bounds for asymptotic expansions of the distributions of the MLE in a GMANOVA model, *Ann. Inst. Statist. Math.*, **39**, 153–161. (383)

Fujikoshi, Y. (1988), Comparison of powers of a class of tests for multivariate linear hypothesis and independence, *J. Multivariate Anal.*, **26**, 48–58. (160)

Fujikoshi, Y. (1988a), Asymptotic power comparison of some tests in MANOVA and canonical correlation models, In *Statistical Theory and Data Analysis, Vol. II*, (K. Matusita, ed.), North-Holland, pp.327–336. (161)

Fujikoshi, Y. (1992), Redundancy of some variables in multivariate analysis, *Japan. J. Behaviormetrics* (in Japanese), **19**, 18–28. (269)

Fujikoshi, Y. (1993), Two-way ANOVA models with unbalanced data, *Discrete Math.*, **116**, 315–334. (153, 154)

Fujikoshi, Y. (1993a), Error bounds for approximations to the distributions of the standardized and studentized estimates in a multivariate linear model, In *J. Statist. Plann. Inference*, **36**, 165–174. (488)

Fujikoshi, Y. (1993b), Error bounds for asymptotic approximations of some distribution functions, *Multivariate Analysis: Future Directions*, (C.R. Rao, ed.), North-Holland, Amsterdam, pp. 181–208. (384)

Fujikoshi, Y. (1994), Error bounds for asymptotic expansions of the distributions of the classification statistic W and related statistics., In *New Trends in Probability and Statistics, Vol. 3.* (E.M. Titt et al., eds.), **3**, pp. 3–15. (468)

Fujikoshi, Y. (1997), An asymptotic expansion of the distribution of Hotelling's T^2-statistic., *J. Multivariate Anal.*, **61**, 187–193. (133, 134)

Fujikoshi, Y. (1997a), A method for improving the large-sample chi-squared approximations to some multivariate tests statistics, *Amer. J. Math. Management Sci.*, **17**, 15–29. (111)

Fujikoshi, Y. (2000), Error bounds for asymptotic approximations of the linear discriminant function when the sample size and dimensionality are large, *J. Multivariate Anal.*, **73**, 1–17. (468)

Fujikoshi, Y. (2002), Asymptotic expansions for the distributions of multivariate basic statistics and one-way MANOVA tests under nonnormality, *J. Statist. Plann. Inference.*, **108**, 263–282. (165)

Fujikoshi, Y. (2002b), Selection of variables for discriminant analysis in a high-dimensional case, *Sankhyā A*, **64**, 256–257. (280)

Fujikoshi, Y. (2002c), Some recent results on asymptotic expansions of multivariate test statistics for mean vectors under nonnormality, *Calcutta Statist. Assoc. Bull.* (Special 4th Triennial Calcutta Proceedings Volume), **52**, 1–46. (165)

Fujikoshi, Y. (2003), Model selection criteria for growth curve model with hierachical within-individuals design matrices, In *New Developments on Psychometrics* (H. Yanai et al. eds.), Springer-Verlag, Tokyo, pp. 433–442. (375)

Fujikoshi, Y. (2004), Multivariate analysis for the case when the dimensionality is large compared to the sample size, *J. Korean Statist. Soc.*, **33**, 1–24. (183)

Fujikoshi, Y. (2009), Selection of components in principal component analysis based on model selection method,, In *Proceedings of International Conference on Multivariate Statistical Modeling and High Dimensional Data Mining*, Kayeseri, June 19–23, 2008. To appear. (304)

Fujikoshi, Y. and Kurata, H. (2008), Information criterion for some conditional independence structures, In *New Trends in Psychometrics* (K. Sigemasu et al. eds.), Universal Academy Press, Tokyo. pp. 69–79. (304)

Fujikoshi, Y. and Rao, C.R. (1991), Selection of covariables in the growth curve model, *Biometrika*, **78**, 779–785. (371)

Fujikoshi, Y. and Sakurai, T. (2008), High-dimensional asymptotic expansions for the distributions of canonical correlations, *J. Multivariate Anal.*, **100**, 231–241. (333, 334)

Fujikoshi, Y. and Satoh, K. (1997), Modified AIC and C_p statistic in multivariate regression models, *Biometrika*, **84**, 707–716. (199)

Fujikoshi, Y. and Seo, T. (1998), Asymptotic approximations for EPMC's of the linear and the quadratic discriminant functions when the sample sizes and the dimension are large, *Random Oper. Stochastic Equations*, **6**, 269–280. (263, 264, 475, 476)

Fujikoshi, Y. and Shimizu, R. (1989), Asymptotic expansions of some mixtures of the multivariate normal distribuion and their error bounds, *Ann. Statist.*, **17**, 1124–1132. (420, 488)

Fujikoshi, Y. and Shimizu, R. (1990), Asymptotic expansions of some distributions and their error bounds: the distributions of sums of independent random variables and scale mixtures, *Sugaku Expositions*, **3**, 75–96. (384)

Fujikoshi, Y. and Ulyanov, V.V. (2006), On accuracy of asymptotic expansion for Wilks' lambda distribution, *J. Multivariate Anal.*, **97**, 1941–1957. (442)

Fujikoshi, Y. and Ulyanov, V.V. (2006a), On accuracy of approximations for location and scale mixture, *J. Math. Sci.*, **138**, 5390–5395. (468)

Fujikoshi, Y. and Veitch, L.G. (1979), Estimation of dimensionality in canonical correlation analysis, *Biometrika*, **66**, 345–351. (274)

Fujikoshi, Y., Himeno, T. and Wakaki, H. (2004), Asymptotic results of a high dimensional MANOVA test and power comparison when the dimension is large, *J. Japan Statist. Soc.*, **34**, 19–26. (183)

Fujikoshi, Y., Himeno, T. and Wakaki, H. (2008), Asymptotic results in canonical discriminant analysis when the dimension is large compared to the sample, *J. Statist. Plann. Inference*, **138**, 3457–3466. (176)

Fujikoshi, Y., Kanda, T. and Ohtaki, M. (1999), Growth curve models with hierachikal within-individuals design, *Ann. Inst. Statist. Math.*, **51**, 707–721. (373)

Fujikoshi, Y., Kanda, T. and Ohtaki, M. (2006), LR tests for some linear hypotheses in an extended growth curve model, *Amer. J. Math. Management Sci.*, **26**, 211–227. (373)

Fujikoshi, Y., Ulyanov, V.V., Shimizu, R. (2005), L_1-norm error bounds for asymptotic expansions of multivariate scale mixtures and their applications to generalized Hotelling's T_0^2, *J. Multivariate Anal.*, **96**, 1–19. (442)

Fujikoshi, Y., Yamada, T., Watanabe, D. and Sugiyama, T. (2007), Asymptotic distribution of LR statistic for equality of the smallest eigenvalues in high-dimensional principalcomponent analysis, *J. Multivariate Anal.*, **98**, 2002-2008. (302)

Fujisawa, H. (1997), Improvement on chi-squared approximation by monotone transformation, *J. Multivariate Anal.*, **60**, 84–89. (134)

Geisser, S. and Eddy, W. F. (1979), A predictive approach to model selection, *J. Amer. Statist. Assoc.*, **74**, 153–160. (194)

Geman, S. (1980), A limit theorem for the norm of random matrices, *Ann. Probab.*, **8**, 252–261. (140)

Ghosh, D. (2003), Penalized discriminant methods for the classification of tumors from gene expression data, *Biometrics*, **59**, 992–1000. (278)

Gleser, I.J. and Olkin, I. (1970), Linear models in multivariate analysis, In *Essays in Probability and Statistics*, (R.C. Bose, ed.), University of North Carolina Press, Chapel Hill, NC, pp. 267–292. (365, 488)

Gleser, L. J. (1966), A note on sphericity test, *Ann. Math. Statist.*, **37**, 464–467. (corrigendum ibid., **39** (1968), p.684). (228)

Gleser, L.J. and Olkin, I. (1972), Estimation for regression model with an unknown covariance matrix, *Proc. Sixth Berkeley Symposium on Mathematical Statistics and Probability I*, University of California, Berkeley, CA, pp. 541–568. (414)

Grizzle, J.E. and Allen, D.M. (1969), Analysis of growth and dose response curves, *Biometrics*, **25**, 357–381. (365, 371)

Gupta, A.K. and Nagar, D.K. (2000), *Matrix Variate Distributions*, Chapman & Hall/CRC, Monographs and Surveys in Pure and Applied Mathematics, Vol. 104, Boca Raton, FL. (35)

Hall, P. (1979), On measures of the distance of a mixture from its parent distribuion, *Stochastic Process Appl.*, **8**, 357–365. (108, 383)

Hall, P. (1983), Inverting an Edgeworth expansion, *Ann. Statist.*, **11**, 569–576. (130)

Hall, P. (1987), Edgeworth expansion for Student's statistic under minimal moment conditions, *Ann. Probab.*, **15**, 920–931. (108)

Hall, P. (1992), *The Bootstrap and Edgeworth Expansion*, Springer-Verlag, New York (134)

Hall, P. (1992a), On the removal of skewness by transformation, *J. R. Statist. Soc. B*, **54**, 221–228. (132)

Hall, P. and Heyde, C.C. (1980), *Multingale Limit Theory and its Applications*, Academic Press, New York. (144)

Hall, P.(1988), Theoretical comparison of bootstrap confidence intervals, *Ann. Statist.*, **16**, 927–953. (135)

Hall, P., Marron, J.S. and Neeman, A. (2005), Geometric representation of high dimension, low sample size data, *J. R. Statist. Soc. B*, **67**, 427–444. (145)

Hastie, T., Buja, A. and Tibshirani, R. (1995), Penalized discriminant analysis, *Ann. Statist.*, **23**, 173–102. (278)

Hastie, T., Tibshirani, R. and Buja, A. (1994), Flexible discriminant analysis by optimal scoring, *Amer. Statist. Assoc.*, **89**, 1255–1270. (278)

Heyde, C.C. (1975), Kurtosis and departure form normality, In *Statistical Distributions in Scientific Work, Vol. 1* (G.P. Patil et al., eds.), Reidel, Dordrecht, The Netherland, pp. 193–201. (383)

Heyde, C.C. and Leslie, J.R. (1976), On moment measures of departure from the normal and exponential laws, *Stochastic Process. Appl.*, **4**, 317–328. (383)

Hill, G.W. and Davis, A.W. (1968), Generalized asymptotic expansions of Cornish-Fisher type, *Ann. Math. Statist.*, **39**, 1264–1273. (126, 127)

Hinkley, D.W. (1988), Bootstrap methods, *J. R. Statist. Soc. B*, **50**, 321–337. (135)

Hoerl, A.E. and Kennard, R.W. (1970), Ridge regression: Biased estimation for nonorthogonal problems, *Technometrics*, **12**, 55–67. (214)

Hotelling, H. (1933), Analysis of a complex of statistical variables into principal components, *J. Educ. Psych.*, textbf24, 417–441, 498–520. (283)

Hotelling, H. (1936), Relationships between two sets of variates, *Biometrika*, **28**, 321–377. (48, 319)

Hsu, P.L. (1940), An algebraic derivation of the distribution of rectangular coordinates, *Proc. Edinb. Math. Soc.*, **6**, 185–189. (30)

Hsu, P.L. (1941), On the limiting distribution of roots of the canonical correlations, *Biometrika*, **32**, 38–45. (327)

Hurvich, C.M. and Tsai, C.L. (1989), Regression and time series model selection in small samples, *Biometrika*, **76**, 297–307. (202)

Ichikawa, M. and Konishi, S. (1999), Model evaluation and information criteria in covariance structure analysis, *British J. Math. & Statist.l Psych.*, **52**, 285–302. (304)

Ito, K. (1956), Asymptotic formulae for the distribution of Hotelling's generalized T_0^2-statistic, *Ann. Math. Statist.*, **27**, 1091–1105. (158)

Ito, K. (1960), Asymptotic formulae for the distribution of Hotelling's generalized T_0^2-statistic-II, *Ann. Math. Statist.*, **31**, 1148-1153. (158)

Iwashita, T. (1997), Asymptotic null and nonnull distributions of Hotelling's T^2-statistic under the elliptical distribution, *J. Statist. Plann. Inference*, **61**, 85–104. (111)

James, A.T. (1954), Normal multivariate analysis and the orthogonal group, *Ann. Math. Statist.*, **25**, 40–75. (31)

James, A.T. (1960), The distribution of the latent roots of the covariance matrix, *Ann. Math. Statist.*, **31**, 151-158. (158, 293)

James, A.T. (1964), Distributions of matrix variates and latent roots derived from normal samples, *Ann. Math. Statist.*, **35**, 475–501. (158)

John, S. (1960), On some classification problem-I, *Sankhyā*, **22**, 301–388. (258)

Johnstone, I. M. and Lu, A. Y. (2009), On consistency and Sparsity for principal components analysis in high dimensions, *J. Amer. Statist. Assoc.*, **104**, 682–693. (302)

Johnstone, I.M. (2001), On the distribution of the largest eigenvalue in principal component analysis, *Ann. Statist.*, **29**, 295–327. (140)

Jolliffe, I.T. (2002), *Principal Component Analysis* (2nd ed.), Springer, New York. (310)

Jonsson, D. (1982), Some limit theorems for eigenvalues of a sample covariance matrix, *J. Multivariate Anal.*, **12**, 1–38. (139, 142)

Kabe, D.G. (1964), A note on the Bartlett decomposition of a Wishart matrix, *J. Roy. Statist. Soc. Ser. B*, **26**, 270–273. (34)

Kano, Y. (1995), An asymptotic expansion of the distribution of Hotelling's T^2-statistic under general conditions, *Amer. J. Math. Management Sci.*, **15**, pp. 317–341. (111)

Kariya, T. and Eaton, M.L.(1977), Rubust tests for sperical symmetry, *Ann. Statist.*, **5**, 206–215. (72)

Kariya, T. and Kurata, H. (2004), *Generalized Least Squares*, Wiley, Hoboken, N.J. (383)

Kariya, T. and Maekawa, K. (1982), A method for approximations to the pdf's and cdf's of GLSE's and its application to the seemingly unrelated regression model, *Ann. Inst. Statist. Math. A*, **34**, 281–297. (383, 491, 492)

Kato, N., Yamada, T. and Fujikoshi, Y. (2009), High-dimensional asymptotic expansion of LR statistic for testing the intraclass correlation structure and its error bound, In *J. Multivariate Anal.* To apperar. (119, 235, 439)

Kendall, M.G. (1966), Discrimination and classification, In *Multivariate Analysis* (P. R. Krichnaiah, ed.), Academic Press, New York. (249)

Kendall, M.G. (1975), *Multivariate Analysis*, Charles Griffin, London (272)

Kenward, M.G. (1986), The distribution of a generalized least squares estimator with covariance adjustment., *J. Multivariate Anal.*, **20**, 244–250. (488)

Khatri, C.G. (1966), A note on a MANOVA model applied to problems in growth curve, *Ann. Inst. Statist. Math.*, **18**, 75–86. (365)

Kleisen, J. and Steutel, F.W. (1974), Mixtures of distributions, moment inequalities and measures of exponentiality and normality, *Ann. Probab.*, **2**, 112–130. (383)

Konishi, S. (1977), Asymptotic expansions for the distribution of a function of latent roots of the covariance matrix, *Ann. Inst. Statist. Math*, **29**, 389–396. (298)

Konishi, S. (1978), An approximation to the distribution of the sample correlation coefficient, *Biometrika*, **65**, pp. 654–656. (75)

Konishi, S. (1991), Normalizing transformations and bootstrap confidence intervals, *Ann. Statist.*, **19**, 2209–2225. (135)

Konishi, S. and Kitagawa, G. (1996), Generalized information criterion in model selection, *Biometrika*, **83**, 875-890. (304)

Kotz, S. and Nadarajah, S. (2004), *Multivariate t-Distributions and Their Applications*, Cambridge University Press, Cambridge, UK. (427)

Kshirsagar, A.M. (1972), *Multivariate Analysis*, Marcel Dekker, New York. (260, 288)

Kudo, A. (1959), The classificatory problem vied as a two-decision problem, *Memoirs of Faculty of Science, Kyusyu Univ. A*, **13**, 96–125. (258)

Kudo, A. (1960), The classificatory problem vied as a two-decision problem-II, *Memoirs of Faculty of Science, Kyusyu Univ. A*, **14**, 63–83. (258)

Kurata, H. and Kariya, T. (1996), Least upper bounds for the covariance matrix of a generalized least squares estimator in regression with applications to a seemingly unrelated regression model and a heteroscedastic model, *Ann. Statist.*, **24**, 1547–1559. (491)

Lachenbruch, P.A. (1968), On expected probabilities of misclassification in discriminant analysis, necessary sample size, and a relation with the multiple correlation coefficients, *Biometrics*, **24**, 823–834. (264, 475)

Lawley, D.N. (1956), Tests of significance for the latent roots of covariance and correlation matrices, *Biometrika*, **43**, 128–136. (123, 300)

Lawley, D.N. (1959), Tests of significance in canonical analysis, *Biometrika*, **46**, 59–66. (179)

Ledoit, O. and Wolf, M. (2002), Some hypothesis tests for the covariance matrix when the dimension is large compared to the sample size, *Ann. Statist.*, **30**, 1081–1102. (140, 142, 226, 230)

Lukacs, E. (1970), *Characteristic Functions*, Charles Griffin, London. (5)

Mallows, C.L. (1973), Some comments on C_p, *Technometrics*, **15**, 661–675. (194, 198)

Mallows, C.L. (1995), More comments on C_p, *Technometrics*, **37**, 362–372. (198)

Marčenko, V.A. and Pastur, L.A.(1967), Distribution for some sets of random matrices, *Math. USSR-Sb*, **1**, 457–483. (139)

Mardia, K.V. (1970), Measures of multivariate skewness and kurtosis with applications, Biometrika, **57**, 519-530. (22)

Mardia, K.V., Kent, J.T. and Bibby, J.M. (1979), *Multivariate Analysis*, Academic Press, New York. (51, 79, 272)

Mathai, A.M., Provost, S.B. and Hayakawa, T. (1995), *Bilinear Forms and Zonal Polynomials*, Lecture Notes in Statistics, **102**, Springer-Verlag, new York. (35, 446)

McLachlan, G.J. (1973), An asymptotic expansion of the expectation of the estimated error rate in discriminant analysis, *Austral. J. Statist.*, **15**, 210–214. (261, 478)

McLachlan, G.J. (1991), *Discriminant Analysis and Statistical Pattern Recognition*, Wiley, Hoboken, NJ. (250)

Memon, A. Z. and Okamoto, M. (1971), Asymptotic expansion of the distribution of the Z statistic in discriminant analysis, *J. Multivariate Anal.*, **1**, 294–307. (262)

Montel, J. (1971), A new analysis of variance model for non-additivedata, *Technometrics*, **13**, 1–18. (309)

Mudholkar, G.S. and Trivedi, M.C. (1980), A normal approximation for the distribution of the likelihood ratio statistic in multivariate analysis of variance, *Biometrika*, **67**, 485–488. (161)

Muirhead, R.J. (1970), Asymptotic distributions of some multivariate tests, *Ann. Math. Statist.*, **41**, 1002–1010. (158)

Muirhead, R.J. (1982), *Aspect of Multivariate Statistical Theory*, Wiley, New York. (15, 16, 37, 223, 288, 293, 323, 414)

Mukaihata, S. and Fujikoshi, Y. (1993), Error bounds for asymptotic expansions of some distributions in a SUR model, In *Stat. Sci. & Data Anal.* (K. Matsusita et al., ed.), VSP, Utrecht, pp.365–374, (491, 492)

Nagao, H. (1973), Some test criteria for covariance matrix, *Ann. Statist.*, **1**, 700–709. (226)

Obenchain, R. L. (1972), Regression optimality of principal componets, *Ann. Math. Statist.*, **43**, 1317–1319. (288)

Okamoto, M. (1963), An asymptotic expansion for the distribution of the linear discriminant function, *Ann. Math. Statist.*, **34**, 1286–1301. (259)

Okamoto, M. (1969), Optimality of principal components, *Multivariate Analysis-II*(P. R. Krishnaiah, ed.), Academic Press, New York, 673–685. (259, 287, 288)

Okamoto, M. and Kanazawa, M. (1968), Minimization 0f eigenvalues of matrix and optimality of principal components, *Ann. Math. Statist.*, **30**, 859–863. (287, 505)

Olkin, I. and Sampson, A. R. (1972), Jacobians of matrix transformationsand induced functional equations, *Linear Algebra and Its Applications*, **5**, 257–276. (509)

Pearson, K. (1901), On lines and planes of closest fit to systems of points in space, *Phil. Mag.*, **2**, 559–572. (283)

Perlman, M.D. (1980), Unbiasedness of the likelihood-ratio tests for equality of several covariance matrices and equality of several multivariate normal populations, *Ann. Statist.*, **8**, 247–263. (242)

Phillips P.C.B. (1977), Approximations to some finite sample distributions asociated with a first order stochastic difference equation, *Econometrica*, **45**, 463–485. (96)

Phillips P.C.B. (1977a), A general theorem in the theory of asymptotic expansions as approximations to the finite sample distributions of econometric estimators, *Econometrica*, **45**, 1517–1534. (96)

Potthoff, R.F. and Roy, S.N. (1964), A generalized multivariate analysis of variance model useful especially for growth curve problems, *Biometrika,*, **51**, 313–326. (xv, 363, 487)

Pynnönen, S. (1987), Selection of variables in nonlinear discriminant analysis by information criteria, *Proceedings of the Second International Tampare Conference in Statistics*(T. Pukkila and S. Puntanen, Ed.), Univ. of Tampere, 627–636. (281)

Rao, C.R. (1948), Tests of significance in multivariate analysis, *Biometrika*, **35**, pp. 58–79. (59)

Rao, C.R. (1951), An asymptotic expansion of the distribution of Wilk's citerion, *Bulletin of the International Statistical Institute*, **33**,Part 2, 177–180 (116)

Rao, C.R. (1959), Some problems involving linear hypotheses in multivariate analysis, *Biometrika*, **46**, 49–58. (351, 365, 487)

Rao, C.R. (1964), The use and interpretation of principal component analysis in applied research, *Sankhyā A*, **26**, 329–358. (269, 287)

Rao, C.R. (1965), Theory of least squares when the parameters are stochastic and its application to the analysis of growth curves, *Biometrika*, **52**, 447–458. (365)

Rao, C.R. (1970), Inference on discriminant function coefficients, In *Essays in Prob. and Statist.* (R. C. Bose, ed.), 587–602. (59, 268, 281)

Rao, C.R. (1973), *Linear Statistical Inference and Its Applications*, 2nd ed., Wiley, New York. (312, 503, 505)

Rao, C.R. and Slater, P. (1949), Multivariate analysis applied to differences between neurotic groups, *British J. Psych.*, **2**, 17–29. (274)

Raudys, S. (1972), On the amount of priori information in designing the classification algorithm (in Russian), *Tech. Cybernet*, **4**, 168–174. (263, 264, 475)

Raudys, S. and Young, D.M. (2004), Results in statistical discriminant analysis: a review of the former Soviet Union Literature, *J. Multivariate Anal.*, **89**, 1–35. (263)

Rencher, A.C. (2002), *Methods of Multivariate Analysis*, 2nd ed., Wiley, Hoboken, NJ. (xv, 202)

Rothenberg, T.J. (1977), Edgeworth expansions for multivariate test statistics, *IP-255*. Center for Research in Management Science, University of California, Berkeley, CA. (159)

Sakurai, T. (2009), Asymptotic expansions of test statistics for dimensionality and additional information in canonical correlation analysis when the dimension is large, *J. Multivariate Anal.*, **100**, pp. 888–901. (342)

Saranadasa, H. (1993), Asymptotic expansion of the misclassification probabilities of D- and A- criteria for discrimination from two high dimensional populations using the theory of large dimensional random matrix, *J. Multivariate Anal.*, **46**, 154–174. (278, 279)

Sargan J.D. (1976), Econometric estimators and the Edgeworth approximation, *Econometrica*, **44**, 421–445. (96)

Schoott, J.R. (2001), Some tests for the equality of covariance matrices, *J. Statist. Plann. Inference*, **94**, 25–36. (245)

Schott, J.R. (2005), Testing for complete independence in high dimensions, *Biometrika*, **92**, 951–956. (143, 144, 239, 240)

Schott, J.R. (2006), Some high-dimensional tests for a one-way MANOVA, *J. Multivariate Anal.*, **98**, pp. 1825–1839. (181, 183)

Schott, J.R. (2006a), A high dimensional test for the equality of the smallest eigenvalues of a covariance matrix, *J. Multivariate Anal.*, **97**, 827–843. (301)

Schott, J.R. (2007), A test for the equality of covariance matrices when the dimension is large relative to the sample sizes, *Comput. Statist. Data Anal.*, **51**, pp. 6535–6542. (245)

Serfling, R.V. (1980), *Approximation Theorems of Mathematical Statistics*, Wiley, New York. (41)

Shimizu, R. (1974), On the remainder term for the central limit theorem, *Ann. Inst. Statist. Math.*, **26**, 195–201. (383)

Shimizu, R. (1987), Error bounds for asymptotic expansion of the scale mixtures of the normal distribution, *Ann. Inst. Statist. Math.*, **39**, 611–622. (383)

Shimizu, R. (1989), Expansion of scale mixtures of the gamma distribution, *J. Statist. Plann. Inference*, **21**, 305–314. (383)

Shimizu, R. (1995), Expansion of the scale mixture of the multivariate normal distribution with error bound evaluated in L_1-norm, *J. Multivariate Anal.*, **53**, 126–138. (384, 488)

Shimizu, R. and Fujikoshi, Y. (1997), Sharp error bounds for asymptotic expansions of the distribution functions of scale mixtures, *Ann. Inst. Statist. Math.*, **49**, 285–297. (384)

Singh, K. (1981), On the asymptotic accuracy of Efron's bootstrap, *Ann. Statist.*, **9**, 1187–1195. (137)

Siotani, M. (1956), On the distribution of the Hotelling's T^2-test, *Ann. Inst. Statist. Math.*, **8**, 1–14. (158)

Siotani, M. (1957), Note on the generalized Student ratio in the analysis of variance or dispersion, *Proc. Inst. Statist. Math.*, **9**, 157–171. (158)

Siotani, M. (1982), Large sample approximations and asymptotic expansions of classification statistic, *Handbook of Statistics*, Vol. 2 (P.R. Krishnaiah and L.N. Kanal, eds.), North-Holand, Amsterdam, pp. 47–60. (258)

Siotani, M. (1990), *Tahenryo-kaiseki Gairon*, Asakura Publishing Company (in Japanese). (334)

Siotani, M. and Wang, R.H. (1977), Asymptotic expansionsfor error rates and comparison of the W-procedure and the Z-procedure in discriminant analysis, *Multivariate Analysis-IV*(P. R. Krishnaiah, ed.), North-Holland, Amsterdam, 523–545. (259, 262)

Siotani, M., Hayakawa, T. and Fujikoshi, Y. (1985), *Modern Multivariate Statistical Analysis: A Graduate Course and Handbook*, American Sciences Press, Columbus, OH. (37, 39, 72, 309, 320, 326, 332, 511)

Srivastava, M.S. (2002), Methods of Multivariate Statistics, Wiley, Hobohen, NJ. (215, 327, 480, 487)

Srivastava, M.S. (1987), Profile analysis of several groups, *Comm. Statist. A*, **16**, 909–926. (480)

Srivastava, M.S. (2007), Multivariate theory for analyzing high dimensional data, *J. Japan Statist. Soc.*, **37**, 53–86. (279, 280)

Srivastava, M.S. and Yau, W.K. (1989), Saddlepoint method for obtaining tail probability of Wilks' likelihood ratio test, *J. Multivariate Anal.*, **31**, 117–126. (117, 161)

Stone, M. (1974), Cross-validatory choice and assesment of statistical predictions (with discussion), *J. Royal Statist. Soc. B*, **36**, 111–147. (260)

Sugiura, N. (1973), Derivatives of the characteristic root of a symmetric or hermitian matrix with two applications in multivariate analysis, *Comm. Statist.*, **1**, 393–417. (122, 296)

Sugiura, N. (1976), Asymptotic expansions of the distributions of the canonical correlations, *Essay in Memory of Establishment of Faculty of Integrated Science*, Hiroshima University, pp. 141–163. (327)

Sugiura, N. (1976a), Asymptotic expansions of the distributions of the latent roots and latent vectors of the Wishart and multivariate F matrices, *J. Multivariate Anal.*, **6**, 500–525. (296)

Sugiura, N. (1978), Further analysis of the data by Akaike's information criterion and the finite corrections, *Comm. Statist. A*, **7**, 500–525. (201, 202)

Sugiura, N. and Fujikoshi, Y. (1969), Asymptotic expansions of the non-null distributions of the likelihood ratio criteria for multivariate linear hypothesis and independence, *Ann. Math. Statist.*, **40**, 942–952. (158, 325)

Sugiura, N. and Nagao, H. (1968), Unbiasedness of some test criteria for the equality of one or two covariance matrices, *Ann. Math. Statist.*, **39**, 1686–1692. (220, 228)

Sugiyama, T. and Tong, H. (1976), On a statistic useful in dimensionality reduction in multivariable linear stochastic systems, *Comm. Statist. A*, **5**, 711–721. (298)

Takeuchi, K. and Takemura, A. (1988), Some results on univariate and multivariate Cornish-Fisher expansion: algebraic properties and validity, *Sankhyā A*, **50**, 111–136. (130)

Taylor, D.E. (1981), Asymptotic inference for eigenvectors, *Ann. Statist.*, **9**, 725–736. (307)

Tonda, T. and Fujikoshi, Y. (2004), Asymptotic expansion of the null distribution of LR statistic for multivariate linear hypothesis when the dimension is large, *Comm. Statist.A̧@A*, **33**, 1205–1220. (161, 163)

Ulyanov, V.V. and Fujikoshi, Y. (2001), On accuracy of improved χ^2-approximat -ions, *Georgian Math. J.*, **8**, 401–414. (135)

Ulyanov, V.V., Wakaki, H. and Fujikoshi.Y. (2006), Berry-Esseen bound for high dimensional asymptotic approximation of Wilks' lambda distribution, *Statist. Probab. Lett.*, **76**, 1191–1200. (130)

Verbyla, A.P. and Venables, W.N. (1988), An extension of the growth curve models, *Biometrika*, **75**, 129–138. (373)

Wakaki, H. (2009), Error bounds for high-dimensional Edgeworth expansions for some tests on covariance matrice, In *Proceedings of International Conference on Multivariate Statistical Modeling and High Dimensional Data Mining*, Kayeseri, June 19–23, 2008. To appear. (119, 120, 225, 230, 244, 439)

Wakaki, H., Eguchi, S. and Fujikoshi, Y. (1990), A class of tests for a general covariance structure, *J. Multivariate Anal.*, **32**, 313–325. ()

Wakaki, H., Fujikoshi, Y. and Ulyanov, V. (2003), Asymptotic expansions of the distributions of MANOVA test statistics when the dimension is large, *Technical Report 02-9*, Statistical Research Group, Hiroshima University, Japan. (57, 162, 163)

Wakaki, H., Yanagihara, H. and Fujikoshi, Y. (2002), Asymptotic expansions of the null distributions of test statistics for multivariate linear hypothesis under nonnormality, *Hiroshima Math. J.*, **32**, 17–50. (165)

Wald, A. (1944), On a statistical problem arising in the classification of an individual into one of two groups, *Ann. Math. Statist.*, **15**, 145–163. (256)

Wallace, D.L. (1958), Asymptotic approximations to distributions, *Ann. Math. Statist.*, **29**, 635–654. (96, 129)

Wallace, D.L. (1959), Bounds on normal approximations to Student's and the chi-square distributions, *Ann. Math. Statist.*, **30**, 1121–1130. (129)

Watamori, Y. (1990), On the moments of traces of Wishart and inverted Wishart matrices, *South African Statist. J.*, **24**, 153–176. (414)

Wishart, J. (1938), Growth rate determination in nutrition studies with the bacon pig, and their analysis, *Biometrika*, **30**, 16–28. (365)

Wyman, F.J., Young, D.M. and Turner, D.W. (1990), A comparison of asymptotic error rate expansions for the sample linear discriminant function, *Pattern Recognition*, **23**, 775–783. (264, 475)

Yin, Y.Q. (1986), LSD' for a class of random matrices, *J. Multivariate Anal.*, **20**, 50–68. (139)

Yin, Y.Q. and Krishnaiah, P.R. (1983), A limit theorem for the eigenvalues of product of two random matrices, *J. Multivariate Anal.*, **13**, 489–507. (142)

Zehna, P.W. (1966), Invariance of maximum likelihood estimators, *Ann. Math. Statist.*, **37**, 348–368. (72)

Zhang, J. and Boos, D.D. (1992), Bootstrap critial values for testing homogeneity of covariance matrices, *J. Amer. Stat. Assoc.*, **87**, 425–429. (223)

Index

WILEY SERIES IN PROBABILITY AND STATISTICS

ESTABLISHED BY WALTER A. SHEWHART AND SAMUEL S. WILKS

The ***Wiley Series in Probability and Statistics*** is well established and authoritative. It covers many topics of current research interest in both pure and applied statistics and probability theory. Written by leading statisticians and institutions, the titles span both state-of-the-art developments in the field and classical methods.

Reflecting the wide range of current research in statistics, the series encompasses applied, methodological and theoretical statistics, ranging from applications and new techniques made possible by advances in computerized practice to rigorous treatment of theoretical approaches.

This series provides essential and invaluable reading for all statisticians, whether in academia, industry, government, or research.

† ABRAHAM and LEDOLTER · Statistical Methods for Forecasting
AGRESTI · Analysis of Ordinal Categorical Data
AGRESTI · An Introduction to Categorical Data Analysis, *Second Edition*
AGRESTI · Categorical Data Analysis, *Second Edition*
ALTMAN, GILL, and McDONALD · Numerical Issues in Statistical Computing for the Social Scientist
AMARATUNGA and CABRERA · Exploration and Analysis of DNA Microarray and Protein Array Data
ANDĚL · Mathematics of Chance
ANDERSON · An Introduction to Multivariate Statistical Analysis, *Third Edition*
* ANDERSON · The Statistical Analysis of Time Series
ANDERSON, AUQUIER, HAUCK, OAKES, VANDAELE, and WEISBERG · Statistical Methods for Comparative Studies
ANDERSON and LOYNES · The Teaching of Practical Statistics
ARMITAGE and DAVID (editors) · Advances in Biometry
ARNOLD, BALAKRISHNAN, and NAGARAJA · Records
* ARTHANARI and DODGE · Mathematical Programming in Statistics
* BAILEY · The Elements of Stochastic Processes with Applications to the Natural Sciences
BALAKRISHNAN and KOUTRAS · Runs and Scans with Applications
BALAKRISHNAN and NG · Precedence-Type Tests and Applications
BARNETT · Comparative Statistical Inference, *Third Edition*
BARNETT · Environmental Statistics
BARNETT and LEWIS · Outliers in Statistical Data, *Third Edition*
BARTOSZYNSKI and NIEWIADOMSKA-BUGAJ · Probability and Statistical Inference
BASILEVSKY · Statistical Factor Analysis and Related Methods: Theory and Applications
BASU and RIGDON · Statistical Methods for the Reliability of Repairable Systems
BATES and WATTS · Nonlinear Regression Analysis and Its Applications
BECHHOFER, SANTNER, and GOLDSMAN · Design and Analysis of Experiments for Statistical Selection, Screening, and Multiple Comparisons

*Now available in a lower priced paperback edition in the Wiley Classics Library.
†Now available in a lower priced paperback edition in the Wiley–Interscience Paperback Series.

BELSLEY · Conditioning Diagnostics: Collinearity and Weak Data in Regression
† BELSLEY, KUH, and WELSCH · Regression Diagnostics: Identifying Influential Data and Sources of Collinearity
BENDAT and PIERSOL · Random Data: Analysis and Measurement Procedures, *Third Edition*
BERRY, CHALONER, and GEWEKE · Bayesian Analysis in Statistics and Econometrics: Essays in Honor of Arnold Zellner
BERNARDO and SMITH · Bayesian Theory
BHAT and MILLER · Elements of Applied Stochastic Processes, *Third Edition*
BHATTACHARYA and WAYMIRE · Stochastic Processes with Applications
BILLINGSLEY · Convergence of Probability Measures, *Second Edition*
BILLINGSLEY · Probability and Measure, *Third Edition*
BIRKES and DODGE · Alternative Methods of Regression
BISWAS, DATTA, FINE, and SEGAL · Statistical Advances in the Biomedical Sciences: Clinical Trials, Epidemiology, Survival Analysis, and Bioinformatics
BLISCHKE AND MURTHY (editors) · Case Studies in Reliability and Maintenance
BLISCHKE AND MURTHY · Reliability: Modeling, Prediction, and Optimization
BLOOMFIELD · Fourier Analysis of Time Series: An Introduction, *Second Edition*
BOLLEN · Structural Equations with Latent Variables
BOLLEN and CURRAN · Latent Curve Models: A Structural Equation Perspective
BOROVKOV · Ergodicity and Stability of Stochastic Processes
BOULEAU · Numerical Methods for Stochastic Processes
BOX · Bayesian Inference in Statistical Analysis
BOX · R. A. Fisher, the Life of a Scientist
BOX and DRAPER · Response Surfaces, Mixtures, and Ridge Analyses, *Second Edition*
* BOX and DRAPER · Evolutionary Operation: A Statistical Method for Process Improvement
BOX and FRIENDS · Improving Almost Anything, *Revised Edition*
BOX, HUNTER, and HUNTER · Statistics for Experimenters: Design, Innovation, and Discovery, *Second Editon*
BOX, JENKINS, and REINSEL · Time Series Analysis: Forcasting and Control, *Fourth Edition*
BOX, LUCEÑO, and PANIAGUA-QUIÑONES · Statistical Control by Monitoring and Adjustment, *Second Edition*
BRANDIMARTE · Numerical Methods in Finance: A MATLAB-Based Introduction
† BROWN and HOLLANDER · Statistics: A Biomedical Introduction
BRUNNER, DOMHOF, and LANGER · Nonparametric Analysis of Longitudinal Data in Factorial Experiments
BUCKLEW · Large Deviation Techniques in Decision, Simulation, and Estimation
CAIROLI and DALANG · Sequential Stochastic Optimization
CASTILLO, HADI, BALAKRISHNAN, and SARABIA · Extreme Value and Related Models with Applications in Engineering and Science
CHAN · Time Series: Applications to Finance
CHARALAMBIDES · Combinatorial Methods in Discrete Distributions
CHATTERJEE and HADI · Regression Analysis by Example, *Fourth Edition*
CHATTERJEE and HADI · Sensitivity Analysis in Linear Regression
CHERNICK · Bootstrap Methods: A Guide for Practitioners and Researchers, *Second Edition*
CHERNICK and FRIIS · Introductory Biostatistics for the Health Sciences
CHILÈS and DELFINER · Geostatistics: Modeling Spatial Uncertainty
CHOW and LIU · Design and Analysis of Clinical Trials: Concepts and Methodologies, *Second Edition*
CLARKE · Linear Models: The Theory and Application of Analysis of Variance

*Now available in a lower priced paperback edition in the Wiley Classics Library.
†Now available in a lower priced paperback edition in the Wiley–Interscience Paperback Series.

CLARKE and DISNEY · Probability and Random Processes: A First Course with Applications, *Second Edition*
* COCHRAN and COX · Experimental Designs, *Second Edition*
COLLINS and LANZA · Latent Class and Latent Transition Analysis: With Applications in the Social, Behavioral, and Health Sciences
CONGDON · Applied Bayesian Modelling
CONGDON · Bayesian Models for Categorical Data
CONGDON · Bayesian Statistical Modelling
CONOVER · Practical Nonparametric Statistics, *Third Edition*
COOK · Regression Graphics
COOK and WEISBERG · Applied Regression Including Computing and Graphics
COOK and WEISBERG · An Introduction to Regression Graphics
CORNELL · Experiments with Mixtures, Designs, Models, and the Analysis of Mixture Data, *Third Edition*
COVER and THOMAS · Elements of Information Theory
COX · A Handbook of Introductory Statistical Methods
* COX · Planning of Experiments
CRESSIE · Statistics for Spatial Data, *Revised Edition*
CSÖRGŐ and HORVÁTH · Limit Theorems in Change Point Analysis
DANIEL · Applications of Statistics to Industrial Experimentation
DANIEL · Biostatistics: A Foundation for Analysis in the Health Sciences, *Eighth Edition*
* DANIEL · Fitting Equations to Data: Computer Analysis of Multifactor Data, *Second Edition*
DASU and JOHNSON · Exploratory Data Mining and Data Cleaning
DAVID and NAGARAJA · Order Statistics, *Third Edition*
* DEGROOT, FIENBERG, and KADANE · Statistics and the Law
DEL CASTILLO · Statistical Process Adjustment for Quality Control
DeMARIS · Regression with Social Data: Modeling Continuous and Limited Response Variables
DEMIDENKO · Mixed Models: Theory and Applications
DENISON, HOLMES, MALLICK and SMITH · Bayesian Methods for Nonlinear Classification and Regression
DETTE and STUDDEN · The Theory of Canonical Moments with Applications in Statistics, Probability, and Analysis
DEY and MUKERJEE · Fractional Factorial Plans
DILLON and GOLDSTEIN · Multivariate Analysis: Methods and Applications
DODGE · Alternative Methods of Regression
* DODGE and ROMIG · Sampling Inspection Tables, *Second Edition*
* DOOB · Stochastic Processes
DOWDY, WEARDEN, and CHILKO · Statistics for Research, *Third Edition*
DRAPER and SMITH · Applied Regression Analysis, *Third Edition*
DRYDEN and MARDIA · Statistical Shape Analysis
DUDEWICZ and MISHRA · Modern Mathematical Statistics
DUNN and CLARK · Basic Statistics: A Primer for the Biomedical Sciences, *Third Edition*
DUPUIS and ELLIS · A Weak Convergence Approach to the Theory of Large Deviations
EDLER and KITSOS · Recent Advances in Quantitative Methods in Cancer and Human Health Risk Assessment
* ELANDT-JOHNSON and JOHNSON · Survival Models and Data Analysis
ENDERS · Applied Econometric Time Series
† ETHIER and KURTZ · Markov Processes: Characterization and Convergence
EVANS, HASTINGS, and PEACOCK · Statistical Distributions, *Third Edition*

*Now available in a lower priced paperback edition in the Wiley Classics Library.
†Now available in a lower priced paperback edition in the Wiley–Interscience Paperback Series.

FELLER · An Introduction to Probability Theory and Its Applications, Volume I, *Third Edition,* Revised; Volume II, *Second Edition*
FISHER and VAN BELLE · Biostatistics: A Methodology for the Health Sciences
FITZMAURICE, LAIRD, and WARE · Applied Longitudinal Analysis
* FLEISS · The Design and Analysis of Clinical Experiments
FLEISS · Statistical Methods for Rates and Proportions, *Third Edition*
† FLEMING and HARRINGTON · Counting Processes and Survival Analysis
FUJIKOSHI, ULYANOV, and SHIMIZU · Multivariate Statistics: High-Dimensional and Large-Sample Approximations
FULLER · Introduction to Statistical Time Series, *Second Edition*
† FULLER · Measurement Error Models
GALLANT · Nonlinear Statistical Models
GEISSER · Modes of Parametric Statistical Inference
GELMAN and MENG · Applied Bayesian Modeling and Causal Inference from Incomplete-Data Perspectives
GEWEKE · Contemporary Bayesian Econometrics and Statistics
GHOSH, MUKHOPADHYAY, and SEN · Sequential Estimation
GIESBRECHT and GUMPERTZ · Planning, Construction, and Statistical Analysis of Comparative Experiments
GIFI · Nonlinear Multivariate Analysis
GIVENS and HOETING · Computational Statistics
GLASSERMAN and YAO · Monotone Structure in Discrete-Event Systems
GNANADESIKAN · Methods for Statistical Data Analysis of Multivariate Observations, *Second Edition*
GOLDSTEIN and LEWIS · Assessment: Problems, Development, and Statistical Issues
GREENWOOD and NIKULIN · A Guide to Chi-Squared Testing
GROSS, SHORTLE, THOMPSON, and HARRIS · Fundamentals of Queueing Theory, *Fourth Edition*
GROSS, SHORTLE, THOMPSON, and HARRIS · Solutions Manual to Accompany Fundamentals of Queueing Theory, *Fourth Edition*
* HAHN and SHAPIRO · Statistical Models in Engineering
HAHN and MEEKER · Statistical Intervals: A Guide for Practitioners
HALD · A History of Probability and Statistics and their Applications Before 1750
HALD · A History of Mathematical Statistics from 1750 to 1930
† HAMPEL · Robust Statistics: The Approach Based on Influence Functions
HANNAN and DEISTLER · The Statistical Theory of Linear Systems
HARTUNG, KNAPP, and SINHA · Statistical Meta-Analysis with Applications
HEIBERGER · Computation for the Analysis of Designed Experiments
HEDAYAT and SINHA · Design and Inference in Finite Population Sampling
HEDEKER and GIBBONS · Longitudinal Data Analysis
HELLER · MACSYMA for Statisticians
HINKELMANN and KEMPTHORNE · Design and Analysis of Experiments, Volume 1: Introduction to Experimental Design, *Second Edition*
HINKELMANN and KEMPTHORNE · Design and Analysis of Experiments, Volume 2: Advanced Experimental Design
HOAGLIN, MOSTELLER, and TUKEY · Fundamentals of Exploratory Analysis of Variance
* HOAGLIN, MOSTELLER, and TUKEY · Exploring Data Tables, Trends and Shapes
* HOAGLIN, MOSTELLER, and TUKEY · Understanding Robust and Exploratory Data Analysis
HOCHBERG and TAMHANE · Multiple Comparison Procedures
HOCKING · Methods and Applications of Linear Models: Regression and the Analysis of Variance, *Second Edition*

*Now available in a lower priced paperback edition in the Wiley Classics Library.
†Now available in a lower priced paperback edition in the Wiley–Interscience Paperback Series.

HOEL · Introduction to Mathematical Statistics, *Fifth Edition*
HOGG and KLUGMAN · Loss Distributions
HOLLANDER and WOLFE · Nonparametric Statistical Methods, *Second Edition*
HOSMER and LEMESHOW · Applied Logistic Regression, *Second Edition*
HOSMER, LEMESHOW, and MAY · Applied Survival Analysis: Regression Modeling of Time-to-Event Data, *Second Edition*
† HUBER and RONCHETTI · Robust Statistics, *Second Edition*
HUBERTY · Applied Discriminant Analysis
HUBERTY and OLEJNIK · Applied MANOVA and Discriminant Analysis, *Second Edition*
HUNT and KENNEDY · Financial Derivatives in Theory and Practice, *Revised Edition*
HURD and MIAMEE · Periodically Correlated Random Sequences: Spectral Theory and Practice
HUSKOVA, BERAN, and DUPAC · Collected Works of Jaroslav Hajek—with Commentary
HUZURBAZAR · Flowgraph Models for Multistate Time-to-Event Data
IMAN and CONOVER · A Modern Approach to Statistics
† JACKSON · A User's Guide to Principle Components
JOHN · Statistical Methods in Engineering and Quality Assurance
JOHNSON · Multivariate Statistical Simulation
JOHNSON and BALAKRISHNAN · Advances in the Theory and Practice of Statistics: A Volume in Honor of Samuel Kotz
JOHNSON and BHATTACHARYYA · Statistics: Principles and Methods, *Fifth Edition*
JOHNSON and KOTZ · Distributions in Statistics
JOHNSON and KOTZ (editors) · Leading Personalities in Statistical Sciences: From the Seventeenth Century to the Present
JOHNSON, KOTZ, and BALAKRISHNAN · Continuous Univariate Distributions, Volume 1, *Second Edition*
JOHNSON, KOTZ, and BALAKRISHNAN · Continuous Univariate Distributions, Volume 2, *Second Edition*
JOHNSON, KOTZ, and BALAKRISHNAN · Discrete Multivariate Distributions
JOHNSON, KEMP, and KOTZ · Univariate Discrete Distributions, *Third Edition*
JUDGE, GRIFFITHS, HILL, LÜTKEPOHL, and LEE · The Theory and Practice of Econometrics, *Second Edition*
JUREČKOVÁ and SEN · Robust Statistical Procedures: Aymptotics and Interrelations
JUREK and MASON · Operator-Limit Distributions in Probability Theory
KADANE · Bayesian Methods and Ethics in a Clinical Trial Design
KADANE AND SCHUM · A Probabilistic Analysis of the Sacco and Vanzetti Evidence
KALBFLEISCH and PRENTICE · The Statistical Analysis of Failure Time Data, *Second Edition*
KARIYA and KURATA · Generalized Least Squares
KASS and VOS · Geometrical Foundations of Asymptotic Inference
† KAUFMAN and ROUSSEEUW · Finding Groups in Data: An Introduction to Cluster Analysis
KEDEM and FOKIANOS · Regression Models for Time Series Analysis
KENDALL, BARDEN, CARNE, and LE · Shape and Shape Theory
KHURI · Advanced Calculus with Applications in Statistics, *Second Edition*
KHURI, MATHEW, and SINHA · Statistical Tests for Mixed Linear Models
KLEIBER and KOTZ · Statistical Size Distributions in Economics and Actuarial Sciences
KLEMELÄ · Smoothing of Multivariate Data: Density Estimation and Visualization
KLUGMAN, PANJER, and WILLMOT · Loss Models: From Data to Decisions, *Third Edition*
KLUGMAN, PANJER, and WILLMOT · Solutions Manual to Accompany Loss Models: From Data to Decisions, *Third Edition*

*Now available in a lower priced paperback edition in the Wiley Classics Library.
†Now available in a lower priced paperback edition in the Wiley–Interscience Paperback Series.

KOTZ, BALAKRISHNAN, and JOHNSON · Continuous Multivariate Distributions, Volume 1, *Second Edition*
KOVALENKO, KUZNETZOV, and PEGG · Mathematical Theory of Reliability of Time-Dependent Systems with Practical Applications
KOWALSKI and TU · Modern Applied U-Statistics
KRISHNAMOORTHY and MATHEW · Statistical Tolerance Regions: Theory, Applications, and Computation
KROONENBERG · Applied Multiway Data Analysis
KVAM and VIDAKOVIC · Nonparametric Statistics with Applications to Science and Engineering
LACHIN · Biostatistical Methods: The Assessment of Relative Risks
LAD · Operational Subjective Statistical Methods: A Mathematical, Philosophical, and Historical Introduction
LAMPERTI · Probability: A Survey of the Mathematical Theory, *Second Edition*
LANGE, RYAN, BILLARD, BRILLINGER, CONQUEST, and GREENHOUSE · Case Studies in Biometry
LARSON · Introduction to Probability Theory and Statistical Inference, *Third Edition*
LAWLESS · Statistical Models and Methods for Lifetime Data, *Second Edition*
LAWSON · Statistical Methods in Spatial Epidemiology
LE · Applied Categorical Data Analysis
LE · Applied Survival Analysis
LEE and WANG · Statistical Methods for Survival Data Analysis, *Third Edition*
LePAGE and BILLARD · Exploring the Limits of Bootstrap
LEYLAND and GOLDSTEIN (editors) · Multilevel Modelling of Health Statistics
LIAO · Statistical Group Comparison
LINDVALL · Lectures on the Coupling Method
LIN · Introductory Stochastic Analysis for Finance and Insurance
LINHART and ZUCCHINI · Model Selection
LITTLE and RUBIN · Statistical Analysis with Missing Data, *Second Edition*
LLOYD · The Statistical Analysis of Categorical Data
LOWEN and TEICH · Fractal-Based Point Processes
MAGNUS and NEUDECKER · Matrix Differential Calculus with Applications in Statistics and Econometrics, *Revised Edition*
MALLER and ZHOU · Survival Analysis with Long Term Survivors
MALLOWS · Design, Data, and Analysis by Some Friends of Cuthbert Daniel
MANN, SCHAFER, and SINGPURWALLA · Methods for Statistical Analysis of Reliability and Life Data
MANTON, WOODBURY, and TOLLEY · Statistical Applications Using Fuzzy Sets
MARCHETTE · Random Graphs for Statistical Pattern Recognition
MARDIA and JUPP · Directional Statistics
MASON, GUNST, and HESS · Statistical Design and Analysis of Experiments with Applications to Engineering and Science, *Second Edition*
McCULLOCH, SEARLE, and NEUHAUS · Generalized, Linear, and Mixed Models, *Second Edition*
McFADDEN · Management of Data in Clinical Trials, *Second Edition*
* McLACHLAN · Discriminant Analysis and Statistical Pattern Recognition
McLACHLAN, DO, and AMBROISE · Analyzing Microarray Gene Expression Data
McLACHLAN and KRISHNAN · The EM Algorithm and Extensions, *Second Edition*
McLACHLAN and PEEL · Finite Mixture Models
McNEIL · Epidemiological Research Methods
MEEKER and ESCOBAR · Statistical Methods for Reliability Data
MEERSCHAERT and SCHEFFLER · Limit Distributions for Sums of Independent Random Vectors: Heavy Tails in Theory and Practice
MICKEY, DUNN, and CLARK · Applied Statistics: Analysis of Variance and

*Now available in a lower priced paperback edition in the Wiley Classics Library.
†Now available in a lower priced paperback edition in the Wiley–Interscience Paperback Series.

Regression, *Third Edition*

* MILLER · Survival Analysis, *Second Edition*

MONTGOMERY, JENNINGS, and KULAHCI · Introduction to Time Series Analysis and Forecasting

MONTGOMERY, PECK, and VINING · Introduction to Linear Regression Analysis, *Fourth Edition*

MORGENTHALER and TUKEY · Configural Polysampling: A Route to Practical Robustness

MUIRHEAD · Aspects of Multivariate Statistical Theory

MULLER and STOYAN · Comparison Methods for Stochastic Models and Risks

MURRAY · X-STAT 2.0 Statistical Experimentation, Design Data Analysis, and Nonlinear Optimization

MURTHY, XIE, and JIANG · Weibull Models

MYERS, MONTGOMERY, and ANDERSON-COOK · Response Surface Methodology: Process and Product Optimization Using Designed Experiments, *Third Edition*

MYERS, MONTGOMERY, and VINING · Generalized Linear Models. With Applications in Engineering and the Sciences

† NELSON · Accelerated Testing, Statistical Models, Test Plans, and Data Analyses

† NELSON · Applied Life Data Analysis

NEWMAN · Biostatistical Methods in Epidemiology

OCHI · Applied Probability and Stochastic Processes in Engineering and Physical Sciences

OKABE, BOOTS, SUGIHARA, and CHIU · Spatial Tesselations: Concepts and Applications of Voronoi Diagrams, *Second Edition*

OLIVER and SMITH · Influence Diagrams, Belief Nets and Decision Analysis

PALTA · Quantitative Methods in Population Health: Extensions of Ordinary Regressions

PANJER · Operational Risk: Modeling and Analytics

PANKRATZ · Forecasting with Dynamic Regression Models

PANKRATZ · Forecasting with Univariate Box-Jenkins Models: Concepts and Cases

* PARZEN · Modern Probability Theory and Its Applications

PEÑA, TIAO, and TSAY · A Course in Time Series Analysis

PIANTADOSI · Clinical Trials: A Methodologic Perspective

PORT · Theoretical Probability for Applications

POURAHMADI · Foundations of Time Series Analysis and Prediction Theory

POWELL · Approximate Dynamic Programming: Solving the Curses of Dimensionality

PRESS · Bayesian Statistics: Principles, Models, and Applications

PRESS · Subjective and Objective Bayesian Statistics, *Second Edition*

PRESS and TANUR · The Subjectivity of Scientists and the Bayesian Approach

PUKELSHEIM · Optimal Experimental Design

PURI, VILAPLANA, and WERTZ · New Perspectives in Theoretical and Applied Statistics

† PUTERMAN · Markov Decision Processes: Discrete Stochastic Dynamic Programming

QIU · Image Processing and Jump Regression Analysis

* RAO · Linear Statistical Inference and Its Applications, *Second Edition*

RAUSAND and HØYLAND · System Reliability Theory: Models, Statistical Methods, and Applications, *Second Edition*

RENCHER · Linear Models in Statistics

RENCHER · Methods of Multivariate Analysis, *Second Edition*

RENCHER · Multivariate Statistical Inference with Applications

* RIPLEY · Spatial Statistics

* RIPLEY · Stochastic Simulation

ROBINSON · Practical Strategies for Experimenting

ROHATGI and SALEH · An Introduction to Probability and Statistics, *Second Edition*

*Now available in a lower priced paperback edition in the Wiley Classics Library.

†Now available in a lower priced paperback edition in the Wiley–Interscience Paperback Series.

ROLSKI, SCHMIDLI, SCHMIDT, and TEUGELS · Stochastic Processes for Insurance and Finance
ROSENBERGER and LACHIN · Randomization in Clinical Trials: Theory and Practice
ROSS · Introduction to Probability and Statistics for Engineers and Scientists
ROSSI, ALLENBY, and McCULLOCH · Bayesian Statistics and Marketing
† ROUSSEEUW and LEROY · Robust Regression and Outlier Detection
* RUBIN · Multiple Imputation for Nonresponse in Surveys
RUBINSTEIN and KROESE · Simulation and the Monte Carlo Method, *Second Edition*
RUBINSTEIN and MELAMED · Modern Simulation and Modeling
RYAN · Modern Engineering Statistics
RYAN · Modern Experimental Design
RYAN · Modern Regression Methods, *Second Edition*
RYAN · Statistical Methods for Quality Improvement, *Second Edition*
SALEH · Theory of Preliminary Test and Stein-Type Estimation with Applications
* SCHEFFE · The Analysis of Variance
SCHIMEK · Smoothing and Regression: Approaches, Computation, and Application
SCHOTT · Matrix Analysis for Statistics, *Second Edition*
SCHOUTENS · Levy Processes in Finance: Pricing Financial Derivatives
SCHUSS · Theory and Applications of Stochastic Differential Equations
SCOTT · Multivariate Density Estimation: Theory, Practice, and Visualization
† SEARLE · Linear Models for Unbalanced Data
† SEARLE · Matrix Algebra Useful for Statistics
† SEARLE, CASELLA, and McCULLOCH · Variance Components
SEARLE and WILLETT · Matrix Algebra for Applied Economics
SEBER · A Matrix Handbook For Statisticians
† SEBER · Multivariate Observations
SEBER and LEE · Linear Regression Analysis, *Second Edition*
† SEBER and WILD · Nonlinear Regression
SENNOTT · Stochastic Dynamic Programming and the Control of Queueing Systems
* SERFLING · Approximation Theorems of Mathematical Statistics
SHAFER and VOVK · Probability and Finance: It's Only a Game!
SILVAPULLE and SEN · Constrained Statistical Inference: Inequality, Order, and Shape Restrictions
SMALL and McLEISH · Hilbert Space Methods in Probability and Statistical Inference
SRIVASTAVA · Methods of Multivariate Statistics
STAPLETON · Linear Statistical Models, *Second Edition*
STAPLETON · Models for Probability and Statistical Inference: Theory and Applications
STAUDTE and SHEATHER · Robust Estimation and Testing
STOYAN, KENDALL, and MECKE · Stochastic Geometry and Its Applications, *Second Edition*
STOYAN and STOYAN · Fractals, Random Shapes and Point Fields: Methods of Geometrical Statistics
STREET and BURGESS · The Construction of Optimal Stated Choice Experiments: Theory and Methods
STYAN · The Collected Papers of T. W. Anderson: 1943–1985
SUTTON, ABRAMS, JONES, SHELDON, and SONG · Methods for Meta-Analysis in Medical Research
TAKEZAWA · Introduction to Nonparametric Regression
TAMHANE · Statistical Analysis of Designed Experiments: Theory and Applications
TANAKA · Time Series Analysis: Nonstationary and Noninvertible Distribution Theory
THOMPSON · Empirical Model Building
THOMPSON · Sampling, *Second Edition*
THOMPSON · Simulation: A Modeler's Approach
THOMPSON and SEBER · Adaptive Sampling

*Now available in a lower priced paperback edition in the Wiley Classics Library.
†Now available in a lower priced paperback edition in the Wiley–Interscience Paperback Series.

THOMPSON, WILLIAMS, and FINDLAY · Models for Investors in Real World Markets
TIAO, BISGAARD, HILL, PEÑA, and STIGLER (editors) · Box on Quality and Discovery: with Design, Control, and Robustness
TIERNEY · LISP-STAT: An Object-Oriented Environment for Statistical Computing and Dynamic Graphics
TSAY · Analysis of Financial Time Series, *Second Edition*
UPTON and FINGLETON · Spatial Data Analysis by Example, Volume II: Categorical and Directional Data
† VAN BELLE · Statistical Rules of Thumb, *Second Edition*
VAN BELLE, FISHER, HEAGERTY, and LUMLEY · Biostatistics: A Methodology for the Health Sciences, *Second Edition*
VESTRUP · The Theory of Measures and Integration
VIDAKOVIC · Statistical Modeling by Wavelets
VINOD and REAGLE · Preparing for the Worst: Incorporating Downside Risk in Stock Market Investments
WALLER and GOTWAY · Applied Spatial Statistics for Public Health Data
WEERAHANDI · Generalized Inference in Repeated Measures: Exact Methods in MANOVA and Mixed Models
WEISBERG · Applied Linear Regression, *Third Edition*
WELSH · Aspects of Statistical Inference
WESTFALL and YOUNG · Resampling-Based Multiple Testing: Examples and Methods for *p*-Value Adjustment
WHITTAKER · Graphical Models in Applied Multivariate Statistics
WINKER · Optimization Heuristics in Economics: Applications of Threshold Accepting
WONNACOTT and WONNACOTT · Econometrics, *Second Edition*
WOODING · Planning Pharmaceutical Clinical Trials: Basic Statistical Principles
WOODWORTH · Biostatistics: A Bayesian Introduction
WOOLSON and CLARKE · Statistical Methods for the Analysis of Biomedical Data, *Second Edition*
WU and HAMADA · Experiments: Planning, Analysis, and Parameter Design Optimization, *Second Edition*
WU and ZHANG · Nonparametric Regression Methods for Longitudinal Data Analysis
YANG · The Construction Theory of Denumerable Markov Processes
YOUNG, VALERO-MORA, and FRIENDLY · Visual Statistics: Seeing Data with Dynamic Interactive Graphics
ZACKS · Stage-Wise Adaptive Designs
ZELTERMAN · Discrete Distributions—Applications in the Health Sciences
* ZELLNER · An Introduction to Bayesian Inference in Econometrics
ZHOU, OBUCHOWSKI, and McCLISH · Statistical Methods in Diagnostic Medicine

*Now available in a lower priced paperback edition in the Wiley Classics Library.
†Now available in a lower priced paperback edition in the Wiley–Interscience Paperback Series.